Sustainable Agricultural, Biological, and Environmental Engineering Applications

Sustainable Agricultural, Biological, and Environmental Engineering Applications

Editors

Muhammad Sultan
Yuguang Zhou
Redmond R. Shamshiri
Aitazaz A. Farooque

MDPI • Basel • Beijing • Wuhan • Barcelona • Belgrade • Manchester • Tokyo • Cluj • Tianjin

Editors

Muhammad Sultan
Department of Agricultural
Engineering
Bahauddin Zakariya University
Multan
Pakistan

Yuguang Zhou
Department of Agricultural
Engineering
College of Engineering
China Agricultural University
Beijing
China

Redmond R. Shamshiri
Leibniz Institute for Agricultural
Engineering and Bioeconomy
Potsdam
Germany

Aitazaz A. Farooque
Sustainable Design Engineering
University of Prince Edward
Island
Charlottetown
Canada

Editorial Office
MDPI
St. Alban-Anlage 66
4052 Basel, Switzerland

This is a reprint of articles from the Special Issue published online in the open access journal *Sustainability* (ISSN 2071-1050) (available at: www.mdpi.com/journal/sustainability/special_issues/ Agricultural_Biological_Environmental_Engineering).

For citation purposes, cite each article independently as indicated on the article page online and as indicated below:

LastName, A.A.; LastName, B.B.; LastName, C.C. Article Title. *Journal Name* **Year**, *Volume Number*, Page Range.

ISBN 978-3-0365-2921-9 (Hbk)
ISBN 978-3-0365-2920-2 (PDF)

Contents

About the Editors

Muhammad Sultan

Muhammad Sultan is an Assistant Professor in the Department of Agricultural Engineering, Bahauddin Zakariya University, Multan (Pakistan). He holds a B.Sc. (2008) and M.Sc. (2010) in Agricultural Engineering from the University of Agriculture, Faisalabad (Pakistan), and a Ph.D. (2015) in Energy and Environmental Engineering from Kyushu University (Japan). He did his postdoctoral research in Energy and Environmental Engineering (2017) at Kyushu University (Japan) and in Mechatronic Systems Engineering (2019) at Simon Fraser University (Canada). He has published more than 200 articles in international journals, conferences, books, and book chapters. He has been a reviewer for more than 80 renowned journals and holds an editor role for five journals with publishers such as SAGE, MDPI, and Frontiers. His research focuses on developing energy-efficient temperature and humidity control systems for agricultural applications including greenhouse, fruits/vegetable storage, livestock, and poultry applications. His research keywords include adsorption heat pumps, desiccant air-conditioning, evaporative cooling, Maisotsenko cycle, adsorption desalination, energy recovery ventilator, atmospheric water harvesting, and wastewater treatment.

Yuguang Zhou

Dr. Zhou Yuguang is currently appointed as an Associate Professor at the College of Engineering, China Agricultural University. His research investigatres biogas engineering technology and the clean combustion of biomass and its emission control, based on platforms such as "National Center for International Research on BioEnergy Science and Technology" (iBEST), Ministry of Science and Technology, and "Key Laboratory of Clean Production and Utilization of Renewable Energy" (CPURE), Ministry of Agriculture and Rural Affairs. He has conducted more than 20 projects that were granted by the National Science Foundation of China, the United Nations Development Programme, Worldbank, etc.

Redmond R. Shamshiri

Dr. Redmond R. Shamshiri holds a Ph.D. in agricultural automation with a focus on control systems and dynamics. He is a scientist at the Leibniz-Institut für Agrartechnik und Bioökonomie working toward the digitization of agriculture for food security. His main research fields include simulation and modeling for closed-field plant production systems, LPWAN sensors, wireless control, and autonomous navigation. His works have appeared in over 110 publications, including peer-reviewed journal papers, book chapters, and conference proceedings. He is the founder of Adaptive AgroTech Consultancy Network and serves as a section editor and reviewer for various peer-reviewed journals in the field of smart farming.

Aitazaz A. Farooque

Dr. Aitazaz A. Farooque is working as an Associate Professor at the Faculty of Sustainable Design Engineering, University of Prince Edward Island. Dr. Farooque's research focuses on the fundamental understanding and development of state-of-the-art precision agriculture (PA) technologies for Eastern Canada's agriculture industry. The development of innovative and novel PA systems requires knowledge of engineering design, development and management, instrumentation, design and evaluation of sensors and controllers, development of hardware and software for automation of machines to sense targets in real time for spot applications of agrochemicals on an as-needed basis to improve farm profitability while maintaining environmental sustainability. Dr. Farooque is actively working on machine vision, the application of multispectral and thermal imagery using drone technology, the delineation of management zones for site-specific fertilization, electromagnetic induction methods, remote sensing, and digital photography techniques for mapping, modeling bio-systems, artificial neural networks, deep learning, and analog and digital sensor integrations into agricultural equipment for real-time soil, plant, and yield mapping. Dr. Farooque has been evaluating the variable rate technologies for potential environmental risks. Dr. Farooque has been very successful in securing research funding from Natural Science and Engineering Council of Canada, Provincial and Federal Governments and agriculture industry. He has been supervising undergraduate and graduate students, research assistants, and a post-doctoral fellow at UPEI and other collaborating institutions. Dr. Farooque's research has been published/shown in peer-reviewed journals, conference proceedings, workshops, industry meetings and farmer's field days.

Preface to "Sustainable Agricultural, Biological, and Environmental Engineering Applications"

Sustainable development in the agriculture sector is crucial to achieve the 2030 Sustainable Development Goals set by the United Nations. There are lots of challenges to developing modern and intelligent agricultural techniques, tools, and systems, by which sustainable agriculture and food security can be satisfied. In addition, carbon-neutral development and clean energy utilization are also associated with the UN-SDGs. Therefore, this book focuses on the recent proliferation and technological advancement in the agricultural, biological, and environmental engineering applications from the viewpoint of the agricultural water–energy–food security nexus. The book presents such engineering technologies and applications in seven categories that include: (i) energy system applications in agriculture, (ii) irrigation and drainage, (iii) biomass, biogas, and biochar, (iv) farm mechanization and soil science, (v) remote sensing and geographical studies, (vi) wastewater and biological studies, and (vii) case studies and societal aspects in agriculture.

In the category "Energy System Applications in Agriculture", three studies are reported. In this regard, an experimental study provides energy-efficient evaporative cooling systems for a poultry farm application. An air heat pump system is discussed for pig farms for sustainable energy efficiency, housing environment, and productivity traits. A CFD study offers a detailed inlet configurations analysis of the microclimate conditions of a novel agricultural greenhouse. The category "Irrigation and Drainage" provides four dedicated studies. An alternate design of a surface drain system is reported for the free discharge of subsurface drainage effluent. A sustainable irrigation system is discussed for small landholdings of the rainfed areas. An assessment of non-conventional irrigation water is conducted for greenhouse cucumber production. Soil erosion and sediment load management strategies are studied for sustainable irrigation in arid regions. The category "Biomass, Biogas, and Biochar" comprises three kinds of research findings. Regional biogas production potential from livestock manure is reviewed and explored. A low-cost biomass furnace development is explained for greenhouse heating. A study describes bioresource transformation to biochar and highlights its importance and potential applications for boosting the circular bioeconomy. Three studies are presented in the category "Farm Mechanization and Soil Science". A double-row pneumatic precision metering device is studied for Brassica Chinensis. The tractive performance of a single grouser shoe is investigated under different soil and moisture conditions. The effect of K and Zn application on the biometric and physiological parameters of different maize genotypes is explored. There are two research studies in the category "Remote Sensing and Geographical Studies". In this regard, the mapping of paddy rice with satellite remote sensing is reviewed. An estimation of regional maize straw resources is given using the Gaofen 6 satellite as the information source which has a high resolution and wide field of view imaging. Two studies are presented in the category "Wastewater and Biological Studies". A dedicated study explores the preparation and application of carbon nanotubes for water and wastewater treatment. One study reports that the compost inoculated with fungi from a mangrove habitat improved the growth and disease defense of vegetable plants. Five kinds of exclusive research studies are offered in the category "Case Studies and Societal Aspects in Agriculture". A state-of-the-art study from Iran presents the effects of the COVID-19 pandemic on food security and agriculture. A study presents the insights from Germany to address an interesting question, i.e., will farmers accept lower gross margins for the sustainable cultivation method of mixed cropping? Similarly, another study reports the empirical evidence from

Polish farm households to address the question: Are farms located in less favored areas financially sustainable? A study from Pakistan predicts the behavioral intentions of rural inhabitants with regard to the economic incentive for deforestation. Another study from Pakistan assesses the contribution of citrus orchards to climate change mitigation through carbon sequestration.

The editors are pleased to share the above-mentioned seven categories comprised of emerging research and applications in the field of agricultural, biological, and environmental engineering. The editors believe that this book *"Sustainable Agricultural, Biological, and Environmental Engineering Applications"* will be useful for agricultural scientists, researchers, and students. The editors would like to acknowledge and thank the dedicated authors from each study for their valuable contributions to this book.

Muhammad Sultan, Yuguang Zhou, Redmond R. Shamshiri, Aitazaz A. Farooque
Editors

sustainability

Article

Experiments on Energy-Efficient Evaporative Cooling Systems for Poultry Farm Application in Multan (Pakistan)

Khawar Shahzad [1], Muhammad Sultan [1,*], Muhammad Bilal [1], Hadeed Ashraf [1], Muhammad Farooq [2], Takahiko Miyazaki [3,4], Uzair Sajjad [5], Imran Ali [6] and Muhammad I. Hussain [7]

[1] Department of Agricultural Engineering, Bahauddin Zakariya University, Multan 60800, Pakistan; khawarshahzad04@gmail.com (K.S.); bilalranauni@gmail.com (M.B.); hadeedashraf15@gmail.com (H.A.)
[2] Department of Mechanical Engineering, University of Engineering and Technology, Lahore 39161, Pakistan; engr.farooq@uet.edu.pk
[3] Faculty of Engineering Sciences, Kyushu University, Fukuoka 816-8580, Japan; miyazaki.takahiko.735@m.kyushu-u.ac.jp
[4] International Institute for Carbon-Neutral Energy Research (WPI-I2CNER), Kyushu University, Fukuoka 819-0395, Japan
[5] Mechanical Engineering Department, National Chiao Tung University, Hsinchu 300044, Taiwan; energyengineer01@gmail.com
[6] Department of Environmental Science and Engineering, College of Chemistry and Environmental Engineering, Shenzhen University, Shenzhen 518060, China; engrimran56@gmail.com
[7] Green Energy Technology Research Center, Kongju National University, Cheonan 122324, Korea; imtiaz@kongju.ac.kr
* Correspondence: muhammadsultan@bzu.edu.pk; Tel.: +92–333–610–8888

Citation: Shahzad, K.; Sultan, M.; Bilal, M.; Ashraf, H.; Farooq, M.; Miyazaki, T.; Sajjad, U.; Ali, I.; Hussain, M.I. Experiments on Energy-Efficient Evaporative Cooling Systems for Poultry Farm Application in Multan (Pakistan). *Sustainability* **2021**, *13*, 2836. https://doi.org/10.3390/su13052836

Academic Editors: Muhammad Sultan, Yuguang Zhou, Redmond R. Shamshiri and Aitazaz A. Farooque

Received: 17 January 2021
Accepted: 25 February 2021
Published: 5 March 2021

Publisher's Note: MDPI stays neutral with regard to jurisdictional claims in published maps and institutional affiliations.

Abstract: Poultry are one of the most vulnerable species of its kind once the temperature-humidity nexus is explored. This is so because the broilers lack sweat glands as compared to humans and undergo panting process to mitigate their latent heat (moisture produced in the body) in the air. As a result, moisture production inside poultry house needs to be maintained to avoid any serious health and welfare complications. Several strategies such as compressor-based air-conditioning systems have been implemented worldwide to attenuate the heat stress in poultry, but these are not economical. Therefore, this study focuses on the development of low-cost and environmentally friendly improved evaporative cooling systems (DEC, IEC, MEC) from the viewpoint of heat stress in poultry houses. Thermodynamic analysis of these systems was carried out for the climatic conditions of Multan, Pakistan. The results appreciably controlled the environmental conditions which showed that for the months of April, May, and June, the decrease in temperature by direct evaporative cooling (DEC), indirect evaporative cooling (IEC), and Maisotsenko-Cycle evaporative cooling (MEC) systems is 7–10 °C, 5–6.5 °C, and 9.5–12 °C, respectively. In case of July, August, and September, the decrease in temperature by DEC, IEC, and MEC systems is 5.5–7 °C, 3.5–4.5 °C, and 7–7.5 °C, respectively. In addition, drop in temperature-humidity index (THI) values by DEC, IEC, and MEC is 3.5–9 °C, 3–7 °C, and 5.5–10 °C, respectively for all months. Optimum temperature and relative humidity conditions are determined for poultry birds and thereby, systems' performance is thermodynamically evaluated for poultry farms from the viewpoint of THI, temperature-humidity-velocity index (THVI), and thermal exposure time (ET). From the analysis, it is concluded that MEC system performed relatively better than others due to its ability of dew-point cooling and achieved THI threshold limit with reasonable temperature and humidity indexes.

Keywords: poultry farms; air-conditioning; evaporative cooling systems; temperature-humidity index; temperature-humidity-velocity index

1. Introduction

1.1. Background

The agriculture sector of Pakistan contributes to about 21% of the gross domestic product (GDP) and absorbs 45.5% of the total labor strength [1]. The total share of livestock sector in agriculture covers about 11.4% of the agriculture gross domestic product and 53.25% of the value-added products [2]. Among the livestock sector, poultry contributed about 1.4% in overall gross domestic product (GDP) during (2017–18) [3]. It also employs directly/indirectly 1.5 million people [4]. Furthermore, the poultry meat production amounted to total 1.43 million tons in 2017–2018 which represented the 32.76% of the total meat production in the country. Keeping in view the economic importance of poultry, it is desirable to monitor the environmental condition for their control sheds where several flocks are brought up on yearly basis. A huge amount of capital is invested to raise the controlled structures. Therefore, minor risks either by labor or machine are vulnerable to poultry. Pakistan is recognized as a tropical country being along the equator on globe.

Climate change causes an increase in frequency, duration, and magnitude of heat events [5,6]. Tropical countries are susceptible to hot and humid weather conditions [7]. In Pakistan, there is a cycle of four seasons giving temporal variations. Summer and winter reach the intense weather conditions. Over this course of time, temperature hits above 40 °C and the corresponding mark of relative humidity drops below 20% in plain areas during summer. Poultry birds are susceptible to environmental conditions. It is advisory to control these factors that adversely affect the production and welfare of broiler chickens. Heat stress is the major contributory force to affect the fate of these broilers. Poultry birds are homoiothermic in nature and have the ability to control the body temperature throughout the year whereas, the thermoregulatory mechanisms are efficient only in the range of thermo-neutral zones (27.5–37.7 °C) [8,9]. The current study consists of the applicability of evaporative cooling systems in the ambient conditions of Multan, Pakistan. To maintain the thermal comfort in a poultry farm, air-conditioning is necessary [10,11]. Figure 1a,b shows the dry-bulb temperature (DBT) and relative humidity (RH) variation for Multan (Pakistan) throughout the year. It is found from the literature that the temperature higher than 25 °C causes heat stress in poultry [12]. This study comprises the poultry thermal comfort under the ambient conditions of Multan, Pakistan. The suitable relative humidity ranges from the efficiency of the poultry farms and the chickens get affected by this temperature-humidity index (THI) [13]. Once these situations reach poorly managed controlled houses for poultry, the mortality rate per flock increases.

Pursuing this trend, growth of chickens is depressed, and heavy economic loss is incurred. High temperatures can be absorbed by the poultry birds to some extent but may go negatively when summer conditions turn severely warm with low humidity in ambient air.

1.2. Heat Stress and Poultry Air-Conditioning

Heat stress is a key problem affecting both the health and performance of the poultry [14]. The chickens try to maintain their body temperature in between the thermo-neutral zone but it is a condition where chickens are unable to maintain the balance between the heat production and heat loss [15]. If the controlled temperature exceeds this zone, heat must be lost in some way by poultry birds. Chickens have no sweat glands. Naturally, a human body has pores on the skin through which moisture loss occurs by specific glands balancing the ambient weather conditions. Unlike humans, chickens are deprived of such sweat glands. Weight gain in chickens gradually goes on with increasing age. During summer, the body heat of poultry birds is also exalted causing raised temperature [16]. At this point, there are two ways: either reduce feed intake by bird or provide optimal weather conditions inside controlled sheds. At 29.4 °C (85 °F), chickens start panting [17]. Figure 2a shows the temperature/humidity heat stress index for chickens which combines the air temperature with the relative humidity to analyze that how increasing humidity affects the thermal comfort zone. Panting is a natural process for heat dissipation in bodies of poultry birds. Analogous to humans, this process maintains metabolic heat balance for chickens. As a result of this phenomenon, water intake is increased to avoid dehydration. Figure 2b illustrates the temperature zones of poultry birds which states that the optimum poultry bird's growth can be obtained by maintaining the desired temperature and humidity zones inside the poultry house. During panting, high values of temperature and humidity pose a serious problem. As the chickens lose moisture heat of the body to their surroundings for attaining thermal comfort. But high humidity in ambient hot air hinders the functioning of this process [18]. It also affects the productive and reproductive performance as well as the economic traits and the welfare of poultry [19,20].

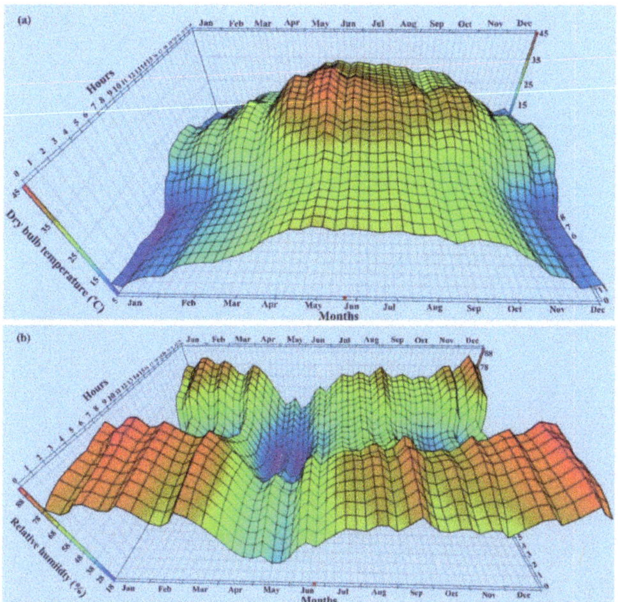

Figure 1. Illustration of (**a**) dry-bulb temperature, and (**b**) relative humidity variation (on hourly basis) for the ambient conditions of Multan, Pakistan.

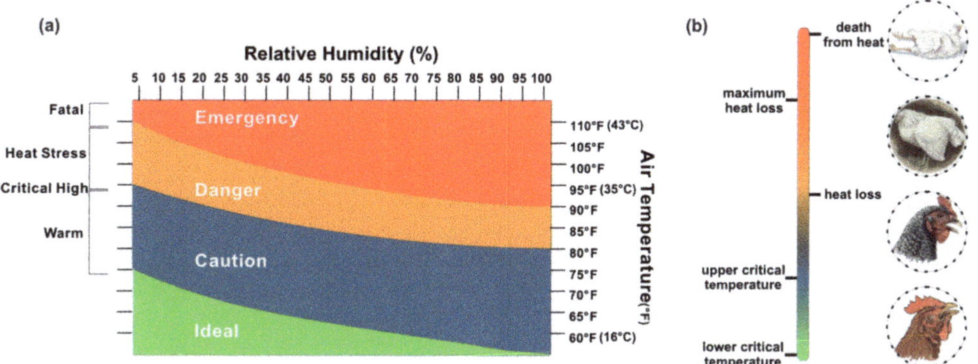

Figure 2. Heat stress effects on poultry birds. (**a**) Illustration of temperature-humidity index (THI) for chickens, reproduced from [21,22], and (**b**) diagram of temperature zones for broiler chickens representing lower, upper, and maximum temperature, reproduced from [23].

The activity and position of broiler chickens in broiler houses if monitored and controlled could potentially lead to ameliorate conditions for health, energy consumption, and welfare of these birds [14]. This research focused on controlling chamber environment for broilers on micro-scale to understand their attributes. Broiler chickens transmit heat flow from their body surface to maintain thermal equilibrium with the environment. The surface temperature of birds can be directly related to the flow of blood in their body. Any change in ambient temperature can be felt through the blood flow in birds near the skin. Climate in poultry houses is a combination of dry air and humidity. Poultry litter is affected when moisture and relative humidity is increased above 70% in room/poultry house. In a result, ammonia (above 70 ppm) production increases which affects bird's health and reduces growth [24–26].

The optimum control of these two parameters guarantees the safety and welfare of broiler chickens. Broiler chickens maintain their body temperatures through sensible (change in body surface temperature) and latent (release of moisture from body in while exhaling) heat emissions. It is suggested that the comprehensive study of metabolic functions be conducted to understand the heat production in poultry birds. Figure 3 illustrates the effects of heat stress on behavioral changes. Chickens under heat stress conditions spend less time in feeding and more in drinking, wings are lifted and less moving. The panting signs are also observed [15]. Physiological changes include the oxidative stress, acid base imbalance, respiratory alkalosis, and changes in cecal microbial profile. Heat stress is correlated with the cellular oxidative stress which causes severe health disorders, lower growth rates and economic losses [16,17]. The heat stress causes production changes by increasing the weight of the chickens and decreasing the quality and quantity of eggs. The mortality rate also increases due to heat stress [18].

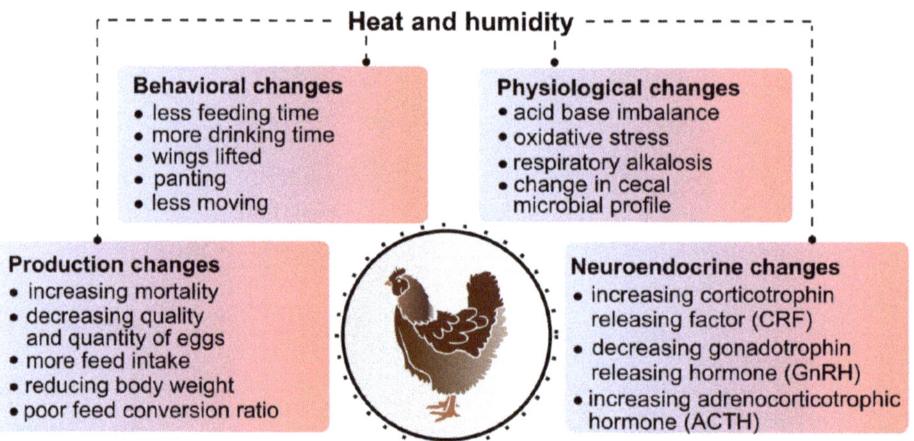

Figure 3. Heat stress effects on the poultry behavioral, physiological, production and neuroendocrine changes, reproduced from [18].

With the increasing ambient temperature, the mortality rate increases. The high levels of temperature not only affect the production performance but also hinders the immune function in poultry [27,28]. To achieve the desired conditions for poultry, many cooling systems have been developed and thus controlled air-conditioning has become necessary. Air conditioning systems specifically evaporative cooling pads alone, or in combination with nozzles are studied in literature [19,20,29,30]. Figure 4 illustrates the schematic of a typical EC-based poultry air-conditioning.

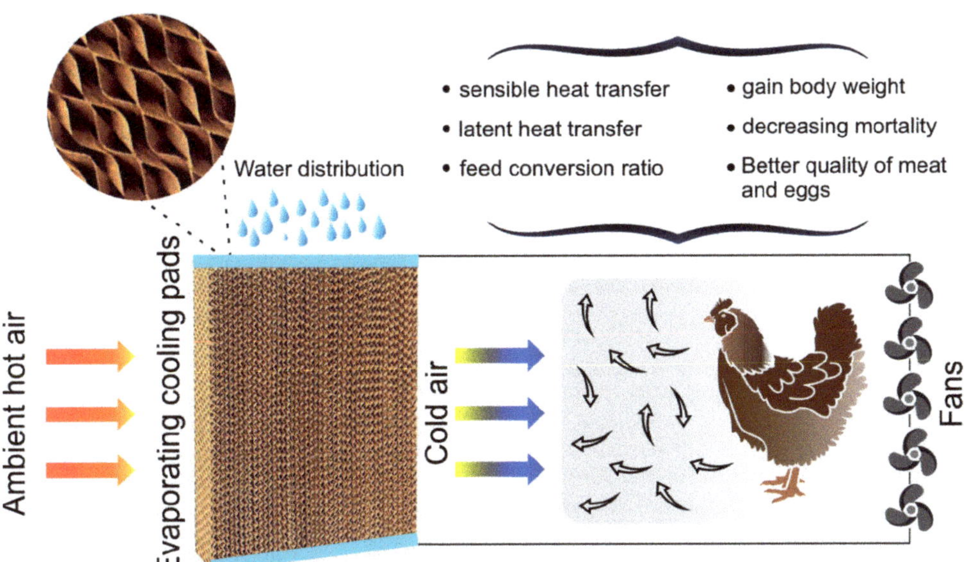

Figure 4. Schematic diagram of typical poultry air-conditioning for thermal comfort.

Air-conditioning happens to be an integrated process which transports the ambient air into the conditioned space with optimal parameters of thermal environment for the occupants [19,20,29,31]. In this way, it controls and maintains the temperature, relative humidity, and air movement, in the conditioned space within the predetermined limits either for thermal comfort or product processing. For poultry birds, thermal environment is of prime significance. The optimal mixture of temperature and relative humidity gives birth to temperature-humidity-index (THI). It is the thermal comfort that enhances health of occupants in any conditioned space [32–34]. DEC, IEC, and MEC systems are performing better as compared to vapor compression systems in terms of saving primary energy and providing desired environmental conditions. Poultry control sheds employ direct evaporative coolers for summer cooling. Evaporative cooling systems are developed and installed to meet air-conditioning requirement in poultry houses. The efficiency of these systems makes them cost effective and acceptable to user end. Poultry control sheds are chambers which are conditioned on the principle of direct evaporative cooling system. Heat production by poultry birds is ejected out of the system with reduction in temperature.

The objectives of this study include the understanding of poultry air-conditioning requirements for poultry birds based on heat and moisture production, calculating heat stress per bird and THI index to assess the desired evaporative cooling systems under the climatic conditions of Multan, understanding the effects of sensible and as well as the latent heat production in poultry birds and evaluation of evaporative cooling systems i.e., direct evaporative cooling (DEC), indirect evaporative cooling (IEC), Maisotsenko-Cycle evaporative cooling (MEC) for the poultry environment in terms of THI and THVI.

2. Evaporative Cooling Systems

Many types of cooling systems are available to provide cool air for commercial or domestic purpose. Since the ancient times, evaporative cooling systems (EC) are used for cooling the ambient air by evaporating water droplets into the air. Evaporation of water is a process in which heat of the ambient air is absorbed and water vapors are imparted to it. In this process, only the latent load is achieved by providing humidity into the ambient air. Whereas the total heat (enthalpy) gets negligible change. Evaporative cooling generally lies on the conversion of sensible heat into the latent one. The main work is accomplished by the water in EC systems. The heat and mass transfer in EC systems occurs on account of temperature and vapor pressure deficits. On the other hand, vapor compression air-conditioning (VCAC) systems employs CFCs or HCFCs which are environmentally harmful being the major exploiters of ozone layer depletion [10,31]. In this regard, increasing research efforts have been made in designing low cost and environmental friendly technologies; specifically EC techniques have been demonstrated effectively [35–37]. Energy consumption is also lower in EC systems as compared to the VCAC [38]. When ambient air is passed through any water steam directly or indirectly, it gets cooled with the effect of water evaporation into the air [20,31,39]. EC system is generally employed in hot and relatively dry climates [7]. On the contrary, humid environment is not suitable for evaporating cooling systems as air is already saturated. The evaporating cooling system can be categorized into DEC and IEC with respect to the interaction of water with air. A new system called M-cycle evaporative cooling (MEC) has also been introduced to get a cool fresh air. Figure 5 shows the laboratory-scale models for DEC, IEC, and MEC systems, respectively. These systems were developed in Agricultural Engineering Department Bahauddin Zakariya University Multan, Pakistan. The experimental setup uses 6.5 L/min water pump for all three developed systems and a standard anemometer for air flow rate (i.e., average 1.7m/s). Standard temperature and moisture sensor (H2) with an experimental uncertainty of ± 1 °C temperature and $\pm 2\%$ RH was used in the experimental setup. Experimental data were collected for a time span of a typical meteorological year and thermodynamic analyses were carried out for poultry air-conditioning. Under the climatic conditions of Multan city, these systems can be employed to achieve certain results which were further optimized to obtain THI and THVI values. Figure 6 illustrates the schematics of typical evaporative cooling with DEC, IEC, and MEC. EC system is an environment friendly and energy saving technology [40]. In terms of thermal comfort, this system can be a suitable option in hot and arid climates as the relative humidity lies in between the 60 and 70%. This system altogether meets the thermal comfort needs of the occupant, being environmentally friendly [41]. The effectiveness of EC system is indicated by wet-bulb and dew-point effectiveness [31]. In this system, ambient air is directly brought in contact with water stream to lower down the temperature and increase relative humidity [19,20,29,31,42].

2.1. Direct Evaporative Cooling (DEC)

It is the easiest and oldest type of EC system in which ambient air is brought in direct contact with air stream to reduce the temperature [38]. The continuous evaporation of water vapors (adiabatic process) causes a cooling effect up to a saturation point in which enthalpy of air remains same, whereas the humidity ratio increases throughout the process [31,44]. These water streams are brought from metal or plastic tubes commonly known as "pads" as their boundary walls. The ambient air is showered with water and thus, it gets cooled and humidified. The water stream is injected with the help of motor and from the top of the wall water droplets are drawn downward with the gravity force and capillary action. The ambient temperature is potentially reduced to its wet bulb temperature at wet-bulb effectiveness of 75–95% [31,45]. Figure 6 shows the working principle of DEC system. In dry climates, DEC system works with 80% efficiency as reported in [38].

2.2. Indirect Evaporative Cooling (IEC)

It is a system in which heat and mass transfer phenomenon takes place without the addition of moisture and works on the principle of sensible cooling [44]. In this system, cooling effect is produced by isenthalpic cooling in the wet channel and sensible heat transfer in the dry channel [44]. In IEC systems, product air passes over the dry side while the working air passes over the wet side. In case of DEC, the conditioned air is obtained but with an increased relative humidity level [46]. To achieve the constant absolute humidity, the IEC systems are desired. Figure 6 shows the working principle of an IEC system. IEC system can reduce the temperature up to the wet-bulb temperature at wet-bulb effectiveness of 50–65% [47]. The humidity ratio of the inlet and outlet air remains same while the enthalpy of the outlet air decreases in an IEC system [44].

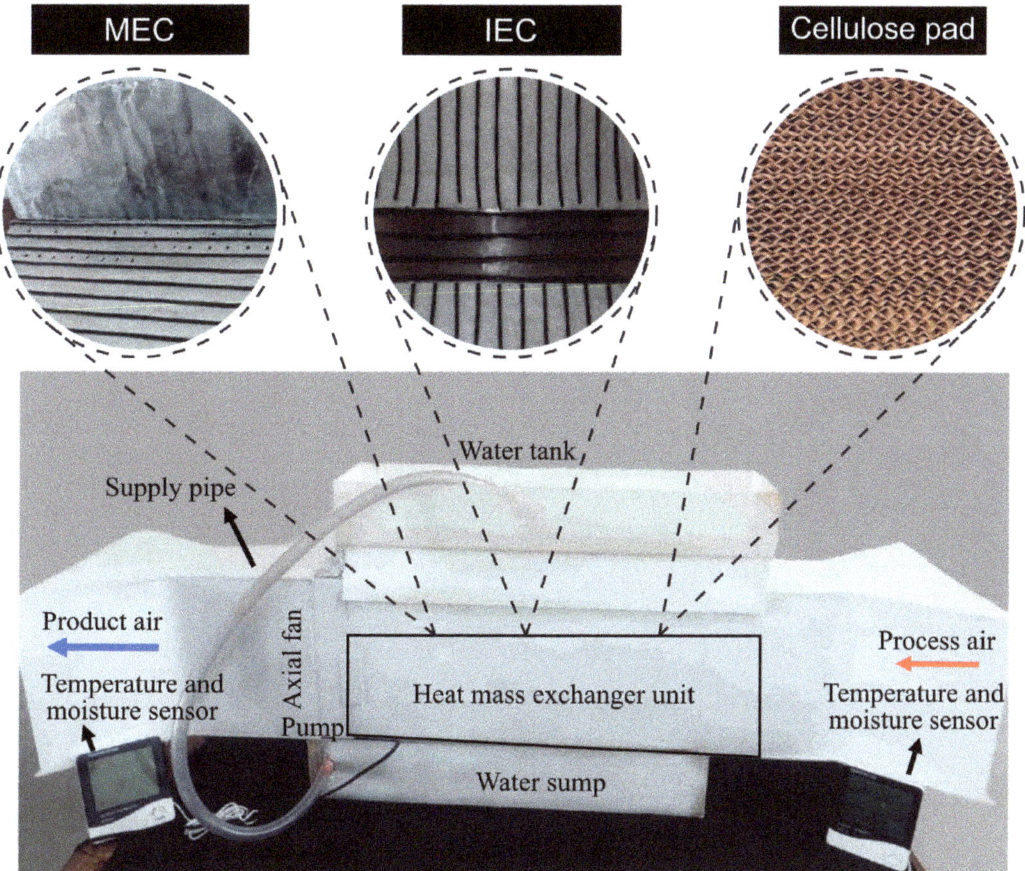

Figure 5. Laboratory-scale models for direct evaporative cooling (DEC), indirect evaporative cooling (IEC), and Maisotsenko-Cycle evaporative cooling (MEC) systems, reproduced from author's work [43].

2.3. M-Cycle Evaporative Cooling (MEC)

Maisotsenko-Cycle (M-Cycle) evaporative cooling technique is an advanced method for achieving the dew point temperature as compared to the other two traditional systems where cooling limit touches the wet bulb temperature [44,48]. In this system, the cooling effect is produced by evaporative cooling and heat transfer where dew-point temperature is achieved instead of wet-bulb temperature [49,50]. Figure 6 illustrates the working principle of MEC system. This system comprises of three channels in which wet channel is sandwiched in between the two dry channels. The ambient air cools down due to the convective heat transfer between the dry and wet channels when it passes through the dry channel [51]. In this system, the humidity ratio of the inlet and outlet air remains same while the enthalpy decreases [45,52]. The studied experimental systems have been developed at lab-scale and analyzed for poultry air-conditioning. However, when developed at large scale, factors like pressure drop, fan power, availability of fresh water for evaporative coolers, availability of the evaporative media, direction of the system installation in the poultry shed, and the energy consumption should be taken into consideration.

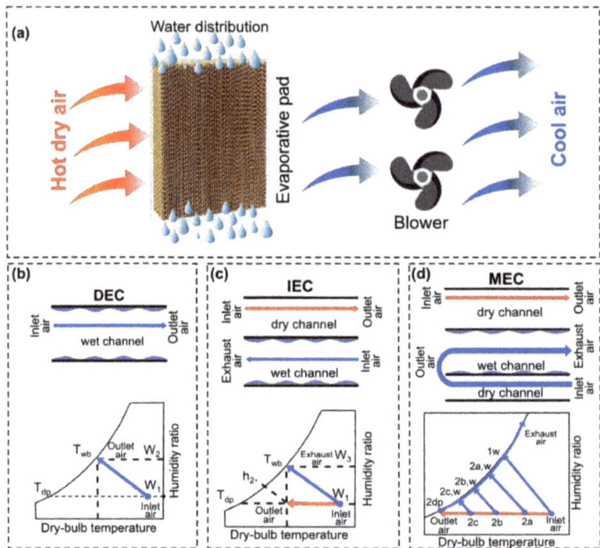

Figure 6. Illustration of evaporative cooling phenomenon. (**a**) Typical evaporative cooling phenomenon using evaporative pads. (**b–d**) DEC, IEC, and MEC systems phenomenon along with air transformations on psychrometric chart.

3. Materials and Methods

Mathematical Models for Poultry Heat Generation

Heat production in poultry birds changes with their body weight and muscle growth due to the consumption of more energy produced from the feed intake [21,22]. Heat production in poultry birds based on live weight is governed by Equation(4) given in literature [15].

$$Q = 60.65 + 0.04W \tag{1}$$

where, Q is the heat production (W/kg) and W is the live weight (kg). Sensible heat production (SHP), and latent heat production (LHP) based on broiler age at various temperatures were calculated using gates model Equations (5)–(12) given in literature [53]. The set of Equations (5)–(12) were replicated from [53] using the self-defined chicken age, and specific temperature ranges (presented below) to investigate the performance of the developed EC systems.

All brooding temperature relations encompass the broiler chickens of age below 15 days. Whereas the rest of relations (i.e., for brooding temperature 15.6 °C, 21.1 °C, and 26.7 °C) are concerned with heat and moisture production of broilers above 15 days and below 48 days.

For all brooding temperatures:

$$SHP = K\exp\left(-6.5194 + 2.9186x - 0.24162x^2\right) \tag{2}$$

$SE = 0.284$ K; $\quad 3 \leq x \leq 5$

$$LHP = K\left(-42.961 + 27.415x - 2.84344x^2\right) \tag{3}$$

$SE = 0.296$ K; $\quad 2 \leq x \leq 5$

For temperature (t = 15.6 °C):

$$SHP = K\left(38.612 - 2.6224x - 0.072047x^2 - 0.00066x^3\right) \tag{4}$$

$$SE = 0.045; \quad 20 \le x \le 41$$

$$LHP = K\left(22.285 - 0.78279x + 0.011503x^2 - 0.000038x^3\right) \tag{5}$$

$$SE = 0.192\ K; \quad 20 \le x \le 43$$

For temperature (t = 21.1 °C):

$$SHP = K\left(36,070 - 2.1307x - 0.058862x^2 - 0.00051x^3\right) \tag{6}$$

$$SE = 0.110\ K; \quad 20 \le x \le 39$$

$$LHP = K\left(11.221 + 0.40495x - 0.02727x^2 - 0.000353x^3\right) \tag{7}$$

$$SE = 0.069\ K; \quad 20 \le x \le 43$$

For temperature (t = 26.7 °C):

$$SHP = K\exp(5.3611 - 0.16177x) \tag{8}$$

$$SE = 0.052 \quad 20 \le x \le 23$$

$$LHP = K\left(20.094 - 0.70318x + 0.015182x^2 - 0.000108x^3\right) \tag{9}$$

$$SE = 0.022\ K; \quad 20 \le x \le 42$$

Where SHP, LHP represent the specific sensible, and latent heat production (W/kg), x represents bird age (days) and SE denotes the standard error of regression.

Temperature-humidity index (THI) is a direct combination of DBT and WBT. The further insights of THI for broilers and layers is given below in Equations (10) and (11) given in literature [54,55].

$$THI_{broilers} = 0.85t_{db} + 0.15t_{wb} \tag{10}$$

$$THI_{layers} = 0.60t_{db} + 0.40t_{wb} \tag{11}$$

where, t_{db}, t_{wb} represent dry-bulb and wet-bulb temperatures (°C) respectively and THI represents temperature-humidity index.

Temperature-humidity-velocity index (THVI) is used to analyze the ability to maintain an internal condition constant by including velocity as one of the factors. The further insights of THVI and ET can be seen in Equations (12)–(15) with normal, alert, danger, and emergency regions of homeostasis for the broilers given in the literature [56].

$$THVI = THI \times V^{-0.058} \quad (0.2 \le V \le 1.2) \tag{12}$$

For 1 °C temperature rise

$$ET = \left(2 \times 10^{29}\right) \times THVI^{-17.68} \tag{13}$$

For 2.5 °C temperature rise

$$ET = \left(4 \times 10^{13}\right) \times THVI^{-7.38} \tag{14}$$

For 4 °C temperature rise

$$ET = \left(3 \times 10^{11}\right) \times THVI^{-5.91} \tag{15}$$

where, THVI represents temperature-humidity-velocity index, and ET stands for exposure time in minutes. Wet-bulb temperature is calculated by Equation (16), as given in literature [57,58].

$$T_{wb} = T_{db} \tan^{-1}\left[0.151977 + (RH + 8.313659)^{\frac{1}{2}}\right] + \tan^{-1}(T_{db} + RH)$$
$$- \tan^{-1}(RH - 1.676331) \tag{16}$$
$$+ 0.00391838\, RH^{\frac{3}{2}} \tan^{-1}(0.023101RH) - 4.686035$$

4. Results and Discussion

Climate control strategy starts with the estimation of ambient weather details for a region. Such kind of analysis makes it visible that what kind of changes occur in the ratio of DBT and RH in a day. Evaporative cooling systems (DEC, IEC, and MEC) were evaluated in the laboratory under summer conditions in Multan. These systems appreciably reduced the ambient temperature and increased relative humidity to meet the threshold THI limit for poultry birds. Figures 7 and 8 show the experimental analysis of DEC, IEC, and MEC systems for the climatic conditions of Multan. Table 1 shows the summary of performance profile of the experimental DEC, IEC, and MEC under the climatic conditions of Multan (Pakistan) for poultry air-conditioning.

Table 1. Summary of performance profile of the experimental DEC, IEC, and MEC under the climatic conditions of Multan (Pakistan) for poultry air-conditioning.

Month	Ambient Temperature	Ambient RH	DEC			IEC			MEC		
			ΔT	ΔRH	ΔTHI	ΔT	ΔRH	ΔTHI	ΔT	ΔRH	ΔTHI
	(°C)	(%)	(°C)	%	(°C)	(°C)	%	(°C)	(°C)	%	(°C)
Apr	36.5	39.5	10	48.5	9	6.5	23	6.5	12	37.5	10
May	37.5	36	10	49	8.5	6.5	20.5	5	11.5	42	9.5
Jun	38	44	7	48	8.5	5	19.5	7	9.5	39	10
Jul	37.5	66	5.5	21	3.5	4.5	12.5	3	7	30.5	5.5
Aug	38.5	62	7	26	4.5	4.5	15.5	3	7.5	31.5	5.5
Sep	35	55	6.5	36	5	3.5	16.5	3.5	7	34.5	6

Δ denotes the gradient/difference in ambient the supply air-conditions.

Broiler heat production increases as its weight increases. It is directly proportional to the physiological growth of birds being grown up under healthy conditions. Broilers need optimal environmental conditions to thrive in tropical regions. In these dry and humid regions, broilers need evaporative cooling effect in hot conditions to maintain their body heat and moisture loss. These birds are sensitive to the slight change in temperature and humidity values and become accustomed to high mortality rates. In Figure 9, it is mentioned through graphical representation that with the increase in broiler age, their body weight keeps on increasing. The way this body weight increases, the need for encountering heat and moisture production arises.

For this purpose, an optimal air-conditioning and ventilation technique needs to be devised. The heat from the broiler started increasing from 60 till 155 W/m² with the corresponding increase in the weight up to 2700 g. Sensible and latent heat production of broilers is studied with respect to its weight under different temperatures according to Gates model. A set of regression equations were developed to study the heat production patterns in broilers. This situation is graphically presented in Figures 10–13, to understand the effects of high temperature and difference in heat production on the overall welfare of poultry birds. These models made it clear that broilers with growing age attain physiological maturity and meanwhile, their sensible and latent heat production is dependent on ambient temperature. These graphs also illustrate that total heat production with varying temperatures increase initially and comes to rest. On the other hand, sensible and latent heat production makes narrower gap at lower temperatures while this gap gets wider at higher temperatures.

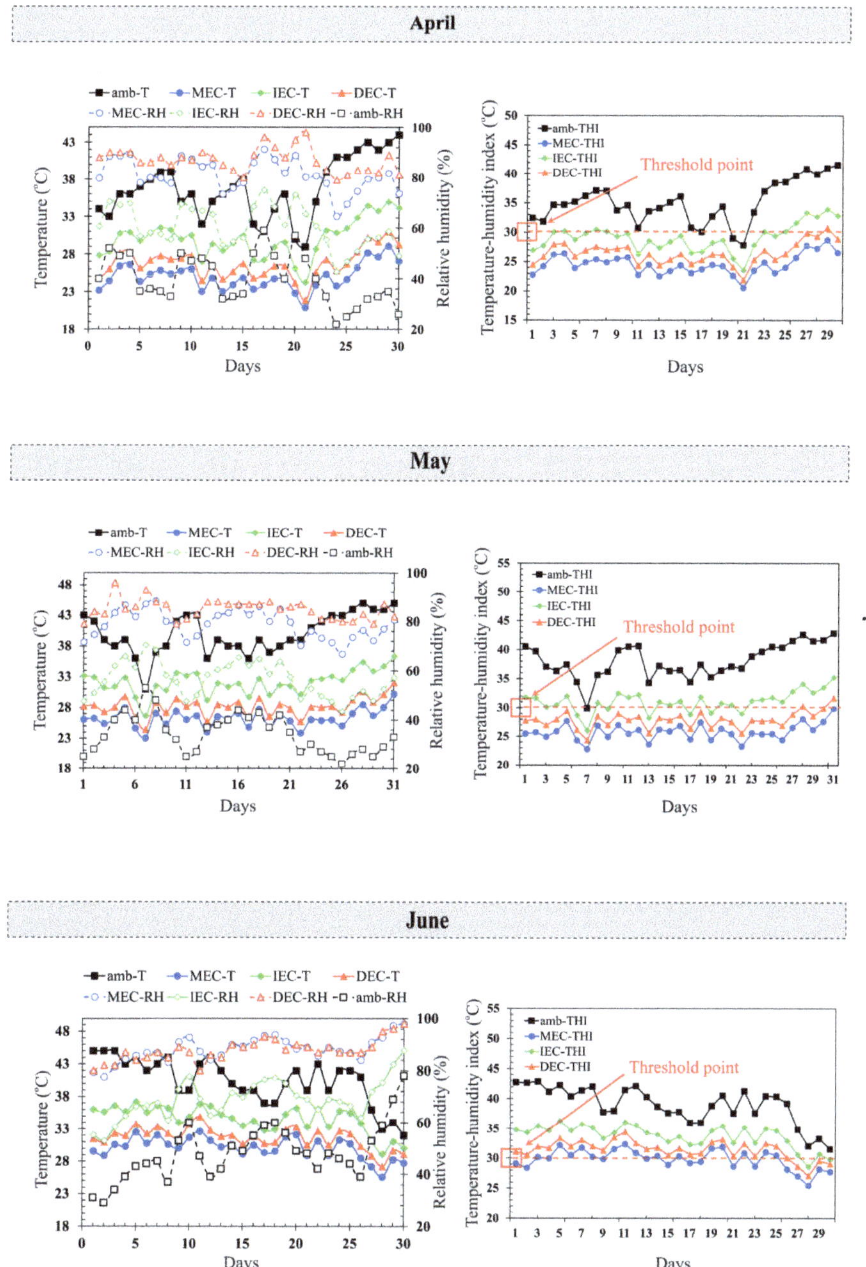

Figure 7. Experimental analysis of DEC, IEC, and MEC systems in the months of April, May, and June for ambient conditions of Multan, Pakistan.

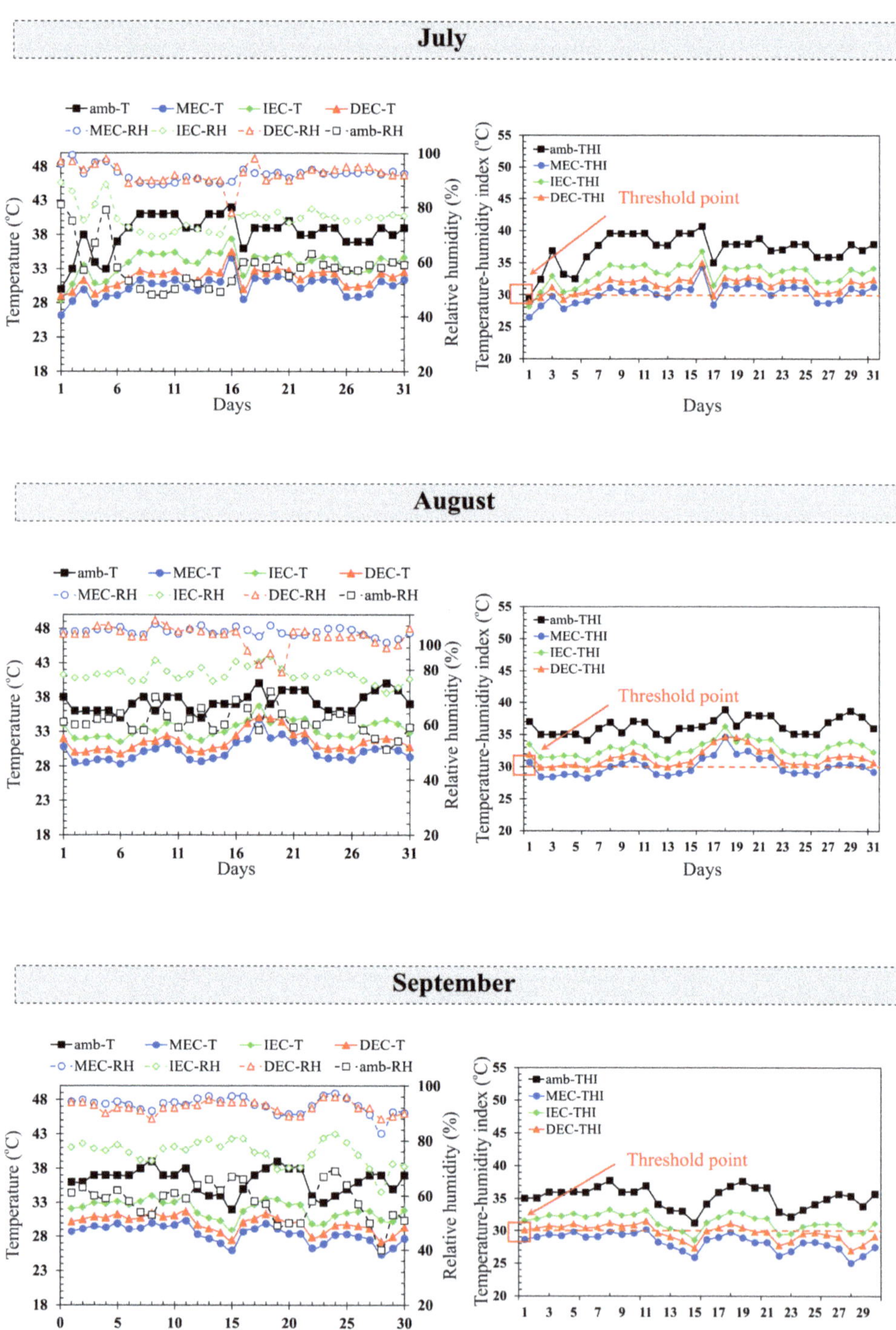

Figure 8. Experimental analysis of DEC, IEC, and MEC systems in the months of July, August, and September for ambient conditions of Multan, Pakistan.

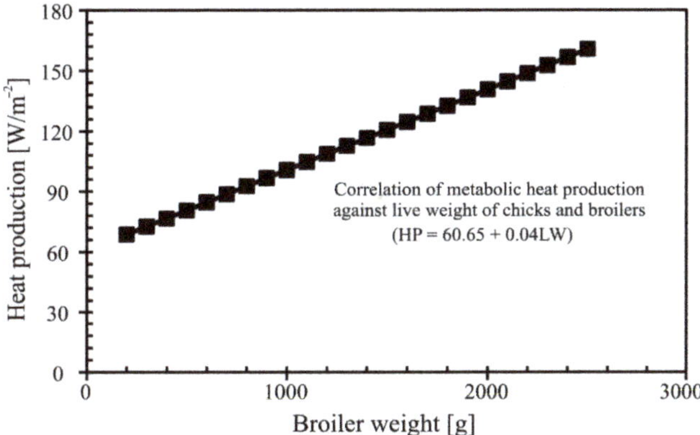

Figure 9. Relationship of broiler weight and heat production of chickens and broilers.

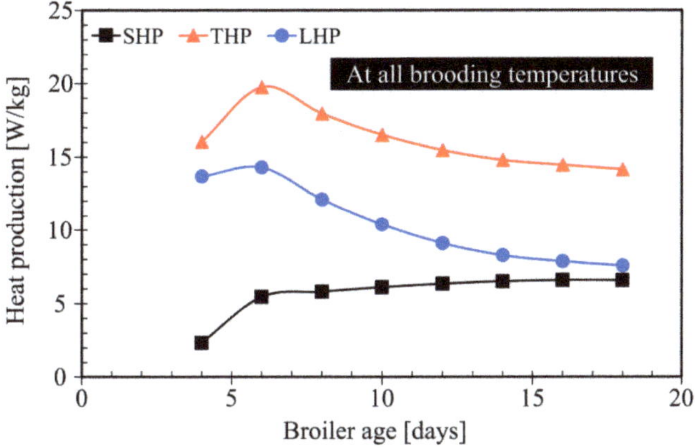

Figure 10. Representation of Broiler age [days] as the function of sensible, latent, and total heat production at all brooding temperatures.

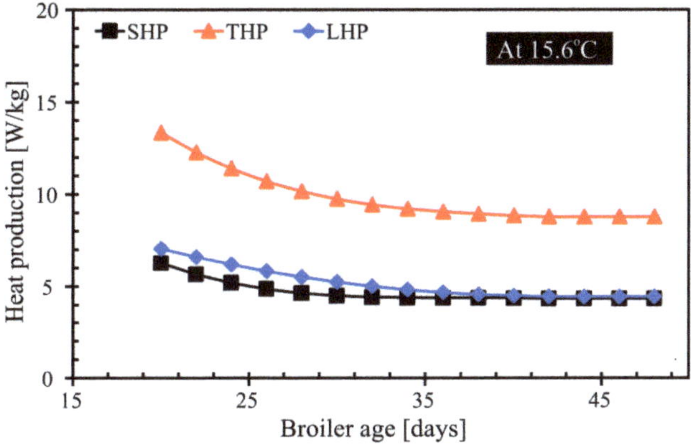

Figure 11. Representation of Broiler age [days] as the function of sensible, latent, and total heat production at 15.6 °C temperature.

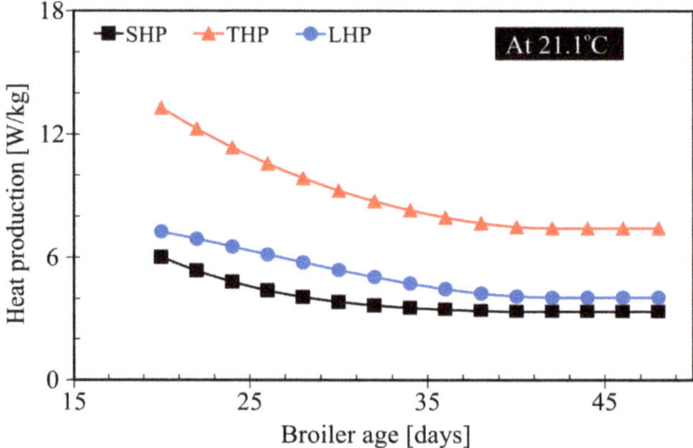

Figure 12. Representation of Broiler age [days] as the function of sensible, latent, and total heat production at 21.1 °C temperature.

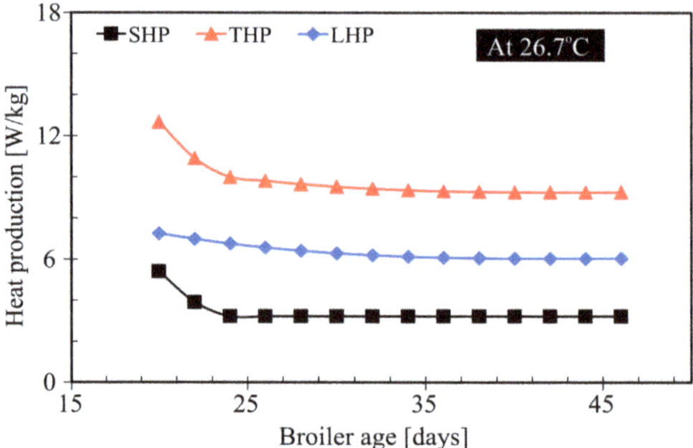

Figure 13. Representation of Broiler age [days] as the function of sensible, latent, and total heat production at 26.7 °C temperature.

This can be controlled with the help of management guide designed for broilers where THI values can resolve this deficiency. The difference between sensible and latent heat production is seriously important to get a know how about the assimilative capacity inside the poultry environment. The wider the gap is, poorer is the resilience of the environment surrounding the poultry birds.

The sustenance capacity of poultry birds to survive heat stress increases with the increasing velocities as depicted in different layouts of THVI with thermal exposure time (ET) in Figure 14. Figure 14 was reproduced from published literature using the tabular data in a research conducted by Tao et al. [56]. These figures categorize exposure time of broiler chickens with THVI and state that it is the air movement that tells the story other way round if not checked properly.

In these figures, it is shown that increasing velocity to some extent makes it easier to achieve thermal comfort zone for poultry birds. THI is the summation of different percentage compositions of dry- and wet-bulb temperatures. Dry-bulb temperature is measured by simple thermometer while wet-bulb temperature is measured with soaked wet cloth wrapped over the measuring segment of the thermometer. In Figure 15a,b, it is stated that there is correlation between T_{ab} and THI to describe the increasing trend from the daily data of weather for broilers from daily weather conditions to calculate THI which is a function of T_{db} and RH.

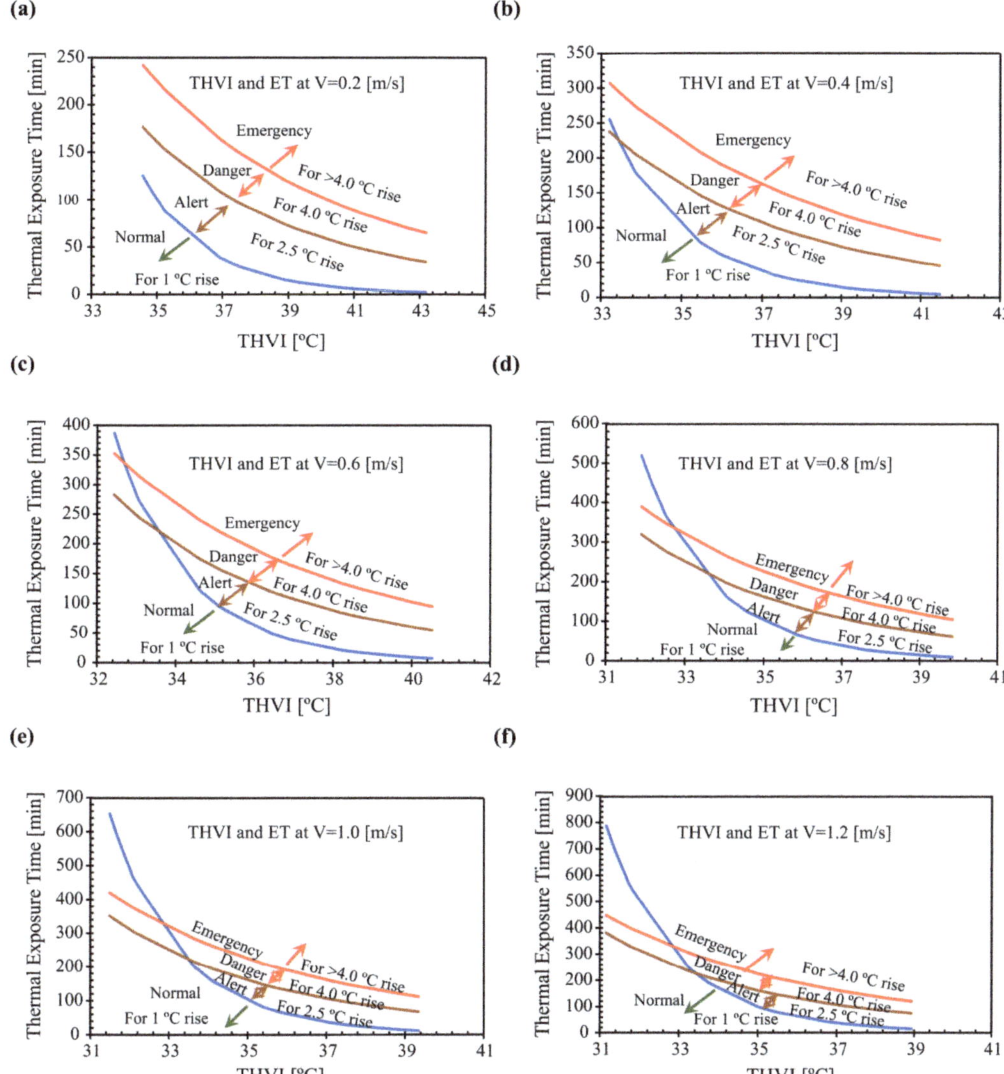

Figure 14. Thermal comfort zone for poultry birds with respect to THVI [°C] and ET [min] at (**a**) 0.2 m/s, (**b**) 0.4 m/s, (**c**) 0.6 m/s, (**d**) 0.8 m/s, (**e**) 1.0 m/s, (**f**) 1.2 m/s velocities, respectively, reproduced from tabular data published by [56].

From Figure 15, an empirical equation is obtained with appreciable R^2 value. Figure 15 concludes that the range of THI variation is less in broiler chickens (Figure 15a) as compared to relatively higher range of THI variation in egg-laying chickens (Figure 15b). This also explains the assimilative capacity of natural environment to resist heat stress to some extent in both cases of broilers and layers, respectively. Figure 16a,b proposes a THI pyramid (i.e., amalgamation of THI) based on daily wet-bulb temperature range. In Figure 16, the green, blue, and red color lines overlaid on top of the regression lines represent boundaries of different zones based on allowed THI of both broilers and egg-laying chickens. In Figure 16, chart area covered underneath the green line represents the threshold zone, chart area covered between green and blue line represents the alert zone, area covered between blue and red line represents the danger zone, and area covered above the red line shows the emergency zone based on the allowed/comfortable THI for both cases. The resilience of layers chicken is more in pyramid (i.e., amalgamation of THI as shown in Figure 16) as compared to broiler chicken with enlarged elliptical trend (in Figure 16). Figure 17a,b explains dry bulb representation with the RH and THI relationship for broilers and egg-laying chickens and states that this trend for broiler chickens was found to be more tolerant to the thermal stress (i.e., heat stress due to ambient air conditions) effectively as compared to layer chickens, which justifies the published literature against the thermal stresses of broilers and layers as shown in Equations (13)–(14). According to Figure 17a, the broiler chicken has higher range of temperature-humidity-index at a specified relative humidity indicating relatively more thermal resilience as compared to layer chicken (Figure 17b) at same relative humidity conditions. Feasibility calendar of EC (DEC, IEC, and MEC) systems for the ambient conditions of Multan (Pakistan) is presented in Figure 18. The ambient conditions of the study area were recorded for a year using standard temperature sensor and were later analyzed for poultry air-conditioning.

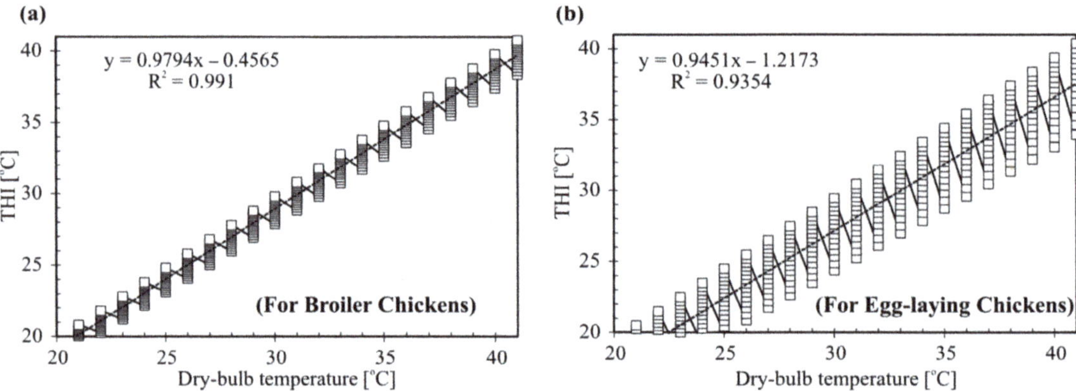

Figure 15. Representation of dry-bulb temperature and THI as a correlating factor for (**a**) broiler chickens, and (**b**) egg-laying chickens.

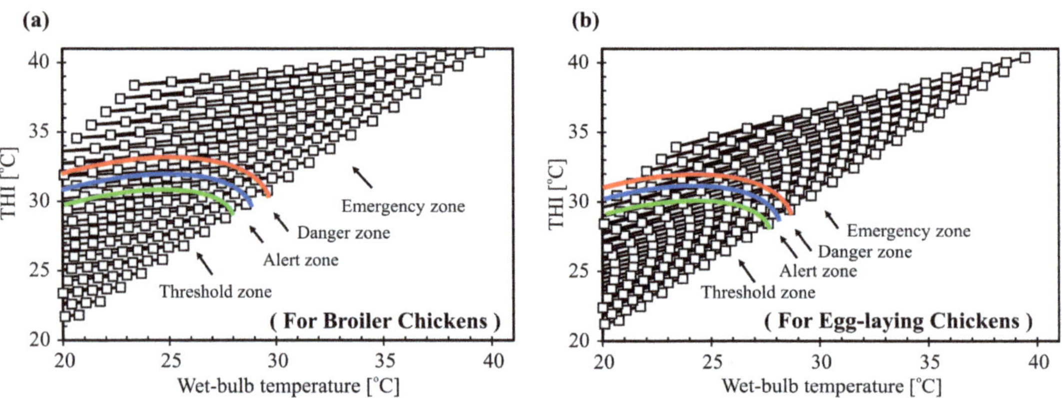

Figure 16. Representation of wet-bulb temperature and THI as correlating factor for (**a**) broiler chickens, and (**b**) egg-laying chickens.

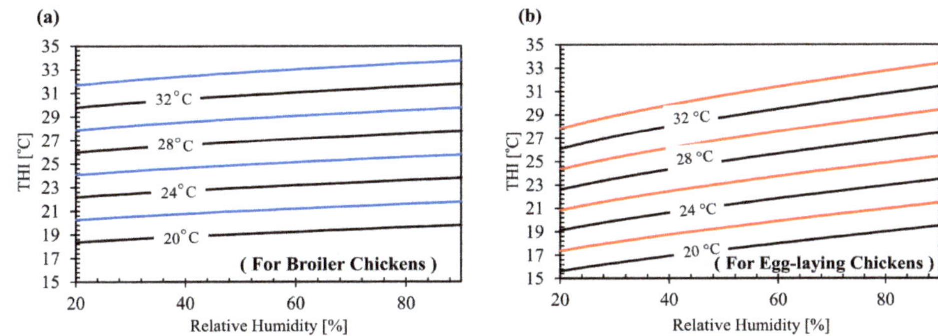

Figure 17. Representation of dry-bulb temperature as a function of relative humidity and THI for (**a**) broiler chickens, and (**b**) egg-laying chickens.

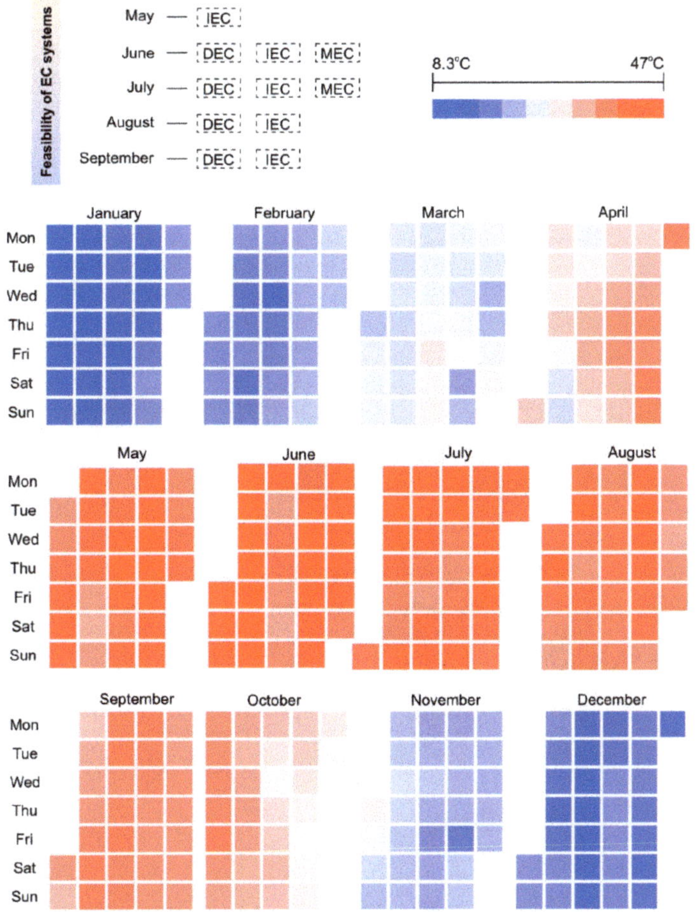

Figure 18. Feasible calendar showing the applicability of evaporative cooling systems for the ambient conditions of Multan, Pakistan.

5. Conclusions

Poultry industry is affected by (sensible/latent) heat stresses and results in substantial economic loss. Heat stress causes severe impacts on poultry health such as mortality rate and body weight increases. Economically, these birds are the cheapest source of proteins in South Asia. Furthermore, poultry farming is catching its momentum with millions of capital investment by many in a quest to gain much more in a short span of time. In view of presented work, the air-conditioning process for broiler chickens carries significant importance. The major issue revolves around is the optimal control of temperature and humidity such as THI. Any minor fluctuations of index wreak havoc for the investor and birds ultimately. For all this there was an open window for the current research to oversee an energy-efficient evaporative cooling system that could condition the air without raising humidity levels. In this regard, the EC systems (DEC, IEC, and MEC) were studied under the weather conditions of Multan in line with the regression equations from Gates model. In fact, the empirical relations of sensible and latent heat production minimized the need to erect a whole new setup to raise poultry birds for studying the heat and moisture production. Moreover, the experimental results of the studied EC systems conclude that the MEC system could be considered as a viable alternate option as compared to the traditional DEC systems used for poultry air-conditioning due to the psychrometric and climatic (i.e., monsoon season) limitations of the DEC system. However, all the studied standalone evaporative cooling systems are still limited by the ambient air conditions. This problem can be resolved by further research on experimental desiccant dehumidification-based evaporative cooling systems for poultry air-conditioning. Therefore, the present study concludes the MEC as the best alternate option to the traditional DEC system used for poultry air-conditioning in Multan, Pakistan.

Author Contributions: Conceptualization, K.S., M.B., and M.S.; data curation, K.S. and M.B.; formal analysis, K.S. and M.B.; funding acquisition, M.S.; investigation, K.S., M.B., H.A., M.F., T.M., U.S., I.A., and M.I.H.; methodology, K.S, M.S., M.B., M.F., and T.M.; project administration, M.S.; resources, M.S. and T.M.; supervision, M.S. and T.M.; validation, M.S. and T.M.; visualization, M.B., H.A., U.S., I.A., and M.I.H.; writing—original draft, K.S.; writing—review and editing, M.S., M.B., H.A., M.F., T.M., U.S., I.A., and M.I.H. All authors have read and agreed to the published version of the manuscript.

Funding: The Bahauddin Zakariya University, Multan-Pakistan funded this research under the Director Research/ORIC grant titled "Development and performance evaluation of prototypes of direct and indirect evaporative cooling-based air-conditioning systems.

Institutional Review Board Statement: Not applicable.

Informed Consent Statement: Not applicable.

Data Availability Statement: Data is contained within the article.

Acknowledgments: All this work is part of the Ph.D. research of Khawar Shahzad (1st Author). This research work has been carried out in the Department of Agricultural Engineering, Bahauddin Zakariya University, Multan-Pakistan. The Bahauddin Zakariya University, Multan-Pakistan funded this research under the Director Research/ORIC grant titled "Development and performance evaluation of prototypes of direct and indirect evaporative cooling-based air-conditioning systems," awarded to Principal Investigator Muhammad Sultan.

Conflicts of Interest: The authors declare no conflict of interest.

Abbreviations

EC	evaporative cooling
GDP	gross domestic product
DBT	dry-bulb temperature
RH	relative humidity
THI	temperature-humidity index

THVI	temperature-humidity-velocity index
DEC	direct evaporative cooling
IEC	indirect evaporative cooling
MEC	M-cycle evaporative cooling
Q	heat produced (Q/kg)
W	Live weight (kg)
SHP	sensible heat production (W/kg)
LHP	latent heat production (W/kg)
x	bird's age (days)
SE	standard error of regression.
ET	thermal exposure time

References

1. Rehman, A.; Jingdong, L.; Du, Y. Last five years Pakistan economic growth rate (GDP) and its comparison with China, India and Bangladesh. *Inter. J. Technol. Enhanc. Emerg. Eng. Res.* **2015**, *4*, 81–86.
2. Rehman, A.; Jingdong, L.; Chandio, A.A.; Hussain, I. Livestock production and population census in Pakistan: Determining their relationship with agricultural GDP using econometric analysis. *Inf. Process. Agric.* **2017**, *4*, 168–177. [CrossRef]
3. FBS Pakistan Statistical Year Book. Available online: http://www.pbs.gov.pk/content/pakistan-statistical-year-book-2017 (accessed on 12 January 2021).
4. Marangoni, F.; Corsello, G.; Cricelli, C.; Ferrara, N.; Ghiselli, A.; Lucchin, L.; Poli, A. Role of poultry meat in a balanced diet aimed at maintaining health and wellbeing: An Italian consensus document. *Food Nutr. Res.* **2015**, *59*, 27606. [CrossRef]
5. Pedrazzi, S.; Allesina, G.; Muscio, A. Indirect evaporative cooling by sub-roof forced ventilation to counter extreme heat events. *Energy Build.* **2020**, *229*, 110491. [CrossRef]
6. Perkins, S.E. A review on the scientific understanding of heatwaves—Their measurement, driving mechanisms, and changes at the global scale. *Atmos. Res.* **2015**, *164*, 242–267. [CrossRef]
7. Poku, R.; Oyinki, T.W.; Ogbonnaya, E.A. The effects of evaporative cooling in tropical climate. *Am. J. Mech. Eng.* **2017**, *5*, 145–150. [CrossRef]
8. Yahav, S. Regulation of body temperature: Strategies and mechanisms. In *Sturkie's Avian Physiology*; Elsevier: Amsterdam, The Netherlands, 2015; pp. 869–905.
9. Van Kampen, M.; Mitchell, B.W.; Siegel, H.S. Thermoneutral zone of chickens as determined by measuring heat production, respiration rate, and electromyographic and electroencephalographic activity in light and dark environments and changing ambient temperatures. *J. Agric. Sci.* **1979**, *92*, 219–226. [CrossRef]
10. Sultan, M.; El-Sharkawy, I.I.; Miyazaki, T.; Saha, B.B.; Koyama, S. An overview of solid desiccant dehumidification and air conditioning systems. *Renew. Sustain. Energy Rev.* **2015**, *46*, 16–29. [CrossRef]
11. Xin, H.; Berry, I.L.; Tabler, G.T.; Barton, T.L. Temperature and humidity profiles of broiler houses with experimental conventional and tunnel ventilation systems. *Appl. Eng. Agric.* **1994**, *10*, 535–542. [CrossRef]
12. Donkoh, A. Ambient temperature: A factor affecting performance and physiological response of broiler chickens. *Int. J. Biometeorol.* **1989**, *33*, 259–265. [CrossRef]
13. Habeeb, A.A.; Gad, A.E.; Atta, M.A. Temperature-Humidity Indices as Indicators to Heat Stress of Climatic Conditions with Relation to Production and Reproduction of Farm Animals. *Int. J. Biotechnol. Recent Adv.* **2018**, *1*, 35–50. [CrossRef]
14. Youssef, A.; Exadaktylos, V.; Berckmans, D.A. Towards real-time control of chicken activity in a ventilated chamber. *Biosyst. Eng.* **2015**, *135*, 31–43. [CrossRef]
15. Lara, L.J.; Rostagno, M.H. Impact of heat stress on poultry production. *Animals* **2013**, *3*, 356–369. [CrossRef]
16. Estévez, M. Oxidative damage to poultry: From farm to fork. *Poult. Sci.* **2015**, *94*, 1368–1378. [CrossRef]
17. Surai, P.F.; Kochish, I.I.; Fisinin, V.I.; Kidd, M.T. Antioxidant defence systems and oxidative stress in poultry biology: An update. *Antioxidants* **2019**, *8*, 235. [CrossRef]
18. Wasti, S.; Sah, N.; Mishra, B. Impact of heat stress on poultry health and performances, and potential mitigation strategies. *Animals* **2020**, *10*, 1266. [CrossRef] [PubMed]
19. Kashif, M.; Niaz, H.; Sultan, M.; Miyazaki, T.; Feng, Y.; Usman, M.; Shahzad, M.W.; Niaz, Y.; Waqas, M.M.; Ali, I. Study on desiccant and evaporative cooling systems for livestock thermal comfort: Theory and experiments. *Energies* **2020**, *13*, 2675. [CrossRef]
20. Sultan, M.; Miyazaki, T.; Mahmood, M.H.; Khan, Z.M. Solar assisted evaporative cooling based passive air-conditioning system for agricultural and livestock applications. *J. Eng. Sci. Technol.* **2018**, *13*, 693–703.
21. Damerow, G. *The Chicken Health Handbook: A Complete Guide to Maximizing Flock Health and Dealing with Disease*; Storey Publishing: North Adams, MA, USA, 2016; ISBN 1603428585.
22. Saeed, M.; Abbas, G.; Alagawany, M.; Kamboh, A.A.; Abd El-Hack, M.E.; Khafaga, A.F.; Chao, S. Heat stress management in poultry farms: A comprehensive overview. *J. Therm. Biol.* **2019**, *84*, 414–425. [CrossRef]
23. Yanagi, T.; Xin, H.; Gates, R.S. A research facility for studying poultry responses to heat stress and its relief. *Appl. Eng. Agric.* **2002**, *18*, 255. [CrossRef]

24. Qian, X.; Yang, Y.; Lee, S.W. Design and Evaluation of the Lab-Scale Shell and Tube Heat Exchanger (STHE) for Poultry Litter to Energy Production. *Processes* **2020**, *8*, 500. [CrossRef]

25. Cui, Y.; Theo, E.; Gurler, T.; Su, Y.; Saffa, R. A comprehensive review on renewable and sustainable heating systems for poultry farming. *Int. J. Low-Carbon Technol.* **2020**, *15*, 121–142. [CrossRef]

26. Yi, B.; Chen, L.; Sa, R.; Zhong, R.; Xing, H.; Zhang, H. High concentrations of atmospheric ammonia induce alterations of gene expression in the breast muscle of broilers (*Gallus gallus*) based on RNA-Seq. *BMC Genomics* **2016**, *17*, 1–11. [CrossRef]

27. Mashaly, M.M.; Hendricks, G.L., 3rd; Kalama, M.A.; Gehad, A.E.; Abbas, A.O.; Patterson, P.H. Effect of heat stress on production parameters and immune responses of commercial laying hens. *Poult. Sci.* **2004**, *83*, 889–894. [CrossRef]

28. Petek, M.; Dikmen, S.; Oğan, M.M. Performance analysis of a two stage pad cooling system in broiler houses. *Turkish J. Vet. Anim. Sci.* **2012**, *36*, 21–26. [CrossRef]

29. Raza, H.M.U.; Ashraf, H.; Shahzad, K.; Sultan, M.; Miyazaki, T.; Usman, M.; Shamshiri, R.R.; Zhou, Y.; Ahmad, R. Investigating Applicability of Evaporative Cooling Systems for Thermal Comfort of Poultry Birds in Pakistan. *Appl. Sci.* **2020**, *10*, 4445. [CrossRef]

30. Noor, S.; Ashraf, H.; Sultan, M.; Khan, Z.M. Evaporative Cooling Options for Building Air-Conditioning: A Comprehensive Study for Climatic Conditions of Multan (Pakistan). *Energies* **2020**, *13*, 3061. [CrossRef]

31. Sultan, M.; Miyazaki, T. Energy-Efficient Air-Conditioning Systems for Nonhuman Applications. In *Refrigeration*; Ekren, O., Ed.; InTech: London, UK, 2017; pp. 97–117.

32. dos Santos, T.C.; Gates, R.S.; Tinôco, I.; de, F.F.; Zolnier, S.; da Baêta, F.C. Behavior of Japanese quail in different air velocities and air temperatures. *Pesqui. Agropecuária Bras.* **2017**, *52*, 344–354. [CrossRef]

33. Bustamante, E.; García-Diego, F.-J.; Calvet, S.; Estellés, F.; Beltrán, P.; Hospitaler, A.; Torres, A.G. Exploring ventilation efficiency in poultry buildings: The validation of computational fluid dynamics (CFD) in a cross-mechanically ventilated broiler farm. *Energies* **2013**, *6*, 2605–2623. [CrossRef]

34. DeShazer, J.A.; Hahn, G.L.; Xin, H. Basic principles of the thermal environment and livestock energetics. In *Livestock Energetics and Thermal Environment Management*; American Society of Agricultural and Biological Engineers: St. Joseph, MI, USA, 2009; pp. 1–22. ISBN 1892769743.

35. Chen, Q.; Pan, N.; Guo, Z.-Y. A new approach to analysis and optimization of evaporative cooling system II: Applications. *Energy* **2011**, *36*, 2890–2898. [CrossRef]

36. Malli, A.; Seyf, H.R.; Layeghi, M.; Sharifian, S.; Behravesh, H. Investigating the performance of cellulosic evaporative cooling pads. *Energy Convers. Manag.* **2011**, *52*, 2598–2603. [CrossRef]

37. De Angelis, A.; Saro, O.; Truant, M. Evaporative cooling systems to improve internal comfort in industrial buildings. *Energy Procedia* **2017**, *126*, 313–320. [CrossRef]

38. Xuan, Y.M.; Xiao, F.; Niu, X.F.; Huang, X.; Wang, S.W. Research and application of evaporative cooling in China: A review (I) - Research. *Renew. Sustain. Energy Rev.* **2012**, *16*, 3535–3546. [CrossRef]

39. Obando, F.A.; Montoya, A.P.; Osorio, J.A.; Damasceno, F.A.; Norton, T. Evaporative pad cooling model validation in a closed dairy cattle building. *Biosyst. Eng.* **2020**, *198*, 147–162. [CrossRef]

40. Kovačević, I.; Sourbron, M. The numerical model for direct evaporative cooler. *Appl. Therm. Eng.* **2017**, *113*, 8–19. [CrossRef]

41. Cuce, P.M.; Riffat, S. A state of the art review of evaporative cooling systems for building applications. *Renew. Sustain. Energy Rev.* **2016**, *54*, 1240–1249. [CrossRef]

42. Panchabikesan, K.; Vellaisamy, K.; Ramalingam, V. Passive cooling potential in buildings under various climatic conditions in India. *Renew. Sustain. Energy Rev.* **2017**, *78*, 1236–1252. [CrossRef]

43. Shahzad, K. Evaluation of Evaporative Cooling Systems for Poultry Air-Conditioning. Master's Thesis, Bahauddin Zakariya University, Multan, Pakistan, 2018.

44. Mahmood, M.H.; Sultan, M.; Miyazaki, T.; Koyama, S.; Maisotsenko, V.S. Overview of the Maisotsenko cycle—A way towards dew point evaporative cooling. *Renew. Sustain. Energy Rev.* **2016**, *66*, 537–555. [CrossRef]

45. Mahmood, M.H.; Sultan, M.; Miyazaki, T. Significance of Temperature and Humidity Control for Agricultural Products Storage: Overview of Conventional and Advanced Options. *Int. J. Food Eng.* **2019**, *15*. [CrossRef]

46. Duan, Z.; Zhan, C.; Zhang, X.; Mustafa, M.; Zhao, X.; Alimohammadisagvand, B.; Hasan, A. Indirect evaporative cooling: Past, present and future potentials. *Renew. Sustain. Energy Rev.* **2012**, *16*, 6823–6850. [CrossRef]

47. Sajjad, U.; Abbas, N.; Hamid, K.; Abbas, S.; Hussain, I.; Ammar, S.M.; Sultan, M.; Ali, H.M.; Hussain, M.; ur Rehman, T.; et al. A review of recent advances in indirect evaporative cooling technology. *Int. Commun. Heat Mass Transf.* **2021**, *122*, 105140. [CrossRef]

48. Zhan, C.; Duan, Z.; Zhao, X.; Smith, S.; Jin, H.; Riffat, S. Comparative study of the performance of the M-cycle counter-flow and cross-flow heat exchangers for indirect evaporative cooling—Paving the path toward sustainable cooling of buildings. *Energy* **2011**, *36*, 6790–6805. [CrossRef]

49. Arun, B.S.; Mariappan, V.; Maisotsenko, V. Experimental study on combined low temperature regeneration of liquid desiccant and evaporative cooling by ultrasonic atomization. *Int. J. Refrig.* **2020**, *112*, 100–109. [CrossRef]

50. Pacak, A.; Worek, W. Review of Dew Point Evaporative Cooling Technology for Air Conditioning Applications. *Appl. Sci.* **2021**, *11*, 934. [CrossRef]

51. Shahzad, M.W.; Lin, J.; Bin Xu, B.; Dala, L.; Chen, Q.; Burhan, M.; Sultan, M.; Worek, W.; Ng, K.C. A spatiotemporal indirect evaporative cooler enabled by transiently interceding water mist. *Energy* **2021**, *217*, 119352. [CrossRef]

52. Sultan, M.; Niaz, H.; Miyazaki, T. Investigation of Desiccant and Evaporative Cooling Systems for Animal Air-Conditioning. In *Refrigeration and Air-Conditioning*; IntechOpen: London, UK, 2019.
53. Gates, R.S.; Overhults, D.G.; Zhang, S.H. Minimum ventilation for modern broiler facilities. *Trans. ASAE* **1996**, *39*, 1135–1144. [CrossRef]
54. Tao, X.; Xin, H. Surface wetting and its optimization to cool broiler chickens. *Trans. ASAE* **2003**, *46*, 483.
55. Zulovich, J.M.; DeShazer, J.A. Estimating egg production declines at high environmental temperatures and humidities. *Pap. Soc. Agric. Eng.* **1990**, *90-4021*, 1–16.
56. Tao, X.; Xin, H. Acute synergistic effects of air temperature, humidity, and velocity on homeostasis of market-size broilers. *Trans. Am. Soc. Agric. Eng.* **2003**, *46*, 491–497. [CrossRef]
57. Bruno, F. On-site experimental testing of a novel dew point evaporative cooler. *Energy Build.* **2011**, *43*, 3475–3483. [CrossRef]
58. Doğramacı, P.A.; Aydın, D. Comparative experimental investigation of novel organic materials for direct evaporative cooling applications in hot-dry climate. *J. Build. Eng.* **2020**, *30*, 101240. [CrossRef]

 *sustainability*

Article

Effect of a Sustainable Air Heat Pump System on Energy Efficiency, Housing Environment, and Productivity Traits in a Pig Farm

Myeong Gil Jeong [1,†], Dhanushka Rathnayake [1,†], Hong Seok Mun [1], Muhammad Ammar Dilawar [1], Kwang Woo Park [2], Sang Ro Lee [2] and Chul Ju Yang [1,*]

[1] Animal Nutrition and Feed Science Laboratory, Department of Animal Science and Technology, Sunchon National University, Suncheon 57922, Korea; wjdaudrlf13@naver.com (M.G.J.); dhanus871@gmail.com (D.R.); mhs88828@nate.com (H.S.M.); ammar_dilawar@yahoo.com (M.A.D.)

[2] WP Co., Ltd., Suncheon 58023, Korea; pkw9872@naver.com (K.W.P.); skylife37@naver.com (S.R.L.)

* Correspondence: yangcj@scnu.kr; Tel.: +82-61-750-3235

† This author contributed equally to this work as co-first author.

Received: 14 October 2020; Accepted: 21 November 2020; Published: 23 November 2020

Abstract: High electricity consumption, carbon dioxide (CO_2), and elevated noxious gas emission in the global livestock sector have a negative influence on environmental sustainability. This study examined the effects of a heating system using an air heat pump (AHP) on the energy saving, housing environment, and productivity traits of pigs. During the experimental period of 16 weeks, the internal temperature was found to be higher ($p < 0.05$) in the AHP house than in the conventional house. Moreover, the average electricity consumption and CO_2 emission decreased by approximately 40 kWh and 19.32 kg, respectively, in the AHP house compared to the house with the conventional heating system. The average NH_3 and H_2S emissions were significantly lower in the AHP house ($p < 0.05$) during the growth stages. The AHP and conventional heating systems did not have a significant influence ($p > 0.05$) on the average ultra-fine dust ($PM_{2.5}$) and formaldehyde level fluctuations. Furthermore, both heating systems did not show a significant difference in the average growth performance of pigs ($p > 0.05$), but the weight gain tended to increase in the AHP house. In conclusion, the AHP system has great potential to reduce energy consumption, greenhouse gas (GHG) emissions, and noxious gas emissions by providing economic benefits and an eco-friendly renewable energy source.

Keywords: air heat pump; carbon dioxide; formaldehyde; electricity consumption; ultra-fine dust

---

1. Introduction

Various energy problems have been identified in the global agriculture sector not only for economic reasons, but also for sustainable ecological persistence [1–3]. This is due to the diminishing fossil fuel reserves and increasing energy prices worldwide [4,5]. In addition, excessive greenhouse gas emissions affect biodiversity degradation through global warming [6]. Furthermore, the increase of global CO_2 emissions into the atmosphere is expected to lead to a temperature increase from 1.1 to 6.4 °C by the end of the 21st century [7]. Fossil-fuel burning is the major contributor of CO_2 emissions, and the atmospheric CO_2 concentration has been enhanced by 31% since 1750, with an average annual increase by 1.5 ppm over the past decades [8]. Beside deforestation and excessive arable land utilization, fossil-fuel combustion is responsible for 90% of CO_2 emissions into the environment [9].

In the global livestock sector, pigs have an inefficient thermoregulation process for dissipating heat from their bodies. Their maximum voluntary feed intake (VFI) ranges from 19 to 25 °C and tends to decrease above 25 °C [10]. NH_3 and methane byproducts released by pigs, together with dust, affect the air quality and are considered important parameters in pig houses [11]. The emission of noxious gas from the livestock sector is one of the major problems; it exerts negative impacts on the environment and accounts for approximately 75–80% of NH_3 emissions in developed nations in the world [12]. Moreover, a combination of both NH_3 and H_2S adversely affects the pig industry [13,14] owing to the direct harmful impact on both animals' and workers' welfare [15]. Dust can penetrate the respiratory organs easily owing to its smaller particle size. Super-fine dust particles less than <1 μm are the most harmful and cause pulmonary diseases [16]. Therefore, essential steps are needed to improve the housing environment by reducing noxious gas emissions, dust concentration, and environmental pollution. Moreover, due to the presence of an abundant renewable resource capacity, South Korea has the potential of utilizing them efficiently to mitigate the problems arising through high energy consumption, thus finding effective solutions for those challenges and the energy distribution process across the various geographical areas [17].

The air heat source pump system has the potential to conserve high-grade energy and allow the effective use of low-grade energy, as well as to provide energy savings and storage [18]. Other than the energy savings, it can reduce CO_2 emissions and is consistent with efficient structural compaction [19].The theoretical and experimental performance and effectiveness of the air heat pump were investigated by previous studies [20–23]. However, there are no publications available on the effects of utilization of air heat pump systems on energy efficiency, housing environments, and productivity traits in livestock sectors. Owing to the environmentally friendly and sustainable source, an air heat pump system can be introduced as an alternative energy system for conventional methods. Therefore, this study compared a conventional electric heating system and an air heat pump (AHP) system for the energy savings, housing environment (NH_3, H_2S, fine dust, formaldehyde), and productivity traits of pigs.

2. Materials and Methods

2.1. Experimental Period and House

The performance of the air pump heating system in a pig house was evaluated for 16 weeks (weaning period, four weeks; growing period, six weeks; finishing period, six weeks) in winter from 2 December 2019 to 2 April 2020 at the Sunchon National University Experimental Farm, South Korea. The pig house consisted of two separate rooms (3 m × 8 m) that were subdivided into 10 pens for individual replication. Two east-facing rooms were contained in the pig house. The room on the south side was considered the control house, which was connected to a conventional electric heating system. The air pump heating system was connected to the other north-facing room (Figure 1). An outdoor unit draws heat in from outside, and thereafter, blows it over a heat exchanger coil. The heat thus generated from the compressor is then transferred through an internal plastic tube with small pores that enable the uniform distribution of the heating pattern inside the house. Finally, the cold liquid vapor coolant mixture enters back into the outdoor unit to be heated once again. The conventional pig house was connected with heating lamps; the heights of these were maintained according to the growth phase of the pigs. The outside walls of the pig house were made from brick plastered on both sides. The floor was installed with a plastic slat, and the slurry was removed daily. The environmentally controlled pig houses' inside temperature ventilation processes were controlled automatically. Moreover, we maintained similar internal temperature settings according to each growing phase, covering both the conventional electric heating house and the AHP installed house to compare the inside temperature fluctuations, energy efficiency, noxious gas emission, ultra-fine dust, and formaldehyde concentration between the two experimental houses. Throughout the experiment, all animals received a commercial basal diet and had access to water ad libitum.

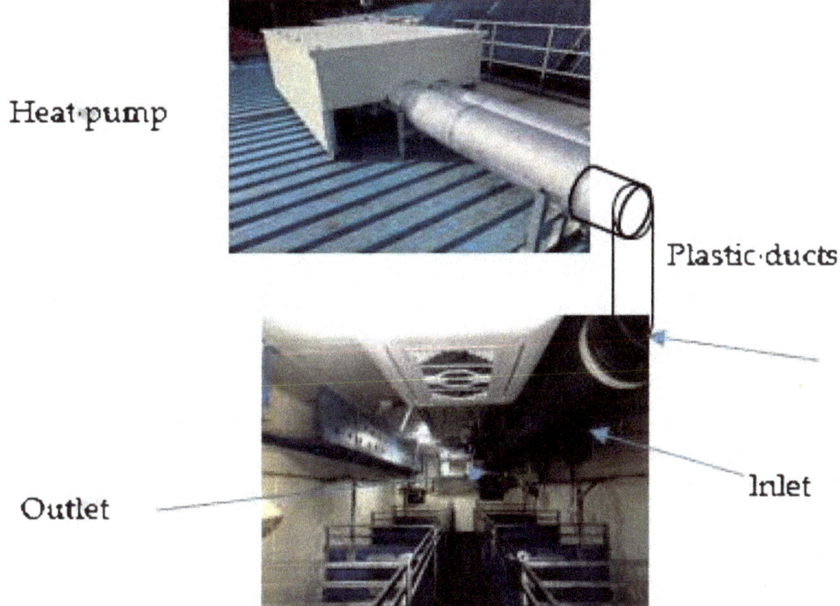

Figure 1. Schematic diagram of the air heat pump (AHP) system for the pig house.

2.2. Description of the Air Heat Pump System

The air heat pump (AHP) (model: BW1450M9S, LG Electronics Inc., Seoul, South Korea) was installed and connected to a pig house according to a slight modification of the procedures recommended by previous studies [24,25]. The major components of the air heat pump system were an air inlet, inhale chamber, air heat pump compressor, discharge chamber, and air-circulating pipes (Figure 2).

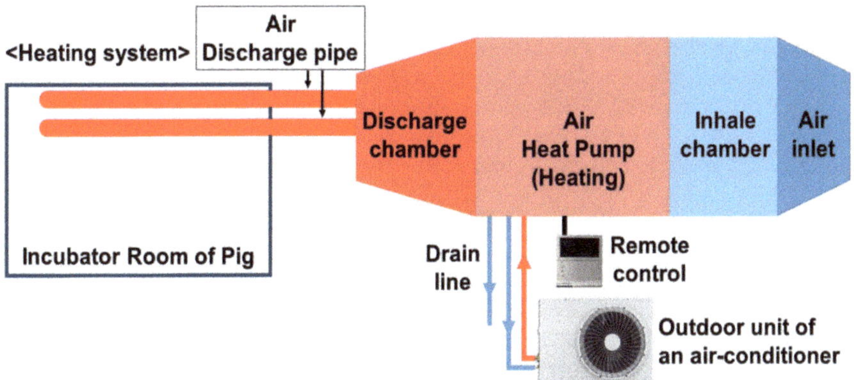

Figure 2. Outline of the air heat pump system.

The power was supplied through a three-phase four-wire system (380 V, 60 Hz). The estimated minimum and maximum heating ability values were 5.2 and 20 kW, respectively. The evaporator coil system of the heat pump (HP) system could dehumidify and cool the extracted hot and wet air. The absorbed fresh and purified air was heated by a condenser, and the circulating fluid was the refrigerant R410A. The inlet fan was controlled thermostatically, and the temperature level was maintained according to its speed. The extraction fan speed was controlled manually. The required power for the operation of the compressor was 1.1 kW; the required level increased to 4.3 kW when the fans were operating. The coefficient of performance was 4.3 for the heating process when the reference temperature values were used: external air at 6.0 °C, evaporation at −4.0 °C, and condensation at 45 °C.

2.3. Measurement and Analysis

The temperatures of the control and air heat pump houses were determined using eight-bit Smart Sensors (model: SMT-75, Seoul, South Korea). Temperature data were taken from the ceiling at the entry (close to door), center and back of the pig houses at 10 cm above the slatted floor (lower point), and 10 cm below the ceiling level. All measuring equipment was connected to a data logger system (CR10X data logger, Campbell Scientific Inc., Edmonton, AB, Canada) to record the data for every hour. The recording equipment was properly designed for an auto-restart process to prevent data losses due to power failures. A digital hygrometer (Electronic Digital Hygrometer HTC-1, Jinggoal International Ltd., Guangdong, China) was used to evaluate the humidity level inside the both pig houses.

The coefficient of performance (COP) of the heat pump was evaluated using the following formula [26]:

$$COP = \frac{\sum \dot{Q}}{\sum \dot{W}} \tag{1}$$

where $\dot{Q}$ is the useful heat extracted from the heat pump (condenser) (kW) of the air heat pump; $\dot{W}$ is the power consumption (kW).

The daily electricity consumption of both the conventional and air heat pump house was measured based on the electricity consumption units recorded by individually installed meters (Model: LD 1210Ra-040, LSis, Seoul, South Korea). The daily electricity cost of each house was calculated according to the current electricity cost in South Korea (Korea Electric Corporation, KEPCO, September 2020 (1 kWh electricity = 39.2 South Korean won, and 39.2 South Korean won = 0.033 USD)). In addition, CO_2 emissions were determined in $kgCO_2e$ (1 kWh = 0.483 kg CO_2 equivalent) [27] according to the electricity consumption in both pig houses.

NH_3 and H_2S gas concentrations were evaluated every day at 8:00 am at the entry, center, and back positions at approximately 30 cm above the slatted floor using a Gastec (model GV-100) gas sampling pump (Gastec Corp., Kanagawa, Japan) and gas detector tubes: No. 3L (0.5–78 ppm, Gastec Corp., Kanagawa, Japan) for NH_3 and 4LT (0.05–4 ppm, Gastec Corp., Kanagawa, Japan) for H_2S. The NH_3 gas emission was expressed in ppm, and the H_2S level was expressed in ppb in both pig houses. The ultra-fine dust concentration and formaldehyde level were measured every day during the experimental period at 8:00 am at the entry, center, and back of each pig house using a Smart Sensor air quality model (model:AR830A-2, Huipu Opto-Electronic Instrument (Zhenjiang) Co., Ltd., Jian, China) at 10 cm above the floor.

The body weight gain, feed intake, and feed conversion ratio (FCR) were measured during the weaning, growing, and finishing periods. The body weight gain was evaluated by dividing the weight difference of the starting and finishing weight by each experimental period. The feed intake was measured every week by weighing the feed weight immediately before the body weight measurement. The FCR was calculated by dividing the feed intake by the average daily gain.

2.4. Statistical Analysis

The inside room temperature, noxious gas emission, ultra-fine dust concentration, and formaldehyde level in the experimental houses were evaluated using PROC GLM of the statistical analysis system (version 9.1, SAS Institute Inc., Cary, NC, USA). The data are reported as the mean ± standard error of the means (SEM). A p-value < 0.05 was considered significant.

3. Results

3.1. Room Temperature and Coefficient of Performance (COP)

As shown in Table 1, the temperature level was increased significantly ($p < 0.05$) in the house with the air heat pump compared to the control house. The highest and lowest temperatures in the AHP house were 26.1 and 19.9 °C, respectively (Figure 3). The COP value was reduced when the external temperature was decreased during the weaning period.

Table 1. Effect of the air heat pump system on inside temperature and coefficient of performance (COP).

Periods	External Temp. (°C)	Control (°C)	AHP (°C)	SEM	*p*-Value	Average COP
Weaning	4.5	24.7 [b]	26.1 [a]	1.84	<0.0001	3.86
Growing	6.1	20.4 [b]	22.8 [a]	1.88	<0.0001	3.98
Finishing	9.7	19.9 [b]	21.1 [a]	1.14	<0.0001	4.12
Average	7.1	21.3 [b]	23.0 [a]	2.59	<0.0001	4.07

[a,b] means that values with different superscripts within same row are significantly different ($p < 0.05$).

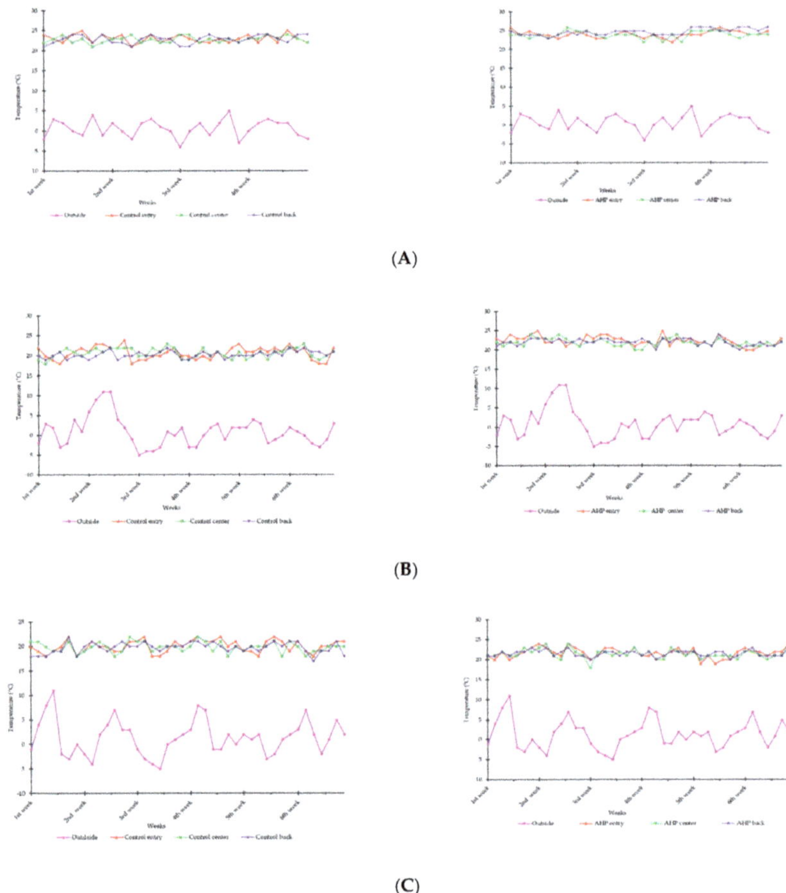

(A)

(B)

(C)

Figure 3. Temperature profile for control (conventional electric heating system) versus air heat pump (AHP) housing system during winter, at the entry, center, and back position of the rooms (average mean value at lower and upper positions). (**A**) Temperature profile during weaning period (four weeks). (**B**) Temperature profile during growing period (six weeks). (**C**) Temperature profile during finishing period (six weeks).

3.2. Electricity Consumption, CO_2 Emissions, and Cost Savings

Table 2 lists the daily electricity consumption and cost analysis per day in both experimental houses. The daily electricity consumption during the weaning, growing, and finishing periods were decreased in the AHP house. The decrease in average daily electricity consumption in the AHP house was 63.5%.

Table 2. Energy consumption and heating costs of the pig houses during the various growth periods.

Period	Electricity Use (kWh/d)			CO$_2$ Emission (kg)			Cost Savings (USD)
	Control	AHP	Reduced	Control	AHP	Reduced	
Weaning	108	33	75	52.16	15.94	36.22	97.02
Growing	60	30	30	28.98	14.50	14.48	38.80
Finishing	35	9	26	16.91	4.35	12.56	33.63
Average	63	23	40	30.43	11.11	19.32	51.74

Electricity consumption was determined using an electric meter installed for each house for every day (8:00–20:00) and night (20:00–8:00), and was summed per day. Carbon dioxide (CO$_2$) emissions were evaluated based on electricity consumption per day. The used conversion factor was 1 kWh = 0.483 kg CO$_2$ emissions [27]. Cost was estimated according to electricity consumption per day. The current value of 1 kWh electricity = 39.2 South Korean Won, and 39.2 South Korean Won = 0.033 USD was used (KEPCO, September 2020).

Consequently, a significant decline of the daily electricity cost was also observed in the AHP house relative to the control house. During the finishing period, the electricity consumption was reduced drastically in the AHP house, and the reduced average daily electricity cost was 63.6% compared to the control pig house. The total CO$_2$ emission was decreased in the AHP-installed house, and the average daily reduction was 63.49% compared to the control house.

3.3. NH$_3$, H$_2$S, Ultra-Fine Dust (PM$_{2.5}$), and Formaldehyde Level

As listed in Table 3, NH$_3$ and H$_2$S emissions during the weaning, growing, and finishing periods were significantly lower ($p < 0.05$) in the AHP house compared to the control house. The average NH$_3$ emissions were reduced by 61%, and the average H$_2$S level was decreased by 45% in the AHP house.

Table 3. Effect of the air heat pump system on the NH$_3$ and H$_2$S emissions in the pig house.

Item	Periods	Control	AHP	SEM	p-Value
NH$_3$ (ppm)	Weaning	0.05 [a]	0.01 [b]	0.02	<0.0001
	Growing	0.63 [a]	0.16 [b]	0.33	<0.0001
	Finishing	2.27 [a]	0.97 [b]	0.45	<0.0001
	Average	1.10 [a]	0.42 [b]	0.78	<0.0001
H$_2$S (ppb)	Weaning	0.00	0.00	0.00	-
	Growing	0.00	0.00	0.00	-
	Finishing	5.14 [a]	2.81 [b]	0.56	<0.0001
	Average	1.93 [a]	1.06 [b]	0.94	<0.0001

[a,b] means that values with different superscripts within the same row are significantly different ($p < 0.05$).

Table 4 lists the ultra-fine dust concentration (PM$_{2.5}$) and formaldehyde concentration due to the air heat pump system. There were no significant differences in the PM$_{2.5}$ dust concentration between the two houses. On the other hand, the dust concentration tended to decrease during all periods, and the average reduction was 6.5% in the AHP house compared to the control house. During the growing and finishing periods, the formaldehyde concentration was significantly lower ($p < 0.05$) in the AHP house.

Table 4. Effect of the air heat pump system on ultra-fine dust ($PM_{2.5}$) and formaldehyde concentration in the pig house.

Item	Periods	Control	AHP	SEM	*p*-Value
Ultra-fine dust ($PM_{2.5}$) ($\mu g/m^3$)	Weaning	28.39	24.74	16.60	0.46
	Growing	29.03	26.33	23.71	0.64
	Finishing	21.14	21.69	14.22	0.87
	Average	25.84	24.14	18.80	0.54
Formaldehyde (ppm)	Weaning	0.06	0.05	0.02	0.26
	Growing	0.08 [a]	0.05 [b]	0.04	0.03
	Finishing	0.22 [a]	0.13 [b]	0.19	0.02
	Average	0.12	0.10	0.12	0.27

[a, b] means that values with different superscripts within the same row are significantly different ($p < 0.05$).

3.4. Effect of the Air Pump Heating System on the Productivity Traits of Pigs

Table 5 lists the results of the growth performances of pigs during their weaning, growing, finishing, and average values. The body weight gain, feed intake, and feed conversion ratio (FCR) did not significantly differ ($p > 0.05$) among the control and AHP-installed house.

Table 5. Effect of the air pump heating system on the productivity parameters of pigs.

Item	Control	AHP	SEM	*p*-Value
Weaning period (0–4 weeks)				
Initial weight (kg)	8.56	8.29	3.17	0.86
Final weight (kg)	25.81	25.87	5.66	0.98
Weight gain (kg)	17.26	17.58	3.01	0.82
Feed intake (kg)	33.48	33.79	7.45	0.93
FCR (Feed/gain)	1.95	2.01	0.60	0.83
Growing period (4–10 weeks)				
Initial weight (kg)	25.81	25.87	5.66	0.98
Final weight (kg)	70.77	66.10	7.80	0.22
Weight gain (kg)	40.23	43.96	4.27	0.03
Feed intake (kg)	100.93	96.66	4.63	0.54
FCR (Feed/gain)	2.25	2.44	0.41	0.36
Finishing period (10–16 weeks)				
Initial weight (kg)	70.77	66.10	7.80	0.22
Final weight (kg)	113.43	107.97	7.18	0.13
Weight gain (kg)	41.87	42.66	4.64	0.73
Feed intake (kg)	150.81	151.48	4.26	0.92
FCR (Feed/gain)	3.53	3.69	0.49	0.51
Average (0–16 weeks)				
Initial weight (kg)	8.56	8.29	2.80	0.96
Final weight (kg)	113.43	107.97	7.18	0.13
Weight gain (kg)	99.68	104.88	5.46	0.06
Feed intake (kg)	285.23	281.93	3.23	0.82
FCR (Feed/gain)	2.71	2.83	0.23	0.30

3.5. Estimation of the Installation and Annual Operational Costs

As shown in Table 6, the initial investment for the air heat pump system was comparatively higher than for the conventional electric heating system. Nevertheless, the AHP system gained a lower annual operational cost, higher life span, and shorter payback period.

Table 6. Installation and operational costs of the air heat pump and conventional electric heating system.

Item	Control	AHP	
Installation cost (USD)	1288	5000	
Life span	5 years	15 years	As per company instruction
Annual operational cost (USD)	4323	394	
Savings (USD)	-	3929	
Payback period (Y)	>useful life	4.1	
Depreciation time	5 years	15 years	

Annual operational cost was evaluated according to annual electricity consumption (131,026 kWh) and price (0.033 USD/kWh).

4. Discussion

Proper temperature maintenance inside a pig house is essential to prevent pigs from cold shock and ensure their optimal growth. In this study, the inside temperature of the AHP house was greater than in the conventional electric heating system. We speculated that the AHP system could distribute a uniform heat pattern more continuously inside the house than the conventional electric heating system due to the high COP value and lower running time period. The calculated average COP of this study was 4.07, which is lower than the values observed by Riva et al. [28] during the heating phase, but it was higher than the experiment conducted by Ji et al. [29] using an air heat pump for domestic heating purposes. In contrast, Zang et al. [30] reported that the COP value tends to increase with decreasing external temperature because the evaporator of the heat pump interacts continuously with hot air circulation during the heating phase. Nevertheless, some studies have reported that the efficiency of the AHP system tends to decrease when exposed to extreme temperature levels [31–33].

Renewable energy sources are abundant, have low cost, and are environmentally safe. In the present study, the AHP system showed lower electricity consumption, CO_2 emissions, and electricity cost relative to the conventional electric heating system. To the authors' knowledge, only one study has evaluated the effects of an air pump system on energy savings and housing environment in pig breeding house [28]. Riva et al. [28] reported that an AHP house could save 11% of the total energy consumption compared to a control house connected to an LPG boiler house. Similarly to the present results, Wu [34] concluded that the AHP system is a more efficient environmental safety system than conventional heating techniques, and can be introduced to minimize the depletion of energy resources. The low electricity consumption in the AHP system might be due to the high COP value, which has the potential to distribute unvarying heat inside the experimental house.

Rabczak et al. [35] reported that the emission of CO_2 to the atmosphere could be reduced by 40% with an air heat pump system compared to a local gas furnace system or particular heating system that is provided for specific geographical areas. Furthermore, decreased CO_2 emissions and energy savings have been investigated in response to the air pump heating system in buildings [36–39]. In addition to the increasing feed cost, energy prices have a huge impact on the productivity of the global pig industry, including in South Korea. In the present study, the electricity cost decreased during each growth period in the AHP-installed house compared to the conventional system.

According to the International Commission of Agricultural and Bio-Systems Engineering, CIGR (2002) [40], the recommended maximum NH_3 concentration is 20 ppm. In the livestock sector, pig growth was slowed by 12% to 30% in intensive swine buildings because of the elevated NH_3 concentration [11]. An improper ventilation system and high concentrations of NH_3, H_2S, and CO_2 lead to poor air quality inside pig houses. In the present study, the concentrations of both NH_3 and H_2S were significantly lower in the AHP house, which is in agreement with the results of a previous study [28]. The lower noxious gas concentration may have occurred due to the increased fresh outdoor air temperature due to the compressor, as well as the subsequent dilution of NH_3 and H_2S levels in the pig house.

Takai et al. [41] reported that the dust concentration in swine houses tends to increase in winter compared to summer. Automotive exhaust and various urea–formaldehyde products are the sources of formaldehyde formation inside houses. Exposure to 0.3 to 50 ppm will depreciate lung compliance [42]. In the present study, during the growing and finishing periods, air contamination with formaldehyde was lower, possibly due to proper air circulation inside the AHP house. On the other hand, the relationship between the dust concentration, formaldehyde level, and installation of an AHP system is unclear. Further research on dust and formaldehyde fluctuations from the utilization of renewable energy sources will be needed.

According to Riva et al. [28], the production parameters, including feed intake, weight gain, and feed conversion ratio, increased significantly in a pig house operating with the AHP system compared to one operating with an LPG gas system. The accumulation of high concentrations of fumes in an LPG house may reduce their voluntary feed intake because of the poor housing environmental conditions. Nevertheless, in our study, there were no significant differences in the productivity parameters, but the weight gain tended to increase in the AHP house during the growth stages. Therefore, further study on the productivity parameters when using the AHP system in the livestock sector will be needed.

Owing to the high COP value of the AHP system, the annual operational cost was reduced by 91% compared to the control heating system. Wu [34] reported that the air heat pump reduced electricity consumption by 46 kWh/m^2. Consequently, it reduces the electricity costs. Islam et al. [43] reported that the installation cost for a renewable geothermal heat pump is considerably more expensive than for a renewable AHP system, and both systems had lower annual operating costs than the electric heating system. The payback period tends to decrease when the COP value is increased. Similarly to our result, the payback period ranges between four and five years when the COP value is 4 [44]. Owing to the high depreciation time, livestock farmers can implement an AHP-based livestock housing system to minimize their electricity costs, and it has the potential to work for a longer period.

5. Conclusions

In global intensive livestock farming systems, higher electricity consumption and inadequate air quality adversely influence the environmental sustainability and slow productivity. Therefore, the implementation of innovative strategies in order to maintain production parameters while reducing energy consumption and providing proper air quality is a current issue and is worthy of being collaboratively investigated. The present study aimed to investigate an AHP system to be utilized for intensive pig farming as an efficient, eco-friendly alternative to the widely used conventional electric heating systems. According to the results of this study, the inside temperature was maintained at a significantly higher level in the AHP house. A significant decrease in average electricity consumption by 40 kWh, overall cost, and CO_2 emissions by 19.32 kg was observed during the experimental period in the AHP house. Furthermore, the NH_3 and H_2S emissions were also lower in the AHP-installed house than in the house with the conventional electric heating system. Although the initial installation cost was high, the investor could obtain long-term benefits with a uniform performance for a longer period (approximately 15 years) while utilizing less electricity and causing less greenhouse gas (GHG) emissions. Therefore, the AHP system is an innovative and sustainable energy source for cost-effective and eco-friendly heating of animal houses in the livestock sector.

Author Contributions: Conceptualization, C.J.Y. and K.W.P.; methodology, M.G.J. and H.S.M.; writing—Original draft preparation, D.R. and M.G.J.; software, D.R. and H.S.M.; data curation, S.R.L. and M.A.D.; formal analysis, C.J.Y. and K.W.P.; investigation, C.J.Y. and H.S.M.; writing—Review and editing, M.A.D., S.R.L., and D.R.; supervision, C.J.Y. and H.S.M., project administration, C.J.Y. and K.W.P.; funding acquisition, S.R.L. and M.G.J. All authors have read and agreed to the published version of the manuscript.

Funding: This study was funded by the Industrial Technology Innovation Business (20194210100020, Development and Demonstration of Renewable Energy Mixed-Use System for the Livestock Industry) and Ministry of Trade, Industry, and Energy, Korea.

Conflicts of Interest: The authors declare no conflict of interest.

References

1. Horne, R.E.; Mortimer, N.D.; Elsayed, M.A. *Energy and Carbon Balances of Biofuels Production: Biodiesel and Bioethanol*, 2nd ed.; International Fertiliser Society: London, UK, 2003; pp. 14–26.
2. Meul, M.; Nevens, F.; Reheul, D.; Hofman, G. Energy use efficiency of specialised dairy, arable and pig farms in Flanders. *Agric. Ecosyst. Environ.* **2007**, *119*, 135–144. [CrossRef]
3. Kythreotou, N.; Florides, G.; Tassou, S.A. A proposed methodology for the calculation of direct consumption of fossil fuels and electricity for livestock breeding, and its application to Cyprus. *Energy* **2012**, *40*, 226–235. [CrossRef]
4. Corrée, W.J.; Jaap, J.S.; Verhagen, A. Energy Use in Conventional and Organic Farming Systems. In *Proceedings of the Open Meeting of the International Fertiliser Society*; International Fertiliser Society: London, UK, 2003.
5. Nakomcic-Smaragdakis, B.; Stajic, T.; Cepic, Z.; Djuric, S. Geothermal energy potentials in the province of Vojvodina from the aspect of the direct energy utilization. *Renew. Sust. Energy Rev.* **2012**, *16*, 5696–5706. [CrossRef]
6. Gardner, G.T.; Stern, P.C. *Environmental Problems and Human Behavior*; Pearson Custom Publishing: Boston, MA, USA, 2002; pp. 79–110.
7. Solomon, S.; Qin, D.; Manning, M.; Chen, Z.; Marquis, M.; Averyt, K.B.; Tignor, M.; Miller, H.L.; IPCC (Intergovernmental Panel on Climate Change). Summary for Policymakers. In *Contribution of Working Group I to the Fourth Assessment Report of the Intergovernmental Panel on Climate Change*; Cambridge University Press: Cambridge, UK, 2007.
8. Apak, R. Alternative solution to global warming arising from CO2 emissions—Partial neutralization of tropospheric H_2CO_3 with NH_3. *Environ. Prog.* **2007**, *26*, 355–359. [CrossRef]
9. Olivier, J.G. *Trends in Global CO_2 Emissions: 2012 Report*; PBL Netherlands Environmental Assessment Agency: Hague, The Netherlands, 2002; p. 40.
10. Collin, A.; van Milgent, J.; Dividich, J.L. Modelling the effect of high, constant temperature on food intake in young growing pigs. *Anim. Sci.* **2001**, *72*, 519–527. [CrossRef]
11. Drummond, J.G.; Curtis, S.E.; Simon, J.; Norton, H.W. Effects of aerial ammonia on growth and health of young pigs. *J. Anim. Sci.* **1980**, *50*, 1085–1091. [CrossRef]
12. Asman, W.A.H. *Ammonia Emissions in Europe: Updated Emission and Emission Variations*; National Institute of Public Health and Environmental Protection: Bilthoven, The Netherlands, 1992; Report 228471008.
13. Blanes-Vidal, V.; Nadimi, E.S.; Ellermann, T.; Andersen, H.V.; Løfstrøm, P. Perceived annoyance from environmental odors and association with atmospheric ammonia levels in non-urban residential communities: a cross-sectional study. *Environ. Health* **2012**, *11*, 27. [CrossRef]
14. Saha, C.K.; Zhang, G.; Kai, P.; Bjerg, B. Effects of a partial pit ventilation system on indoor air quality and ammonia emission from a fattening pig room. *Biosyst. Eng.* **2010**, *105*, 279–287. [CrossRef]
15. Ni, J.Q.; Heber, A.J.; Lim, T.T. Ammonia and hydrogen sulphide in swine production. In *Air Quality and Livestock Farming*; CRC Press: Florida, FA, USA, 2018; pp. 69–88.
16. Nõu, T.; Viljasoo, V. The effect of heating systems on dust, an indoor climate factor. *Agron. Res.* **2011**, *9*, 165–174.
17. Alsharif, M.; Kim, J.; Kim, J. Opportunities and challenges of solar and wind energy in South Korea: A Review. *Sustainability* **2018**, *10*, 1822. [CrossRef]
18. Zhao, X.; Long, E.; Zhang, Y.; Liu, Q.; Jin, Z.; Liang, F. Experimental study on heating performance of air - source heat pump with water tank for thermal energy storage. *Proc. Eng.* **2017**, *205*, 2055–2062. [CrossRef]
19. Gajewski, A.; Gładyszewska-Fiedoruk, K.; Krawczyk, D.A. Carbon dioxide emissions during Air, Ground, or Groundwater heat pump performance in Białystok. *Sustainability* **2019**, *11*, 5087. [CrossRef]
20. Mei, V.C.; Chen, F.C.; Domitrovic, R.E.; Kilpatrick, J.K.; Carter, J.A. A Study of a Natural Convection Immersed Condenser Heat Pump Water Heater/Discussion. *ASHRAE Trans.* **2003**, *109*, 3–8.
21. Wang, Y.; Jiang, H.; Ma, Z.; Jiang, Y.; Yao, Y. Experiment and analysis of delaying frost of air source heat pump water chiller heater units by increasing the evaporator area. *HV&AC* **2006**, *36*, 83–88.
22. Xu, G.; Zhang, X.; Deng, S. A simulation study on the operating performance of a solar–air source heat pump water heater. *Appl. Therm. Eng.* **2006**, *26*, 1257–1265. [CrossRef]
23. Jie, Z. *Study on Seasonal Performance Optimization and Refrigerant Flux Characters of Air Source Heat Pump Water Heater*; Shanghai Jiaotong University: Shanghai, China, 2007; pp. 12–38.

24. Huang, L.; Zheng, R.; Piontek, U. Installation and Operation of a Solar Cooling and Heating System Incorporated with Air-Source Heat Pumps. *Energies* **2019**, *12*, 996. [CrossRef]

25. Lun, Y.H.V.; Tung, S.L.D. *Heat Pumps for Sustainable Heating and Cooling*, 1st ed.; Springer: Cham, Switzerland, 2019; pp. 35–48.

26. Han, D.; Chang, Y.S.; Kim, Y. Performance analysis of air source heat pump system for office building. *J. Mech. Sci. Technol.* **2016**, *30*, 5257–5268. [CrossRef]

27. D'Agostino, D.; Mele, L.; Minichiello, F.; Renno, C. The Use of Ground Source Heat Pump to Achieve a Net Zero Energy Building. *Energies* **2020**, *13*, 3450. [CrossRef]

28. Riva, G.; Pedretti, E.F.; Fabbri, C. Utilization of a heat pump in pig breeding for energy saving and climate and ammonia control. *J. Agric. Eng.* **2000**, *77*, 449–455. [CrossRef]

29. Ji, J.; Chow, T.; Pei, G.; Dong, J.; He, W. Domestic air-conditioner and integrated water heater for subtropical climate. *Appl. Therm. Eng.* **2003**, *23*, 581–592. [CrossRef]

30. Zhang, L.; Jiang, Y.; Dong, J.; Yao, Y. Advances in vapor compression air source heat pump system in cold regions: A review. *Renew. Sust. Energy Rev.* **2018**, *81*, 353–365. [CrossRef]

31. Zheng, N.; Song, W.; Zhao, L. Theoretical and experimental investigations on the changing regularity of the extreme point of the temperature difference between zeotropic mixtures and heat transfer fluid. *Energy* **2013**, *55*, 541–552. [CrossRef]

32. Zhang, Q.; Zhang, L.; Nie, J.; Li, Y. Techno-economic analysis of air source heat pump applied for space heating in northern China. *Appl. Energy* **2017**, *207*, 533–542. [CrossRef]

33. The Annual Report Korea Energy Agency. Available online: http://www.energy.or.kr/web/kem_home_new/ energy_issue/mail_vol22/pdf/publish_05_201507.pdf (accessed on 22 May 2018).

34. Wu, R. Energy efficiency technologies–air source heat pump vs. ground source heat pump. *J. Sust. Dev.* **2009**, *2*, 14–23. [CrossRef]

35. Rabczak, S.; Proszak-Miąsik, D. The impact of selected heat pumps on CO_2 emissions. *E3S Web Conf.* **2018**, *45*, 71. [CrossRef]

36. Patteeuw, D.; Reynders, G.; Bruninx, K.; Protopapadaki, C.; Delarue, E.; D'haeseleer, W.; Saelens, D.; Helsen, L. CO_2-abatement cost of residential heat pumps with active demand response: demand- and supply-side effects. *Appl. Energy* **2015**, *156*, 490–501. [CrossRef]

37. Hedegaard, K.; Münster, M. Influence of individual heat pumps on wind power integration – Energy system investments and operation. *Energy Convers. Manag.* **2013**, *75*, 673–684. [CrossRef]

38. Hedegaard, K.; Mathiesen, B.V.; Lund, H.; Heiselberg, P. Wind power integration using individual heat pumps—Analysis of different heat storage options. *Energy* **2012**, *47*, 284–293. [CrossRef]

39. Johansson, D. The life cycle costs of indoor climate systems in dwellings and offices taking into account system choice, airflow rate, health and productivity. *Build. Environ.* **2009**, *44*, 368–376. [CrossRef]

40. Søren, P.; Krister, S. *International Commission of Agricultural and Biosystems Engineering*, 4th ed.; Research Centre Bygholm, Danish Institute of Agricultural Sciences: Horsens, Denmark, 2002; pp. 75–92.

41. Takai, H.; Pedersen, S.; Johnsen, J.O.; Metz, J.H.M.; Groot Koerkamp, P.W.G.; Uenk, G.H.; Phillips, V.R.; Holden, M.R.; Sneath, R.W.; Short, J.L. Concentrations and Emissions of Airborne Dust in Livestock Buildings in Northern Europe. *J. Agric. Eng. Res.* **1998**, *70*, 59–77. [CrossRef]

42. National Research Council. *Formaldehyde—An Assessment of Its Health Effects*; National Academies Press (US): Washington, DC, USA, 1980. Available online: https://www.ncbi.nlm.nih.gov/books/NBK217655/ (accessed on 5 March 1980).

43. Islam, M.M.; Mun, H.-S.; Bostami, A.B.M.R.; Ahmed, S.T.; Park, K.-J.; Yang, C.-J. Evaluation of a ground source geothermal heat pump to save energy and reduce CO_2 and noxious gas emissions in a pig house. *Energy Build.* **2016**, *111*, 446–454. [CrossRef]
44. Masa, V.; Havlasek, M. Integration of air to water heat pumps into industrial district heating substations. *Chem. Eng. Trans.* **2016**, *52*, 739–744. [CrossRef]

Publisher's Note: MDPI stays neutral with regard to jurisdictional claims in published maps and institutional affiliations.

sustainability

MDPI

Article

Analysis of Inlet Configurations on the Microclimate Conditions of a Novel Standalone Agricultural Greenhouse for Egypt Using Computational Fluid Dynamics

Mohammad Akrami [1,*], Can Dogan Mutlum [1], Akbar A. Javadi [1], Alaa H. Salah [2], Hassan E. S. Fath [3], Mahdieh Dibaj [1], Raziyeh Farmani [1], Ramy H. Mohammed [4] and Abdelazim Negm [5,*]

1 Department of Engineering, University of Exeter, Exeter EX4 4QF, UK; dm637@exeter.ac.uk (C.D.M.); a.a.javadi@exeter.ac.uk (A.A.J.); md529@exeter.ac.uk (M.D.); r.farmani@exeter.ac.uk (R.F.)
2 City of Scientific Research and Technological Applications (SRTA), Alexandria 21934, Egypt; alaa.h.salah@gmail.com
3 Ex-Environmental Engineering Department, School of Energy Resources, Environment, Chemical and Petrochemical Engineering, Egypt-Japan University of Science and Technology, Alexandria 21934, Egypt; h_elbanna_f@yahoo.com
4 Department of Mechanical Power Engineering, Zagazig University, Zagazig 44519, Egypt; rhamdy@knights.ucf.edu
5 Water and Water structures Engineering Department, Faculty of Engineering, Zagazig University, Zagazig 44519, Egypt
* Correspondence: m.akrami@exeter.ac.uk (M.A.); amnegm@zu.edu.eg or amnegm85@yahoo.com (A.N.)

check for **updates**

Citation: Akrami, M.; Mutlum, C.D.; Javadi, A.A.; Salah, A.H.; Fath, H.E.S.; Dibaj, M.; Farmani, R.; Mohammed, R.H.; Negm, A. Analysis of Inlet Configurations on the Microclimate Conditions of a Novel Standalone Agricultural Greenhouse for Egypt Using Computational Fluid Dynamics. *Sustainability* **2021**, *13*, 1446. https://doi.org/10.3390/su13031446

Academic Editors: Muhammad Sultan and Marc A. Rosen

Received: 10 December 2020
Accepted: 27 January 2021
Published: 30 January 2021

Publisher's Note: MDPI stays neutral with regard to jurisdictional claims in published maps and institutional affiliations.

Abstract: Water shortage, human population increase, and lack of food resources have directed societies towards sustainable energy and water resources, especially for agriculture. While open agriculture requires a massive amount of water and energy, the requirements of horticultural systems can be controlled to provide standard conditions for the plants to grow, with significant decrease in water consumption. A greenhouse is a transparent indoor environment used for horticulture, as it allows for reasonable control of the microclimate conditions (e.g., temperature, air velocity, rate of ventilation, and humidity). While such systems create a controlled environment for the plants, the greenhouses need ventilation to provide fresh air. In order to have a sustainable venting mechanism, a novel solution has been proposed in this study providing a naturally ventilating system required for the plants, while at the same time reducing the energy requirements for cooling or other forced ventilation techniques. Computational fluid dynamics (CFD) was used to analyse the ventilation requirements for different vent opening scenarios, showing the importance of inlet locations for the proposed sustainable greenhouse system.

Keywords: greenhouse; computational fluid dynamics; airflow; temperature; humidity; sustainable agriculture; horticulture; Zagazig; Egypt

1. Introduction

Production of fresh horticultural crops in greenhouses is an essential agricultural practice. The use of greenhouses results in increased harvest, water and nutrients, higher fruit yield, longer production times, and the capacity to grow off-season [1]. They are being used to protect the plants from severe climate conditions such as high wind rates, intense sunshine, and high levels of temperature and humidity [2,3]. Such different parameters may be managed simply by opening/closing vents automatically or manually to control wind speeds, or even by choosing proper covering materials [4,5]. Using dyed glass can prevent high solar irradiance from impacting plant growth by shielding the greenhouse (GH)'s translucent walls and roof [3].

Although a mono-span known as a walk-in greenhouse is a common greenhouse structure in Egypt, many other types of greenhouse have been developed over the last

25 years, such as the double-span, the Parron system, wooden greenhouses, and the multi-span. However, double-span greenhouses are the most widely used, with their sufficient ventilation and simple management, according to growers in Egypt [1]. Greenhouses covered with screen nets or shade nets are frequently used in Egypt, especially during hot summer days, to reduce the radiation intensity in the greenhouse. Additionally, shading screens used inside the glasshouses caused the reduction of photosynthetically active radiation leading to the better quality of agricultural production [6–9]. A screen shade net can be placed outside on top of the greenhouse (using proper construction) and would be effective in minimising heat load from crops grown in the greenhouse [10,11]. The design and sustainability of the greenhouse in Egypt must consider both high temperatures on hot days in summer, low air temperatures at night in winter, and insufficient humidity levels, especially in the south of the country, throughout the whole year [12]. The natural greenhouse ventilation is built with netting on the edges, as well as one- or two-sided openings on the top floor. The top openings can be versatile to enable it to open or close, depending on the environment. Fogging can be applied inside the greenhouse for cooling and increasing relative air humidity. Applying a fogging system may also reduce crop evapotranspiration, but total water use may be the same because water is required for fogging itself [13]. Egypt has imported a range of greenhouses from countries that have highly developed greenhouse technology, such as the Netherlands, Spain, China, and Hungary, as part of its national project for 100,000 greenhouses. The Egyptian decision-makers evaluated the suitability of these greenhouses for conditions in Egypt based on the agricultural sector's experience with manufactured greenhouse systems [1].

One of the most significant issues in Mediterranean greenhouses is that from early spring to the end of autumn, there are extremely high interior temperatures during the day. These have negative impacts on the yield and quality of nearly every greenhouse crop. The main reason for those high temperatures is generally insufficient ventilation [14]. In semi-arid regions, control of the inner temperature and relative humidity is crucial in order to maintain the photosynthetic and transpiration rates of plants [15]. Forced ventilation is not economical due to its energy consumption and maintenance costs. Natural ventilation is a cheaper and more reasonable method and is very commonly used in both summer and winter to ensure a nearly optimal greenhouse climate [16]. Indoor microclimate regulation is thus a central concern in analysing the greenhouses, and natural ventilation plays a key role in indoor climate control as it directly influences heat and mass exchanges between the outside environment and the greenhouse. Vent measurements and locations are important elements in the design of natural ventilation. The correlations between ventilation rates and environmental parameters were evaluated with various approaches, including wind speed and direction [17]. Decay-rate tracer nitrate techniques were used in a single-span greenhouse with a circular arch roof and vertical walls to experimentally examine the vent form and screening effect on airflow and temperature distribution. It was observed that the indoor air velocity exhibited a rapid flow near the ground and low velocity near the roof in the case of side openings alone, while the combination of roof and side openings resulted in increase in air velocity and a reduction in indoor temperature, together with a higher microclimate heterogeneity [18]. The basic energy balance of a large greenhouse in a hot climate is calculated based on values of indoor and outdoor air temperature and humidity, outside global solar radiation, and measured wind speed and direction over significant periods, thus determining ventilation fluxes. Measurements of airspeed through vents and inside the greenhouse were also conducted to determine patterns of air movement [19]. Convection within greenhouses has been researched experimentally and numerically [20,21]. Higher indoor air temperatures are needed during cold weather for optimum plant growth and can be achieved by retaining the greenhouse effect or using some effective heating technology. On the other hand, in relatively hot climates, the greenhouse effect is needed only for a limited period of time spanning from around two to three months [22,23]. Many types of greenhouse have been used at different latitudes to grow off-season vegetables in different regions [24].

The greenhouse effect is not sufficient for all months of the year [25]. During summer, the mean air temperature in the Arabian Peninsula typically reaches 45 °C with insufficient relative humidity [26]. For instance, in Iraq during summer, ambient air temperatures can reach nearly 50 °C, making the solar greenhouse in this period unworkable. To attempt to resolve this situation, a system consisting of one indirect evaporative heat exchanger and three pads as a direct evaporative refrigeration using groundwater has been suggested [22] as an efficient technique for decreasing the temperature of the air and increasing its humidity to meet the climatic conditions needed for agriculture [23]. Greenhouse crops are mainly warm-season crops that are suited to maximum air temperatures between 17 and 27 °C, with minimum and maximum nominal temperatures between 10 °C and 35 °C. The GH indoor temperature without a climate controller could be 20–30 °C higher than the outside under hot and humid tropical climate conditions, while the air temperature may rise to 38 °C. A temperature above 26 °C is identified as a failure value and indicated that values over 25 °C would most likely reduce the yield of tomatoes. In addition, the maximum temperature of the greenhouse air should not exceed 30–35 °C [27].

Specific methodologies have been utilised in the study of natural greenhouse ventilation. Quantitative models were initially used to research natural ventilation in urban and industrial buildings and were used to establish realistic methodologies for quantifying greenhouse ventilation levels [28]. One of the methodologies commonly used to research this process is computational fluid dynamics (CFD), which, as opposed to laboratory experiments, can provide fast and reliable simulations at lower cost [29]. This simulation method has allowed a comprehensive explanation of the flow fields and thermal distribution in several greenhouses. It should be noted that relatively few transient CFD experiments are able to simulate the complex scenarios of shifts in wind speed and distance, as well as temperature [30].

Structures in agriculture such as greenhouses and ventilation mechanisms play a crucial role in climate and environmental control. Ventilation not only induces transfer of heat and humidity between the greenhouse and the ambient environment, but also leads to supplying fresh air to prevent the shortage of indoor carbon dioxide. Nowadays, there are many ventilation strategies in operation. However, due to its low cost and reduced energy consumption, natural ventilation is becoming more and more popular in the field. Nonetheless, there are several factors that influence inherent ventilation efficiency, with wind speed and wind direction having dominant effects [31]. Despite the significant amount of research undertaken to estimate the effect of wind speed on ventilation, the characteristics of airflow through the roof openings of a multi-span greenhouse are not adequately documented, especially concerning two important characteristics: namely, the effects of wind direction and magnitude on the flow patterns on the greenhouse opening planes and the detailed flow pattern at the crop level [32].

One study evaluated the efficiency of single-span commercial greenhouse ventilation according to the wind characteristics of reclaimed coastal lands, showing that the external wind patterns, along with the ratio of side vent area to greenhouse length, have a significant effect on the greenhouse's natural ventilation [29]. This study also demonstrated that ventilation rates increase as the wind speed rises. Wind towers can be used for solar greenhouses to improve natural airflow and provide higher rates of airflow. Wind towers work based on pressure gradient (the difference in pressure between the windward and leeward sides). The windward side is characterised by positive pressure, which guides air into the structure, while the negative pressure on the leeward side leads air outwards [33]. An insect screen can substantially decrease indoor wind speed and increase the temperature and humidity inside the greenhouse. Their simulation results also showed that within the canopy region, the wind speed above the canopy is higher than its below [34]. Greenhouse conditions were simulated considering the fact that while the wind force existed, the ventilation rate of the naturally ventilated greenhouse was directly proportional to the scale of the sidewall opening and the wind speed. They also reported that insect screens and dense crop rows perpendicular to the airflow would significantly hinder greenhouse ventilation by the wind [35]. The effect of wind direction on the rate of ventilation of the Spanish "parral" greenhouse with two styles of roof openings was investigated; the findings revealed that in some situations, differences in wind direction of only 10 °C could improve ventilation by up to 50% [36]. The effects of wind direction on flow patterns and ventilation efficiency, compared with a single-span pitched-roofed greenhouse, were studied, and it was reported that the speed of ventilation and flow patterns in a single-span greenhouse with continuous roof vents depended on the wind direction and opening angles of windward and leeward wind [37].

Previous studies have mainly concentrated on greenhouses ventilated by roof and side vents and presented experimental results for the key factors of the greenhouse environment (including air velocity and air temperature), culminating in the compilation of a database for validating greenhouse ventilation analysis strategies for computational fluid dynamics (CFD) [38]. A discussion was presented on the efficiency of the various discretisation methods used as CFD solvers for simulating GH's natural ventilation [39]. Ventilation is one of the main challenges of greenhouse construction, and it is possible to create a reasonable balance between ventilation and airspeed by means of careful CFD experiments [40]. The significance of analysing greenhouse air movements, caused by ventilation, and their effects on the uniformity of indoor microclimates are highly important. In reality, growers have increasingly managed to utilise any greenhouse region for high-quality yields, owing to the increasingly stronger global market rivalry [41]. At the same time, ventilation rates were lower when the wind speed increased, with subsequently more reduced air-exchange output due to inadequate matching of the supply air with the greenhouse air [42]. The sophistication and accuracy in both scientific experiments and simulations have been gradually improved, as recent tests compensate for all important greenhouse system variables [43]. One study used a 2D CFD analysis to examine the impact of a Chinese solar greenhouse segment resulting in longer sections producing higher internal temperatures than the shorter parts [44]. A two-dimensional analysis was performed on an Italian greenhouse and found that an open sidewall with a closed windward roof was the most vigorous airflow arrangement available, which eliminated 64% of the sun's rays [45]. The main elements of greenhouse design are cladding material, and shape and faces of the greenhouse [46,47]. Therefore, in a realistic physical model, precise measurements of solar radiation, and mass and heat transfer coefficients are vital because these parameters have a direct effect on greenhouse energy and temperature [48,49]. The greenhouse interior space where microclimate conditions for plant growth should be adjusted is known as the greenhouse cavity. The key parameters for the cavity of the greenhouse need to be monitored: namely, temperature, relative humidity, concentration of carbon dioxide and photosynthetic photon flux in the air inside. Plants often require a 10–30 °C temperature range and 60–90% range of relative humidity [50]. If the temperature is above 30 °C, water stress will occur due to the amount of water loss through the leaves of the plants [51]. The same effect occurs by rapid transpiration at low relative humidity. On the other hand, water and nutrients are not transferred from the root zone due to the high relative humidity, which reduces the evaporation and transpiration levels of plants [52,53]. For greenhouses in a cold environment, during the daytime, the transmitted solar radiation inside the GH is absorbed and re-emitted during the night time to be captured by the GH cover, heating the air within the GH and thereby minimising or removing the heating power needed for GH operation. Unlike in cold climates, the solar radiation of the GH in a hot environment is higher than the comfort zone of the plants. This means that a cooling system should remove that extra solar radiation from the GH environment. The appropriate GH heating or cooling system usually depends on the location (ambient conditions) of the site [54]. If the outside temperature average is less than 10 °C, GH is likely to require heating, especially at night. If the average outside temperature is below 27 °C, during the day, ventilation will prevent excessive internal temperatures; however, if the average temperature exceeds 27–28 °C, then artificial cooling may be required [52]. In this paper, a CFD model was built to test the microclimate of the GH and to the test the different inlet ventilation scenarios on the indoor microclimate of the GH under construction in Egypt.

2. Materials and Methods

2.1. The GH Model

The conceptual greenhouse (GH) model in Figure 1 was modelled as a solar-powered desalination greenhouse in Egypt. In this GH, the solar energy is used to desalinate seawater using translucent solar still units mounted on the roof of the GH [55]. The plant transpiration, which is partly extracted utilizing a condenser that serves as a dehumidifier at the GH exit, is another outlet for water output. Salah et al. [55] have developed a mathematical model focused on mass and heat transfer equations to estimate GH efficiency based on the Clear Sky Day [56,57] solar radiation experiment. The condenser shown in Figure 1 is bypassed by 75% of the cavity air (i.e., just 25% of the GH cavity air moves through the condenser), and 90% of the GH cavity air is recirculated through the down-cup, which is blended with 10% fresh air before re-entering the GH cavity. The meteorological data used for that day were determined on the basis of average values for a period of 10 years (2004–2014) [58]. The input parameters used for the model are explained in Figure 1. In Egypt, the sun shines for 12h a day in the spring season, with an average strength of around 1000 W/m^2. The solar stills (SS) can be used for desalination and generate a Zero Liquid Discharge (ZLD) model [59].

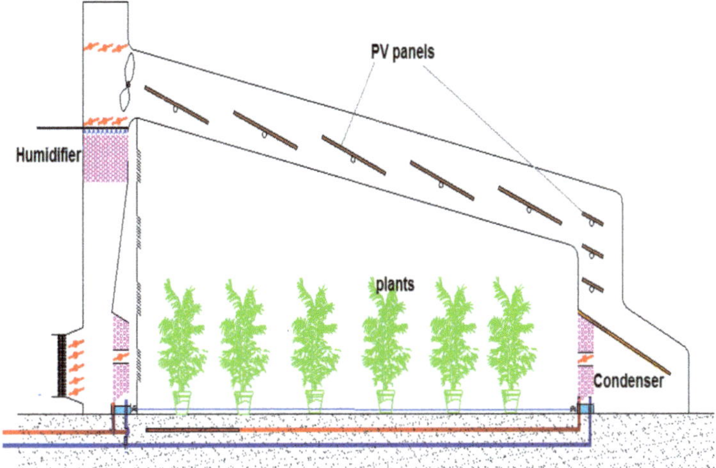

Figure 1. Conceptual model of a naturally ventilated greenhouse (GH) in Egypt [60].

2.2. Building the GH Model with Ansys

The model of the greenhouse (GH) was designed with Ansys Fluent 19.3 as a 2D model. Primary analysis was conducted on the 2D GH, as the literature [61–63] found that 2D and 3D tests on wind perpendicular to the GH ridge provided comparable results for the cross-section perpendicular to the ridge. The simulations were conducted using a pressure-based solver, and steady-state analysis was completed in such a way that the results provided were time independent with a constant wind speed [64]. Gravity was allowed, with the gravity acceleration set at -9.81 ms^{-2} on the y-axis (vertical). The full GH dimensions can be found in Table 1, with a cross-section shown in Figure 2. The internal vent, number 7 in Figure 2, is being moved up and down for the CFD simulations to analyse how its position can influence the microclimate conditions within the GH.

Table 1. The dimensions of the greenhouse structure.

Name of Dimensions	Value
Height of the main inlet	1.00 m
Height of the left-side external wall from top point to ground	5.80 m
Width of the main outlet	0.90 m
Length of the external roof	7.21m
Height of the right-side external wall from the roof to ground	2.88 m
Length of the greenhouse	8.25 m
Width of the internal vent	0.50 m
Height of the left-side internal wall	4.26 m
Length of the internal roof	5.56 m
Height of the right-side internal wall from no. 9 to ground	2.50 m
Width of the rear vents	0.50 m
Length of the rear blade	2.00 m
Length of the solar stills	1.60 m
Length of the baffles	1.00 m

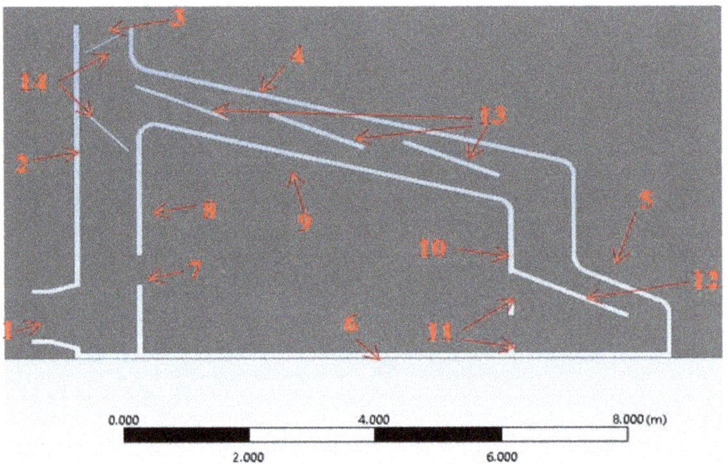

Figure 2. Cross-sectional view of the greenhouse.

To analyse the mesh sensitivity on the built CFD model, a mesh convergence analysis was performed on the geometry in Figure 3. Lower cell density was needed where strong gradients exist to minimise computational requirements without sacrificing precision, so controls of cell size were applied close to the GH walls in the model. The results were observed to converge at about 7514 triangular cells, with a maximum global element size of 5 mm and a maximum size of 50 mm for each edge of the geometry.

In this study, solar radiation was modelled using temperature and heat flux boundary conditions. The fluid properties were left as the default settings for both air and water. As the Boussinesq approximation was used in the model, the Boussinesq density was estimated using the same value as the constant fluid density, and the thermal expansion coefficient was also measured. The coefficient of thermal expansion of air was found to be 0.0034 K^{-1} at 25 °C [65], and the coefficient of thermal expansion of water was 0.000257 K^{-1} at 25 °C [66]. Throughout the simulations, the solids used were glass and soil. The material properties impacting the fluid movement and temperature calculations are the liquid pressure, and heat and thermal conductivity (see Table 2). The glass was added to both of the GH walls and roof edges, and the soil was introduced to the GH base. Nonetheless, these are just surface properties, as the wall thickness was not calculated to minimise the computational necessity; hence, the influence of the solid properties on the construct is minimal. The boundary conditions are chosen from the experimental measurements of Aiz et al.'s study for the 2D model (see Table 3) [39].

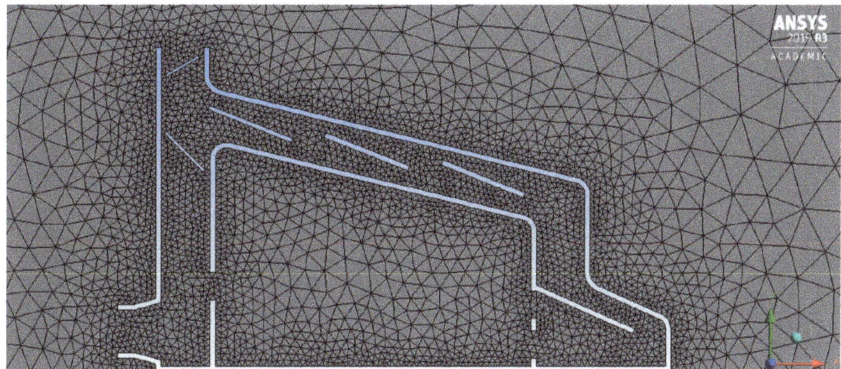

Figure 3. Final discretised model of thethe GH.

Table 2. The material properties used in the simulation for glass and soil.

Material	Density (kg m^{-3})	Specific Heat (J kg^{-1} K^{-1})	Thermal Conductivity (W m^{-1} K^{-1})
Glass	2400	753	1.0
Soil	2200	871	0.5

Table 3. The boundary conditions used for 2D design of the GH.

Named Selection	Boundary Type	Boundary Condition(s)
GH Walls	Wall (glass)	T = 310 K
GH Roof	Wall (glass)	T = 310 K
GH Floor	Wall (soil)	T = 320 K
External Floor	Wall (soil)	T = 300 K
Inlet	Velocity inlet	U = 2 ms^{-1}, 5 ms^{-1}, 10 ms^{-1} T = 290 K, 300 K, 310 K
Outlet	Pressure outlet	N/A
Internal Roof	Symmetry	N/A

The original 2D model was used as the basis for the research, and in the study conducted, each model simulation had one parameter differing from the initial model, either in the model geometry or limit conditions. Different locations for the internal vents and different configurations [opening the lower vent, opening the upper vent, having both vents open, having the third and fourth vents open] were generated to investigate the effects on the air velocity and the temperature. The internal vents were studied for heights of 0.25 m and 1 m above ground at intervals of 0.25. There are 27 different scenarios for vent configurations with three temperature and three velocity values. The external wind speed was assigned at 2 m/s, 5 m/s, and 10 m/s to analyse the effect of the wind speed on the airflow patterns and temperature contours in the GH, while the air temperature was analysed for 290 K, 300 K, and 310 K in different scenarios.

3. Results and Discussion

The air-flow patterns, velocity pathlines, and temperature contours produced by the simulation are in very close agreement with Sase et al.'s experimental work [63]. The temperature contours are distinctive, and this is possibly due to the particular form of ventilator used in the current study. Concerning quantitative validation against the previous studies, the maximum temperature for the current study is 320 K, while for the previous studies, a maximum value of 343 K was reported [55] with less than seven per cent deviation. For this analysis, the temperature range defined for plant growth is approximately 300–306 K, which is the maximum acceptable range for cultivation (average 308 K) [67]. As stated by Bartzanas et al. [68], the consequence of using pivoting ventilation is that the air flowing towards the ventilation will first flow around the ventilation itself, which allows the air to slow. The numerous airflow patterns caused by this type of vent can also be found in Shklyar and Arbel [37], which indicates similar trends to this paper.

The results of the 2D model (see Figure 4) demonstrate that the ambient air came through the main inlet on the left windward side of the greenhouse (GH) and then split into two parts. While the biggest part of it entered into the internal vent, the rest went directly to the main outlet. The internal air then moved up through the top of the GH and generated a central loop, while the air followed pathlines through the bottom of the vent, where it exited the GH main outlet at the top-left, reaching the solar stills.

The temperature contours in Figure 4a,g show that even though there was a loop in the centre of the greenhouse, smaller vortices occurred in the upper left corners. These caused sudden changes in temperature patterns in the GH. As a result of these changes, the natural ventilation conditions required for plants may not be met, and their growth may be negatively affected. On the other hand, as shown in Figure 4d, the temperature contours were more evenly distributed in the greenhouse in the scenario where the lower vent was open, which created a big loop in the centre. While the temperature of the air in the GH's centre caused a difference of 1.5 K in Figure 4a, it was 0.75 K and 0.42 K in Figure 4d,g, respectively. It is seen that in the scenario where both vents were open, more air entered the greenhouse, causing an increase in temperature.

Although the air velocity was below 1.5 m/s in the greenhouse centre for all three images (Figure 4c,f,i), in Figure 4b,h, it did not exceed 1 m/s. According to the literature, the desired natural ventilation velocity should be between 0.5–1.0 m/s. However, it can be said that in Figure 4e the velocity vectors are more homogeneously and evenly distributed in the greenhouse. It was clearly seen that the airflow accelerated between the inner and outer roofs, and the velocity vectors became more prominent in these regions. This can be explained by the fact that the air hitting the surfaces accelerated in smooth corners. Inspection of the velocity paths revealed that the barriers placed under the main outlet prevented the air from coming out directly into the greenhouse.

It can be seen from Figure 5b,e,h that the increase in wind speed generally caused the temperature to rise in the greenhouse. Vortices still occured in the scenarios (see Figure 5a,g). However, the increase in the ambient air velocity caused the increase in the velocity of air entering the greenhouse and the continuity of the air circulation. The temperature difference remained between 288.227–289.780 K for the upper-ventilation-open scenario (see Figure 5a), while for the "lower vent open" and "two vents open" scenarios (see Figure 5a,d,g), it was 288.456–289.001 K and 288.620–289.565 K, respectively. However, the increase in velocity values had a considerable effect onthe changes of the velocity streamlines (see Figure 5b,e,h) and vectors (see Figure 5c,f,i). Additionally, the results showed that by increasing the inlet air velocity, the air speed within the greenhouse could reach 3.75 m/s. The effect of sudden and continuous velocity changes on the development of plants may be harmful. Additionally, just like the scenario in Figure 4g, the higher-velocity air flows through the two vents in Figure 5g caused the temperature to rise even more.

In the scenarios in Figure 6, the initial temperature was determined as 290 K, and the speed as 10 m/s. In this case, according to the temperature contours in the experiments, the temperature cycle formed in the greenhouse centre became smaller as compared with the previous two studies (see Figure 6a,d,g). While the temperature value was in the range of 288.724–289.941 K in Figure 5a, it was 288.330–289.798 K and 289.404–290.025 K in Figure 6d,g, respectively.

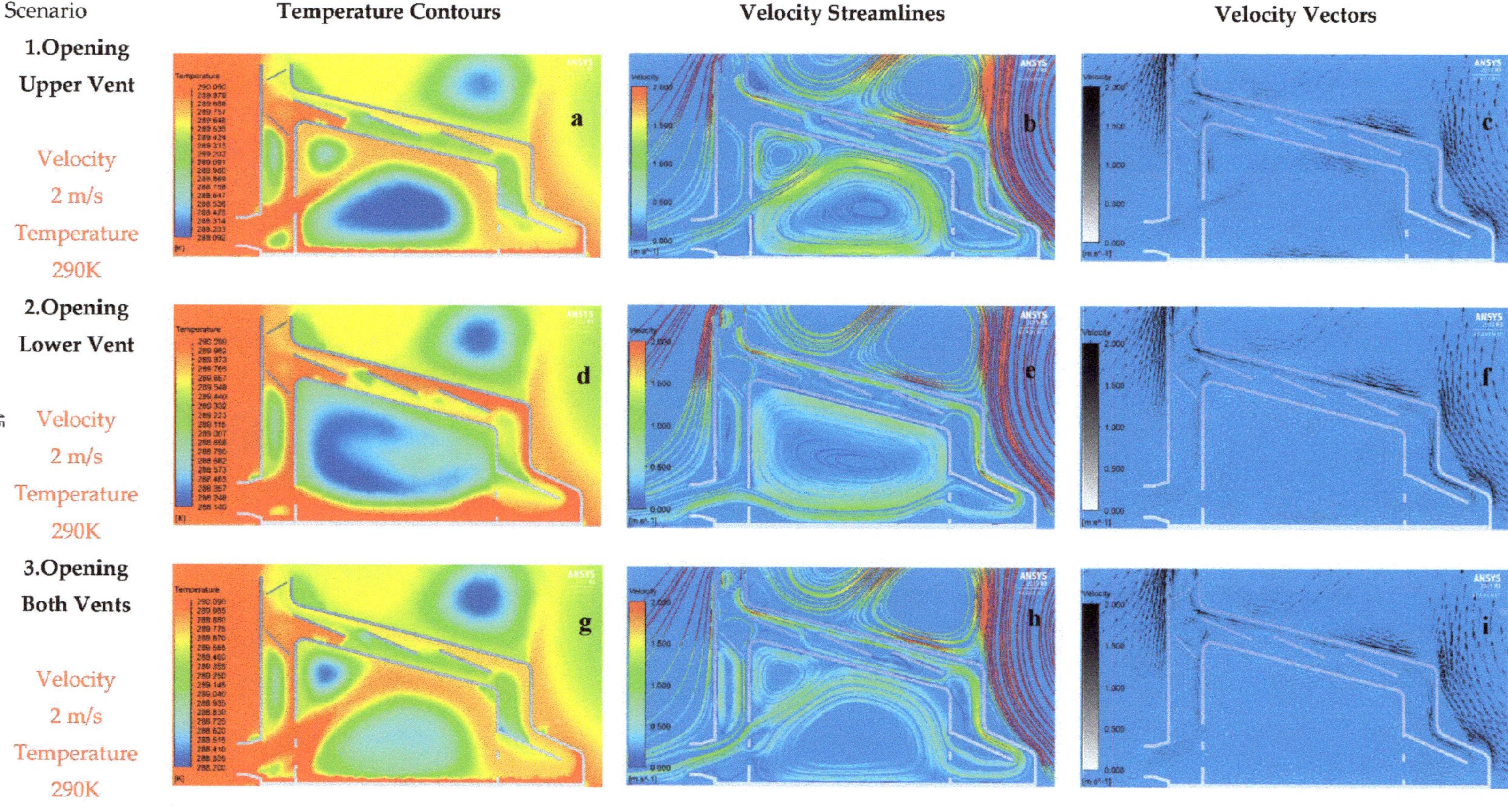

Figure 4. Temperature contours, velocity streamlines, and velocity vectors for three different vent scenarios with the initial conditions of velocity 2 m/s and temperature 290 K.

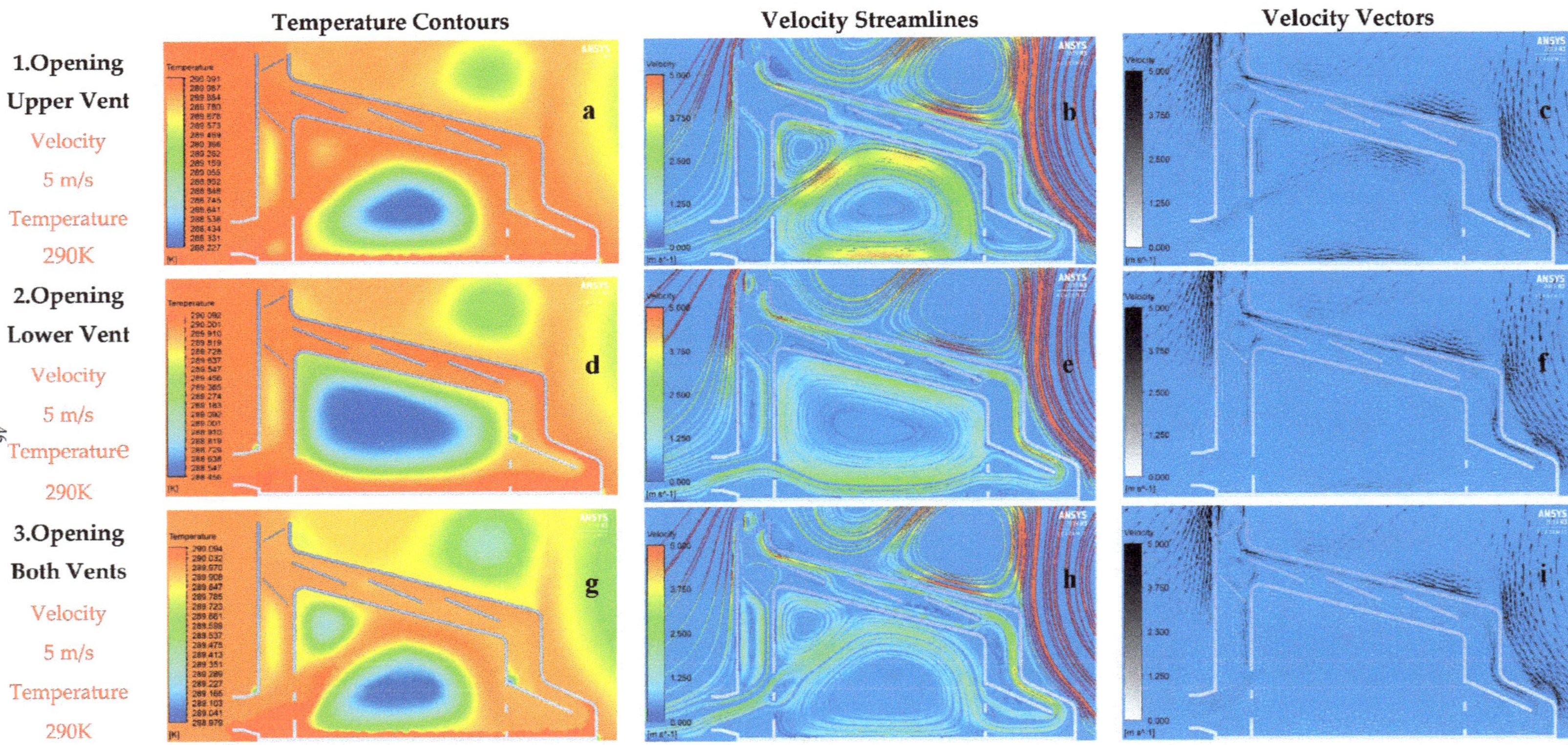

Figure 5. Temperature contours, velocity streamlines, and velocity vectors for three different vent scenarios with the initial conditions of velocity 5 m/s and temperature 290 K.

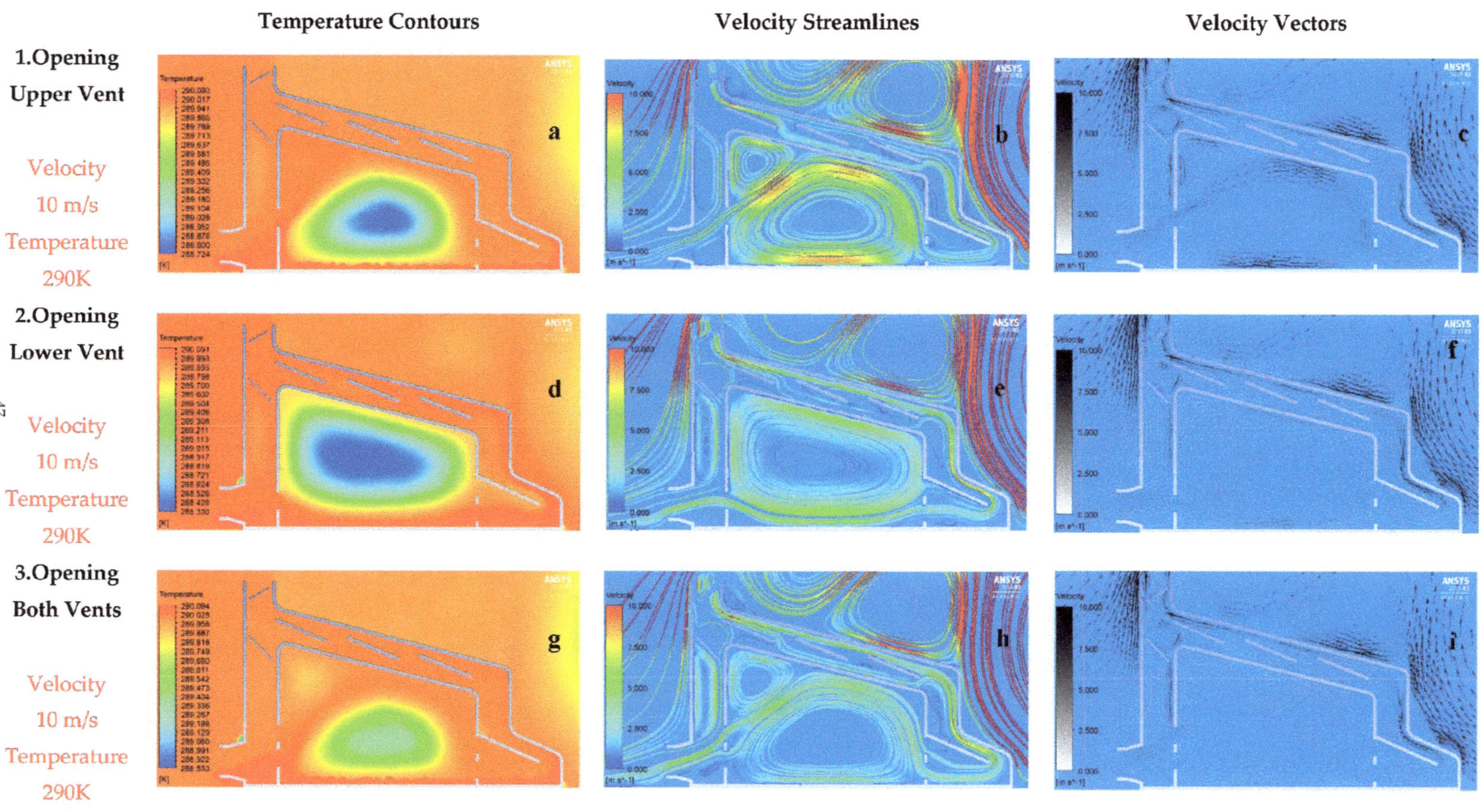

Figure 6. Temperature contours, velocity streamlines, and velocity vectors for three different vent scenarios with the initial conditions of velocity 10 m/s and temperature 290 K.

The remarkable situation here is that the optimum temperature distribution required for growing plants compared to the previous study (see Figure 6d–f). Almost all cases have a temperature change with clear contour lines. However, the temperature formed in the roof of the greenhouse reached almost the maximum level. In the scenario where both vents were open, relatively higher temperature values were obtained as compared with other cases. As this may affect the greenhouse's natural cooling system, forced ventilation may be required.

Although there is not a noticeable difference in the velocity streamline scenarios (Figure 6b,e,h) and velocity vectors (Figure 6c,f,i), it can be said that the air joining the loop accelerated, and these scenarios will not be suitable for sustainable horticulture. Besides, although a visual change in velocity streamlines and vectors was not observed in these cases, this situation had an impact on the temperature distribution in the greenhouse. However, the velocity values in the greenhouse centre varied between 2.5 and 7.5 m/s in Figure 6.

Figures 7–9 represent the data for the initial conditions of the "lower vent open", "upper vent open", and" both vents open" scenarios for 2 m/s, 5 m/s, 10 m/s velocity and 300 K temperature, respectively. It was clearly seen (see Figure 7a,d,g) that although the temperature values for all three cases gave inhomogeneous patterns in the greenhouse centre, the temperature distribution was different in the "opening lower vent" scenario (see Figure 7d) and was more suitable for plant growth. Additionally, in the same figure, in the "upper vent open" (Figure 7a) and "lower vent open" (Figure 7g) scenarios, the air entering the main inlet can be observed through the cooled temperature curves as a result of circulation. However, the temperature in the center of GH for the three scenarios ranged between 288–295 K, 288–292 K, and 293–296 K, respectively (Figure 7a,d,g). The air temperature was higher for the "both vents open" scenario (Figure 7g). The main reason for this is that after the air passed through the main inlet, the two vents took in more air (than a single vent) to the plant area in the GH. It was also seen in Figure 7 that apart from the main loop for the 1st (Figure 7b,) and 3rd (Figure 7h,) cases, there was a vortex in the upper left corners that would trigger irregularity. Inspecting the velocity paths (see Figure 7b,e,h) and vectors (see Figure 7c,f,i) in the same figure, it was seen that the streams were more regular in the 1st and 3rd figures, but in the 2nd scenario, the incoming air showed a more balanced distribution. Finally, the velocity values were between 0.1 and 0.5 m/s in the central region of the greenhouse.

Figure 8 shows that the air temperature had increased slightly in the greenhouse for all three cases. Although the temperature had dropped slightly due to the air circulation in the centre of the greenhouse, it had reached the maximum point in the remaining regions. Another significant change was that the vortices formed in Figure 8 for the 1st and 3rd scenarios dod not change the air temperature or affect it very slightly. However, when looking at velocity paths, it can be said that these irregularities still exist. However, the temperatures remained more suitable in the centre for the second scenario, and these values were between 289–296 K, 290–293 K, and 293–297 K in the centre for the "upper vent open" (Figure 8a), "lower vent open" (Figure 8d) and "both vents open" (Figure 8h) scenarios. Lastly, there was no noticeable change in velocity streamlines (see Figure 8b,e,h) and vectors (see Figure 8c,f,i), and the velocity was above 1.25 m/s for the greenhouse centre.

When the initial velocity of 10 m/s was selected according to Figure 9, the air temperature in the greenhouse reached the maximum level (see Figure 9a,h), except for the 2nd scenario (see Figure 9d). This could have a detrimental effect on the development of plants. For the 2nd scenario, the temperature remained within the desired values in the central region, but at the bottom of the greenhouse, it was higher. Although there was no noticeable difference in the velocity streamlines (Figure 9b,e,h) and velocity vectors (see Figure 9c,f,i) between the different scenarios, it can be said that the air entering the cycle accelerates and hence these scenarios will not be suitable for sustainable horticulture. Furthermore, although there was no huge change in velocity streamlines and vectors in comparison with other scenarios, it can be said that this situation had an impact on the temperature distribution in the greenhouse. However, the velocity values in the greenhouse centre varied between 2.5 and 7.5 m/s and there was still vortex in the upper left corner (see Figure 9).

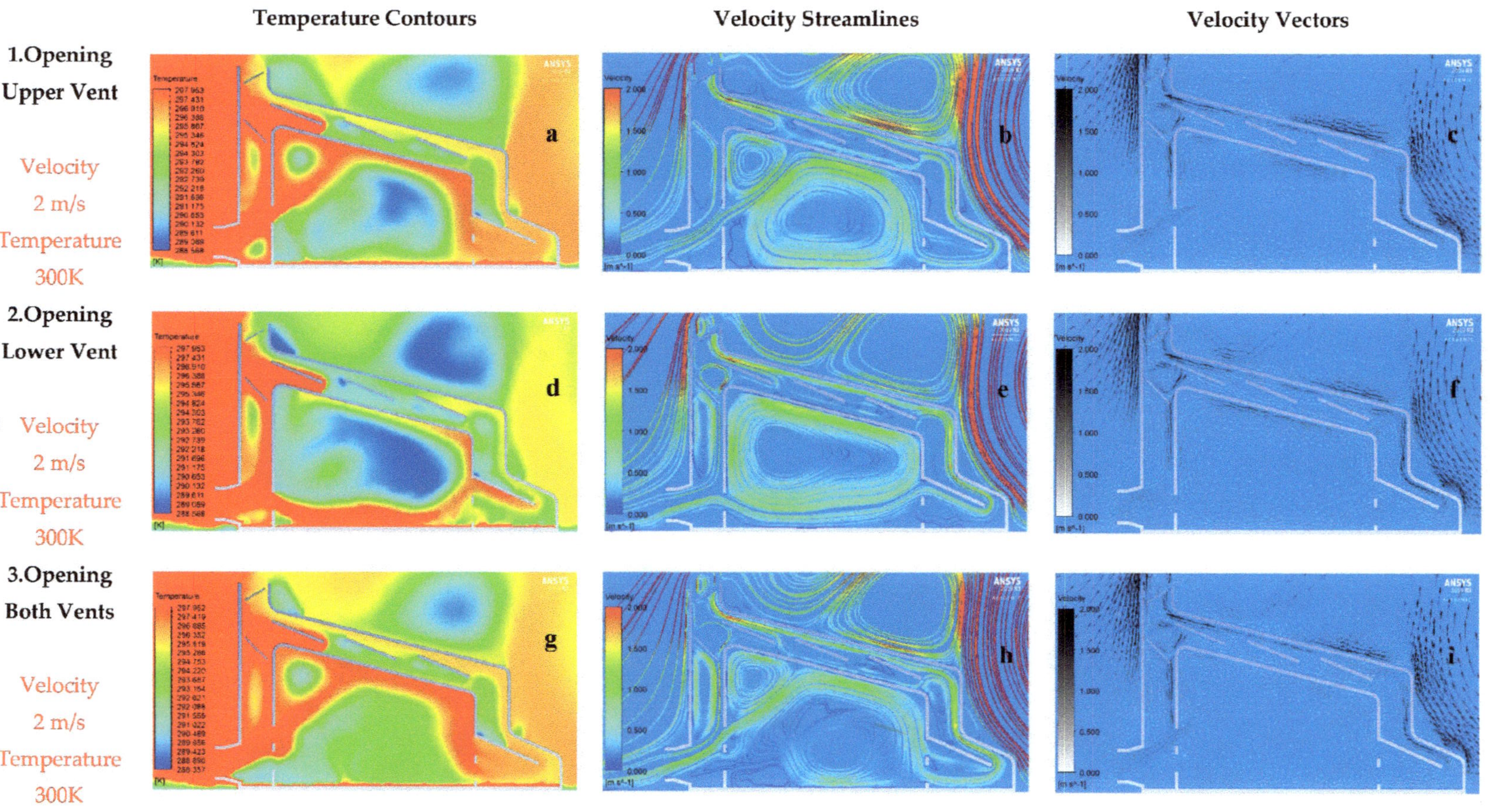

Figure 7. Temperature contours, velocity streamlines, and velocity vectors for three different vent scenarios with the initial conditions of velocity 2 m/s and temperature 300 K.

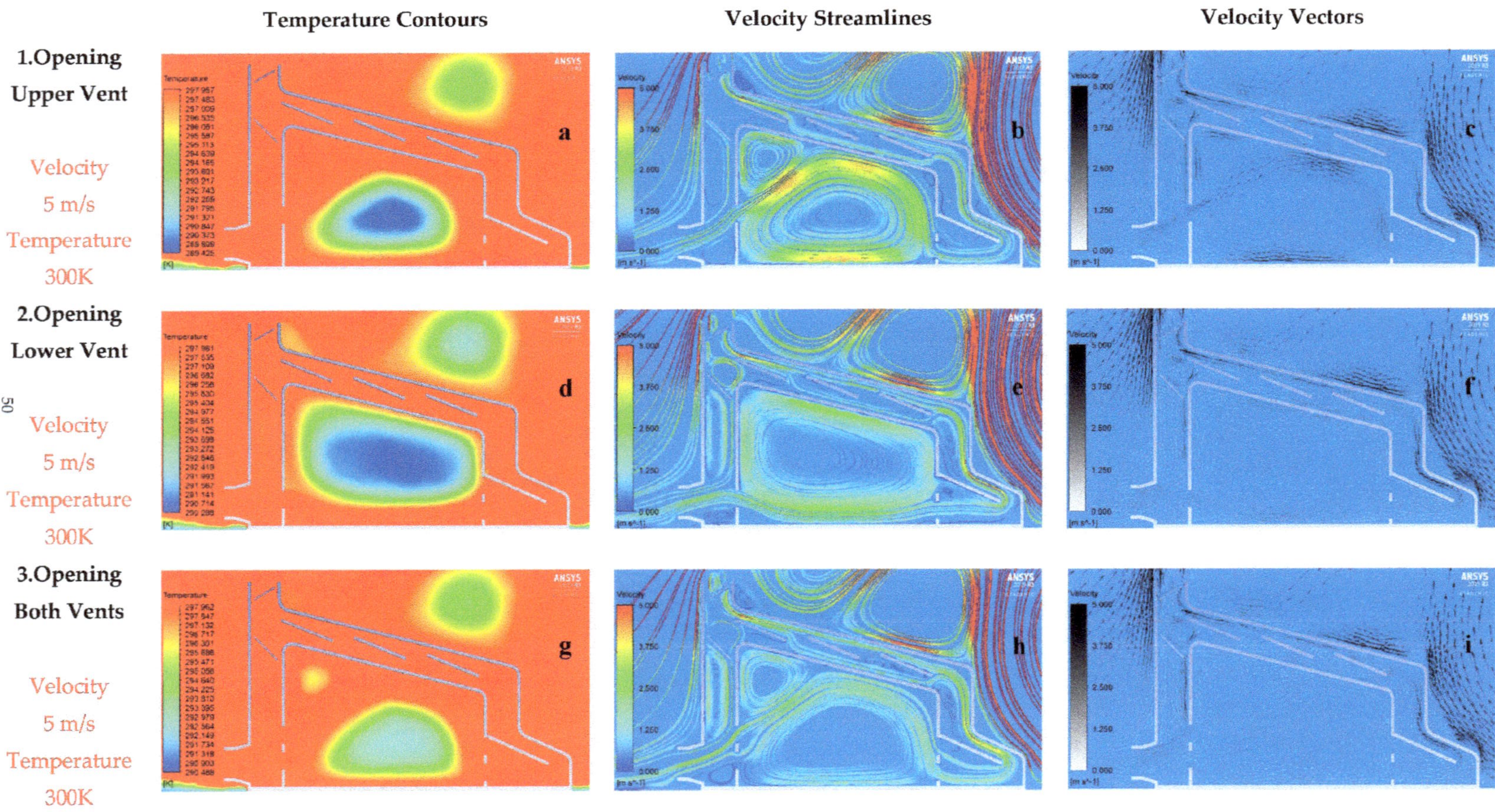

Figure 8. Temperature contours, velocity streamlines, and velocity vectors for three different vent scenarios with the initial conditions of velocity 5 m/s and temperature 300 K.

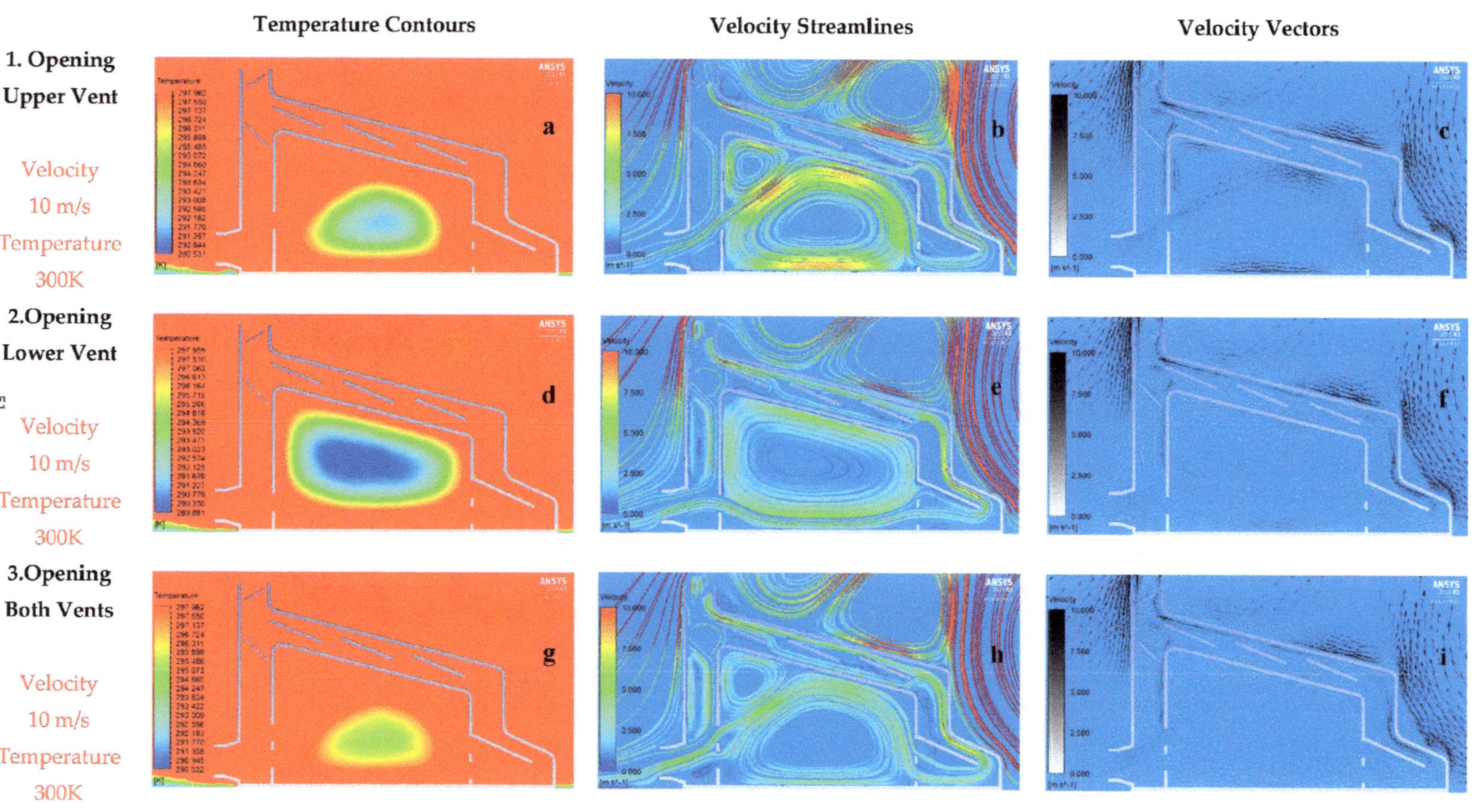

Figure 9. Temperature contours, velocity streamlines, and velocity vectors for three different vent scenarios with the initial conditions of velocity 10 m/s and temperature 300 K.

Figures 10–12 show the results of the CFD analysis for the initial temperature of 310 K and wind speeds of 2, 5, and 10 m/s, respectively. The ambient temperature was considerably higher than the temperature required for a typical plant to grow. The purpose of this last study was to study the distribution of heat and wind in the greenhouse. According to Figure 10, temperature values were not dispersed in the greenhouse in a distinct profile for the three different scenarios (see Figure 10a,d,g). The irregular temperature distribution was not beneficial for efficient and sustainable horticulture. At the same time, the temperature values were higher for the "both vents open" (Figure 10g) scenario compared to the other scenarios, and the lowest initial temperature was about 299 K. It can be said that the irregularity in this temperature distribution was caused by the formed vortices (see Figure 10b,e,h). Many large and small vortices, especially in the velocity streamlines of the 1st (Figure 10b) and 3rd (Figure 10h) scenarios, were observed. This situation arises when the high-temperature air enters the greenhouse, encounters other boundary conditions, and consequently, undergoes sudden changes. Therefore, the air cannot be circulated entirely. Additionally, inspection of the temperature contours revealed that the air hitting the greenhouse floor had cooled while going up, while the air hitting the greenhouse ceiling had warmed up.

Figure 11b,e,h show that up to 5 m/s, the air created a big loop in the greenhouse centre. However, for the 1st scenario in the same figure, there was air at the desired temperature in the greenhouse centre, while the loop was smaller and most of the greenhouse was exposed to a high temperature (see Figure 11a,d,g). In the 2nd scenario (Figure 11d), although the desired temperature condition was seen more clearly, in the 3rd scenario (Figure 11g), the minimum temperature value was even higher. In Figure 12, where the initial velocity was 10 m/s, the temperature values (see Figure 12a,d,g) increased significantly and were above the desired values. However, for the 2nd scenario, the values were more acceptable (Figure 12d). Finally, there were no substantial changes for Figures 11 and 12 in the shape of velocity streamlines and velocity vectors as compared with other studies.

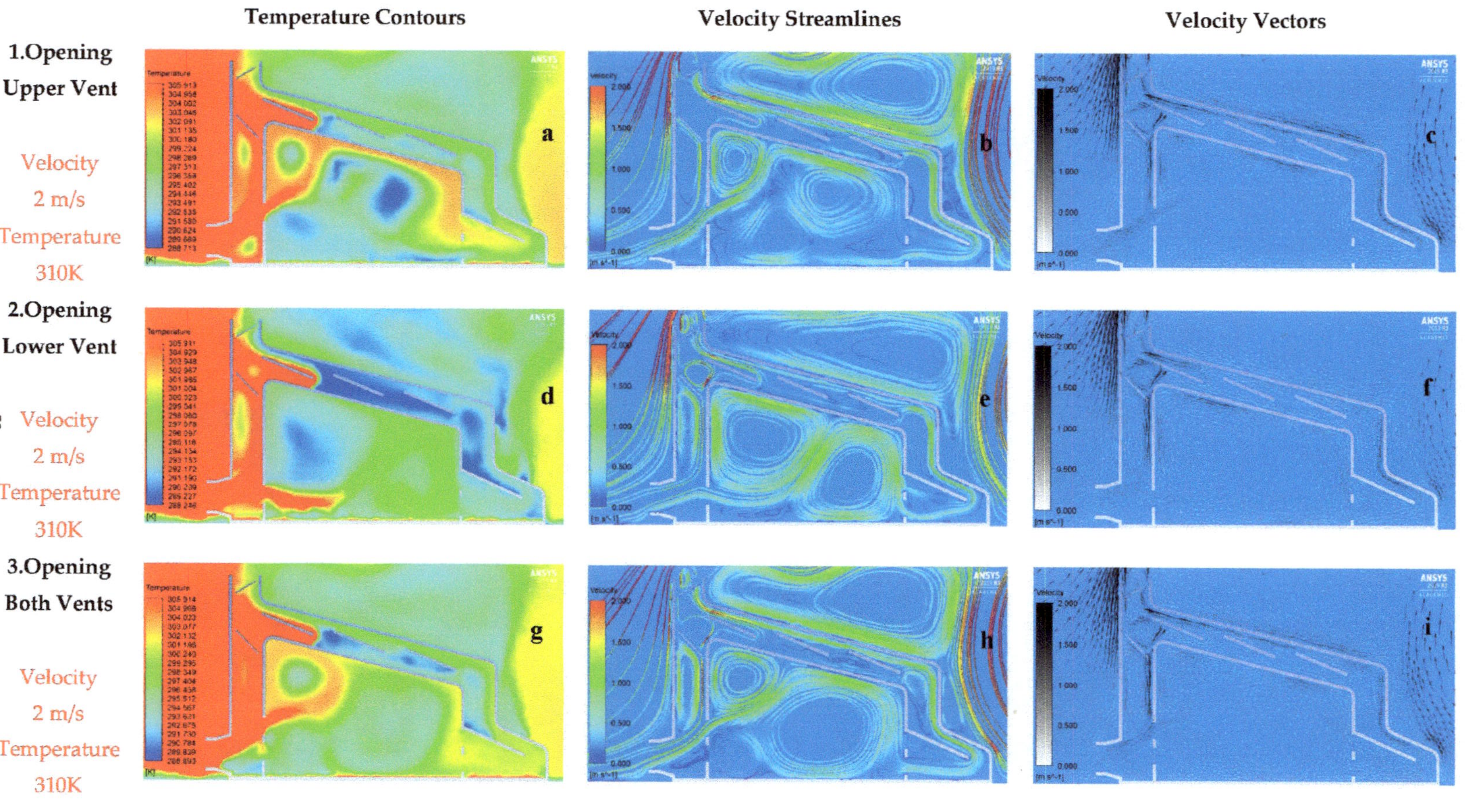

Figure 10. Temperature contours, velocity streamlines, and velocity vectors for three different vent scenarios with the initial conditions of velocity 2 m/s and temperature 310 K.

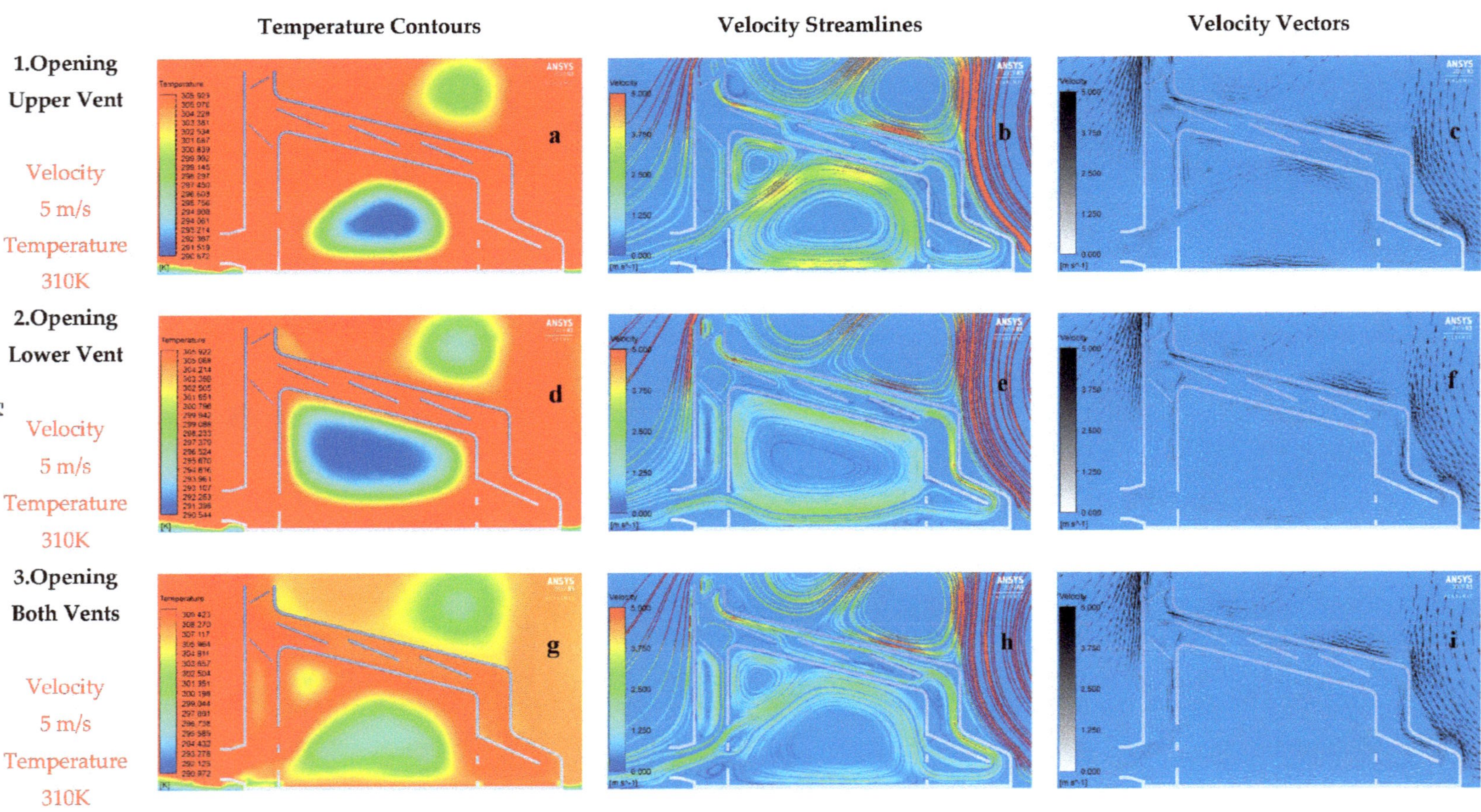

Figure 11. Temperature contours, velocity streamlines, and velocity vectors for three different vent scenarios with the initial conditions of velocity 5 m/s and temperature 310 K.

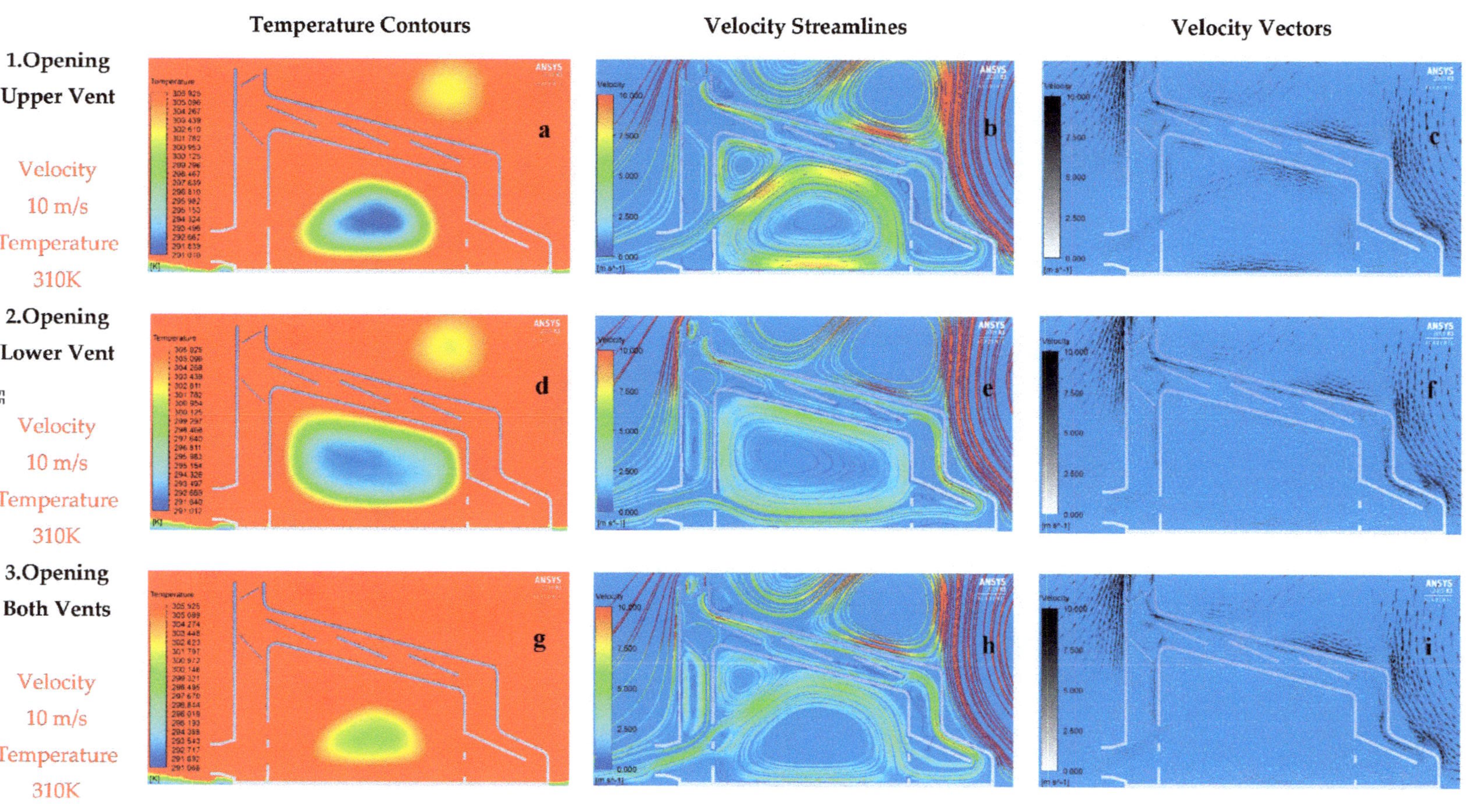

Figure 12. Temperature contours, velocity streamlines, and velocity vectors for three different vent scenarios with the initial conditions of velocity 10 m/s and temperature 310 K.

4. Conclusions

In this study, a CFD model of a solar-powered desalination greenhouse model was developed. Airflow patterns, streamlines, and the contours of temperature were found in order to analyse the microclimate conditions within the developed greenhouse. The velocity streamlines had a similar shape in all scenarios. The design parameters of the greenhouse had a significant effect on the distribution of air. In both "lower vent open" and "both vents open" scenarios, the temperature contours revealed the formations of vortices around the upper left corner. This issue triggered dramatic fluctuations in GH temperature trends. The natural ventilation requirements needed for plants may not be satisfied because of the fluctuations, and their growth may be negatively impacted. On the other hand, the "upper vent open" scenario revealed that the distribution of the air had created a vortex in the centre of the greenhouse. Furthermore, the temperature was high in the outer part of the cycle caused by the airflow formed in the greenhouse cavity. Almost all scenarios had a sudden increase in temperature. Relatively higher temperature values were obtained in the case where both vents were open. This can impact the natural cooling mechanism in the greenhouse, adding to the requirements for forced ventilation, or using fans or coolers. The case with the open lower vent works as a better option, while other scenarios can be used for different seasonal changes. The presented model can be used to develop a sustainable, naturally ventilated standalone greenhouse for countries facing water–energy shortages. Moreover, different scenarios with a change of position of the lower and upper vents and the addition of motility to the vents may be set for future works. These changes could likely affect not only the microclimate conditions (velocity of wind and temperature) inside the GH, but also its structural design.

Author Contributions: Conceptualization, M.A., C.D.M., A.H.S., H.E.S.F., A.A.J., R.F., A.N.; methodology, M.A., C.D.M., M.D., A.H.S., H.E.S.F., A.A.J.; software, M.A., C.D.M.; formal analysis, M.A., C.D.M.; investigation, M.A., A.A.J., A.N.; resources, M.A., A.A.J.; data curation, M.A., C.D.M.; writing—original draft preparation, M.A., C.D.M.; writing—review and editing, M.A., C.D.M., M.D., A.H.S., H.E.S.F., R.H.M., A.A.J., R.F., A.N.; visualization, C.D.M., M.A.; supervision, M.A., A.A.J.; project administration, M.A.; funding acquisition, A.A.J., A.N., R.F. All authors have read and agreed to the published version of the manuscript.

Funding: This paper is based on work supported by the British Council (BC) of UK, Grant No. (332435306) and Science, Technology, and Innovation Funding Authority (STIFA) of Egypt, Grant No. (30771), through the project titled "A Novel Standalone Solar-Driven Agriculture Greenhouse-Desalination System: That Grows its Energy and Irrigation Water" via the Newton-Musharafa funding scheme.

Institutional Review Board Statement: Not applicable.

Informed Consent Statement: Not applicable.

Data Availability Statement: Not applicable.

Conflicts of Interest: The authors declare no conflict of interest.

References

1. Abdrabbo, M.; Negm, A.; Fath, H.E.; Javadi, A. Greenhouse management and best practice in Egypt. *Int. Water Technol. Assoc.* **2019**, *9*, 118–201.
2. Fath, H.E.-B.S. Desalination and Greenhouses. In *Unconventional Water Resources and Agriculture in Egypt*; Negm, A.M., Ed.; Springer International Publishing: Cham, Switzerland, 2019; pp. 455–483. [CrossRef]
3. Hassan, G.E.; Salah, A.H.; Fath, H.; Elhelw, M.; Hassan, A.; Saqr, K.M. Optimum operational performance of a new stand-alone agricultural greenhouse with integrated-TPV solar panels. *Sol. Energy* **2016**, *136*, 303–316. [CrossRef]
4. Maraveas, C. Environmental Sustainability of Greenhouse Covering Materials. *Sustainability* **2019**, *11*, 6129. [CrossRef]
5. Rabbi, B.; Chen, Z.-H.; Sethuvenkatraman, S. Protected Cropping in Warm Climates: A Review of Humidity Control and Cooling Methods. *Energies* **2019**, *12*, 2737. [CrossRef]
6. Abdel-Mawgoud, A.; El-Abd, S.; Singer, S.; Abou-Hadid, A.; Hsiao, T. Effect of shade on the growth and yield of tomato plants. *Strateg. Mark. Oriented Greenh. Prod.* **1995**, *434*, 313–320. [CrossRef]

7. Matsoukis, A.; Kamoutsis, A. Studies on the growth of Lantana camara L. subsp. camara in relation to glasshouse environment and paclobutrazol. *Adv. Hortic. Sci.* **2003**, *17*, 153–158.

8. Gent, M.P. Effect of degree and duration of shade on quality of greenhouse tomato. *HortScience* **2007**, *42*, 514–520. [CrossRef]

9. Aroca-Delgado, R.; Pérez-Alonso, J.; Callejón-Ferre, Á.J.; Velázquez-Martí, B. Compatibility between crops and solar panels: An overview from shading systems. *Sustainability* **2018**, *10*, 743. [CrossRef]

10. Medany, M.; Abdrabbo, M.; Awny, A.; Hassanien, M.; Abou-Hadid, A. Growth and productivity of mango grown under greenhouse conditions. *Egypt. J. Hort* **2009**, *36*, 373–382.

11. Hasanein, N.; Abdrabbo, M.; El-Khulaifi, Y. The effect of bio-fertilizers and amino acids on tomato production and water productivity under net-house conditions. *Arab Univ. J. Agric. Sci.* **2014**, *22*, 43–54.

12. El Afandi, G.; Abdrabbo, M. Evaluation of reference evapotranspiration equations under current climate conditions of Egypt. *Turk. J. Agric. Food Sci. Technol.* **2015**, *3*, 819–825. [CrossRef]

13. Abul-Soud, M.; Emam, M.; Abdrabbo, M. Intercropping of some brassica crops with mango trees under different net house color. *Res. J. Agric. Biol. Sci.* **2014**, *10*, 70–79.

14. Lorenzo, P.; Maroto, C.; Castilla, N. Carbon dioxide air levels in plastic greenhouse in Almeria. *Agricultura* **1990**, *697*, 696–698.

15. Mistriotis, A.; Bot, G.P.A.; Picuno, P.; Scarascia-Mugnozza, G. Analysis of the efficiency of greenhouse ventilation using computational fluid dynamics. *Agric. For. Meteorol.* **1997**, *85*, 217–228. [CrossRef]

16. Papadakis, G.; Mermier, M.; Meneses, J.F.; Boulard, T. Measurement and Analysis of Air Exchange Rates in a Greenhouse with Continuous Roof and Side Openings. *J. Agric. Eng. Res.* **1996**, *63*, 219–227. [CrossRef]

17. Ganguly, A.; Ghosh, S. Model development and experimental validation of a floriculture greenhouse under natural ventilation. *Energy Build.* **2009**, *41*, 521–527. [CrossRef]

18. Kittas, C.; Boulard, T.; Bartzanas, T.; Katsoulas, N.; Mermier, M. Influence of an Insect Screen on Greenhouse Ventilation. *Trans. ASAE* **2002**, *45*, 1083. [CrossRef]

19. Demrati, H.; Boulard, T.; Bekkaoui, A.; Bouirden, L. SE—Structures and Environment: Natural Ventilation and Microclimatic Performance of a Large-scale Banana Greenhouse. *J. Agric. Eng. Res.* **2001**, *80*, 261–271. [CrossRef]

20. Shukla, A.; Tiwari, G.; Sodha, M. Energy conservation potential of inner thermal curtain in an even span greenhouse. *Trends Appl. Sci. Res* **2006**, *1*, 542–552.

21. Rico-Garcia, E.; Lopez-Cruz, I.; Herrera-Ruiz, G.; Soto-Zarazua, G.; Castaneda-Miranda, R. Effect of temperature on greenhouse natural ventilation under hot conditions: Computational Fluid Dynamics simulations. *J. Appl. Sci.* **2008**, *8*, 4543–4551. [CrossRef]

22. Esen, M.; Yuksel, T. Experimental evaluation of using various renewable energy sources for heating a greenhouse. *Energy Build.* **2013**, *65*, 340–351. [CrossRef]

23. Aljubury, I.M.A.; Ridha, H.D.a. Enhancement of evaporative cooling system in a greenhouse using geothermal energy. *Renew. Energy* **2017**, *111*, 321–331. [CrossRef]

24. Von Elsner, B.; Briassoulis, D.; Waaijenberg, D.; Mistriotis, A.; von Zabeltitz, C.; Gratraud, J.; Russo, G.; Suay-Cortes, R. Review of Structural and Functional Characteristics of Greenhouses in European Union Countries: Part I, Design Requirements. *J. Agric. Eng. Res.* **2000**, *75*, 1–16. [CrossRef]

25. Choab, N.; Allouhi, A.; El Maakoul, A.; Kousksou, T.; Saadeddine, S.; Jamil, A. Review on greenhouse microclimate and application: Design parameters, thermal modeling and simulation, climate controlling technologies. *Sol. Energy* **2019**, *191*, 109–137. [CrossRef]

26. Abdel-Ghany, A.M.; Al-Helal, I.M. Solar energy utilization by a greenhouse: General relations. *Renew. Energy* **2011**, *36*, 189–196. [CrossRef]

27. Kittas, C.; Karamanis, M.; Katsoulas, N. Air temperature regime in a forced ventilated greenhouse with rose crop. *Energy Build.* **2005**, *37*, 807–812. [CrossRef]

28. Foster, M.P.; Down, M.J. Ventilation of livestock buildings by natural convection. *J. Agric. Eng. Res.* **1987**, *37*, 1–13. [CrossRef]

29. Lee, S.-Y.; Lee, I.-B.; Kim, R.-W. Evaluation of wind-driven natural ventilation of single-span greenhouses built on reclaimed coastal land. *Biosyst. Eng.* **2018**, *171*, 120–142. [CrossRef]

30. Aich, W.; Kolsi, L.; Borjini, M.N.; Aissia, H.B.; Öztop, H.; Abu-Hamdeh, N. Three-dimensional CFD Analysis of Buoyancy-driven Natural Ventilation and Entropy Generation in a Prismatic Greenhouse. *Therm. Sci.* **2016**, *52*, 1–12.

31. Boulard, T.; Baille, A. Modelling of air exchange rate in a greenhouse equipped with continuous roof vents. *J. Agric. Eng. Res.* **1995**, *61*, 37–47. [CrossRef]

32. Teitel, M.; Ziskind, G.; Liran, O.; Dubovsky, V.; Letan, R. Effect of wind direction on greenhouse ventilation rate, airflow patterns and temperature distributions. *Biosyst. Eng.* **2008**, *101*, 351–369. [CrossRef]

33. Pakari, A.; Ghani, S. Airflow assessment in a naturally ventilated greenhouse equipped with wind towers: Numerical simulation and wind tunnel experiments. *Energy Build.* **2019**, *199*, 1–11. [CrossRef]

34. Majdoubi, H.; Boulard, T.; Fatnassi, H.; Bouirden, L. Airflow and microclimate patterns in a one-hectare Canary type greenhouse: An experimental and CFD assisted study. *Agric. For. Meteorol.* **2009**, *149*, 1050–1062. [CrossRef]

35. Bournet, P.-E.; Boulard, T. Effect of ventilator configuration on the distributed climate of greenhouses: A review of experimental and CFD studies. *Comput. Electron. Agric.* **2010**, *74*, 195–217. [CrossRef]

36. Campen, J.; Bot, G. Determination of greenhouse-specific aspects of ventilation using three-dimensional computational fluid dynamics. *Biosyst. Eng.* **2003**, *84*, 69–77. [CrossRef]

37. Shklyar, A.; Arbel, A. Numerical model of the three-dimensional isothermal flow patterns and mass fluxes in a pitched-roof greenhouse. *J. Wind Eng. Ind. Aerodyn.* **2004**, *92*, 1039–1059. [CrossRef]
38. Kittas, C.; Katsoulas, N.; Bartzanas, T.; Mermier, M.; Boulard, T. The impact of insect screens and ventilation openings on the greenhouse microclimate. *Trans. ASABE* **2008**, *51*, 2151–2165. [CrossRef]
39. Molina-Aiz, F.D.; Fatnassi, H.; Boulard, T.; Roy, J.C.; Valera, D.L. Comparison of finite element and finite volume methods for simulation of natural ventilation in greenhouses. *Comput. Electron. Agric.* **2010**, *72*, 69–86. [CrossRef]
40. Ayuga, F. Present and future of the numerical methods in buildings and infrastructures areas of biosystems engineering. *J. Agric. Eng.* **2015**, *46*, 1–12. [CrossRef]
41. Teitel, M.; Liran, O.; Tanny, J.; Barak, M. Wind driven ventilation of a mono-span greenhouse with a rose crop and continuous screened side vents and its effect on flow patterns and microclimate. *Biosyst. Eng.* **2008**, *101*, 111–122. [CrossRef]
42. Teitel, M.; Wenger, E. Air exchange and ventilation efficiencies of a monospan greenhouse with one inflow and one outflow through longitudinal side openings. *Biosyst. Eng.* **2014**, *119*, 98–107. [CrossRef]
43. Lee, I.-B.; Bitog, J.P.P.; Hong, S.-W.; Seo, I.-H.; Kwon, K.-S.; Bartzanas, T.; Kacira, M. The past, present and future of CFD for agro-environmental applications. *Comput. Electron. Agric.* **2013**, *93*, 168–183. [CrossRef]
44. Tong, G.; Christopher, D.M.; Zhang, G. New insights on span selection for Chinese solar greenhouses using CFD analyses. *Comput. Electron. Agric.* **2018**, *149*, 3–15. [CrossRef]
45. Benni, S.; Tassinari, P.; Bonora, F.; Barbaresi, A.; Torreggiani, D. Efficacy of greenhouse natural ventilation: Environmental monitoring and CFD simulations of a study case. *Energy Build.* **2016**, *125*, 276–286. [CrossRef]
46. Afou, Y.E.; Msaad, A.A.; Kousksou, T.; Mahdaoui, M. Predictive control of temperature under greenhouse using LQG strategy. In Proceedings of the 2015 3rd International Renewable and Sustainable Energy Conference (IRSEC), Ouarzazate, Morocco, 10–13 December 2015; pp. 1–5.
47. Bot, G.P. Physical modeling of greenhouse climate. *IFAC Proc. Vol.* **1991**, *24*, 7–12. [CrossRef]
48. Su, Y.; Xu, L. Towards discrete time model for greenhouse climate control. *Eng. Agric. Environ. Food* **2017**, *10*, 157–170. [CrossRef]
49. Iga, J.L.; García, E.A.; Fuentes, H.R. Modeling and Validation of a Greenhouse Climate Model. *IFAC Proc. Vol.* **2005**, *38*, 173–178. [CrossRef]
50. Paull, R. Effect of temperature and relative humidity on fresh commodity quality. *Postharvest Biol. Technol.* **1999**, *15*, 263–277. [CrossRef]
51. Sharkey, T.D.; Loreto, F. Water stress, temperature, and light effects on the capacity for isoprene emission and photosynthesis of kudzu leaves. *Oecologia* **1993**, *95*, 328–333. [CrossRef]
52. Kittas, C.; Katsoulas, N.; Bartzanas, T. Greenhouse climate control in mediterranean greenhouses. *Cuad. Estud. Agroaliment.* **2012**, *3*, 89–114.
53. Yohannes, T.; Fath, H. Novel agriculture greenhouse that grows its water and power: Thermal analysis. In Proceedings of the 24th Canadian Congress of Applied Mechanics CANCAM, Saskatoon, SK, Canada, 2–6 June 2013.
54. Akrami, M.; Salah, A.H.; Javadi, A.A.; Fath, H.E.S.; Hassanein, M.J.; Farmani, R.; Dibaj, M.; Negm, A. Towards a Sustainable Greenhouse: Review of Trends and Emerging Practices in Analysing Greenhouse Ventilation Requirements to Sustain Maximum Agricultural Yield. *Sustainability* **2020**, *12*, 2794. [CrossRef]
55. Salah, A.H.; Hassan, G.E.; Fath, H.; Elhelw, M.; Elsherbiny, S. Analytical investigation of different operational scenarios of a novel greenhouse combined with solar stills. *Appl. Therm. Eng.* **2017**, *122*, 297–310. [CrossRef]
56. Amarananwatana, P.; Sorapipatana, C. An assessment of the ASHRAE clear sky model for irradiance prediction in Thailand Nuntiya. *Asian J. Energy Env.* **2007**, *8*, 523–532.
57. *ASHRAE Handbook—Fundamentals*; American Society of Heating, Refrigerating and Air-Conditioning Engineers, Inc.: Atlanta, GA, USA, 2009.
58. Wu, W.-Y.; Lan, C.-W.; Lo, M.-H.; Reager, J.T.; Famiglietti, J.S. Increases in the annual range of soil water storage at northern middle and high latitudes under global warming. *Geophys. Res. Lett.* **2015**, *42*, 3903–3910. [CrossRef]
59. Akrami, M.; Salah, A.H.; Dibaj, M.; Porcheron, M.; Javadi, A.A.; Farmani, R.; Fath, H.E.S.; Negm, A. A Zero-Liquid Discharge Model for a Transient Solar-Powered Desalination System for Greenhouse. *Water* **2020**, *12*, 1440. [CrossRef]
60. Salah, A.; Fath, H.; Negm, A.; Akrami, M.; Javadi, A. Simulation of agriculture greenhouse integrated with on-roof Photo-Voltaic panels: Case study for a winter day. In Proceedings of the International Conference on Innovative Applied Energy (IAPE'20), Cambridge, UK, 15–16 September 2020.
61. Mistriotis, A.; Arcidiacono, C.; Picuno, P.; Bot, G.P.A.; Scarascia-Mugnozza, G. Computational analysis of ventilation in greenhouses at zero- and low-wind-speeds. *Agric. For. Meteorol.* **1997**, *88*, 121–135. [CrossRef]
62. Okushima, L.; Sase, S.; Nara, M. A support system for natural ventilation design of greenhouses based on computational aerodynamics. In Proceedings of the International Symposium on Models for Plant Growth, Environmental Control and Farm Management in Protected Cultivation 248, Hanover Germany, 28 August–2 September 1988; pp. 129–136.
63. Sase, S.; Takakura, T.; Nara, M. Wind tunnel testing on airflow and temperature distribution of a naturally ventilated greenhouse. In Proceedings of the III International Symposium on Energy in Protected Cultivation 148, Columbus, OH, USA, 21–26 August 1983; pp. 329–336.
64. Akrami, M.; Javadi, A.A.; Hassanein, M.J.; Farmani, R.; Dibaj, M.; Tabor, G.R.; Negm, A. Study of the Effects of Vent Configuration on Mono-Span Greenhouse Ventilation Using Computational Fluid Dynamics. *Sustainability* **2020**, *12*, 986. [CrossRef]

65. Hu, J.; Cai, W.; Li, C.; Gan, Y.; Chen, L. In situ X-ray diffraction study of the thermal expansion of silver nanoparticles in ambient air and vacuum. *Appl. Phys. Lett.* **2005**, *86*, 151915. [CrossRef]
66. Tang, J.C.; Lin, G.L.; Yang, H.C.; Jiang, G.J.; Chen-Yang, Y.W. Polyimide-silica nanocomposites exhibiting low thermal expansion coefficient and water absorption from surface-modified silica. *J. Appl. Polym. Sci.* **2007**, *104*, 4096–4105. [CrossRef]
67. Fernández, M.; Bonachela, S.; Orgaz, F.; Thompson, R.; López, J.; Granados, M.; Gallardo, M.; Fereres, E. Measurement and estimation of plastic greenhouse reference evapotranspiration in a Mediterranean climate. *Irrig. Sci.* **2010**, *28*, 497–509. [CrossRef]
68. Bartzanas, T.; Boulard, T.; Kittas, C. Effect of Vent Arrangement on Windward Ventilation of a Tunnel Greenhouse. *Biosyst. Eng.* **2004**, *88*, 479–490. [CrossRef]

 *sustainability*

Article

Free Discharge of Subsurface Drainage Effluent: An Alternate Design of the Surface Drain System in Pakistan

Muhammad Ali Imran [1], Jinlan Xu [1], Muhammad Sultan [2,*], Redmond R. Shamshiri [3,*], Naveed Ahmed [4], Qaiser Javed [5], Hafiz Muhammad Asfahan [2], Yasir Latif [6], Muhammad Usman [7] and Riaz Ahmad [8]

1   School of Environmental and Municipal Engineering, Xi'an University of Architecture and Technology, Xi'an 710055, China; maimran@xauat.edu.cn (M.A.I.); xujinlan@xauat.edu.cn (J.X.)
2   Department of Agricultural Engineering, Bahauddin Zakariya University, Multan 60800, Pakistan; hmasfahan@gmail.com
3   Department of Engineering for Crop Production, Leibniz Institute for Agricultural Engineering and Bioeconomy, 14469 Potsdam-Bornim, Germany
4   Key Laboratory of Mountain Surface Process and Ecological Regulations, Institute of Mountain Hazards and Environment, Chinese Academy of Sciences, Chengdu 610041, China; naveedahmed@imde.ac.cn
5   School of the Environment and Safety Engineering, Jiangsu University, Zhenjiang 212013, China; qjaved@ujs.edu.cn
6   Institute of Computer Sciences, Czech Academy of Sciences, 18200 Prague, Czech Republic; latif@cs.cas.cz
7   Institute for Water Resources and Water Supply, Hamburg University of Technology, Am Schwarzenberg-Campus 3, 20173 Hamburg, Germany; muhammad.usman@tuhh.de
8   College of Engineering, Nanjing Agricultural University, Nanjing 210031, China; riaz@cau.edu.cn
*   Correspondence: muhammadsultan@bzu.edu.pk (M.S.); rshamshiri@atb-potsdam.de (R.R.S.); Tel.: +92-333-6108888 (M.S.)

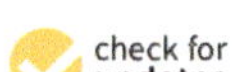
check for updates

Citation: Imran, M.A.; Xu, J.; Sultan, M.; Shamshiri, R.R.; Ahmed, N.; Javed, Q.; Asfahan, H.M.; Latif, Y.; Usman, M.; Ahmad, R. Free Discharge of Subsurface Drainage Effluent: An Alternate Design of the Surface Drain System in Pakistan. *Sustainability* **2021**, *13*, 4080. https://doi.org/10.3390/su13074080

Academic Editor: Mohammad Valipour

Received: 13 March 2021
Accepted: 1 April 2021
Published: 6 April 2021

Publisher's Note: MDPI stays neutral with regard to jurisdictional claims in published maps and institutional affiliations.

Abstract: In Pakistan, many subsurface (SS) drainage projects were launched by the Salinity Control and Reclamation Project (SCARP) to deal with twin problems (waterlogging and salinity). In some cases, sump pumps were installed for the disposal of SS effluent into surface drainage channels. Presently, sump pumps have become dysfunctional due to social and financial constraints. This study evaluates the alternate design of the Paharang drainage system that could permit the discharge of the SS drainage system in the response of gravity. The proposed design was completed after many successive trials in terms of lowering the bed level and decreasing the channel bed slope. Interconnected MS-Excel worksheets were developed to design the L-section and X-section. Design continuity of the drainage system was achieved by ensuring the bed and water levels of the receiving drain were lower than the outfalling drain. The drain cross-section was set within the present row with a few changes on the service roadside. The channel side slope was taken as 1:1.5 and the spoil bank inner and outer slopes were kept as 1:2 for the entire design. The earthwork was calculated in terms of excavation for lowering the bed level and increasing the drain section to place the excavated materials in a specific manner. The study showed that modification in the design of the Paharang drainage system is technically admissible and allows for the continuous discharge of SS drainage effluent from the area.

Keywords: surface drain system; design; drainage effluent; Pakistan

1. Introduction

Waterlogging and salinization issues (typically referred to as twin problems) elevate due to seepage from unlined canals, flooding of inferior quality groundwater, and practicing poor techniques for irrigating the land [1,2]. Because of the aforementioned twin problem, a significant proportion of the crop yield decreases in waterlogged areas worldwide due to the uplifting of anaerobic conditions and the growth of hydrophilic weeds, which develop nutrient deficiency in the root zone [3–5].

Unfortunately, in Pakistan along the Indus Basin Flood Plains (IBFP), approximately 7 million hectares (ha) of the land are affected by waterlogging and salinization. Nearly half of the waterlogged area (~3 million ha) is located in Punjab, mainly responsible for regulating ~90% of the country's food chain [1,2,6]. This might be due to this area comprising the largest irrigation structure run by gravity, which is regrettably unlined. However, the unlined system contributes a significant portion of groundwater replenishment by seepage, which is the second most crucial freshwater source. However, in some areas of Pakistan, the groundwater level is already close to the soil bed, especially the areas close to the IBFP. This leads to the conversion of precious fertile land into toxic saline unculturable land. Some of the problems associated with waterlogging and salinity in the farms and lands of Pakistan are demonstrated in Figure 1. However, keeping in mind that various efforts and approaches have been actively employed to eliminate the twin evils. For instance, management practices related to soil and crop [7] drainage networks in either the surface or subsurface [8–11] and raising the soil bed [12–14]. A significant improvement has been documented in the literature due to the consideration of management practices, as mentioned earlier [8–11].

Figure 1. Example of farm and land problems associated with waterlogging and salinity in Pakistan showing: (**a**) improper control of waterlogging and salinity in low-lying areas of Sindh, (**b**) the reduction of essential nutrients including N, P, and K, and an increase of minerals such as Fe and Mn in the soil, and (**c,d**) the mismanagement of surface water resources and salinity causing water with the total dissolved solids (TDS) over 1500 in the Thatta and Sindh districts. (Source: Adaptive AgroTech Consultancy).

Similarly, the government of Pakistan (GOP) initiated mega-projects, such as the Salinity Control and Reclamation Project (SCARP) wells between the 1960s and 2000s, whereby the vast number of tubewells (vertical drainage system) were installed to maintain the sustainable groundwater table for crops [15]. The purpose of SCARP tubewells was to lower the water table, leaching the salts by intensified irrigation, and to ensure the comfort zone for crop production [16,17] by simply pumping the water from the ground and disposing of it into a well-lined drainage channel. From here, the gravity force was employed for conveying the water into nearby streams. It was reported that about 8 million ha of land were reclaimed at a cost of USD 2 billion [18]. However, later, it was recognized that SCARP tubewells were not a promising option due to the massive operational costs, maintenance costs, and the country's socioeconomic instability. An alternative option preferred by farmers is the privatization of tubewells to obtain good-quality shallow groundwater [1,18].

Another sustainable solution for reclaiming the wetlands is developing surface or subsurface drainage structures for carrying away the water from the waterlogged areas. In this regard, the World Bank releases the funds to line the water conveying channels, particularly watercourses and farmers' field channels, for enhancement of the culturable land [2,19]. Surface or subsurface drains merely collect the water below the root zone through collectors/laterals and drain it to nearby sumps or reservoirs. The sumps or reservoirs are constructed deep below the surface or subsurface channels in order to carry a significant amount of water. When the sumps get filled, they are emptied by employing a pumping unit to pump the water from the sumps or reservoirs and outfalls into the surface drains, usually constructed bottom-up from the subsurface drains. In this regard, subsurface drains and laterals work as water collecting entities from waterlogged areas, whereas pumping units and surface drains operate as water conveying entities from one locality to another locality [3,20–22]. Fourth Drainage Project (FDP) is one of the leading projects of the World Bank's twin evil removal initiative, which was installed in Faisalabad. In this regard, the literature documented a significant improvement of groundwater quality and reclamation of salinity from the agriculturable land [23,24]. However, electricity shortfalls and extensive pumping make it dysfunctional overall. Therefore, researchers need to further explore alternative options that could be energy-efficient and economical. In order to do so, there would be technical modifications required on the FDP site. It could be done by redesigning the surface drains and employing gravity for conveying the water from the sumps to the surface drains. Alternatively, the latest techniques of GIS and geophysics have been considered for selecting suitable basins with freshwater aquifers for the formation of an efficient exploration strategy [2]. However, redesigning the surface drainage network still seems necessary.As such, various design parameters need to be taken into account, including permissible velocity, side slope, and bedslope. Permissible velocity for clear and muddy water was calculated by Fortier et al. [25] at 0.45 and 0.75 m/s, respectively. Ritzema et al. [26] stated that for fine sand and loam, the velocity ranges from 0.1 to 0.3 m/s and from 0.3 to 0.6 m/s, respectively. The government of Pakistan (GOP, 1993) [27,28] concluded that for fine sand, the maximum velocity is 0.45 m/s, and for sand and sandy loam the maximum velocity is 0.75 m/s. Deshmukh et al. [29] concluded that the time required for water to flow from a section of channel (time of flow) and the flow velocity can be affected by a resectioning of the drain, and overland flow time can be increased by raising the cross-section of the drain. The side slope should be checked for stability against erosiveness, under wet conditions, and against the sudden lowering of water levels on the cessation of flows. Steep side slopes are desirable to save excavation as well as the land area occupied. However, the steeper side slope can only be used in cohesive and well-aggregated soils or where bank protection is provided. The GOP (1993) [27] have recommended following side slope ratios also called horizontal to vertical distance (H:V) for sand, loam, and clay, which are 3:1, 2:1, and 1:1, respectively. According to the 1984 United States Bureau of Reclamation (USBR) [30] guide, the value of Manning's roughness coefficient 'n' lies between 0.025 and0.030. The value of Manning's roughness coefficient 'n' reflects the net effect of all factors causing the reduction of flow. As a general guide, berm width is taken as twice the channel depth with a minimum of 5 to 10 feet. Tipton and Kalmbach et al. [31–33] recommended a berm width of 4, 6, and 10 feet for channels having a design depth (D) of <4, 4–6, and >6 feet, respectively. The minimum berm width is taken equal to the drain design depth (D) or the depth of cut (H). The berm may be increased when higher velocities tend to enlarge the section, or the drain passes through marshy or unstable soils.

At present, the subsurface drains outfall their effluent into the sump via drainage collector pipes, from which water is pumped and carried through shallow surface disposal channels, and ultimately outfalls into surface drains. Lackey K. A. [34] reported that pumping stations involve a complex balancing of several critical priorities, including reliable long-term service, operations and maintenance considerations, capital cost, and constructability. An investigation of the current situation reveals that pumping units, electricity transmission lines, transformer devices, and other instrumentation have been pilfered, and that there is a shortage of budget and electricity for operating such systems. Okarwy et al. [35] also investigated the impact of shallow surface drainage, rather than a conventional design. The design is more economical and adoptable for economically unstable countries such as Pakistan because it excludes the requirement of sump and pump accessories.

In the present study, we proposed a redesigned consideration for the Paharang drainage network that could receive drainage effluents under the gravity flow from sumps to surface drains. The idea is to lower down the beds of surface drains from the corresponding sump beds to develop the hydraulic gradient. In this regard, active surface drain beds are excavated further to achieve the prerequisite depth, so that drainage collectors directly dispose the collected water (via perforated pores) from waterlogged areas into the surface drains under the action of gravity. In addition, the study also determines the additional design requirements for the surface drains' cross-section and subsurface drainage units. In order to achieve the subsurface gravity flow, the modified bed level of surface drains was set below the drainage collector. The required surface drain bed level is achieved by deepening the existing bed level and/or decreasing the bed slope. The proposed design was completed after many successive trials. Interconnected MS-Excel worksheets were developed to optimize and redesign the L-section and the X-section of the proposed approach.

2. Data and Methods

2.1. Study Area

The Paharang drainage network (PDN) lies in the southwestern part of Faisalabad city (longitude 73°05′1.68″ to 73°11′27.27″ N Latitude 31°34′38.50″ to 31°42′00″ E). It has a drainage area of 560 km^2. The project area contains a wide range of coarse to medium soil textures. It has an overall length of 63 km (39.15 miles). The uniform slope of 1:5000, which is generally prevalent in the area, was adopted from the source to the receiving body, the Chakbandi main drain. Moreover, eight branch/tributary drains contribute to the main drain. Figure 2 demonstrates the map and schematic of the FDP and the PDN, respectively. Existing blueprints of the PDN, which include maps, L-sections, X-sections, sump pumps along the collector's outfall, discharge in drains, hydraulic structures, water table depth, and soil type along the drain, were collected from the Punjab Irrigation Department (PID) in Faisalabad.

2.2. Computation of Required Bed Level

To discharge SS outflow into the surface drain network by the action of gravity, it is mandatory to maintain a downward slope between the SS and the surface drainage channels. It is merely done by lowering the bed of the surface drain from the level of the SS drainage collector pipe. Elevation of the SS drainage collector pipe at the surface drain (EL_C) is calculated as:

$$EL_c = CL - (D \times S) \tag{1}$$

where D is the distance from the sump to the surface drain; S is the slope of the collector drain, and CL is the existing SS drainage collector level at sump. EL_C is calculated to determine the required bed level of the surface drain. The minimum water level in the surface drain is calculated as:

$$WL_{LF} = EL_C - SF \tag{2}$$

$$BL_{RV} = WL_{LF} - d_{LF} \tag{3}$$

where WL_{LF} is low-flow water level (minimum level) in the surface drain; SF is the safe margin (0.3 to 0.5 m); BL_{RV} is the revised drain bed level, and d_{LF} is the flow depth at low flow. The existing and proposed design of the drain section, along with the water level for both the high/maximum (WL_{HF}) and low/minimum (WL_{LF}) flow conditions, are described in Figure 3. An extended SS drainage collector pipe receives water from the existing SS drainage network and directly disposes of into the surface drain.

All the sumps have the potential to operate with the proposed extended drainage collector pipe facility to outflow the SS water into the surface drainage channel because of gravity. Drainage collector pipes discharge SS water freely when the surface drain is operating at a low/minimum flow. However, this free outfall is hampered temporarily when the water level in the surface drain reaches the level of the drainage collector pipe, particularly during a storm or a high-level surface runoff. In turn, the drainage collector pipe ceases to dispose of the SS outflow into the surface drainage for a few hours/days. A self-closing valve/lid is thus mounted at the outlet of the drainage collector pipe to block/stop the backward flow of the surface drain water into the conventional sump via the drainage collector pipe.

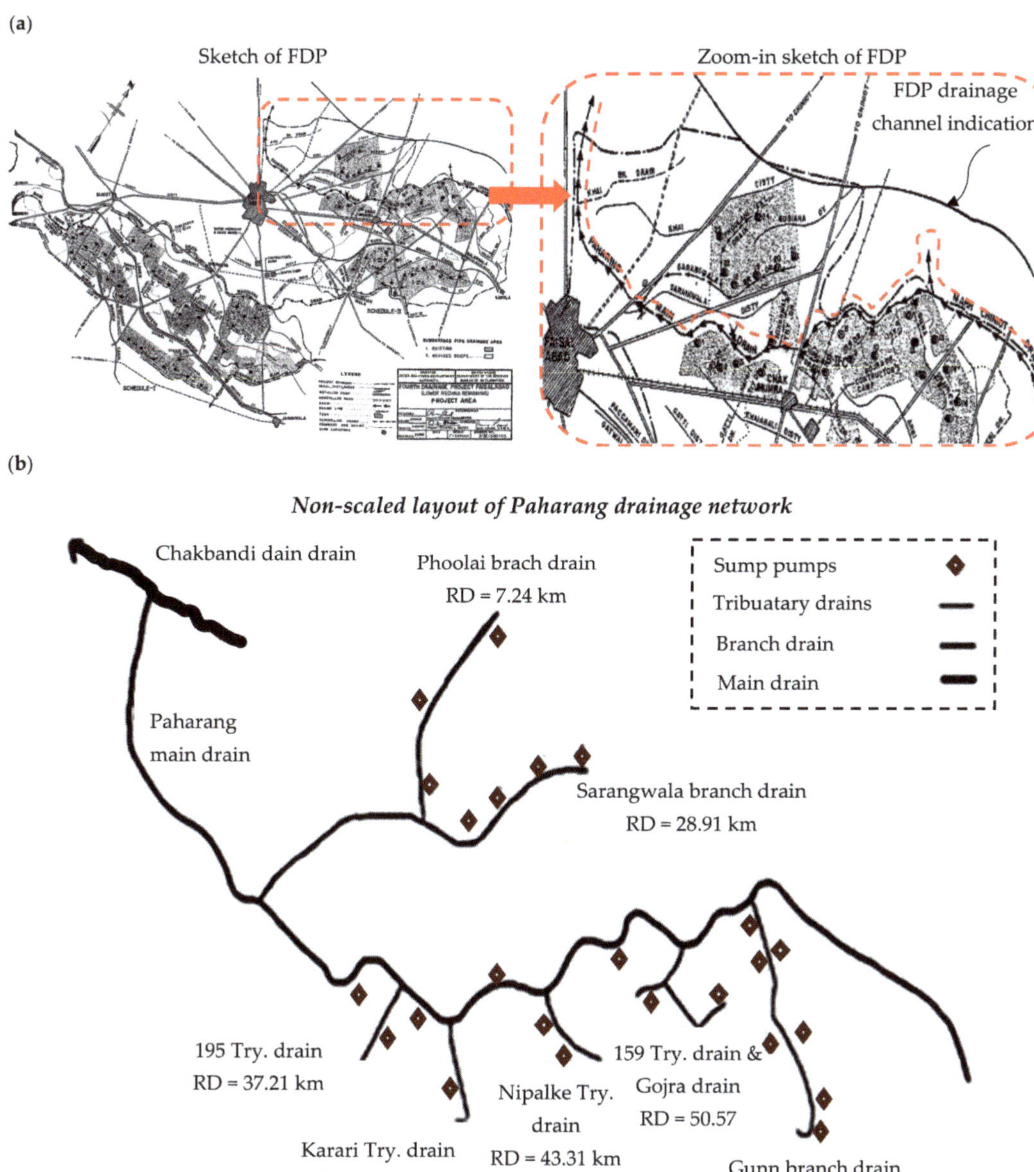

Figure 2. (**a**) Layout map of the Fourth Drainage Project (FDP), and (**b**) non-scaled layout of drainage and location of sump pumps in the Paharang drainage network (PDN).

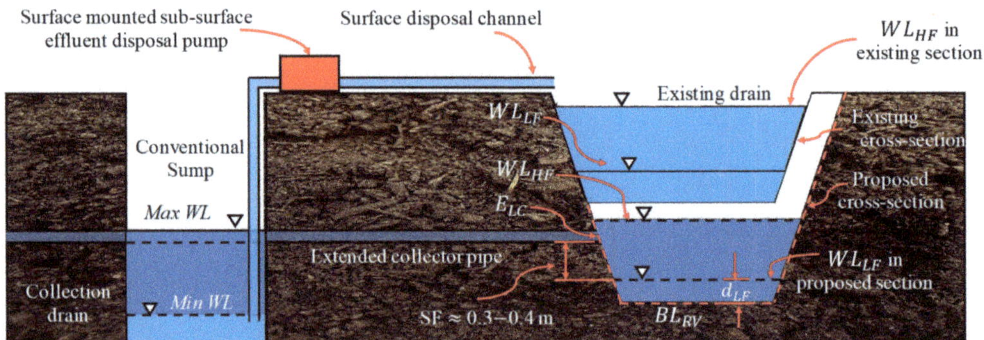

Figure 3. Schematic arrangement of extended pipe collector from sump to surface drain.

2.3. Redesign of Surface Drain L-Section

To ensure the gravity flow of SS effluent; surface drainage channels needed to be redesigned; lowering the surface drain bed and water level by 1 to 2 m. This was achieved by changing the existing bed slope of drains S_1:1 (H:V) to a smaller value S_2:1 ($S_2 > S_1$) by such an extent that the requisite bed level was achieved over the selected distance D, as shown in Figure 4.

$$\Delta BL_B = EB_{LB} - \frac{D}{(S_2 - S_1)} \tag{4}$$

$$BL_B = BL_A + \frac{D}{S_2} \tag{5}$$

where BL_A and BL_B are bed levels at A and B, respectively, D is the distance from point A to point B, EB_{LB} is the existing bed level at B, and ΔBL_B is the change in the bed level at B. The slope beyond point B may be either kept the same as the existing slope or changed to accomplish further lowering. This was repeated until the bed level of all main/branch/tributary drains was achieved as required at all the collector outfall locations.

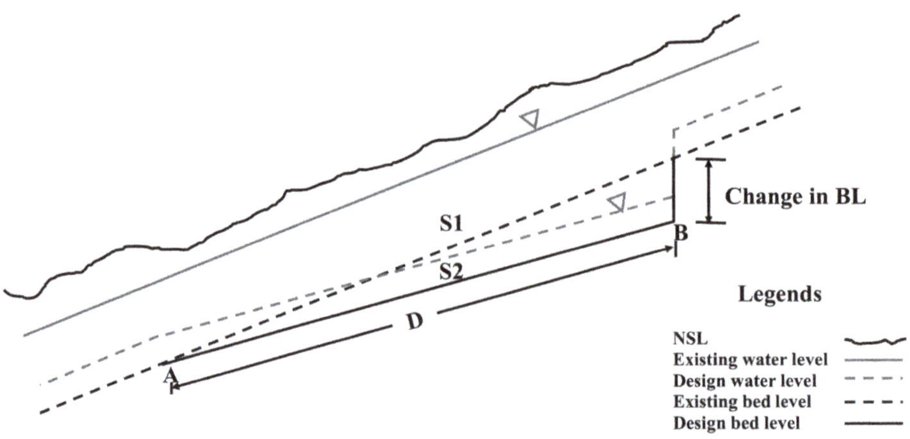

Figure 4. Typical schematic diagram of longitudinal section.

2.4. Flow Section Design

Manning's formula was used to redesign the cross-section of the proposed drains using the existing drains roughness coefficient (n) and side slope values. The drain cross-section at low flow and high flow was achieved by changing the values of bed width (B), and revised bed slope (S_2). The point where the bed slope of the surface drainage channel changes is represented as point A in Figure 4. A compromise design was achieved for all main, branch, and tributary drains and all their reaches by endorsing the prime efforts on the redesigning of the bed level of the Paharang main drain (surface drain). The simultaneous design was carried out using the Excel workbook. Interconnected worksheets were designed for each main, branch, or tributary drain. During the redesign, the drain top width was taken as a design constraint, such that the row for the revised cross-section is not varied too much from the existing row of the drain.

2.5. Flow Section Design

Design continuity of the outfalling drain (i.e., branch drain) and receiving drain (i.e., main drain) was ensured by employing the following conditions:

$$BL_O \geq BL_R \text{ and } WL_{LFO} \geq WL_{LFR} \text{ and } WL_{HFO} \geq WL_{HFR}$$

where BL is the bed level and WL is the water level. The subscripts O and R stand for the outfalling and receiving drains, respectively, whilst LFO and HFR stand for the low flow in the outfalling drains and the high flow in the receiving drains, respectively.

3. Results and Discussion

3.1. Required Bed Lowering and Design Continuity

Extended SS drainage collector pipe levels were calculated with slope ranges between 0.0007 and 0.0013. Low-flow depth was measured anddepends upon the corresponding channel section. Figure 4 shows that the outfall levels of the proposed extended drainage collector pipes are well below the existing bed level. Hence, it was required to lower the bed level of the surface drainage channel from 2.29 to 4.1 m, which includes the level of the extended drainage collector pipe, depth of discharge, and safe margin for the free disposal of the sumps effluent. Overall, 11 out of 24 reaches of the surface drainage channels were required to lower the bed level by about 3 m or more from the outfall of extended drainage collector pipe.

The required bed level of the surface drain (main drain) depends upon outfall levels of the SS drainage pipe collectors as well as the outfall levels of branch drains. Contrastingly, the bed levels of the branch and tributary drains depend upon the SS drainage pipe collector levels and their outfall, the bed level of the surface drain (main drain), and the receiving body. Thus, the selection of any parameter for the upstream drain affects the design parameters of all other higher-/lower-order surface, branch, and tributary drains accordingly. This requirement was resolved by many successive trials. The results of the final selected trial are presented in Table 1.

Table 1. Design continuity of the Paharang main drain.

Drain	RD (km)	BL_O (m)	BL_R (m)	WL_{HFO} (m)	WL_{HFR} (m)	WL_{LFO} (m)	WL_{LFR} (m)
Sarangwala Branch	28.91	178.5	178.2	180.6	180.6	179.7	179.2
195 Try. Drain	37.21	181.7	180.4	183.1	182.7	182.0	181.2
Karari Try. Drain	39.81	181.6	180.7	183.2	183.1	181.9	181.7
NIPALKE Try.	43.31	181.5	181.1	183.8	183.5	182.3	182.0
159 Try. Drain	50.57	184.0	182.8	184.8	183.5	184.5	183.6
Gojra Drain	50.57	184.0	182.8	184.7	184.5	184.3	183.6
Gunna Branch	53.91	183.5	183.2	184.7	184.4	184.2	184.0

3.2. Redesigned Longitudinal Sections

The whole length of the Paharang main drain was divided into 17 reaches, in which 7 branches/tributary drains and 4 SS drainage pipe collectors were directly outfalling into the main drain. Based on the proposed configuration of the longitudinal section, this needed to be redesigned. In the first trial, it was ensured that the bed level of the main drain remained lower than the outfall levels of the extended drainage collector pipes. From this trial, it was observed that the bed level of Paharang main drain at the outfalling point was lower than the receiving Chakbandi main drain, which resulted in non-uniformity and free flow of the hydraulic structure. Alternatively, it seemed that the design did not need any alteration up to the reduced distance (RD) of 7.16 km because there was no interconnected branch/tributary drain and no drainage pipe collector outfall. Another significant purpose to keep the section as per the existing design was the large aqueduct of Waghwal (Jhang BC and Mudhuana) laid in this section at the RD of 7.097 km. Undoubtedly, lowering the bed level of the aqueduct could be too expensive and problematic. In the second trial, the bed level of the main drain was started lower from the RD of 7.160 km. However, at the bed slope of 1:5000, the required level could not be achieved; also, it was not possible to increase the depth more than the downstream portion, as the bed level was lowered more than the first reach. In the third trial, from the reach at the RD of 7.160 to the RD of 53.950 km, the channel bed slope was decreased up to 1:8000. The required bed level was achieved, but the capacity to carry the discharge of the channel reduced as the bed slope decreased. Therefore, the channel bed width was increased to pass the design discharge. The complete proposed design of Paharang main drain along with existing design are described in Figure 5. All sumps can outfall at gravity flow during the low-flow conditions and the reach-wise design of the parameters with low-flow water levels are given in Table 2.

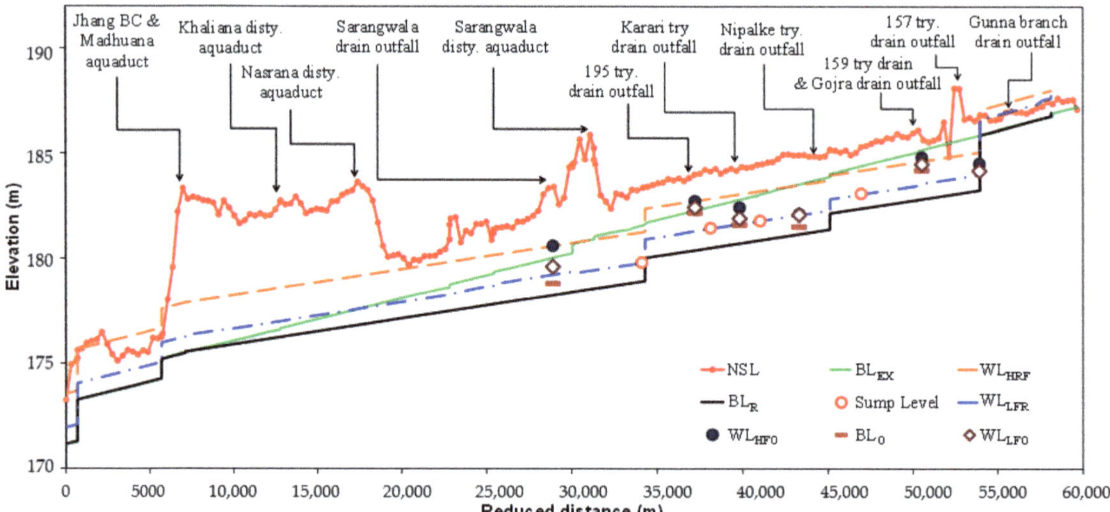

Figure 5. Proposed L-section of the Paharang main drain.

Table 2. Reach-wise design for Paharang main drain.

Reach (km)	B (m)	dHF (m)	BS H:1V	V (m/s)	High Q (m³/s)	dLF (m)	Low Q (m³/s)
0–7.16	11.28	7.9	5000	0.52	18.58	0.74	2.72
7.16–9.14	15.1	7.7	8000	0.42	18.39	0.74	2.69
9.14–12.80	14.32	7.7	8000	0.42	17.50	0.76	2.60
12.80–17.04	14.02	7.7	8000	0.42	17.16	0.77	2.58
17.04–22.55	13.72	7.7	8000	0.41	16.82	0.78	2.58
22.55–25.42	10.82	7.7	8000	0.40	13.59	0.90	2.55
25.42–30.05	9.30	7.7	8000	0.40	11.92	0.98	2.49
30.05–31.09	7.77	7.7	8000	0.39	10.25	0.88	1.81
31.09–34.14	7.62	7.7	8000	0.38	10.08	0.89	1.81
34.14–40.84	5.33	7.7	8000	0.36	7.61	0.93	1.38
40.84–45.11	4.72	7.7	8000	0.36	6.96	0.91	1.22
45.11–50.29	4.72	6	8000	0.32	4.30	0.68	0.74
50.29–53.95	3.66	6	8000	0.30	3.57	0.68	0.57
53.95–57.30	6.10	4	4500	0.35	3.40	0.76	0.57
57.30–58.12	4.57	4	4500	0.34	2.63	0.87	0.57
58.12–59.65	3.05	3.5	4500	0.30	1.47	0.87	0.57

3.3. Redesigned Cross-Section

There were six sites selected randomly to represent the different reach of length for showing the existing and proposed design of the cross-section of the Paharang main drain, presented in Figure 6. In the proposed design, the right bank of the channel was kept as the same design in most of the reaches, because it would be difficult and expensive to shift the service road and dowel. Therefore, the left side of the channel was used to increase the bed width, because extra berm width was available on the left side for future extension. However, for some reaches after the RD of 34.140 km, it became necessary to shift the dowel and service road to maintain the berm width required for the proposed design. However, there was no paved service road in these reaches. In the typical cross-section design of both the main and sub drains, the side slope of the channel was used at 1:1.5, and for the spoil bank, 1:2 inner and outer bank slope was used. In the plots of the cross-section channel (Figure 6), the side slope may not look perfect because of the unequal horizontal and vertical scaling, but in the whole design, its value was taken as 1:1.5.

In the remodeled section from the RD of 7.160 km to the RD of 53.950 km, a mild channel slope of 1:8000 was used, which reduced the discharge capacity of the drain. Therefore, passing the design discharge at the mild bed slope required a larger channel section. Channel capacity can be increased by increasing depth and bed width. Depth was increased to attain the required bed level. Thus, keeping in mind the relation of $B/d = 2$, the bed width was increased to increase the discharge capacity of the drain. As the bed width increased, the top width also increased. At the start of the tail side, increase in the top width was managed with extra width available for future extension. However, in the next section towards the head side, the extra berm was not enough. Therefore, an additional row was required for the proposed design. Similarly, the bed level difference was also increasing from tail to head, which caused an increase in the row. Important features for the proposed design cross-sections presented in the Table 3.

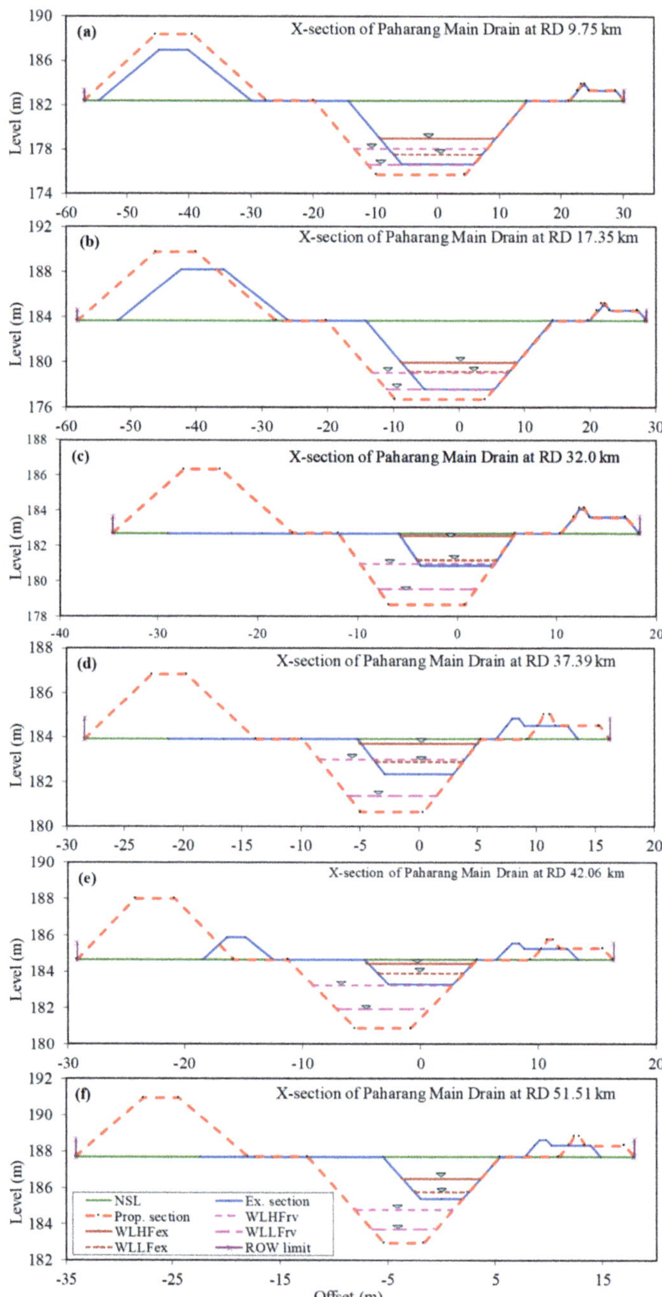

Figure 6. Cross-section design of (**a**) relative distance (RD) of 9.750 km, (**b**) RD of 17.350 km, (**c**) RD of 32.000 km, (**d**) RD of 37.390 km, (**e**) RD of 42.060 km, and (**f**) RD of 51.510 km on the Paharang main drain.

Table 3. Features of design cross-section of the Paharang main drain.

RD (km)	NSL (m)	Existing Bed Level (m)	Design Bed Level (m)	WLHF$_{RV}$ (m)	WLLF$_{RV}$ (m)	WLHF$_{EX}$ (m)	WLHF$_{EX}$ (m)	Additional ROW (m)	
								Left	Right
7.62	182.91	175.98	175.47	177.82	176.39	178.32	177.07	0	0
9.75	182.41	176.65	175.73	178.08	179.68	179.0	177.54	2.34	0
13.72	182.93	177.08	176.23	178.58	177.13	179.43	177.64	3.59	0
17.35	183.65	177.59	176.69	179.04	177.60	179.94	179.15	60.32	0
19.81	180.21	178.15	176.99	179.34	177.60	179.88	179.17	6.34	0
26.82	181.75	179.62	178.15	180.35	178.98	181.08	181.06	1.06	0
32.00	182.71	180.89	178.65	181.0	189.05	182.56	181.19	4.09	0
37.39	183.91	182.34	180.64	182.99	178.17	183.69	182.90	7.27	2.68
42.06	184.65	183.29	180.87	183.22	181.36	184.41	183.86	10.77	2.99
49.07	185.95	184.91	182.66	184.49	18.32	185.82	185.21	13.85	0
51.51	187.69	185.39	182.95	184.78	183.71	186.49	185.76	11.69	10.3

To lower the bed level and redesign the cross-section of the PDN, it was necessary to calculate the earthwork in terms of excavation, and to dump the excavated materials in a specific manner. For the proposed design, 1435 m^3 and 1030 m^3 of earthwork were required for the Paharang main drain and its branch/tributary drain, respectively.

3.4. Modification in Hydraulic Structures

Mainly, there are three types of hydraulic structures along the main drain that are bridges, aqueducts, and water coarse crossings. Because of the excavation, in terms of lowering the bed level of the drain, the stability of the structure reduces as the foundations of the supporting piers are exposed. To protect the structure and foundation's stability, one must either increase the depth of the foundation by providing a pad (expensive solution) or line the section around the structure to increase the stability and remove the risk of erosion. Both solutions can be adopted, as the proposed L-section design of the main drain shows that the change in bed level is increasing towards the source point. Thus, the proposed design can be divided in two sections at the RD of 25.910 km; at this point, the difference between the existing and proposed bed levels was 1.57 m. In this section, the exposure of the foundation was less, so the section was lined around the structure. In the second section after the RD of 25.910 km, the bed level difference was more. Therefore, the foundation depth can be increased by providing a pad below the foundation, as shown schematically in Figure 7 (left). There were three aqueducts lying in the redesigned section of the Paharang main drain. Table 4 describes the details of the changes made with respect to the bed level. In the cross-section design at the proposed bed level, the top width increases because of an increase in the bed width. Therefore, it causes an increase in aqueduct length and change in bed level and water level that are explained by the schematic, proposed design of the aqueduct in Figure 7 (right).

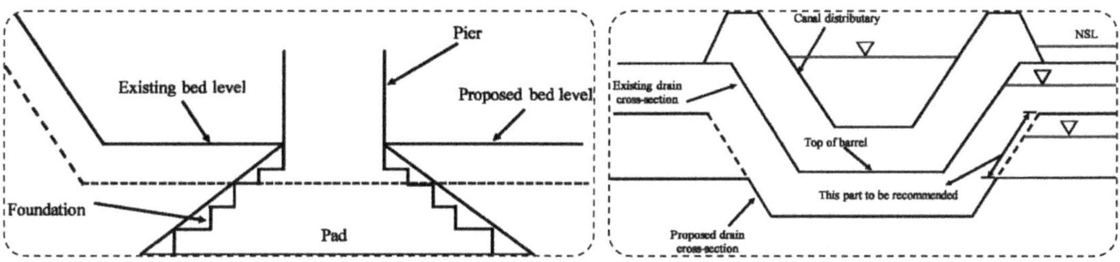

Figure 7. Schematic diagram of the pad design (**left**) and the aqueduct with the proposed design (**right**).

Table 4. Difference in design bed level (DBL) of the aqueduct.

Name	RD (km)	Existing DBL (m)	Proposed DBL (m)	Difference in DBL (m)
Aquaduct of Khaliana	12.809	176.68	176.26	0.43
Aquaduct of Nasrana	17.034	177.52	176.78	0.74
Aquaduct of Sarangwala Disty.	31.320	180.93	178.56	2.36

4. Conclusions

An alternate design of the Paharang drainage system was evaluated that can receive gravitation discharge of subsurface drainage effluent. A model based on Microsoft Excel was developed to complete the design. The proposed design was achieved by lowering the bed level and decreasing the drain bed slope. As the bed level and low-flow water level of the surface drain become lower than the subsurface drainage collector pipes, the water can discharge freely into the surface drain under gravity at all times. The present constraints related to budget, electricity, law and order, theft of sump pump parts, etc. would thus become irrelevant. Therefore, the subsurface drainage system can operate uninterrupted throughout the year. The cost related to operating the pumping unit in terms of electricity/fuel, required labor, and maintenance can also be reduced. The redesign of the Paharang main drain to admit free flow may be adopted after further economic and structural analysis. In the future, subsurface drainage projects should be designed to ensure gravity outflow into surface drains by deepening the flow section of surface drains.

Author Contributions: Conceptualization, M.A.I.; Data curation, M.A.I., and J.X.; Formal analysis, M.A.I. and J.X.; Funding acquisition, M.S. and R.R.S.; Investigation, M.S., R.R.S., Q.J., H.M.A., and M.U.; Methodology, M.A.I., N.A., and M.U.; Project administration, J.X., R.A., and M.S.; Resources, N.A. and R.A.; Software, M.A.I., Q.J., and Y.L.; Supervision, J.X., M.S., and R.R.S.; Validation, J.X.; Visualization, H.M.A., Y.L., and M.S.; Writing—original draft, M.A.I.; Writing—review and editing, J.X., M.S., R.R.S., and H.M.A. All authors have read and agreed to the published version of the manuscript.

Funding: This research received no external funding.

Institutional Review Board Statement: Not applicable.

Informed Consent Statement: Not applicable.

Data Availability Statement: Data is contained within the article.

Acknowledgments: The present research was conducted at the School of Environmental and Municipal Engineering, Xi'an University of Architecture and Technology, 710055 Shaanxi, Xi'an, China. The first author acknowledges the financial support for postdoctoral studies from University of Architecture and Technology, 710055 Shaanxi, Xi'an, China. The authors acknowledge the editorial and financial supports by the Open Access Publication Fund of Adaptive AgroTech Consultancy International.

Conflicts of Interest: The authors declare no conflict of interest.

References

1. Qureshi, A.S.; McCornick, P.G.; Qadir, M.; Aslam, Z. Managing salinity and waterlogging in the Indus Basin of Pakistan. *Agric. Water Manag.* **2008**, *95*, 1–10. [CrossRef]
2. Qureshi, A.S. Groundwater Governance in Pakistan: From Colossal Development to Neglected Management. *Water* **2020**, *12*, 3017. [CrossRef]
3. Cannell, R.Q.; Belford, R.K.; Gales, K.; Dennis, C.W.; Prew, R.D. Effects of waterlogging at different stages of development on the growth and yield of winter wheat. *J. Sci. Food Agric.* **1980**, *31*, 117–132. [CrossRef]
4. Liu, H.; Li, M.; Zheng, X.; Wang, Y.; Anwar, S. Surface Salinization of Soil under Mulched. *Water* **2020**, *12*, 3031. [CrossRef]
5. Mdrpk, D. *Operation Updates Report Pakistan: Monsoon Floods*; IFRC: Islamabad, Pakistan, 2020; pp. 1–21. Available online: https://reliefweb.int/sites/reliefweb.int/files/resources/MDRPK019du1.pdf (accessed on 9 March 2021).
6. Hocking, P.J.; Reicosky, D.C.; Meyer, W.S. Effects of intermittent waterlogging on the mineral nutrition of cotton. *Plant Soil* **1987**, *101*, 211–221. [CrossRef]
7. Manik, S.M.N.; Pengilley, G.; Dean, G.; Field, B.; Shabala, S.; Zhou, M. Soil and Crop Management Practices to Minimize the Impact of Waterlogging on Crop Productivity. *Front. Plant Sci.* **2019**, *10*, 140. [CrossRef]

8.   Estimating Long-Term Regional Groundwater Recharge for the Evaluation of Potential Solution Alternatives to Waterlogging and Salinisation-ScienceDirect. Available online: https://www.sciencedirect.com/science/article/pii/S0022169411004434?casa_token=LQtlR1vdwh0AAAAA:bZwfRpx1CZgPAxG-b4RC3iAX4cNEuAPcVrZ7QNZoEnVTfOAwlGBrHa_uxFLVnBRWS-2D13b4Wg (accessed on 9 March 2021).

9.   Chandio, A.S.; Lee, T.S.; Mirjat, M.S. Simulation of horizontal and vertical drainage systems to combat waterlogging problems along the Rohri Canal in Khairpur District, Pakistan. *J. Irrig. Drain. Eng.* **2013**, *139*, 710–717. [CrossRef]

10.   Ritzema, H.P.; Satyanarayana, T.V.; Raman, S.; Boonstra, J. Subsurface drainage to combat waterlogging and salinity in irrigated lands in India: Lessons learned in farmers' fields. *Agric. Water Manag.* **2008**, *95*, 179–189. [CrossRef]

11.   Gupta, S.K. Sub-surface drainage for waterlogged saline soils. *Water Energy Int.* **1985**, *42*, 335–344.

12.   Velmurugan, A.; Swarnam, T.P.; Ambast, S.K.; Kumar, N. Managing waterlogging and soil salinity with a permanent raised bed and furrow system in coastal lowlands of humid tropics. *Agric. Water Manag.* **2016**, *168*, 56–67. [CrossRef]

13.   Bakker, D.M.; Hamilton, G.J.; Houlbrooke, D.J.; Spann, C. The effect of raised beds on soil structure, waterlogging, and productivity on duplex soils in Western Australia. *Soil Res.* **2005**, *43*, 575–585. [CrossRef]

14.   Sayre, K.D.; Moreno Ramos, O.H. *Applications of Raised-Bed Planting Systems to Wheat*; CIMMYT: El Batan, Mexico, 1997; ISBN 9706480005.

15.   Johnson, S. Large scale irrigation and drainage schemes in Pakistan: A study of rigidities in public decision making. *Food Res. Inst. Stud.* **1982**, *18*, 149–180.

16.   Tariq, A.-R. *Drainage System Engineering*; Center of Excellence in Water Resources Engineering, University of Engineering and Technology: Lahore, Pakistan, 2008.

17.   Smedema, L.K.; Abdel-Dayem, S.; Ochs, W.J. Drainage and agricultural development. *Irrig. Drain. Syst.* **2000**, *14*, 223–235. [CrossRef]

18.   Qureshi, A.S.; McCornick, P.G.; Sarwar, A.; Sharma, B.R. Challenges and prospects of sustainable groundwater management in the Indus Basin, Pakistan. *Water Resour. Manag.* **2010**, *24*, 1551–1569. [CrossRef]

19.   Kelleners, T.J.; Chaudhry, M.R. Drainage water salinity of tubewells and pipe drains: A case study from Pakistan. *Agric. Water Manag.* **1998**, *37*, 41–53. [CrossRef]

20.   Waller, P.; Yitayew, M. *Subsurface Drainage Design and Installation BT-Irrigation and Drainage Engineering*; Waller, P., Yitayew, M., Eds.; Springer International Publishing: Cham, Germany, 2016; pp. 531–544, ISBN 978-3-319-05699-9.

21.   Hamid, Y. *Experience Gained from Interceptor Drains Installed in LBOD Stage-1 Project*; Water and Agriculture Division NESPAK: Lahore, Pakistan, 1998. Available online: https://pecongress.org.pk/images/upload/books/578.pdf (accessed on 9 March 2021).

22.   Llor, M.; Bel, A.; Ortuño, J.F.; Meseguer, F. Oxygen Consumption in Two Subsurface Wastewater Infiltration Systems under Continuous Operation Mode. *Water* **2020**, *12*, 3007.

23.   Huang, P.; Han, S. Assessment by multivariate analysis of groundwater-surface water interactions in the Coal-mining Exploring District, China. *Earth Sci. Res. J.* **2016**, *20*, G1–G8. [CrossRef]

24.   Sohag, S. Groundwater Investigation in Awlad Salameh. *Earth Sci. Res. J.* **2010**, *14*, 63–75.

25.   Fortier, S.; Scobey, F.C. Permissible Canal Velocities. *Trans. Am. Soc. Civ. Eng.* **1926**, *89*, 940–949. [CrossRef]

26.   Ritzema, H.P.; Nijland, H.J.; Croon, F.W. Subsurface drainage practices: From manual installation to large-scale implementation. *Agric. Water Manag.* **2006**, *86*, 60–71. [CrossRef]

27.   Government of Pakistan. *Surface Drainage Manual for Pakistan: Irrigation System Management Project: II, USAID Project Number 391-0467*; Government of Pakistan: Islamabad, Pakistan, 1993.

28.   Rilinger, G. Toward a Sociology of Economic Engineering: The Creation and Collapse of California's Electricity Markets between 1993 and 2001. Ph.D. Thesis, The University of Chicago, Chicago, IL, USA, 2020.

29.   Deshmukh, S.; Pansare, K.; Balsane, V.; Borse, K.; Samtani, B.K. Analytical study of quantitatvie chracheterists of watershed between Tapi and Wansuki distry River of Kakarapar and Ukai Command Area. *Ecol. Environ. Conserv.* **2015**, *21*, 345–354.

30.   Fisher, L. Minnesota Water Management Law and Section 404 Permits: A Practitioner's Perspective. *Hamline L. Rev.* **1984**, *7*, 249.

31.   Simons, D.B.; Richardson, E.V. A study of variables affecting flow characteristics and sediment transport in alluvial channels. In *Proceedings of the Federal Interagency Sedimentation Conference, Agricultural Research Service*; U.S. Department of Agriculture: Jackson, MI, USA, 1963; pp. 193–206.

32.   Witaschek, F.V. International Control of River Water Pollution. *Denver J. Int. Law Policy* **2020**, *2*, 6.

33.   Simons, D.B.; Rlchardson, E.V.; Nordin, C.F., Jr. Sedimentary structures generated by flow in alluvial channels. *AAPG Bull.* **1964**, *48*, 547.

34.   Lackey, K.A. Innovation in large capacity wastewater pumping station design. In Proceedings of the NC AWWA-WEA, WInston-Salem, NC, USA, 14–17 November 2010.

35.   Okwany, R.O.; Prathapar, S.; Bastakoti, R.C.; Mondal, M.K. Shallow Subsurface Drainage for Managing Seasonal Flooding in Ganges Floodplain, Bangladesh. *Irrig. Drain.* **2016**, *65*, 712–723. [CrossRef]

sustainability

A Sustainable Irrigation System for Small Landholdings of Rainfed Punjab, Pakistan

Marjan Aziz [1,*], Sultan Ahmad Rizvi [2], Muhammad Azhar Iqbal [3], Sairah Syed [4], Muhammad Ashraf [5], Saira Anwer [6], Muhammad Usman [6], Nazia Tahir [7], Azra Khan [8], Sana Asghar [9] and Jamil Akhtar [1]

[1] Department of Agricultural Engineering, Barani Agricultural Research Institute, Chakwal 48800, Pakistan; jamilakhtar8231@gmail.com

[2] Water Conservation Division, Soil and Water Conservation Research Institute, Chakwal 48800, Pakistan; engrsultan68@yahoo.com

[3] Centre of Excellence for Olive Research and Training (CEFORT), Barani Agricultural Research Institute, Chakwal 48800, Pakistan; azhar.horticulture@gmail.com

[4] Department of Agronomy, Barani Agricultural Research Institute, Chakwal 48800, Pakistan; sairah81@gmail.com

[5] Pakistan Council of Research in Water Resources, Islamabad 44000, Pakistan; muhammad_ashraf63@yahoo.com

[6] Faculty of Agricultural Engineering and Technology, Pir Mehr Ali Shah Arid Agriculture University, Rawalpindi 46300, Pakistan; saira.anwer791@gmail.com (S.A.); us.usman791@uaar.edu.pk (M.U.)

[7] Department of Agriculture, Abdul Wali Khan University, Mardan 23200, Pakistan; naziatahir@awkum.edu.pk

[8] Department of Agronomy, Soil and Water Conservation Research Institute, Chakwal 48800, Pakistan; azrasawcri33@gmail.com

[9] Department of Horticulture, Horticultural Research Station, Sahiwal 57000, Pakistan; sanaasghar14@yahoo.com

* Correspondence: marjan_aziz19@hotmail.com; Tel.: +92-332-677-4140

Citation: Aziz, M.; Rizvi, S.A.; Iqbal, M.A.; Syed, S.; Ashraf, M.; Anwer, S.; Usman, M.; Tahir, N.; Khan, A.; Asghar, S.; et al. A Sustainable Irrigation System for Small Landholdings of Rainfed Punjab, Pakistan. *Sustainability* **2021**, *13*, 11178. https://doi.org/10.3390/su132011178

Academic Editor: Michael S. Carolan

Received: 3 September 2021
Accepted: 30 September 2021
Published: 11 October 2021

Publisher's Note: MDPI stays neutral with regard to jurisdictional claims in published maps and institutional affiliations.

Abstract: Drip irrigation has long been proven beneficial for fruit and vegetable crops in Pakistan, but the only barrier in its adoption is the high cost of installation for small landholders, which is due to overdesigning of the system. In the present study, the cost of a conventional drip irrigation system was reduced by redesigning and eliminating the heavy filtration system (i.e., hydrocyclon, sand media, disc filters (groundwater source), pressure gauges, water meters, and double laterals).Purchasing the drip system from local vendors also reduced the cost. Field trials were conducted during 2015 and 2016 to observe the productive and economic effects of low-cost drip irrigation on vegetables (potato, onion, and chilies) and fruits (olive, peach, and citrus). The low-cost drip irrigation system saved 50% cost of irrigation and increased 27–54% net revenue in comparison with the furrow irrigation system. Further, water use efficiency (WUE) was found from 3.91–13.30 kg/m^3 and 1.28–4.89 kg/m^3 for drip irrigation and furrow irrigation systems, respectively. The physical and chemical attributes of vegetables and fruits were also improved to a reasonably good extent. The present study concluded that low-cost drip irrigation increased the yield by more than 20%, as compared with traditional furrow irrigation, and thus, it is beneficial for the small landholders (i.e., less than 2 hectares).

Keywords: agricultural economy; drip irrigation system; net revenue; small landholders; sustainable irrigation

1. Introduction

Irrigation, along with other quality inputs, is crucial for the livelihood and food security of Pakistan [1]. Land and water management practices are two very important components to outstrip the water use efficiency and livelihood of rainfed areas [2,3]. In the present system of irrigation, low water use efficiency, and low agricultural productivity are the topmost concerns of the Government of Pakistan [4]. Two possible ways to enhance agricultural productivity include either bringing more area under farming (horizontal expansion) or increasing the production per hectare (vertical expansion) [5]. Historically farmers of Pakistan have been using conventional irrigation methods comprising basin, border, and furrow to irrigate the crops, in which the entire fieldis watered without considering the actual crop water requirement. These traditional methods of irrigation have created immense issues such as waterlogging and salinity, and on the other hand, their application efficiency is very low [6]. There are numerous substitute strategies to improve the water application efficiency such as using drip-and-sprinkler irrigation, considering climatic and land parameters, as well as altering the cropping pattern or varieties [4].

Punjab is Pakistan's agro-economic hub that contributes to about 80 percent of the country's food needs [7]. During the last some decades, climate change has had a crucial effect on the country's water resources. In response, progressive farmers started using high-efficiency irrigation systems. Due to their high initial and operational costs, small landholders are constrained to employ these modern technologies due to their poor economic conditions and low potential returns. Modern technologies are necessary to address water scarcity and enhance crop performance and water productivity. The use of high efficiency and low-cost irrigation system is one of many options to overcome the water losses caused by conventional methods [8]. Drip irrigation system, when compared with the furrow irrigation system, gives the optimum potential to enhance yields and irrigation water use efficiency [9]. Evidently, the furrow irrigation system involves a little initial cost, and it appears to be beneficial, but in reality, furrow irrigation systems require vigorous labor for their establishment and need regular maintenance due to having low application efficiency (45%), as indicated in a study by [10].

Efficient systems such as drip irrigation have been tested in various crops and found to be beneficial in water resources conservation and water productivity enhancement. Many farmers have limited financial resources to install this system. Pakistan is an agricultural country that is currently facing the problem of water scarcity to fulfill different crop requirements. Drip/trickle irrigation technology was introduced in Pakistan during the early 21st century. After years of research and promotion of high-efficiency irrigation systems through subsidized schemes, drip irrigation technology has become available for easy adoption by farmers. Due to high installation costs, less awareness, and training of farmers for its use, this technology still needs to be tested and evaluated at farmer's fields to achieve large-scale farmer adoption [11,12]. Although the subsidized schemes of the government have promoted drip-and-sprinkler technologies, training and knowledge support to farmers for shifting toward high-value cash crops are limiting factors [13]. Farmers worldwide have been using drip irrigation systems since the 1990s, but the trend of adoption is quite moderate in Pakistan for small landholders because of (1) excessive designing, which makes this system very costly for small landholdings and (2) poor management of drip irrigation system.

Keeping in mind the adoption constraints by the small landholders, the current study planned to redesign the system by setting up the simpler parts without the involvement of companies. Hence, the main objectives of this research study were to (1) redesign the system and examine the economics of a low-cost drip irrigation system for small landholders (farmers with lands less than 2 hectares) and (2) compare the drip irrigation system with furrow irrigation in terms of water saving and yield improvement.

2. Materials and Methods

The experiment was conducted at Barani Agricultural Research Institute (BARI), which is located at 72°43.4′ longitude, 32°55.5′ latitude, having an altitude of 522 m. The weather conditions of Chakwal are arid to semiarid with annual rainfall varying from 500 to 1000 mm (1979–2016) [14]. The soil of the experimental site is piedmontalluvial (plains order: ALFISOL belongs to Therpal/Satwal/Kotli series). The physical and chemical properties of the soil as reported [15] are presented in Table 1.

Table 1. Soil physical and chemical properties of the experimental site.

Physical and Chemical Properties of Soil	Depth below Ground Surface	
	0–15 cm	16–30 cm
Clay (%)	10	10
Silt (%)	30	30
Sand (%)	60	60
Nitrogen (%)	0.8	2.0
Phosphorus (ppm)	5.0	3.4
Potassium (ppm)	138.0	132
Organic matter (%)	0.6	0.33
Electric conductivity (dS/m)	0.3	0.25
pH	7.68	7.79

The trials were set up in a completely randomized block design (RBCD) with two treatments T1 (low-cost drip irrigation) and T2 (conventional furrow irrigation), each having five replications, as shown in Figure 1a–d. Furrows and ridges were prepared by means of a ridger, keeping the maximum length of furrow as 30 m to avoid deep percolation losses.

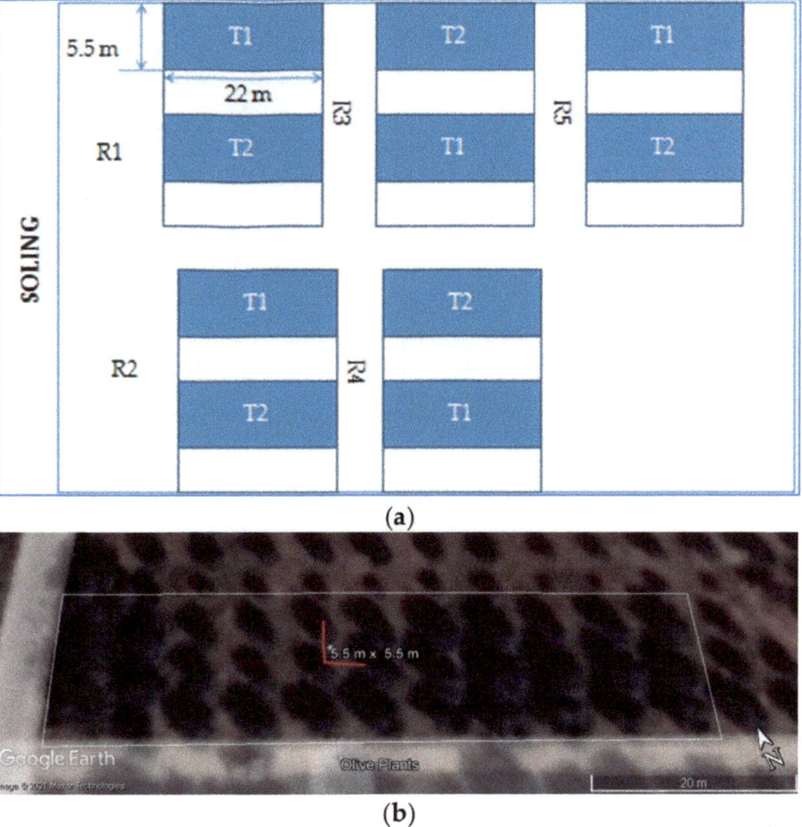

Figure 1. *Cont.*

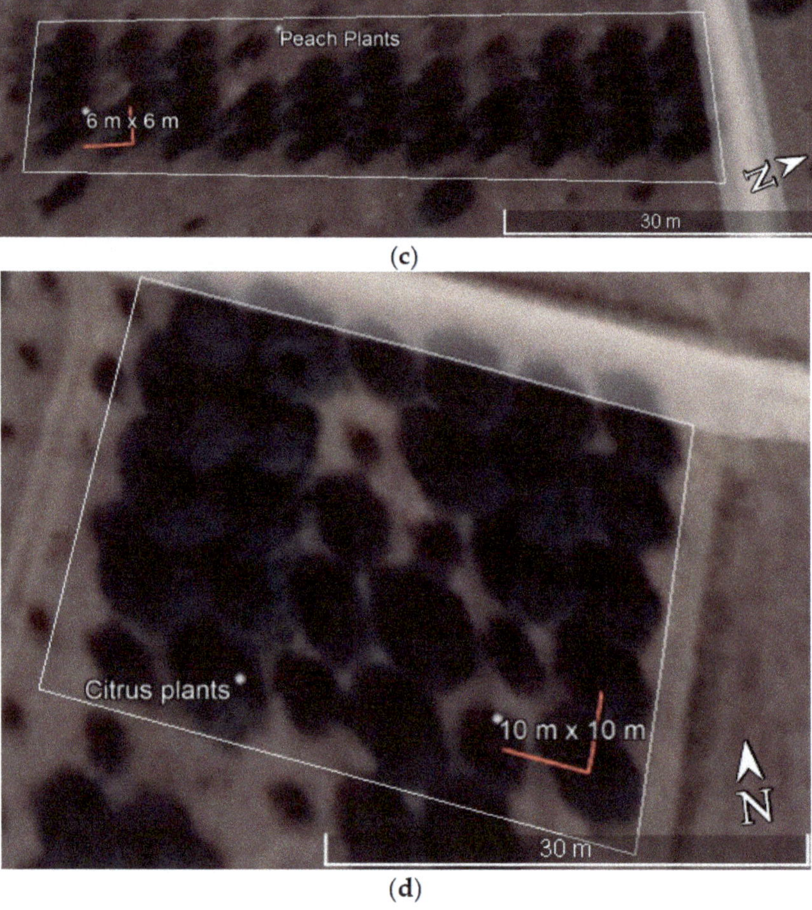

Figure 1. (**a**) Block diagram of vegetables (potato, onion, and chilies) plots; (**b**) the layout of olive plant; (**c**) the layout of peach plant; (**d**) the layout of citrus plant.

In Figure 1a, R1, R2, R3, R4, and R5 refer to replications, and T1 and T2 refer to treatments.

Vegetables were sown as per conventional farmers' practice, and plants having age of six years were selected from the existing orchards of BARI, as depicted in Figure 1b–d. The planting geometry of vegetables (potato, onion, and chilies) and plants (olive, citrus, and peach) is presented in Table 2.

A low-cost drip irrigation system was designed and installed in the fields manually. This system comprised a main and sub main lines for each set having 38 mm dia pipe made of polyvinyl chloride (PVC), further attached to lateral lines having 16 mm dia made of low-density polyethylene (LDPE) fitted with 0.006 m^3/h drippers (Figure 2). In all crops, lateral lines were placed parallel to the plant lines. Lateral lines with built-in drippers were used for row crops (onion, potato, and chilies), while two (2) drippers/plants were placed on the lateral line for fruit plants (olive, peach, and citrus). The parts of the drip irrigation system were purchased from local vendors (local market) and installed manually (without the involvement of a company). Testing of drippers was performed to check the pressure and flow variations by using the standard method described in [16]. Pressure and flow rates were maintained and recorded as given in Table 3.

Table 2. Planting geometry of the crops (vegetables and plants).

Crops (Variety Name)	Age of Plant	Row-Row Distance (m)	Plant-Plant Distance (m)	Area/Plant (m^2)	Total Area under Crop (m^2)
Potato (Desirie)	1 season	0.61	0.204	0.124	1220
Onion (Phulkara)	1 season	0.69	0.101	0.070	1220
Chilies (Ghotki)	1 season	0.735	0.46	0.338	1220
Olive (BARI Zaitoon1)	6 years	5.5	5.5	30	990
Peach (Early Grand)	6 years	6	6	36	1584
Citrus (Musambi)	6 years	10	10	100	3000

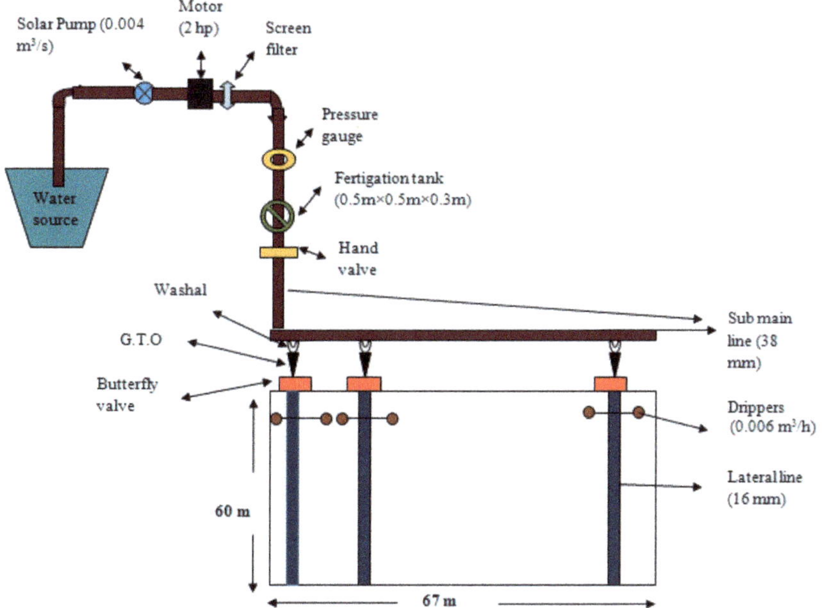

Figure 2. Schematic diagram of the low-cost drip irrigation system.

Table 3. Pressure and flow variation in the low-cost drip irrigation system.

Emitter	Pressure (kPa)	Flow Rate (m^3/h)
1	215	0.0062
2	210	0.0063
3	210	0.0012
4	195	0.0095
5	120	0.0064
6	190	0.0064
7	200	0.007
8	230	0.0063
9	210	0.0064
Average	197.778	0.006
Midpoint	175	0.0101
Variation calculation (%)	−9	−11
Acceptable range	<±10%	<±5%

A typical drip system is normally equipped with a venturi injector; including a heavy filtration unit (hydrocyclon filter, sand media filter, and disc filter). However, in the present study, only a screen filter was used at the inlet point of the water source. A simple drum (0.5 m × 0.5 m × 0.3 m) was placed for fertigation instead of a venturi injector to reduce the cost of the system (Figure 2). The life span of the low-cost drip irrigation system was considered to be 10 years, as adopted by [17]. The solar pump with a flow rate of

0.004 m^3/s was installed for pumping water. The effective life span of the solar pump was assumed to be 30 years, as adopted in [18].

For irrigation scheduling, vacuum-gauge-type tensiometers were installed down to effective root depths. Irrigation applications were scheduled on 60% soil moisture depletion (SMD) after accounting for effective rainfall. The irrigation requirements of crops were calculated using the moisture retention curve (Figure 3). The effective rainfall was calculated using CROPWAT 8.0 model, as shown in [19].

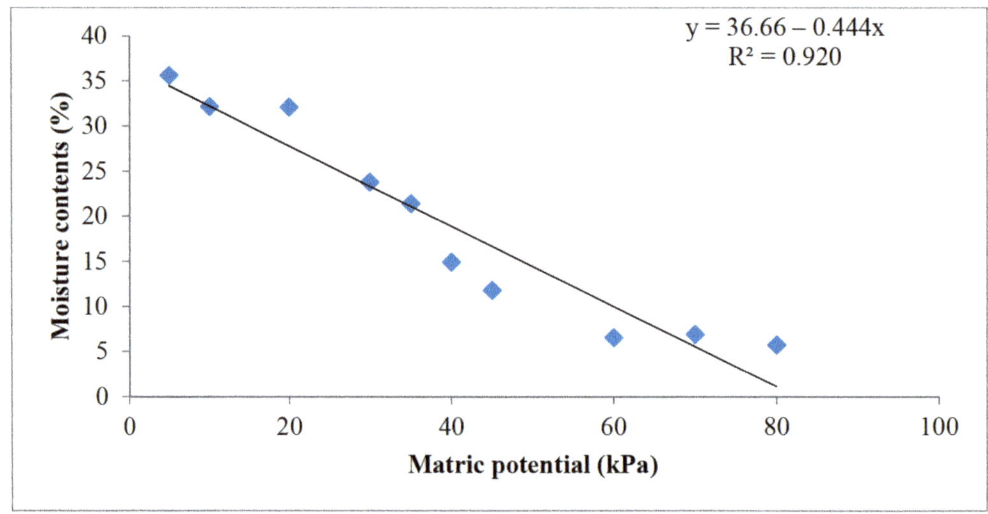

Figure 3. Moisture retention curve.

2.1. Experimental Data Collection and Analysis

Pre sowing moisture contents were determined gravimetrically from a depth of 15 cm to 90 cm, with an interval of 15 cm. To schedule irrigation, soil moisture contents were taken using tensiometers (Figure 3), after 7-day intervals from each of the experimental sets. The seasonal crop water requirements were assessed with the CROPWAT 8.0 model for which input data comprising climate data (maximum and minimum temperatures (°C), relative humidity (%), sunshine hours (hours), wind speed (km/day), and rainfall (mm)) were acquired from the nearest weather station installed at the campus, while crop data (planting and harvesting dates, Kc values at each growth stage, root depth (m), plant height (m)) and soil data (soil type, total available water (mm/meter), maximum rain infiltration rate (m/day), and initial soil moisture depletion (%)) were recorded on-site.

2.1.1. Vegetative Growth

Crops (onion, potato, and chilies) attributes, i.e., plant height (m), root depth (m), and leaf area (m^2) were measured at the time of harvest. Plant height and root depth were measured from randomly selected 20 plants/replication with the help of a measuring tape. Leaf area (m^2) was calculated by selecting 5 plants/treatment by separating the leaves from the plant, washed with plain water, and drying them in the open air, using a portable leaf area meter. For fruit trees (olive, peach, and citrus), the plant height (m) and canopy volume (m^3) were calculated according to the formula: 0.536 × tree height × crown diameter, as proposed in [20].

2.1.2. Yield

Yield data of row crops (onion, potato, and chilies) were recorded on each picking from each trial. Similarly, for fruit trees (olive, peach, and citrus), yield data of each fruit tree were measured in (kg)/tree at the time of each picking.

2.1.3. Fruit Quality

Fruit quality was assessed by selecting 20 fruits per treatment at random and determining physical and chemical characteristics of fruit, including fruit length and diameter (mm), fruit weight (kg)/plant, and its health, with visual observation. Fruit length and diameter were calculated by digital Vernier caliper in the laboratory. A total of 10fruits per replication were selected to record juice quality of citrus and peach such as total soluble solids (TSS) by hand refractometer, titrable acidity (%), as citric acid according to [21], and juice contents (%), as proposed in [22]. For all selected crops, the cross-sectional data of fixed costs, variable costs, depreciation costs, and the net return attained during the experimental period 2015–2016 for both drip and furrow irrigation systems were determined.

3. Results

3.1. Water Application

Water application to a rainfed crop depends on the water availability at the time of sowing and the amount of precipitation received throughout the growing season. For this purpose, long-term rainfall analysis was very important. The weather data for the last 37 years (1979–2016) were collected at the weather station of Soil and Water Conservation Research Institute (SAWCRI), Chakwal, located adjacent to the experimental field, and were analyzed to use in CROPWAT for estimation of crop water requirements. Rainfall data of 2015 and 2016 are shown in Figure 4.

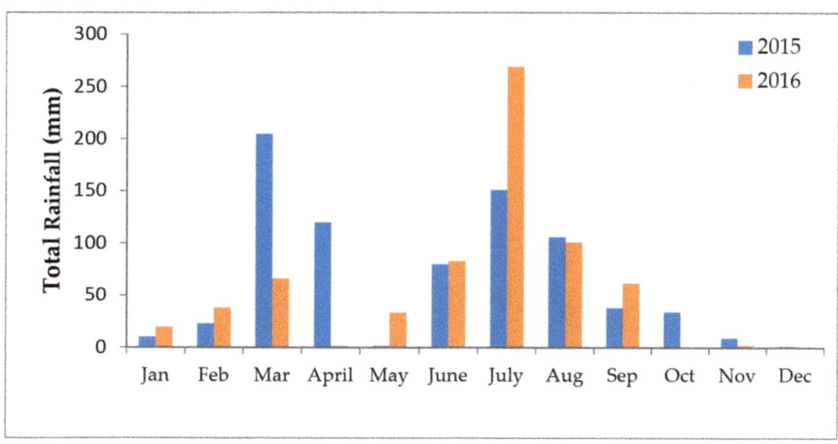

Figure 4. Monthly rainfall (mm) for the experimental period.

Total rainfall during the years 2015 and 2016 was 779 and 675 mm, and effective rainfall was 580 and 502 mm, respectively. The comparison of monthly climatic data with long-term means climatic data showed that total rainfall received during 2015 was higher than in 2016, and 62 % of yearly rainfall was received during the months of July to September in both years. Table 4 shows the amount of effective rainfall and irrigation (m^3) applied to each crop through drip and furrow irrigation techniques, along with the consequent yield (kg/ha) and water use efficiency (WUE) values during cropping seasons of 2015 and 2016.

Table 4. Water and yield data of different crops averaged over two years (2015–2016) at BARI.

Crop	Water Requirement (m³/ha)	Effective Rain Fall (m³)	Water Applied (m³)		Water Saving (%)	Yield (kg/ha)		Yield Increase (%)	Water Use Efficiency (WUE) (kg/m³)	
			Drip	Furrow		Drip	Furrow		Drip	Furrow
Potato	1500	280	1350	2440	45	10,930	7287	33	8.10	2.99
Onion	2000	1060	1040	1880	45	13,832	9201	33	13.30	4.89
Chilies	5040	3550	1660	2980	44	13,049	9077	30	7.86	3.05
Olive	5940	4790	1280	2870	55	5000	3667	27	3.91	1.28
Peach	7370	4790	2780	6450	57	25,676	19,270	25	9.24	2.99
Citrus	8480	4790	4100	9220	56	35,135	26,027	26	8.57	2.82

The amount of water applied to each crop was calculated by subtracting the effective rainfall from the total water requirement. Effective rainfall was calculated by using the CROPWAT model, which has the built-in function that uses various parameters, along with total rainfall. Table 4 shows that drip irrigation required 50% less water, as compared with furrow irrigation, to achieve the required SMD. Moisture levels were kept at an optimal range (60% SMD), which improved the plant production and quality. Drip irrigation allowed the rows between plants to remain dry, reduce weed growth, and reduce leaching of water and nutrients below the root zone. The water use efficiency (WUE) values under the drip irrigation system and furrow irrigation system ranged from 3.91 to 13.30 kg/m³ and 1.28–4.89 kg/m³, respectively. It was observed that water use efficiency was maximum in onion under drip irrigation (13.30 kg/m³). The results showed that drip irrigation gave three times more yield per unit of water applied in all vegetables and fruit crops when compared with furrow irrigation. Water use efficiency was exceptionally low in the furrow irrigation system due to conveyance, deep percolation, and evaporation losses. The results of this study are in line with [12], who reported that a low-cost drip system used 30–40% less water, as compared with the furrow irrigation method. Water savings were also higher (55%, 57% and 56%) in water-intensive crops such as olive, peach, and citrus, respectively (Figure 5).

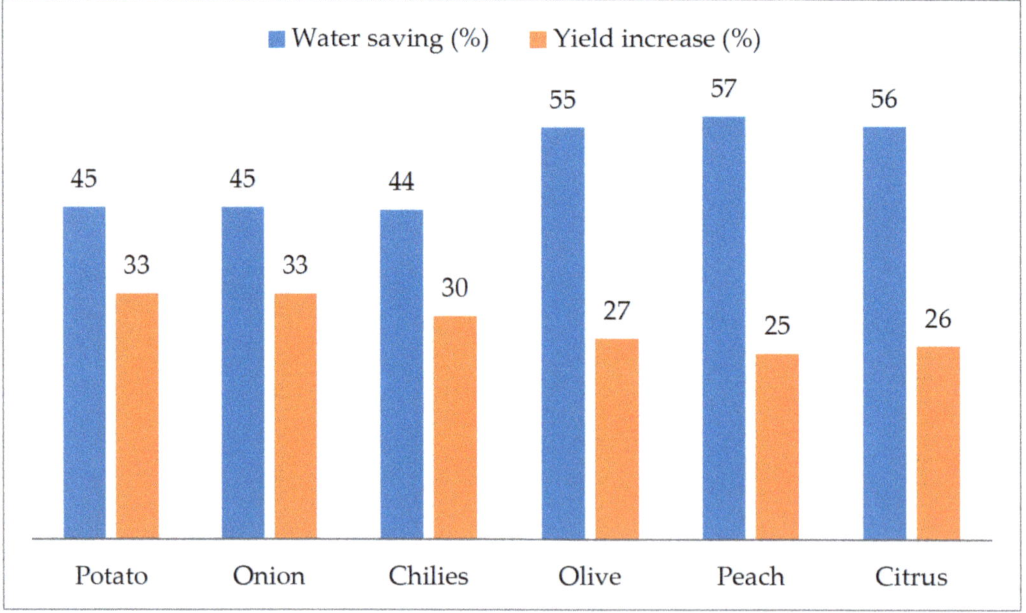

Figure 5. Increase in yield (%) and water saving (%) under drip irrigation system, averaged over 2 years (2015 and 2016).

3.2. Effect of Irrigation on Physical and Chemical Properties of Fruit

Values obtained from the treatments related to plant height (m), root depth (m), fruit weight/plant (kg), leaf area (m^2) for row crops (onion, potato, and chilies) and fruit plants (olive, peach, and citrus) are shown in Table 5, and values of plant height (m), fruit weight/plant (g), canopy volume (m^3), fruit length (mm), fruit diameter (mm), fruit weight/plant (kg) for fruit plants (olive, peach, and citrus) are given in Table 6.

Table 5. Effect of irrigation treatments on plant attributes of vegetables and fruit plants (averaged over 2 years).

Parameters/Crops		Plant Height (m)	Root Depth (m)	Fruit wt./Plant (kg)	Leaf Area (m^2)	Fruit Length (mm)	Fruit Diameter (mm)
Potato	Drip	0.6	0.33	0.391	0.2241	88	55
	Furrow	0.52	0.36	0.348	0.2012	75	48
Onion	Drip	0.51	0.27	0.136	0.0425	65.5	70.9
	Furrow	0.44	0.31	0.11	0.0385	59.2	62.4
Chilies	Drip	0.92	0.39	0.438	0.0475	55.5	
	Furrow	0.85	0.42	0.347	0.0398	50.4	44.4
Olive	Drip	2.1	15	4.3	19.3	14.9	2.03
	Furrow	1.9	12	3.2	17.4	12.8	1.75
Peach	Drip	3.5	130	26.48	87	6.31	116
	Furrow	3.0	100	25.2	74	5.94	105.5
Citrus	Drip	2.0	95	9.7	66	77	120
	Furrow	1.85	75	8.4	62	7	115

Table 6. Effect of irrigation treatments on chemical parameters of fruit juice (averaged over 2 years).

Parameters	TSS (°Brix)		Titratable Juice Acidity (%)		Juice Contents (%)	
Fruits \ Treatments	Drip	Furrow	Drip	Furrow	Drip	Furrow
Peach	5.3	3.5	0.5	0.36	48.5	46.3
Citrus	10.3	9.56	0.45	0.3	57.5	55.2

From the data in Table 5, it is obvious that vegetative growth parameters of all crops (plant height, leaf area, and canopy volume, and fruit wt. (kg) per plant) increased in the treatment of drip irrigation system. The drip irrigation system maintained soil moisture around the plant roots by maintaining the soil physical properties, which could be a possible reason for the enhanced plant growth and yield under drip irrigation. Similar results were reported in [23–25] for potato, onion, and chilies, respectively, and in [26–28] for olive, peach, and citrus, respectively.

The chemical properties of peach and citrus juice were also recorded, as shown in Table 6, which included total soluble salts (TSSs), Brix, titratable juice acidity (%), and juice contents (%). Some studies [20,27] reported an increase in TSS and titratable acidity under drip irrigation treatment as the amount of water applied decreased, and in the furrow irrigation system, plants received ample water; thus, the values of fruit juice quality parameters were reduced. The comparative wet conditions that enhanced the fruit size may be conducive for the production of higher total soluble salts (TSSs). High soil moisture levels helped in increasing titratable acidity and juice contents.

3.3. Economic Evaluation

The drip irrigation method requires an initial fixed cost for installation, and the cost depends on the crop nature, plant spacing, water requirement, discharge of the dripper, and distance from the water source. The crops with more plant to plant and row to row distance require a relatively low capital cost. Moreover, the fixed cost also depends on the quality of the materials used for the system. In Pakistan, the adoption of drip irrigation systems is quite slow mainly because of overdesigning of the system. In Government-sponsored subsidized schemes, high-efficiency irrigation systems are generally equipped with solar-powered groundwater pumps, heavy filters, fertigation chambers, etc. Further, many companies are involved in designing (who overdesign in their interest) without good experience, installing the drip system by adding large and unnecessary parts such as filters, gauges, fertigation tanks, water meters, etc., which make the system costly. The management of such systems is difficult for common farmers; therefore, they are reluctant to install the system. In this study, the authors proposed a low-cost/economical drip irrigation system by eliminating unwanted parts, purchasing the parts from local vendors, and installing them manually. All costs involved in making furrows and designing/installing the low-cost drip irrigation system are listed in Tables 7 and 8.

Table 7. Fixed costs and depreciation costs of furrow irrigation system.

Crops	Fixed Cost			Depreciation Cost (Labor Involved in Irrigation and Furrow Repairing)
	Pumping Cost	Laser Leveling Cost + 8 Daily Paid Labor @ 365 Rs/Day	Ridge Making through Tractor + 8 Daily Paid Labor @ 365 Rs/Day	
	Rs/ha	Rs/ha	Rs/ha	Rs/ha
Potato	10,033	8788	7518	14,834
Onion	10,033	8788	7518	14,834
Chilies	10,033	8788	7518	22,250
Olive	10,033	8788	0	44,501
Peach	10,033	8788	0	44,501
Citrus	10,033	8788	0	44,501

Table 8. Fixed costs for the drip irrigation system.

Crops	Total Cost of the System (Rs/ha)	Life of the Drip System (Years)	Pumping Cost of the System (Rs/ha)	Life of the Solar Pump (Years)	Fixed Cost [(A/B) + (C/D)]
	A	B	C	D	Rs/ha
Potato	874,090	10	118,500	30	91,359
Onion	779,468	10	118,500	30	81,897
Chilies	729,211	10	118,500	30	76,871
Olive	106,175	10	118,500	30	14,567
Peach	96,576	10	118,500	30	13,608
Citrus	99,342	10	118,500	30	13,884

4. Discussion

Irrigation was scheduled with respect to effective rainfall events during crop growing seasons. Irrigation scheduling devices (tensiometers) were installed to monitor the soil moisture to schedule the irrigation events. Water saving in drip irrigation (Figure 5) was high because the furrow system is less efficient (50%), excess amount of water leached down to the groundwater, and consequently, a large amount of irrigation had to be applied to meet the crop water requirement. The findings of this study are in close agreement with [29], for vegetables, and [30] for fruit crops. The data presented in Figure 5 show the percent increase in yield and water saving in the drip irrigation system. Drip irrigation increased production and, at the same time, increased the quality of fruit, reducing shoot growth, as was reported in [28]. Numerous research studies suggested that wetting only 20% to 50% of the effective rooting depth of full-grown deciduous fruit trees is adequate to maximize yield, provided enough water is available to meet water requirements during critical periods of fruit development, as proposed in [31]. Plant growth is badly affected when using the furrow irrigation method because after irrigation, soil moisture contents change from saturation to field capacity to dryness, and therefore, plants bear moisture stress before the next irrigation. A minimum interval of irrigation throughout the crop growing season creates water and nutrient balance and ensures optimum growth of the crop.

Figure 6a–e presents the complete comparison of all costs, i.e., fixed costs, variable costs, and depreciation costs, involved in the establishment and operation of furrow and drip irrigation systems. The total cost of installation of drip irrigation per hectare was calculated as Rs 50,000–150,000, assuming 10 years of its useful life, with a payback period of 1–2 years for fruit plants and 3–6 years for vegetables. The fixed capital costs varied for all crops due to variation in plant spacing of the respective crops (Table 2); it included the cost of installation of drip system, along with pumping cost using a solar pump. Fixed costs in furrow irrigation comprised the cost of land leveling through laser leveler and the tractor expenses to make ridges (Table 7). Laser leveling required 4–4.15 h/acre to level 10 cm to 15 cm deep layers of soil. Short-level furrows required accurate field grading, which was performed by machines. The plowing and furrowing were also performed by machines. These operations required skilled labor, fuel, and machinery tools, and all these cots added to the fixed costs of the furrow irrigation method (Table 7). The variation in variable costs was mainly due to incurred expenses with the purchase of seeds, fertilizers, pesticides, weedicides, and labor involved in field operations. In the drip irrigation system, the labor cost was half, as compared to the furrow irrigation system, because in furrow irrigation, more labor was required for hoeing, weeding, and watering operations. The drip irrigation system required lower field operations, which also reduced the cost of the system. Depreciation costs in drip irrigation systems include the repair and maintenance of drip parts such as damage or leakage in lateral lines, drip emitter clogging, etc., which was fixed for all crops (Figure 6a). In the furrow irrigation system, the depreciation cost comprised the cost of labor for the repair and maintenance of furrows after every irrigation or a high-rainfall event. Every month during the crop season eight (8) persons were deployed for these operations for a hectare.

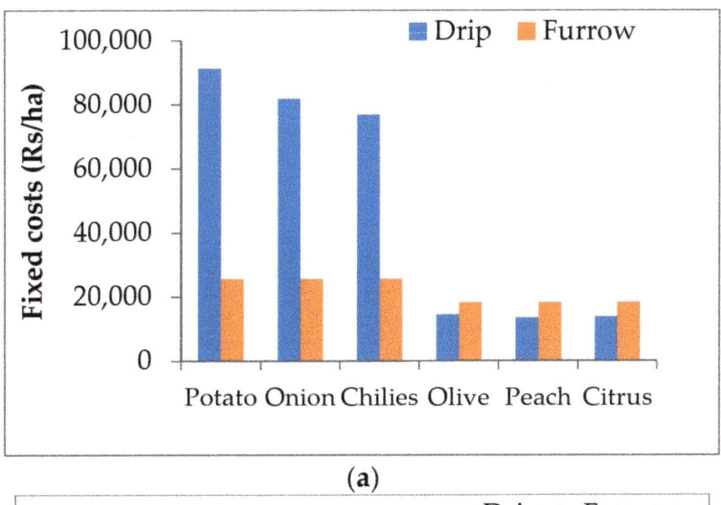

(**a**)

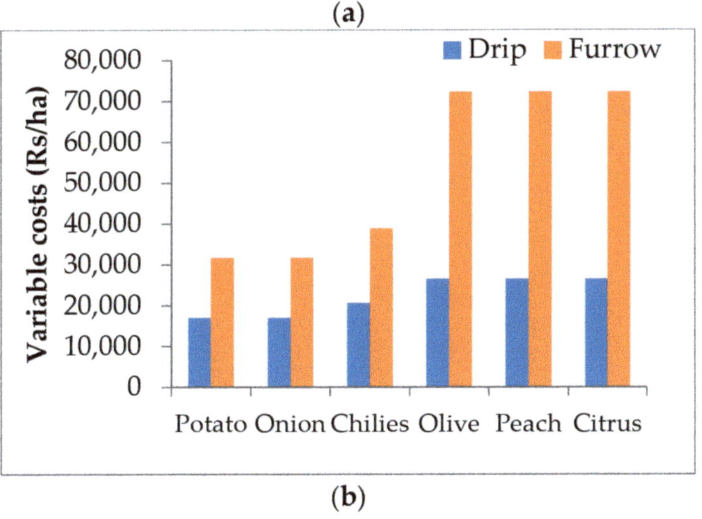

(**b**)

Figure 6. *Cont.*

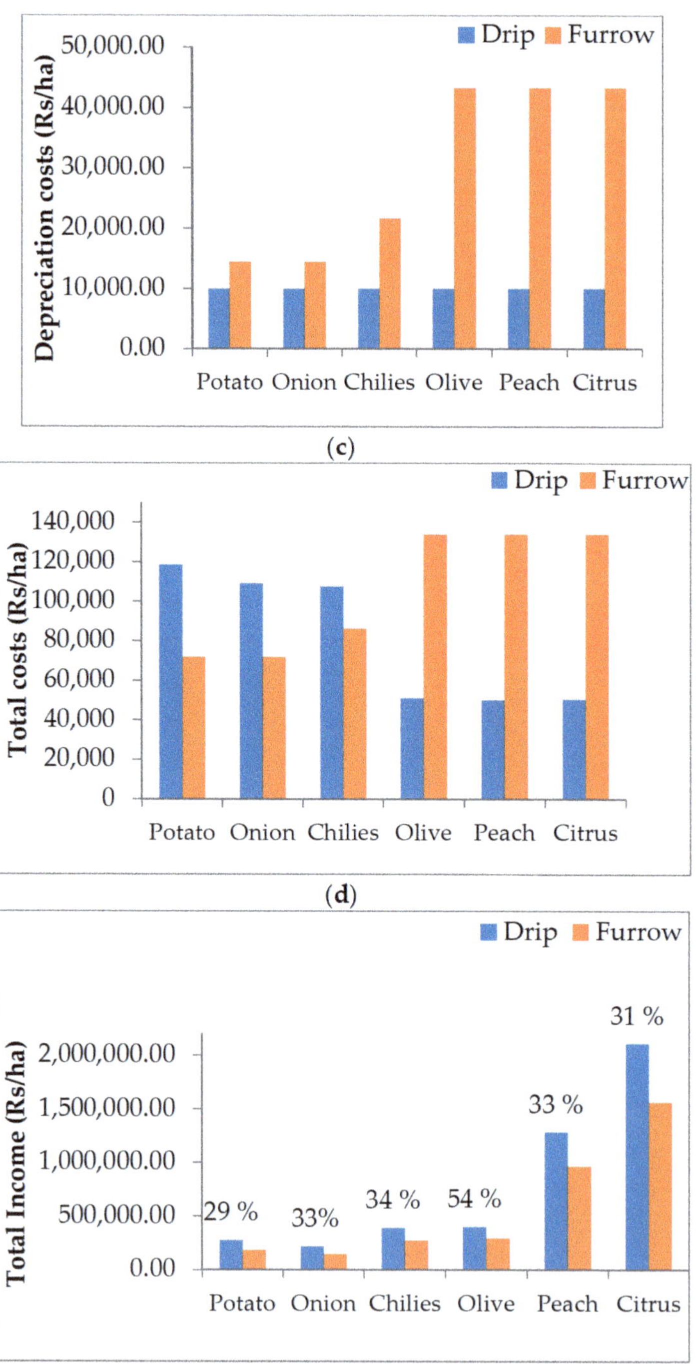

Figure 6. (**a**) Comparison of fixed costs incurred in T1 and T2; (**b**) comparison of variable costs incurred in T1 and T2; (**c**) comparison of depreciation costs incurred in T1 and T2; (**d**) comparison of total costs incurred in T1 and T2; (**e**) comparison of total income received from T1 and T2.

The gross returns were computed by multiplying the average market rate with the yield of respective vegetables and fruits during the crop harvesting period. The seasonal gross expenditure, gross return, net return, and percentage increase in net return for drip irrigation and furrow irrigation systems for all the selected crops are also depicted in Figure 6d,e. The results revealed that the highest percentage net return per hectare under drip irrigation system was recorded for olive (54%) and the lowest percentage net returns recorded for potato (29%), as shown in Figure 6e.

Financial viability analysis was performed by computing the net present value of crops and fruit plants by discounting both the costs and the returns at the prevailing rate of interest (10%), which is shown in Table 9. From the table, it is clear that net present values of crops and fruit plants were computed for the entire life of the drip system (10 years). Net present values in Table 9 showed that the low-cost drip system discounted cash flows over the entire life of the drip set (10 years). The tear-wise net present worth was estimated to calculate the number of years required to recover the capital cost of the drip system. The payback period for olive, peach, and citrus was 2 years, 1 year, and 1 year, respectively, and for potato, onion, and chilies, the payback period was 5 years, 6 years, and 3 years, respectively. The cost incurred on the drip irrigation system was Rs 118,451 for potato and Rs 108,989 for onion; thus, the payback period for both crops is maximum. Due to narrow plant spacing in potato and onion, the initial costs of drip sets were high.

Table 9. Net present worth and payback periods of drip irrigated crops.

Crops	Net Present Value (Rs/ha) @ 10% Discount Rate	Pay Back Period (Years)
Potato	65,247	5
Onion	38,185	6
Chilies	139,032	3
Olive	202,235	2
Peach	746,163	1
Citrus	1,252,415	1

Gross expenditures in the drip irrigation system were higher because of the high initial investment. However, the gross income in the drip irrigation system was high because of the good quality of produce and high yield. Furrow irrigation system consistently underperformed in the case of all the vegetables and fruit crops.

5. Conclusions

From the results of this study, it was concluded that the low-cost drip irrigation system applies water near the roots of the plant, as per requirement, and therefore produces more vegetables and fruits with less water. The low-cost drip irrigation under rainfed conditions saved up to 86% of irrigation water and increased yield by 26–33%, as compared with the furrow irrigation method. Reduced cost of labor in irrigation, fertilizer application, and weeding, combined with increased economic returns, leads to higher economics of vegetable production under the drip system. Fruits and vegetables performed well in the drip irrigation system; however, as observed in this study, the performance of vegetables (potato, onion and chilies) was far low, as compared with fruit plants (olive, peach, and citrus). Based on the present research findings, the average cost of drip sets was calculated to be Rs 50,000–150,000 per hectare for all given crops. It was also concluded that the gross expenditures of the low-cost drip irrigation set can be fully recovered in the second year of crops and orchards. The low-cost drip irrigation was found efficient and economically viable, gave long-term benefits for small landholders, and is feasible/suitable for those areas where the capital costs of existing drip systems are the main barrier to their adoption. There is considerable potential for farmers to grow their orchards and vegetables by installing a low-cost drip irrigation system in their farms/fields.

Author Contributions: Conceptualization, methodology, execution, data curation, writing—original draft preparation, M.A. (Marjan Aziz); formal data analysis, interpretation of results, review and final revision, S.A.R.; investigation, M.A.I.; resources, S.S.; software, S.A. (Saira Anwer) and S.A. (Sana Asghar); validation, N.T., M.U. and A.K.; acquisition of parts and repair maintenance, J.A.; writing—review and editing, M.A. (Muhammad Ashraf). All authors have read and agreed to the published version of the manuscript.

Funding: This research received no external funding.

Institutional Review Board Statement: Not Applicable.

Informed Consent Statement: Not Applicable.

Data Availability Statement: Not applicable.

Acknowledgments: The authors are indebted to Abdul Majid, ICARDA office in Pakistanfor providing technical guidance and moral help, and Bareerah Fatima from Pakistan Council of Research in Water Resources (PCRWR), Islamabad, is highly acknowledged for her help in completing this study.

Conflicts of Interest: The authors declare no conflict of interest.

References

1. Government of Pakistan (GoP). Available online: https://www.finance.gov.pk/survey1920.html (accessed on 4 January 2021).
2. Albeyi, A.; Ustun, H.; Oweis, T.; Pala, M.; Benli, B. Wheat water productivity and yield in a cool highland environment: Effect of early sowing with supplemental irrigation. *Agric. Water Manag.* **2006**, *82*, 399–410. [CrossRef]
3. Passioura, J. Increasing crop productivity when water is scarce—From breeding to field management. *Agric. Water Manag.* **2006**, *8*, 176–196. [CrossRef]
4. Yu, W.; Yi-Chen, Y.; Savitsky, A.; Alford, D.; Brown, C.; Wescoat, J.; Debowicz, D.; Sherman, R. *The Indus Basin of Pakistan: The Impacts of Climate Risks on Water and Agriculture*; World Bank: Washington, DC, USA, 2013. [CrossRef]
5. Qureshi, S.A. Water management in the Indus basin in Pakistan: Challenges and opportunities. *Mt. Res. Dev.* **2011**, *31*, 252–260. [CrossRef]
6. Tagar, A.; Chandio, F.A.; Mari, I.A.; Wagan, B. Comparative study of drip and furrow irrigation methods at farmer's field in Umarkot. *Int. J. Agric. Biosyst. Eng.* **2012**, *6*, 788–792.
7. Government of Pakistan GoP. Available online: https://www.finance.gov.pk/survey_1213.html (accessed on 24 November 2019).
8. Mahmood, A.; Oweis, T.; Ashraf, M.; Majid, A.; Aftab, M.; Aadal, N.K.; Ahmad, I. Performance of improved practices in farmers' fields under rainfed and supplemental irrigation systems in a semi-arid area of Pakistan. *Agric. Water Manag.* **2015**, *155*, 1–10. [CrossRef]
9. Hanson, B.R.; Hutmacher, R.B.; May, D.M. Drip irrigation of tomato and cotton under shallow saline ground water conditions. *Irrig. Drain. Syst.* **2006**, *20*, 155–175. [CrossRef]
10. Irmak, S.; Odhiambo, L.O.; Kranz, W.L.; Eisenhauer, D.E. Irrigation efficiency and uniformity and crop water use efficiency. *Biol. Syst. Eng.* **2011**, *2011*, 451.
11. Asif, M.; Ahmad, M.; Mangrio, A.G.; Akbar, G.; Memon, A.H. Design, Evaluation and Irrigation Scheduling of Drip Irrigation System on Citrus Orchard. *Pak. J. Meteorol.* **2015**, *12*, 37–48.
12. Razzaq, A.; Rehman, A.; Qureshi, A.H.; Javed, I.; Saqib, R.; Iqbal, M.N. An economic analysis of high efficiency irrigation systems in Punjab, Pakistan. *Sarhad J. Agric.* **2018**, *34*, 818–826. [CrossRef]
13. Muhammad, M.; Zahid, A.; Breue, L. Water resources management strategies for irrigated agriculture in the Indus basin of Pakistan. *Water* **2020**, *12*, 1429. [CrossRef]
14. Aziz, M.; Tariq, M. Assessing the Potential of Rain-Water Harvesting (in situ) for Sustainable Olive (*Oleaeuropaea* L.) Cultivation in Water-Scarce Rain-Fed Areas. *Irrig. Drain. Syst. Eng.* **2018**, *7*, 212. [CrossRef]
15. Abdul, H.; Yousaf, M. Detailed Soil Survey of Barani Agricultural Research Institute. *Chakwal Dir. Soil Surv. Interpret.* **1989**, (Institutional internal report, unpublished).
16. Zellman, P. Drip Irrigation System Evaluations: How to Measure & Use Distribution Uniformity Tests Importance of Distribution Uniformity 1–9. 2016. Available online: https://www.sustainablewinegrowing.org/docs/DUArticle.pdf (accessed on 10 May 2018).
17. Reddy, M.; Ayyanagowdar, M.S.; Patil, M.G.; Polisgowdar, B.S.; Nemichandrappa, M.; Anantachar, M.; Balanagoudar, S.R. Water use efficiency and economic feasibility of drip irrigation for watermelon (*Citrulluslunatus*). *Int. J. Pure App. Biosci.* **2017**, *5*, 1058–1064. [CrossRef]
18. Hossain, M.A.; Hassan, M.S.; Ahmmed, S.; Islam, M.S. Solar pump irrigation system for green agriculture. *Agric. Eng. Int. CIGR J.* **2014**, *16*, 1–15.
19. FAO. Irrigation water requirements. In Irrigation Potential in Africa: A Basin Approach, Chapter, 5. 2005. Available online: http://www.fao.org/docrep/WRome./w4347e.htm (accessed on 10 March 2017).
20. Verma, N.K.; Lamb, D.W.; Reid, N.; Wilson, B. Comparison of canopy volume measurements of scattered eucalypt farm trees derived from high spatial resolution imagery and lidar. *Remote Sens.* **2016**, *8*, 388. [CrossRef]
21. Heising, J.K.; Dekker, M.; Bartels, P.V.; Van Boekel, M.A. Monitoring the quality of perishable foods: Opportunities for intelligent packaging. *Crit. Rev. Food Sci. Nutr.* **2014**, *54*, 645–654. [CrossRef]

22. OECD. Guidance on Objective Tests to Determine Quality of Fruits and Vegetables and Dry and Dried Produce. *International Standardization of Fruit and Vegetables*. 2005. Available online: https://www.oecd.org/agriculture/fruit-vegetables/publications/guidelines-on-objective-tests.pdf (accessed on 19 May 2018).
23. Trifonov, P.; Lazarovitch, N.; Arye, G. Water and nitrogen productivity of potato growth in desert areas under low-discharge drip irrigation. *Water* **2018**, *10*, 970. [CrossRef]
24. Bagali, A.N.; Patil, H.B.; Guled, M.B.; Patil, V.R. Effect of scheduling of drip irrigation on growth, yield and water use efficiency of onion (*Allium cepa* L.). *Karnataka J. Agric. Sci.* **2012**, *25*, 116–119.
25. Maind, M.M.; Yadload, S.S.; Bhalerao, R.V.; Thalkari, G.N. Effect of Irrigation and Fertilizer Levels on Growth and Yield of Chilli (*Capsicum annuum* L.). *Int. J. Curr. Microbiol. Appl. Sci.* **2018**, *6*, 1192–1199.
26. Karrou, M.; Nangia, V.; Oweis, T. Effect of deficit irrigation on vegetative growth and fruit yield parameters of young olive trees (*Oleaeuropaea* L.) in semi-arid area of Morocco. In Proceedings of the 7th International Conf. on Water Resources in the Mediterranean Basin, Marrakech, Morocco, 10–12 October 2014.
27. El-sayed, S.A.; Ennab, H.A. Effect of different levels of irrigation water and nitrogen fertilizer on vegetative growth, yield and fruit quality of nitrogen fertilizer on vegetative growth, yield and fruit quality of valencia orange trees. *Minufiya J. Agric. Res.* **2013**, *38*, 761–773.
28. Yildirim, O.; Dumanoğlu, H.; Güneş, N.T.; Yildiri, M. Effect of wetted soil area on trunk growth, yield, and fruit quality of drip-irrigated sour cherry trees. *Turk. J. Agric. For.* **2012**, *36*, 439–450. [CrossRef]
29. Kumar, R.; Khanna, R. Comparative study of different irrigation methods on tomato crop (*Lycopersiconesculentum*) in western Uttar Pradesh, India. *Int. J. Chem. Stud.* **2019**, *7*, 59–64.
30. Elsayed, M.; Mohammed, A.; Alhajhoj, M.R.; Ali-dinar, H.M.; Munir, M. Impact of a novel water-saving subsurface irrigation system on water productivity, photosynthetic characteristics, yield, and fruit quality of date palm under arid conditions. *Agronomy* **2020**, *10*, 1265.
31. Mezghani, M.A.; Mguidiche, A.; Khebour, F.A.; Zouari, I. Water Status and Yield Response to Deficit Irrigation and Fertilization of Three Olive Oil Cultivars under the Semi-Arid Conditions of Tunisia. *Sustainability* **2019**, *11*, 4812. [CrossRef]

Article

Assessment of Non-Conventional Irrigation Water in Greenhouse Cucumber (*Cucumis sativus*) Production

Muhammad Mumtaz Khan [1,*], Mansour Hamed Al-Haddabi [2], Muhammad Tahir Akram [3], Muhammad Azam Khan [3], Aitazaz A. Farooque [4,*] and Sajjad Ahmad Siddiqi [5]

1   Department Plant Sciences, College of Agriculture and Marine Science, Sultan Qaboos University, P.O. Box 34, Al-khoud 123, Oman
2   Department Soils, Water and Agricultural Engineering, College of Agriculture and Marine Science, Sultan Qaboos University, P.O. Box 34, Al-khoud 123, Oman; mans99@squ.edu.om
3   Department of Horticulture, PMAS-Arid Agriculture University, Rawalpindi 46300, Pakistan; tahiruaf786@gmail.com (M.T.A.); drazam1980@uaar.edu.pk (M.A.K.)
4   Faculty of Sustainable Design Engineering, School of Sustainable Design Engineering, University of Prince Edward Island, Charlottetown, PE C1A 4P3, Canada
5   Department of Civil and Architectural Engineering, College of Engineering, Sultan Qaboos University, P.O. Box 33, Al-khoud 123, Oman; s110153@student.squ.edu.om
*   Correspondence: mumtaz@squ.edu.om (M.M.K.); afarooque@upei.ca (A.A.F.)

Abstract: Climate change, urbanization and subsequent environmental changes are depleting freshwater resources around the globe. The reuse of domestic, industrial and agricultural wastewater is an alternative approach to freshwater that can be used for irrigation purposes. However, these wastewaters may contain hazardous and toxic elements, such as heavy metals that are hazardous for human health and the environment. Therefore, an experiment was conducted to evaluate the concentration of macro, micro and heavy metals in cucumber irrigated with different resources (tap water, greywater, dairy water and wastewater). The results showed that the use of different irrigation resources has increased the level of macro (sodium (Na), potassium (K), calcium (Ca), magnesium (Mg)), microelements (zinc (Zn), iron (Fe), manganese (Mn)), and heavy metals (copper (Cu), barium (Ba), lead (Pb) and cadmium (Cd)) in cucumber leaves and fruits. However, their levels were in the range that is safe for human health and the environment was as recommended by FAO maximum values of trace elements (Zn, 2.0; Fe 1.0; Mn, 0.2; Cu, 0.2; Pb, 5.0, and Cd, 0.01 mgL^{-1}). Based on observations, it was also revealed that among different irrigation resources, the use of dairy water in cucumber improved its agronomic attributes and maximum plant yield (1191.02 g), while the different irrigation resources showed a non-significant impact on fruit diameter. However, total soluble solid contents (TSS) were more significant in cucumber fruits treated with wastewater (2.26 °brix) followed by dairy water (2.06 °brix), while the least TSS contents (1.57 °brix) were observed in cucumber plants treated with tap water. The significance of non-conventional irrigation water use in agriculture, particularly greenhouse cucumber (*Cucumis sativus*) production, is discussed.

Keywords: climate change; environment; heavy metals; human health; resources; wastewater

Citation: Khan, M.M.; Al-Haddabi, M.H.; Akram, M.T.; Khan, M.A.; Farooque, A.A.; Siddiqi, S.A. Assessment of Non-Conventional Irrigation Water in Greenhouse Cucumber (*Cucumis sativus*) Production. *Sustainability* **2022**, *14*, 257. https://doi.org/10.3390/su14010257

Academic Editor: Agostina Chiavola

Received: 1 October 2021
Accepted: 20 December 2021
Published: 27 December 2021

Publisher's Note: MDPI stays neutral with regard to jurisdictional claims in published maps and institutional affiliations.

1. Introduction

Worldwide, climate change has an adverse impact on water quality, water availability, food security, and human health. Globally, about 40% of the earth's total area is comprised of arid, semiarid, and range lands [1] and nearly 50% of European countries are already facing water scarcity [2]. In previous decades, the amount of water required in agriculture has been tripled while the available freshwater resources are depleting, and the agriculture sector is experiencing water shortages. Under existing climatic conditions, almost half of the world population will be confronted with water scarcity by 2030 as the resources of freshwater are depleting day by day, and Middle East/Gulf countries will suffer severe water scarcity, with future conflicts over scarce water resources due to climatic and socioeconomic issues [3]. In the past few decades, drought and desertification cycles have affected particularly the semiarid zone. Therefore, under these circumstances, water is the most critical ingredient for a sustainable ecosystem and, even more importantly, for economic development. Irrigation-dependent agriculture comprises 20% of the worldwide cultivated land, accounts for approximately 40% of global food production, and agriculture is also responsible for 70% of global water usage [4]. Hydrological poverty is also being caused by population growth and poor management of water supplies as 70% of available freshwater is being consumed in agriculture. In arid and semiarid regions, water demand for domestic, industrial, and agricultural uses is steadily rising and it is estimated that more than 40% of the world's population could face water scarcity by 2050 [5].

Domestic water would be of high quality, but industrial and agricultural water should be of lower quality. A huge volume of fresh water is consumed in households and industries, out of which 50–80% is wasted in households. The re-evaluation or reuse of this wastewater is a viable, alternative, and sustainable approach to conserve freshwater [6]. Besides household wastewater, municipal, and livestock slurries may be used in agriculture as these are affordable and appealing resources of irrigation [7]. The use of wastewater in agriculture minimizes aquatic degradation by reducing sewage sludge.

Wastewater contains a high concentration of nutrients and it has a significant potential for application in agricultural irrigation as it provides soil organic carbon (SOC), nutrients (NPK), minerals, organic matter and inorganic micronutrients to crops [8]. The treating systems do not remove nitrogen and potassium from the wastewater and are harmful to aquatic life. However, they are important from an agronomic view as they improve soil fertility, crop yield, and minimize fertilizer use and input cost. The studies highlighted that the use of wastewater, particularly for crop irrigation, has enhanced crop production as it is enriched with nutrients [9,10], and the reuse of urban wastewater has fulfilled the phosphorus (P) and potassium (K) demands for maize crops in Saudi Arabia [11].

Like Saudi Arabia, Oman is an arid country with severe water problems, and the practice of wastewater irrigation may be an affordable and appealing resource of irrigation as this approach is continuously increasing throughout the world for water security [12,13]. The availability of different resources of wastewater has also emphasized the attention for farmers of crop selection and different resources of wastewater have increased the production of lettuce, spinach, onion, tomato, potato, carrot, cucumber, and other different vegetables [14–16]. The wastewater is enriched with macro and micronutrients required for plant growth and it may be a resource to enhance soil productivity and fertility level [17].

However, the improper use of wastewater may cause environmental issues and the application of wastewater may increase the uptake of heavy metals in vegetable crops [18]. The continuous use of wastewater may affect soil physical and chemical characteristics [19]. It has been argued that high electrical conductivity (EC) values in irrigation water, e.g., 2900, 3900, and 2400 S/cm, resulted in a 50% drop in production in onions, potatoes, and dry bean crops [20]. Similarly, it is reported that irrigation water with high biological oxygen demand (BOD) or chemical oxygen demand (COD) inhibits plant growth, while water carrying chlorine and fluorine can harm plant tissues severely [21]. However, the mung bean (*Vigna radiate*) growth and yield were found to be lower with raw sewage irrigation water compared to biologically treated wastewater irrigation water [11]. The investigations on the application of wastewater irrigation showed significantly enhanced trace metal concentrations (mg kg^{-1}) in cucumber crops, with Fe (393.2) > Pb (145.1) > Cu (92.3) > Cr (84.8) > Zn (46.6) > Ni (48.2) > Mn (46.6) > Cd (14.1) > Co (11.1) [22]. The use of wastewater in spinach, radish, cauliflower, and mustard may increase the uptake of heavy metals (Cd, Cr, Cu, Pb and Cd) [18]. However, the climatic conditions, rooting media and plant species selection may alter the adsorption rate of heavy metals, such as in spinach, there is a high accumulation of Pb and Ni during summer, while in winter the accumulation of Cd increases [23]. Moreover, the highest values of Hg and U were observed during the wet season as compared to the dry season [24]. Likewise, carrots favored the build-up of Zn and Cu while spinach and mint have shown the highest uptake of Mn and Fe [25]. Similarly, there were 20 times more concentrations of heavy metals observed in crops treated with sewage water as compared to European countries [26]. The studies have also shown the absence of health risk factors when contaminated heavy metal vegetables were consumed by humans [27,28]. However, it should be kept in mind that these trials were conducted for short period. It is also observed that wastewater irrigation has increased eggplant production and its nutritional status without contaminating it with heavy metals [29]. A similar finding has described that wastewater application in the arid regions increased the soil organic matter (SOM), electrical conductivity, nitrogen, and heavy metal concentrations in soil [30].

Despite having been studied, there is still a need to explore the impact of different wastewater irrigation resources on soil properties, vegetables, and fruit qualities, as the reuse of wastewater for irrigation will be greater in the future, especially in water deficit areas due to increased population and ever increasing demands for food and fresh water Therefore, non-conventional water resources such as wastewater should be tested for agriculture to supply food and careful management should be adopted to ensure long-term agricultural productivity. By considering all the above factors, an experiment was designed in Sultan Qaboos University, Oman, with the aim to re-utilize wastewater, greywater and dairy water in cucumber production and to investigate their environmental impact on growing media and plant nutrient uptake.

2. Materials and Methods

2.1. Experimental Conditions

The experiment was conducted in the greenhouse conditions located at the Agriculture Experiment Station facility, Sultan Qaboos University, Oman. Greenhouse cucumber "Beit Alpha" cultivar seeds were procured by Mr. Waleed Al-Busaidi (technician) Department of Plant Sciences, from the Island harvest trading LLC, Barka, Oman. In this experiment, the environmental conditions of the greenhouse, such as temperature (27/20 °C) and light (240 μmol), were maintained throughout. Greenhouse cucumber seeds were sown in nursery plug trays (50 holes) having compost as a growing substrate and were watered uniformly. After one month of sowing, uniform seedlings of 4 inches in length were transplanted in plastic pots of size 7 × 10″, each having an equal volume of compost. Pots were supplemented uniformly with Hoagland's nutrient solution (H2395 Sigma-Aldrich, St. Louis, MO, USA) once a week. The experiment was laid out under a completely randomized design (CRD) and four treatments of varied water, with freshwater as control, treated wastewater, greywater, and dairy cleansing water, were applied uniformly with 500 mL concentration according to the crop requirement. There were three replicates used in the experiment and in each replication eight plants were used.

2.2. Water Analysis

Before irrigation, the water quality of all water resources of freshwater, treated wastewater, greywater and dairy water was analyzed to determine the concentrations of macronutrients (sodium (Na), potassium (K), calcium (Ca), and magnesium (Mg), micronutrients (zinc (Zn), iron (Fe) and manganese (Mn), and heavy metals (copper (Cu), barium (Ba), chromium (Cr), lead (Pb) and cadmium (Cd) by using inductively coupled plasma atomic emission spectroscopy (ICP AES, Interpid Il XDL) [31].

2.3. Growing Substrate Analysis

The analysis of plant growing substrate (compost) was conducted before the experiment to analyze the nutrient status. Therefore, three samples of each replicate were collected randomly at 10 cm pot depth and were taken to the soil chemistry laboratory at the Department of Soil, Water and Agricultural Engineering, College of Agricultural and Marine Sciences, Sultan Qaboos University, Oman. The soil organic matter content, total N content, macroelements, microelements, and heavy metals were determined by ICP after nitric-perchloric acid (2:1) digestion.

2.4. Leaf and Fruit Analysis

After eight weeks of transplanting, six mature leaves from individual cucumber plants were collected and were immediately transported to the Department of Plant Sciences, College of Agriculture and Marine Science, Sultan Qaboos University, Oman, for laboratory analysis. Before analysis, samples were washed thoroughly with special detergent (Alconox 0.1%) and rinsed in tap water, after each sample was cleaned with a diluted solution of 0.005% HCL and was finally rinsed in distilled water. To dry out leaf samples were left on filter paper for 2 h and were oven-dried for 48 h at 65 °C. After that macro/microelements and heavy metals were determined by ICP [31].

2.5. Fruit Yield and Quality Analysis

The yield of each plant was taken at each picking and the total yield of the plant was calculated at final maturity and the unit was expressed in grams (g). For fruit quality analysis, fruit size was measured by using Vernier caliper (model CP33659-00, Company VWR, Radnor, PA, USA) and values were taken in millimeters (mm), while fruit total soluble solid content (TSS, °Brix) was determined by hand refractometer (model MASTER-53α, ATAGO, Bellevue, WA, USA). Macro/microelements and heavy metals were determined by ICP [31].

2.6. Statistical Analysis

The experiment was laid out under a completely randomized design (CRD) and eight plants were taken as per single replicate and a total of twenty-four plants were used in three replications. The data were statistically analyzed by using analysis of variance (ANOVA) and differences among treatments were compared at 5% level of probability by applying Tukey's HSD.

3. Results

3.1. Availability of Essential Elements and Heavy Metals in Irrigation Water Resources

The data regarding the assessment of water resources are shown in Table 1. The results revealed that the concentration of macro, micro and heavy metals was significantly different in diverse water resources. Among macro essential elements, the highest Na was observed in wastewater (5370 mg/L) followed by a dairy water resource (1.50 mg/g). The amount K was highest in dairy water (1430 mg/L) followed by wastewater (380 mg/L). Similarly, Ca (1960 mg/L) and Mg (940 mg/L) were observed at maximum in wastewater, while Mg was not observed in fresh water and grey water resources. Regarding micro essential elements, the highest amount of Zn (9.89 mg/L), Fe (17.63 mg/L) and Mn (0.81 mg/L) was observed in dairy water. Likewise, the number of heavy metals significantly varied with water resources. The maximum concentration of Cu (3.16 mg/L) was observed in greywater while its least amount (1.60 mg/L) was observed in freshwater. Similarly, the concentrations of Pb (4.25 mg/L), Cr (1.64 mg/L) and Cd (0.08 mg/L) were noted highest in greywater. Meanwhile Ba was found highest in wastewater (5 mg/L) at par with a freshwater source (4.62 mg/L) and its quantity was observed lowest (0.25 mg/L) in dairy water.

Table 1. Nutrients and heavy metal analysis of irrigation resources before application.

	Freshwater	Greywater	Dairy Water	Wastewater
Macroelements				
Na (mg/L)	320 ± 100 c	530 ± 90 c	1500 ± 80 b	5370 ± 90 a
K (mg/L)	190 ± 20 b	310 ± 90 b	1430 ± 80 a	380 ± 50 b
Ca (mg/L)	550 ± 100 b	100 ± 40 c	670 ± 90 b	1960 ± 80 a
Mg (mg/L)	0.00 ± 0.00 c	0.00 ± 0.00 c	590 ± 20 b	940 ± 20 a
Microelements				
Zn (mg/L)	6.40 ± 0.4 d	9.18 ± 0.9 b	9.89 ± 0.9 a	7.07 ± 0.2 c
Fe (mg/L)	2.95 ± 0.1 d	12.97 ± 0.2 b	17.63 ± 0.3 a	4.85 ± 0.1 c
Mn (mg/L)	0.20 ± 0.09 c	0.61 ± 0.10 ab	0.81 ± 0.08 a	0.40 ± 0.10 c
Heavy metals				
Cu (mg/L)	1.60 ± 0.18 c	3.16 ± 0.06 a	1.96 ± 0.01 b	1.85 ± 0.10 bc
Ba (mg/L)	4.62 ± 0.08 a	3.71 ± 0.20 b	3.76 ± 0.09 b	5.00 ± 0.20 a
Cr (mg/L)	0.40 ± 0.09 b	1.46 ± 0.10 a	1.33 ± 0.18 a	0.40 ± 0.04 b
Pb (mg/L)	0.70 ± 0.01 c	4.25 ± 0.10 a	0.25 ± 0.02 d	1.20 ± 0.20 b
Cd (mg/L)	0.05 ± 0.01 a	0.08 ± 0.01 a	0.05 ± 0.01 a	0.05 ± 0.02 a

Any two means in a column having similar letters indicate a non-significant relationship ($p \leq 0.05$).

3.2. Concentrations of Macro–Micro and Heavy Metal Elements Available in the Compost When Treated with Different Irrigation Water Resources

The nutrient analysis of compost after final crop harvesting is presented in Table 2. The results exhibited that among macro essential elements, the highest amount of Na (690 mg/L) was observed in compost treated with wastewater followed by the compost treated with dairy water (530 mg/L) and greywater (450 mg/L). Similarly, the highest contents of Ca (18,400 mg/L) were observed in compost treated with wastewater followed by dairy water (17,870 mg/L). While the highest amount of K was observed in compost treated with dairy water (800 mg/L) at par with wastewater (780 mg/L). Likewise, the concentration of all micro essential elements Zn (91.30 mg/L), Fe (862 mg/L), and Mn (52.57 mg/L) was found at maximum in dairy water while these micronutrients were observed in minimum quantity in compost treated with fresh water. Regarding heavy metals, the highest amount of Cu (172 mg/L), Cr (11.53 mg/L), Pb (12.22 mg/L) and Cd (1.82 mg/L) was noticed in compost treated with greywater. Meanwhile, the maximum amount of Ba (8.25 mg/L) was recorded in compost treated with fresh water at par with composts treated with wastewater (8.5 mg/L) and dairy water (8.11 mg/L), respectively.

Table 2. Nutrients and heavy metal contents of compost (growing medium) after irrigation applications.

	Freshwater	Greywater	Dairy Water	Wastewater
Macroelements				
Na (mg/L)	430 ± 40 b	450 ± 50 b	530 ± 110 b	690 ± 30 a
K (mg/L)	770 ± 20 a	780 ± 30 a	800 ± 50 a	780 ± 100 a
Ca (mg/L)	17,180 ± 170 c	17,070 ± 120 c	17,870 ± 100 b	18,400 ± 150 a
Mg (mg/L)	1390 ± 30 b	1370 ± 100 b	1640 ± 100 a	1710 ± 100 a
Microelements				
Zn (mg/L)	41.85 ± 5 d	62.00 ± 0.89 b	91.30 ± 20 a	51.55 ± 17 c
Fe (mg/L)	727.00 ± 100 b	746.67 ± 40 b	862.00 ± 30 a	728.67 ± 89 b
Mn (mg/L)	38.82 ± 0.05 d	41.96 ± 0.04 c	52.57 ± 0.03 a	42.65 ± 0.09 b
Heavy metals				
Cu (mg/L)	16.26 ± 0.45 d	172.0 ± 17 a	59.77 ± 0.12 b	20.77 ± 0.10 c
Ba (mg/L)	8.25 ± 0.10 a	7.71 ± 0.20 b	8.11 ± 0.11 a	8.15 ± 0.09 a
Cr (mg/L)	11.25 ± 0.22 a	11.53 ± 0.20 a	7.82 ± 0.3 b	3.63 ± 0.27 c
Pb (mg/L)	7.83 ± 0.13 b	12.22 ± 0.18 a	2.93 ± 0.03 d	7.46 ± 0.8 c
Cd (mg/L)	1.31 ± 0.10 b	1.82 ± 0.10 a	1.01 ± 0.09 c	0.73 ± 0.09 d

Any two means in a column having similar letters indicate a non-significant relationship ($p \leq 0.05$).

3.3. Accumulation of Essential Minerals and Heavy Metals in Plant Leaves

The application through different water resources has significantly affected the accumulation level of nutrients in plant leaf tissues as shown in Table 3. The results revealed that in macro elements, the highest concentration of Na (1940 mg/L) was observed in leaves that were supplied wastewater irrigation followed by dairy water irrigation application (1220 mg/L). The macronutrients, K (9470 mg/L) and Ca (45,080 mg/L) were found maximum in plant leaves that were treated with dairy water resources while the least quantity of these nutrients K (6730 mg/L) and Ca (42,830 mg/L) was observed in plants that were treated with a freshwater resource. For micronutrients, Zn (48.37 mg/L) was found highest in plant leaves that were treated with dairy water followed by grey water (44.58 mg/L) as the resource of irrigation, while the least amount of Zn (34.68 mg/L) was observed in plants treated with wastewater. Similarly, other micro essentials Fe (95 mg/L) and Mn (41 mg/L) were also found to be highest in plant leaves that were treated with dairy water. Regarding heavy metals, the concentrations of Cu (5.38 mg/L), Cr (5.24 mg/L) and Pb (4.34 mg/L) were noticed in leaves that were treated with greywater resources. Meanwhile, the concentrations of Cr (5.28 mg/L) and Pb (4.34 mg/L) were the least in plant leaves that were treated with dairy water. However, the concentration of Cd (0.46 mg/L) was found to be a minimum in plant leaves treated with fresh water.

Table 3. Distribution status of macroelements, microelements, and heavy metals in cucumber leaves irrigated with different irrigations resources.

	Freshwater	Greywater	Dairy Water	Wastewater
Macroelements				
Na (mg/L)	710 ± 20 c	690 ± 10 c	1220 ± 90 b	1940 ±80 a
K (mg/L)	6730 ± 160 c	7550 ± 190 b	9470 ± 120 a	6130 ± 90 d
Ca (mg/L)	42,830 ± 1000 c	44,830 ± 900 a	45,080 ± 800 a	44,120 ± 700 b
Mg (mg/L)	5460 ± 110 c	5430 ± 70 c	6580 ± 100 a	6030 ± 80 b
Microelements				
Zn (mg/L)	38.05 ± 0.10 c	44.58 ± 0.8 b	48.37 ± 0.9 a	34.68 ± 0.78 d
Fe (mg/L)	78.80 ± 1.78 b	50.56 ± 1.09 c	95.00 ± 3.00 a	76.18 ± 0.9 b
Mn (mg/L)	23.70 ± 1.80 b	31.10 ± 0.9 b	41.00 ± 0.10 a	31.16 ± 0.88 b
Heavy metals				
Cu (mg/L)	5.07 ± 0.08 b	5.38 ± 0.12 a	4.06 ± 0.08 d	4.63 ± 0.09 c
Ba (mg/L)	11.85 ± 1.79 a	7.40 ± 0.40 b	8.38 ± 0.08 b	6.72 ± 0.09 b
Cr (mg/L)	4.46 ± 0.04 b	5.28 ± 0.10 a	3.86 ± 0.12 c	4.28 ± 0.06 b
Pb (mg/L)	2.11 ± 0.09 c	4.34 ± 0.020 a	1.81 ± 0.09 c	3.13 ± 0.08 b
Cd (mg/L)	0.46 ± 0.04 c	0.78 ± 0.03 ab	0.73 ± 0.02 b	0.81 ± 0.01 a

Any two means in a column having similar letters indicate a non-significant relationship ($p \leq 0.05$).

3.4. Accumulation of Essential Minerals and Heavy Metals in Cucumber Fruit

The results regarding the minerals and heavy metals are shown in Table 4. The results indicated that the use of different irrigation resources has a significant effect on the accumulation of minerals and heavy metals in fruits. The highest amount of macro element accumulation was observed in the fruits that were irrigated with wastewater (Na (1360 mg/L) followed by the dairy water (790 mg/L). Regarding K, it was found highest (16,780 mg/L) in fruits treated with dairy water while the least amount of K (12,530 mg/L) was observed in fruits treated with wastewater. While the maximum amount of Ca (2420 mg/L) and Mg (1550 mg/L) was observed in dairy water, however, the minimum amount of Ca (1400 mg/L) and Mg (1090 mg/L) was detected in fruits irrigated with fresh water.

Table 4. Distribution status of macro, microelements, and heavy metals in cucumber fruit irrigated with different irrigation resources.

	Freshwater	Greywater	Dairy Water	Wastewater
Macroelements				
Na (mg/L)	620 ± 10 c	660 ± 20 bc	790 ± 40 b	1360 ± 80 a
K (mg/L)	14,330 ± 100 b	13,770 ± 160 c	16,780 ± 180 a	12,530 ± 190 d
Ca (mg/L)	1400 ± 90 c	1920 ± 30 b	2420 ± 17 a	1420 ± 50 c
Mg (mg/L)	1090 ± 40 c	1260 ± 10 bc	1550 ± 90 a	1370 ± 10 ab
Microelements				
Zn (mg/L)	24.11 ± 0.11 c	25.27 ± 0.07 b	28.13 ± 0.12 a	20.18 ± 0.05 d
Fe (mg/L)	33.88 ± 0.06 c	61.20 ± 3.58 b	67.96 ± 0.25 a	33.90 ± 0.11 c
Mn (mg/L)	10.35 ± 0.20 c	13.95 ± 0.27 b	15.68 ± 0.18 a	8.91 ± 0.04 d
Heavy metals				
Cu (mg/L)	5.31 ± 0.18 a	5.68 ± 0.16 a	4.67 ± 0.20 b	5.26 ± 0.08 a
Ba (mg/L)	2.25 ± 0.35 a	2.62 ± 0.51 a	0.71 ± 0.10 b	2.81 ± 0.09 a
Cr (mg/L)	3.00 ± 0.08 b	5.02 ± 0.02 a	2.73 ± 0.03 c	3.03 ± 0.03 b
Pb (mg/L)	0.86 ± 0.07 c	1.80 ± 0.27 a	1.15 ± 0.11 bc	1.50 ± 0.09 ab
Cd (mg/L)	0.13 ± 0.18 b	0.22 ± 0.02 a	0.15 ± 0.03 b	0.20 ± 0.02 a

Any two means in a column having similar letters indicate a non-significant relationship ($p \leq 0.05$).

For micronutrients, the highest amount of Zn (28.13 mg/L), Fe (67.96 mg/L) and Mn (15.68 mg/L) were found in fruits treated with dairy water while the least quantity Zn (20.18 mg/L), and Mn (8.91 mg/L) were found in fruits treated with wastewater, and the Fe (33.88 mg/L) minimum amount was observed in freshwater. In heavy metals, the highest amount of Cu (5.68 mg/L), Ba (2.62 mg/L), Cr (5.02 mg/L), Pb (1.80 mg/L) and Cd (0.22 mg/L) were observed in fruits treated with greywater.

3.5. Effect of Irrigation Water on Fruit Production and Quality

The results regarding fruit production and quality are shown in Figure 1. The results showed that the highest yield in cucumber (1191.02 g) was achieved through dairy water irrigation followed by wastewater (976.65 g) and freshwater irrigation (938.14 g), while the least amount of fruit yield (872.01 g) was obtained in plants irrigated with greywater as shown in Figure 1a. However, the different irrigations have shown a non-significant impact on cucumber fruit diameter as indicated in Figure 1b. With regard to the quality of the total soluble solids, maximum TSS (2.26 °brix) was observed in cucumber fruits that were irrigated with wastewater followed by the dairy water (2.06 °brix). However, the least amount of TSS (1.61 °brix) was observed in fruits that were irrigated with fresh water and grey water (1.57 °brix) as shown in Figure 1c.

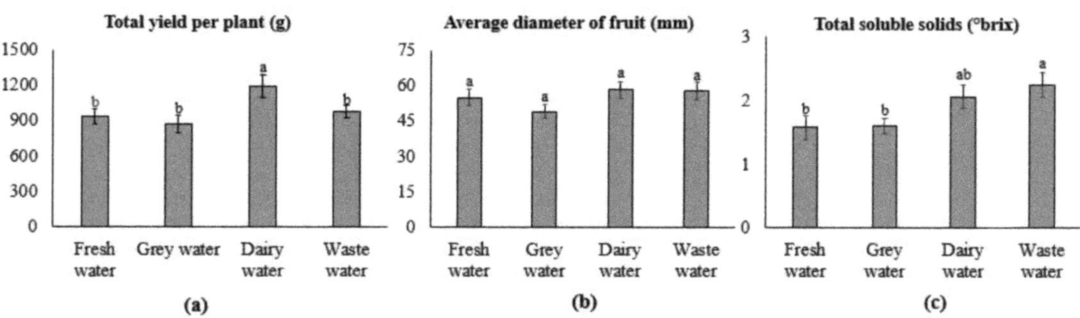

Figure 1. Effect of irrigation resources on (**a**) cucumber yield, (**b**) fruit diameter and (**c**) total soluble solids contents.

4. Discussion

Worldwide, the availability of fresh water is becoming a major concern, and the recycling of wastewater for crops irrigation has received a lot of interest. However, untreated or inadequately treated wastewater may harm soil, plants, and the environment, as well as cause serious risks to human health [32]. The presence of macro, micro, and heavy metals at high levels in wastewater disturbs plant physiological processes, metabolic processes and enzyme activities as the combination of different elements does not always increase growth and fruit yield [33,34]. However, the toxicity of these elements may affect plant yield, growth and may induce abnormal anatomical changes [35]. Furthermore, it is argued that heavy metals can accumulate in plants when municipal wastewater is used for irrigation, and an excessive amount can affect residents who consume crops and/or vegetables grown in these contaminated locations [36]. As a result, this could be the first food safety indicator.

In irrigation water, sodium (Na), magnesium (Mg) and calcium (Ca) are dominantly present and play an important role in plant growth and development. The wastewater carries significant amounts of nitrogen, phosphorus, potassium, calcium, magnesium, organic matter, and trace minerals, which are regarded as good sources of nutrients for plant growth and development [37]. After irrigation, some of the water is up taken and consumed by plants or evaporates directly from the soil. Among these salts, Ca and Mg often precipitate in carbonates while Na remains dominant in soil [38]. In our findings, a higher concentration of Na was observed in compost treated with wastewater. This higher accumulation in media may be due to wastewater as it already contains a higher amount of Na. However, the higher level of Na in soil may depress the plants' nutrient uptake and may increase the salts accumulation, particularly sodium chlorides (NaCl) that are most common in causing soil salinity. It has been reported that agriculture practices are increasingly pressurizing fertilizer demand; as a result, it is critical to find alternatives that maximize nutrient uptake by crop plants while minimizing the associated environmental impacts, and wastewater streams can be used as an alternate irrigation water supply [39]. The accumulation of Na generates osmotic potential and prevents water uptake, which creates similar conditions as drought [40]. Moreover, there may be accelerated drought and salinity due to an increase in low-quality water for irrigation, inadequate drainage and climate change [41]. Therefore, the increased level in Na may lead plants towards drought conditions and the application of wastewater to arid soils may enhance salinity and drought. The excess accumulation of other essential elements K, Ca, Cl and NO_3 in plant root zones may develop osmotic pressure in plant roots and imbalance cell ion homeostasis that leads towards photosynthesis inhibition, inactivates certain enzyme functions and damage to chloroplast and other organelles [42]. However, the presence of these elements up to optimal level promotes plant growth, development, photosynthesis and transpiration. Our results presented in Table 4 showed that the plant content of micronutrients and heavy metals surpassed (in most cases) the WHO permitted limits as shown in Table 5. However, Zn and Fe levels in cucumber fruit (Table 4) irrigated with freshwater and wastewater was lower compared to the threshold values.

Nevertheless, it was observed that heavy metals taken up by the vegetables produced under wastewater irrigation prefer to persist in the roots, a small percentage of the heavy metals translocated into the shoots, and an even less percentage reaches the fruit [47]. These elements also play a pivotal role in human health and their deficiency or toxicity may affect the metabolic functions occurring in the human body. Zinc is one of the essential elements required for the human body as it promotes cell growth, cell division, carbohydrates breakdown, wound healing, plays an important role in the defensive system and helps in developing immunity [48]. It is an omnipotent metal having amphoteric nature and its daily requirement is about 15–20 mg/day [49]. However, the high uptake of zinc may interfere with Cu uptake and may induce trauma due to zinc accumulation [50]. The researchers also reported that heavy metals in irrigation wastewater tend to collect in soils, where they may become bioavailable to crops and enhance bacterial populations linked with plant growth [51]. Besides, Iron (Fe) is a micronutrient that is vital for plant growth, it is a critical component in energy conversions required for syntheses and other cell life processes [52]. However, the Cu (5.26 mg/L) and Ca (1420 mg/L) were found in the highest concentrations in cucumber fruit from various irrigation resource water applications (Table 4), earlier studies indicate that minerals were extensively translocated from leaves to fruits in greenhouse tomato and cumbers [53]. It has also been reported that Cu plays an important role as a redox transition element in photosynthesis, respiration, and the metabolism of carbon and nitrogen, as well as protecting from oxidative stress [54].

Table 5. The World Health Organization (WHO), the United States Environmental Protection Agency (USEPA), Oman, the United Arab Emirates, Egypt, and Algerian Standards have established wastewater threshold levels for reuse applications [43–46].

Parameters	Units	WHO	US-EPA	Oman	UAE	Jordan	Egypt	Algeria
Na	mg/L	250		70	150	230		-
K	mg/L	100		-	12	-		-
Ca	mg/L	450		-	-	230		-
Mg	mg/L	80		30	0.4	100		-
Zn	mg/L	20	2	2	5	5	0.01	10
Fe	mg/L	50	5	1	0.2	5	0.5	-
Mn	mg/L	0.2	0.2	0.2	-	0.2	0.2	-
Cu	mg/L	0.2	0.2	0.2	1	0.2	0.01	0.5
Cr	mg/L	0.1	0.1	0.1	0.5	0.1	0.05	5
Pb	mg/L	5	5	0.1	0.1	5	0.01	10
Cd	mg/L	0.01	0.01	0.01	0.003	0.01	0.001	-

In our findings, the highest concentrations of K, Zn, Fe and Mn were observed in dairy water resources and the maximum yield was also obtained in cucumber in a similar resource of irrigation. However, heavy metals (Cr and Pb) were lower in all irrigation water types compared to the threshold values (Table 5). The highest yield may be due to the presence of K as it regulates plant stomata conductance and ensures optimal plant growth [55]. Further, it activates certain enzymes that trigger protein synthesis, sugar transport, and photosynthesis. Our finding has a resemblance with other researchers who stated that K is essential for cell growth and increases fruit yield and quality [56]. It has been also argued that K, on the other hand, is the most abundant cationic inorganic nutrient, with a large role in activating enzymatic reactions in plants and reducing heavy metal intake due to its availability as a nutrient in irrigation water and soil [57]. A similar finding was observed by [58] who stated that K is involved in stomatal opening, cell elongation and certain physiological processes that enhance plant yield. The studies have shown that Mn is an essential element; nevertheless, high quantities of Mn in the soil might smother plant growth. The current study found slightly elevated Mn levels in cucumber fruit, ranging from 8.91 mg/L to 15.68 mg/L, which could be hazardous to plants and soil media [52]. Likewise, Zn is an essential micronutrient that is required for plant growth and is involved in various enzymatic reactions as a catalyst, metabolism and energy transfer [59]. Therefore, it promoted plant yield attributes such as yield and fruit diameter in our study. Zn is equally essential for human health as it improves the immune system, prevents hair loss, muscle weakness, and memory loss. About 50% of the world population is suffering from Zn deficiency, and the ideal dose of 15 mg/day is required every day [60]. Likewise, according to FAO, 30% of world land is deficient in Zinc, therefore such cultivated areas are deficient in providing required range to the plants [61]. In our study, the range of Zn obtained in cucumber fruits treated with different irrigation resources was 20.18 to 28.13 mg/L, whereas in irrigation resources the level of Zn was less in fresh water, wastewater, greywater and followed by dairy water, respectively. Therefore, irrigation through these water resources may be used in some Zn deficient soils. It has been reported that Zn is a mineral that is required for plant nutrition, it serves structural and/or catalytic roles in a range of enzymes, including Cu–Zn superoxide dismutase, alcohol dehydrogenase, RNA polymerase, and DNA-binding proteins, and is important in carbohydrate metabolism [62]. However, it has also revealed that plants, on the other hand, are unlikely to absorb the excess Zn and Cu, which could be damaging to food consumers and hazardous to plants long before reaching a level that is toxic to humans [46].

In our study, the highest yield of cucumber was obtained in plants treated with dairy water. This might be due to the highest availability of essential macro and microelements in dairy water that promoted plant growth and fruit cell division and expansion. Our findings were in harmony with other researchers who reported that the application of wastewater boosted tomato yield compared to normal irrigation due to the presence of nutrients [63,64]. In our findings, TSS in cucumber fruits was within the range (1.57–2.26 °brix) that showed non-toxicity or non-accumulation of heavy metals. The highest TSS (2.26 °brix) was observed in fruits treated with wastewater while the least TSS (1.57 °brix) was observed in freshwater. The stress conditions and the abundance of elements might be a reason for the increase in soluble solids. Similar results were observed by other researchers who suggested that higher TSS concentrations provide a mechanism in plants to maintain osmotic pressure during stress conditions by metals [18,65]. However, high total soluble solids levels in fruits are also linked with consumer acceptance and are an important trait for crop harvest [66].

The reuse of wastewater in agriculture may be a greater source of saving freshwater consumption, as non-availability of fresh water and degraded water quality is a major issue for agriculture production [67]. Therefore, the reuse of wastewater in agriculture offers an ideal alternative source of irrigation to conserve precious freshwater resources particularly at a time of ever-changing climate scenarios [31]. Moreover, wastewater has been used as a source of fertilizer for different vegetable and fruit crops [68]. However, the use of wastewater requires consideration as it has an impact on soil physical, chemical, and microbial properties. Further, it has an influence on crop productivity and human health [69]. Therefore, precise monitoring and adequate strategies are required to minimize waste water risks for human and environment to ensure sustainable reuse practice [70].

5. Conclusions

The resources of freshwater are depleting rapidly due to climate change and urbanization. The utilization of wastewater is becoming popular in the agriculture sector, particularly in the arid and semiarid regions of the world because of the severe water scarcity. The irrigation of cucumber plants with different water resources, such as tap water, greywater, dairy water and wastewater enhanced the quantity of essential and non-essential elements. However, the level of these elements was within the range that was non-toxic to human consumption and the environment. Further, the use of dairy water revealed advantages in enhancing sustainable greenhouse cucumber production compared to other irrigation water resources. The application of alternate water resources offered marked opportunities for sustainable agriculture and conservation of ever-depleting precious water reserves.

Author Contributions: M.M.K., has conceptualized, collected material for the research and experiment conducted, and planned the manuscript structure; M.H.A.-H., has assisted in sample analysis; M.T.A., has assisted in gathering information, manuscript formatting and data analysis. M.A.K., has reviewed the manuscript and helped in manuscript formatting; A.A.F., has reviewed the manuscript extensively and suggested suitable changes. S.A.S., has edited the manuscript and has provided the related literature info. All authors have read and agreed to the published version of the manuscript.

Funding: This research received no external funding.

Institutional Review Board Statement: Not applicable.

Informed Consent Statement: Not applicable.

Data Availability Statement: Not applicable.

Acknowledgments: The authors greatly acknowledge the College of Agricultural and Marine Sciences, Sultan Qaboos University, Oman, for their continuous support throughout the study.

Conflicts of Interest: The authors declare that they have no conflict of interest.

References

1. Becerra-Castro, C.; Lopes, A.; Vaz-Moreira, I.; Silva, M.E.F.; Manaia, C.M.; Nunes, O.C. Wastewater reuse in irrigation: A microbiological perspective on implications in soil fertility and human and environmental health. *Environ. Int.* **2015**, *75*, 117–135. [CrossRef]
2. Bixio, D.; Thoeye, C.; De Koning, J.; Joksimovic, D.; Savic, D.; Wintgens, T.; Melin, T. Wastewater reuse in Europe. *Desalination* **2006**, *187*, 89–101. [CrossRef]
3. UNDESA. International Decade for Action 'WATER FOR LIFE' 2005–2015. 2014. Available online: https://www.un.org/waterforlifedecade/ (accessed on 30 September 2021).
4. FAO. Growing More Food—Using less water. In Proceedings of the Fifth World Water Forum, Istanbul, Turkey, 16–22 March 2009. Available online: https://www.fao.org/land-water/news-archive/news-detail/en/c/267311/ (accessed on 30 September 2021).
5. Boretti, A.; Rosa, L. Reassessing the projections of the World Water Development Report. *NPJ Clean Water* **2019**, *2*, 1–6. [CrossRef]
6. Sato, T.; Qadir, M.; Yamamoto, S.; Endo, T.; Zahoor, A. Global, regional, and country level need for data on wastewater generation, treatment, and use. *Agric. Water Manag.* **2013**, *130*, 1–13. [CrossRef]
7. Sgroi, M.; Vagliasindi, F.G.; Roccaro, P. Feasibility, sustainability and circular economy concepts in water reuse. *Curr. Opin. Environ. Sci. Health* **2018**, *2*, 20–25. [CrossRef]
8. Alcalde-Sanz, L.; Gawlik, B. Minimum Quality Requirements for Water Reuse in Agricultural Irrigation and Aquifer Recharge. Towards a Legal Instrument on Water Reuse at EU Level. 2017. Available online: https://publications.jrc.ec.europa.eu/repository/bitstream/JRC109291/jrc109291_online_08022018.pdf (accessed on 30 September 2021).
9. Maaß, O.; Grundmann, P. Governing Transactions and Interdependences between Linked Value Chains in a Circular Economy: The Case of Wastewater Reuse in Braunschweig (Germany). *Sustainability* **2018**, *10*, 1125. [CrossRef]
10. Hanjra, M.A.; Blackwell, J.; Carr, G.; Zhang, F.; Jackson, T.M. Wastewater irrigation and environmental health: Implications for water governance and public policy. *Int. J. Hyg. Environ. Health* **2012**, *215*, 255–269. [CrossRef] [PubMed]
11. Chojnacka, K.; Witek-Krowiak, A.; Moustakas, K.; Skrzypczak, D.; Mikula, K.; Loizidou, M. A transition from conventional irrigation to fertigation with reclaimed wastewater: Prospects and challenges. *Renew. Sustain. Energy Rev.* **2020**, *130*, 109959. [CrossRef]
12. Al-Rashed, M.F.; Sherif, M.M. Water Resources in the GCC Countries: An Overview. *Water Resour. Manag.* **2000**, *14*, 59–75. [CrossRef]
13. Mohammad, M.J.; Mazahreh, N. Changes in Soil Fertility Parameters in Response to Irrigation of Forage Crops with Secondary Treated Wastewater. *Commun. Soil Sci. Plant Anal.* **2003**, *34*, 1281–1294. [CrossRef]
14. Hussain, A.; Priyadarshi, M.; Dubey, S. Experimental study on accumulation of heavy metals in vegetables irrigated with treated wastewater. *Appl. Water Sci.* **2019**, *9*, 1–11. [CrossRef]
15. Woldetsadik, D.; Drechsel, P.; Keraita, B.; Itanna, F.; Erko, B.; Gebrekidan, H. Microbiological quality of lettuce (Lactuca sativa) irrigated with wastewater in Addis Ababa, Ethiopia and effect of green salads washing methods. *Int. J. Food Contam.* **2017**, *4*, 49. [CrossRef]
16. Lonigro, A.; Rubino, P.; Lacasella, V.; Montemurro, N. Faecal pollution on vegetables and soil drip irrigated with treated municipal wastewaters. *Agric. Water Manag.* **2016**, *174*, 66–73. [CrossRef]
17. Rusan, M.J.M.; Hinnawi, S.; Rousan, L. Long term effect of wastewater irrigation of forage crops on soil and plant quality parameters. *Desalination* **2007**, *215*, 143–152. [CrossRef]
18. Gupta, S.; Satpati, S.; Nayek, S.; Garai, D. Effect of wastewater irrigation on vegetables in relation to bioaccumulation of heavy metals and biochemical changes. *Environ. Monit. Assess.* **2010**, *165*, 169–177. [CrossRef] [PubMed]
19. Tabatabaei, S.-H.; Najafi, P.; Amini, H. Assessment of Change in Soil Water Content Properties Irrigated with Industrial Sugar Beet Wastewater. *Pak. J. Biol. Sci.* **2007**, *10*, 1649–1654. [CrossRef]
20. Ding, X.; Jiang, Y.; Zhao, H.; Guo, D.; He, L.; Liu, F.; Zhou, Q.; Nandwani, D.; Hui, D.; Yu, J. Electrical conductivity of nutrient solution influenced photosynthesis, quality, and antioxidant enzyme activity of pakchoi (Brassica campestris L. ssp. Chinensis) in a hydroponic system. *PLoS ONE* **2018**, *13*, e0202090. [CrossRef]
21. Chen, C.-Y.; Wang, S.-W.; Kim, H.; Pan, S.-Y.; Fan, C.; Lin, Y.J. Non-conventional water reuse in agriculture: A circular water economy. *Water Res.* **2021**, *199*, 117193. [CrossRef]
22. Shehata, H.S.; Galal, T.M. Trace metal concentration in planted cucumber (Cucumis sativus L.) from contaminated soils and its associated health risks. *J. Consum. Prot. Food Saf.* **2020**, *15*, 205–217. [CrossRef]
23. Sharma, R.K.; Agrawal, M.; Marshall, F. Heavy Metal Contamination in Vegetables Grown in Wastewater Irrigated Areas of Varanasi, India. *Bull. Environ. Contam. Toxicol.* **2006**, *77*, 312–318. [CrossRef] [PubMed]
24. Ricolfi, L.; Barbieri, M.; Muteto, P.V.; Nigro, A.; Sappa, G.; Vitale, S. Potential toxic elements in groundwater and their health risk assessment in drinking water of Limpopo National Park, Gaza Province, Southern Mozambique. *Environ. Geochem. Health* **2020**, *42*, 2733–2745. [CrossRef]

25. Arora, M.; Kiran, B.; Rani, S.; Rani, A.; Kaur, B.; Mittal, N. Heavy metal accumulation in vegetables irrigated with water from different sources. *Food Chem.* **2008**, *111*, 811–815. [CrossRef]
26. Muchuweti, M.; Birkett, J.; Chinyanga, E.; Zvauya, R.; Scrimshaw, M.; Lester, J. Heavy metal content of vegetables irrigated with mixtures of wastewater and sewage sludge in Zimbabwe: Implications for human health. *Agric. Ecosyst. Environ.* **2006**, *112*, 41–48. [CrossRef]
27. Khan, S.; Cao, Q.; Zheng, Y.; Huang, Y.; Zhu, Y. Health risks of heavy metals in contaminated soils and food crops irrigated with wastewater in Beijing, China. *Environ. Pollut.* **2008**, *152*, 686–692. [CrossRef] [PubMed]
28. Avci, H. Trace metals in vegetables grown with municipal and industrial wastewaters. *Toxicol. Environ. Chem.* **2012**, *94*, 1125–1143. [CrossRef]
29. Al-Nakshabandi, G.; Saqqar, M.; Shatanawi, M.; Fayyad, M.; Al-Horani, H. Some environmental problems associated with the use of treated wastewater for irrigation in Jordan. *Agric. Water Manag.* **1997**, *34*, 81–94. [CrossRef]
30. Mojiri, A.; Aziz, H.A.; Aziz, S.Q.; Gholami, A.; Aboutorab, M. Impact of Urban Wastewater on Soil Properties and Lepidium sativum in an Arid Region. *Int. J. Sci. Res. Environ. Sci.* **2013**, *1*, 7–15. [CrossRef]
31. Pedrero, F.; Kalavrouziotis, I.; Alarcón, J.J.; Koukoulakis, P.; Asano, T. Use of treated municipal wastewater in irrigated agriculture—Review of some practices in Spain and Greece. *Agric. Water Manag.* **2010**, *97*, 1233–1241. [CrossRef]
32. Bañón, S.; Miralles, J.; Ochoa, J.; Franco, J.; Sánchez-Blanco, M. Effects of diluted and undiluted treated wastewater on the growth, physiological aspects and visual quality of potted lantana and polygala plants. *Sci. Hortic.* **2011**, *129*, 869–876. [CrossRef]
33. Rengel, Z.; Bose, J.; Chen, Q.; Tripathi, B.N. Magnesium alleviates plant toxicity of aluminium and heavy metals. *Crop. Pasture Sci.* **2015**, *66*, 1298–1307. [CrossRef]
34. Cole, J.C.; Smith, M.W.; Penn, C.J.; Cheary, B.S.; Conaghan, K.J. Nitrogen, phosphorus, calcium, and magnesium applied individually or as a slow release or controlled release fertilizer increase growth and yield and affect macronutrient and micronutrient concentration and content of field-grown tomato plants. *Sci. Hortic.* **2016**, *211*, 420–430. [CrossRef]
35. Hajihashemi, S.; Mbarki, S.; Skalicky, M.; Noedoost, F.; Raeisi, M.; Brestic, M. Effect of Wastewater Irrigation on Photosynthesis, Growth, and Anatomical Features of Two Wheat Cultivars (*Triticum aestivum* L.). *Water* **2020**, *12*, 607. [CrossRef]
36. Cao, C.; Zhang, Q.; Ma, Z.-B.; Wang, X.-M.; Chen, H.; Wang, J.-J. Fractionation and mobility risks of heavy metals and metalloids in wastewater-irrigated agricultural soils from greenhouses and fields in Gansu, China. *Geoderma* **2018**, *328*, 1–9. [CrossRef]
37. Taghipour, M.; Jalali, M. Impact of some industrial solid wastes on the growth and heavy metal uptake of cucumber (*Cucumis sativus* L.) under salinity stress. *Ecotoxicol. Environ. Saf.* **2019**, *182*, 109347. [CrossRef]
38. Taalab, A.S.; Ageeb, G.W.; Siam, H.S.; Mahmoud, S.A. Some Characteristics of Calcareous soils. A review. *Middle East J.* **2019**, *8*, 96–105.
39. Santos, F.M.; Pires, J.C. Nutrient recovery from wastewaters by microalgae and its potential application as bio-char. *Bioresour. Technol.* **2018**, *267*, 725–731. [CrossRef] [PubMed]
40. Jogawat, A. Osmolytes and their Role in Abiotic Stress Tolerance in Plants. *Molecular Plant Abiotic Stress* **2019**, 91–104. [CrossRef]
41. Machado, R.M.A.; Serralheiro, R.P. Soil salinity: Effect on vegetable crop growth. Management practices to prevent and mitigate soil salinization. *Horticulturae* **2017**, *3*, 30. [CrossRef]
42. Paranychianakis, N.; Chartzoulakis, K. Irrigation of Mediterranean crops with saline water: From physiology to management practices. *Agric. Ecosyst. Environ.* **2005**, *106*, 171–187. [CrossRef]
43. Cherfi, A.; Achour, M.; Cherfi, M.; Otmani, S.; Morsli, A. Health risk assessment of heavy metals through consumption of vegetables irrigated with reclaimed urban wastewater in Algeria. *Process. Saf. Environ. Prot.* **2015**, *98*, 245–252. [CrossRef]
44. Gabr, M. Wastewater Reuse Standards for Agriculture Irrigation in Egypt. 2019. Available online: https://www.researchgate. net/profile/Mohamed-Gabr-7/publication/333676905_WASTEWATER_REUSE_STANDARDS_FOR_AGRICULTURE_ IRRIGATION_IN_EGYPT/links/5cfe991d4585157d15a1f2f3/WASTEWATER-REUSE-STANDARDS-FOR-AGRICULTURE-IRRIGATION-IN-EGYPT.pdf (accessed on 30 September 2021).
45. MEWR. Sultanate of Oman on the Conservation of the Environment and Prevention of Pollution. Regulations for Wastewater Re-Use and Discharge, Ministreial Decision 5/86 Dated 17th May, 1986. 1986; pp. 1–6. Available online: http://www. sustainableoman.com/wp-content/uploads/2016/05/MD-5-86.pdf (accessed on 30 September 2021).
46. Shakir, E.; Zahraw, Z.; Al-Obaidy, A.H.M. Environmental and health risks associated with reuse of wastewater for irrigation. *Egypt. J. Pet.* **2017**, *26*, 95–102. [CrossRef]
47. Emongor, V.; Ramolemana, G. Treated sewage effluent (water) potential to be used for horticultural production in Botswana. *Phys. Chem. Earth Parts A/B/C* **2004**, *29*, 1101–1108. [CrossRef]
48. Roohani, N.; Hurrell, R.; Kelishadi, R.; Schulin, R. Zinc and its importance for human health: An integrative review. *J. Res. Med Sci.* **2013**, *18*, 144–157.
49. Prashanth, L.; Kattapagari, K.K.; Chitturi, R.T.; Baddam, V.R.R.; Prasad, L.K. A review on role of essential trace elements in health and disease. *J. Dr. NTR Univ. Health Sci.* **2015**, *4*, 75–85. [CrossRef]
50. Plum, L.M.; Rink, L.; Haase, H. The Essential Toxin: Impact of Zinc on Human Health. *Int. J. Environ. Res. Public Health* **2010**, *7*, 1342–1365. [CrossRef]
51. Chaoua, S.; Boussaa, S.; El Gharmali, A.; Boumezzough, A. Impact of irrigation with wastewater on accumulation of heavy metals in soil and crops in the region of Marrakech in Morocco. *J. Saudi Soc. Agric. Sci.* **2019**, *18*, 429–436. [CrossRef]
52. Hajar, E.W.I.; Bin Sulaiman, A.Z.; Sakinah, A.M. Assessment of Heavy Metals Tolerance in Leaves, Stems and Flowers of Stevia Rebaudiana Plant. *Procedia Environ. Sci.* **2014**, *20*, 386–393. [CrossRef]

53. Ekinci, M.; Esringu, A.; Dursun, A.; Yildirim, E.; Turan, M.; Karaman, M.R.; Arjumend, T. Growth, yield, and calcium and boron uptake of tomato (*Lycopersicon esculentum* L.) and cucumber (*Cucumis sativus* L.) asaffected by calcium and boron humate application in greenhouse conditions. *Turk. J. Agric. For.* **2015**, *39*, 613–632. [CrossRef]

54. Ramírez-Pérez, L.J.; Morales-Díaz, A.B.; De Alba-Romenus, K.; González-Morales, S.; Benavides-Mendoza, A.; Juárez-Maldonado, A. Determination of Micronutrient Accumulation in Greenhouse Cucumber Crop Using a Modeling Approach. *Agronomy* **2017**, *7*, 79. [CrossRef]

55. White, P.J.; Karley, A.J. Potassium. In *Cell Biology of Metals and Nutrients*; Springer: Berlin/Heidelberg, Germany, 2010; pp. 199–224. Available online: https://link.springer.com/book/10.1007/978-3-642-10613-2 (accessed on 30 September 2021).

56. Oosterhuis, D.M.; Loka, D.A.; Kawakami, E.M.; Pettigrew, W.T. The Physiology of Potassium in Crop Production. *Adv. Agron.* **2014**, *126*, 203–233. [CrossRef]

57. Farahat, E.; Linderholm, H.W. The effect of long-term wastewater irrigation on accumulation and transfer of heavy metals in Cupressus sempervirens leaves and adjacent soils. *Sci. Total. Environ.* **2015**, *512–513*, 1–7. [CrossRef]

58. Xu, X.; Du, X.; Wang, F.; Sha, J.; Chen, Q.; Tian, G.; Zhu, Z.; Ge, S.; Jiang, Y. Effects of Potassium Levels on Plant Growth, Accumulation and Distribution of Carbon, and Nitrate Metabolism in Apple Dwarf Rootstock Seedlings. *Front. Plant Sci.* **2020**, *11*, 904. [CrossRef] [PubMed]

59. ChitraMani, P.K. Evaluation of antimony induced biochemical shift in mustard. *Plant Arch.* **2020**, *20*, 3493–3498.

60. Hafeez, B.; Khanif, Y.; Saleem, M. Role of zinc in plant nutrition—A review. *J. Exp. Agric. Int.* **2013**, 374–391. [CrossRef]

61. Sillanpaa, M. *Micronutrient Assessment at the Country Level: An International Study*; Food and Agriculture Organization of the United Nations (FAO): Rome, Italy, 1990.

62. Lee, S.R. Critical Role of Zinc as Either an Antioxidant or a Prooxidant in Cellular Systems. *Oxidative Med. Cell. Longev.* **2018**, *2018*, 1–11. [CrossRef]

63. Khan, M.J.; Jan, M.T.; Farhatullah, N.; Khan, M.A.; Perveen, S.; Alam, S.; Jan, A. The effect of using waste water for tomato. *Pak. J. Bot.* **2011**, *43*, 1033–1044.

64. Jahan, K.; Khatun, R.; Islam, M. Effects of wastewater irrigation on soil physico-chemical properties, growth and yield of tomato. *Progress. Agric.* **2020**, *30*, 352–359. [CrossRef]

65. Verma, S.; Dubey, R.S. Effect of Cadmium on Soluble Sugars and Enzymes of their Metabolism in Rice. *Biol. Plant.* **2001**, *44*, 117–124. [CrossRef]

66. Petousi, I.; Daskalakis, G.; Fountoulakis, M.; Lydakis, D.; Fletcher, L.; Stentiford, E.; Manios, T. Effects of treated wastewater irrigation on the establishment of young grapevines. *Sci. Total. Environ.* **2019**, *658*, 485–492. [CrossRef]

67. Hoekstra, A.Y.; Mekonnen, M. Reply to Ridoutt and Huang: From water footprint assessment to policy. *Proc. Natl. Acad. Sci. USA* **2012**, *109*, E1425. [CrossRef]

68. Cooper, P. Historical aspect of wastewater treatment. In *Decentralised Sanitation Reuse: Concepts, System and Implementation*; IWA Publishing: London, UK, 2001; pp. 11–38.

69. Ibekwe, A.M.; Gonzalez-Rubio, A.; Suarez, D.L. Impact of treated wastewater for irrigatioon soil microbial communities. *Sci. Total Environ.* **2018**, *622–623*, 1603–1610. [CrossRef] [PubMed]

70. Maaß, O.; Grundmann, P. Added-value from linking the value chains of wastewater treatment, crop production and bioenergy production: A case study on reusing wastewater and sludge in crop production in Braunschweig (Germany). *Resour. Conserv. Recycl.* **2016**, *107*, 195–211. [CrossRef]

sustainability

MDPI

Soil Erosion and Sediment Load Management Strategies for Sustainable Irrigation in Arid Regions

Muhammad Tousif Bhatti [1,*], **Muhammad Ashraf** [2,*] **and Arif A. Anwar** [1]

[1] International Water Management Institute, Lahore 53700, Pakistan; a.anwar@cgiar.org
[2] Department of Agricultural Engineering, Khwaja Fareed University of Engineering & Information Technology, Rahim Yar Khan 64200, Pakistan
* Correspondence: tousif.bhatti@cgiar.org (M.T.B.); dr.ashraf@kfueit.edu.pk (M.A.)

Abstract: Soil erosion is a serious environmental issue in the Gomal River catchment shared by Pakistan and Afghanistan. The river segment between the Gomal Zam dam and a diversion barrage (~40 km) brings a huge load of sediments that negatively affects the downstream irrigation system, but the sediment sources have not been explored in detail in this sub-catchment. The analysis of flow and sediment data shows that the significant sediment yield is still contributing to the diversion barrage despite the Gomal Zam dam construction. However, the sediment share at the diversion barrage from the sub-catchment is much larger than its relative size. A spatial assessment of erosion rates in the sub-catchment with the revised universal soil loss equation (RUSLE) shows that most of the sub-catchment falls into very severe and catastrophic erosion rate categories (>100 t $h^{-1}y^{-1}$). The sediment entry into the irrigation system can be managed both by limiting erosion in the catchment and trapping sediments into a hydraulic structure. The authors tested a scenario by improving the crop management factor in RUSLE as a catchment management option. The results show that improving the crop management factor makes little difference in reducing the erosion rates in the sub-catchment, suggesting other RUSLE factors, and perhaps slope is a more obvious reason for high erosion rates. This research also explores the efficiency of a proposed settling reservoir as a sediment load management option for the flows diverted from the barrage. The proposed settling reservoir is simulated using a computer-based sediment transport model. The modeling results suggest that a settling reservoir can reduce sediment entry into the irrigation network by trapping 95% and 25% for sand and silt particles, respectively. The findings of the study suggest that managing the sub-catchment characterizing an arid region and having steep slopes and barren mountains is a less compelling option to reduce sediment entry into the irrigation system compared to the settling reservoir at the diversion barrage. Managing the entire catchment (including upstream of Gomal Zam dam) can be a potential solution, but it would require cooperative planning due to the transboundary nature of the Gomal river catchment. The output of this research can aid policy and decision-makers to sustainably manage sediment erosion issues of the irrigation network.

Keywords: soil erosion; sediment yield; RUSLE; sediment transport modeling; Gomal River; arid regions

Citation: Bhatti, M.T.; Ashraf, M.; Anwar, A.A. Soil Erosion and Sediment Load Management Strategies for Sustainable Irrigation in Arid Regions. *Sustainability* **2021**, *13*, 3547. https://doi.org/10.3390/su13063547

Academic Editors: Muhammad Sultan, Yuguang Zhou, Redmond R. Shamshiri and Aitazaz A. Farooque

Received: 6 February 2021
Accepted: 17 March 2021
Published: 23 March 2021

Publisher's Note: MDPI stays neutral with regard to jurisdictional claims in published maps and institutional affiliations.

1. Introduction

Soil erosion in catchments occurs in various forms such as sheet, rill, gully, river bed and bank erosion, and landslides that contribute sediments to the water bodies. The rate of erosion is primarily determined by the erosive events (e.g., short duration and high-intensity rainfall events), soil type, and characteristics of the terrain [1]. The impacts of accelerated soil erosion processes can be severe, not only through land degradation and fertility loss but through a conspicuous number of off-site effects such as sedimentation, siltation, and eutrophication of waterways or enhanced flooding [2]. Soil erosion rates are exacerbated for the arid and semi-arid regions due to barren mountains with scattered vegetation that provide direct exposure to heavy rainfall. For example, 50% of the soil loss occurs in Pakistan during the monsoon season [3] causing huge sediment load diverted into the irrigation canals. Also, landslides and debris flow from the catchment increase the sediment load in the river flows. The eroded sediments finally deposit in reservoirs, stream channels, irrigation canals, and water conveyance structures and reduce the capacities of these water bodies to perform their prime functions and often requires costly treatments [1,4].

Soil erosion and the fate of eroded sediments are widely recognized as one of the major environmental concerns worldwide. The results of soil erosion rate for the year 2012 published in a global study [5] show that South America, Africa, and Asia had respectively 8.3%, 7.7%, and 7.6% of the continental area in high erosion rate classes (>10 t $ha^{-1}yr^{-1}$). In the global map of soil erosion rates in [5], the northern parts of Pakistan (where the study area of this research lies) fall in the range of vulnerable to high erosion rates (>50 t $ha^{-1}yr^{-1}$).

Sustainable irrigation requires appropriate soil erosion and sediment load management options and strategies both at the catchment level and the diversion structures. The catchment management options such as improved vegetation, slope stabilization, etc. are often considered preferred solutions to limit soil erosion rates as opposed to the structural measures for its control. However, in some cases, it becomes impractical to achieve anticipated benefits of the catchment management interventions amid factors such as; steep terrain, transboundary nature, urgency and severity of the challenges, etc. Another critical factor to consider is the feasibility of implementing such interventions.

Significant sediment entry from river flows may reduce the discharge capacity of irrigation canals up to 40% due to the settlement of coarse particles [6]. Many remedial measures have been suggested by different researchers to reduce the sediment load in canals, e.g., 41% using silt excluder [7], up to 40% using submerged vanes [8], and 40–45% using the vortex tube sediment extractor [9]. These structural measures only remove the coarser sediment in the canals; therefore, settling basins are assumed more appropriate options of sediment control for both coarser and finer particles [10]. Dredging of deposited sediment from irrigation networks also requires considerable effort and cost that farmers often have to put in. Similarly, the control of sediment load at the diversion structure depends on the feasibility of the proposed solution [11]. Hence, before making investment decisions, it is more practical to identify the targeted areas (i.e., hotspot analysis) in the case of opting for catchment management options. In the case of remedial measures (e.g., a hydraulic structure proposed to control sediment entry) the simulation model can be useful to predict the impacts of the proposed intervention on sustainable irrigation supplies.

Pakistan and Afghanistan share three main rivers, namely the Kabul, Kurram, and Gomal rivers. The Gomal river is the smallest among these three in terms of average annual inflow, but it is an essential source of livelihood for the downstream users in the Khyber Pakhtunkhwa (KP) province of Pakistan. The combined average flow of Kabul, Kurram, and Gomal rivers is 23.58 Gm^3 yr^{-1}, out of which the Gomal river contributes 4.1% (0.974 Gm^3 yr^{-1}). The physical infrastructures in the Gomal river catchment on the Pakistan side comprise a storage dam, two powerhouses, a diversion barrage, and a canal irrigation system to irrigate more than 77,000 ha of agricultural land.

Afghanistan is the upstream riparian on all shared rivers with Pakistan, including the Gomal river. The eroded sediment from the catchment in Afghanistan finds its fate in the reservoir of the Gomal Zam (GZ) dam and periodically sluiced in the downstream river reach. In the absence of a joint river management agreement between Pakistan and Afghanistan, integrated catchment management intervention cannot be introduced effectively. There is also no data-sharing mechanism between the two catchment-sharing countries.

The Gomal river brings a considerable amount of sediments to the diversion barrage, as shown in Figure 1. This sediment is a combination of soil eroded by rainfall in the sub-catchment below the GZ dam and the sediment generated in the catchment upstream of the GZ dam. Hydro-meteorological monitoring is very limited in the catchment area. Furthermore, the sediment flushing operation at the GZ dam is ad hoc, and its concentration in the outflows is not monitored. All this makes the catchment of the Gomal river a data-sparse catchment. When sediment-laden flow enters into the irrigation canal network, the sediments settle down in the channels and impede the conveyance capacity of the canals. This deprives the downstream users of their due share and gives rise to social unrest and conflicts.

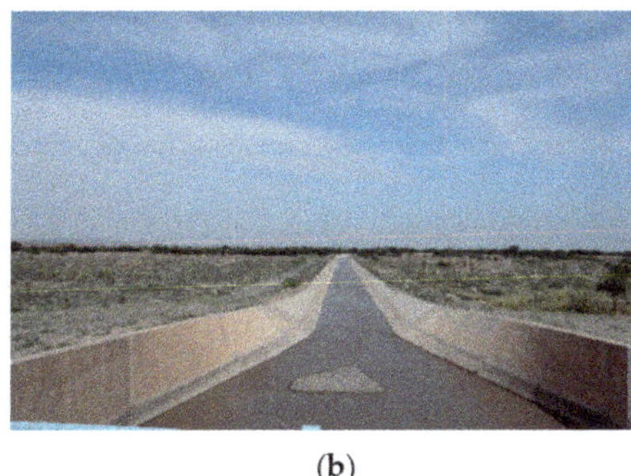

(a)   (b)

Figure 1. Sediment conditions at the diversion barrage. (**a**) High sediments concentration at downstream of the under sluice sections of the barrage as a result of flushing from GZ dam during 9–10 May 2019; (**b**) View of the main canal from canal head regulator on 10 May 2019.

The decisions to manage erosion rates in the catchment and eroded sediments in the water conveyance system are informed by the erosion/sediment assessment studies. These assessments help identify catchment areas with high soil erosion susceptibility and a clear understanding of the system's operation. The revised universal soil loss equation (RUSLE) is commonly used to assess soil erosion rates. Soil loss predictions are frequently unrealistic because the methods used to estimate the soil loss equation's factors are empirically derived. Therefore, the methodology section emphasizes the selection of suitable methods to determine RUSLE's factors. Few studies have identified the impact of RUSLE factors on soil loss. For example, both ref [12,13] concurred in their findings that the topographic factor is the most significant factor among others, affecting the soil erosion rates. On the other hand, ref [14] has found that improving the support practice factor by adopting terracing showed a reduction in sediment yields. Previous studies show that no one rule fits all when investing in catchment management interventions.

Many studies that employ the RUSLE equation limit their analysis to improve the estimation of potential soil erosion. The input factors in RULSE can be estimated by using values from the literature or adapted for empirical and statistical data in combination with geographic information system (GIS) software [15], and this remains the focus of most of the studies.

This paper contributes to the science and application related to soil erosion as it is a critical problem worldwide, and especially in dryland areas. The paper first identifies the sources of sediment in a segment of the Gomal river (between GZ dam and the diversion barrage) in Pakistan. The incoming sediment limits supply in the irrigation system downstream and creates huge maintenance problems. A novel contribution of this study is its solution-oriented approach to the prevailing problem of sedimentation in the water conveyance system. There could be many possible solutions, but this paper has analyzed in detail a catchment management option (preventive) and a structural option (remedial). The analysis consists of erosion rates estimation and their severity level in the sub-catchment of the Gomal river using a computer-based empirical model. The study is focused on the sub-catchment because it is not transboundary and, therefore, management options can be introduced, monitored, and evaluated with relative ease. The analysis expands on predicting the impact of improved vegetation in the sub-catchment (i.e., crop management factor of RUSLE) as a catchment management option. Finally, a sediment transport model is used to determine the sediment trapping efficiency of a hypothetical settling reservoir at the diversion barrage.

2. Materials and Methods

2.1. The Study Area

Afghanistan and Pakistan share the catchment of the Gomal river. The Gomal river originates in the mountains of Ghazni province of Afghanistan and enters the South Waziristan district of KP, Pakistan. The total catchment area at Gomal Zam (GZ) dam is 33,950 km^2, about 70% of the area lies within Pakistan, and the remaining lies in Afghanistan.

The Government of Pakistan constructed the GZ dam (32°5′55″ N 69°52′53″ E) in South Waziristan district (formerly the South Waziristan Agency of the Federally Administered Tribal Area) with the financial assistance of the United States Agency for International Development (USAID). The construction of the dam started in August 2001 and was completed in April 2011. The dam impounds the Gomal River in a reservoir with 1.41 km^3 storage capacity. The primary purpose of the dam is to supply water to irrigate 77,295 ha of land through a newly built irrigation system in the Tank and Dera Ismail Khan districts of Khyber Pakhtunkhwa province of Pakistan. The additional benefits of the dam are flood control and a modest amount of hydroelectric power generation (17.4 MW).

The Gomal river traverses around 40 km downstream of the dam site where a barrage is constructed near the town of Kot Murtaza, to divert the river flows into the irrigation system. The sub-catchment of this river segment (herein referred to as sub-catchment) is shown in Figure 2. The sub-catchment undergoes erosion and imparts sediments into the river stream, which find their way to the diversion barrage and enters into the canal irrigation system. This not only challenges the management of the irrigation system but also places significant demands on the operations and maintenance budget of the irrigation system.

Figure 2 shows the location and pictures of the GZ dam and diversion barrage. The diversion barrage is 140 m long and 7 m high. Three gates divert water into the main canal; in addition to that, four under sluice gates are also provided as shown in Figure 1a. The barrage is designed to pass a flood of 5200 m^{3}s^{-1} safely. The primary purpose of the GZ dam is to irrigate 77,295 ha of land in two districts of Khyber Pakhtunkhwa province i.e., D I Khan and Tank districts. For this purpose, an irrigation system consisting of the diversion barrage, 60.5 km long main canal, and 206 km of distribution canals have been constructed. The irrigation system is operational since November 2014, except for the portion fed by a branch canal named Waran Canal. A flow and sediment monitoring station is located approximately 2.5 km upstream of the diversion barrage, as shown in Figure 2.

At the diversion barrage, the river inflow is a combination of the flows released from the GZ dam and runoff from the sub-catchment. The area of this sub-catchment is ~450 km^2 constituting only 1.3% of the entire catchment of the Gomal river. This sub-catchment constitutes the study area for detailed analysis in this research.

The sub-catchment is mostly barren, with no essential utilities for human settlements. The community living in the command area of the irrigation system depends entirely on the water from the Gomal river not only for irrigated agriculture but in some parts also for domestic uses.

The existing dam operation is such that water is released from the tailrace of two hydroelectric power units at the GZ dam into the Gomal River. The main spillway has never been operated because water levels in the reservoir have not yet reached the spillway crest level. A low-level outlet is provided in the main dam to flush the sediments, which is operated on an ad hoc basis. This bottom outlet at an elevation of 680 m above mean sea level (dead storage level is 711 m) was designed for flushing out deposited sediments in the reservoir during monsoon, but one of the hydroelectric power units has never worked and, therefore, part of the water demand (8–10 m^{3}s^{-1}) is supplied through this bottom outlet.

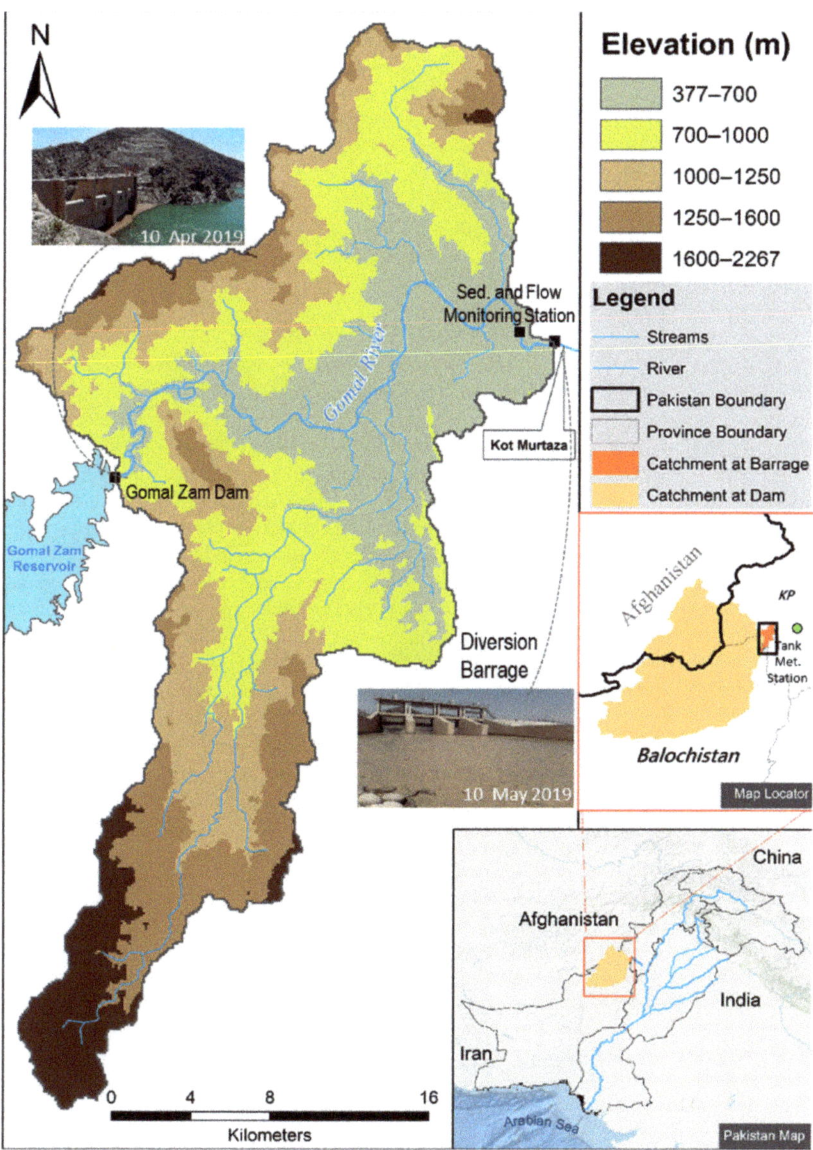

Figure 2. Map of the Gomal river sub-catchment.

The mean annual rainfall in the sub-catchment is 276 mm yr^{-1}. The rainfall events are typically of high intensity, generate significant runoff, and bring with them eroded sediments. Erosion in this sub-catchment is often considered as a leading source of sediment [16] and, therefore, efforts for watershed management and conservation practices are emphasized as a potential solution to this issue of soil erosion [17]. The sediments from the catchment upstream of the GZ dam are trapped in the reservoir behind the dam and flushed through the low-level outlet. This sediment contribution is not quantified and monitored systematically; therefore, the only data available for analysis are at the monitoring station near the diversion barrage.

2.2. Data and Methods

2.2.1. Annual Sediment Yield in Gomal River

The long-term annual sediment yield data (1960–2015) was analyzed to find the cumulative trend at the monitoring station before and after the construction of the GZ dam. The monitoring station is located 2.5 km upstream of the diversion barrage site, and hence the sediment yield at this location includes the contribution from the entire catchment of the Gomal river, including the sub-catchment. The monitoring station is maintained by the Water and Power Development Authority (WAPDA) of Pakistan, which is responsible for the collection and analysis of sediment data. Moreover, it is noteworthy that only suspended sediment samples are collected by WAPDA using depth-integrated samplers (DH-48).

2.2.2. Assessment of Erosion Rates

The universal soil loss equation (USLE) was developed in 1965 [18] to estimate the rate of soil loss. This empirical equation was improved and revised in 1995, since then it has been known as the RUSLE [1] and has been applied in numerous catchments worldwide. RUSLE is not intended to estimate the sediment yield (as the amount of sediment reaching or passing a point of interest in a given period) instead; its application is limited to calculate the annual soil loss rate. However, the sediment delivery ratio (SDR) method can be used to link soil erosion with sediment delivery to the stream. With the RUSLE model, the average annual rate of soil loss for a catchment of interest can be predicted for any number of scenarios in association with cropping systems, management techniques, and erosion control practices [19]. Therefore, many studies have applied RUSLE to estimate soil loss rates and identification of high-risk areas, e.g., [20–22]. It has become the most widely used approach during an 80-year history of erosion modeling applied in 109 countries [23]. Alewell et al. [23] point out the limitations of USLE-type modeling that it does not address larger rills or gully erosion but is restricted to sheet/interrill and small rill erosion only. Due to these limitations, some researchers, e.g., reference [24] have criticized RUSLE for using it in predicting sediment delivery ratios.

In this research, RUSLE is applied to the sub-catchment of the Gomal river to assess erosion rate and test an improved vegetation scenario.

$$A = RKLSCP \tag{1}$$

where A denotes average annual soil loss (t ha^{-1} y^{-1}), R is rainfall-runoff erosivity factor (MJ mm ha^{-1} h^{-1} yr^{-1}), K is soil erodibility factor (Mg ha h ha^{-1} MJ^{-1} mm^{-1}), LS is topographic factor, C is crop management factor (ranges from zero to 1), and P is conservation support practice factor (ranges from zero to 1).

Soil loss predictions using RUSLE yield varying results because the methods used to estimate the factors in Equation (1) are empirically derived. The regional applicability of the RUSLE requires the sub-factors to be adjusted and modified based on the specific characteristics of the study site [25]. Some studies have attempted to evaluate the impact of different RUSLE factors on soil loss. For example, reference [12] found the land slope (LS) as the factor most significant factor affecting soil erosion. The authors in [12] conclude that severe erosion occurs along the drainage channels due to steep bank slopes. Similarly, ref [13] found that the land slope (LS) factor is that with most influence on soil erosion, followed by P, K, C, and R. The support practice P may also be an important factor when the management of soil erosion from the catchment is considered [14].

Assessment of soil loss using RUSLE approach requires a careful selection of methods determining its factors. A recent study by [25] reviews the different factors of USLE and RUSLE, and analyses how various studies around the world have adapted the equations to local conditions.

Rainfall-Runoff Erosivity Factor *R*

The factor *R* considers the erosion due to rainfall and runoff, which is the function of long-term rainfall kinetic energy and maximum 30-min intensity [1,26]. For this study, precipitation data were collected from Tank meteorological station (32°12′23″ N, 70°23′30″ E) maintained by the Water and Power Development Authority (WAPDA). The meteorological station does not have an automatic recording rain-gauge, and hence high-intensity rainfall data of considerable length was not available. The rainfall data for the period of 2014–2016 was collected to estimate *R* using an appropriate empirical equation for the study area. Many regional studies have suggested empirical equations (see in [25]) that can be adopted to estimate *R* using monthly rainfall data. The advantage of considering these empirical equations is that they do not necessitate long-term and high-resolution rainfall data and are suitable for use in data-sparse catchments like the Gomal river catchment. Table 1 lists down a few of these empirical equations, which we considered to calculate *R* but finally adopted the empirical equation by [27] in this research. A comparison of *R* calculated with the empirical equations (as shown in Table 1) with the *R* from the global rainfall erosivity dataset is presented in the Results and Discussion section. The global rainfall erosivity dataset is collected from the Joint Research Centre, European Soil Data Centre (ESDAC) (https://esdac.jrc.ec.europa.eu/, accessed on 30 June 2019).

Table 1. Different empirical relationships for calculation of rainfall-runoff erosivity factor *R*.

Serial Number	Equation	Geographic Location	Author(s)
1	$R = 4.79\text{MFI} - 142$	Morocco and other locations West Africa	Arnoldus (1980) [28]
2	$R = 4.79\text{MFI} - 143$		
3	$R = 0.66\text{MFI} - 3$		
4	$R = 0.5 \times p$	Africa	Roose (1975) [27]
5	$R = -823.8 + 5.213p$	USA	Fernandez et al. (2003) [29] originally developed by the USDA-ARS
6	$R = 0.1215 \times \text{MFI}^{2.2421}$	Turkey	İrvem et al. (2007) [30]
7	$R = 839.1 \times e^{0.0008p}$	India	Nakil (2014) [31]

In Table 1, MFI is the modified Fournier index calculated as in Equation (2).

$$\text{MFI} = \sum_{i=1}^{12} \frac{P_i^2}{P} \tag{2}$$

where p_i is the monthly rainfall in mm, and p is the annual rainfall in mm.

Soil Erodibility Factor K

The soil erodibility factor *K* represents the susceptibility of soil to erosion under standard plot conditions. Soil erodibility is scaled from 0–1, depending on the soil texture, e.g., high values are used for silt to fine sand and low values for coarse sand due to its resistance to erosion. To calculate soil erodibility factor *K*, ref [32] has proposed a relationship shown in Equation (3) that considers the soil textural information, organic matter, information about the soil structure, and permeability. Although Equation (3) is based on the soil data of regional conditions in the United States, in many studies outside the United States researchers have used Equation (3) for soil erodibility factor calculations [25].

$$K = \frac{0.1317\left[2.1 \times 10^{-4} M^{1.14}(12 - a) + 3.25(b - 2) + 2.5(c - 3)\right]}{100} \tag{3}$$

where M is a parameter based upon soil texture and estimated using Equation (4) as adopted from [33], a is the organic matter content (%), b is a code related to soil structure (1 for very fine granular, 2 for fine granular, 3 for coarse granular, 4 for massive), and c is a code related to soil permeability (1 to 6, fast to very slow draining characteristics).

$$M = (m_{silt} + m_{vfs}) \times \left(100 - m_{clay}\right) \tag{4}$$

where m_{clay} is a fraction of clay contents with particle size <0.002 mm, m_{silt} is a fraction of silt contents with particle size ranging from 0.002 to 0.05 mm, and m_{vfs} is a fraction of very fine sand contents with particle size ranging from 0.05 to 0.1 mm. All fractions are expressed as a percentage.

Soil maps help to define the codes related to soil structure and soil permeability. Soil erodibility can also be identified based on the regional studies, e.g., reference [34] have reported the erodibility factor K values for the Potohar region, Pakistan. In Pakistan, the Soil Survey of Pakistan and the Geological Survey of Pakistan prepare maps that can be used to define erodibility classes. Moreover, some satellite products are also available at a coarser scale for harmonized soil type information. Satellite data of different soil fractions are also available on 250 m grid resolution. However, the erodibility factor for the current study was calculated using the soil texture information extracted from the Food and Agriculture Organization (FAO) soil datasets (www.fao.org, accessed on 23 May 2019). Soil erodibility values were assigned based on the texture and organic matter of the soil. Three soil layers were extracted from the FAO soil database. Soil classes were identified from clay loam to loam with organic matter of more than 2% because the soils in the catchment are very rich in organic carbon. The values assigned for clay loam and loam were 0.26 and 0.28, respectively.

Topographic Factor (LS)

The topographic factor LS reflects the effect of slope length and steepness on soil erosion from the catchment. It considers sheet, rill, and inter-rill erosion by water, and is a ratio of expected soil loss from a field slope relative to the original USLE unit plot [32]. In RUSLE, this relationship was extended to the more complex hill slopes for better estimation of the topographic effect by modifying the steepness factor that is more sensitive than the length factor [1]. Some studies extend the LS factor to topographically complex terrains using a method that incorporates contributing area and flow accumulation [35]. The flow accumulation tool built into a GIS allows the estimation of the upslope contributing area using a digital elevation model (DEM) that, in turn, is used to calculate LS factor to account for more topographically complex landscapes [35,36]. The flow accumulation method to calculate slope length and steepness explicitly accounts for convergence and divergence of flow, which is important while considering soil erosion over a complex landscape [37]. However, a high-resolution DEM is needed for an accurate representation of the topography to calculate LS factor. Benavidez et al. [25] reported that the method of flow accumulation performed significantly better at the sub-watershed or field scale. We have used Equation (5) in this research to determine the LS factor, developed by [36] and widely used globally as well as for regional studies.

$$LS = (m+1)\left(\frac{U}{L_0}\right)^m \left(\frac{Sin\beta}{S_0}\right)^n \tag{5}$$

where U is upslope contributing area per unit width (m^2/m), L_0 is the length of the unit plot (m), S_0 is the slope of the unit plot (%), β is the slope (%), m is a factor for sheet erosion and n is a factor for rill erosion. The value of m ranges from 0.4 to 0.6 whereas that of n ranges from 1.0 to 1.3, depending on the prevailing erosion types. For this study, the value of m and n were used as 0.4 and 1.3, respectively as suggested by [36]. L_0 and S_0 are constants, and their values are 22.1 m and 9% in Equation (5).

The DEM of the Shuttle Radar Topography Mission (SRTM) with a 30 m resolution was used to determine catchment and sub-catchment boundaries and calculate slope length and slope steepness which are used in the RUSLE topographic factor *LS* (https://earthexplorer.usgs.gov/, accessed on 6 May 2019). ArcGIS tool was used for spatial analysis for *LS* factor and catchment delineation using appropriate tools.

Crop Management Factor C and Conservation Support Practice Factor P

The crop management factor *C* reflects the effects of cropping and management practices on erosion. In RUSLE, its value is calculated as a product of five subfactors: prior land use, canopy cover, surface cover, surface roughness, and soil moisture. Factor *C* is expressed as a ratio of soil loss of a parcel with certain land use and a fallow condition [32,38]. Generally, *C* ranges between 1 and 0; higher values (≤ 1) are used for less vegetation, thus considering the land as barren whereas lower values (near to zero) indicate very strong cover and well-protected soil. Lanorte et al. [39] have reported that many authors have adopted simplified approaches to estimate *C*, e.g., by using land cover maps and assigning a *C* value to each class [40] or by applying remote-sensing techniques such as image classification [41,42] and vegetation indices [43]. Various studies [44–46] report mathematical functions to calculate *C* using the normalized difference vegetation index (NDVI), which is positively correlated with the amount of green biomass and indicates differences in green vegetation coverage [45]. In this study, the mathematical function developed by [45] has been used to estimate the crop management factor as shown in Equation (6). The NDVI calculations were made from Band 5 (NIR) and 4 (Red) of Landsat 8 image (https://earthexplorer.usgs.gov/, accessed on 29 May 2019) for July 2018, as shown in Figure 3. Most of the precipitation in the study area is received during the monsoon season (July to September) and the scattered vegetation dependent on the monsoon rainfall. Therefore, single imagery was used to calculate NDVI; otherwise, average values of different periods may be used where different cropping patterns prevail, as suggested by [13].

$$C = -exp[l(\text{NDVI}/r - \text{NDVI})] \tag{6}$$

where *C* is crop management factor, NDVI is normalized difference vegetation index, *l* and *r* are constants, and their values are taken as 2 and 1, respectively.

Conservation support practice factor *P* represents the ratio of soil loss for a unit area with a given erosion control method compared to the soil under the same conditions without any erosion control measures [25,26]. The value of the *P* factor ranges from 0 to 1 (0 indicates good conservation practice and 1 represents poor conservation practice). In this research, *P* was set equal to one across the sub-catchment because there are no erosion-control works in the sub-catchment to prevent soil erosion.

Finally, the annual soil erosion rate was calculated by applying RUSLE. Morgan [26] developed a coding system for soil erosion appraisal in the field where classes have been defined based on erosion rate. We use the coding system by [26] to explain erosion rates calculated in the sub-catchment.

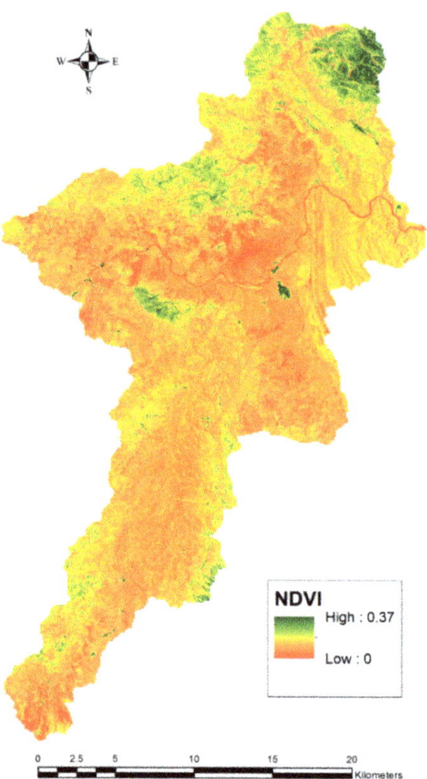

Figure 3. Normalized difference vegetation index (NDVI) for the sub-catchment.

2.3. Sediment Transport Modeling

About 70% of annual sediment load is diverted into the canals only during the *monsoon* season [9]. In flashy streams, significant proportions of sediment move as bedload, which is challenging to account for with conventional sampling methods for suspended sediments. The bed load fraction creates more problems when it enters irrigation canals due to the limited sediment transport capacity of the irrigation canals [10]. In our case, sediment enters into the irrigation system through a diversion barrage build across the Gomal river. The description of the diversion barrage is provided in the Study Area section. In this research, we simulated flows in a settling reservoir proposed immediately downstream of the diversion barrage and the main canal feeding from this settling reservoir rather than from a head regulator at the diversion barrage. The size of this hypothetical settling reservoir is taken as $100 \times 80 \times 2$ m^3. Sediment Simulation In Intakes with Multiblock option (SSIIM) 3D model [47] is used for sediment transport modeling to compute sediment trapped in the hypothetical settling reservoir. The simulation informs how efficient this structure could be in reducing sediment entry into the canal irrigation system. The SSIIM program solves the Navier Stokes equations over a structured mesh, using the k-epsilon model to represent the turbulent viscosities. The main strengths of SSIIM are the capability of modeling sediment transport with a movable bed in complex geometry.

Once the flow field is computed, the convection-diffusion 3D sediment equations are solved for each of the sediment fractions considered. From these, the trapping efficiencies, as well as the depositional sediment pattern, are obtained.

For the hypothetical settling reservoir, a grid of $22 \times 6 \times 6$ (792 cells) was used. Inflow into the basin was taken as 24 m^3 s^{-1}; sediment discharge was considered as 10, 156, and

151 kg s^{-1} for sand silt and clay particle sizes, respectively. The water level of 2 m from the bed of the settling reservoir was considered as an outflow boundary condition. The flows and sediment concentration data of 2016 was used to calculate the amount of sediment deposition in the proposed settling reservoir. For this purpose, a sediment rating curve was developed to calculate the sediment concentration during missing days (Figure 4). The amount of bedload in river flows during rainfall events assumed as 10% of suspended sediment load as suggested by [48]. For improved estimates of the sediment load, the continuous records approach applied by [49] can be used, but it requires the hourly or daily sediment concentration data.

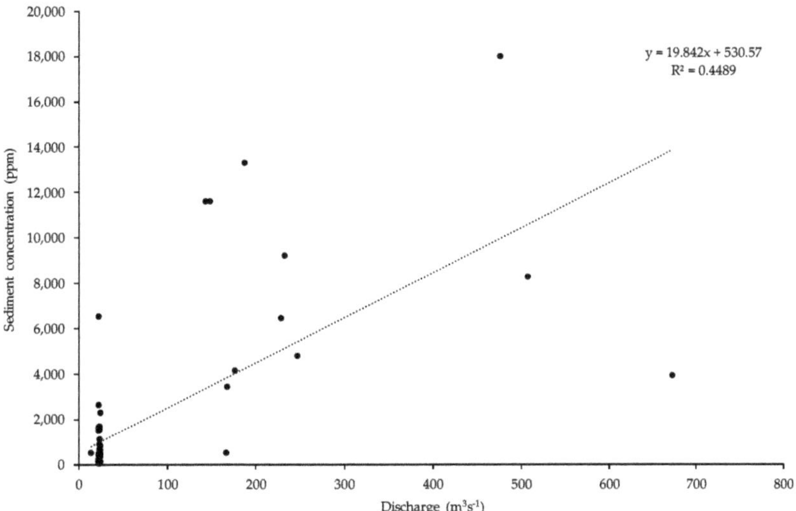

Figure 4. Sediment rating curve of the Gomal river at monitoring station during the year 2016.

3. Results and Discussion

3.1. Annual Sediment Yield

Figure 5 shows the cumulative sediment yield at the monitoring station near the diversion barrage (see location in Figure 2). At this location, the sediment yield includes sediment contribution from the entire GZ catchment (including the sub-catchment). Before the GZ dam construction in 2011, the sediment load from the entire catchment was contributing to the monitoring station. There was no canal irrigation system, and the community had historic rights to use the river water through spate irrigation. Flow and sediment monitoring remained discontinued from 1996 to 2002. The trend lines for the periods of 1960–2010 and 2003–2010 are almost parallel, showing no substantial difference in the pattern of sediment load in the Gomal river. The trend line for the period of 2011–2015 shows a significant change in the sediment load. This is intuitive because the GZ reservoir has been filling during this period, and the sediment from the upstream catchment used to be trapped in the reservoir. The trend line pre and post-dam constructions show an 85% decrease in the annual sediment yield, suggesting that significant sediment yield is still contributing to the diversion barrage although the sub-catchment area comprised 1.3% of the total catchment area. This high sediment contribution can be attributed to the sediment releases from the dam during the flushing operation, and the sub-catchment as well during rainfall events.

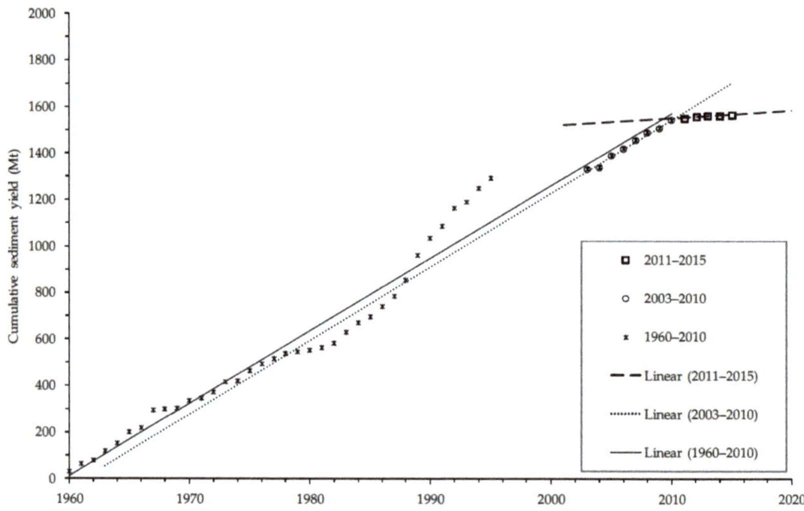

Figure 5. Cumulative sediment yield from Gomal river catchment during 1960–2015.

Figure 6 presents a sediment rating curve for the selected years from 1994 to 2016 due to the availability of limited data. Suspended sediment concentration observed at the monitoring station ranges from 11 to 82,500 parts per million (ppm). The data after dam operation (2015–2016) gives more insight into the sediment and discharge at the monitoring station. Figure 7 presents the histogram of discharge and sediment concentration during 2015–2016. In Figure 7a, it is evident that the discharge remained 82% of the time around 24 m^{3}s^{-1}, which is about one-quarter of the irrigation demand (indent) of 100 m^{3}s^{-1}. On the other hand, Figure 7b shows that 50% of the time, sediment concentration remained in the zero to 250 ppm range.

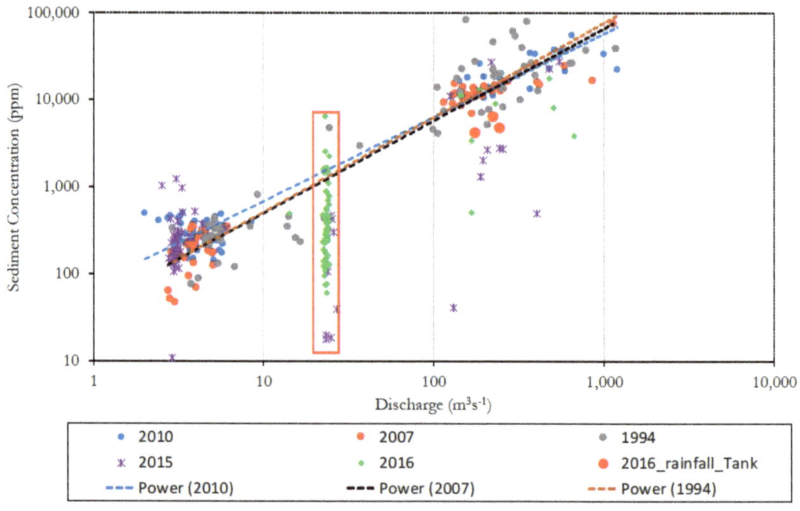

Figure 6. Sediment rating curve of the Gomal river at the monitoring station.

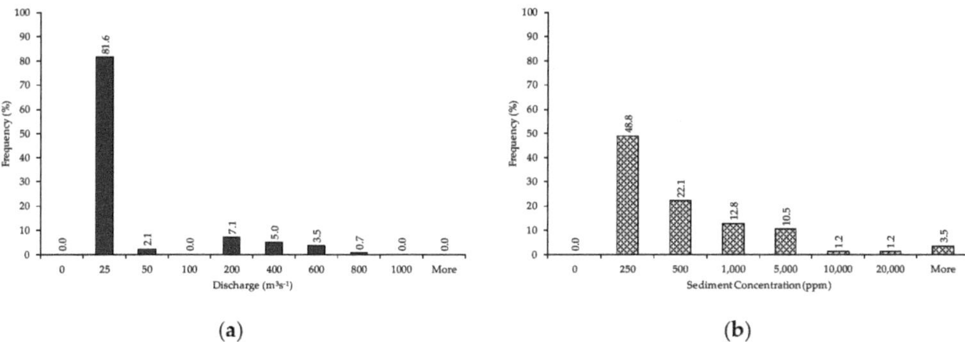

Figure 7. Histogram for data during 2015–2016; (**a**) discharge; (**b**) sediment concentration.

A few high sediment concentrations were also observed at high discharges, i.e., >100 m^{3}s^{-1}, as shown in Figure 6. The figure shows high sediment concentration in 2016 as large red circles which correspond to the days when rainfall was observed at the Tank meteorological station. This confirms that the sub-catchment contributes high sediments in response to rainfall events as the sediment concentration was observed more than 4000 ppm during all rainfall events. However, sediment concentrations ranging from 61 to 6540 ppm in the same year 2016 (shown in green color) are clustered corresponding to 24 m^{3}s^{-1}. This sediment concentration is likely to be coming from water releases from the GZ dam, which is controlled to 24 m^{3}s^{-1}. It suggests that the dam also contributes to high sediment load during the flushing operation. Similar sediment concentration scatter was observed in the Upper Chenab Canal (UCC) and Marala Link Canal (MRLC) in Pakistan by [49], where average sediment concentrations were 450 and 500 ppm, respectively. However, the sediment concentration observed at the Gomal river monitoring station is much higher than UCC and MRLC. The above analysis is based on the suspended sediment concentration data only because the bed load at the monitoring station is not observed.

3.2. Erosion Rate Estimation

Table 2 shows calculated values of rainfall-runoff erosivity factor R by using various empirical equations and the rainfall data on which calculations are based. The table also shows the range of R in the sub-catchment from the ESDAC global data set. The variation in calculated values of R is too wide to adopt a particular empirical equation. We have preferred the Roose equation [27] because researchers have used this equation to calculate R in recent studies to estimate soil erosion rate in Pakistan's Potohar catchments [34] and India's Pravara catchment [50]. The adopted value of R in this research is, therefore, 95 MJ mm ha^{-1} h^{-1} yr^{-1} (Table 2) which is based on rainfall data from the Tank meteorological station from 2014 to 2016.

Ideally, R is calculated as a sum of individual storm erosivity values (the rainfall intensity–kinetic energy relationship), as suggested by [1], whereby data of rainfall events should be of considerable length and high frequency. Such high temporal resolution rainfall (pluviograph) data needed for accurate computation of R is available only in a few regions of the world [51]. The ESDAC global dataset is based on the rainfall intensity–kinetic energy relationship and uses high-resolution rainfall data collected from 3540 stations across the globe [52]. The extensive list of stations, however, does not include many countries, and Pakistan is not an exception to this. In the case of the Gomal river sub-catchment, the limitation of low temporal resolution of rainfall data steered the analysis to consider empirical equations from regional studies. Table 2 shows significant variation amongst the computed R values using empirical equations and the global data set. The comparison in Table 2 is presented for the benefit of future studies in data-sparse catchments, where an empirical equation may yield a more realistic R than a value from the global data set. It may be particularly useful to consider empirical equations in those regions where high-resolution rainfall data was not available for global rainfall erosivity assessment by [52].

Table 2. Rainfall-runoff erosivity factor R using different approaches.

	Annual Data			Average
	2014	2015	2016	
Rainfall at Tank station (mm)	226	212	134	191
Rainfall-runoff erosivity factor R (MJ mm ha $^{-1}$ h $^{-1}$ yr $^{-1}$)				
Arnoldus (1980) [28], West Africa	38	59	8	35
Arnoldus (1980) [28], Western USA	37	58	7	34
Arnoldus (1980) [28], Northwest USA	22	25	18	22
Roose (1975) [27], Africa	113	106	67	95
Fernandez (2003) [29], USA	356	281	−128	170
İrvem (2007) [30], Turkey	412	531	272	405
Nakil (2014) [31], India	1006	994	934	978
The European Soil Data Centre (ESDAC), Global dataset	538–1300			919

The sub-catchment of the Gomal river is mostly hilly; thus, slopes vary from gentle to very steep. The average slope of the catchment is 33%; therefore, terrain may be considered steep. However, a significant portion of the sub-catchment exhibits less than a 40% slope. The topographic factor LS varies from 0 to more than 20, with a substantial proportion of the area (about 80%) having LS values less than 20 (Figure 8a).

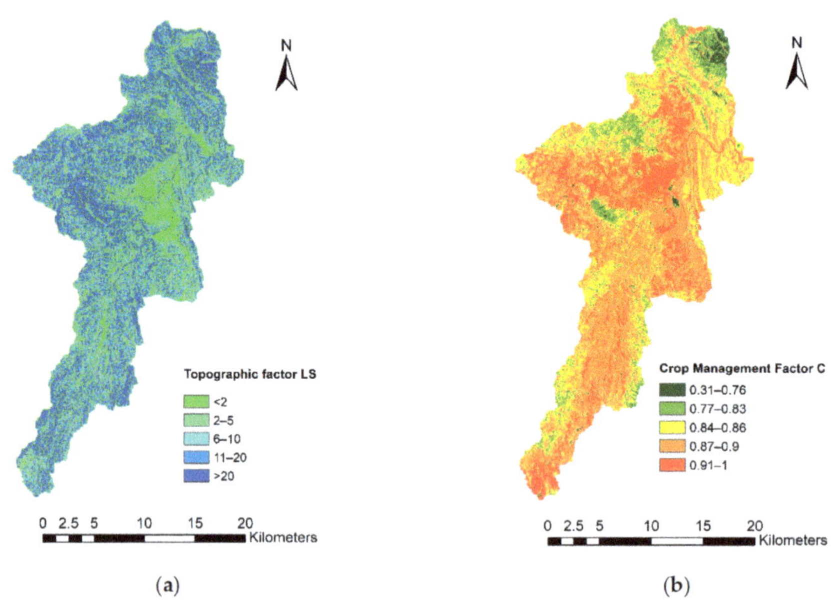

Figure 8. Revised universal soil loss equation (RUSLE) factors estimation in the sub-catchment; (**a**) topographic factor LS; (**b**) crop management factor C.

Higher values >100 indicate more risk of soil erosion [53]. Moreover, the *LS* factor has been found to affect soil erosion rates more than any other RUSLE factor in the Appalachian hills of the United States [12]. In the Dobrov river catchment, Romania, researchers [54] have used different *LS* factors calculation methods and reported that the equations by [36] produced comparable erosion rates with the measured values; therefore, it was used to calculate the topographic factor for this study.

The computed crop management factor *C* (as shown in Figure 8b) is close to the published values reported by [34] for the regional catchment areas. The factor *C* > 0.83 indicates minimal vegetation cover, predominating in the sub-catchment and hence the soil is very vulnerable to erosion (Figure 8b).

The results (Figure 9a) show that 46% of the sub-catchment area shows a very severe annual erosion rate (100–500 t h^{-1}) while 16% shows a catastrophic annual erosion rate (>500 t ha^{-1}). Very severe and catastrophic erosion rates have also been reported by different researchers for Pakistan, India, Ethiopia, and Brazil [34,55–57]. The main reasons for such high erosion rates are steep topography, sparse vegetative cover, and highly erodible soil type. The sub-catchment can be characterized as highly prone to erosion as compared with other catchments in Pakistan, e.g., the Simly dam catchment [58] or the Potohar Plateau [34,59]. Similarly, the soil erosion rates in the sub-catchment are much higher than reported catchments in the other region of the world, for example, in Nepal [60] and Ethiopia [61].

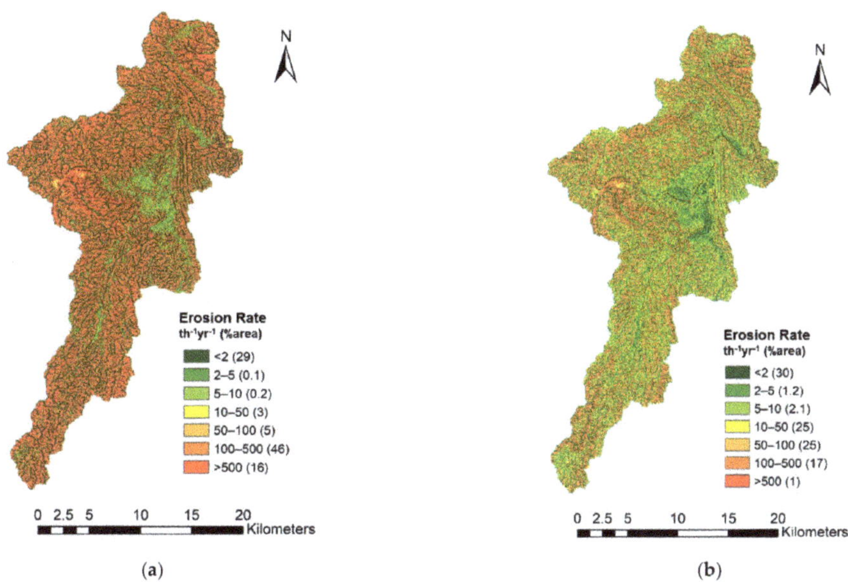

Figure 9. Annual soil erosion rate with RUSLE; (**a**) with calculated parameters; (**b**) with improved C factor (scenario testing).

3.3. Scenario Testing in Revised Universal Soil Loss Equation (RUSLE)

Figure 8b shows computed *C* values in the Gomal river sub-catchment. The factor *C* can range from zero (for a non-erodible surface) to 1 for a bare plot (no vegetation). The sub-catchment represents mostly barren land with values >0.8. The minimum value of *C* in the sub-catchment is 0.3 that prevails sparsely in small pockets. We have tested a scenario to understand the impact of improved *C* on erosion rates. For this scenario, the value of *C* was set to 0.18 for the entire sub-catchment rather than computing *C* values with Equation (6) as shown in Figure 8a, and keeping all other factors of the RUSLE unaltered. The *C* used in the scenario is quite low, representing very well-covered and managed land.

To draw a parallel with the real cases, researchers [62] have mapped C at the European Union level and found the lowest mean values in Denmark and Hungary (0.178, and 0.188, respectively).

In practical terms, achieving the C value of 0.18 in the sub-catchment by improving vegetation would be extremely difficult amid steep topography and less precipitation. However, we assume a meager value of C to see how well this potential catchment management solution can work in reducing erosion rates.

The result in Figure 9b showed a decrease in the proportion of sub-catchment under catastrophic erosion rate, but the majority of the sub-catchment remained in the categories of high and severe erosion rates. A comparison of the erosion rates in various classes is shown in Table 3, which confirms that the crop management factor alone does not achieve a substantial reduction in soil erosion rates in the sub-catchment. Overall, the study area exhibits very high erosion rates. Global studies [5] show that management and the related land-use changes affect the spatial patterns and magnitude of accelerated soil erosion. The estimates of soil loss reduction derived from the implementation of conservation agriculture are encouraging in 40 countries, as reported by [5]. In contrast, the study area in this research has not shown a substantial reduction in soil erosion in response to a vegetation management scenario. These results also mirror the findings of [12], who reported that the Hocking river basin of the United States is covered with deciduous forests (about 50%) but is vulnerable to soil erosion due to steep slopes. Similarly, in a study on the Mangla River Basin [63], the authors reported that heavy rainfall increased the sediment load during the year 1992 despite increased vegetation cover compared to the previous years. Hence, vegetation alone in the study area would not help to significantly reduce erosion.

Table 3. Area of Gomal river sub-catchment in various erosion rate classes.

Erosion Rate (t/ha)	Category [1]	RUSLE		RUSLE with Improved C Factor		Change in the Area with Improved C Factor	
		Area (km^2)	Proportion (%)	Area (km^2)	Area (%)	km^2	Percent
<2	Very slight	133.3	29.3	135.1	29.6	1.8	1.3
2–5	Slight	0.4	0.1	5.3	1.2	4.9	1247.8
5–10	Moderate	1.0	0.2	9.7	2.1	8.7	848.2
10–50	High	15.4	3.4	111.9	24.6	96.5	625.6
50–100	Severe	21.8	4.8	114.6	25.1	92.8	425.6
100–500	Very severe	209.6	46.1	76.0	16.7	−133.6	−63.7
>500	Catastrophic	73.4	16.1	3.1	0.7	−70.3	−95.8
Total		454.9	100.0	455.6	100.0		

[1] Erosion rates categorization is used after Morgan [26].

Limiting management intervention to the sub-catchment can result in a partial reduction in sediments at the most, and even this may not turn out to be a viable option due to arid climatic conditions in this sub-catchment. However, the plantation campaign should be encouraged in the catchment upstream of the Gomal Zam dam without singling out any portion of the catchment that could be beneficial for environmental reasons. The geographic nature of the catchment is such that 30% of the catchment area is in Afghanistan, and two provinces share the part in Pakistan. Therefore, the catchment management option would be more effective if the two catchment-sharing countries jointly prepare a cooperative catchment management plan.

Moreover, check-dams can reduce the sediment load in the river basins with high erosion rates, as suggested by [64]. Check-dams in gullies could be one of the more effective ways to conserve soil and water, as observed in the Loess Plateau of China, where afforestation methods were not successful due to the arid climate and barren soils [65].

However, a few torrential storms could fill up the check dams due to high sediment load, as concluded by [66]. Also, the terracing at the catchment level could reduce the sediment yield by 4 to 5 times [14]. However, the use of appropriate management methods requires that whether the soil and climate conditions are feasible for the area or not. It could be inferred from the above discussion that any management option at the catchment scale would be challenging and have short-term impacts to control the sediment load in the river.

3.4. Sediment Transport in a Settling Reservoir

A hypothetical settling reservoir provided immediately downstream of the existing diversion barrage is modeled. On the ground, a considerable area (256,691 m^2) is available downstream of the diversion barrage where a settling reservoir can be constructed. This area is currently not designated for productive uses, e.g., settlements, agriculture, etc. This makes the proposed settling reservoir a feasible option in the short term.

The sediment-laden flow would first enter into the settling reservoir where the sediments would settle down, and relatively clear water would outflow into the canal. The SSIIM model was set up using the hypothetical geometry and boundary conditions, as defined in the Data and Methods section. The results of the SSIIM model verify that the settling reservoir would be able to trap 16% of the incoming sediments. More precisely, it would be able to trap sand particles (coarse particles) almost entirely while the silt and clay would be trapped up to 25% and 0.5%, respectively. The overall trap efficiency is not as apparent as that of sand particles. However, the hopeful fact is that the settling reservoir is very efficient in preventing the entry of coarse particles, which are the most problematic particles among others. The model results reveal that 60,265 m^3 sediment will deposit in the settling reservoir every year that would otherwise have been entering the irrigation system.

The settling reservoir can provide an expeditious solution compared to catchment management options, e.g., improving vegetation, slope stabilization, etc. It would also minimize the dredging requirement from the pond area upstream of the existing diversion barrage. The modeling results also suggest that the settling reservoir should be operated such that a flow depth of 1.5 m should be maintained in the settling reservoir, which in turn would require regular dredging of the deposited sediments inside it.

The initial results of the model are encouraging and provide the basis for a quick solution. This modeling exercise was limited to only one geometry of the settling reservoir and a commonly prevailing flow condition. Prospective studies are needed to optimize settling reservoir size and use the transient sediment transport modeling approach.

4. Conclusions

The irrigation system fed by the Gomal river is challenged by the massive amounts of sediment flowing into it. Limited sediment monitoring arrangement and hydroclimatic data make it a data-sparse catchment. Therefore, the sources of sediment are not clearly known. The soil erosion estimations and management option focuses on a sub-catchment of 450 km^2 (~1.2% of total catchment area) between the Gomal Zam dam and a diversion barrage ~40 km downstream of the dam.

The trends of cumulative sediment load pre and post-dam construction suggest an 85% decrease in the annual load. It can be inferred that the sediment share from the sub-catchment is ~15% that is larger compared to its relative size (1.3%) to the entire catchment. It means the sub-catchment could be considered as a hotspot within the Gomal river catchment. Moreover, results of the suspended sediment scatter also indicate that outflows during the flushing operation of the dam substantially contribute to the sediment load of the Gomal River that calls for a monitoring mechanism to compute the proportion of this contribution.

Soil erosion rates estimated using the revised universal soil loss equation (RUSLE) reveals that the different approaches and available data sets significantly impact the erosion rates. In data-sparse catchments like the Gomal river, the options become limited to select a factor estimation method. In the case of rainfall-runoff erosivity factor R, a comparison of empirical equations and global data sets revealed colossal variation. The European Soil Data Centre (ESDAC) data set resulted in much higher values as compared with Roose's [27] relationship which is adopted for this study. In our case, the values of R adopted from a global dataset yield unrealistic erosion rates of catastrophic severity. Similarly, estimation of the cropping management factor with Knijff et al. [45] approach resulted in better values than the Durigon et al. (2014) [44] estimations. Therefore, the parameter estimation method for RUSLE empirical approach should be selected through careful consideration of regional literature rather than relying solely on global data sets, which in turn may give misleading results.

The results from RUSLE showed that most of the Gomal river sub-catchment area falls in very severe and catastrophic erosion rate categories (>100 t h^{-1}y^{-1}), and hence there are no apparent hotspots within the sub-catchment. High erosion rates can be attributed to steep topography, low vegetation, and highly erodible soil, as explained by RUSLE factors estimation.

A scenario was tested to understand the effectiveness of improved vegetation in the sub-catchment as an option to limit soil erosion. The results showed a small reduction in erosion rates. Half of the sub-catchment area still showed a high and severe erosion rate (10–100 t h^{-1}y^{-1}), and 17% of the area showed a very severe erosion rate (100–500 t ha^{-1}y^{-1}). This implies that improving vegetation in the sub-catchment can be beneficial for environmental reasons, but it may not be a single solution to prevent soil erosion from the catchment and consequently reducing sediment load in the river and that finds its way into the downstream irrigation system. For any catchment management option, e.g., improving vegetation, the entire catchment should be considered in the planning.

This study reveals that for catchments with such high erosion rates, complexities, and data limitations, improving one factor of RUSLE might not play a definitive role in managing land degradation. Sensitivity analysis of computed soil loss to changes in multiple factors of RUSLE and conditions could give better insight and poses an area of future work.

Reducing immediate negative impacts of high erosion rates remains a priority for catchment managers. In this context, the study finds a relatively quick solution by building a settling reservoir immediately downstream of the diversion barrage. The results from a 3D sediment transport model (SSIIM) suggest that 95% of sand particles and 25% silt particles can be trapped if the required water depth is maintained in the settling reservoir. The settling reservoir option, if adopted, would substantially reduce the sediment entry in the irrigation network. However, dredging would be required on an annual basis for the efficient operation of the settling reservoir.

Future research work should, therefore, seek to optimize the geometry of the settling reservoir and compare the results of various sediment transport models. It is also recommended to improve the sediment monitoring network at the Gomal river and the canal network for making informed management decisions. To better understand the sediment budget, future studies may use sediment yield assessment models for a better understanding of the sediment load contribution from the catchment area. The findings of this study are helpful for practitioners and to inform future investments to resolve the sedimentation problem. This research also provides the comparison of different methods to estimate the RUSLE factors applied in other parts of the world. The discussion around the selection of the methods can be helpful for the global audience to choose the appropriate methods of parameter calculation for catchments with similar characteristics.

Author Contributions: Conceptualization, M.T.B. and M.A.; Data curation, M.T.B.; Formal analysis, M.T.B. and M.A.; Funding acquisition, A.A.A.; Investigation, M.T.B. and M.A.; Methodology, M.T.B. and M.A.; Project administration, A.A.A.; Resources, A.A.A.; Software, M.A.; Supervision, A.A.A.; Validation, M.T.B. and A.A.A.; Writing—original draft, M.T.B.; Writing—review and editing, M.T.B., M.A. and A.A.A. All authors have read and agreed to the published version of the manuscript.

Funding: U.S. Agency for International Development's Mission to Pakistan (USAID/Pakistan) under the terms of Award No. 72039118 IO 00003.

Institutional Review Board Statement: Not applicable.

Informed Consent Statement: Not applicable.

Data Availability Statement: The data used for the study e.g., digital elevation model (DEM), satellite images, global rainfall-runoff erosivity factor, and soil type classification are available in the public domain and the web links are mentioned in the manuscript.

Acknowledgments: The International Water Management Institute (IWMI) is in receipt of financial support from the United States Agency for International Development (USAID) through Cooperative Agreement #72039118 IO 00003 which was used in part to support this study. The study design, data collection, analysis, and interpretation of the results are exclusively those of the authors and do not reflect the views or opinions of USAID, IWMI, or the CGIAR Research Program on Water, Land, and Ecosystems. The authors wish to acknowledge the contribution of Muhammad Humza in preparing the GIS maps used in this manuscript. The authors acknowledge NASA and USGS for providing Landsat images and SRTM DEM data free of cost. The authors also acknowledge the European Soil Data Centre (ESDAC) and FAO for providing free Global R and soil data, respectively.

Conflicts of Interest: The authors declare no conflict of interest.

Notation

a	organic matter content (%)
A	average annual soil loss
b	code related to soil structure
c	code related to soil permeability
C	crop management factor
K	soil erodibility factor
LS	topographic factor
M	percent silt and very fine sand contents
MFI	modified Frontier Index
NDVI	normalized Difference Vegetation Index
NIR	near Infra-Red
p	annual rainfall
P	Conservation support practice factor
p_i	monthly rainfall
R	rainfall-runoff erosivity factor
U	upslope contributing area per unit width
l	constant in Equation (6)
m	exponent in Equation (5) representing sheet erosion
m_{clay}	clay fraction content (<0.002 mm);
m_{silt}	silt fraction content (0.002–0.05 mm);
m_{vfs}	very fine sand fraction content (0.05–0.1 mm)
n	exponent in Equation (5) representing rill erosion
r	constant in Equation (6)
β	slope
L_0	length of the unit plot
S_0	slope of unit plot

References

1. Renard, K.; Foster, G.; Weesies, G.; McCool, D.; Yoder, D. *Predicting Soil Erosion by Water: A Guide to Conservation Planning with the Revised Universal Soil Loss Equation (RUSLE)*; United States Government Printing: Washington, DC, USA, 1997.
2. Boardman, J.; Poesen, J. Soil Erosion in Europe: Major Processes, Causes and Consequences. In *Soil Erosion in Europe*; John Wiley & Sons, Ltd.: Hoboken, NJ, USA, 2006; pp. 477–487, ISBN 978-0-470-85920-9.
3. Zokaib, S.; Naser, G. Impacts of Land Uses on Runoff and Soil Erosion A Case Study in Hilkot Watershed Pakistan. *Int. J. Sediment Res.* **2011**, *26*, 343–352. [CrossRef]
4. National Research Council. *Soil and Water Quality: An Agenda for Agriculture*; National Academies Press: Washington, DC, USA, 1993; p. 2132, ISBN 978-0-309-04933-7.

5. Borrelli, P.; Robinson, D.A.; Fleischer, L.R.; Lugato, E.; Ballabio, C.; Alewell, C.; Meusburger, K.; Modugno, S.; Schütt, B.; Ferro, V.; et al. An Assessment of the Global Impact of 21st Century Land Use Change on Soil Erosion. *Nat. Commun.* **2017**, *8*, 2013. [CrossRef]

6. Munir, S. Role of Sediment Transport in Operation and Maintenance of Supply and Demand Based Irrigation Canals: Application to Machai Maira Branch Canals. Ph.D. Thesis, UNESCO-IHE Institute for Water Education, Delft, The Netherlands, January 2011.

7. Sarwar, M.K.; Anjum, M.N.; Mahmood, S. Impact of Silt Excluder on Sediment Management of an Irrigation Canal: A Case Study of D.G. Khan Canal, Pakistan. *Arab. J. Sci. Eng.* **2013**, *38*, 3301–3307. [CrossRef]

8. Allahyonesi, H.; Omid, M.H.; Haghiabi, A.H. A Study of the Effects of the Longitudinal Arrangement Sediment Behavior near Intake Structures. *J. Hydraul. Res.* **2008**, *46*, 814–819. [CrossRef]

9. Atkinson, E. Vortex-Tube Sediment Extractors. I: Trapping Efficiency. *J. Hydraul. Eng.* **1994**, *120*, 1110–1125. [CrossRef]

10. Melone, A.M. Canals and Waterways, Sediment Control. In *General Geology*; Springer: Boston, MA, USA, 1988; pp. 55–63, ISBN 978-0-387-30844-9.

11. Jones, K.R.; Berney, O.; Carr, D.P.; Barrett, E.C. *Arid Zone Hydrology for Agricultural Development*; FAO Irrigation and Drainage Paper 37; Food and Agriculture Organization of the United Nations: Rome, Italy, 1981; ISBN 92-5-101079-X.

12. Chang, T.J.; Zhou, H.; Guan, Y. Applications of Erosion Hotspots for Watershed Investigation in the Appalachian Hills of the United States. *J. Irrig. Drain Eng.* **2016**, *142*, 4015057. [CrossRef]

13. Pham, T.G.; Degener, J.; Kappas, M. Integrated Universal Soil Loss Equation (USLE) and Geographical Information System (GIS) for Soil Erosion Estimation in A Sap Basin: Central Vietnam. *Int. Soil Water Conserv. Res.* **2018**, *6*, 99–110. [CrossRef]

14. Hussain, F.; Nabi, G.; Wu, R.-S.; Hussain, B.; Abbas, T. Parameter Evaluation for Soil Erosion Estimation on Small Watersheds Using SWAT Model. *Int. J. Agric. Biol. Eng.* **2019**, *12*, 96–108. [CrossRef]

15. Chuenchum, P.; Xu, M.; Tang, W. Estimation of Soil Erosion and Sediment Yield in the Lancang–Mekong River Using the Modified Revised Universal Soil Loss Equation and GIS Techniques. *Water* **2019**, *12*, 135. [CrossRef]

16. HALCROW. *Gomal Zam Irrigation Project-Siltation Rapid Assessment Overview*; Halcrow Pakistan (Pvt) Limited: Islamabad, Pakistan, 2017.

17. Pakistani and, U.S. Experts Conserve Water in Gomal Zam Dam Area. Available online: https://pk.usembassy.gov/pakistani-and-u-s-experts-conserve-water-in-gomal-zam-dam-area/ (accessed on 12 July 2020).

18. Wischmeier, W.H.; Smith, D.D. *Predicting Rainfall-Erosion Losses from Cropland East of the Rocky Mountains: Guide for Selection of Practices for Soil and Water Conservation*; USDA Agriculture Handbooks; Agricultural Research Service, U.S. Department of Agriculture in Cooperation with Purdue Agricultural Experiment Station: Washington, DC, USA, 1965; p. 47.

19. Lee, G.-S.; Lee, K.-H. Scaling Effect for Estimating Soil Loss in the RUSLE Model Using Remotely Sensed Geospatial Data in Korea. *Hydrol. Earth Syst. Sci. Discuss.* **2006**, *3*, 135–157. [CrossRef]

20. Panditharathne, D.L.D.; Abeysingha, N.S.; Nirmanee, K.G.S.; Mallawatantri, A. Application of Revised Universal Soil Loss Equation (RUSLE) Model to Assess Soil Erosion in "Kalu Ganga" River Basin in Sri Lanka. Available online: https://www.hindawi.com/journals/aess/2019/4037379/ (accessed on 6 January 2020).

21. Ganasri, B.P.; Ramesh, H. Assessment of Soil Erosion by RUSLE Model Using Remote Sensing and GIS—A Case Study of Nethravathi Basin. *Geosci. Front.* **2016**, *7*, 953–961. [CrossRef]

22. Terranova, O.; Antronico, L.; Coscarelli, R.; Iaquinta, P. Soil Erosion Risk Scenarios in the Mediterranean Environment Using RUSLE and GIS: An Application Model for Calabria (Southern Italy). *Geomorphology* **2009**, *112*, 228–245. [CrossRef]

23. Alewell, C.; Borrelli, P.; Meusburger, K.; Panagos, P. Using the USLE: Chances, Challenges and Limitations of Soil Erosion Modelling. *Int. Soil Water Conserv. Res.* **2019**, *7*, 203–225. [CrossRef]

24. Trimble, S.W.; Crosson, P.U.S. Soil Erosion Rates—Myth and Reality. *Science* **2000**, *289*, 248–250. [CrossRef] [PubMed]

25. Benavidez, R.; Jackson, B.; Maxwell, D.; Norton, K. A Review of the (Revised) Universal Soil Loss Equation ((R)USLE): With a View to Increasing Its Global Applicability and Improving Soil Loss Estimates. *Hydrol. Earth Syst. Sci.* **2018**, *22*, 6059–6086. [CrossRef]

26. Morgan, R.P.C. *Soil Erosion and Conservation*, 3rd ed.; John Wiley & Sons, Inc.: Hoboken, NJ, USA, 2005; ISBN 978-1-4051-4467-4.

27. Roose, E.J. *Erosion et Ruissellement En Afrique de l'Ouest: Vingt Années de Mesures En Petites Parcelles Expérimentales*; Laboratoire de Pédologie de l'ORSTOM: Abidjan, Ivory Coast, 1975.

28. Arnoldus, H.M.J. *An Approximation of the Rainfall Factor in the Universal Soil Loss Equation*; John Wiley and Sons Ltd: Hoboken, NJ, USA, 1980.

29. Fernandez, C.; Wu, J.Q.; McCool, D.K.; Stockle, C.O. Estimating Water Erosion and Sediment Yield with GIs, RUSLE, and SEDD. *J. Soil Water Conserv.* **2003**, *58*, 128–136.

30. İrvem, A.; Topaloğlu, F.; Uygur, V. Estimating Spatial Distribution of Soil Loss over Seyhan River Basin in Turkey. *J. Hydrol.* **2007**, *336*, 30–37. [CrossRef]

31. Nakil, M. *Analysis of Parameters Causing Water Induced Soil Erosion*; Fifth Annual Progress Seminar; Indian Institute of Technology: Bombay, India, 2014; unpublished.

32. Wischmeier, W.H.; Smith, D.D. *Predicting Rainfall Erosion Losses: A Guide to Conservation Planning*; Department of Agriculture, Science and Education Administration: Washington, DC, USA, 1978.

33. Panagos, P.; Meusburger, K.; Ballabio, C.; Borrelli, P.; Alewell, C. Soil Erodibility in Europe: A High-Resolution Dataset Based on LUCAS. *Sci. Total Environ.* **2014**, *479–480*, 189–200. [CrossRef] [PubMed]

34. Ullah, S.; Ali, A.; Iqbal, M.; Javid, M.; Imran, M. Geospatial Assessment of Soil Erosion Intensity and Sediment Yield: A Case Study of Potohar Region, Pakistan. *Environ. Earth Sci.* **2018**, *77*, 705. [CrossRef]
35. Desmet, P.J.J.; Govers, G. A GIS Procedure for Automatically Calculating the USLE LS Factor on Topographically Complex Landscape Units. *J. Soil Water Conserv.* **1996**, *51*, 427–433.
36. Moore, I.D.; Burch, G.J. Physical Basis of the Length-Slope Factor in the Universal Soil Loss Equation 1. *Soil Sci. Soc. Am. J.* **1986**, *50*, 1294–1298. [CrossRef]
37. Wilson, J.P.; Gallant, J.C. (Eds.) *Terrain Analysis: Principles and Applications*; Wiley: New York, NY, USA, 2000; ISBN 978-0-471-32188-0.
38. Kinnell, P.I.A. Event Soil Loss, Runoff and the Universal Soil Loss Equation Family of Models: A Review. *J. Hydrol.* **2010**, *385*, 384–397. [CrossRef]
39. Lanorte, A.; Cillis, G.; Calamita, G.; Nolè, G.; Pilogallo, A.; Tucci, B.; De Santis, F. Integrated Approach of RUSLE, GIS and ESA Sentinel-2 Satellite Data for Post-Fire Soil Erosion Assessment in Basilicata Region (Southern Italy). *Geomat. Nat. Hazards Risk* **2019**, *10*, 1563–1595. [CrossRef]
40. Borrelli, P.; Märker, M.; Panagos, P.; Schütt, B. Modeling Soil Erosion and River Sediment Yield for an Intermountain Drainage Basin of the Central Apennines, Italy. *CATENA* **2014**, *114*, 45–58. [CrossRef]
41. Karydas, C.G.; Sekuloska, T.; Silleos, G.N. Quantification and Site-Specification of the Support Practice Factor When Mapping Soil Erosion Risk Associated with Olive Plantations in the Mediterranean Island of Crete. *Environ. Monit. Assess.* **2009**, *149*, 19–28. [CrossRef]
42. Lazzari, M.; Gioia, D.; Piccarreta, M.; Danese, M.; Lanorte, A. Sediment Yield and Erosion Rate Estimation in the Mountain Catchments of the Camastra Artificial Reservoir (Southern Italy): A Comparison between Different Empirical Methods. *CATENA* **2015**, *127*, 323–339. [CrossRef]
43. Vatandaşlar, C.; Yavuz, M. Modeling Cover Management Factor of RUSLE Using Very High-Resolution Satellite Imagery in a Semiarid Watershed. *Environ. Earth Sci.* **2017**, *76*, 65. [CrossRef]
44. Durigon, V.L.; Carvalho, D.F.; Antunes, M.A.H.; Oliveira, P.T.S.; Fernandes, M.M. NDVI Time Series for Monitoring RUSLE Cover Management Factor in a Tropical Watershed. *Int. J. Remote Sens.* **2014**, *35*, 441–453. [CrossRef]
45. Van der Knijff, J.M.; Jones, R.J.A.; Montanarella, L. *Soil Erosion Risk Assessment in Europe*; European Commission: Brussels, Belgium, 2000.
46. Ma, H.L.; Wang, Z.L.; Zhou, X. Estimation of Soil Loss Based on RUSLE in Zengcheng, Guangdong Province. *Yangtze River* **2010**, *41*, 90–93.
47. Olsen, N.R.B. *SSIIM User's Manual*; The Norwegian University of Science and Technology: Trondheim, Norway, 2018.
48. Ali, K.F.; De Boer, D.H. Spatial Patterns and Variation of Suspended Sediment Yield in the Upper Indus River Basin, Northern Pakistan. *J. Hydrol.* **2007**, *334*, 368–387. [CrossRef]
49. Ashraf, M.; Bhatti, M.T.; Shakir, A.S.; Tahir, A.A.; Ahmad, A. Sediment Control Interventions and River Flow Dynamics: Impact on Sediment Entry into the Large Canals. *Environ. Earth Sci.* **2015**, *74*, 5465–5474. [CrossRef]
50. Sinha, D.; Joshi, V.U. Application of Universal Soil Loss Equation (USLE) to Recently Reclaimed Badlands along the Adula and Mahalungi Rivers, Pravara Basin, Maharashtra. *J. Geol. Soc. India* **2012**, *80*, 341–350. [CrossRef]
51. Sholagberu, A.T.; Mustafa, M.R.U.; Yusof, K.W.; Ahmad, M.H. Evaluation of rainfall-runoff erosivity factor for Cameron highland, Pahang, Malaysia. *J. Ecol. Eng.* **2016**, *17*, 1–8. [CrossRef]
52. Panagos, P.; Borrelli, P.; Meusburger, K.; Yu, B.; Klik, A.; Jae Lim, K.; Yang, J.E.; Ni, J.; Miao, C.; Chattopadhyay, N.; et al. Global Rainfall Erosivity Assessment Based on High-Temporal Resolution Rainfall Records. *Sci. Rep.* **2017**, *7*, 4175. [CrossRef] [PubMed]
53. Kouli, M.; Soupios, P.; Vallianatos, F. Soil Erosion Prediction Using the Revised Universal Soil Loss Equation (RUSLE) in a GIS Framework, Chania, Northwestern Crete, Greece. *Environ. Geol.* **2009**, *57*, 483–497. [CrossRef]
54. Patriche, C.V.; Pirnau, R.; Grozavu, A.; Rosca, B. A Comparative Analysis of Binary Logistic Regression and Analytical Hierarchy Process for Landslide Susceptibility Assessment in the Dobrov River Basin, Romania. *Pedosphere* **2016**, *26*, 335–350. [CrossRef]
55. Marques, V.S.; Ceddia, M.B.; Antunes, M.A.H.; Carvalho, D.F.; Anache, J.A.A.; Rodrigues, D.B.B.; Oliveira, P.T.S. USLE K-Factor Method Selection for a Tropical Catchment. *Sustainability* **2019**, *11*, 1840. [CrossRef]
56. Tesfaye, G.; Debebe, Y.; Fikirie, K. Soil Erosion Risk Assessment Using GIS Based USLE Model for Soil and Water Conservation Planning in Somodo Watershed, South West Ethiopia. *IJOEAR* **2018**, *4*, 9.
57. Sujatha, E.R.; Sridhar, V. Spatial Prediction of Erosion Risk of a Small Mountainous Watershed Using RUSLE: A Case-Study of the Palar Sub-Watershed in Kodaikanal, South India. *Water* **2018**, *10*, 1608. [CrossRef]
58. Abuzar, M.K.; Shakir, U.; Ashraf, M.A.; Khan, S.; Shaista, S.; Pasha, A.R. GIS Based Risk Modeling of Soil Erosion under Different Scenarios of Land Use Change in Simly Watershed of Pakistan. *J. Himal. Earth Sci.* **2018**, *51*, 132–143.
59. Bashir, S.; Baig, M.A.; Ashraf, M.; Anwar, M.M.; Bhalli, M.N.; Munawar, S. Risk Assessment of Soil Erosion in Rawal Watershed Using Geoinformatics Techniques. *Sci.Int.* **2013**, *25*, 583–588.
60. Koirala, P.; Thakuri, S.; Joshi, S.; Chauhan, R. Estimation of Soil Erosion in Nepal Using a RUSLE Modeling and Geospatial Tool. *Geosciences* **2019**, *9*, 147. [CrossRef]
61. Lencha, B.K.; Moges, A. Identification of Soil Erosion Hotspots in Jimma Zone (Ethiopia) Using GIS Based Approach. *Ethiop. J. Environ. Stud. Manag.* **2016**, *8*, 926. [CrossRef]
62. Panagos, P.; Borrelli, P.; Meusburger, K.; Alewell, C.; Lugato, E.; Montanarella, L. Estimating the Soil Erosion Cover-Management Factor at the European Scale. *Land Use Policy* **2015**, *48*, 38–50. [CrossRef]

63. Butt, M.J.; Mahmood, R.; Waqas, A. Sediments Deposition Due to Soil Erosion in the Watershed Region of Mangla Dam. *Environ. Monit. Assess.* **2011**, *181*, 419–429. [CrossRef] [PubMed]
64. Boix-Fayos, C.; de Vente, J.; Martínez-Mena, M.; Barberá, G.G.; Castillo, V. The Impact of Land Use Change and Check-Dams on Catchment Sediment Yield. *Hydrol. Process.* **2008**, *22*, 4922–4935. [CrossRef]
65. Xiang-Zhou, X.; Hong-Wu, Z.; Ouyang, Z. Development of Check-Dam Systems in Gullies on the Loess Plateau, China. *Environ. Sci. Policy* **2004**, *7*, 79–86. [CrossRef]
66. Castillo, V.M.; Mosch, W.M.; García, C.C.; Barberá, G.G.; Cano, J.A.N.; López-Bermúdez, F. Effectiveness and Geomorphological Impacts of Check Dams for Soil Erosion Control in a Semiarid Mediterranean Catchment: El Cárcavo (Murcia, Spain). *CATENA* **2007**, *70*, 416–427. [CrossRef]

 *sustainability*

Article

Biogas Production Potential from Livestock Manure in Pakistan

Muhammad U. Khan [1,*], Muhammad Ahmad [2], Muhammad Sultan [3,*], Ihsanullah Sohoo [4,*], Prakash C. Ghimire [5], Azlan Zahid [6,7], Abid Sarwar [8], Muhammad Farooq [9], Uzair Sajjad [10], Peyman Abdeshahian [11] and Maryam Yousaf [2,12]

[1] Department of Energy Systems Engineering, Faculty of Agricultural Engineering and Technology, University of Agriculture, Faisalabad 38040, Pakistan

[2] School of Chemistry and Chemical Engineering, Beijing Institute of Technology, Beijing 102488, China; engineer_nust@yahoo.com (M.A.); maryamyousaf79@yahoo.com (M.Y.)

[3] Department of Agricultural Engineering, Bahauddin Zakariya University, Multan 60800, Pakistan

[4] Institute of Environmental Technology and Energy Economics, Hamburg University of Technology, Blohmstr. 15, 21073 Hamburg, Germany

[5] Energy Sector Leader at S.N.V. Netherlands Development Organization, 2514 JG Den Haag, The Netherlands; prakashchgh@gmail.com

[6] Department of Agricultural and Biological Engineering, Penn State University, University Park, PA 16802, USA; axz264@psu.edu

[7] Department of Farm Machinery and Power, University of Agriculture, Faisalabad 38040, Pakistan

[8] Department of Irrigation and Drainage, University of Agriculture Faisalabad, Punjab 38040, Pakistan; abidsarwar@uaf.edu.pk

[9] Department of Mechanical Engineering (New Campus-KSK), University of Engineering and Technology, Lahore 54890, Pakistan; engr.farooq@uet.edu.pk

[10] Mechanical Engineering Department, National Chiao Tung University, Hsinchu 30010, Taiwan; energyengineer01@gmail.com

[11] Department of Microbiology, Masjed Soleiman Branch, Islamic Azad University, Masjed Soleiman, Iran; peyman_137@yahoo.com

[12] Department of Chemistry, Faculty of Sciences, University of Agriculture Faisalabad, Punjab 38040, Pakistan

* Correspondence: usman.khan@uaf.edu.pk (M.U.K.); muhammadsultan@bzu.edu.pk (M.S.); sohoo.ihsanullah@tuhh.de (I.S.); Tel.: +92-311-730-2162 (M.U.K.)

Citation: Khan, M.U.; Ahmad, M.; Sultan, M.; Sohoo, I.; Ghimire, P.C.; Zahid, A.; Sarwar, A.; Farooq, M.; Sajjad, U.; Abdeshahian, P.; et al. Biogas Production Potential from Livestock Manure in Pakistan. *Sustainability* **2021**, *13*, 6751. https://doi.org/10.3390/su13126751

Academic Editor: Alessio Siciliano

Received: 3 May 2021
Accepted: 26 May 2021
Published: 15 June 2021

Publisher's Note: MDPI stays neutral with regard to jurisdictional claims in published maps and institutional affiliations.

Abstract: Pakistan is facing a severe energy crisis due to its heavy dependency on the import of costly fossil fuels, which ultimately leads to expansive electricity generation, a low power supply, and interruptive load shedding. In this regard, the utilization of available renewable energy resources within the country for production of electricity can lessen this energy crisis. Livestock waste/manure is considered the most renewable and abundant material for biogas generation. Pakistan is primarily an agricultural country, and livestock is widely kept by the farming community, in order to meet their needs. According to the 2016–2018 data on the livestock population, poultry held the largest share at 45.8%, followed by buffaloes (20.6%), cattle (12.7%), goats (10.8%), sheep (8.4%), asses (1.3%), camels (0.25%), horses (0.1%), and mules (0.05%). Different animals produce different amounts of manure, based upon their size, weight, age, feed, and type. The most manure is produced by cattle (10–20 kg/day), while poultry produce the least (0.08–0.1 kg/day). Large quantities of livestock manure are produced from each province of Pakistan; Punjab province was the highest contributor (51%) of livestock manure in 2018. The potential livestock manure production in Pakistan was 417.3 million tons (Mt) in 2018, from which 26,871.35 million m^3 of biogas could be generated—with a production potential of 492.6 petajoules (PJ) of heat energy and 5521.5 MW of electricity. Due to its favorable conditions for biodigester technologies, and through the appropriate development of anaerobic digestion, the currently prevailing energy crises in Pakistan could be eliminated.

Keywords: renewable energy; biogas production; livestock manure; anaerobic digestion

1. Introduction

The production of cheap, green energy has been considered a prime objective for a country heading towards sustainable development. Pakistan, as a developing country, needs an enormous amount of energy, around 25,000 megawatts (MW), for its industrial, agricultural, and household needs (Figure 1). However, this energy demand has not been met, which has led to electricity crises [1–3]. Due to these severe energy crises, Pakistan is currently facing tremendous electricity load shedding (10–14 h/day) [4–7]. Energy consumption has increased due to the growth in industrialization and the increasing urban population. For example, the per capita energy consumption of Pakistan has shown an increasing trend from the year 2000, from 373.13 to 484.45 kWh [8].

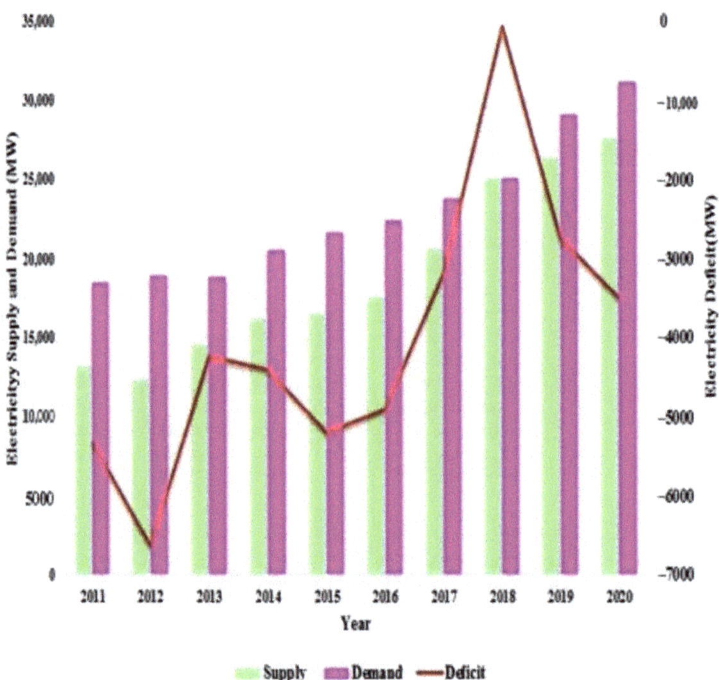

Figure 1. Electricity supply, demand, and deficit of the country.

The current determined capacity of Pakistan is 22,000 MW of electricity (Figure 1). Mismanagement at transmission and distribution networks and high discharges have resulted in high losses that consequently increase load shedding. This is the foremost reason that Pakistan is facing energy shortfalls of 4000–6000 MW, as presented in Figure 1 [9–14]. The industrial, agricultural, and domestic sectors are suffering badly due to the ongoing energy shortfall [6]. A recent survey—conducted by the Private Power and Infrastructure Board, Government of Pakistan, Ministry of Energy (Power Division)—indicated that electricity consumption of 90.36 terawatt-hours (TWh) was recorded in Pakistan during 2015–2016, with 6.01% electricity export and 0.49% electricity import.

The electricity demand of the country is increasing at an annual rate of 11–13% [15] because of the increase in growth centers and the industrialization process. Pakistan is likely to follow the same trend in the future, as well. The energy demand for Pakistan is projected to rise to 54,000 MW in 2020 and 113,000 MW in 2030 [16–18].

It was found that 70% of Pakistan's population lives in rural areas, and that 96.6% of rural people have no access to modern energy facilities. Thus, they are facing energy poverty. In this regard, it was also found that about 45% of their energy expenditure is spent on solid biomass such as dung cakes, firewood, and crop residues, with an additional 12% spent on natural gas, LPG, kerosene, and candles used for lighting and cooking purposes in rural areas [19].

Natural gas, firewood, kerosene, livestock dung, liquid petroleum gas (LPG), firewood, and electricity are the most common fuels currently being used for cooking purposes in Pakistan [20]. Natural gas (supplied through pipes) is the cheapest fuel for cooking purposes. However, due to the limited reserves and insufficient supply systems, it cannot be a promising fuel for cooking [21]. The limited reserves and high prices of fossil fuels have resulted in the fact that kerosene and LPG are not viable options for cooking purposes in Pakistan. On the other hand, the most common cooking fuels such as firewood, crop residues, and animal dung have lower efficiency with higher heating values as compared to the other fuels [22].

Raising livestock is one of the major agricultural activities in Pakistan which contributes to the agricultural economy of Pakistan. In this context, the agriculture-based economy has a 24.5% share of the gross domestic product (GDP) and provides 60% of export earnings in Pakistan. Likewise, 55.6% of the economy is from the livestock sector and contributes 11.8% of Pakistan's GDP. Cattle raising is one of the major agricultural activities in Pakistan, meaning that a large quantity of livestock waste is produced in Pakistan which could be utilized as an appropriate source of sustainable energy. Cattle manure in most villages is used to prepare dried dung cakes that are burned for cooking energy.

People living in hilly areas of Pakistan are in difficult conditions to fulfill their energy demands, and they spend a lot of time collecting animal dung and woody biomass. In this regard, the use of livestock waste for energy production will be a worthwhile approach for providing an energy supply to the rural areas which in turn is beneficial for the economic development of the country, with a reduction in environment concerns [23,24].

Moreover, the use of biogas as a clean energy source will also reduce the utilization of conventional fossil fuels which in turn will lower GHG, and other hazardous gas emissions which are detrimental to the environment [25,26]. In this view, the development of biodigester technology will provide a manure management facility for dairy farms as well as for poultry farms. Digested manure is a natural fertilizer which can be applied to crops as a cost-effective alternative to synthetic fertilizers [26]. On the other hand, the development of biodigester technology will result in the conservation of resources and protection of the environment [27]. Considering the large quantity of livestock manure production in rural areas of Pakistan, it has potential to be utilized for energy production in order to overcome prevailing energy crises. The biogas produced from cattle manure is a unique sustainable energy supply due to its high availability as a decentralized energy source [28]. However, currently, the main issue with anaerobic digestion of livestock manure is the ammonia toxicity due to the higher concentrations of nitrogen as well as lower degradation during anaerobic digestion due to the higher concentrations of lignocellulosic materials [29,30]. This problem can be tackled by co-digestion of manure with material having a lower concentration of nitrogen [31,32].

Some studies have already been carried out to show the role of agro-industrial waste for biogas generation as an important source of sustainable energy in Pakistan [10,33,34]. Currently, about 8000 biogas plants are operative in Pakistan [35]. However, there is a lack of scientific study to evaluate the potential of livestock manure as a pivotal bioresource and the potential of biogas generation via anaerobic digestion of the available livestock manure in different provinces of Pakistan. Furthermore, it is also not clear how the potential of biogas from animal manure can contribute to the heat and electrical energy supply in Pakistan. Hence, it is essential to find out the potential of biogas, methane, and electricity generation using animal manure for enhancing biodigester technology in the country as well as for overcoming the prevailing energy and environmental issues.

Herein, we studied the potential of renewable energy production (e.g., biogas, methane, electricity, heat energy) from livestock manure in Pakistan by spatially analyzing and characterizing the data (from 1960 to 2018) that were collected from the Pakistan Bureau of

Statistics and Ministry of National Food Security & Research. The results of this study will be useful for developing biogas-based electricity projects in all provinces of Pakistan which will not only be helpful in overcoming the ongoing energy crises but will also create employment opportunities, particularly in rural areas. This analysis will also be useful to the policymakers of developing countries that can change the lives of many villagers.

2. Methodology

2.1. Calculation of Livestock Population

The livestock population and density records were extracted from the archives of livestock census data (collected by the Pakistan Bureau of Statistics from 1960 to 2018) and arrayed provincially [36,37]. However, in this study, livestock populations were estimated for 2016, 2017, and 2018, using the annual growth rate of 8–10% [38].

2.2. Calculation and Measurement of Total Amount of Livestock Manure in Pakistan

The amount of manure produced by an animal depends on many parameters, including body weight, size, age, amount of feed, and type of animal [39]. The reference study found that the amount of manure produced by cattle and camels is 10–20 kg/day and 15–17 kg/day, respectively [39]. For sheep/goats, it is 2 kg/day, whereas, for mules, horses, and asses, it ranges from 10 to 15 kg/day [39]. Similarly, for poultry, daily manure generation is estimated to be 0.08–0.1 kg [39]. Keeping in view the effect of influential parameters, in this study, the average manure production for cattle/buffalo, goats/sheep, camels, and mules/horses/asses was considered 10 kg/day, 2 kg/day, 15 kg/day, 10 kg/day, and 0.1 kg/day, respectively.

2.3. Calculation of Total Potential of Biogas Production from Livestock Manure

The potential of biogas generation from livestock manure in the country was calculated using manure produced annually. The biogas production from animal manure can be affected by various factors such as the amount of manure, the availability of manure, and the total solids content in animal manure [39]. A variable coefficient of manure availability was introduced to concede the manure collection and transportation losses in the calculation. Table 1 summarizes the numeric values of influential parameters that were considered in this study. The theoretical potential of biogas (TPB) generated from animal manure was determined by the following Equation (1).

$$TPB = M \times AC \times TS \times \frac{BY}{kgTS} \tag{1}$$

where TPB is the theoretical potential of biogas (million m^3 year^{-1}), M is the quantity of livestock manure/year/province (million kg year^{-1}), AC denotes the availability coefficient of animal manure for selected species, TS is the total solids content of animal manure, and BY is the biogas yield of animal manure for each kilogram of total solids (m^3 kg^{-1} TS).

Table 1. Amount of animal manure produced, manure availability coefficient, biogas yield, and ratio of the total solids of animal manure for selected species [25,40,41].

Species	Manure Yield (kg/Day)	Manure Availability Coefficient (%)	Biogas Yield (m^{3}kg^{-1} TS)	Ratio of the TS (TS%)
Cattle	10–20	50	0.6–0.8	25–30
Buffalo	10–20	50	0.6–0.8	25–30
Sheep	2	33	0.3–0.4	18–25
Goat	2	33	0.3–0.4	18–25
Camel	15–17	50	0.6–0.8	25–30
Horse	10–15	50	0.6–0.8	25–30
Ass/Donkey	10–15	50	0.6–0.8	25–30
Mule	10–15	50	0.6–0.8	25–30
Poultry	0.08–0.1	99	0.3–0.8	10–29

In this study, the biogas potential determined for the manure obtained from the selected animal species was calculated by considering AC and BY values of 50% and 0.6 m^3 kg^{-1} TS, respectively, for cattle, buffaloes, camels, horses, asses, and mules. Moreover, AC and BY values of 33% and 0.30 m^3 kg^{-1} TS were considered for sheep/goats, whereas for poultry, 99% and 0.15 m^3 kg^{-1} TS, respectively, were considered. Similarly, the TS value was considered 25% for cattle, buffaloes, camels, horses, asses, and mules, whereas 20% was considered for sheep/goats and 29% for poultry [42].

2.4. Calculation of Potential of Methane and Electricity Production from Livestock Manure

In this section, a few assumptions were considered to estimate the methane and electricity production potential from the available livestock manure. However, it has been well documented that the proportion of methane content in goat/sheep manure ranges between 40 and 50%, whereas it ranges between 50 and 70% for poultry [43]. Biogas production has been significantly dependent upon the amount of methane production. It has been found that approximately 50–70% of the methane content transforms into biogas [44]. For this study, the biogas generation through anaerobic digestion of manure for the specified livestock was assumed to be 60% of methane, while methane was considered to form 50% of the biogas content for poultry manure. The heating value of methane was calculated by considering a conversion efficiency of 85% in the boiler, and the calorific value of methane was considered as 36 MJ/m^3. The annual electricity generation potential using biogas was determined by Equation (2):

$$e_{biogas} = E_{biogas} \times \eta \tag{2}$$

where e_{biogas} = amount of electricity generated using biogas (kWh year^{-1}), E_{biogas} = total amount of energy in biogas which has not been converted, and η = efficiency of the power plant for conversion of biogas to electricity (~30%). The unconverted energy content of the biogas was determined by the following Equation (3):

$$E_{biogas} = C.V_{biogas} \times m_{biogas} \tag{3}$$

where $C.V_{biogas}$ = caloric value of the biogas, ~6 kWh m^{-3} [45], and m_{biogas} = annual amount of biogas produced from the selected species of livestock.

3. Results and Discussion

3.1. Livestock Population and Potential of Biodigester Technology

The livestock growth rate was calculated, and influential parameters were evaluated accordingly. Table 2 shows the provincial livestock population record of Pakistan from 1960 to 2018. From Table 2, it is summarized that the total livestock population achieved the highest number of 362,111,000 in 2018. Poultry exhibited the largest share, i.e., 45.8%, followed by goats, cattle, buffaloes, sheep, asses, camels, horses, and mules, with shares of 20.6%, 12.7%, 10.8%, 8.4%, 1.3%, 0.25%, 0.1%, and 0.05%, respectively. Punjab ranked at the top with a livestock population share of 39.7% on the regional scale, and Balochistan had the lowest population share, i.e., 13.6%. The temporal increment in the livestock population (4.9 times from 1960) emphasizes the potential of biogas origination and, consequently, biodigester technology development in the country.

Table 2. Livestock population in different provinces of Pakistan for the years 1960–2018 (×1000 heads) [36,37].

Animal Type	1960	1972	1976	1986	1996	2006	2016	2017	2018
Punjab									
Cattle	9673	8226	8108	8817	9382	14,412	20,826	21,607	22,417
Buffaloes	6129	7413	7979	11,150	13,101	17,747	23,850	24,566	25,302
Sheep	5583	6280	8037	6686	6142	6362	7168	7254	7341
Goats	2973	5943	7767	10,755	15,301	19,831	26,011	26,726	27,461
Camels	266	365	338	321	187	199	199	199	199
Horses	226	264	286	245	181	163	163	163	163
Asses	897	1063	1139	1657	1948	2232	2465	2490	2515
Mules	23	20	29	36	57	63	63	63	63
Poultry	6440	8688	13,783	27,848	24,511	25,906	50,961	54,528	58,345
Total	32,210	38,262	47,466	67,515	70,810	86,915	131,706	137,596	143,806
Sindh									
Cattle	2936	2800	2854	3874	5664	6925	10,007	10,382	10,771
Buffaloes	1353	1522	1834	3220	5615	7340	9684	10,160	10,465
Sheep	1590	840	1829	2616	3710	3959	4460	4514	4568
Goats	2201	2275	4237	6755	9734	12,572	16,490	16,943	17,409
Camels	62	80	144	218	225	278	278	278	278
Horses	40	71	94	76	63	45	45	45	45
Asses	159	242	373	500	694	1004	1109	1120	1131
Mules	1	2	3	5	12	20	22	22	23
Poultry	1250	2743	6295	8798	11,549	14,136	27,807	29,754	31,836
Total	9592	10,575	17,663	26,062	37,266	46,279	69,902	73,218	76,526
KPK									
Cattle	3206	2962	3000	3285	4237	5968	8624	8947	9283
Buffaloes	651	791	762	1271	1395	1928	2591	2668	2748
Sheep	2432	2455	3675	1599	2821	3363	3789	3834	3880
Goats	3035	3737	4686	2899	6764	9599	12,590	12,936	13,292
Camels	76	101	95	70	65	64	65	66	66
Horses	23	31	29	34	47	76	81	83	85
Asses	306	408	381	446	534	560	618	624	631
Mules	19	32	28	23	60	67	74	75	76
Poultry	4190	4939	9708	17,203	22,501	27,695	54,480	58,294	62,374
Total	13,938	15,456	22,364	26,830	38,424	49,320	82,912	87,527	92,435
Balochistan									
Cattle	643	482	684	1157	1341	2254	3257	3379	3505
Buffaloes	26	22	33	63	161	320	430	442	456
Sheep	2564	3859	5075	11,111	10,841	12,804	14,426	14,599	14,774
Goats	1596	3238	4441	7299	9369	11,785	15,457	15,882	16,319
Camels	86	185	212	349	339	380	380	380	380
Horses	10	19	23	29	43	60	60	60	60
Asses	99	171	244	370	383	472	521	526	531
Mules	0.4	1	1	4	6	6	6	6	7
Poultry	454	1183	1958	3295	4637	5911	11,628	12,441	13,312
Total	5478.4	9160	12,671	23,677	27,120	33,992	46,165	47,715	49,344

3.2. Suitability of Livestock Manure as a Potential Substrate for Biodigester Technology

The livestock manure potential of the country in 2018 increases approximately 2.6 times from 1960 due to accretion in the livestock population. The gradual increase in livestock manure indicates that waste management through anaerobic digestion could be a viable solution, which also assists in overcoming the prevailing energy crises of the country. Moreover, manure management through anaerobic digestion will also reduce the consumption of synthetic fertilizers and increase crop yields due to the utilization of organic fertilizer, resulting in revenue generation.

Table 3 shows the temporal increment in animal manure production from 1960 to 2018. Based on calculations, ~417.3 million tons (Mt) of animal manure was produced in 2018. At the regional level, Punjab manifested the highest livestock manure potential with a 51% share of the total manure in 2018, followed by Sindh, KPK, and Balochistan, with shares of 24.1%, 14.85%, and 10.04%, respectively, whereas among animals species, cattle showed the highest contribution of 40.21% to the total manure produced in 2018, followed by buffaloes, goats, sheep, asses, poultry, camels, horses, and mules, with shares of 34.08%, 13.02%, 5.34%, 4.19%, 1.44%, 1.20%, 0.3%, and 0.14%, respectively.

Table 3. Animal manure production potential in Pakistan from 1960 to 2018 (Mt/year).

Animal Type	1960	1972	1976	1986	1996	2006	2016	2017	2018
Cattle	35.30	30.02	29.59	32.18	34.24	52.60	76.01	78.86	81.82
Buffaloes	22.37	27.05	29.12	40.69	47.81	64.77	87.05	89.66	92.35
Sheep	4.07	4.58	5.86	4.88	4.48	4.64	5.23	5.29	5.35
Goats	2.17	4.33	5.66	7.85	11.16	14.47	18.98	19.50	20.04
Camels	1.45	1.99	1.85	1.75	1.02	1.08	1.08	1.08	1.08
Horses	0.82	0.96	1.04	0.89	0.66	0.59	0.59	0.59	0.59
Asses	3.27	3.87	4.15	6.04	7.11	8.14	8.99	9.08	9.17
Mules	0.08	0.07	0.10	0.13	0.20	0.22	0.22	0.22	0.22
Poultry	0.23	0.31	0.50	1.01	0.89	0.94	1.86	1.99	2.12
Total	69.79	73.23	77.91	95.45	107.61	147.50	200.05	206.33	212.80
Sindh									
Cattle	10.71	10.22	10.41	14.14	20.67	25.27	36.52	37.89	39.31
Buffaloes	4.93	5.55	6.69	11.75	20.49	26.79	35.34	37.08	38.19
Sheep	1.16	0.61	1.33	1.90	2.70	2.89	3.25	3.29	3.33
Goats	1.60	1.66	3.09	4.93	7.10	9.17	12.03	12.36	12.70
Camels	0.33	0.43	0.78	1.19	1.23	1.52	1.52	1.52	1.52
Horses	0.14	0.25	0.34	0.27	0.22	0.16	0.16	0.16	0.16
Asses	0.58	0.88	1.36	1.82	2.5331	3.66	4.04	4.08	4.12
Mules	0.003	0.007	0.01	0.01	0.04	0.07	0.08	0.08	0.08
Poultry	0.04	0.10	0.22	0.32	0.42	0.51	1.01	1.08	1.16
Total	19.53	19.73	24.27	36.36	55.44	70.07	93.99	97.58	100.6
KPK									
Cattle	11.70	10.81	10.95	11.99	15.46	21.78	31.47	32.65	33.88
Buffaloes	2.37	2.88	2.78	4.63	5.09	7.03	9.45	9.73	10.03
Sheep	1.77	1.79	2.68	1.167	2.05	2.45	2.76	2.79	2.83
Goats	2.21	2.72	3.42	2.11	4.93	7.00	9.19	9.44	9.70
Camels	0.41	0.55	0.52	0.38	0.35	0.35	0.35	0.36	0.36
Horses	0.08	0.11	0.10	0.12	0.17	0.27	0.29	0.30	0.31
Asses	1.11	1.48	1.39	1.62	1.94	2.04	2.25	2.27	2.30
Mules	0.06	0.11	0.10	0.08	0.21	0.24	0.27	0.27	0.27
Poultry	0.15	0.18	0.35	0.62	0.82	1.01	1.98	2.12	2.27
Total	19.90	20.67	22.30	22.76	31.07	42.20	58.05	59.98	61.97
Balochistan									
Cattle	2.34	1.75	2.49	4.22	4.89	8.22	11.88	12.33	12.79
Buffaloes	0.09	0.08	0.12	0.22	0.58	1.16	1.56	1.61	1.66
Sheep	1.87	2.81	3.70	8.11	7.91	9.34	10.53	10.65	10.78
Goats	1.16	2.36	3.24	5.32	6.83	8.60	11.28	11.59	11.91
Camels	0.47	1.01	1.16	1.91	1.85	2.08	2.08	2.08	2.08
Horses	0.03	0.06	0.08	0.10	0.15	0.2	0.21	0.21	0.21
Asses	0.36	0.62	0.89	1.35	1.39	1.72	1.90	1.91	1.93
Mules	0.001	0.003	0.003	0.014	0.021	0.021	0.021	0.021	0.025
Poultry	0.01	0.04	0.07	0.12	0.16	0.21	0.42	0.45	0.48
Total	6.36	8.77	11.77	21.39	23.83	31.60	39.91	40.89	41.90

3.3. Potential of Biogas Production from the Utilization of Biodigester Technology

Table 4 shows the regional increase in biogas production from 1960 to 2018. It is found that 417.3 Mt of manure possesses the potential of producing 26,871.35 Mm^3 of biogas. Due to the province having the highest population and manure production, Punjab is leading in biogas generation with a 53.92% share, followed by Sindh, KPK, and Balochistan, with 24.64%, 14.45%, and 6.97% shares in total biogas generation, respectively. Moreover, large animals such as cattle and buffaloes showed the highest biogas production potential with 46.83% and 39.70% shares of the total biogas generation, respectively. The other large animals such as camels, horses, asses, and mules revealed 1.41%, 0.35%, 4.89%, and 0.17% shares in the total biogas production, respectively. At the same time, smaller animals such as goats and sheep were found to have 4% and 1.64% shares, respectively. Similarly, poultry contributes 0.97%. In comparison to other agricultural countries, Pakistan leads in the biogas production potential (26,871.35 Mm^3/year), followed by Iran (16,146.35 Mm^3/year), Malaysia (4589.5 Mm^3/year), and Turkey (2180 Mm^3/year), as shown in Figure 2 [25,39,46–48].

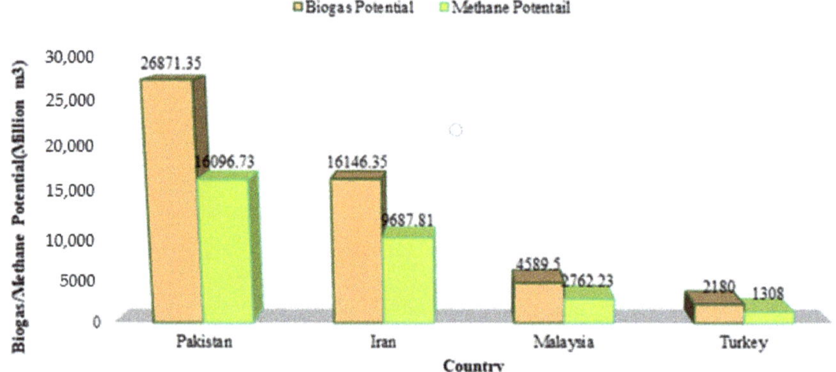

Figure 2. Biogas and methane potential of Pakistan in comparison to Iran, Turkey, and Malaysia.

3.4. Potential of Methane Production from the Utilization of Biodigester Technology

The methane production potential using farm animal manure in Pakistan is shown in Table 5. The results proclaim that the total methane production potential in 2018 showed the highest amount of 16,096.73 Mm^3. The methane production potential in 2018 was estimated to be 2.45, 2.35, 2.12, 1.41, 0.96, 0.45, 0.06, and 0.03 times higher than the methane production potential in 1960, 1972, 1976, 1986, 1996, 2006, 2016, and 2017, respectively. Punjab had the highest methane potential with a 53.95% share, while Sindh, KPK, and Balochistan had 24.65, 14.41, and 6.97% shares in the total methane production potential, respectively. In comparison, it was found from cited studies that the potential of methane generation from livestock manure in Iran, Canada, Malaysia, Turkey, and Indonesia was 5160, 2310, 2289, 1308, and 5758 Mm^3 $year^{-1}$, respectively [25,39,46–48].

Table 4. Potential of biogas generation from livestock manure in Pakistan from 1960 to 2018 (Mt/year).

Animal Type	1960	1972	1976	1986	1996	2006	2016	2017	2018
Punjab									
Cattle	2647.98	2251.86	2219.56	2413.65	2568.32	3945.28	5701.11	5914.91	6136.65
Buffaloes	1677.81	2029.30	2184.25	3052.31	3586.39	4858.24	6528.93	6724.94	6926.42
Sheep	80.69	90.77	116.16	96.63	88.77	91.95	103.60	104.84	106.10
Goats	42.97	85.90	112.26	155.45	221.16	286.63	375.96	386.29	396.92
Camels	109.22	149.87	138.79	131.81	76.78	81.71	81.71	81.71	81.71
Horses	61.86	72.27	78.29	67.06	49.54	44.62	44.62	44.62	44.62
Asses	245.55	290.99	311.80	453.60	533.26	611.01	674.79	681.63	688.48
Mules	6.29	5.47	7.93	9.85	15.60	17.24	17.24	17.24	17.24
Poultry	10.12	13.65	21.66	43.77	38.52	40.72	80.10	85.71	91.71
Total	4882.53	4990.12	5190.73	6424.17	7178.39	9977.43	13,608.1	14,041.94	14,489.88
Sindh									
Cattle	803.73	766.5	781.28	1060.5	1550.52	1895.71	2739.41	2842.07	2948.56
Buffaloes	370.38	416.64	502.05	881.47	1537.10	2009.32	2650.99	2781.3	2864.79
Sheep	22.98	12.14	26.43	37.81	53.62	57.22	64.46	65.24	66.02
Goats	31.81	32.88	61.24	97.63	140.69	181.71	238.34	244.89	251.62
Camels	25.45	32.85	59.13	89.51	92.39	114.15	114.15	114.15	114.15
Horses	10.95	19.43	25.73	20.80	17.24	12.31	12.31	12.31	12.31
Asses	43.52	66.24	102.10	136.87	189.98	274.84	303.58	306.6	309.61
Mules	0.27	0.54	0.82	1.36	3.28	5.47	6.02	6.02	6.29
Poultry	1.96	4.31	9.89	13.82	18.15	22.21	43.70	46.76	50.04
Total	1311.08	1351.56	1568.70	2339.82	3603	4572.99	6173.01	6419.37	6623.43
KPK									
Cattle	877.64	810.84	821.25	899.26	1159.87	1633.74	2360.82	2449.24	2541.22
Buffaloes	178.21	216.53	208.59	347.93	381.88	527.79	709.28	730.36	752.26
Sheep	35.15	35.48	53.11	23.11	40.77	48.60	54.76	55.41	56.08
Goats	43.86	54.01	67.73	41.90	97.76	138.74	181.97	186.97	192.12
Camels	31.20	41.47	39	28.74	26.69	26.28	26.69	27.10	27.10
Horses	6.29	8.48	7.93	9.30	12.86	20.80	22.17	22.72	23.26
Asses	83.76	111.69	104.29	122.09	146.18	153.3	169.17	170.82	172.73
Mules	5.20	8.76	7.66	6.29	16.42	18.34	20.25	20.53	20.80
Poultry	6.58	7.76	15.25	27.04	35.36	43.53	85.63	91.63	98.04
Total	1267.93	1295.05	1324.86	1482.58	1917.83	2611.14	3630.78	3754.80	3883.64
Balochistan									
Cattle	176.02	131.94	187.24	316.72	367.09	617.03	891.60	925	959.49
Buffaloes	7.11	6.02	9.03	17.24	44.07	87.6	117.71	120.99	124.83
Sheep	37.06	55.77	73.35	160.59	156.69	185.06	208.51	211.01	213.54
Goats	23.06	46.80	64.19	105.49	135.41	170.34	223.41	229.55	235.87
Camels	35.31	75.96	87.05	143.30	139.20	156.03	156.03	156.03	156.03
Horses	2.73	5.20	6.29	7.93	11.77	16.42	16.42	16.42	16.42
Asses	27.10	46.81	66.79	101.28	104.84	129.21	142.62	143.99	145.36
Mules	0.10	0.27	0.27	1.09	1.64	1.64	1.64	1.64	1.91
Poultry	0.71	1.85	3.07	5.17	7.28	9.29	18.27	19.55	20.92
Total	309.24	370.66	497.31	858.88	968.03	1372.64	1776.25	1824.22	1874.40

Table 5. Potential of methane generation from livestock manure in Pakistan from 1960 to 2018 (Mt/year).

Animal Type	1960	1972	1976	1986	1996	2006	2016	2017	2018
					Punjab				
Cattle	1588.79	1351.12	1331.73	1448.19	1540.99	2367.17	3420.67	3548.95	3681.99
Buffaloes	1006.68	1217.58	1310.55	1831.38	2151.83	2914.94	3917.36	4034.96	4155.85
Sheep	48.41	54.46	69.70	57.98	53.26	55.17	62.16	62.90	63.66
Goats	25.78	51.54	67.35	93.27	132.6	171.98	225.57	231.77	238.15
Camels	65.53	89.92	83.27	79.08	46.07	49.02	49.02	49.02	49.02
Horses	37.12	43.36	46.97	40.24	29.72	26.77	26.77	26.77	26.77
Asses	147.33	174.59	187.08	272.16	319.95	366.60	404.87	408.98	413.08
Mules	3.77	3.28	4.76	5.91	9.36	10.34	10.34	10.34	10.34
Poultry	5.06	6.82	10.83	21.88	19.26	20.36	40.05	42.85	45.85
Total	2928.50	2992.70	3112.27	3850.12	4303.18	5982.38	8156.85	8416.59	8684.75
					Sindh				
Cattle	482.23	459.9	468.76	636.30	930.31	1137.43	1643.65	1705.24	1769.13
Buffaloes	222.23	249.98	301.23	528.88	922.26	1205.59	1590.59	1668.78	1718.87
Sheep	13.78	7.28	15.86	22.68	32.17	34.33	38.67	39.14	39.61
Goats	19.08	19.72	36.74	58.58	84.41	109.02	143.00	146.93	150.97
Camels	15.27	19.71	35.47	53.70	55.43	68.49	68.49	68.49	68.49
Horses	6.57	11.66	15.43	12.48	10.34	7.39	7.39	7.39	7.39
Asses	26.11	39.74	61.26	82.12	113.98	164.90	182.15	183.96	185.76
Mules	0.16	0.32	0.49	0.82	1.97	3.28	3.61	3.61	3.77
Poultry	0.98	2.15	4.94	6.91	9.07	11.10	21.85	23.38	25.02
Total	786.45	810.50	940.23	1402.51	2159.98	2741.57	3699.43	3846.94	3969.05
					KPK				
Cattle	526.58	486.50	492.75	539.56	695.92	980.24	1416.49	1469.54	1524.73
Buffaloes	106.92	129.92	125.15	208.76	229.12	316.67	425.57	438.21	451.35
Sheep	21.09	21.29	31.87	13.86	24.46	29.16	32.85	33.24	33.64
Goats	26.32	32.40	40.63	25.14	58.66	83.24	109.18	112.18	115.27
Camels	18.72	24.88	23.40	17.24	16.01	15.76	16.01	16.26	16.26
Horses	3.77	5.09	4.76	5.58	7.71	12.48	13.30	13.63	13.96
Asses	50.26	67.01	62.57	73.25	87.70	91.98	101.50	102.49	103.64
Mules	3.12	5.25	4.59	3.77	9.85	11.00	12.15	12.31	12.48
Poultry	3.29	3.88	7.62	13.52	17.68	21.76	42.81	45.81	49.02
Total	760.10	776.25	793.39	900.71	1147.16	1562.33	2169.90	2243.72	2320.38
					Balochistan				
Cattle	105.61	79.16	112.34	190.03	220.25	370.21	534.96	555.00	575.69
Buffaloes	4.27	3.61	5.42	10.34	26.44	52.56	70.62	72.59	74.89
Sheep	22.23	33.46	44.01	96.35	94.01	111.04	125.10	126.60	128.12
Goats	13.84	28.08	38.51	63.29	81.25	102.20	134.04	137.73	141.52
Camels	21.18	45.57	52.23	85.98	83.52	93.62	93.62	93.62	93.62
Horses	1.64	3.12	3.77	4.76	7.06	9.85	9.85	9.85	9.85
Asses	16.26	28.08	40.07	60.77	62.90	77.52	85.57	86.39	87.21
Mules	0.065	0.16	0.16	0.65	0.98	0.98	0.98	0.98	1.14
Poultry	0.35	0.92	1.53	2.58	3.64	4.64	9.13	9.77	10.46
Total	185.47	222.21	298.08	514.81	580.09	822.65	1063.92	1092.57	1122.55

The gradual increase in methane production from livestock manure will reduce the energy imports of the country, which are currently at 34%, and the government is spending about USD 1.27 billion annually on these imports.

3.5. Potential of Heat Energy and Electricity Production from Biodigester Technology

Figure 3 shows that the potential of heat energy acquired from the burning of methane in 2018 was 492.6 PJ. Compared to 1960, heat energy increased 245.46% in 2018 due to the higher livestock manure production, thereby exalting methane production. On the regional scale, Punjab yielded more heating energy (266 PJ/year), followed by Sindh (121.71 PJ/year), KPK (510.5 PJ/year), and Balochistan (34.4 PJ/year). Consequently, Pakistan is leading in heat energy production from methane as compared to Malaysia, Turkey, and Iran due to the high livestock population [25,39,46–49]. In addition, the potential of heat energy produced in Pakistan is higher than in Canada [25]. Similarly, the potential of electricity generation from biogas was computed and is showcased in Figure 4. The highest potential of electricity generation by manure-based biogas obtained was 5521.5 MW in 2018. This value accounts for ~22% of the country's electricity requirement which is an indication of the considerable energy share from livestock waste. A similar study was conducted in Canada which showed that biogas electricity could fulfill ~22% of the country's electricity demands using agricultural waste such as wood waste and municipal solid waste [25].

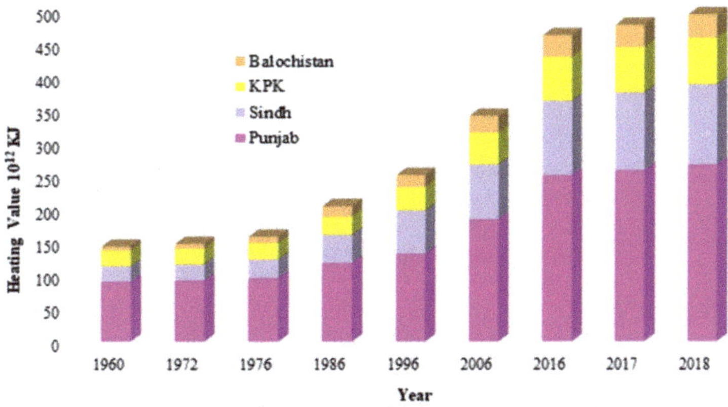

Figure 3. Potential of heat energy obtained from the methane produced by livestock manure in different provinces of Pakistan.

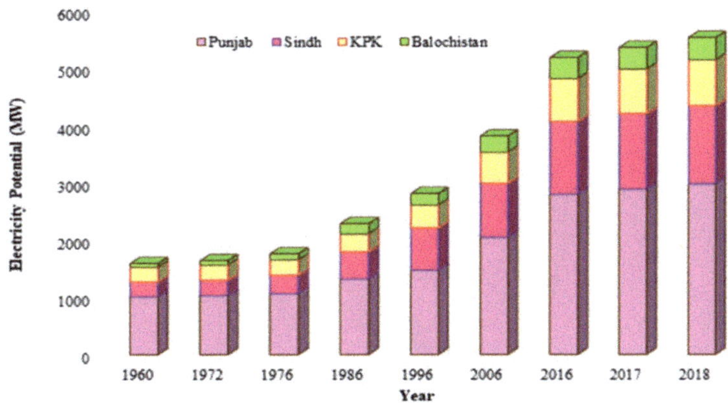

Figure 4. Potential of electricity generation from manure-based biogas in different provinces of Pakistan.

Punjab province had the highest electricity generation potential, with a value of 2977.3 MW, followed by Sindh, KPK, and Balochistan, with values of 1360.9, 798.0, and 385.1 MW, respectively. Furthermore, Punjab province had the highest electricity generation potential in 2018, contributing 54% of the total electricity generation, followed by Sindh, KPK, and Balochistan, with 25%, 14%, and 7% shares in electricity generation, respectively, as shown in Figure 5.

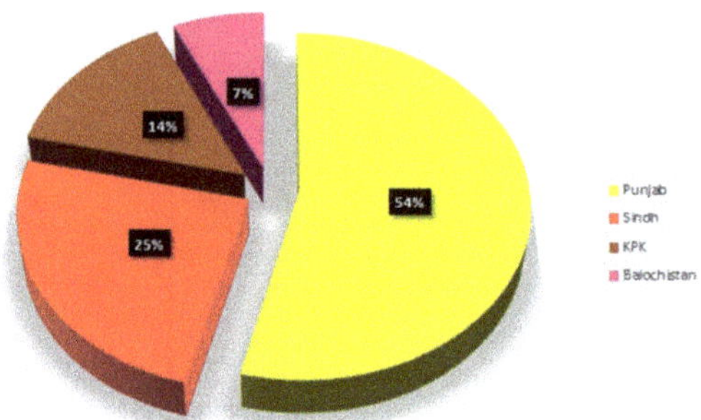

Figure 5. Province-wise share of biogas-based electricity.

Figures 6 and 7 illustrate the potential of electricity generation by manure-based biogas from different livestock animals. It is found that large ruminants, namely, cattle and buffaloes, had the highest potential for electricity generation, followed by asses and goats. The potential of electricity generated from cattle manure-based biogas had the maximum share, meaning that in 2018, it had a percentage value of 47%, followed by buffaloes, asses, goats, sheep, camels, poultry, horses, and mules, with percentage values of 40%, 5%, 4%, 2%, 1%, 1%, 0.35%, and 0.17%, respectively. It has previously been found that the potential of electricity generation by manure-based biogas in Malaysia, Turkey, and Iran could be 944 MW year^{-1}, 448 MW year^{-1}, and 3317 MW year^{-1}, respectively [25,39,46–48].

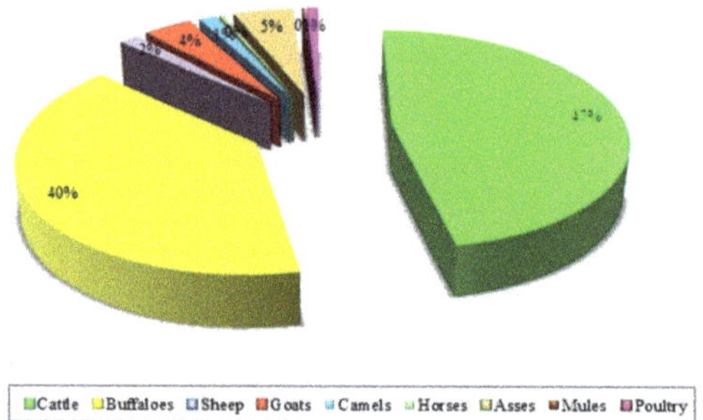

Figure 6. Share of selected livestock species in biogas-based electricity.

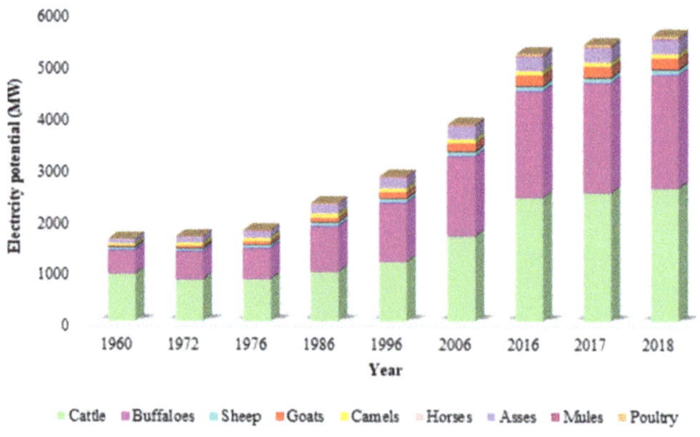

Figure 7. Potential of electricity generation from manure-based biogas by different farm animals in Pakistan.

3.6. Feasibility and Suitability of Biodigester Technology in Pakistan

Studies conducted in many countries such as China [50], Malaysia [47], Turkey [39], Brazil [51], Serbia [52], Ecuador [42], Nepal [53], Indonesia [54], and Ethiopia [55] have indicated that biodigester technology is becoming popular because of its user-friendliness, cost-effectiveness, and robustness.

The prevailing energy crises of Pakistan can be eliminated by the appropriate development of biodigester technology. Biodigester technology in Pakistan has been considered over recent decades. In this regard, the first biogas plant was installed in Sindh in 1959 [19,56]. The government of Pakistan focused on the development of biodigester technology during the year 1974; the Pakistan Council for Appropriate Technology (PCAT) constructed 31 fixed dome digesters in different areas of Pakistan. Figure 8 depicts the biodigesters installed by different organizations in Pakistan during 1974–2015 [56].

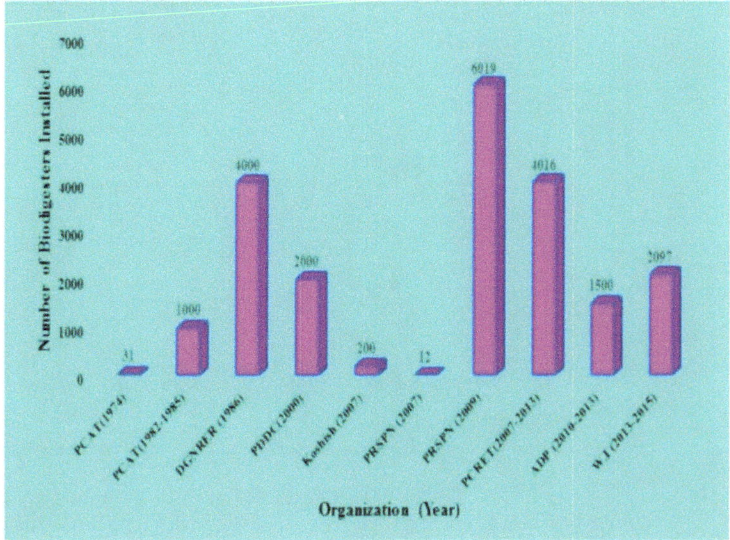

Figure 8. The number of biodigesters that have already been installed by different organizations in Pakistan between 1974 and 2015.

According to Ghimire and Nepal, 2009 [56], many factors can drastically affect the potential of biodigesters, including technical factors, economic and financial factors, social factors, and institutional factors. Figure 9 shows all the main factors mentioned along with their classification. These inhibiting factors could be minimized if special attention is paid during the program implementation phase.

Most parts of Pakistan have favorable conditions for biodigesters. It is clear from the country's livestock population that most of these animals are found in Punjab, Sindh, and KPK. It has been estimated that 10 million households are involved in raising livestock. In most parts of Punjab and Sindh, the temperature is favorable for the production of biogas. Construction materials and the labor force are easily available. Moreover, the land for installing biogas plants is not a problem for most farmers in Pakistan. However, about 30% of farmers in the country do not have favorable conditions for installing biodigesters due to the non-availability of land or harsh temperature conditions [56,57].

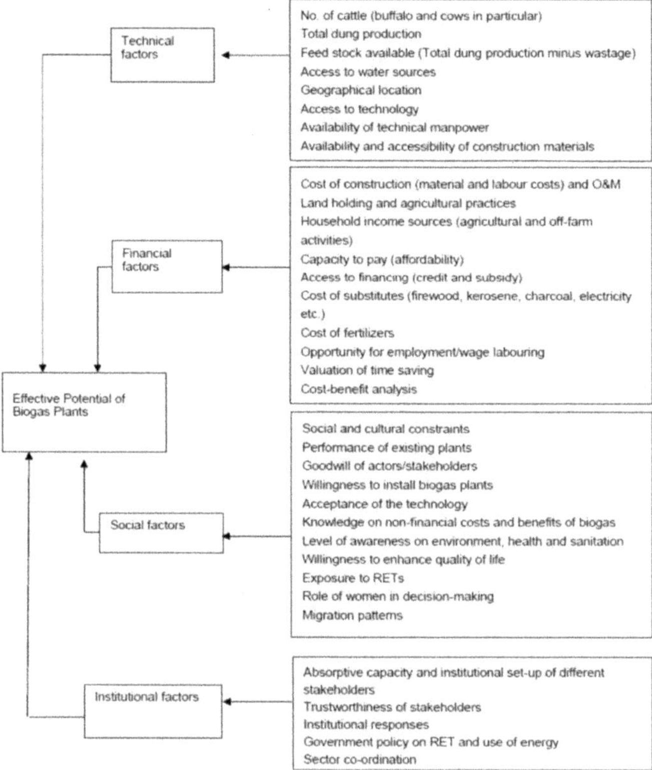

Figure 9. Factors affecting biodigester technology development in Pakistan.

From Table 6, it is found that Pakistan has a capacity of 5 million biodigesters which can be easily installed in different farming areas. In this regard, the annual increase in the livestock population (as presented in Table 2) indicates a promising technology for biodigester development, especially in rural areas of Pakistan.

Table 6. Potential for biogas plants in Pakistan [56].

Particulars	Number of Households
Total number of households that have livestock animals	10 million
Households with only one cattle or buffalo which are technically not feasible for installing biodigesters	2 million
Households having no potential for biodigesters due to various factors such as temperature and competitiveness of biogas	3 million
Total number of households having potential for biodigester installation	5 million

4. Conclusions

This study accentuates the livestock manure production potential and its utilization in different areas of Pakistan. This study found that livestock manure is a sustainable bioresource for energy generation in Pakistan. The highest population of livestock in Pakistan is found in Punjab province, followed by Sindh, KPK, and Balochistan. Livestock is mainly managed in almost 10 million households. The total potential of animal manure in the country for 2018 was 417.3 Mt, and 26,871.35 million m^3 of biogas, 492.6 PJ of heat energy, and 5521.5 MW of electricity could potentially be produced from animal manure in 2018 to reduce the ongoing energy crises in Pakistan. Moreover, there are ample opportunities to harness biodigester technologies in Pakistan because of the space available for installing 5 million biodigesters in different farming areas. Considering the huge potential of biodigester technology, the country has a high need for the development and implementation of national programs focusing on disseminating domestic biodigesters in Pakistan.

Author Contributions: Conceptualization, M.U.K. and M.S.; data curation, M.U.K., M.A., and A.Z.; formal analysis, M.U.K., M.A., I.S., and P.C.G.; funding acquisition, M.S. and I.S.; investigation, P.C.G., A.Z., A.S., M.F., U.S., and M.Y.; methodology, M.U.K., M.A., M.S., I.S., A.Z., U.S., and P.A.; project administration, M.U.K. and M.S.; resources, M.U.K., P.C.G., M.F., U.S., and P.A.; software, M.U.K., M.A., A.S., M.F., and M.Y.; supervision, M.U.K. and M.S.; validation, M.U.K., M.A., M.S., I.S., and A.Z.; visualization, I.S., P.C.G., A.Z., A.S., P.A., and M.Y.; writing—original draft, M.U.K.; writing—review and editing, M.A., M.S., I.S., P.C.G., A.Z., A.S., M.F., U.S., P.A., and M.Y. All authors have read and agreed to the published version of the manuscript.

Funding: This research received no external funding.

Institutional Review Board Statement: This study was conducted according to the guidelines of the Declaration of Helsinki and approved by the Institutional Review Board of the University of Agriculture, Faisalabad, Pakistan.

Data Availability Statement: Not applicable.

Acknowledgments: This work was supported by the Higher Education Commission (HEC), Pakistan [No:21-1830/SRGP/R&D/HEC/2018]. Publishing fees were supported by the funding program *Open Access Publishing* of Hamburg University of Technology (TUHH), Hamburg, Germany. Moreover, the authors express their gratitude towards Washington State University, Peking University, China, and the University of Agriculture Faisalabad, Pakistan, for providing all opportunities to conduct this research.

Conflicts of Interest: The authors declare no conflict of interest.

References

1. Uddin, W.; Zeb, K.; Haider, A.; Khan, B.; Islam, S.U.; Ishfaq, M.; Khan, I.; Adil, M.; Kim, H.J. Current and future prospects of small hydro power in Pakistan: A survey. *Energy Strat. Rev.* **2019**, *24*, 166–177. [CrossRef]
2. Shakeel, S.R.; Takala, J.; Shakeel, W. Renewable energy sources in power generation in Pakistan. *Renew. Sustain. Energy Rev.* **2016**, *64*, 421–434. [CrossRef]
3. Younas, U.; Khan, B.; Ali, S.; Arshad, C.; Farid, U.; Zeb, K.; Rehman, F.; Mehmood, C.A.; Vaccaro, A. Pakistan geothermal renewable energy potential for electric power generation: A survey. *Renew. Sustain. Energy Rev.* **2016**, *63*, 398–413. [CrossRef]
4. Kessides, I.N. Chaos in power: Pakistan's electricity crisis. *Energy Policy* **2013**, *55*, 271–285. [CrossRef]
5. Grainger, C.A.; Zhang, F. Electricity shortages and manufacturing productivity in Pakistan. *Energy Policy* **2019**, *132*, 1000–1008. [CrossRef]
6. Baloch, M.H.; Chauhdary, S.T.; Ishak, D.; Kaloi, G.S.; Nadeem, M.H.; Wattoo, W.A.; Younas, T.; Hamid, H.T. Hybrid energy sources status of Pakistan: An optimal technical proposal to solve the power crises issues. *Energy Strat. Rev.* **2019**, *24*, 132–153. [CrossRef]
7. Arshad, N.; Ali, U. An analysis of the effects of residential uninterpretable power supply systems on Pakistan's power sector. *Energy Sustain. Dev.* **2017**, *36*, 16–21. [CrossRef]
8. Hou, Y.; Iqbal, W.; Shaikh, G.M.; Iqbal, N.; Solangi, Y.A.; Fatima, A. Measuring Energy Efficiency and Environmental Performance: A Case of South Asia. *Processes* **2019**, *7*, 325. [CrossRef]
9. Aslam, W.; Soban, M.; Akhtar, F.; Zaffar, N.A. Smart meters for industrial energy conservation and efficiency optimization in Pakistan: Scope, technology and applications. *Renew. Sustain. Energy Rev.* **2015**, *44*, 933–943. [CrossRef]
10. Naqvi, S.R.; Jamshaid, S.; Naqvi, M.R.; Farooq, W.; Niazi, M.B.; Aman, Z.; Zubair, M.; Ali, M.; Shahbaz, M.; Inayat, A.; et al. Potential of biomass for bioenergy in Pakistan based on present case and future perspectives. *Renew. Sustain. Energy Rev.* **2018**, *81*, 1247–1258. [CrossRef]
11. Khan, M.A.; Abbas, F. The dynamics of electricity demand in Pakistan: A panel cointegration analysis. *Renew. Sustain. Energy Rev.* **2016**, *65*, 1159–1178. [CrossRef]
12. Shahzad, K.; Bajwa, S.U.; Ansted, R.B.; Mamoon, D.; Rehman, K.-U. Evaluating human resource management capacity for effective implementation of advanced metering infrastructure by electricity distribution companies in Pakistan. *Util. Policy* **2016**, *41*, 107–117. [CrossRef]

13. Sher, H.; Murtaza, A.F.; Addoweesh, K.E.; Chiaberge, M. Pakistan's progress in solar PV based energy generation. *Renew. Sustain. Energy Rev.* **2015**, *47*, 213–217. [CrossRef]

14. Wasti, S. *Economic Survey of Pakistan 2014–2015*; Government of Pakistan: Islamabad, Pakistan, 2015.

15. Rehman, S.A.U.; Cai, Y.; Fazal, R.; Das Walasai, G.; Mirjat, N.H. An integrated modeling approach for forecasting long-term energy demand in Pakistan. *Energies* **2017**, *10*, 1868. [CrossRef]

16. Qazi, U.; Jahanzaib, M.; Ahmad, W.; Hussain, S. An institutional framework for the development of sustainable and competitive power market in Pakistan. *Renew. Sustain. Energy Rev.* **2017**, *70*, 83–95. [CrossRef]

17. Zuberi, M.J.S.; Torkmahalleh, M.A.; Ali, S.H. A comparative study of biomass resources utilization for power generation and transportation in Pakistan. *Int. J. Hydrogen Energy* **2015**, *40*, 11154–11160. [CrossRef]

18. Kamran, M. Current status and future success of renewable energy in Pakistan. *Renew. Sustain. Energy Rev.* **2018**, *82*, 609–617. [CrossRef]

19. Uddin, W.; Khan, B.; Shaukat, N.; Majid, M.; Mujtaba, G.; Mehmood, C.A.; Ali, S.; Younas, U.; Anwar, M.; Almeshal, A.M. Biogas potential for electric power generation in Pakistan: A survey. *Renew. Sustain. Energy Rev.* **2016**, *54*, 25–33. [CrossRef]

20. Rahut, D.B.; Ali, A.; Mottaleb, K.A.; Aryal, J.P. Wealth, education and cooking-fuel choices among rural households in Pakistan. *Energy Strat. Rev.* **2019**, *24*, 236–243. [CrossRef]

21. Malik, S.N.; Sukhera, O.R. Management of natural gas resources and search for alternative renewable energy resources: A case study of Pakistan. *Renew. Sustain. Energy Rev.* **2012**, *16*, 1282–1290. [CrossRef]

22. Imran, M.; Özçatalbaş, O.; Bakhsh, K. Rural household preferences for cleaner energy sources in Pakistan. *Environ. Sci. Pollut. Res.* **2019**, *26*, 22783–22793. [CrossRef]

23. Yasmin, N.; Grundmann, P. Adoption and diffusion of renewable energy—The case of biogas as alternative fuel for cooking in Pakistan. *Renew. Sustain. Energy Rev.* **2019**, *101*, 255–264. [CrossRef]

24. Ashraf, S.; Luqman, M.; Hassan, Z.Y.; Yaqoob, A. Determinants of Biogas Technology Adoption in Pakistan. *Pak. J. Sci. Ind. Res. Ser. A Phys. Sci.* **2019**, *62*, 113–123. [CrossRef]

25. Levin, D.B.; Zhu, H.; Beland, M.; Cicek, N.; Holbein, B.E. Potential for hydrogen and methane production from biomass residues in Canada. *Bioresour. Technol.* **2007**, *98*, 654–660. [CrossRef]

26. Kafle, G.K.; Kim, S.H. Effects of chemical compositions and ensiling on the biogas productivity and degradation rates of agricultural and food processing by-products. *Bioresour. Technol.* **2013**, *142*, 553–561. [CrossRef]

27. TR, Y.S.S.; Kohli, S.; Rana, V. Enhancement of biogas production from solid substrates using different techniques. *J. Bioresour. Technol.* **2004**, *95*, 1–10. [CrossRef]

28. Mushtaq, K.; Zaidi, A.A.; Askari, S.J. Design and performance analysis of floating dome type portable biogas plant for domestic use in Pakistan. *Sustain. Energy Technol. Assess.* **2016**, *14*, 21–25. [CrossRef]

29. Khan, M.U.; Ahring, B.K. Improving the biogas yield of manure: Effect of pretreatment on anaerobic digestion of the recalcitrant fraction of manure. *Bioresour. Technol.* **2021**, *321*, 124427. [CrossRef]

30. Khan, M.U.; Ahring, B.K. Anaerobic Digestion of Digested Manure Fibers: Influence of Thermal and Alkaline Thermal Pretreatment on the Biogas Yield. *BioEnergy Res.* **2020**, 1–10. [CrossRef]

31. Lee, J.T.; Khan, M.U.; Dai, Y.; Tong, Y.W.; Ahring, B.K. Influence of wet oxidation pretreatment with hydrogen peroxide and addition of clarified manure on anaerobic digestion of oil palm empty fruit bunches. *Bioresour. Technol.* **2021**, *332*, 125033. [CrossRef] [PubMed]

32. Khan, M.U.; Ahring, B.K. Anaerobic digestion of biorefinery lignin: Effect of different wet explosion pretreatment conditions. *Bioresour. Technol.* **2020**, *298*, 122537. [CrossRef]

33. Amjid, S.S.; Bilal, M.Q.; Nazir, M.S.; Hussain, A. Biogas, renewable energy resource for Pakistan. *Renew. Sustain. Energy Rev.* **2011**, *15*, 2833–2837. [CrossRef]

34. Raheem, A.; Hassan, M.Y.; Shakoor, R. Bioenergy from anaerobic digestion in Pakistan: Potential, development and prospects. *Renew. Sustain. Energy Rev.* **2016**, *59*, 264–275. [CrossRef]

35. Shaukat, N.; Khan, B.; Khan, T.; Younis, M.N.; ul Faris, N.; Javed, A.; Iqbal, M.N. A comprehensive review of biogas sector for electric power generation in Pakistan. *PSM Biol. Res.* **2016**, *1*, 43–48.

36. Pakistan Livestock Census. Pakistan Bureau of Statistics, Government of Pakistan. 2006. Available online: https://www.pbs.gov.pk/content/pakistan-livestock-census-2006 (accessed on 10 June 2019).

37. Livestock Census Report 2006, Ministry of National Food Security & Research, Ismamabad. 2006. Available online: https://phkh.nhsrc.pk/sites/default/files/2019-06/All%20Pakistan%20Report%20Livestock%20Census%2006.pdf (accessed on 9 June 2019).

38. Ministry of Finance, Government of Pakistan. Pakistan Economic Survey 2016–2017. Available online: http://www.finance.gov.pk/survey_1617.html (accessed on 10 June 2019).

39. Avcioğlu, A.O.; Türker, U. Status and potential of biogas energy from animal wastes in Turkey. *Renew. Sustain. Energy Rev.* **2012**, *16*, 1557–1561. [CrossRef]

40. Vedrenne, F.; Béline, F.; Dabert, P.; Bernet, N. The effect of incubation conditions on the laboratory measurement of the methane producing capacity of livestock wastes. *Bioresour. Technol.* **2008**, *99*, 146–155. [CrossRef]

41. Liu, G.; Zhang, R.; El-Mashad, H.M.; Dong, R. Effect of feed to inoculum ratios on biogas yields of food and green wastes. *Bioresour. Technol.* **2009**, *100*, 5103–5108. [CrossRef]

42. Cornejo, C.; Wilkie, A.C. Greenhouse gas emissions and biogas potential from livestock in Ecuador. *Energy Sustain. Dev.* **2010**, *14*, 256–266. [CrossRef]
43. Nasir, I.M.; Ghazi, T.I.M.; Omar, R.; Idris, A. Anaerobic digestion of cattle manure: Influence of inoculums concentration. *Int. J. Eng. Technol.* **2013**, *10*, 22–26.
44. Igliński, B.; Buczkowski, R.; Cichosz, M. Biogas production in Poland—Current state, potential and perspectives. *Renew. Sustain. Energy Rev.* **2015**, *50*, 686–695. [CrossRef]
45. Rahimnejad, M.; Adhami, A.; Darvari, S.; Zirepour, A.; Oh, S.-E. Microbial fuel cell as new technology for bioelectricity generation: A review. *Alex. Eng. J.* **2015**, *54*, 745–756. [CrossRef]
46. Maghanaki, M.M.; Ghobadian, B.; Najafi, G.; Galogah, R.J. Potential of biogas production in Iran. *Renew. Sustain. Energy Rev.* **2013**, *28*, 702–714. [CrossRef]
47. Abdeshahian, P.; Lim, J.S.; Ho, W.S.; Hashim, H.; Lee, C.T. Potential of biogas production from farm animal waste in Malaysia. *Renew. Sustain. Energy Rev.* **2016**, *60*, 714–723. [CrossRef]
48. Khalil, M.; Berawi, M.A.; Heryanto, R.; Rizalie, A. Waste to energy technology: The potential of sustainable biogas production from animal waste in Indonesia. *Renew. Sustain. Energy Rev.* **2019**, *105*, 323–331. [CrossRef]
49. Saidmamatov, O.; Rudenko, I.; Baier, U.; Khodjaniyazov, E. Challenges and Solutions for Biogas Production from Agriculture Waste in the Aral Sea Basin. *Processes* **2021**, *9*, 199. [CrossRef]
50. Cheng, S.; Li, Z.; Mang, H.-P.; Huba, E.-M. A review of prefabricated biogas digesters in China. *Renew. Sustain. Energy Rev.* **2013**, *28*, 738–748. [CrossRef]
51. Coimbra-Araújo, C.H.; Mariane, L.; Júnior, C.B.; Frigo, E.P.; Frigo, M.S.; Araújo, I.R.C.; Alves, H.J. Brazilian case study for biogas energy: Production of electric power, heat and automotive energy in condominiums of agroenergy. *Renew. Sustain. Energy Rev.* **2014**, *40*, 826–839. [CrossRef]
52. Cvetkovic, S.; Radoičić, T.K.; Vukadinović, B.; Kijevčanin, M. Potentials and status of biogas as energy source in the Republic of Serbia. *Renew. Sustain. Energy Rev.* **2014**, *31*, 407–416. [CrossRef]
53. Katuwal, H.; Bohara, A.K. Biogas: A promising renewable technology and its impact on rural households in Nepal. *Renew. Sustain. Energy Rev.* **2009**, *13*, 2668–2674. [CrossRef]
54. Putra, A.; Liu, Z.; Lund, M. The impact of biogas technology adoption for farm households—Empirical evidence from mixed crop and livestock farming systems in Indonesia. *Renew. Sustain. Energy Rev.* **2017**, *74*, 1371–1378. [CrossRef]
55. Mengistu, M.; Simane, B.; Eshete, G.; Workneh, T. A review on biogas technology and its contributions to sustainable rural livelihood in Ethiopia. *Renew. Sustain. Energy Rev.* **2015**, *48*, 306–316. [CrossRef]
56. Ghimire, P.C. *Final Report on Technical Study of Biogas Plants Installed in Pakistan*; Prepared by: Asia/Africa Biogas Programme, Netherlands Development Organisation (SNV). 2007, pp. 1–74. Available online: https://bibalex.org/baifa/Attachment/Documents/172360.pdf (accessed on 14 June 2021).
57. Farooq, M.K.; Kumar, S. An assessment of renewable energy potential for electricity generation in Pakistan. *Renew. Sustain. Energy Rev.* **2013**, *20*, 240–254. [CrossRef]

sustainability

Article

Development of a Low-Cost Biomass Furnace for Greenhouse Heating

Asif Ali [1,2], Tahir Iqbal [1,2,*], Muhammad Jehanzeb Masud Cheema [1,2], Arslan Afzal [1], Muhammad Yasin [1], Zia ul Haq [1], Arshad Mahmood Malik [2] and Khalid Saifullah Khan [3]

[1] Faculty of Agricultural Engineering and Technology, PMAS-Arid Agriculture University, Shamsabad 46000, Pakistan; miraniabei@yahoo.com (A.A.); mjm.cheema@gmail.com or mjm.cheema@uaar.edu.pk (M.J.M.C.); arslanafzal@uaar.edu.pk (A.A.); yasindga@gmail.com (M.Y.); ziaulhaquaf@gmail.com (Z.u.H.)

[2] National Center of Industrial Biotechnology, PMAS-Arid Agriculture University, Rawalpindi, Shamsabad 46000, Pakistan; arshadmm@uaar.edu.pk

[3] Institute of Soil Science, PMAS-Arid Agriculture University, Shamsabad 46000, Pakistan; khalidsaifullah@uaar.edu.pk

* Correspondence: tahir.iqbal@uaar.edu.pk; Tel.: +92-51-9292163

check for **updates**

Citation: Ali, A.; Iqbal, T.; Cheema, M.J.M.; Afzal, A.; Yasin, M.; Haq, Z.u.; Malik, A.M.; Khan, K.S. Development of a Low-Cost Biomass Furnace for Greenhouse Heating. *Sustainability* **2021**, *13*, 5152. https://doi.org/10.3390/su13095152

Academic Editors: Muhammad Sultan, Yuguang Zhou, Redmond R. Shamshiri and Aitazaz A. Farooque

Received: 18 March 2021
Accepted: 27 April 2021
Published: 5 May 2021

Publisher's Note: MDPI stays neutral with regard to jurisdictional claims in published maps and institutional affiliations.

Abstract: The energy crisis and increasing fossil fuel prices due to increasing demands, controlled supplies, and global political unrest have adversely affected agricultural productivity and farm profitability across the globe and Pakistan is not an exception. To cope with this issue of energy deficiency in agriculture, the best alternate strategy is to take advantage of biomass and solid waste potential. In low-income countries such as Pakistan, the greenhouse heating system mostly relies on fossil fuels such as diesel, gasoline, and LPG. Farmers are reluctant to adopt greenhouse farming due to the continuously rising prices of the fossil fuels. To reduce reliance on fossil fuel energy, the objective of this study was to utilize biomass from crop residues to develop an efficient and economical biomass furnace that could heat greenhouses to protect the crop from seasonal temperature effects. Modifications made to the biomass furnace, such as the incorporation of insulation around the walls of the furnace, providing turbulators in fire tubes, and a secondary heat exchanger (heat recovery system) in the chimney, have increased the thermal efficiency of the biomass furnace by about 21.7%. A drastic reduction in hazardous elements of flue gases was observed due to the addition of a water scrubber smoke filter in the exit line of the flue. The efficiency of the biomass furnace ranged from 50.42% to 54.18%, whereas the heating efficiency of the diesel-fired heater was 71.19%. On the basis of the equal heating value of the fuels, the unit material and operating costs of the biomass furnace for wood, cotton stalks, corn cobs, and cow dung were USD 2.04, 1.86, 1.78, and 2.00 respectively against USD 4.67/h for the diesel heater. The capital and operating costs of the biomass furnace were about 50% and 43.7% of the diesel heater respectively, resulting in a seasonal saving of about 1573 USD. The produced smoke was tested as environmental friendly under the prescribed limits of the National Environmental Quality Standards (NEQS), which shows potential for its large-scale adoption and wider applications.

Keywords: biomass; furnace; greenhouse; efficiency; economics

1. Introduction

Energy requirements in the future are certain to increase drastically with the ever-growing global population. In the coming years, more people will require the excess of energy that is presently available from different sources [1,2] Generally, energy is considered the main pillar of the economic growth of a country as most of the industries run on energy, which is playing remarkable role in socio-economic growth [3]. Energy is first and foremost a requirement for sustainable development. In terms of energy mix, Pakistan's energy sector heavily depends on thermal energy which consists of imported coal, local coal, re-gasified liquid natural gas, and natural gas that constitute about 58.4% of the total energy mix. The share of different sources such as hydroelectric, thermal, nuclear, and renewable is 30.9, 58.4, 8.2, and 2.4%, respectively. From the energy mix, the contribution of renewable energy needs to be enhanced as uncontrolled burning of fossil fuels is leading to increased environmental pollution by the release of greenhouse gases [4]. Moreover, a substantial potential for renewable energy is present in the country. All these aspects demand switching to renewable energy resources, which will help reduce the gap between energy supply and demand in Pakistan [5,6]. Table 1 shows that there is a large potential for biomass-based renewable energy in Pakistan. Renewable energy sources like solar, geothermal, and biomass are commonly used in greenhouse heating [7]. During the winter season, the increased demand for heat energy to keep the temperature of the greenhouses at the desirable level for crop production is essential. Hence, a proper greenhouse heating system is unavoidable for healthy and optimum crop production [8].

Table 1. Calculation of estimated annual surplus biomass production [9].

Crop Type	Estimated Crop Production (000 tons/year)	Crop Residue Ratio (CRR)	Estimated Biomass Production (000 tons/year)	Estimated Surplus Biomass (000 ton/year)
Sugarcane	65,257	0.12	7831	2552
Cotton	14,531	3.40	49,405	5039
Wheat	34,581	1.00	34,581	5689
Rice	16,754	1.00	16,754	6534
Maize	4260	1.25	5325	680
Total	135,383		113,896	20,494

Renewable energy sources like solar, geothermal, and biomass are commonly used in greenhouse heating [7]. To cater for the increasing energy demands, developed nations are continuously making efforts to explore alternate energy sources and coin new methods and technological innovations for energy conservation and efficiency improvement [10,11]. The literature has provided information about the application of heating and cooling devices for use in food and agriculture sectors in Portugal [12]. However, these studies do not provide the latest research and development framework and modalities that could help towards the design of a low-cost furnace or hot air generators for the agriculture sector, especially, when talking about remote locations. However, these studies discussed the utilization of renewable energy to provide hot air for the agriculture sector, in detail. Some studies suggested a methodology for analyzing the regional potential for developing biomass district heating systems based on forestry biomasses [13–15].

In consideration of a holistic and cost-effective approach, the overall energy price tag on greenhouses comes around 10–15% of its total cost of production. The profit margins of greenhouse farming have decreased due to rising energy costs, which have doubled over the last two decades [16]. Most greenhouse heating systems in Pakistan rely on electricity or fossil fuels, the prices of which have remained volatile and are continuously rising. Therefore, farmers are reluctant to adopt greenhouse farming due to decreasing profit margins. Higher energy costs in greenhouse farming have motivated the farmers to explore alternative means to reduce energy costs. Many growers use firewood for the heating of greenhouses, the cost of which is also rising. However, being an agrarian economy, Pakistan produces a large mass of crop residues annually [17,18]. The estimated production of major crop residues of cotton stalks, wheat straw, rice straw, sugarcane trash, and corn stalk in Pakistan are 49.4, 34.581, 16.75, 7.83, and 5.325 million tons per annum, respectively [2]. These crop residues are abundantly available in the country and require a viable strategy to be utilized in an efficient way, as compared to direct land filling and open air burning which is the current practice [2,19,20].

Generally, farmers are very cautious in exploring new sources of energy for their agricultural operations. Most farmers understand that firewood is the best option and low-cost source for thermal heating at the farm level [16]. Two types of biomass boilers and furnaces are currently used in the world on the basis of manual fuel feeding and automated feeding. Manually loaded boilers and furnaces are mostly run on waste wood whereas automatically fueled boilers run on different biomass sources like wood chips, biomass pellets, wood biomass, grain, and bagasse [21,22].

As per author information, there is no comprehensive study in Pakistan that has developed an indigenous waste fuel-based furnace in the agricultural sector for better and improved thermal applications. In this study, we developed a biomass/solid waste fuel furnace that can be helpful for thermal applications in the agriculture sector for better and improved production purpose. The objective of the study was to utilize biomass/solid waste in an efficient and economical way as a greenhouse heating system. The produced heat can be utilized for greenhouse heating to maintain optimal temperature during winter season.

2. Materials and Methods

2.1. Design Parameters

The biomass furnace for greenhouse heating was designed and developed at the Faculty of Agricultural Engineering, Pir Mehr Ali Shah Arid Agriculture University, Rawalpindi, Pakistan. The important design considerations included simple design, local manufacturing, light weight, portability, economics, and ease of operation. The 1st prototype model of the biomass furnace was installed in a 30.48×12.19 m^2 greenhouse tunnel for further testing and evaluation. Biomass furnaces work on the principle of a boiler, where the direct burning of biomass takes place in the burning chamber. The clean and hot air moves in a separate enclosure surrounding the hot air tubes. A typical biomass furnace consists of a combustion chamber, primary heat exchanger, chimney, secondary heat exchanger, water scrubber smoke filter, air distribution system, automatic air temperature control system, ash chamber, axial fan, blower for combustion, and temperature gauges. The basic design considerations for the design and development of the biomass furnace for greenhouse heating included:

1.   It should be simple, light weight, portable, and easy to operate.
2.   It can be manufactured using indigenous material and local technology.
3.   It should be affordable (economical) and efficient.

The important parameters for the design of a biomass furnace for greenhouse heating are described hereunder.

2.1.1. Equation (1): Volume of the Targeted Tunnel

$$\text{Volume} = \text{Length} \times \text{Width} \times \text{Height} \tag{1}$$

2.1.2. Equation (2): Energy Required to Heat the Targeted Tunnel

$$Q = m \times C_p \times \Delta t \tag{2}$$

where m = mass of air in the tunnel, Cp = specific heat of air (1.008 kJ kg^{-1} K^{-1}), and Δt = the difference in final and initial temperature inside the tunnel.

2.1.3. Equation (3): Biomass Required to Maintain the Required Heat in the Targeted Tunnel

$$\text{Biomass required} = \frac{\text{Heat Required}}{\text{Calorific value of biamass}} \tag{3}$$

2.1.4. Heat Exchanger Design

Equations (4) and (5): The heating surface required is computed by:

$$A = \frac{Q}{U \times \Delta tm} \tag{4}$$

$$Q = U \times A \times \Delta t_m \tag{5}$$

where A = heat transfer area (m^2), Q = heat transfer rate, kJ h^{-1}, U = overall heat transfer coefficient, kJ h^{-1} m^2 °C (for air 28.58 kJ h^{-1} m^2 °C), Δt_m = Log mean temperature difference (°C), which is given by Equation (6):

$$\Delta tm = \frac{(T1 - t2) - (T2 - t1)}{\ln \frac{(T1-t2)}{(T2-t1)}} \tag{6}$$

where T_1 = inlet fire tube temperature (°C), T_2 = outlet fire tube temperature (°C), t_1 = inlet shell side air temperature (°C), and t_2 = outlet shell side air temperature (°C).

Equation (7): The heat transfer rate can be measured as:

$$Q = m \times C_p \times \Delta t \tag{7}$$

where m = mass flow rate of air (kg hr^{-1}), and C_p and Δt are as specified above.

The schematic diagram of the biomass furnace is presented in Figure 1 and its isometric view is presented in Figure 2, whereas the connectivity of biomass furnace with the greenhouse tunnel is shown in Figure 3. The design final parameters of the biomass furnace are outlined in Table 2.

Table 2. Specification of biomass furnace.

Parameters	Values
Length of furnace	168.0 cm
Width of furnace	108.0 cm
Height of furnace	183.0 cm
Fuel loading capacity	50 kg/batch
Construction material	MS steel
Volume flow rate	0.17 m^3/s
Heat exchanger area	5 m^2
Cross-sectional area of exhaust	0.0046 m^2
Volume of tunnel	1303.6 m^3
Mass of air in tunnel	1469.16 kg
Efficiency	54%
Total weight	400 kg
Price USD	1562.5

2.2. Fabrication of Biomass Furnace

Apart from design parameters, fabrication material is the most important aspect of the biomass furnace that directly affects its thermal efficiency and capital cost. A multitude of fabrication materials (silver, copper, brass, iron, steel) with varying thermal conductivities (406, 385, 109, 80, 50 W/mK) and melting points (962, 1085, 930, 1538, 1450 °C), respectively, are in use across the world. However, their manufacturing industry and capital costs limit their wider-scale adoption. Pakistan is a low-income agrarian economy and the manufacturing sector is still at its infancy but has tremendous growth potential. For fabrication of the prototype type biomass furnace designed in this study, we selected easily available and the lowest cost material, i.e., mild steel. A local vendor (M/S Malik Engineering and Works, Rawalpindi) was hired for the fabrication of the biomass furnace in a precise manner to ensure the precision of all the design parameters. Figure 4 displays the different views of the fabricated biomass furnace.

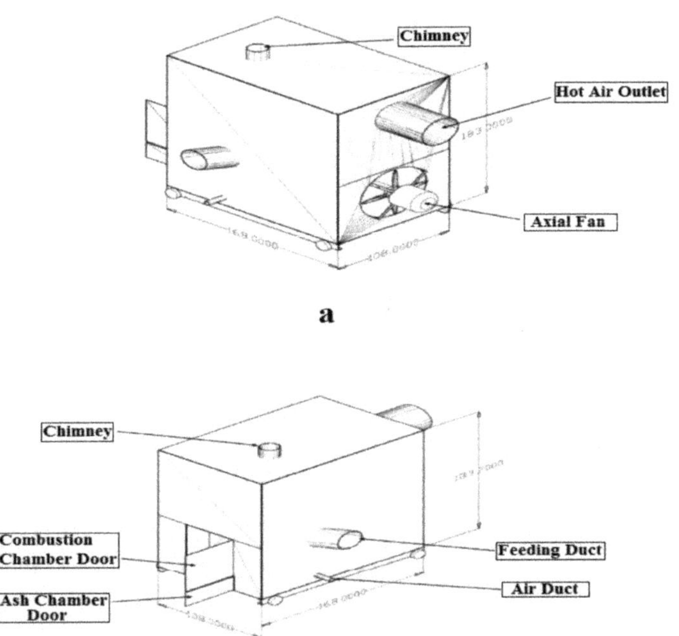

a

b

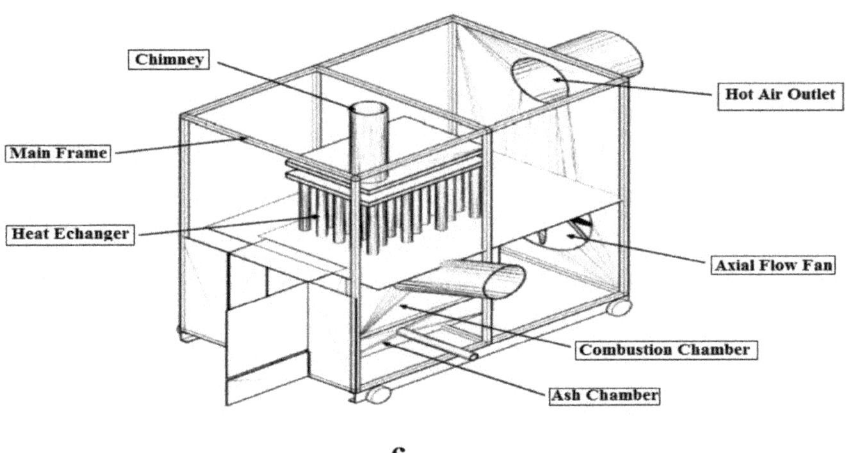

c

Figure 1. Schematic diagram of biomass furnace (**a**) back view (**b**) front view, and (**c**) cross-sectional view.

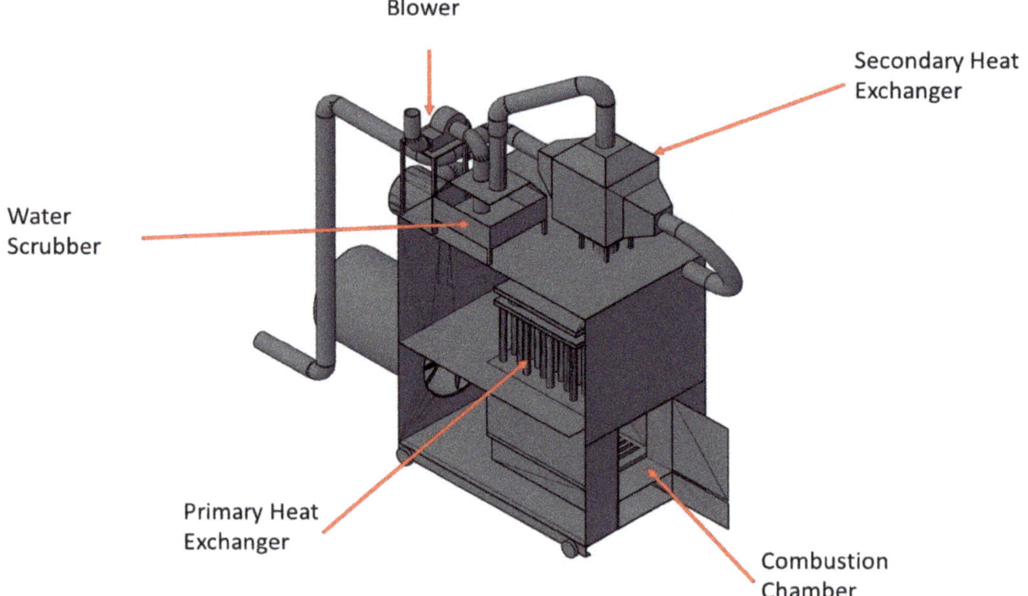

Figure 2. Isometric view of biomass furnace.

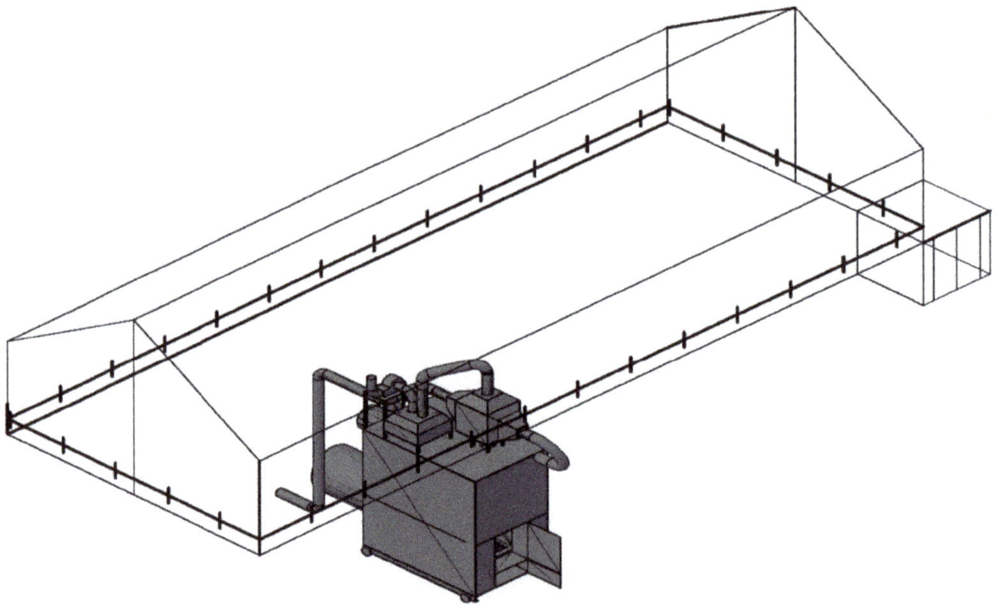

Figure 3. Connectivity of the biomass furnace to the greenhouse tunnel.

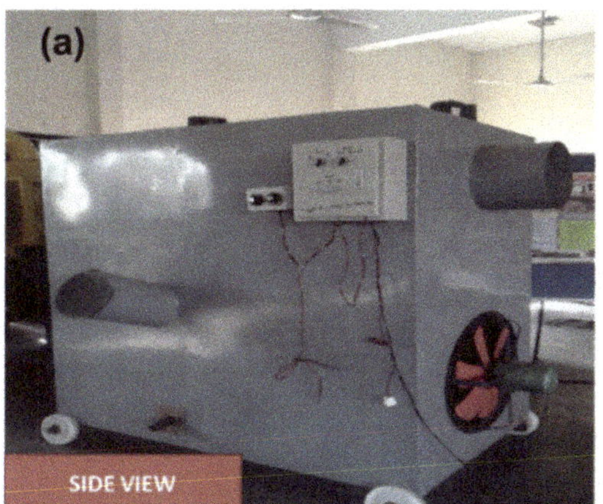

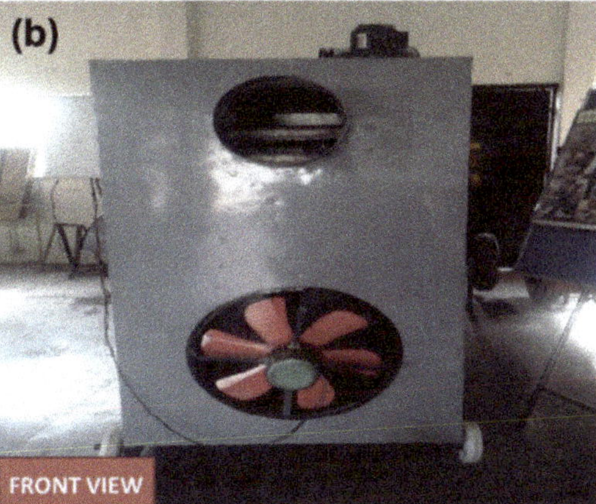

Figure 4. Different views of fabricated biomass furnace. (**a**) side view. (**b**) front view. (**c**) back view.

A variable flow blower is provided in the combustion chamber for the mixing of the stoichiometric air-fuel ratio for optimum combustion. The combustion gases move in a network of parallel vertical tubes provided with spiral baffles for delaying the passage of gases in the tubes. The heat is transferred from biomass flames to fire tubes through the radiation and convection principles of heat transfer, while from inner surfaces to the outer surfaces of tubes heat is transferred through the conduction principle. The hot air passing from all fire tubes gets very hot and exits from one end, which directly opens into the greenhouse or tunnel. Processed air passes through a zigzag having different baffles for getting maximum heat transfer from the hot air tubes to the processed air. An auxiliary suction fan is provided at the exit of the hot air for getting the maximum amount of hot air for drying purposes.

2.3. Biomass Collection and Preparation

Biomass residue samples (wood, cotton stalk, corn cob, and cow dung) were collected from farmers' fields in the surrounding areas. The moisture content affects the heating value of a biomass fuel. For a moist fuel, the heating value decreases because a portion of the heat is used to evaporate the water present in the biomass. The collected biomass was sun-dried to lower the moisture content (10%). Thereafter, the biomass was converted into pellets, chips, and briquettes for greenhouse heating using a biomass furnace. Therefore, almost no cost was incurred to dry the biomass used for this study. However, the conversion of biomass to pellets, chips, and briquettes were prepared manually with a labor cost of USD 0.375/day.

2.4. Evaluation of the Biomass Furnace

The prototype biomass furnace unit was evaluated at the National Agricultural Research Centre, Islamabad using different feed rates of wood. The biomass furnace was operated continuously for 10 h, while the air flow rate of the blower was fixed at 0.17 m^3/s during the entire testing period. The thermal efficiency of the furnace was recorded for each feed rate of the firwood. The pretesting indicated considerable heat losses in the walls of the combustion chamber and through the hot flue gases releasing from the exhaust vent. Therefore, the design and fabrication of the prototype biomass furnace unit was further modified to increase its thermal efficiency and make it environmentally friendly.

3. Results

3.1. Pretesting of the Biomass Furnace

The prototype biomass furnace was pretested using firewood feed rates of 8, 10, and 12 kg/h. The prototype biomass furnace was transported to the engineering workshop of the National Agricultural Research Centre (NARC), Islamabad for its pretesting, evaluation, and further modification (Figure 5). Data were collected on ambient air temperature, furnace temperature, exhaust temperature, heated air temperature, tunnel temperature, output heat flow rate, efficiency of the furnace, and heating time during the different tests (Table 2). The efficiency, input and output power, and operational cost of the furnace were determined to make the comparison between different feed rates of wood (8, 10, and 12 kg/h). Based on input and output powers, the average efficiency of the furnace increased linearly with the feed rate of the firewood (Table 3). Thermal efficiency was calculated with the ratio of output heat flow rate to input heat flow rate. The output heat flow rate was calculated using Equation (8):

$$Q_{output} = m \times (kg) \times Cp. \times \Delta T \tag{8}$$

where Q = energy flow rate (Kw); m (fan speed) = mass flow rate (0.1008 kg/s); Cp = specific heat of air (1.012 KJ/kg·k); ΔT = temperature difference ($T_2 - T_1$); T_2 = final temperature; T_1 = initial temperature.

Table 3. Temperature, flow rate, and efficiency of wood at different feed rates during pretesting of the biomass furnace.

Biomass Type and Feeding Rate (kg/h)	Amb. Air Temp. (°C)	Furnace Temp. (°C)	Exhaust Temp. (°C)	Heated Air Temp. (°C)	Tunnel Temp. (°C)	Output Heat Flow Rate (kW)	Thermal Efficiency η (%)
Wood 8	10.2	247.7	191.3	98.5	26.4	14.30	42.89
Wood 10	13.8	300.1	211.8	126.8	33.5	18.30	43.91
Wood 12	9.5	318.8	229.4	146.1	34.1	22.12	44.24

The input heat flow rate is the product of a biomass burnt per hour and the calorific heating value of that biomass, i.e., Q_{input} = m, the calorific or heating value of biomass multiplied with quantity used per hour. However, the increase in efficiency was disproportionate with the increase in the feed rate of wood. Nevertheless, pretesting of the prototype biomass furnace suggested its larger suitability for greenhouse heating in cold environments.

Figure 5. Pretesting of the biomass furnace at NARC.

The design of a diesel heater is relatively better due to its better engineering, and copper material having good thermal conductivity (385 W/m K). However, the material of the biomass furnace was mild steel, which possesses relatively lower thermal conductivity (64.8 W/m K). The design temperature of exhaust gas was assumed to be 373.15 K, but the actual values always differ from the design values because in theoretical design, mostly ideal conditions are assumed while the actual conditions are always different from the theoretical. This fact leads to lower actual efficiency of the machine than the theoretical efficiency.

3.2. Modification of the Prototype Biomass Furnace

The major deficiencies identified during the pre-testing of the prototype biomass furnace unit were its low thermal efficiency due to considerable heat losses from furnace walls as well as from the vent. The risk of environmental pollution due to the uncontrolled emission of toxic flue gases was another drawback of the prototype furnace. To cater for these issues, the design and fabrication of the furnace was modified by providing insulation work inside, installing turbulators in the heat exchanger tubes, adding a secondary heat recovery unit, and installing a water scrubber smoke filter in the furnace to reduce environmental pollution.

The insulation work was carried out by providing a thick layer of glass wool across the walls of the furnace as a barrier to reduce thermal losses during the operation of the biomass furnace. Hot air turbulators are commonly used for enhancing the thermal efficiency of boilers, air heaters, and heat exchangers. They retain hot air for longer durations inside the heat exchanger, resulting in saving fuel and increasing thermal efficiency. The air turbulators in the modified biomass furnace were provided inside the heat exchanger and heat recovery unit. A significant volume of precious hot air was wasted through the vent located at the top of furnace. To address this issue, a secondary heat exchanger was provided on top of the vent as a heat recovery unit. The flue gases again passed through this secondary heat exchanger unit or heat recovery unit, which contained tubes and a convection chamber for further recovery of heat from the flue gases. The addition of this heat recovery unit considerably improved the thermal efficiency of the furnace and reduced exhaust temperature. The emission of raw flue gases from the furnace into the atmosphere is detrimental to environment. A wide variety of exhaust emission control devices such as venturi wet scrubbers, packed tower wet scrubber, impingement wet scrubbers, and catalytic converters are used in industrial applications to control environment pollutions, but these devices are very costly and not affordable to the common farmer. To address this problem, a simple and economical water scrubber smoke filter device was designed, developed, and tested for the cleaning of flue gases emitting from the biomass furnace developed in this study. This modification added 312.5 USD more in capital cost of the furnace. It was provided on the top of the furnace from where flue gases pass through before mixing into the environment. The water scrubber smoke filter controls air pollution and removes particulate matter by dissolving it in liquid. The smoke filtering device is very useful for reducing air pollution and is also used for the reduction of many exhaust gases, which include CO, NO, NO2, H_2S and SO_2 [23,24]. To monitor the exhaust flue gases, the US-EPA and PAK-EPA certified TESTO-350 flue gas analyzers were used. The modified biomass furnace is shown in Figure 6.

Figure 6. Side view of the modified biomass furnace.

3.3. Testing of the Modified Biomass Furnace

The modified biomass furnace was shifted to the field and installed outside an existing greenhouse/tunnel with the dimensions 30.48 m × 12.19 m × 4.26 m. The heating efficiency of the biomass furnace was evaluated by varying three feeding rates of wood biomass, e.g., 8, 10, and 12 kg/h. The efficiency and economics of the modified biomass furnace was also evaluated for common types of crop residue-based biomasses (cotton stalks, corn cobs, and cow dung), as well as diesel, which is the standard fuel for greenhouse heating in Pakistan. Data were collected on furnace parameters, such as ambient air temperature, furnace inside temperature, heated air temperature, furnace exhaust temperature, tunnel temperature, heat flow rate, and furnace efficiency. The furnace was operated continuously for 10 h, while the air flow rate of the furnace was kept constant at 0.17 m^3/s during the entire testing period. The initial performance of the prototype biomass furnace was poor due to considerable heat loss from the furnace walls and the vent. However, a considerable reduction in heat energy dissipated from the exhaust vent and significant increases in the furnace temperature, heated air temperature, output heat flow rate, and thermal efficiency were observed after modification of the biomass furnace (Tables 3 and 4).

Table 4. Temperature, flow rate, and efficiency of wood at different feed rates after modification of the biomass furnace.

Biomass Type and Feeding Rate (kg/h)	Amb. Air Temp. (°C)	Furnace Temp. (°C)	Exhaust Temp. (°C)	Heated Air Temp. (°C)	Tunnel Temp. (°C)	Output Heat Flow Rate (kW)	Thermal Efficiency η (%)
Wood 8	6.0	371.4	149.0	122.0	15.0	17.33	51.98
Wood 10	5.8	394.5	158.3	153.0	15.9	22.20	53.28
Wood 12	7.1	404.8	174.4	183.7	16.4	27.09	54.18
Cotton Stalks 13.0	7.2	389.2	155.4	174.6	15.0	25.84	51.71
Corn Cobs 13.0	6.3	388.9	158.9	178.7	15.3	26.46	52.33
Cow Dung 23	5.4	364.2	161.3	169.8	14.1	25.21	50.42
Diesel fuel 4 L/h	7.1	477.4	107.5	129.4	27.5	36.09	72.19

4. Discussion

4.1. Efficiency of the Modified Biomass Furnace

Modification in the design and fabrication of the biomass furnace resulted in significant improvements in its performance under different feed rates of firewood. On average, the furnace temperature increased by 36.1%, heated air temperature by 23.4%, output heat flow rate by 21.7%, and thermal efficiency by 21.7%, and exhaust temperature reduced by 23.8% over the prototype biomass furnace (Figure 7). These improvements were achieved with a nominal increase in the capital cost of the modified biomass furnace.

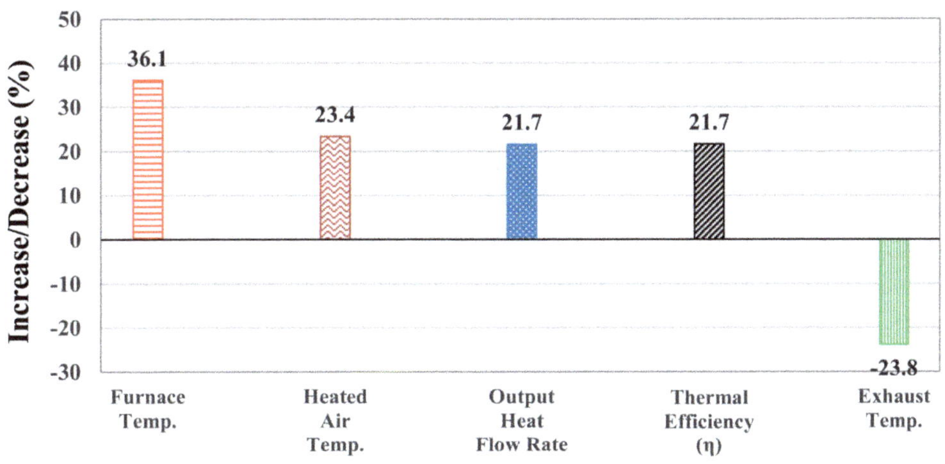

Figure 7. Improvements in various parameters after the modification of the biomass furnace.

4.2. Efficiency of Different Biomasses Relative to Diesel

Diesel is the standard and most widely used fuel for greenhouse heating in Pakistan. However, due to its increasing price, greenhouse growers are continuously searching for viable alternatives for heating of their greenhouses. As such, the efficiency of different biomass fuels used in this study was compared with respect to the heating value of the diesel fuel. The average heating value of 4 L diesel fuel is about 45 MJ/kg. The feed rates of the selected biomass fuels were adjusted to meet the reference heating value of diesel fuel. During the tests, three feed rates of wood viz. 8.0, 10.0, and 12.0 kg/h were used, whereas one feed rate for cotton stalks (13.0 kg/h), corn cobs (13.0 kg/h), and cow dung (22.5 kg/h) was used for testing purpose. The biomass furnace was operated continuously for 10 h under all diesel and biomass fuel tests. The idea behind this was to explore economic and viable solutions for greenhouse heating so that the growers could use their crop residues for this purpose.

Table 4 shows average efficiencies of the modified biomass furnace for different biomasses and diesel fuel. Based on input and output powers, the efficiency of the diesel fuel was 72.19%, which is higher than the efficiency of different biomasses. The reason is that there were still many heat losses from the biomass furnace. In contrast, the diesel-fired heater had minimal heat loss from the heating surface.

4.3. Economics of Biomass Furnace for Greenhouse Heating

The economic analysis of the biomass furnace for greenhouse heating is the most important factor for farmers, as well as end users, in order to understand the cost of greenhouse heating that they have to pay by adopting this innovative technology. Therefore, an economic comparison of the biomass-based heating of greenhouse tunnel with a commercially available diesel-fired heater was carried out. The following assumptions were made in order to make an economic comparison of the two system:

1. It was assumed that both systems would be operated for 600 h per annum.
2. Labor cost as well as man-hours were assumed to be equal for both systems.
3. The life of the system was assumed to be equal for both systems.
4. The feeding rate of biomass (wood) for the furnace was fixed with the fuel consumption of the diesel heater. For example, 4 L/h was the fuel consumption of diesel heater. The calorific value of diesel was 45 MJ/Kg. This makes 12 kg/h of wood equal to the heating value of diesel.

Following Kepner et al. [25], the cost analysis based on fixed and variable costs of the biomass furnace for greenhouse heating is presented in Table 4. The purchase price (capital cost) of the new biomass furnace for greenhouse heating was estimated as USD 1562.50, whereas the market price of the diesel-fired heater was assumed to be USD 3125, and the useful life of the both systems was taken as 15 years. The annual fixed cost and variable cost of the biomass furnace for greenhouse heating was calculated to be USD 398.44 and USD 828.11, respectively. This made the total cost (fixed + variable) equal USD 1226.55. The cost of wood fuel consumption in an hour of furnace operation was USD 0.75, whereas for diesel heater it was USD 2.78. The repair and maintenance costs were USD 0.13/h for the biomass furnace and USD 0.42 for diesel heater. The electric load of the furnace was calculated as 1000 watts per hour whereas that of the diesel heater was 420 watts.

Apart from the capital cost, the operating cost of the biomass furnace is the most important economic consideration for the selection of greenhouse heating system. Table 5 shows the detail comparison of operational costs per day and per season of the biomass furnace using different biomass sources. The economic cost analysis presented in Table 6 indicates that the adoption of the biomass furnace as a greenhouse heating system could save about 1573.09 USD annually for each greenhouse grower. This means that the diesel heater is 2–3 times more expensive than the biomass furnace. Thus, the biomass furnace is economical and environmentally friendly as compared with the diesel-fired heater.

Table 5. Comparison of operational cost of biomass furnace using different biomass fuels with the diesel heater.

Fuel Source	Unit Calorific Value MJ/kg	Estimated Unit Cost (USD)	Biomass Feeding Rates Equivalent to 4 L Diesel Heating	Operational Cost (USD/h)	Operational Cost (USD/Season)	Net Seasonal Saving over Diesel Heating (USD)
Diesel	45.00	0.63	4.0 lit	4.67	2799.64	NIL
Wood	15.00	0.06	12.0 kg	2.04	1226.55	1573.09
Cotton stalks	14.00	0.04	13.0 kg	2.04	1117.80	1681.84
Corn cobs	14.00	0.04	13.0 kg	1.78	1069.05	1730.59
Cow dung	8.00	0.03	22.5 kg	2.00	1198.43	1601.21
Rice husk	15.00	0.05	12.0 kg	1.89	1136.55	1663.09
Wood chips	18.00	0.05	10.0 kg	1.79	1076.55	1723.09

Operational cost is determined by following Kepner et al. [25].

Table 6. Cost analysis of a biomass furnace for greenhouse heating.

Item	Diesel Heater	Biomass Furnace
Basic information		
Purchase price (USD)	3125	1562.5
Annual usage (hr)	600	600
Life (yrs)	15	15
Life (hrs)	9000	9000
Salvage value (USD)	312.5	165.25
Fixed cost		
Depreciation (USD/hr)	0.31	0.16
Interest (USD/hr)	0.86	0.43
Insurance (USD/hr)	0.05	0.03
Tax (USD/hr)	0.05	0.03
Shelter (USD/hr)	0.05	0.03
Sub-total (USD/hr)	1.33	0.66
Variable cost		
Electricity cost (USD/hr)	0.13	0.13
Cost of diesel/wood (USD/hr)	2.78	0.75
Labor cost (USD/hr)	0.38	0.38
Repair and maintenance (USD/hr)	0.42	0.13
Sub-total (USD/hr)	3.34	1.38
Total Cost (USD/hr)	4.67	2.04
Operating cost (USD/day)	46.66	20.44
Saving over diesel heater (USD/hr)		2.62
Saving over diesel heater (USD/day)		26.22
Saving over diesel heater (USD/Season)		1573.13

Conversion rate is based on 1 USD being equal to PKR 160.

4.4. Emissions of Flue Gases

One of the major drawbacks of the conventional greenhouse heating systems is the emission of a higher concentration of greenhouse gases into the atmosphere. The emission of four flue gases (carbon monoxide (CO), sulfur dioxide (SO_2), nitrogen dioxide (NO_2), and carbon dioxide (CO_2)) was monitored before and after the installation of the water scrubber smoke filter over the top of the furnace. In case of no water scrubber, excessive (greater than National Environmental Quality Standards-NEQS limits) carbon monoxide (CO) was only detected for cotton stalks and cow dung. The concentrations of the remaining three flue gases remained well below the NEQS limits, which are specified as 698, 649, and 638 ppm for CO, SO_2, and NO_2, respectively, whereas such a limit is not specified for

CO_2. Nevertheless, the provision of a water scrubber smoke filter significantly reduced the concentrations of the emitting flue gases into the atmosphere (Figure 8).

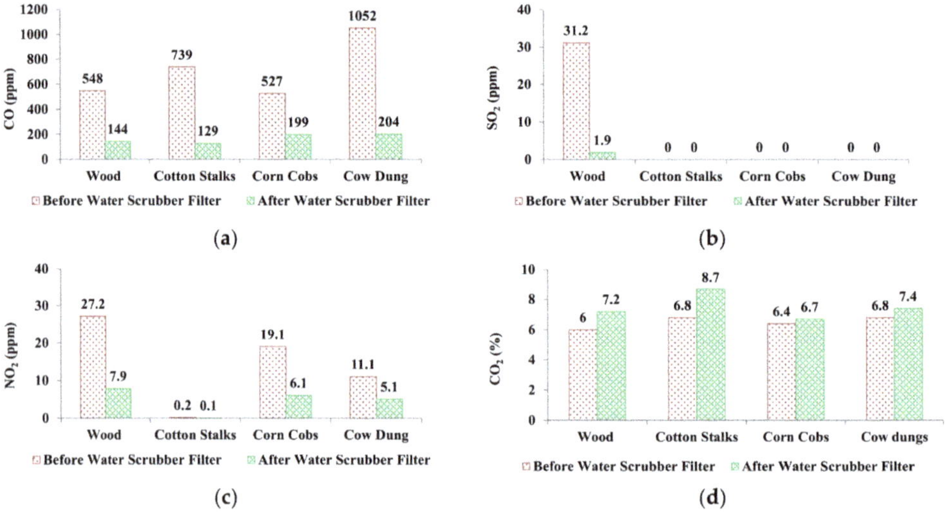

Figure 8. Flue gas concentrations of different biomass fuels before and after the water scrubber. (**a**) Carbon monoxide (CO). (**b**) Sulphur dioxide (SO_2). (**c**) Nitrogen dioxide (NO_2). (**d**) Carbon dioxide (CO_2).

The provision of the water scrubber smoke filter reduced CO concentrations by 73.7% from wood, 73.9% from cotton stalks, 62.2% from corn cobs, and 80.6% from cow dung, while 99.4% of the SO_2 was removed from wood. Sulfur dioxide was in considerable concentration only in wood, whereas in cotton stalks, corn cobs, and cow dung, it could not be detected. Similarly, the removal of NO_2 concentration was about 71.0% for wood, 50.0% for cotton stalks, 68.1% for corn cobs, and 54.1% for cow dung. It is worth noting that the addition of the water scrubber smoke filter did not reduce the emission of CO_2; rather its concentration was slightly increased due to the reaction of CO with water to form CO_2. The elevated levels of CO emissions from cotton stalks and cow dung were efficiently lowered to meet the NEQS.

5. Conclusions

A biomass furnace was successfully designed and developed at the Faculty of Agricultural Engineering and Technology, PMAS-AAUR, with the aim of utilizing biomass (crop residues) in an efficient and economical way as an alternative energy source to fossil fuels for greenhouse heating. Based on the experimental results, the following conclusions were drawn:

i. A biomass furnace is an efficient and attractive heating system for greenhouse heating and has great potential for similar uses like the heating of farmhouses, poultry sheds, and water; and the drying of grains, fruits, and vegetables.

ii. The designed biomass furnace is lightweight and portable, which enhances its practical utility.

iii. Modifications made to the biomass furnace, such as the insulation of the outer walls of the furnace, the provision of turbulators in fire tubes, and the addition of a secondary heat exchanger (heat recovery unit) in vent/chimney increased the thermal efficiency of the biomass furnace by 21.7% (from 43.68% to 53.15%).

iv. The thermal efficiency of the biomass furnace can be increased considerably by using fabrication materials with greater thermal conductivities (e.g., silver, copper, brass, etc.) and by installing the furnace inside the tunnel.

v. The thermal efficiencies of the biomass furnace with different biomasses were lower than the diesel-fired heater and ranged from 50.42% to 54.18% against 71.9% by diesel fuel. In terms of equal calorific value of 4 L diesel, the thermal efficiencies of different biomasses vary slightly with the highest efficiency for wood followed by corn cobs, cotton stalks, and cow dung.

vi. The designed biomass furnace is significantly more economical as compared to a commonly used diesel heater. Its capital cost is only 50% and operational cost is about 43.7% of the traditional diesel fuel heater. Hence, a seasonal saving of 1573 USD over the diesel heater can be achieved by using a biomass furnace.

vii. The seasonal operating cost of the biomass furnace is about 50% of the diesel heater (1562.5 USD against 3125 USD).

viii. The produced smoke was tested as environmentally friendly under the prescribed limits of the National Environmental Quality Standards (NEQS), which shows potential for its large-scale adoption, and wider applications can be a source of safe disposal of agricultural wastes.

Keeping in view the increasing rates of fossil fuels and easy availability of crop residues at the farms, the designed biomass furnace displays a very high potential for its large-scale adoption in the heating of various systems. However, the lack of policy framework, adequate research and development, market development, commercial services, farmer awareness, trainings, demonstration, and legal and regularity issues are the major bottlenecks in the utilization of these biomass resources within the country.

Author Contributions: Conceptualization, A.A. (Asif Ali) and T.I.; methodology, T.I.; and M.Y.; validation, A.A. (Asif Ali), T.I. and A.A. (Arslan Afzal); formal analysis, A.A. (Asif Ali); and Z.u.H.; investigation, A.A. (Asif Ali); resources, A.M.M.; data curation, A.A. (Asif Ali); writing—original draft preparation, A.A. (Asif Ali) and M.Y.; writing—review and editing, T.I. and M.J.M.C.; visualization, A.A. (Arslan Afzal); and K.S.K.; supervision, T.I.; project administration, K.S.K.; and M.J.M.C.; funding acquisition, M.J.M.C.; All authors have read and agreed to the published version of the manuscript.

Funding: This research was funded by the National Center of Industrial Biotechnology, PMAS-Arid Agriculture University, Rawalpindi, Shamsabad 46000, Pakistan.

Institutional Review Board Statement: Not applicable.

Informed Consent Statement: Not applicable.

Data Availability Statement: Not applicable.

Acknowledgments: The authors are thankful to the Faculty of Agricultural Engineering and Technology, PMAS Arid Agriculture University Rawalpindi, Pakistan and National Agriculture Research Council (NARC) for providing the platform for this study.

Conflicts of Interest: There is no conflict of interest.

References

1. Javed, M.S.; Raza, R.; Hassan, I.; Saeed, R.; Shaheen, N.; Iqbal, J.; Shaukat, S.F. The energy crisis in Pakistan: A possible solution via biomass-based waste. *J. Renew. Sustain. Energy* **2016**, *8*, 043102. [CrossRef]
2. Iqbal, T.; Dong, C.Q.; Lu, Q.; Ali, Z.; Khan, I.; Hussain, Z.; Abbas, A. Sketching Pakistan's energy dynamics: Prospects of biomass energy. *J. Renew. Sustain. Energy* **2018**, *10*, 023101. [CrossRef]
3. Naseem, I.; Khan, J. Impact of energy crisis on economic growth of Pakistan. *Int. J. Afr. Asian Stud. J.* **2015**, *7*, 33–42.
4. Government of Pakistan. Pakistan Economic Survey 2019–2020. *Econ. Surv. Islamabad Econ. Advis. Wing Minist. Financ.* **2020**.
5. Malkani, M.S. A review of coal and water resources of Pakistan. *Sci. Technol. Dev.* **2012**, *31*, 202–218.
6. Latif, A.; Ramzan, N. A review of renewable energy resources in Pakistan. *J. Glob. Innov. Agric. Soc. Sci.* **2014**, *2*, 127–132. [CrossRef]
7. Vourdoubas, J. Overview of heating greenhouses with renewable energy sources a case study in Crete-Greece. *J. Agric. Environ. Sci.* **2015**, *4*, 70–76. [CrossRef]
8. Abdellatif, S.; Ashmawy, N.; El-Bakhaswan, M.; Tarabye, H. Hybird, solar and biomass energy system for heating greenhouse sweet coloured pepper. *Adv. Res.* **2016**, *8*, 1–21. [CrossRef]

9.   Mhlanga, A.; Yasir, A.; Khan, A. Policy advisory services in biomass technology in Pakistan, Policy on Biomass Energy Technology. 2016. Available online: https://www.google.com/url?sa=t&rct=j&q=&esrc=s&source=web&cd=&ved=2ahUKEwj0x4 yVuqDwAhUOXsAKHe-IAQQQFjAAegQIAhAD&url=http%3A%2F%2Faedb.org%2Fimages%2FUNIDOBiomassPolicyDocumentSecondDraftV1.pdf&usg=AOvVaw0TjhYR8v5SRERyX4VS0G3G (accessed on 20 April 2021).

10. Saghir, M.; Zafar, S.; Tahir, A.; Ouadi, M.; Siddique, B.; Hornung, A. Unlocking the potential of biomass energy in Pakistan. *Front. Energy Res.* **2019**, *7*, 24. [CrossRef]

11. Güney, T.; Kantar, K. Biomass energy consumption and sustainable development. *Int. J. Sustain. Dev. World Ecol.* **2020**, *27*, 1–10. [CrossRef]

12. Nunes, J.; Silva, P.D.; Andrade, L.P.; Gaspar, P.D. Key points on the energy sustainable development of the food industry: Case study of the Portuguese sausages industry. *Renew. Sustain. Energy Rev.* **2016**, *57*, 393–411. [CrossRef]

13. Lizana, J.; Ortiz, C.; Soltero, V.M.; Chacartegui, R. District heating systems based on low-carbon energy technologies in Mediterranean areas. *Energy* **2017**, *120*, 397–416. [CrossRef]

14. Soltero, V.M.; Chacartegui, R.; Ortiz, C.; Lizana, J.; Quirosa, G. Biomass District Heating Systems Based on Agriculture Residues. *Appl. Sci.* **2018**, *8*, 476. [CrossRef]

15. Soltero, V.M.; Chacartegui, R.; Ortiz, C.; Velázquez, R. Potential of biomass district heating systems in rural areas. *Energy* **2018**, *156*, 132–143. [CrossRef]

16. Sanford, S. *Biomass Energy for Heating Greenhouses Biomass Boilers*; University of Wisconsin-Extension, Cooperative Extension: Madison, WI, USA, 2010.

17. Naqvi, S.R.; Jamshaid, S.; Naqvi, M.; Farooq, W.; Niazi, M.B.K.; Aman, Z.; Zubair, M.; Ali, M.; Shahbaz, M.; Inayat, A.; et al. Potential of biomass for bioenergy in Pakistan based on present case and future perspectives. *Renew. Sustain. Energy Rev.* **2018**, *81*, 1247–1258. [CrossRef]

18. Mirza, U.K.; Ahmad, N.; Majeed, T. An overview of biomass energy utilization in Pakistan. *Renew. Sustain. Energy Rev.* **2008**, *17*, 1988–1996. [CrossRef]

19. Iqbal, T.; Zulfiqar, A.; Abbas, A.; Hussain, Z.; Rafique, S.F.; Dong, C.Q.; Lu, Q. Fast pyrolysis of agricultural residues: A sustainable way to produce clean energy. *Fresenius Environ. Bull.* **2011**, *2*, 1253–1261.

20. Zuberi, M.J.S.; Hasany, S.Z.; Tariq, M.A.; Fahrioglu, M. Assessment of Biomass Energy Resources Potential in Pakistan for Power Generation. In Proceedings of the International Conference on Power Engineering, Energy and Electrical Drives, Turkey, Istanbul, 13–17 May 2013; pp. 1301–1306.

21. Saidur, R.; Abdelaziz, E.A.; Demirbas, A.; Hossain, M.S.; Mekhilef, S. A review on biomass as a fuel for boilers. *Renew. Sustain. Energy Rev.* **2011**, *15*, 2262–2289. [CrossRef]

22. Chau, J.; Sowlati, T.; Sokhansanj, S.; Preto, F.; Melin, S.; Bi, X. Techno-economic analysis of wood biomass boilers for the greenhouse industry. *Appl. Energy* **2009**, *88*, 364–371. [CrossRef]

23. Pakistan=Environmental-Protection-Act-1997. Available online: https://www.google.com/url?sa=t&rct=j&q=&esrc=s& source=web&cd=&ved=2ahUKEwiV0vj8_6LwAhVRolwKHUDjBPcQFjAAegQIAhAD&url=https%3A%2F%2Fwww.elaw. org%2Fsystem%2Ffiles%2FLaw-PEPA-1997.pdf&usg=AOvVaw3Ycel4ucKW_w1rzen4MGRh (accessed on 20 April 2021).

24. Erdohelyi, A.; Fodor, K.; Suru, G. Reaction of carbon monoxide with water on supported iridium catalysts. *Appl. Catal. A Gen.* **1996**, *139*, 131–147. [CrossRef]

25. Kepner, A.; Bainer, R.; Barger, E.L. *Principles of Farm Machinery*; AVI Publishing Company Inc.: West Port, CT, USA, 1978; Volume 3, pp. 112–134.

 *sustainability*

Article

Transforming a Valuable Bioresource to Biochar, Its Environmental Importance, and Potential Applications in Boosting Circular Bioeconomy While Promoting Sustainable Agriculture

Farhat Abbas [1,*,†], Hafiz Mohkum Hammad [2,†], Farhat Anwar [2], Aitazaz Ahsan Farooque [3,*], Rashid Jawad [4], Hafiz Faiq Bakhat [2], Muhammad Asif Naeem [2], Sajjad Ahmad [2] and Saeed Ahmad Qaisrani [2]

[1] School of Climate Change and Adaptation, University of Prince Edward Island, Charlottetown, PE C1A 4P3, Canada
[2] Department of Environmental Sciences, COMSATS University Islamabad, Vehari 61100, Pakistan; mohkum@ciitvehari.edu.pk (H.M.H.); farhatch3@gmail.com (F.A.); faiqsiddique@ciitvehari.edu.pk (H.F.B.); asif.naeem@ciitvehari.edu.pk (M.A.N.); sajjad.ahmad@ciitvehari.edu.pk (S.A.); saeed.qaisrani@ciitvehari.edu.pk (S.A.Q.)
[3] Faculty of Sustainable Design Engineering, University of Prince Edward Island, Charlottetown, PE C1A 4P3, Canada
[4] Department of Horticulture, Ghazi University, Dera Ghazi Khan 32260, Pakistan; rashidjawad74@gmail.com
* Correspondence: fabbas@upei.ca (F.A.); afarooque@upei.ca (A.A.F.)
† These authors have equal contribution.

Citation: Abbas, F.; Hammad, H.M.; Anwar, F.; Farooque, A.A.; Jawad, R.; Bakhat, H.F.; Naeem, M.A.; Ahmad, S.; Qaisrani, S.A. Transforming a Valuable Bioresource to Biochar, Its Environmental Importance, and Potential Applications in Boosting Circular Bioeconomy While Promoting Sustainable Agriculture. *Sustainability* **2021**, 13, 2599. https://doi.org/10.3390/su13052599

Academic Editor: Elio Dinuccio

Received: 29 January 2021
Accepted: 25 February 2021
Published: 1 March 2021

Publisher's Note: MDPI stays neutral with regard to jurisdictional claims in published maps and institutional affiliations.

Abstract: Biochar produced from transforming bioresource waste can benefit sustainable agriculture and support circular bioeconomy. The objective of this study was to evaluate the effect of the application of biochar, produced from wheat straws, and a nitrification inhibitor, sourced from neem (*Azadirachta indica*), in combinition with the recommended synthetic fertilizer on soil properties, maize (*Zea mays* L.) plant growth characteristics, and maize grain yield and quality paramters. The nitrification inhibitor was used with the concentrations of 5 and 10 mL pot^{-1} (N_1 and N_2, respectively) with four levels of biochar ($B_0 = 0$ g, $B_1 = 35$ g, $B_2 = 70$ g, $B_3 = 105$ g, $B_4 = 140$ g pot^{-1}), one recommended nitrogen, phosphorous, and potassium syntactic fertilizer (250, 125, and 100 kg ha^{-1}, respectively) treatment, and one control treatment. The results showed that the nitrification inhibitor enhanced crop growth while the application of biochar significantly improved soil fertility. The application of biochar significantly enhanced soil organic matter and soil nitrogen as compared with nitrogen–phosphorus–potassium treatment. The highest root length (65.43 cm) and root weight (50.25 g) were observed in the maize plants treated with B_4 and N_2 combinedly. The grain yield, total biomass production, protein content from biochar's B_4, and nitrogen–phosphorus–potassium treatments were not significantly different from each other. The application of 140 g biochar pot^{-1} (B_4) with nitrification inhibitor (10 mL pot^{-1}) resulted in higher crop yield and the highest protein contents in maize grains as compared to the control treatments. Therefore, the potential of biochar application in combination with nitrification inhibitor may be used as the best nutrient management practice after verifying these findings at a large-scale field study. Based on the experimental findings, the applied potential of the study treatments, and results of economic analysis, it can be said that biochar has an important role to play in the circular bioeconomy.

Keywords: bioresources; circular bioeconomy; economic analysis; Nitrification inhibitor; smog; wheat straw

---

1. Introduction

Developing countries in South Asia face serious environmental problems from poor management of waste materials such as the burning of crop residues [1]. The antienvironmental burning of crop residues takes place to get ready for the next cropping cycle.

Through such burning, although agricultural fields are cleared and get quickly ready for next sowing yet the adverse impacts of the release of greenhouse gases [2] on public health offsets the personal gains of individual farmers. Avoiding the burning of crop residues can help reduce smog-based public issues originating from poor air quality (including diseases and traffic accidents) that have been reported for more than 2 decades in India and Pakistan particularly between October and November every year [3]. The circular bioeconomy finds the best place to play its role in such conditions with options for transforming crop residues through recycling this valuable bioresource to biochar for sustainable agriculture [4]. Circular bioeconomy benefits from the enhanced circularity of bioresources (wheat and/or rice straws) as its agriculture-based waste feedstock [1].

Biochar application to agricultural soils has been identified as a low-cost approach with an environmentally sound option in the wake of the global depletion of clean environment. It has attracted attentiveness in recent years mainly due to importance of soil carbon sequestration [4,5]. Biochar application is viable in enhancing crop growth [6–8] through improving soil chemical and physical properties [9,10] such as its extremely porous interior structure [11,12]. It acts as a soil conditioning mediator thereby improving soil water holding capacity by altering the soil pore size distribution [13] thus preventing nutrient loss from agricultural fields [14–16].

Feedstock for biochar ranges from a variety of raw materials including agricultural waste. Figueredo et al. [17] reported that the raw material and the pyrolysis temperature impact the nutrient concentration of biochar. They characterized and reported the release of nutrients and contaminants from types of biochar made from sugarcane bagasse, eucalyptus bark, and sewage sludge on 350–500 °C pyrolysis temperature. Biochar is an enriched carbon-based material and is the product of biomass pyrolysis and has profound impacts on improving soil carbon storage [18]. An important attribute of biochar is its cation exchange capacity (CEC) due to its large surface area and porosity which impact the soil biota and nutrient dynamics [6,19]. It enhances the soil nutrient availability to plants [20,21], flourishes the soil microbial population [19,22,23], and reduces greenhouse gas emissions through carbon sequestration [24]. Eventually, it increases the crop yield [25]. For example, Peng et al. [26] stated that 1% application of biochar increased 64% total biomass (above and below ground) of the maize in ultisol soils. Henceforth, it might play a positive role against climate change [27–29]. By active carbon sequestration, biochar has the potential to gain carbon credits [4]. The positive response of crop productivity against biochar application is attributed to its nutrients such as Ca, mg, K, and unintended fertility. These indirect and direct fertility aspects of biochar are categorized as a soil conditioner and soil fertilizer, respectively [6,26,30] that improve soil fertility [31]. The soil pH is also improved by the alkalinity of biochar [12] and it also facilitates the availability of phosphorous [32].

Biochar had a major and significant effect on different characters like a seedling, stem girth, number of roots, length of roots, and percentage germination [33]. Among the positive effects of biochar on plant development, the nitrogen use efficiency (NUE) has also been moderately recognized [10,34]. Laird et al. [35] found better N retention in soil hence, preventing approximately 11% N loss following 2% biochar application. Similarly, Clough et al. [36] reported that the biochar amendment had great agronomic advantages including changes soil nitrogen dynamics.

Nitrogen losses, precisely in agricultural soils are a widespread problem and are categorized into denitrification, leaching down with water as well as transformations into gaseous components [37]. In the case of anthropogenic N supplementation to agricultural soils, Zhang et al. [38] and others [39] found that about 30–80% of this N is taken up and incorporated by crops with loss of the remaining N proportion. Reactive N is effectively conserved through intrinsic soil N dynamics within natural environments [40–42]. Nitrate (NO_3^-) losses in soils of subtropical regions are more characterized by the leaching or runoff due to high rainfall patterns [40]. Nitrogen losses through nitrification are common in unsaturated N soils particularly upon the application of ammonium sulfate;

nonetheless, in saturated agricultural soils, N immobilization and mineralization into NH_4 are more frequent [43].

Nitrification inhibitors (NIs) are commonly employed in agricultural soils for enhancing the N retention by preventing its loss in the N_2O form and reducing the leaching of N [44–46]. Hence, to overcome N losses, NIs are distinguished in cropping systems [45,47] for enhancing crop production and decreasing the N_2O emission [46] hence improving the NUE in agricultural soils [48,49]. The NIs have shown a reduction in leaching of ammonium and urea-based fertilizers [50]. These inhibitors encourage the N retention in the soil in NH_4 by inhibiting the activity of ammonium monooxygenase (AMO). This AMO is recognized as a broad-spectrum efficiency for substrates [51]. The NIs compete with the active sites of this enzyme and aids in preventing the NH_4-enzyme complex and in this way, delay the rate-limiting step of nitrification [52]. A variety of NIs are used in agricultural biochar-amended soils. For example, 3,4-dimethyl pyrazole phosphate is viable for reducing N losses even at low application rates [53] with little adverse effects on soil ecology [48]. Another important NI is the dicyandiamide that is useful in reducing soil N losses [54]. These NIs have profoundly reduced N losses for example potassium thiosulfate is also characterized as a good NI [55]. Moreover, Cai et al. [56] in a laboratory experiment, found that dicyandiamide can reduce N_2O emissions up to 70% and predicted that these substances might be performed excellently at field scale as well [57,58].

Besides, recent studies have suggested that NIs correlate with biochar, explaining that the sorption of NIs is influenced by applied biochar [36,59–63]. The soil amendment of biochar, regardless of its feedstock, adds up new binding sites, thereby altering soil attributes such as pH and hydrophobicity and ultimately affecting the sorption of applied NIs resulting in the high productivity of cropping systems [64–66].

We hypothesized that the wheat crop residues would make nutrient-rich biochar and that such a soil amendment (biochar mixed with NI and NPK) will benefit soil health, plant growth, and crop yield and quality leading a way to circular bioeconomy. The hypothesis was tested by evaluating the effect of the application of biochar, produced from wheat straws, and NI, sourced from neem (*Azadirachta indica*), in combinition with recommended doeses of NPK on soil properties, maize (*Zea mays* L.) plant growth characteristics and maize grain yield and quality paramters. The use of neem as a NI in combination with NPK and biochar produced from wheat crop residues accounts for novelty of this work. Another novelty component of this work is the economic analysis that could not be found in biochar mixed with other fertilizers literature.

2. Materials and Methods

This experimental study was carried out at COMSATS University Islamabad, Vehari Campus Pakistan located at latitude 32°03′ N longitude 72°31′ E and with an altitude of 184 m. Long-term mean annual rainfall and reference evapotranspiration were approximately 231 mm and 1790 mm, respectively, while the annual mean daily maximum and minimum temperature were 28.0 °C and 13.7 °C, respectively as the experiment (Figure 1).

2.1. Preparation of Biochar, Neem Extract, and Experimental Pots

Biochar for this study was prepared with wheat straw via the pyrolysis method, which is also known as the thermal decomposition under oxygen-free conditions. The feedstock (wheat straw) of biochar were first heated at 105 °C for 30 min to remove the moisture from the raw materials. During the processing of biochar production, the temperature of the biochar pyrolysis apparatus was between 450 and 550 °C in a perpendicular oven. The gas produced from biochar preparation was condensed in the plant and collected as a liquid bio-oil for the safety of environmental pollution. The final biochar product was milled to pass through a 1 mm filter before its use. Selective properties of the produced biochar are given in Table 1.

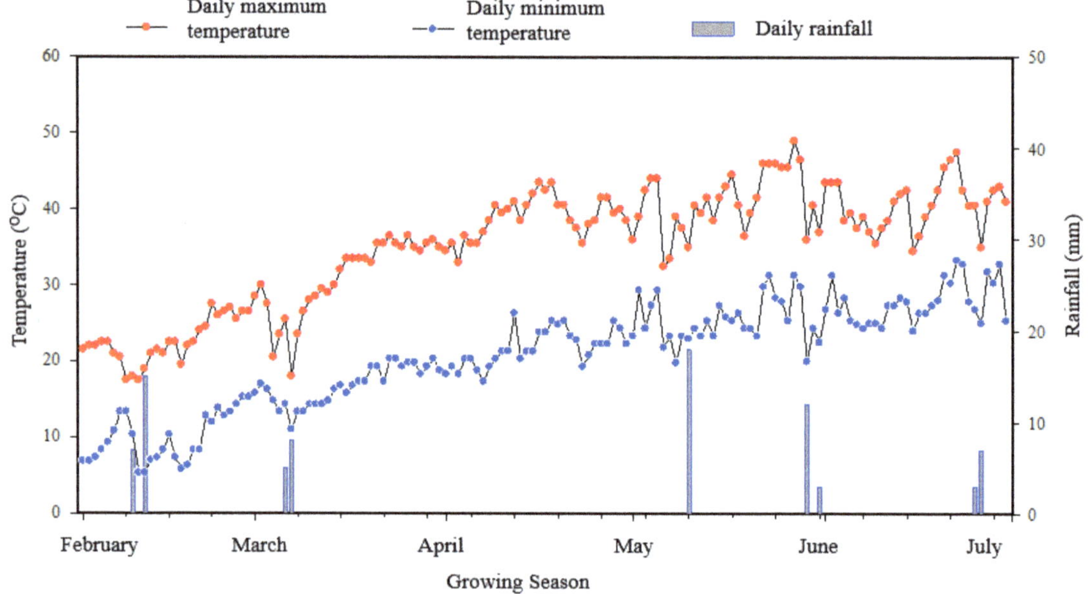

Figure 1. Daily temperature and rainfall data of experimental site during the growing season.

Table 1. Physio-chemical characteristics of biochar and soil used in the experiment.

Characteristics	Biochar	Soil
Organic matter (%)	45.5	0.74
Total nitrogen (g kg^{-1})	0.35	0.04
Total phosphorous (g kg^{-1})	1.34	6.5
Total potassium (g kg^{-1})	9.40	14.0
Electrical conductivity (dSm^{-1})	—	1.41
pH	8.8	7.5
Ash content (g kg^{-1})	120	—
Moisture (%)	31	—
Cation exchange capacity (cmolc kg^{-1})	93	6.5

As per local practice of preparing neem extract for kitchen/backyard gardening, the neem leaves plus seeds were soaked in water overnight with 1:2 neem to water ratio (5 kg of neem leaves/seeds in 10 L of water). The same material was then boiled on the next day to the point when approximately 50% of the water was evaporated and/or left in the boiling pan. The boiled solution was then sieved to collect neem extract to be used as NI in this experiment.

The soil made pots (30-cm height, 15-cm radius from the bottom, and 20-cm radius from the neck) were used during this experiment to grow maize under the experimental treatments. Each pot had a filling capacity of 15 kg of soil. All the pots were filled with 5 kg of non-sterilized soil collected from a nearby agricultural field that was sieved by using a 4.5-mm sieve to remove plant roots and other debris. A small hole was permitted at the bottom of each pot to let the excess water drain out in case of excessive rain. The properties of experimental soil are given in Table 1.

2.2. Experimental Design and Treatments

The experimental design for this was a factorial split-plot design with three replications. Four levels of biochar (B_0 = 0 g, B_1 = 35 g, B_2 = 70 g, B_3 = 105 g, B_4 = 140 g per pot), one treatment of recommended the N, P, and K (250, 125 and 100 kg ha^{-1}, respectively) and one control treatment were used to make the experimental treatments. A treatment of one selected NI (neem extract solution; N_1 = 5 mL, N_2 = 10 mL pot^{-1}) was applied to each of the four biochar levels, one NPK level, and one control. Resultantly, the set of four biochar treatments separately existed with 5 mL NI and with 10 mL NI. Therefore, the total experimental units were twelve as given below.

- T1 = N_1B_0 *(control)*: 5 mL neam + 0 g biochar
- T2 = N_1NPK: 5 mL neam + N, P, and K added @ 250, 125 and 100 kg ha^{-1}, respectively
- T3 = N_1B_1: 5 mL neam + 35 g biochar
- T4 = N_1B_2: 5 mL neam + 70 g biochar
- T5 = N_1B_3: 5 mL neam + 105 g biochar
- T6 = $N_1 B_4$: 5 mL neam + 140 g biochar
- T7 = N_2B_0 *(control)*: 10 mL neam + 0 g biochar
- T8 = N_2NPK: 5 mL neam + N, P, and K added @ 250, 125 and 100 kg ha^{-1}, respectively
- T9 = $N_2 B_1$: 10 mL neam + 35 g biochar
- T10 = $N_2 B_2$: 10 mL neam + 70 g biochar
- T11 = $N_2 B_3$: 10 mL neam + 105 g biochar
- T12 = $N_2 B_4$: 10 mL neam + 140 g biochar

2.3. Sample Analysis

The experimental soil (collected from the field) and soils from each experimental pot were analyzed for various soil properties. Soil organic matter was determined by the dichromate oxidation method [67]. Soil electrical conductivity (EC) and pH were determined in a 1:5 soil/water extract. Plant available-N in the soil was determined by the methods defined by Hesse [68] and available-P was determined by using the method as described by Olsen [69]. Available soil potassium (K) was determined by the method described by Junsomboon and Jakmunee [70].

The experiment started on February 12, 2018 and the maize variety Pioneer 31R88 was sown in experimental pots on the same day right after fertilization and crop sowing. At maturity, one plant was randomly extracted from each replication and washed with water. Root length was measured from plant base to root tip with the help of scale. The plant roots were oven-dried separately at 70 °C till constant weight and their dry weight was recorded. The number of days to tasseling, silking, and maturity were noted in each plant and the mean number of days taken to tasseling, silking, and crop maturity was calculated from the sowing date. A sample for thousand grains was taken from each pot and sundried up to standard moisture content in the grains and weighed by an electrical balance. At maturity, grain yield was calculated. The harvested plants were threshed manually, and grain yield was recorded on a g plant^{-1} basis. For biological yield whole plant was harvested and weighed. At harvest, the grains were taken from each plant and nitrogen contents of the seeds were calculated by using the micro-Kjeldahl method [71], and then crude protein contents were calculated by using the following formula.

$$\text{Crude protein} = \text{Nitrogen} \times 6.25$$

2.4. Statistical and Economic Analysis

The treatment effects on the studied variables were analyzed by constructing an analysis of variance (ANOVA) using SAS [72]. When F-values were significant, the least significant difference test was used for comparing means of treatments. The difference in treatment means was considered significant at $p < 0.05$. An economic analysis of the crop inputs (expenses) and output was performed on the basis of costs that varied in different treatments and by adding fixed cost following the procedure devised by Byerlee [73]. For economic analysis, the yeild was converted from plant pot^{-1} to Ton ha^{-1} by considering 666,666 plants per ha as reported by Hammad et al. [74]. All the input and output prices were made based on numbers obtained from consulting growers and the 2018 Economic Survey of Pakistan.

3. Results

Basis for presenting the study findings were made from the ANOVA results for the study variables. Sample ANOVA results for selective variables (root length, grain yield, total biomass, and protein content) are presented in Table 2. If the interaction of NIs and biochar levels were non-significant, the results were presented individually for each treatment. For example, the interaction of NIs and biochar levels were non-significant for root length (p = 0.9343). Therefore, results of such variables are discussed separately (see Tables 3 and 4). However, if interactions of the NIs and biochar levels were significant; for example, for grain yield (p = 0.0029), total biomass (p = 0.0031), and protein contents (p = 0.0030), the results of these variables are discussed for the combined effects of experimental treatments (see Table 5).

Table 2. Sample analysis of variance (ANOVA) values for selective variables (root length, grain yield, total biomass, and protein content) to base method for presenting study results.

Source of Variation	DF	SS	MS	F	p
Root Length					
Replication	2	12.77	6.384		
NI	1	81.60	81.601	42.2	0.0229
Error Replication×NI	2	3.87	1.934		
Biochar	5	2547.1	509.420	61.39	0.0000
NI×Biochar	5	10.42	2.084	0.25	0.9343
Error Replication×NI×Biochar	20	165.97	8.299		
Total	35	2821.73			
Grain Yield					
Replication	2	4.39	2.19		
NI	1	348.44	348.44	33.1	0.0289
Error Replication×NI	2	21.06	10.53		
Biochar	5	6050.56	1210.11	89.45	0.0000
NI× Biochar	5	358.22	71.64	5.3	0.0029
Error Replication×NI×Biochar	20	270.56	13.53		
Total	35	7053.22			
Total Biomass					
Replication	2	30.2	15.08		
NI	1	584	584.03	1617.31	0.0006
Error Replication×NI	2	0.7	0.36		
Biochar	5	34,868.3	6973.65	124.04	0.0000
NI×Biochar	5	1477.1	295.43	5.25	0.0031
Error Replication×NI×Biochar	20	1124.4	56.22		
Total	35	38,084.8			
Protein Content					
Replication	2	0.574	0.2869		
NI	1	6.588	6.5878	765.03	0.0013
Error Replication×NI	2	0.017	0.0086		
Biochar	5	280.939	56.1878	213.55	0.0000
NI×Biochar	5	6.922	1.3844	5.26	0.0030
Error Replication×NI×Biochar	20	5.262	0.2631		
Total	35	300.302			

DF: Degree of freedom, SS: Some of squares, MS: Mean squares, NI: Nitrification inhibitor.

3.1. Soil Properties

Soil organic matter is an important characteristic that plays a key role in maize grain yield. The result showed that the maximum soil organic matter (1.03%) was observed in the N_2 treatment of NI (Table 3). The soil organic matter increased with increase of biochar application levels. The application of biochar level B_4 (140 g pot^{-1}) resulted in soil organic matter of 1.30% for this treatment, which was 65% (0.84 vs. 1.30) greater from the soil organic matter content of control treatment.

Table 3. Effect of biochar and nitrogen inhibitor application on soil physic-chemical properties.

Treatments	Soil Organic Matter (%)	Soil pH	Soil EC (dSm^{-1})	N in the Soil $(mg\ g^{-1})$	P in the Soil $(mg\ kg^{-1})$	K in the Soil $(mg\ kg^{-1})$
N_1	1.02 a	7.61 a	1.58 a	0.046 b	6.96 a	15.60 b
N_2	1.03 a	7.59 a	1.56 a	0.049 a	6.97 a	15.83 a
Significance	<0.07	<0.03	<0.08	<0.001	<0.01	<0.01
LSD 5%	0.05	0.15	0.053	0.0011	0.56	0.15
Control	0.74 c	7.51 a	1.41 b	0.045 b	6.48 c	13.97 b
NPK (Recommended)	0.83 c	7.65 a	1.54 ab	0.045 b	4.46 a	16.73 a
B_1	0.92 bc	7.51 a	1.50 ab	0.046 ab	6.68 bc	14.96 ab
B_2	0.98 b	7.61 a	1.59 ab	0.046 ab	6.82 abc	15.18 ab
B_3	1.26 a	7.63 a	1.66 a	0.047 ab	7.05 abc	16.27 a
B_4	1.30 a	7.69 a	1.69 a	0.049 a	7.26 ab	16.88 a
Mean	1.02	7.60	1.57	0.046	6.96	15.67
Significance	<0.02	<0.3	<0.03	<0.021	<0.03	<0.01
LSD 5%	0.13	0.64	0.20	0.003	0.67	1.97
CV	7.45	4.65	7.02	3.57	5.28	6.96

EC: Electrical conductivity, N: Nitrogen, P: Phosphorous, K: Potassium, NPK: Nitrogen–phosphorus–potassium treatment. Means values that share different homogeneous group letters (a, b, or c) in a column vary significantly at $p \leq 0.05$, CV: coefficient of variance, LSD: Least significance difference, N_1 and N_2 are neem extract solutions (5 mL, 10 mL pot^{-1}, respectively) B_1 = 35, B_2 = 70, B_3 = 105, B_4 = 140 g biochar pot^{-1}.

The maximum soil pH (7.59) was observed in the N_2 of NI treatment, which was was improved with biochar applications; however, the effect biochar levels on soil pH was non-significance. The application of biochar level B_4 at the rate of 140 g pot^{-1} resulted in the highest soil pH 7.69. The lowest soil pH (7.51) was observed in unfertilized treatment; i.e., control treatment. Soil electrical conductivity (EC) is another important characteristic that plays a key role in plant growth. The result showed that the maximum soil EC was attained at the N_1 of NI which was 1.58 dSm^{-1}. The results showed that soil EC was improved with increasing of the biochar application rate. The application of biochar level B_4 resulted in the highest soil EC (1.69 dSm^{-1}). The lowest soil EC (1.41 dSm^{-1}) was observed in control treatment.

The result showed that the maximum N in the soil was observed at the N_2 level of NI which was 0.049 mg N g^{-1} (Table 3) and it increased with increase in biochar application reaching to its higest value of 0.049 mg g^{-1} in B_4 treatment and the lowest value (i.e., 0.045 mg N g^{-1}) in the soil of control treatment. Similarly, the highest P concentration (6.97 mg kg^{-1}) was determined in the soil of N_2 application (Table 3). Like N, the concentraiton of P also increased with increasing biochar application rate. The application of biochar at the rate of 140 g pot^{-1} (level B_4) resulted in the highest P (7.26 mg kg^{-1}) concentration in the soil of B_4 treatment and the lowest concentration of P (6.48 mg kg^{-1}) was observed in the soil of control treatment. In addition to N and P, the K in the soil is also an important characteristic that plays a key role in growth and yield quality. The concentration of K was the highest in soil of the N_2 level of NI (15.83 mg kg^{-1}). Its concentration was significantly affected by levels of biochar application (Table 3). The K had the increasing trends with increasing the biochar application rate also as its highest value (16.88 mg kg^{-1}) was from the biochar application at the 140 g pot^{-1} (B_4) and the lowest value was (13.97 mg kg^{-1}) in the soil of control treatment.

3.2. Plant Growth Characteristics

The results showed that the maximum root length (55.35 cm) was observed at N_2 level (Table 4). Besides, root length was significantly also affected by levels of biochar application

($p < 0.0001$). Among the biochar treatment levels, maximum root length (65.43 cm) was noted at B_4 treatment (140 g biochar pot^{-1}) followed by NPK treatment which had the root length equavalent to 61.36 cm that was 38% greater (40.7 vs. 65.4) than the root length of plants of control treatment. The lowest root length (40.70 cm) was observed in the control treatment. Like root length, there was also a substantial difference between the root weight of maize treated with two different treatment levels of NIs. The maximum root weight (42.13 g plant^{-1}) was observed for N_2 level that was significantly different from N_1 treatment ($p < 0.01$).

Table 4. Effect of biochar and nitrogen inhibitor applications on maize growth parameters.

Treatments	Root Length (cm)	Root Weight (g)	Days to Tasseling (Day)	Days to Silking (Day)
N_1	52.34 b	38.51 b	45 a	51 a
N_2	55.35 a	42.13 a	47 a	52 a
Significance (P)	<0.022	<0.02	<0.10	<0.09
LSD 5%	1.99	3.58	2.53	1.95
Control	40.70 d	30.60 d	42 c	47 c
NPK (Recommended)	61.36 ab	41.83 b	47 ab	53 ab
B_1	47.15 c	35.25 cd	44 bc	49 bc
B_2	51.40 c	39.90 bc	46 ab	52 abc
B_3	57.01 b	44.13 b	48 a	54 a
B_4	65.43 a	50.25 a	50 a	56 a
Mean	53.84	40.33	46	52
Significance (P)	<0.01	<0.01	<0.01	<0.02
LSD 5%	5.22	5.57	3.99	4.88
CV	5.35	7.62	4.78	5.23

Means values that share different homogeneous group letters (a, b, c, or d) in a column vary significantly at $p \leq 0.05$, CV: coefficient of variance, LSD: Least significance difference, Control: A treatment without fertilizer, N_1 and N_2 are Neem extract solutions (5 mL, 10 mL pot^{-1}, respectively) and B1: 35, B2: 70, B3: 105, and B4: 140 g biochar pot^{-1}.

Similarly, maize treated with N_2 took non-significantly lesser days (47 days) for tasselling and silking (52 days) as compared to N_1 treatment in which the tasselling and silking took place after 45 and 51 days, respectively (Table 4). However, there was significant differences in the tasselling and silking days of maize treated with different biochar levels ($p < 0.05$). Maximum days to tasselling (50 days) and silking (56 days) were reported at biochar level B_4 while tasselling occurred after 42 days and silking after 47 days in the control treatment. The onset of tasselling (47 days) and silking (53 days) was also a bit earlier in maize treated with NPK. Furthermore, the results from B_4 were statistically similar to the NPK treatment level while silking in B_4 treatment was statistically at par with NPK application treatment. In both NI treatment levels; i.e., N_1 and N_2, crop maturity occured after 102 days and maturity during N_1 and N_2 was statistically similar. However, maturity was significantly affected by different levels of biochar application; i.e., maturity was delayed with increasing level of biochar application. Maize treated with B_4 reached maturity after 108 days which was 11 days later than that in B_1 (97 days). Maturity in B_4 was statistically at par with NPK application treatment and the mean number of days to maturity was 102 days.

3.3. Yield and Quality Parameters

Grain yield and total biomass were also significantly affected by levels of biochar and NPK applications and significantly ($p < 0.01$) increased with an increasing level of biochar application (Table 5). The maximum grain yield (84.00 g plant^{-1}) and biomass (266.67 g plant^{-1}) were resulted from the application of recommended NPK with the combination of N_2. However, the application of biochar level B_4 with N_1 resulted in grain yield of 76.67 g plant^{-1} and total biomass of 256 g plant^{-1}. The highest grain yield (43.00 g plant^{-1}) and the lowest total biomass (172. 33 g plant^{-1}) were observed in N_2B_0. In the case of total biomass, N_1B_4 and N_2B_4 were statistically similar to NPK treatment as represented by similar LSD letters. In the case of protein content, the highest protein content (14.3%) in maize grain was observed in the grains of NPK treatment in combination with N_2. Besides, significantly increasing protein content percentage with increasing biochar applications was also observed. At the largest biochar application level, the protein content was 12.37%, which was slightly lower than that of N_1NPK treatment (12.97%). The lowest protein content was observed in the N_2B_1 treatment (4.90%).

Table 5. Effect of biochar and nitrogen inhibitor application on maize yields and quality.

Treatments	Grain Yield (g Plant^{-1})	Total Biomass (g Plant^{-1})	Protein Content (%)
N_1B_0 *(control)*	43.33 f	183.33 de	5.67 g
N_1NPK	73.33 b	252.67 a	12.97 ab
N_1B_1	44.67 f	181.00 de	6.27 fg
N_1B_2	51.33 e	198.00 cd	7.77 ef
N_1B_3	59.67 cd	221.33 b	9.80 cd
$N_1 B_4$	76.67 b	256.00 a	10.30 c
N_2B_0 *(control)*	43.00 f	172.33 e	4.90 g
N_2NPK	84.00 a	266.67 a	14.30 a
N_2B_1	53.00 de	193.67 cde	7.53 ef
N_2B_2	61.33 c	214.67 bc	8.43 de
N_2B_3	72.67 b	245.00 a	10.37 c
$N_2 B_4$	72.33 b	248.33 a	12.37 b
Mean	61.27	219.42	9.22
Significance (P)	<0.003	<0.003	<0.003
LSD 5%	6.26	22.45	1.54
CV	6.00	3.42	5.56

Means values that share different homogeneous group letters (a–g) in a column vary significantly at $p \leq 0.05$, CV: Coefficient of variance, LSD: Least significance difference, Control: A treatment without fertilizer, N_1 and N_2 are Neem extract solutions (5 mL, 10 mL pot^{-1}, respectively) and B_1: 35, B_2: 70, B_3: 105, and B_4: 140 g biochar pot^{-1}.

3.4. Economic Analysis Results

The highest net returns were calculated for N_2NPK treatment ($759.4 ha^{-1}) followed by N_2B_3 ($664.7 ha^{-1}) and N_1B_4 ($587.7 ha^{-1}) treatments (Table 6). Although the NPK treatment had the highest returns but it is argued that the difference between its and biochars treatments' profit may be traded of with the long-term treasure of soil health with a wealth of sequestered soil organic carbon and the bioremediation role of biochar for soil health [4]. With improvements in biochar production technologies, inclusion of biochar in the best nutrient management pracitces, and reduction of its cost due to higher commercial production, circulation, and demand, its market price is anticipated to drop down. Hence, the economical availability of biochar will reduce the costs of crop inputs and will increase the farm profitability. The mixed use of biochar with compost or with synthetic fertilizers can also be argued for its importance in farmer's income. Numerous studies have highlighted [75,76] that the crop growth is affected precisely due to biochar made changes in soil nutrient cycles, specifically the cycling of P and K.

Table 6. Economic analysis of input and output costs on maize cultivated with the experimental treatments.

Treatment	Total Yield, Ton ha^{-1}	Adjusted Yield after Considering 10% Yield Lost during Field Harvest, Ton ha^{-1}	Gross Income Based on Maize Grain Price in Pakistan ($360 ton^{-1})	Biochar Cost Based on Biochar Price ($140 ton^{-1})	Fertilizer per Hectare Cost, $ ton^{-1}	Variable Cost, $ ha^{-1}	Fixed Cost, $ ha^{-1}	Net Benefit, $ ha^{-1}
N_0B_0	2.389	2.15	773.9	—	—	0	415	358.9
N_1NPK	4.889	4.4	1583.9	—	694.0	640.0	415	528.9
N_1B_1	2.978	2.68	964.9	163.3	—	163.33	415	386.5
N_1B_2	3.422	3.08	1108.7	326.7	—	326.66	415	367.1
N_1B_3	3.978	3.58	1288.9	490.0	—	490.0	415	383.7
N_1B_4	5.111	4.60	1656.1	653.3	—	653.33	415	587.7
N_0B_0	2.367	2.13	766.8	—	—	0	415	351.8
N_2NPK	5.600	5.04	1814.4	—	694.0	640.0	415	759.4
N_2B_1	3.533	3.18	1144.8	163.3	—	163.33	415	566.5
N_2B_2	4.089	3.68	1324.7	326.7	—	326.66	415	583.1
N_2B_3	4.845	4.36	1569.7	490.0	—	490	415	664.7
$N_2 B_4$	4.822	4.34	1562.3	653.3	—	653.33	415	494.0

All calculations are based on numbers obtained from consulting growers and the 2018 Economic Survey of Pakistan.

4. Discussion

The role of biochar and NIs on growth and yield of maize has been reported in literature [7,26]. Among the direct and indirect effects of biochar, the latter are more distinguished as reported by Glaser et al. [6]. According to Genesio et al. [77] the biochar application to the soils changes the natural state and thermal dynamics of the soil thereby promoting crop growth. They further reported that biochar supplementation with the NI had a promising role in the germination and phenology of plants.

Slow-release of N from synthetic fertilizers is achieved by coating the fertilizer grains with hydrophobic chemicals to provide a physical barrier against water for minimizing N losses and improving N uptake by crops; however, such alternatives may harm the soil health and crop growth [78]. In contrast, the natural NIs are soil environment friendly and plant growth stimulators. The nature-based inhibitors have been exhaustively investigated as alternatives [79]; these include powder of *Azadirachta indica* seed [62] and bark of *Acacia* caven [63]. Such alternatives promote the slow release of N to soil solution [80]. In our experimentl treatments involving higher concentration of NI sourced from naturally occurig neem significantly reduced N loss from soil. These reults are in agreement with the finding of Mohanty et al. [62] who used neem seed powder, and found that the difference in urea content of treated and untreated samples was less significant at the start but became more profound with time, pointing to an inhibitory mechanism of neem whereby it takes some time for the bio inhibitor to be activated [78].

In our study, better root length and root weight were reported for the application of 140 g biochar pot^{-1} that is linked with better nutrient accessibility to roots after the biochar application to soils [26,32]. Besides, maize root growth was increased with increasing biochar application because the biochar hold a slight ratio of labile carbon [5], which either improves root growth or facilitates the root contact to available P [81,82].

In the case of N, biochar application increased the quantity of N reserved in the soils that is not according to the earlier conclusions that biochar expands the absorption capacity of the soil but decreases leakage of nitrate and ammonium because of its great surface area and absorbent structure [35,83] as found in the soils tested by Zhang et al. [38] with biochar adjustment. The results of the current study showed K availability was also increased through the biochar amendment resulting in enhanced K content in the soil. This K content then increased the maize total biomass and grain yield in treatments that received greater biochar application. However, at this point, the fertilizing effect of biochar is more characterized because K availability to maize was increased due to the high content of K in biochar along with its reduced leaching [35,84]. Martinsen et al. [84] argued that K is the major nutrient supplied by biochar which helps in delaying the tasseling and silking days and also helps in alleviating the nutrient stress conditions.

The treatments with increasing level of biochar application also enhanced the P availability to maize, which also improved maize grain yield, total biomass, and protein content For example, the B_4 treatment relative to other treatments, made P available for plants by increasing the soil pH [13,85] that helps in reducing P sorption [86,87]. DeLuca et al. [88] further elucidated that the biochar amendment ensures better P availability to crops, with the ability of biochar to retain exchangeable P ions due to its positively charged sites. The increase in maize total biomass and grain yield in B_4 treatment is also attributed to biochar's role in increasing the total soil organic carbon as reported by Trupiano et al. [89]. Likewise, Pandit et al. [30] mentioned increased maize biomass production with increasing biochar supplementation.

Better soil water retention is governed by biochar amendments as found by Hagemann et al. [90] suggesting that biochar influences in forming organic coatings of soils by reducing pore spaces (resulting in increased capillary rise) and enhanced hydrophilicity. This leads to better soil health and enhanced crop yield [7,74,91]. From our study results, it can be assumed that the application of biochar to agricultural soils is thoughtful and can be used as an alternative option to lime materials in raising the pH, especially in acidic soils because it is noted that approximately 30% world's soils are acidic and 50% of them have the arable potential [92].

The impact of applying biochar as a soil amendment is for approximately 30% world's soils that are acidic and 50% of which have arable potential. The application of biochar to agricultural soils can alternate soil liming, which is used to raising the pH of acidic soils [93]. This leads to the potential of improving acidic soils of Atlantic Canada, to make them suitable for potato cultivation. Soil liming is a common practice in potato fields where the pH is either too acidic or too alkaline. The lime application in Canadian soils varies from province to province; as 11.3 and 20.2% of the croplands of New Brunswick and Prince Edward Island were treated with lime for making them suitable for potato cultivation [94]. Overall, the use of biochar improves soil health especially in poor soils of arid and semiarid regions [95]. Based on the experimental findings, the applied potential of the study treatments, and results of economic analysis, it can be said that biochar has an important role to play in the circular bioeconomy in future.

5. Conclusions

Biochar amendment to agricultural soils is environmentally safe and a sustainable approach relative to synthetic fertilization. Besides, it is also helpful in increasing the fertilizer use efficiency as well as reducing soil pollution. Like biochar supplementation, applying the nitrification inhibitor (neem extract) revealed better maize growth and yield. The maize had the best growth parameters namely the maximum root length, root weight, tasseling, silking, and crop maturity under the treatment of 140 g of biochar applied per pot/plant and 10 mL pot^{-1} application of neem extract. Therefore, the potential of biochar application in combination with nitrification inhibitors should be further exploited for sustainable crop production. It is therefore concluded that the circular bioeconomy seems one of the solutions to transform wheat straw biowastes into a useful bioproduct (biochar) that can ensure agricultural sustainability in terms of a closed-loop sustainability framework involving biomass. With attributes of success of circular bioeconomy at small as well as at large scales, farmers can recycle their crop residues and benefit from a circular resource economy. Biochar can be synthesized by farmers themselves at a low cost instead of spending on the purchase of commercially produced biochar that has the same fertility components. Waste to biochar is a sustainable partway to the circular bioeconomy.

Author Contributions: Conceptualization, F.A. (Farhat Abbas) and H.M.H.; methodology, F.A. (Farhat Abbas); H.M.H. and F.A. (Farhat Anwar); software, H.M.H.; validation, A.A.F.; formal analysis, F.A. (Farhat Anwar); investigation, H.M.H., and F.A. (Farhat Anwar); resources, H.M.H.; data curation, F.A. (Farhat Anwar); writing—original draft preparation, F.A. (Farhat Abbas); H.M.H.

and A.A.F.; writing—review and editing, R.J.; H.F.B.; M.A.N.; S.A.; and S.A.Q.; visualization, R.J.; H.F.B.; M.A.N.; S.A.; and S.A.Q.; supervision, H.M.H.; project administration, F.A. (Farhat Abbas); H.M.H.; funding acquisition, A.A.F. All authors have read and agreed to the published version of the manuscript.

Funding: This research received no external funding.

Institutional Review Board Statement: Not applicable.

Informed Consent Statement: Not applicable.

Acknowledgments: The Natural Sciences and Engineering Research Council of Canada (NSERC).

Conflicts of Interest: The authors declare no conflict of interest.

References

1. Abbas, F.; Aini, Q.; Farooque, A.A. Sustainable, Safe, and Economical Bioresource Management: Basis for Circular Bioeconomy. *Sustainability* **2021**, *13*, under review.
2. Haider, G.; Steffens, D.; Moser, G.; Müller, C.; Kammann, C.I. Biochar reduced nitrate leaching and improved soil moisture content without yield improvements in a four-year field study. *Agric. Ecosyst Environ.* **2017**, *237*, 80–94. [CrossRef]
3. Singh, R.P.; Kaskaoutis, D.G. Crop residue burning: A threat to South Asian air quality. *Eos Trans. Am. Geophys Union* **2014**, *95*, 333–334. [CrossRef]
4. Lehmann, J. A handful of carbon. *Nature* **2007**, *447*, 143–144. [CrossRef] [PubMed]
5. Luo, Y.; Durenkamp, M.; De Nobili, M.; Lin, Q.; Brookes, P. Short term soil priming effects and the mineralisation of biochar following its incorporation to soils of different pH. *Soil Biol. Biochem.* **2011**, *43*, 2304–2314. [CrossRef]
6. Glaser, B.; Lehmann, J.; Zech, W. Ameliorating physical and chemical properties of highly weathered soils in the tropics with charcoal–a review. *Biol. Fertil. Soils* **2002**, *35*, 219–230. [CrossRef]
7. Van Zwieten, L.; Kimber, S.; Morris, S.; Chan, K.; Downie, A.; Rust, J.; Joseph, S.; Cowie, A. Effects of biochar from slow pyrolysis of papermill waste on agronomic performance and soil fertility. *Plant. Soil* **2010**, *327*, 235–246. [CrossRef]
8. Zhu, J.; Ingram, P.A.; Benfey, P.N.; Elich, T. From lab to field, new approaches to phenotyping root system architecture. *Curr. Opin. Plant. Biol.* **2011**, *14*, 310–317. [CrossRef] [PubMed]
9. Güereña, D.; Lehmann, J.; Hanley, K.; Enders, A.; Hyland, C.; Riha, S. Nitrogen dynamics following field application of biochar in a temperate North American maize-based production system. *Plant. Soil* **2013**, *365*, 239–254. [CrossRef]
10. Karer, J.; Wimmer, B.; Zehetner, F.; Kloss, S.; Soja, G. Biochar application to temperate soils: Effects on nutrient uptake and crop yield under field conditions. Agric. *Food Sci.* **2013**, *22*, 390–403. [CrossRef]
11. Chan, K.Y.; Van Zwieten, L.; Meszaros, I.; Downie, A.; Joseph, S. Agronomic values of greenwaste biochar as a soil amendment. *Soil Res.* **2008**, *45*, 629–634. [CrossRef]
12. Oguntunde, P.G.; Abiodun, B.J.; Ajayi, A.E.; van de Giesen, N. Effects of charcoal production on soil physical properties in Ghana. J. Plant Nutr. *Soil Sci.* **2008**, *171*, 591–596.
13. Asai, H.; Samson, B.K.; Stephan, H.M.; Songyikhangsuthor, K.; Homma, K.; Kiyono, Y.; Inoue, Y.; Shiraiwa, T.; Horie, T. Biochar amendment techniques for upland rice production in Northern Laos: 1. Soil physical properties, leaf SPAD and grain yield. *Field Crops Res.* **2009**, *111*, 81–84. [CrossRef]
14. Sohi, S.P.; Krull, E.; Lopez-Capel, E.; Bol, R. A review of biochar and its use and function in soil. In *Advances in Agronomy*; Elsevier: Amsterdam, The Netherlands, 2010; Volume 105, pp. 47–82.
15. Chien, C.C.; Huang, Y.P.; Wang, W.C.; Chao, J.H.; Wei, Y.Y. Efficiency of moso bamboo charcoal and activated carbon for adsorbing radioactive iodine. *Clean Soil Air Water* **2011**, *39*, 103–108. [CrossRef]
16. Mukherjee, A.; Lal, R. Biochar impacts on soil physical properties and greenhouse gas emissions. *Agronomy* **2013**, *3*, 313–339. [CrossRef]
17. Figueredo, N.; Costa, L.; Melo, L.; Siebeneichler, E.; Tronto, J. Characterization of biochars from different sources and evaluation of release of nutrients and contaminants. *Rev. Cienc. Agron.* **2017**, *48*, 3–403. [CrossRef]
18. Xu, G.; Lv, Y.; Sun, J.; Shao, H.; Wei, L. Recent advances in biochar applications in agricultural soils: Benefits and environmental implications. *Clean Soil Air Water* **2012**, *40*, 1093–1098. [CrossRef]
19. Lehmann, J.; Rillig, M.C.; Thies, J.; Masiello, C.A.; Hockaday, W.C.; Crowley, D. Biochar effects on soil biota–a review. *Soil Biol. Biochem.* **2011**, *43*, 1812–1836. [CrossRef]
20. Jeffery, S.; Verheijen, F.G.; van der Velde, M.; Bastos, A.C. A quantitative review of the effects of biochar application to soils on crop productivity using meta-analysis. *Agric. Ecosyst. Environ.* **2011**, *144*, 175–187. [CrossRef]
21. Biederman, L.A.; Harpole, W.S. Biochar and its effects on plant productivity and nutrient cycling: A meta-analysis. *Gcb Bioenergy* **2013**, *5*, 202–214. [CrossRef]
22. Quilliam, R.S.; Glanville, H.C.; Wade, S.C.; Jones, D.L. Life in the 'charosphere'–Does biochar in agricultural soil provide a significant habitat for microorganisms? *Soil Biol. Biochem.* **2013**, *65*, 287–293. [CrossRef]

23. Jaafar, N.M.; Clode, P.L.; Abbott, L.K. Microscopy observations of habitable space in biochar for colonization by fungal hyphae from soil. *J. Integr. Agric.* **2014**, *13*, 483–490. [CrossRef]
24. Crombie, K.; Mašek, O. Pyrolysis biochar systems, balance between bioenergy and carbon sequestration. *GCB Bioenergy* **2015**, *7*, 349–361. [CrossRef]
25. Zhu, Q.; Peng, X.; Huang, T. Contrasted effects of biochar on maize growth and N use efficiency depending on soil conditions. *Int. Agrophys.* **2015**, *29*, 257–266. [CrossRef]
26. Peng, X.; Ye, L.; Wang, C.; Zhou, H.; Sun, B. Temperature-and duration-dependent rice straw-derived biochar: Characteristics and its effects on soil properties of an Ultisol in southern China. *Soil Tillage Res.* **2011**, *112*, 159–166. [CrossRef]
27. Ennis, C.J.; Evans, A.G.; Islam, M.; Ralebitso-Senior, T.K.; Senior, E. Biochar: Carbon sequestration, land remediation, and impacts on soil microbiology. *Crit. Rev. Env. Sci. Technol.* **2012**, *42*, 2311–2364. [CrossRef]
28. Malghani, S.; Gleixner, G.; Trumbore, S.E. Chars produced by slow pyrolysis and hydrothermal carbonization vary in carbon sequestration potential and greenhouse gases emissions. *Soil Biol. Biochem.* **2013**, *62*, 137–146. [CrossRef]
29. Stewart, C.E.; Zheng, J.; Botte, J.; Cotrufo, M.F. Co-generated fast pyrolysis biochar mitigates green-house gas emissions and increases carbon sequestration in temperate soils. *GCB Bioenergy* **2013**, *5*, 153–164. [CrossRef]
30. Pandit, N.R.; Mulder, J.; Hale, S.E.; Martinsen, V.; Schmidt, H.P.; Cornelissen, G. Biochar improves maize growth by alleviation of nutrient stress in a moderately acidic low-input Nepalese soil. *Sci. Total Environ.* **2018**, *625*, 1380–1389. [CrossRef]
31. Widowati, W.U.; Guritno, B.; Soehono, L. The effect of biochar on the growth and N fertilizer requirement of maize (Zea mays L.) in green house experiment. *J. Agric. Sci.* **2012**, *4*, 255.
32. Qiao-Hong, Z.; Xin-Hua, P.; HUANG, T.-Q.; Zu-Bin, X.; Holden, N. Effect of biochar addition on maize growth and nitrogen use efficiency in acidic red soils. *Pedosphere* **2014**, *24*, 699–708.
33. Ndor, E.; Amana, S.; Asadu, C. Effect of biochar on soil properties and organic carbon sink in degraded soil of Southern Guinea Savanna Zone, Nigeria. *Int. J. Plant. Soil Sci.* **2015**, *4*, 252–258. [CrossRef]
34. Clough, T.J.; Condron, L.M. Biochar and the nitrogen cycle: Introduction. *J. Environ. Qual.* **2010**, *39*, 1218–1223. [CrossRef]
35. Laird, D.A.; Fleming, P.; Davis, D.D.; Horton, R.; Wang, B.; Karlen, D.L. Impact of biochar amendments on the quality of a typical Midwestern agricultural soil. *Geoderma* **2010**, *158*, 443–449. [CrossRef]
36. Clough, T.; Condron, L.; Kammann, C.; Müller, C. A review of biochar and soil nitrogen dynamics. *Agronomy* **2013**, *3*, 275–293. [CrossRef]
37. Cameron, K.; Di, H.J.; Moir, J. Nitrogen losses from the soil/plant system: A review. *Ann. Appl. Biol.* **2013**, *162*, 145–173. [CrossRef]
38. Zhang, Y.; Ding, H.; Zheng, X.; Cai, Z.; Misselbrook, T.; Carswell, A.; Müller, C.; Zhang, J. Soil N transformation mechanisms can effectively conserve N in soil under saturated conditions compared to unsaturated conditions in subtropical China. *Biol. Fertil. Soils* **2018**, *54*, 495–507. [CrossRef]
39. Van Drecht, G.; Bouwman, A.; Knoop, J.; Beusen, A.; Meinardi, C. Global modeling of the fate of nitrogen from point and nonpoint sources in soils, groundwater, and surface water. *Glob. Biogeochem. Cycles* **2003**, *17*. [CrossRef]
40. Huygens, D.; Rütting, T.; Boeckx, P.; Van Cleemput, O.; Godoy, R.; Müller, C. Soil nitrogen conservation mechanisms in a pristine south Chilean Nothofagus forest ecosystem. *Soil Biol. Biochem.* **2007**, *39*, 2448–2458. [CrossRef]
41. Rütting, T.; Müller, C. Process-specific analysis of nitrite dynamics in a permanent grassland soil by using a Monte Carlo sampling technique. *Eur. J. Soil Sci.* **2008**, *59*, 208–215. [CrossRef]
42. Rütting, T.; Huygens, D.; Müller, C.; Van Cleemput, O.; Godoy, R.; Boeckx, P. Functional role of DNRA and nitrite reduction in a pristine south Chilean Nothofagus forest. *Biogeochemistry* **2008**, *90*, 243–258. [CrossRef]
43. Choi, W.-J.; Ro, H.-M.; Lee, S.-M. Natural 15N abundances of inorganic nitrogen in soil treated with fertilizer and compost under changing soil moisture regimes. *Soil Biol. Biochem.* **2003**, *35*, 1289–1298. [CrossRef]
44. Di, H.; Cameron, K. Reducing environmental impacts of agriculture by using a fine particle suspension nitrification inhibitor to decrease nitrate leaching from grazed pastures. *Agric.; Ecosyst. Environ.* **2005**, *109*, 202–212. [CrossRef]
45. Cai, Z.; Gao, S.; Xu, M.; Hanson, B.D. Evaluation of potassium thiosulfate as a nitrification inhibitor to reduce nitrous oxide emissions. *Sci. Total Environ.* **2018**, *618*, 243–249. [CrossRef]
46. Zhang, M.; Fan, C.; Li, Q.; Li, B.; Zhu, Y.; Xiong, Z. A 2-yr field assessment of the effects of chemical and biological nitrification inhibitors on nitrous oxide emissions and nitrogen use efficiency in an intensively managed vegetable cropping system. *Agric. Ecosyst. Environ.* **2015**, *201*, 43–50. [CrossRef]
47. Maeda, M.; Zhao, B.; Ozaki, Y.; Yoneyama, T. Nitrate leaching in an Andisol treated with different types of fertilizers. *Environ. Pollut.* **2003**, *121*, 477–487. [CrossRef]
48. Yang, M.; Fang, Y.; Sun, D.; Shi, Y. Efficiency of two nitrification inhibitors (dicyandiamide and 3, 4-dimethypyrazole phosphate) on soil nitrogen transformations and plant productivity: A meta-analysis. *Sci. Rep.* **2016**, *6*, 22075. [CrossRef]
49. Friedl, J.; Scheer, C.; Rowlings, D.W.; Mumford, M.T.; Grace, P.R. The nitrification inhibitor DMPP (3, 4-dimethylpyrazole phosphate) reduces N2 emissions from intensively managed pastures in subtropical Australia. *Soil Biol. Biochem.* **2017**, *108*, 55–64. [CrossRef]
50. Monaghan, R.; Smith, L.; Ledgard, S. The effectiveness of a granular formulation of dicyandiamide (DCD) in limiting nitrate leaching from a grazed dairy pasture. *N. Z. J. Agric. Res.* **2009**, *52*, 145–159. [CrossRef]

51. Marsden, K.A.; Marín-Martínez, A.J.; Vallejo, A.; Hill, P.W.; Jones, D.L.; Chadwick, D.R. The mobility of nitrification inhibitors under simulated ruminant urine deposition and rainfall: A comparison between DCD and DMPP. *Biol. Fertil. Soils* **2016**, *52*, 491–503. [CrossRef]

52. Zerulla, W.; Barth, T.; Dressel, J.; Erhardt, K.; von Locquenghien, K.H.; Pasda, G.; Rädle, M.; Wissemeier, A. 3, 4-Dimethylpyrazole phosphate (DMPP)–a new nitrification inhibitor for agriculture and horticulture. *Biol. Fertil. Soils* **2001**, *34*, 79–84. [CrossRef]

53. Benckiser, G.; Christ, E.; Herbert, T.; Weiske, A.; Blome, J.; Hardt, M. The nitrification inhibitor 3, 4-dimethylpyrazole-phosphat (DMPP)-quantification and effects on soil metabolism. *Plant. Soil* **2013**, *371*, 257–266. [CrossRef]

54. Wissemeier, A.; Linzmeier, W.; Gutser, R.; Weigelt, W.; Schmidhalter, U. The new nitrification inhibitor DMPP (ENTEC®)—comparisons with DCD in model studies and field applications. In *Plant Nutrition*; Springer: Berlin/Heidelberg, Germany, 2001; pp. 702–703.

55. Abbasi, M.K.; Hina, M.; Tahir, M.M. Effect of *Azadirachta indica* (neem), sodium thiosulphate and calcium chloride on changes in nitrogen transformations and inhibition of nitrification in soil incubated under laboratory conditions. *Chemosphere* **2011**, *82*, 1629–1635. [CrossRef]

56. Cai, Z.; Gao, S.; Hendratna, A.; Duan, Y.; Xu, M.; Hanson, B.D. Key factors, soil nitrogen processes, and nitrite accumulation affecting nitrous oxide emissions. *Soil Sci. Soc. Am. J.* **2016**, *80*, 1560–1571. [CrossRef]

57. McGeough, K.; Laughlin, R.J.; Watson, C.; Müller, C.; Ernfors, M.; Cahalan, E.; Richards, K.G. The effect of cattle slurry in combination with nitrate and the nitrification inhibitor dicyandiamide on in situ nitrous oxide and dinitrogen emissions. *Biogeosciences* **2012**, *9*, 4909–4919. [CrossRef]

58. Misselbrook, T.; Cardenas, L.; Camp, V.; Thorman, R.; Williams, J.; Rollett, A.; Chambers, B. An assessment of nitrification inhibitors to reduce nitrous oxide emissions from UK agriculture. *Environ. Res. Lett.* **2014**, *9*, 115006. [CrossRef]

59. Artola, E.; Cruchaga, S.; Ariz, I.; Moran, J.F.; Garnica, M.; Houdusse, F.; Houdusse, F.; Mina, J.M.G.; Irigoyen, I.; Lasa, B.; et al. Effect of N-(n-butyl) thiophosphoric triamide on urea metabolism and the assimilation of ammonium by Triticum aestivum L. *Plant. Growth Regul.* **2011**, *63*, 73–79. [CrossRef]

60. Du, N.; Chen, M.; Liu, Z.; Sheng, L.; Xu, H.; Chen, S. Kinetics and mechanism of jack bean urease inhibition by Hg^{2+}. *Chem. Cent. J.* **2012**, *6*, 1–7. [CrossRef] [PubMed]

61. Prakash, O.; Vishwakarma, D.K. Inhibition of urease from seeds of the water melon (*Citrullus vulgaris*) by heavy metal ions. *J. Plant. Biochem.* **2001**, *10*, 147–149. [CrossRef]

62. Mohanty, S.; Patra, A.K.; Chhonkar, P.K. Neem (*Azadirachta indica*) seed kernel powder retards urease and nitrification activities in different soils at contrasting moisture and temperature regimes. *Bioresour. Technol.* **2008**, *99*, 894–899. [CrossRef] [PubMed]

63. Suescun, F.; Paulino, L.; Zagal, L.; Ovalle, C.; Munoz, C. Plant extracts from the Mediterranean zone of Chile potentially affect soil microbial activity related to N transformations: A laboratory experiment. *Acta Agric. Scand. Sect. B Soil Plant. Sci.* **2012**, *62*, 556–564. [CrossRef]

64. Kumari, K.; Moldrup, P.; Paradelo, M.; de Jonge, L.W. Phenanthrene sorption on biochar-amended soils: Application rate, aging, and physicochemical properties of soil. *Water Air Soil Pollut.* **2014**, *225*, 2105. [CrossRef]

65. Rechberger, M.V.; Kloss, S.; Rennhofer, H.; Tintner, J.; Watzinger, A.; Soja, G.; Lichtenegger, H.; Zehetner, F. Changes in biochar physical and chemical properties: Accelerated biochar aging in an acidic soil. *Carbon* **2017**, *115*, 209–219. [CrossRef]

66. Mandal, S.; Thangarajan, R.; Bolan, N.S.; Sarkar, B.; Khan, N.; Ok, Y.S.; Naidu, R. Biochar-induced concomitant decrease in ammonia volatilization and increase in nitrogen use efficiency by wheat. *Chemosphere* **2016**, *142*, 120–127. [CrossRef] [PubMed]

67. Walkley, A.; Black, I.A. An examination of the Degtjareff method for determining soil organic matter, and a proposed modification of the chromic acid titration method. *Soil. Sci.* **1934**, *37*, 29–38. [CrossRef]

68. Hesse, P.R. *A Textbook of Soil Chemical Analysis (No. 631.41 H4)*; Chemical Publishing Company: Gloucester, MA, USA, 1972.

69. Olsen, S.R. *Estimation of Available Phosphorus in Soils by Extraction with Sodium Bicarbonate (No. 939)*; US Department of Agriculture: Washington, DC, USA, 1954.

70. Junsomboon, J.; Jakmunee, J. Determination of potassium, sodium, and total alkalies in portland cement, fly ash, admixtures, and water of concrete by a simple flow injection flame photometric system. *J. Anal. Meth. Chem.* **2011**, *2011*, 742656. [CrossRef]

71. Helrich, K. *Nitrogen in Meat Kjeldahl Method*, 15th ed.; Association of Official Analytical Chemists, Inc.: Rockville, MD, USA, 1990; p. 935.

72. SAS Institute. *SAS/STAT 9.1 User's Guide*; SAS Inst.: Cary, NC, USA, 2004.

73. Byerlee, D. *From Agronomic data to Farmer's Recommendation: An Economics Training Manual*; CIMMYT: Mexico City, Mexico, 1988; pp. 31–33.

74. Hammad, H.M.; Ahmad, A.; Abbas, F.; Farhad, W.; Cordoba, B.C.; Hoogenboom, G. Water and Nitrogen Productivity of Maize under Semiarid Environments. *Crop. Sci.* **2015**, *55*, 877–888. [CrossRef]

75. Dempster, D.; Jones, D.; Murphy, D. Organic nitrogen mineralisation in two contrasting agro-ecosystems is unchanged by biochar addition. *Soil Biol. Biochem.* **2012**, *48*, 47–50. [CrossRef]

76. Taghizadeh-Toosi, A.; Clough, T.J.; Sherlock, R.R.; Condron, L.M. Biochar adsorbed ammonia is bioavailable. *Plant. Soil* **2012**, *350*, 57–69. [CrossRef]

77. Genesio, L.; Miglietta, F.; Lugato, E.; Baronti, S.; Pieri, M.; Vaccari, F. Surface albedo following biochar application in durum wheat. *Env. Res. Lett.* **2012**, *7*, 014025. [CrossRef]

78. Mathialagan, R.; Mansor, N.; Al-Khateeb, B.; Mohamad, M.H.; Shamsuddin, M.R. Evaluation of Allicin as Soil Urease Inhibitor. *Procedia Eng.* **2017**, *184*, 449–459. [CrossRef]

79. Juszkiewicz, A.; Zaborska, A.; Aptas, Z. Olech. A study of the inhibition of jack bean urease by garlic extract. *Food Chem* **2004**, *85*, 553–558. [CrossRef]

80. Akiyama, H.; Yan, X.; Yagi, K. Evaluation of effectiveness of enhanced-efficiency fertilizers as mitigation options for N_2O and NO emissions from agricultural soils: Meta-analysis. *Glob. Chang. Biol.* **2010**, *16*, 1837–1846. [CrossRef]

81. Jaiswal, A.K.; Elad, Y.; Graber, E.R.; Frenkel, O. Rhizoctonia solani suppression and plant growth promotion in cucumber as affected by biochar pyrolysis temperature, feedstock and concentration. *Soil Biol. Biochem.* **2014**, *69*, 110–118. [CrossRef]

82. Olmo, M.; Villar, R.; Salazar, P.; Alburquerque, J.A. Changes in soil nutrient availability explain biochar's impact on wheat root development. *Plant. Soil* **2016**, *399*, 333–343. [CrossRef]

83. Kookana, R.S.; Sarmah, A.K.; Van Zwieten, L.; Krull, E.; Singh, B. Biochar application to soil: Agronomic and environmental benefits and unintended consequences. In *Advances in Agronomy*; Elsevier: Amsterdam, The Netherlands, 2011; Volume 112, pp. 103–143.

84. Martinsen, V.; Mulder, J.; Shitumbanuma, V.; Sparrevik, M.; Børresen, T.; Cornelissen, G. Farmer-led maize biochar trials: Effect on crop yield and soil nutrients under conservation farming. *J. Plant. Nutr. Soil Sci.* **2014**, *177*, 681–695. [CrossRef]

85. Hale, S.; Alling, V.; Martinsen, V.; Mulder, J.; Breedveld, G.; Cornelissen, G. The sorption and desorption of phosphate-P, ammonium-N and nitrate-N in cacao shell and corn cob biochars. *Chemosphere* **2013**, *91*, 1612–1619. [CrossRef] [PubMed]

86. Nigussie, A.; Kissi, E.; Misganaw, M.; Ambaw, G. Effect of biochar application on soil properties and nutrient uptake of lettuces (Lactuca sativa) grown in chromium polluted soils. *Am. Eurasian J. Agric. Env. Sci.* **2012**, *12*, 369–376.

87. Agegnehu, G.; Bass, A.M.; Nelson, P.N.; Muirhead, B.; Wright, G.; Bird, M.I. Biochar and biochar-compost as soil amendments: Effects on peanut yield, soil properties and greenhouse gas emissions in tropical North Queensland, Australia. *Agric. Ecosyst. Environ.* **2015**, *213*, 72–85. [CrossRef]

88. DeLuca, T.H.; Gundale, M.J.; MacKenzie, M.D.; Jones, D.L. Biochar effects on soil nutrient transformations. *Biochar Environ. Manag. Sci. Technol. Implement.* **2015**, *2*, 421–454.

89. Trupiano, D.; Cocozza, C.; Baronti, S.; Amendola, C.; Vaccari, F.P.; Lustrato, G.; Di Lonardo, S.; Fantasma, F.; Tognetti, R.; Scippa, G.S. The effects of biochar and its combination with compost on lettuce (*Lactuca sativa* L.) growth, soil properties, and soil microbial activity and abundance. *Int. J. Agron.* **2017**, *2017*, 3158207. [CrossRef]

90. Hagemann, N.; Joseph, S.; Conte, P.; Albu, M.; Obst, M.; Borch, T.; Orsetti, S.; Subdiaga, E.; Behrens, S.; Kappler, A. Composting-derived organic coating on biochar enhances its affinity to nitrate. *Egu Gen. Assem. Conf. Abstr.* **2017**, *19*, 10775.

91. Schmidt, H.; Pandit, B.; Martinsen, V.; Cornelissen, G.; Conte, P.; Kammann, C. Fourfold increase in pumpkin yield in response to low-dosage root zone application of urine-enhanced biochar to a fertile tropical soil. *Agriculture* **2015**, *5*, 723–741. [CrossRef]

92. Mensah, A.K.; Frimpong, K.A. Biochar and/or Compost Applications Improve Soil Properties, Growth, and Yield of Maize Grown in Acidic Rainforest and Coastal Savannah Soils in Ghana. *Int. J. Agron.* **2018**, *2018*, 6837404. [CrossRef]

93. Farooque, A.A.; Zare, M.; Abbas, F.; Bos, M.; Esau, T.; Zaman, Q. Forecasting potato tuber yield using a soil electromagnetic induction method. *Eur. J. Soil Sci.* **2019**, *71*, 880–897. [CrossRef]

94. Dorff, E.; Beaulieu, M.S. Canadian Agriculture at a Glance—Feeding the Soil Puts Food on Your Plate. Analytical paper, Agriculture Division. Catalogue no. 96-325-X—No. 004; ISSN 0-662-35659-4. 2011. Available online: https://www150.statcan.gc.ca/n1/en/pub/96-325-x/2014001/article/13006-eng.pdf?st=hwBVfcxx (accessed on 28 February 2021).

95. Xiong, J.; Yu, R.; Islam, E.; Zhu, F.; Zha, J.; Sohail, M.I. Effect of Biochar on Soil Temperature under High Soil Surface Temperature in Coal Mined Arid and Semiarid Regions. *Sustainability* **2020**, *12*, 8238. [CrossRef]

 *sustainability*

Article

Design Evaluation and Performance Analysis of a Double-Row Pneumatic Precision Metering Device for *Brassica chinensis*

Bohong Li [1,2], Riaz Ahmad [2,3], Xindan Qi [1,*], Hua Li [2,3,*], Samuel Mbugua Nyambura [2,3], Jufei Wang [2,3], Xi Chen [1,2] and Shengbing Li [1,2]

1 College of Mechanical and Power Engineering, Nanjing Tech University, Nanjing 211816, China; li_bohong319@163.com (B.L.); njut_chenxi@163.com (X.C.); lishengbing1996@163.com (S.L.)
2 Key Laboratory of Intelligent Agricultural Equipment in Jiangsu Province, Nanjing Agricultural University, Nanjing 210031, China; Riazahmad@njau.edu.cn (R.A.); accessmbugua@gmail.com (S.M.N.); wangjufei@stu.njau.edu.cn (J.W.)
3 College of Engineering, Nanjing Agricultural University, Nanjing 210031, China
* Correspondence: xindanqi@njtech.edu.cn (X.Q.); lihua@njau.edu.cn (H.L.)

Abstract: In view of the low seeding efficiency and precision of seeders used for *Brassica chinensis* in China, a new double-row pneumatic precision metering device for *Brassica chinensis* was designed, fabricated, and evaluated. With the characteristics of small size and high sphericity of *Brassica chinensis* seeds in mind, the structure and key dimensions of the metering plate were determined, and a force analysis of the seed-filling process was carried out. The negative pressure (NP), angular velocity (AV) of the metering plate, and cone angle (CA) of the suction hole were selected as the main influencing factors of the experiment. In order to explore the influence of each single factor and the interaction between factors on the seeding performance, a single factor experiment and a central composite design (CCD) experiment were designed, respectively, and the experimental results were analyzed by analysis of variance (ANOVA). After optimizing the main influencing factors such that the target of the qualified index (QI) was greater than 94% and the miss index (MI) was less than 2.5%, it was found that when CA was 60°, NP was 1.55–1.72 kPa, and AV was 1.1–1.9 rad/s, the seeding performance was excellent. The bench verification results of seeding performance ($94\% \leq Q \leq 100\%$, $0 \leq M \leq 2.5\%$) and the coefficient of variation (CV) of seed mass (CV of seed mass in outer and inner circle: 5.15%; CV of total seed mass: 8.60%) under the condition of parameter optimization were analyzed; as a result, the accuracy of the parameter optimization was confirmed.

Keywords: ANOVA; *Brassica chinensis*; coefficient of variation; double-row; metering device; pneumatic

Citation: Li, B.; Ahmad, R.; Qi, X.; Li, H.; Nyambura, S.M.; Wang, J.; Chen, X.; Li, S. Design Evaluation and Performance Analysis of a Double-Row Pneumatic Precision Metering Device for *Brassica chinensis*. *Sustainability* **2021**, *13*, 1374. https://doi.org/10.3390/su13031374

Academic Editor: Muhammad Sultan
Received: 10 December 2020
Accepted: 12 January 2021
Published: 28 January 2021

Publisher's Note: MDPI stays neutral with regard to jurisdictional claims in published maps and institutional affiliations.

1. Introduction

Brassica chinensis (BC) is rich in vitamins and minerals, with a high plant cellulose content. It is widely cultivated in China because its vegetable is favored by people in the North and South [1]. The planting mode of BC is mainly individual planting in China, and artificial seeding is also the main method for the plantation process. Because this kind of seeding method is relatively inefficient, the implementation of mechanized precision seeding technology is a primary requirement for BC cultivation.

The term "small seeds" usually refer to seeds with an average diameter of less than 3 mm, and includes most of the seeds of vegetables and flowers [2]. BC seeds are small in size, but their regular shape and high average sphericity index are suitable for high-speed precision seeding. Precision seeding is defined as the process of using a precision seeder to make a single seed fall accurately into a reserved position in the soil according to certain agronomic requirements [3]. The seed metering device is a core component for the realization of high-speed precision seeding, and its performance is one of the most important factors that affect the quality of the seeding [4]. A pneumatic metering device has the advantages of having a low seed size requirement, does not damage the seed, is suitable for single-grain precision seeding, and is suitable for high-speed seeding [5]. In order to meet the agronomic requirements of narrow-row dense planting of small sized seeds, and to achieve the goal of high-speed precision seeding, it is necessary to design a double-row precision metering device with a simple structure, strong adaptability, and good seeding performance [6].

At present, many experts and scholars have carried out research on pneumatic seed metering devices, but research on multi-row precision metering devices is not extensive. Research carried out has included an air-suction potato seed metering device that improved the seeding performance of large-size seed crops [7]. After theoretical analysis, the main structure and operating parameters of the device were determined. Two orthogonal tests (conventional tuber test and mini-tuber test) were then carried out to analyze the influence of the operating parameters on the seeding quality, which were then evaluated with corresponding indicators (multiple-seeding index (MTI), missing-seeding index (MI), and qualified index (QI)). The results of the conventional tuber test (mini-tuber test) indicated that the MTI was 1.1% (0.5%), MI was 0.8% (0.6%), and QI was 98.1% (98.9%) under the conditions of a 30 (35) r/min rotating speed, 25 (17) cm seed height, and 10 (3.5) kPa pickup vacuum pressure. A novel combination vacuum and spoon belt metering device to improve the efficiency and precise seeding of potatoes was designed by [8]. The structure and dimensions of the seed metering device are key components that have to be included in the experimental design to verify the seeding performance of the seed metering device. The experiment results found that, when the seeding belt speed was 0.43 m·s^{-1}, the spoon aperture was 15.72 mm, and the cleaning air pressure was 2.94 kPa, the seeding effect of the metering device was high (the missing seed index was 3.97%, the multiple seed index was 4.65%, and the qualified seed index was 91.38%). A six-row air-blowing centralized precision seed-metering device for the realization of precision seeding of *Panax notoginseng* was designed by [9]. A mechanical model of the movement process for the seed metering device was constructed based on a method combining theoretical calculations and simulation analysis. The outlet pressure of the air nozzle and the forward velocity and cone angle of the hole were selected as the test factors for carrying out the quadratic rotation orthogonal combination test. After parameter optimization, it was found that when the cone angle of the hole was 50°, the forward velocity was less than 0.73 m/s, and the outlet pressure of the air nozzle was 0.32–0.52 kPa, the qualified index of grain spacing was higher than 94%, the miss index was less than 3%, the multiple index was less than 5%, and the coefficient of variation of the row displacement consistency was less than 5%.

A pneumatic disk with four rows for planting rapeseed was designed by Elebaid et al. [10], and its performance under several rotating speeds and vacuum pressure values was investigated. Subsequently, the seed mass of the four-row design was measured and analyzed under the influence of rotational speed and negative pressure. It was found that the seed mass of row 1 and the seed mass of row 4 were significantly different under high-speed conditions (25 and 30 r/min). Taghinezhad et al. [11] designed and modeled a new mechanism for a sugarcane metering device with Catia software. The effect of metering device tooth length and the speed of the sugarcane billet metering device were studied in order to find the best combination for improving the distance uniformity and filling performance of the metering device cells. The analytical hierarchy process was used to select the best combination, and it was found that a 2 cm tooth length and a 0.75 m/s forward speed was the best-suited combination for the metering device, and the consistency ratio was computed as being lower than 0.1. Mandal et al. [12] designed a pneumatic seed metering mechanism for a power tiller-operated three-row precision planter. The optimum design and operating parameters of the modular seed metering device were determined by conducting experiments on the sticky belt test stand, with various performance indexes being considered. The optimum design and operation parameters were determined as follows: the number of holes for the seed metering disc was eight, the diameter of the hole was 3.5 mm, the pitch circle diameter of the disc was 116 mm, operational speed was 0.11 m·s^{-1}, and negative pressure was 6 kPa.

The objectives of this study are to design, fabricate, and evaluate a new double-row pneumatic precision metering device for BC. Additionally, the effects of the cone angle (CA) of the suction hole, the angular velocity (AV) of the metering plate, and the negative pressure (NP) on the seeding performance of the metering device are investigated. Additionally, the range of optimal working parameters is determined and verified.

2. Materials and Methods

2.1. Structure and Working Principle of Seed Metering Device

A new double-row pneumatic precision metering device for BC was designed. The power input mechanism of the seed metering device consisted of a sprocket which transferred power to the shaft for the rotation of the metering plate. The overall structure of the seed metering device is shown in Figure 1. The seeds in the seed box become adsorbed in the suction hole of the metering plate under the effect of NP. The stirring wheel located in the seed-filling zone rotates with the metering plate under the periodic collision by the striking column. As a result, the seeds which were originally accumulated in the seed-filling zone achieved a flowing state. The seeds adsorbed by the suction hole in the seed-filling zone are transferred using the metering plate to the seed-cleaning zone. The seed cleaning devices of the inner and outer circles scrape off the multiple seeds absorbed by one hole to ensure that one seed is absorbed by one hole. The seed rotates with the metering plate to the seed-throwing zone, and due to the cut-off of the negative pressure airway, the seed breaks away from the adsorption force of NP and falls under the action of gravity and centrifugal force. The diversion channel separates the seeds from the inner and outer circles; after that, the seeds enter the diversion tube to form a double-row seed flow.

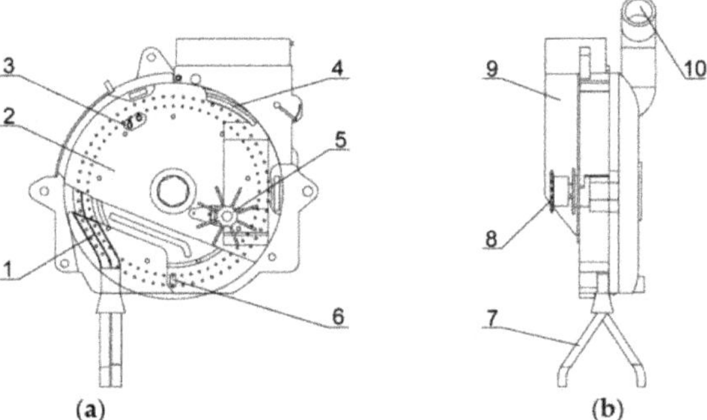

1. Diversion channel; 2. Metering plate; 3. Seed cleaning device in inner circle; 4. Seed cleaning device in outer circle; 5. Stirring wheel; 6. Anti-blocking channel; 7. Diversion tube; 8. Sprocket wheel; 9. Seed box; 10. Negative pressure outlet

Figure 1. Integral structure of seed metering device: (**a**) Front view; (**b**) Side view.

2.2. Physical Properties of Brassica Chinensis

"*Shanghai Ai Ji*" is a common variety of BC in China, and the seeds of this variety were used for the experiment. The physical properties of BC are shown in Table 1. Fifty samples were randomly selected, and the triaxial dimensions were measured using a digital vernier caliper with an accuracy of 0.01 mm. One thousand samples were weighed five times using an electronic balance with an accuracy of 0.001 g, and the mean was then calculated.

Table 1. Physical properties of *Brassica chinensis*.

Physical Properties	Maximum Value, mm	Minimum Value, mm	Mean, mm	Standard Error
Length l	2.04	1.39	1.74	0.15
Width w	1.91	1.30	1.65	0.13
Thickness h	1.97	1.33	1.63	0.16
Thousand Seed mass, g	-	-	2.410	-

According to the results of the triaxial dimensions of seeds in Table 1, the average equivalent diameter $(\overline{De})$ and the average spherical rate $(\overline{Sp})$ of the seeds can be calculated by the following formulas:

$$\overline{De} = \sqrt[3]{\bar{l} \cdot \bar{w} \cdot \bar{h}} \tag{1}$$

$$\overline{Sp} = \frac{\sqrt[3]{\bar{l} \cdot \bar{w} \cdot \bar{h}}}{\bar{l}} \times 100\% \tag{2}$$

where $\bar{l}$ is the average length of the seeds, in mm; $\bar{w}$ is the average width of the seeds, in mm; $\bar{h}$ is the average thickness of the seeds, in mm; $\overline{De}$ is the average equivalent diameter of the seeds, in mm; and $\overline{Sp}$ is the average spherical rate of the seeds, in %.

According to the above equations, the average equivalent diameter of seeds $(\overline{De})$ was 1.67 mm, and the average spherical ratio $(\overline{Sp})$ was 96%.

2.3. Structural Design and Theoretical Analysis of the Metering Plate

2.3.1. Determination of Key Parameters for the Metering Plate

As key parts of the seed metering device, the structure parameters of the metering plate have a significant effect on seed filling performance [13,14]. The integral structure and key dimensions of the metering plate are shown in Figure 2. The position of the outer circle hole is low at the seed-filling zone, and the pressure exerted by the upper seeds on the lower seeds is large, resulting in an increase in the seed-filling resistance of the lower seeds. Moreover, the linear velocity at the center of the outer circle hole is larger than that at the inner circle hole. The seed-filling time is relatively shorter, so the outer circle hole is more difficult to fill under the same conditions. The diameter of the outer circle hole is determined according to the size of the seeds using Equation (3):

$$d_2 = (0.64 \sim 0.66)\overline{De} \tag{3}$$

where d_2 is the diameter of the outer circle hole, in mm.

From Equation (1), the average equivalent diameter $(\overline{De})$ of seeds was 1.67 mm. The range of the diameter of the outer circle hole can be obtained by substituting $\overline{De} = 1.67$ mm into Equation (3): $d_2 = (1.07\sim1.10$ mm). For this design, the diameter (d_2) of the outer circle hole was selected to be 1.1 mm. It is more difficult to fill the outer circle hole than the inner circle, so the diameter (d_2) of the outer circle hole was larger than the diameter (d_1) of the inner circle hole (1.0 mm).

The number of holes on the inner and outer circles determines the size of the double-row metering plate. With the increase in the number of holes of the inner and outer circles, the diameter of the metering plate increased accordingly. Similarly, the linear velocity at the center of the suction hole of the metering plate decreased. As a result, the seed-filling performance of the hole increased with the increase in the seed-filling time [15]. The number of holes was inversely proportional to the product of the rotation speed of the double-row metering plate (r/min) and the planting space of the seeds (mm), and directly proportional to the forward speed of the planter (m/s). The equation for calculating the number of holes is as follows:

$$N = \frac{60v_d}{nl} \tag{4}$$

where N is the number of holes; v_d is the forward speed of the planter, in m/s; n is the rotation speed of the double-row metering plate, in r/min; l is the planting space of the seeds, in mm.

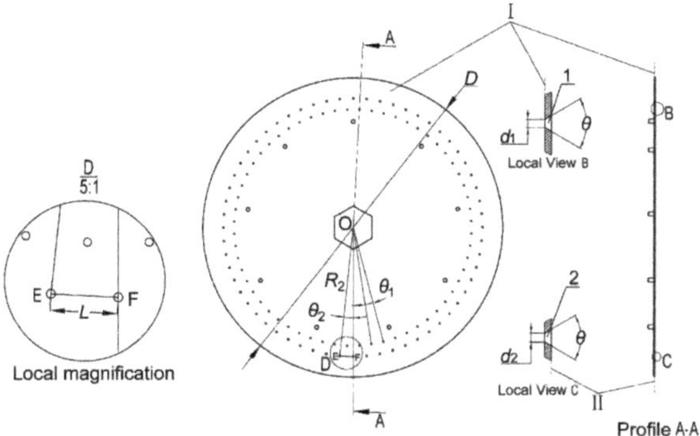

Notes: I. The side in contact with the seed; II. The side in contact with the NP; O. The center of metering plate; E. The center of outer circle hole; F. The center of outer circle adjacent hole.

θ is the cone angle of the suction hole, °; θ_1 is the angle between two adjacent holes in the inner circle, °; θ_2 is the angle between two adjacent holes in the outer circle, °; D is the diameter of the metering plate, mm; L is the distance between point E and point F, mm; R_2 is the radius of the center of outer circle hole, mm.

Figure 2. Integral structure and key dimensions of metering plate of the seed metering device in front view and side view planes.

According to the agronomic requirements, the dense planting spacing of BC is 4~5 cm with a row spacing of 10~11 cm. As such, the planting space for the seeds was $l \leq 5$ cm = 0.05 m. According to the *Design Manual for Agricultural Machinery* [16], the linear velocity at the center of holes of the metering plate was $v \leq 0.35$ m/s. Considering the actual installation size of the seed metering device, the radius of the metering plate with double-row holes was determined to be $R \geq 100$ mm. According to the equation of rotation speed ($n = v/2\pi R$), it can be calculated that the rotation speed of the metering plate was $n \leq 0.557$ r/s = 33.42 r/min. At present, the pneumatic precision metering device can generally adapt to high-speed seeding ($v_d \geq 6$ km/h = 1.67 m/s). Therefore, it can be calculated from Equation (4) that $N \geq 60$, and the number of holes of the inner and outer circles was finally determined as $N_1 = N_2 = 60$.

Since the number of holes (N_1, N_2) in the inner and outer circles was 60, the angle between the two adjacent holes in the inner and outer circles is as follows: $\theta_1 = \theta_2 = 360°/60 = 6°$. As shown in Figure 2, three points (OEF) constitute an isosceles triangle. The distance L satisfies the following: $L > d_2 + 2l_{max} = 1.1$ mm + 2 × 2.04 mm = 5.18 mm, where l_{max} is the maximum length of seeds, mm. From the cosine theorem of isosceles triangle OEF, the equation for calculating R_2 is as follows:

$$R_2 = \sqrt{\frac{L^2}{2(1 - \cos(\theta_2))}} \tag{5}$$

It can be calculated from the above equation that $R_2 > 49.39$ mm. Considering that sufficient space was reserved for the inner circle hole of the metering plate and the space of the metering plate was fully utilized, the radius of the center of the outer circle hole was determined to be 75 mm ($R_2 = 75$ mm). The design of the metering plate diameter should be greater than $2R_2$ and a sufficient margin should be left. Combined with the *Design Manual for Agricultural Machinery* [16], the diameter of the metering plate was finally determined to be 200 mm ($D = 200$ mm).

The space between the inner and outer circles holes is an important factor to ensure the seeding quality of a double-row metering device. Furthermore, the space of the holes directly affects the stability of seed movement. The relationship between the inner and outer holes is shown in Figure 3.

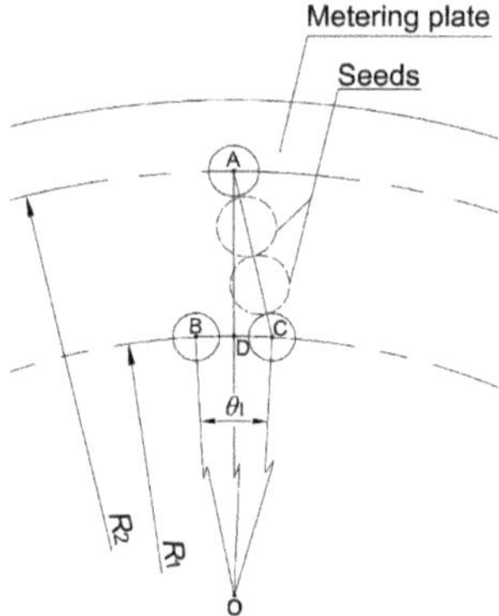

Notes: A. The center of outer circle hole; B. The center of inner circle hole; C. The center of inner circle adjacent hole; D. The center of arc length BC; O. The center of the metering plate

Figure 3. Geometry relationship of holes of inner and outer circles.

According to Figure 3, the following geometric relationships are determined:

$$\begin{cases} \text{Pythagorean theorem}: l^2_{AD} + l^2_{CD} = l^2_{AC} \\ \text{Restrictions on length AC}: l_{AC} > \frac{d_1+d_2}{2} + 2\overline{De} \end{cases} \tag{6}$$

$$\begin{cases} l_{CD} = R_1 \sin\left(\frac{\theta_1}{2}\right) \\ R_1 = R_2 - l_{AD} \end{cases} \tag{7}$$

where R_1 is the radius of the center of the inner circle hole, in mm.

The spacing (l_{AD}) between the inner and outer holes can be obtained from Equations (6) and (7):

$$l_{AD} > \sqrt{\left(\frac{d_1+d_2}{2} + 2\overline{De}\right)^2 - \left[(R_2 - l_{AD})\sin\left(\frac{\theta_1}{2}\right)\right]^2} \tag{8}$$

The spacing between the inner and outer holes can be obtained ($l_{AD} > 2.18$ mm) by substituting the known parameters into Equation (8). Taking into account the reasonable use of the space of the metering plate and the non-interference of the inner and outer holes, the spacing between the holes of the inner and outer circles (l_{AD}) was determined to be 8 mm.

2.3.2. Design of Diversion Tube for Seed Metering Device

The diversion tube guides the seeds placed on the metering plate of double-row holes to form double-rows and the seeds flow and fall smoothly along the surface of the wall. The seeds drop from the outer circle holes through the No. 1 diversion tube, and the inner circle holes through the No. 2 diversion tube. The diversion tube is shown in Figure 4.

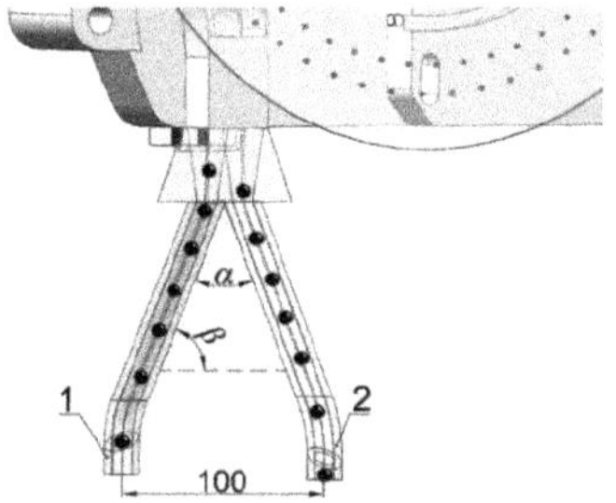

α is the diversion angle, °; β is the angle between diversion tube and horizontal plane, °.

Figure 4. The structure of diversion tube.

After analyzing the process of seed-dropping, the diversion angle α of the diversion tube was finally determined to be 60°. According to the experiment measurements, the static friction coefficient was completed earlier and the average value of the static friction angle between the seed and the diversion tube (Material: DSM IMAGE8000) was 18.95°. When β is greater than the static friction angle, the relative movement occur between the solid surfaces. The angle β can be calculated as: $\beta = (180° - \alpha)/2 = 60°$. The seeds can fall smoothly in the diversion tube and form double-row seed flows at $\beta > 18.95°$. In this study, narrow rows were considered for planting BC, and row space was set to 100 mm (10 cm) according to agronomic requirements.

2.3.3. Force Analysis of Seed-Filling Process

Owing to high average spherical ratio, the seeds of BC can be regarded as a sphere during the force analysis of the seed-filling process. The force analysis is aimed at the outer circle seeds that are difficult to be adsorbed by the suction hole in the seed-filling zone. During the seed-filling process, the force acting on the adsorbed seeds was divided into component forces in three directions (x, y, z), as shown in Figure 5.

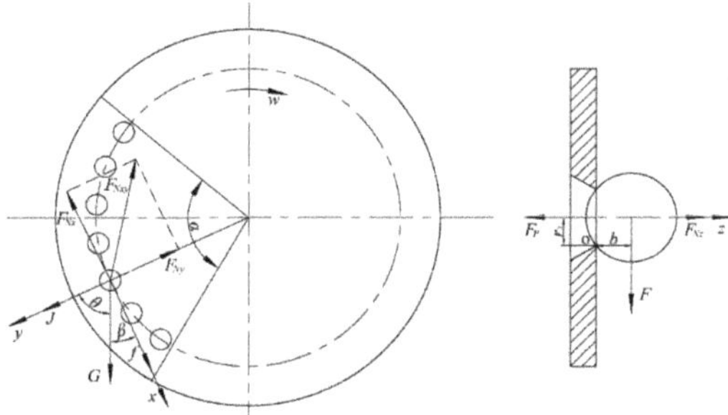

Note: f is the friction resistance of air and seed population to adsorbed seed, N; G is gravity, N; J is the centrifugal force of the seed; F_{Nx} is the support of the suction hole to the seed in the x axis direction, N; F_{Ny} is the support of the suction hole to the seed in the y axis direction, N; F_{Nz} is the support of the suction hole to the seed in the z axis direction, N; F_{Nxy} is the resultant force of F_{Nx} and F_{Ny}, N; F_P is the adsorption force on the seed, N; F is the resultant force of J, G, and f, N; α is the seed-filling zone; β is the angle between G and x axis, °; θ is the angle between G and y axis, °; b is the distance between F and o, m; r_2 is the distance between F_P and o, m;

Figure 5. Force analysis of seeds on suction holes.

A suction hole adsorbs a single seed, which must meet the following moment equilibrium conditions:

$$F_b \leq F_P r_2 \tag{9}$$

For the seeds to be steadily adsorbed by the suction hole, the equilibrium condition of forces must be satisfied in the xy plane:

$$\begin{cases} x \text{ axis} : f + G\cos(\beta) = F_{Nx} \\ y \text{ axis} : J + G\cos(\theta) = F_{Ny} \end{cases} \tag{10}$$

The resultant force F_{Nxy} of F_{Nx} and F_{Ny} on the xy plane are obtained according to Equation (10):

$$F_{Nxy} = \sqrt{F_{Nx}^2 + F_{Ny}^2} = \sqrt{(f + G\cos(\beta))^2 + (J + G\cos(\theta))^2} \tag{11}$$

Since F is numerically equal to F_{Nxy} ($F = F_{Nxy}$), Equation (9) is rewritten as follows:

$$F_P \geq \frac{\sqrt{(f + G\cos(\beta))^2 + (J + G\cos(\theta))^2}\, b}{r_2} \tag{12}$$

According to the pressure formula ($P = F/S$), the adsorption pressure of the suction hole is as follows:

$$P = \frac{F_P}{S} \geq \frac{\sqrt{(f+G\cos(\beta))^2+(J+G\cos(\theta))^2}\,b}{r_2 S}$$
$$\geq \frac{\sqrt{(f+G\cos(\beta))^2+(J+G\cos(\theta))^2}\,b}{\pi r_2^3} \tag{13}$$

where P is the adsorption pressure of the suction hole, in Pa; S is the area of suction hole, in mm^2; $f = mg\lambda$, $\lambda = (6{\sim}10)\tan(\varepsilon)$, ε is the angle of repose of seeds, in $°$; $J = mv^2/R_2$, v_h is the linear velocity at the center of the suction hole, in m/s.

The 1000-grain weight of the seed measured in the early stage was 2.41 g, so the average mass of each seed was 2.41×10^{-6} kg ($m = 2.41 \times 10^{-6}$ kg). The angle $\theta = 60°$ and $\beta = 30°$ were measured when the hole was at the optimum seed-filling position. In the early stage, the angle of repose of seeds was measured to be $25.46°$ ($\varepsilon = 25.46°$), so the value of λ can be calculated as 3.809 ($\lambda = 8\tan(25.46°) = 3.809$). According to the *Design Manual for Agricultural Machinery* [16], the linear velocity at the center of the holes of the metering plate was $v \leq 0.35$ m/s. The range of distance b was determined to be 1.21–1.83 mm, so the average value of b was 1.52 mm ($b = 0.00152$ m). Substituting the above parameters into Equation (13), the P obtained is greater than 324.4 Pa ($P \geq 324.4$ Pa).

2.4. Experimental Materials and Equipment

The seeds of "*Shanghai Ai Ji*" were used as experimental material for the study. A self-built double-row metering device bench (Figure 6) was used for the experiments.

1. Conveyor belt; 2. Digital display governor; 3. Vacuum tube; 4. Bench; 5. Piezometric tube; 6. Positive pressure tube

Figure 6. Experimental object and devices: (**a**) Seeds of "*Shanghai Ai Ji*"; (**b**) Experimental devices used for seeding performance.

2.5. Experimental Methods and Evaluation Indicators

Combined with the research results of relevant scholars and previous experimental research, the main parameters affecting seeding performance were determined to be negative pressure, angular velocity of the metering plate, and cone angle of the suction hole. Therefore, NP, AV, and CA were selected as the main experimental factors of this experiment [17,18].

A suitable NP value can adsorb the seeds and ensure that only one seed is adsorbed by one suction hole. According to the theoretical calculation results, the minimum value of NP was ($P \geq 324.4$ Pa) and based on the design of this study, a combination of a double-row metering plate coupled with the existence of pressure loss resulted in the selection requirements for the NP value being relatively strict [19]. After the pre-experiment, the NP value was selected to be 0.5~2.5 kPa. It was found that when the rotation speed of the metering plate exceeded 35 r/min (converted to AV of 3.67 rad/s), the seeding performance of the metering device decreased sharply. As such, the AV of the metering plate was changed at 1.5–3.5 rad/s. The change in the CA of the suction hole directly affected the change in the flow field at the suction hole, which led to a change in adsorption force of the suction hole. According to the pre-experiment, when the CA of the suction hole changed from 45° to 75°, it was observed that the adsorption situation of the suction hole of the inner and outer circles was good. Therefore, three kinds of metering plate with different cone angles of the suction hole were custom-machined, having cone angles of 45°, 60°, and 75°.

Each group of experiments was repeated three times, and its average value was taken. According to the National Standard of P.R.C (GB/T 6973-2005 Testing Methods of Single Seed Drills (Precision Drills)) [20], the QI and the MI of the inner and outer circles of the metering device were determined as the indexes of seeding performance in this experiment. This experiment measured 180 samples of planting spacing. The dense planting spacing of BC was 4~5 cm according to the agronomic requirements, so the theoretically qualified planting spacing (L) of this experiment was set to 4.5 cm. According to the requirements where the planting spacing was within the range of ($0.5L = 2.25$ cm, and $1.5L = 6.75$ cm), it was qualified spacing; where the planting spacing was greater than 6.75 cm, it was miss spacing; and where the planting spacing was less than 2.25 cm, it was multiple spacing. Qualified spacing and miss spacing coupled with the total sample number (180) were used to calculate the qualified index (QI) and miss index (MI), respectively, as percentages. The experimental scheme is shown in Table 2.

Table 2. Experiment factors and levels.

Level	NP [a] x_1, kPa	AV [b] x_2, rad·s^{-1}	CA [c] x_3, °
−1	0.5	1.5	45
0	1.5	2.5	60
1	2.5	3.5	75

Note: [a] NP = Negative pressure; [b] AV = Angular velocity; [c] CA = Cone angle.

3. Results and Discussion

3.1. Single Factor Experiment

3.1.1. Effect of Negative Pressure on Seeding Performance

From Figure 7, it can be seen that when the AV of the metering plate was 2.5 rad/s and the CA was 60°, the influence trend of NP on the seeding performance of the inner and outer circles was essentially similar. With increase in NP, the QI of the inner and outer circles first increased and then decreased, and the MI of the inner and outer circles decreased continuously. The qualified index of the outer circle (Q_O) reached the maximum value ($Q_{Omax} = 95.45\%$) and the qualified index of the inner circle (Q_I) reached the maximum value ($Q_{Imax} = 94.58\%$) when the NP was 1.71 and 1.63 kPa, respectively.

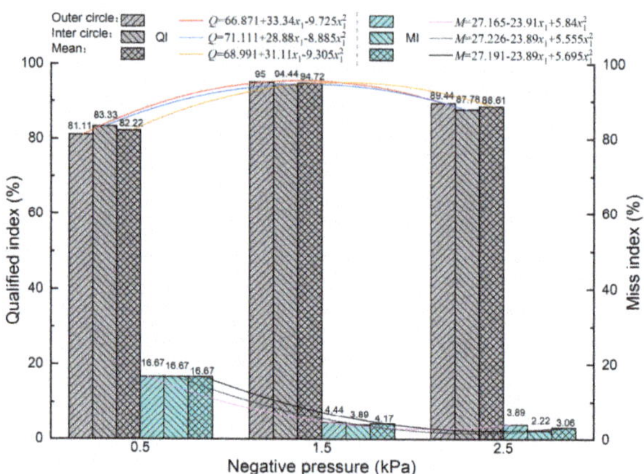

Figure 7. Effect of NP on seeding performance at AV of 2.5 rad/s and CA of 60°.

When the NP value was low, the adsorption force of the suction holes on the seeds was relatively small such that the phenomenon of missed suctioning occurred in the suction hole of the inner and outer circles. This resulted in the MI being high ($M \geq 7\%$). At this time, owing to the large radius of the outer circle, the high linear velocity at the center of the suction hole of the outer circle, and the smaller time of seed-filling, the Q_I was higher than the Q_O. When the NP gradually increased, the adsorption force of the suction holes on the seeds also increased, and the MI of the inner and outer circles decreased sharply. When the NP was too high, there were some suction holes in the inner and outer circles that adsorbed multiple seeds, which led to an increase in the multiple index of the inner and outer circles. As a result, the QI was reduced at this time.

From Figure 7, it can also be seen that the QI of inner and outer circles was high ($Q \geq 92\%$) and the MI of inner and outer circles was low ($M \leq 4\%$) when the NP was in the range of 1.57–2.16 kPa.

3.1.2. Effect of Angular Velocity on Seeding Performance

The influence trend of AV of the metering plate on the seeding performance of the inner and outer circles was fundamentally similar (Figure 8) at 1.5 kPa negative pressure and 60° cone angle.

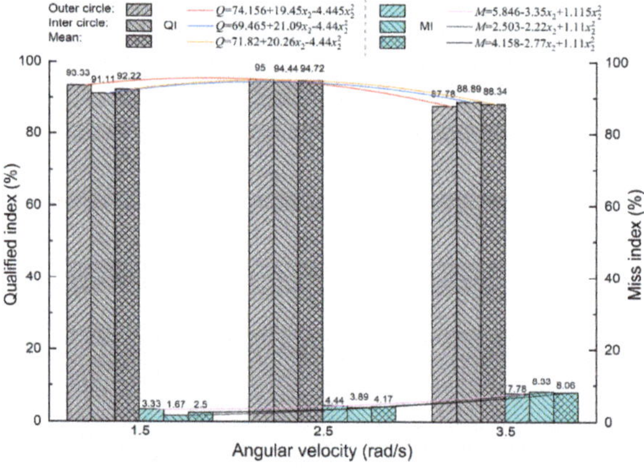

Figure 8. Effect of AV on seeding performance at NP of 1.5 kPa and CA of 60°.

When the AV of the metering plate was low, the seed cleaning device of the inner circle had weaker collision strength for the seeds which then were reabsorbed by the suction hole. As a result, there was a higher multiple index for the inner circle, such that the Q_O was higher than the Q_I. When the AV of the metering plate gradually increased, the QI of inner and outer circles first increased and reached the peak (Q_{Imax} = 94.58%, Q_{Omax} = 95.45%), and then decreased. Where the AV of the metering plate was too large, the seed-filling time of the suction hole was too short when the metering plate passed through the seed-filling zone. As a result, there was missed suctioning of the suction hole of the inner and outer circles, so that the MI of the inner and outer circles was high ($M \geq 7\%$).

From Figure 8, it can also be seen that the QI of the inner and outer circles was high ($Q \geq 92\%$) and the MI of the inner and outer circles was low ($M \leq 4\%$) when the AV was in the range of 1.62–2.28 rad/s.

3.1.3. Effect of Cone Angle on Seeding Performance

The influence trend of CA of the suction hole on the seeding performance of the inner and outer circles was fundamentally similar (Figure 9) at 1.5 kPa negative pressure and 2.5 rad/s angular velocity.

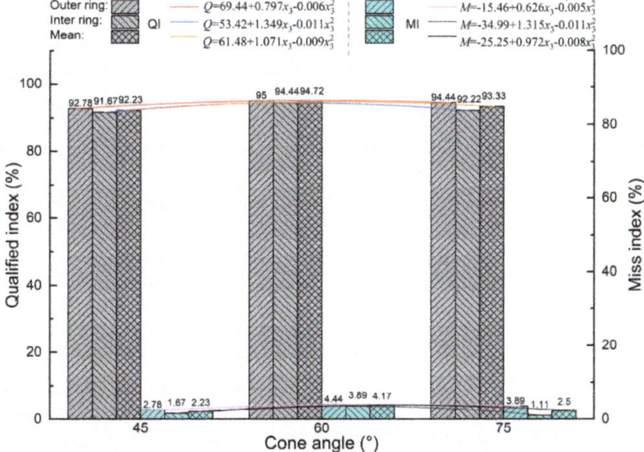

Figure 9. Effect of CA on seeding performance at NP of 1.5 kPa and AV of 2.5 rad/s.

When the CA of the suction hole was small, the adsorption force of the suction hole acting on the seed was relatively concentrated, which led to the phenomenon of some suction holes in the inner and outer circles adsorbing multiple seeds, so that the QI of the inner and outer circles was not high ($Q \leq 93\%$). When the CA of the suction hole increased, the QI of the inner and outer circles first increased and reached the peak (Q_{Imax} = 94.78%, Q_{Omax} = 95.91%), and then decreased.

3.2. Central Composite Design Experiment

3.2.1. Experimental Design and Results

The experimental scheme design and coding of influencing factors was performed according to the Central Composite Design in Design-Expert 8.0.6 software. This software was also used to process and analyze the experimental data. The experiment scheme and results are shown in Table 3.

Table 3. Experimental scheme and results.

NO.	NP [a] x_1, kPa	AV [b] x_2, rad·s^{-1}	CA [c] x_3, °	QI [d], %	MI [e], %	QI, %	MI, %
				OC [f]		IC [g]	
1	−1	−1	−1	85.56	13.33	91.11	8.33
2	1	−1	−1	92.78	3.33	88.89	2.22
3	−1	1	−1	78.33	21.67	74.44	25.56
4	1	1	−1	91.11	4.44	88.33	6.11
5	−1	−1	1	87.22	9.44	91.67	7.22
6	1	−1	1	92.78	3.89	87.78	1.67
7	−1	1	1	77.22	22.78	75.00	25.00
8	1	1	1	92.22	4.44	87.22	3.33
9	−1	0	0	81.11	16.67	83.33	16.67
10	1	0	0	89.44	3.89	87.78	2.22
11	0	−1	0	93.33	3.33	91.11	1.67
12	0	1	0	87.78	7.78	88.89	8.33
13	0	0	−1	92.78	2.78	91.67	1.67
14	0	0	1	94.44	3.89	92.22	1.11
15	0	0	0	95.00	4.44	94.44	3.89
16	0	0	0	92.78	3.33	93.33	2.78
17	0	0	0	93.33	5.00	92.78	4.44
18	0	0	0	92.78	4.44	91.67	3.89
19	0	0	0	94.44	3.89	94.44	3.33
20	0	0	0	93.89	4.44	92.78	3.89

Note: [a] NP = Negative pressure; [b] AV = Angular velocity; [c] CA = Cone angle; [d] QI = Qualified index; [e] MI = Miss index; [f] OC = Outer circle; [g] IC = Inner circle.

3.2.2. Analysis of Variance

Multiple regression fitting of the experimental results was performed using ANOVA in Design-Expert 8.0.6 software. This is shown in Table 4. After multiple regression analysis, the regression equations of the QI (Q) of the inner and outer circles, the MI (M) of the inner and outer circles, and various influencing factors can be obtained as follows:

$$Q_O = 101.399 + 19.622x_1 + 1.951x_2 - 0.893x_3 + 1.875x_1x_2 + 0.005x_1x_3 \\ - 0.014x_2x_3 - 6.567x_1{}^2 - 1.287x_2{}^2 + 0.008x_3{}^2 \tag{14}$$

$$M_O = 10.811 - 19.610x_1 - 1.740x_2 + 0.397x_3 - 2.503x_1x_2 + 0.028x_1x_3 \\ + 0.037x_2x_3 + 5.935x_1{}^2 + 1.210x_2{}^2 - 0.004x_3{}^2 \tag{15}$$

$$Q_I = 93.042 + 12.154x_1 - 1.755x_2 - 0.151x_3 + 4.028x_1x_2 - 0.028x_1x_3 \\ + 1.591 \times 10^{-15}x_2x_3 - 6.036x_1{}^2 - 1.591x_2{}^2 + 0.002x_3{}^2 \tag{16}$$

$$M_I = -15.778 - 15.550x_1 + 1.872x_2 + 0.961x_3 - 3.683x_1x_2 - 0.014x_1x_3 \\ -0.014x_2x_3 + 6.288x_1{}^2 + 1.843x_2{}^2 - 0.008x_3{}^2 \tag{17}$$

From Table 4, it can be seen that the *P*-value of each evaluation indicator (Q_O, M_O, Q_I, M_I) is less than 0.01, indicating that the regression model established in this paper is highly significant. Furthermore, the lack of fit for each model was greater than 0.05, which indicates that the four regression equations are highly fitted. The coefficients of determination (R^2) of the four models are 0.965, 0.990, 0.947, and 0.991, respectively. These coefficients of determination are all close to 1, indicating that the four models had a high fitting degree to the experimental data. The non-significant factors with *p*-value > 0.05 were eliminated according to the significant level *p*-value of different influencing factors in each model, and the following optimized regression equations were obtained:

$$Q_O = 84.880 + 19.036x_1 - 5.314x_2 + 1.875x_1x_2 - 6.278x_1{}^2 \tag{18}$$

$$M_O = 15.121 - 16.804x_1 + 2.373x_2 - 2.503x_1x_2 + 5.557x_1{}^2 + 0.832x_2{}^2 \tag{19}$$

$$Q_I = 97.688 + 12.710x_1 - 9.709x_2 + 4.028x_1x_2 - 6.778x_1{}^2 \tag{20}$$

$$M_I = 9.248 - 14.478x_1 + 4.202x_2 - 3.683x_1x_2 + 5.654x_1{}^2 + 1.209x_2{}^2 - 0.0003x_3{}^2 \tag{21}$$

Table 4. ANOVA for response surface quadratic model.

EI [a]	Source	SS [b]	DF [c]	F Value	p-Value	Significance
Q_O [d]	Model	538.09	9	30.60	<0.0001	**
	x_1-NP	239.02	1	122.33	<0.0001	**
	x_2-AV	62.55	1	32.01	0.0002	**
	x_3-CA	1.10	1	0.56	0.4699	
	$x_1 x_2$	28.13	1	14.39	0.0035	**
	$x_1 x_3$	0.04	1	0.02	0.8902	
	$x_2 x_3$	0.34	1	0.18	0.6835	
	x_1^2	118.59	1	60.69	<0.0001	**
	x_2^2	4.55	1	2.33	0.1578	
	x_3^2	8.60	1	4.40	0.0623	
	Residual	19.54	10			
	Lack of Fit	15.44	5	3.76	0.0861	
	Pure Error	4.10	5			
	Cor Total	557.63	19			
M_O [e]	Model	728.02	9	109.13	<0.0001	**
	x_1-NP	408.32	1	550.86	<0.0001	**
	x_2-AV	77.23	1	104.19	<0.0001	**
	x_3-CA	0.12	1	0.17	0.6921	
	$x_1 x_2$	50.10	1	67.59	<0.0001	**
	$x_1 x_3$	1.39	1	1.88	0.2002	
	$x_2 x_3$	2.46	1	3.32	0.0982	
	x_1^2	96.88	1	130.70	<0.0001	**
	x_2^2	4.03	1	5.44	0.0419	*
	x_3^2	2.80	1	3.78	0.0805	
	Residual	7.41	10			
	Lack of Fit	5.77	5	3.50	0.0976	
	Pure Error	1.65	5			
	Cor Total	735.44	19			
Q_I [f]	Model	562.37	9	19.84	<0.0001	**
	x_1-NP	59.78	1	18.98	0.0014	**
	x_2-AV	134.54	1	42.71	<0.0001	**
	x_3-CA	0.03	1	0.01	0.9239	
	$x_1 x_2$	129.77	1	41.19	<0.0001	**
	$x_1 x_3$	1.39	1	0.44	0.5209	
	$x_2 x_3$	1.14E-13	1	3.61E-14	1.0000	
	x_1^2	100.19	1	31.80	0.0002	**
	x_2^2	6.96	1	2.21	0.1680	
	x_3^2	0.34	1	0.11	0.7476	
	Residual	31.50	10			
	Lack of Fit	25.73	5	4.45	0.0634	
	Pure Error	5.78	5			
	Cor Total	593.87	19			
M_I [g]	Model	1000.81	9	120.17	<0.0001	**
	x_1-NP	451.99	1	488.43	<0.0001	**
	x_2-AV	222.97	1	240.95	<0.0001	**
	x_3-CA	3.09	1	3.34	0.0975	
	$x_1 x_2$	108.49	1	117.23	<0.0001	**
	$x_1 x_3$	0.34	1	0.37	0.5554	
	$x_2 x_3$	0.35	1	0.38	0.5507	
	x_1^2	108.72	1	117.49	<0.0001	**
	x_2^2	9.34	1	10.09	0.0099	**
	x_3^2	8.59	1	9.28	0.0123	*
	Residual	9.25	10			
	Lack of Fit	7.61	5	4.65	0.0586	
	Pure Error	1.64	5			
	Cor Total	1010.07	19			

Note: [a] EI = Evaluation Indicators; [b] SS = Sum of Squares; [c] DF = Degree of Freedom; [d] Q_O = Qualified index of outer circle; [e] M_O = Miss index of outer circle; [f] Q_I = Qualified index of inner circle; [g] M_I = Miss index of inner circle; * ($0.01 < p < 0.05$); ** ($p < 0.01$).

It can be seen from Equations (18)–(21) that the significant factors that affect the Q_I of the inner and outer circles are NP (x_1), AV (x_2), the interaction term of NP and AV ($x_1 x_2$), and the quadratic term of NP (x_1^2). The significant factors affecting the MI of the inner and outer circles are NP (x_1), AV (x_2), the interaction term of NP and AV (x_1x_2), the quadratic term of NP (x_1^2), and the quadratic term of AV (x_2^2). In addition to the above, other factors affecting the miss index of the inner circle (M_I) include quadratic term of the CA (x_3^2).

3.2.3. Effect of Interaction Factors on Seeding Performance

The descending dimension method was used to adjust the coding of any of the three influencing factors of NP, AV, and CA to zero [21], and the response surface diagram of the interaction of the other two influencing factors on the seeding performance of the inner and outer circle was plotted, as shown in Figures 10 and 11.

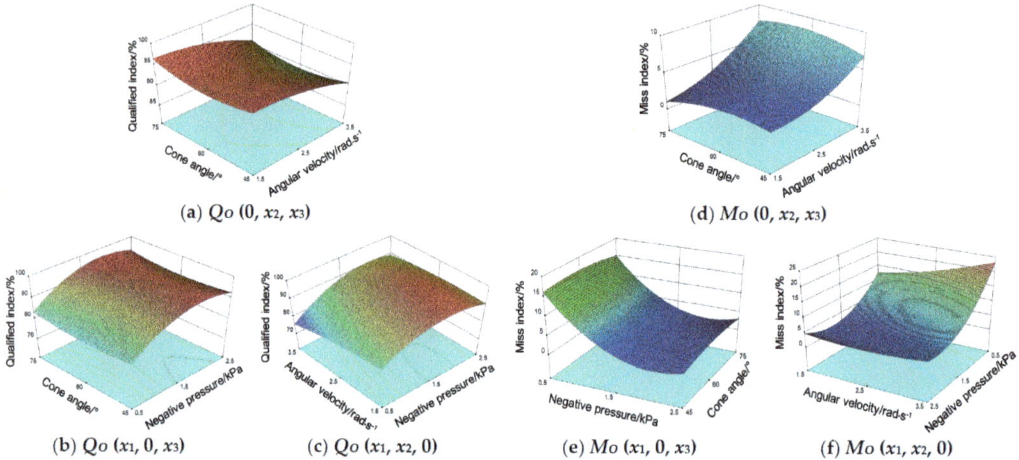

Figure 10. Effect of interaction factors on seeding performance of outer circle when NP = 1.5 kPa, AV = 2.5 rad/s, and CA = 60°, respectively

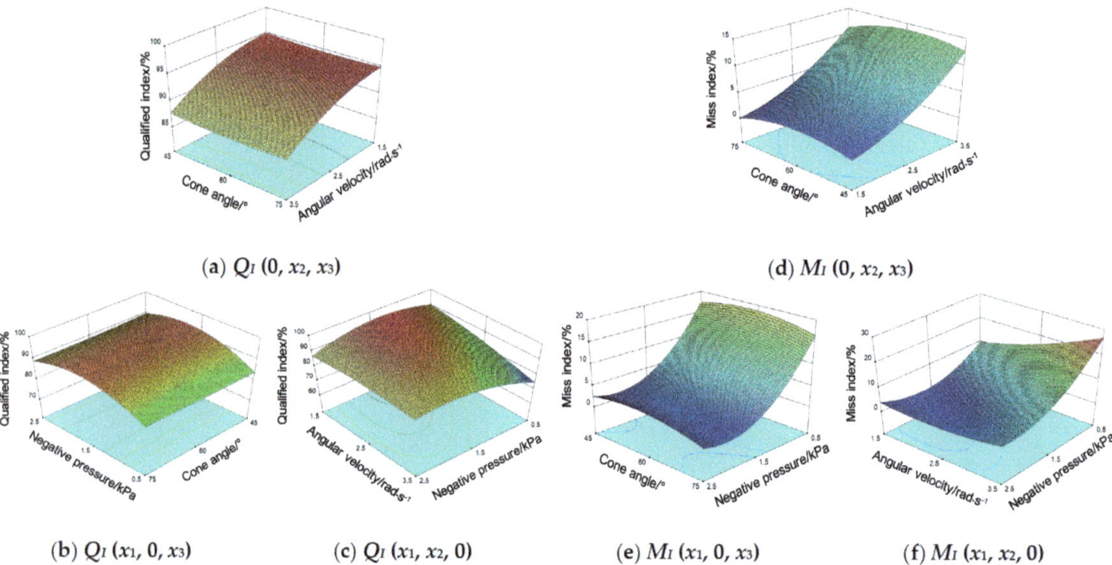

Figure 11. Effect of interaction factors on seeding performance of inner circle when NP = 1.5 kPa, AV = 2.5 rad/s, and CA = 60°, respectively.

It can be seen from Figure 10a at 1.5 kPa negative pressure under similar CA, the Q_O decreased with an increase in AV, but the trend for M_O was the opposite. This can be attributed to the shortening of the residence time of the metering plate in the seed-filling zone with the increase in the AV of the metering plate, resulting in missed suctioning of the suction hole. Under the condition of similar AV, the change in CA had no significant effect on the Q_O and M_O.

Figure 10b shows that when the AV of the metering plate was 2.5 rad/s, under the condition of similar CA, the Q_O increased with the increase in NP, and then decreased, while the M_O was the opposite. At 0.5–1.8 kPa negative pressure, the suction force of the holes for seeds gradually increased with the increase in NP, so that the QI increased significantly and the MI decreased suddenly. When the NP was in the range of 1.8–2.5 kPa, one suction hole adsorbed multiple seeds because of the excessive adsorption force, which led to an increase in multiple index and decrease in QI. Under the condition of similar NP, the change in CA had no significant effect on the Q_O and M_O.

It can be seen from Figure 10c that when the CA was 60°, under the condition of similar AV, the Q_O first increased and then decreased with the increase in NP, while the M_O was completely opposite. Under the condition of similar NP, the Q_O first increased and then decreased with the increase in AV. When the AV was in the range of 1.5–2.2 rad/s, the residence time of the metering plate in the seed-filling zone was lengthy, and this was beneficial to the suction hole for seed-filling, since it increased the Q_O. When the AV was greater than 2.2 rad/s, the residence time of the metering plate in the seed-filling zone was shortened, and the phenomenon of missed suctioning of the suction hole increased. As a result, the M_O increased rapidly.

Figure 11a shows that when the NP was 1.5 kPa, under the condition of similar AV, the change in CA had no significant effect on the Q_I, while the M_I first increased and then decreased with the increase in CA. Under the condition of similar CA, Q_I decreased with the increase in AV, but the trend for M_I was the opposite. It can be seen from Figure 11b that when the AV of the metering plate was 2.5 rad/s, under the condition of similar NP, the change in CA had no significant effect on the Q_I and M_I. Under the condition of similar CA, when the NP was in the range of 0.5–1.7 kPa, Q_I increased with the increase in NP, but the M_I decreased sharply with the increase in NP. When NP was greater than 1.7 kPa, Q_I decreased with the increase in NP. From Figure 11c, it is evident that when CA was 60°, under the condition of similar AV, the Q_I first increased and then decreased with the increase in NP, while the trend for M_O was the opposite. When the NP was low (NP $\leq$ 1.4 kPa), the effect of AV on the Q_I and M_I was significant. Q_I decreased with the increase in AV, while the trend for M_I was the opposite. When NP was large (NP > 1.4 kPa), the effect of AV on Q_I and M_I was not significant. Overall, it was determined from the figure that Q_I was high (Q_I > 92%) and M_I was low (M_I < 5.5%) when NP was in the range of 1.2–1.6 kPa and AV was in the range of 1.65–2.35 rad/s.

3.2.4. Parameter Optimization

In order to explore the optimal parameter combination, the parameter optimization design was carried out. According to the National Standard of P.R.C (GB/T 6973-2005 Testing Methods of Single Seed Drills (Precision Drills)) [20], under the premise of ensuring a high QI and low MI of seeding, parameter optimization was carried out with the goal that the QI of the inner and outer circle was greater than 94% and the MI was less than 2.5%. This is shown in Figure 12. According to the previous single factor experiment on CA, when the other two factors were at zero level and CA was 60°, the performance of seeding was better. There, CA was set at 60°. The parameter optimization conditions are as follows:

$$\begin{cases} 94\% \leq (Q_O(x_1, x_2, x_3), Q_I(x_1, x_2, x_3)) \leq 100\% \\ 0 \leq (M_O(x_1, x_2, x_3), M_I(x_1, x_2, x_3)) \leq 2.5\% \\ \text{s.t.} \begin{cases} 0.5\,\text{kPa} \leq x_1 \leq 2.5\,\text{kPa} \\ 1.5\,\text{rad/s} \leq x_2 \leq 3.5\,\text{rad/s} \\ x_3 = 60° \end{cases} \end{cases} \qquad (22)$$

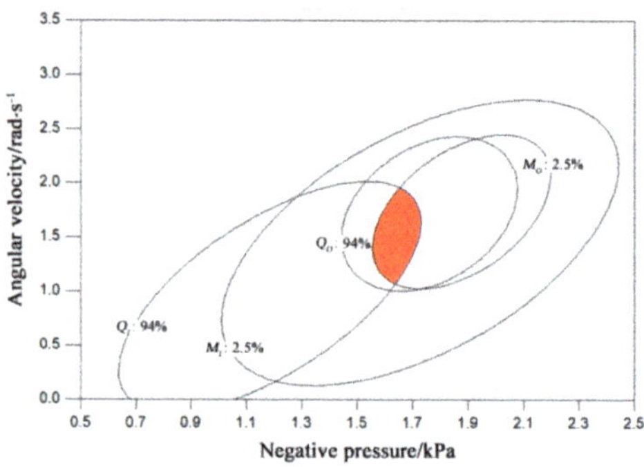

Figure 12. Parameter optimization area.

The graph intersection area is the parameter optimization area as described in Figure 12. When NP was 1.55–1.72 kPa and AV was 1.1–1.9 rad/s, the seeding performance was excellent.

3.2.5. Bench Verification Experiment

A bench verification experiment was used to verify whether the expected seeding performance [94% $\leq (Q_O, Q_I) \leq$ 100%, 0 $\leq (M_O, M_I) \leq$ 2.5%] and excellent CV of the seed mass could be achieved under the condition of optimized parameters, and the experiment was repeated three times and five times, respectively, under the median value of the range of parameter optimization (NP = 1.63 kPa, AV = 1.5 rad/s, CA = 60°). The results of the bench verification experiments are shown in Tables 5 and 6. It was observed from the experiment results that the seeding performance was excellent {$(Q_O, Q_I) >$ 94%, $(M_O, M_I) <$ 2.5%} and CV was good (CV of the seed mass in the outer and inner circle: 5.15%; CV of total seed mass: 8.60%) under the condition of optimized parameters. Therefore, the parameter optimization results were accurate.

Table 5. The results of seeding performance.

Experiment Indices		Number of Experiments			Mean	Standard Error
		NO.1	NO.2	NO.3		
QI, %	OC	94.58	94.17	95.42	94.72	0.52
	IC	94.17	94.58	94.58	94.44	0.19
MI, %	OC	2.08	2.5	2.5	2.36	0.20
	IC	2.08	1.25	1.67	1.67	0.34

Table 6. The results of coefficient of variation of seed mass.

Experiment Indices	Number of Experiments					Mean	TSM [a]	RE [b], %	CV [c], %	CV of Total Seed Mass, %
	NO.1	NO.2	NO.3	NO.4	NO.5					
Seed mass of outer circle, g	1.838	1.808	1.769	1.824	1.816	1.811	2.068	12.43	5.15	8.60
Seed mass of inner circle, g	2.072	1.897	1.972	1.910	1.889	1.948	2.068	5.80		
Total seed mass, g	3.910	3.705	3.741	3.734	3.705	3.759	4.136	9.12		

[a] TSM = Theoretical seed mass; [b] RE = Relative error; [c] CV = Coefficient of variation.

4. Conclusions

1. Aiming at the characteristics of small size and high sphericity of BC seeds and meeting the row spacing of seeding by agronomic requirements, a double-row pneumatic metering device that can be used for high-speed and precise double-row seeding was developed. The metering device can complete double-row seeding with a double-row metering plate, and can meet different row spacing of seeding requirements by replacing different diversion tubes. Furthermore, the structure and key dimensions of the double-row metering plate were determined, the other key components of metering device were designed, and the force analysis of the seed-filling process was carried out.

2. The NP, the AV of the metering plate, and the CA of the suction hole were selected as the main influencing factors of the experiment. A single factor experiment was carried out to investigate the influence of each factor on the seeding performance of the inner and outer circles.

3. For further exploring the relationship between the seeding performance and the NP, AV, and CA, the CCD scheme was applied in the experiment. Moreover, the experimental results were analyzed by ANOVA, and the regression equation between the evaluation indicators and the influencing factors was established. It was found from ANOVA that the NP, AV, the interaction term of NP and AV, the quadratic term of NP, the quadratic term of AV, and the quadratic term of CA were significant factors affecting the seeding performance of the inner and outer circles. Furthermore, the effect of interaction factors on the seeding performance of the inner and outer circle was also studied.

4. To explore the best combination of parameters, the parameters were optimized with the goal that the QI of the inner and outer circle was greater than 94% and the MI was less than 2.5%. After parameter optimization, it was found that when the NP was 1.55–1.72 kPa, the AV was 1.1–1.9 rad/s, and the CA was 60°, the seeding performance of the metering device and the coefficient of variation (CV) of the seed mass were good.

Author Contributions: Conceptualization, B.L., H.L. and X.Q.; methodology, B.L. and R.A.; software, B.L. and S.M.N.; validation, B.L., J.W. and S.L.; formal analysis, B.L. and X.C.; investigation, B.L. and H.L.; resources, B.L. and R.A.; data curation, B.L.; writing—original draft preparation, B.L.; writing—review and editing, B.L., H.L., X.Q., R.A., S.M.N., J.W., X.C. and S.L.; supervision, H.L. and X.Q.; project administration, H.L.; funding acquisition, J.W. All authors have read and agreed to the published version of the manuscript.

Funding: This research was supported by the National Key R&D Program of China "Vegetable Intelligent Fine Production Technology and Equipment R&D" (Grant No. 2017YFD0701302).

Data Availability Statement: The data presented in this study are available on request from the corresponding author. The data are not publicly available due to privacy.

Acknowledgments: We would like to thank "College of Engineering, Nanjing Agricultural University" and "College of Mechanical and Power Engineering, Nanjing Tech University"

Conflicts of Interest: The authors declare no conflict of interest.

Abbreviations

NP	Negative pressure (kPa)
AV	Angular velocity (rad/s)
CA	Cone angle (°)
CCD	Central composite design
ANOVA	Analysis of variance
QI	Qualified index (%)
MI	Miss index (%)
CV	Coefficient of variation (%)
BC	*Brassica chinensis*
MTI	Multiple index (%)
Q_O	Qualified index of outer circle (%)
Q_I	Qualified index of inner circle (%)
OC	Outer circle
IC	Inner circle

References

1. Han, W.; Sun, C.X.; Zhao, H.L.; Hu, Q.; Zheng, Q.Z.; Song, X.L. Compensatory Ability and Defense Mechanism of Chinese Cabbage under High Temperature Stress. *Chin. J. Agrometeorol.* **2018**, *39*, 119–128. [CrossRef]
2. Jin, X.; Li, Q.W.; Zhao, K.X.; Zhao, B.; He, Z.T.; Qiu, Z.M. Development and test of an electric precision seeder for small-size vegetable seeds. *Int. J. Agric. Biol. Eng.* **2019**, *12*, 75–81. [CrossRef]
3. Singh, R.C.; Singh, G.; Saraswat, D.C. Optimisation of Design and Operational Parameters of a Pneumatic Seed Metering Device for Planting Cottonseeds. *Biosyst. Eng.* **2005**, *92*, 429–438. [CrossRef]
4. Yang, L.; Yan, B.X.; Cui, T.; Yu, Y.M.; He, X.T.; Liu, Q.W.; Liang, Z.J.; Yin, X.W.; Zhang, D.X. Global overview of research progress and development of precision maize planters. *Int. J. Agric. Biol. Eng.* **2016**, *9*, 9–26. [CrossRef]
5. Chen, J.; Li, Y.M.; Wang, X.Q.; Zhao, Z. Finite Element Analysis for the Sucking Nozzle Air Field of Air-suction Seeder. *Trans. Chin. Soc. Agric. Mach.* **2007**, *9*, 59–62.
6. Li, Y.H.; Yang, L.; Zhang, D.X.; Cui, T.; Ding, L.; Wei, Y.N. Design and Experiment of Pneumatic Precision Seed-metering Device with Single Seed-metering Plate for Double-row. *Trans. Chin. Soc. Agric. Mach.* **2019**, *50*, 61–73. [CrossRef]
7. Lü, J.Q.; Yang, Y.; Li, Z.H.; Shang, Q.Q.; Li, J.C.; Liu, Z.Y. Design and experiment of an air-suction potato seed metering device. *Int. J. Agric. Biol. Eng.* **2016**, *9*, 33–42. [CrossRef]
8. Zhang, W.Z.; Liu, C.L.; Lv, Z.Q.; Qi, X.T.; Lv, H.Y.; Hou, J.L. Optimized Design and Experiment on Novel Combination Vacuum and Spoon Belt Metering Device for Potato Planters. *Math. Probl. Eng.* **2020**, *2020*, 1–12. [CrossRef]
9. Lai, Q.H.; Sun, K.; Yu, Q.Y.; Qin, W. Design and experiment of a six-row air-blowing centralized precision seed-metering device for Panax notoginseng. *Int. J. Agric. Biol. Eng.* **2020**, *13*, 111–122. [CrossRef]
10. Elebaid, J.I.; Liao, Q.X.; Wang, L.; Liao, Y.T.; Yao, L. Design and experiment of multi-row pneumatic precision metering device for rapeseed. *Int. J. Agric. Biol. Eng.* **2018**, *11*, 116–123. [CrossRef]
11. Taghinezhad, J.; Alimardani, R.; Jafari, A. Development and Evaluation of a New Mechanism for Sugarcane Metering Device Using Analytical Hierarchy Method and Response Surface Methodology. *Sugar Tech* **2015**, *17*, 258–265. [CrossRef]
12. Mandal, S.; Kumar, G.V.P.; Tanna, H.; Kumar, A. Design and Evaluation of a Pneumatic Metering Mechanism for Power Tiller Operated Precision Planter. *Curr. Sci.* **2018**, *115*, 1106. [CrossRef]
13. Yan, B.X.; Zhang, D.X.; Cui, T.; He, X.T.; Ding, Y.Q.; Yang, L. Design of pneumatic maize precision seed-metering device with synchronous rotating seed plate and vacuum chamber. *Trans. Chin. Soc. Agric. Eng.* **2017**, *33*, 15–23. [CrossRef]
14. Jia, H.L.; Zhang, S.W.; Chen, T.Y.; Zhao, J.L.; Guo, M.Z.; Yuan, H.F. Design and Experiment of Self-suction Precision Mung-bean Metering Device. *Trans. Chin. Soc. Agric. Mach.* **2020**, *51*, 51–60. [CrossRef]
15. Cheng, X.P.; Lu, C.Y.; Meng, Z.J.; Yu, J.Y. Design and parameter optimization on wheat precision seed meter with combination of pneumatic and type hole. *Trans. Chin. Soc. Agric. Eng.* **2018**, *34*, 1–9. [CrossRef]
16. Chinese Academy of Agricultural Mechanization Sciences. *Design Manual for Agricultural Machinery*; China Agricultural Science and Technology Press: Beijing, China, 2007; Volume 1, pp. 343–368.
17. Zhai, J.B.; Xia, J.F.; Zhou, Y. Design and Experiment of Pneumatic Precision Hill-drop Drilling Seed Metering Device for Hybrid Rice. *Trans. Chin. Soc. Agric. Mach.* **2016**, *47*, 75–82. [CrossRef]
18. Shi, L.R.; Sun, B.G.; Zhao, W.Y.; Yang, X.P.; Xin, S.L.; Wang, J.X. Optimization and Test of Performance Parameters of Elastic Air Suction Type Corn Roller Seed-metering Device. *Trans. Chin. Soc. Agric. Mach.* **2019**, *50*, 88–95, 207. [CrossRef]
19. Chen, M.Z.; Diao, P.S.; Zhang, Y.P.; Gao, Q.M.; Yang, Z.; Yao, W.Y. Design of pneumatic seed-metering device with single seed-metering plate for double-row in soybean narrow-row-dense-planting seeder. *Trans. Chin. Soc. Agric. Eng.* **2018**, *34*, 8–16. [CrossRef]

20. National Standards of the P.R.C National Standard of the People's Republic of China. *GB/T 6973-2005: Testing Methods of Single Seed Drills (Precision Drills)*; Standardization Administration of the P.R.C: Beijing, China, 2005.
21. Chen, H.T.; Li, T.H.; Wang, H.F.; Wang, Y.; Wang, X. Design and parameter optimization of pneumatic cylinder ridge three-row close-planting seed-metering device for soybean. *Trans. Chin. Soc. Agric. Eng.* **2018**, *34*, 16–24. [CrossRef]

sustainability

MDPI

Article

Effect of Grouser Height on the Tractive Performance of Single Grouser Shoe under Different Soil Moisture Contents in Clay Loam Terrain

Sher Ali Shaikh [1,2], Yaoming Li [1,*], Ma Zheng [1], Farman Ali Chandio [2], Fiaz Ahmad [1,3], Mazhar Hussain Tunio [1,2] and Irfan Abbas [1]

[1] School of Agricultural Equipment Engineering, Jiangsu University, Zhenjiang 212013, China; sashaikh@sau.edu.pk (S.A.S.); mazheng123@ujs.edu.cn (M.Z.); engrfiaz@yahoo.com (F.A.); mazharhussaintunio@sau.edu.pk (M.H.T.); dr.iabbas@yahoo.com (I.A.)
[2] Faculty of Agricultural Engineering, Sindh Agriculture University, Tando Jam 70060, Pakistan; farman_chandio@hotmail.com
[3] Department of Agricultural Engineering, Bahauddin Zakariya University, Multan 60800, Pakistan
* Correspondence: ymli@ujs.edu.cn

Abstract: The grouser height and soil conditions have a considerable influence on the tractive performance of single-track shoe. A soil bin-based research was conducted to assess the influence of grouser height on the tractive performance of single-track shoe at different moisture contents of clay loam soil. Eight moisture contents (7.5, 12, 16.7, 21.5, 26.2, 30.7, 35.8, and 38%) and three grouser heights (45, 55, and 60 mm) were comprised during this study. The tractive performance parameters of (thrust, running resistance, and traction) were determined by penetration test. A sensor-based soil bin was designed for penetration tests, which was included penetration system (AC motor, loadcell, and displacement sensor). The test results revealed that soil cohesion was decreased, and adhesion was increased after 16.7% moisture content. Soil thrust at lateral sides and bottom of grouser were increased before 16.7%, and then decreased for all the three heights but the major decrease was observed at 45 mm height. The motion resistance was linearly decreased, the more reduction was on 45 mm at 38% moisture content. The traction of the single-track shoe was decreased with a rise in moisture content, the maximum decrease was on 45 mm grouser height at 38% moisture content. It could be concluded that an off-road tracked vehicle (crawler combine harvester) with 45 mm grouser height of single-track shoe could be operated towards a moderate moisture content range (16.7–21.5%) under paddy soil for better traction.

Keywords: traction; soil; sinkage; single-track shoe; penetration; cohesion; adhesion

Citation: Shaikh, S.A.; Li, Y.; Zheng, M.; Chandio, F.A.; Ahmad, F.; Tunio, M.H.; Abbas, I. Effect of Grouser Height on the Tractive Performance of Single Grouser Shoe under Different Soil Moisture Contents in Clay Loam Terrain. *Sustainability* **2021**, *13*, 1156. https://doi.org/10.3390/su13031156

Received: 20 November 2020
Accepted: 29 December 2020
Published: 22 January 2021

Publisher's Note: MDPI stays neutral with regard to jurisdictional claims in published maps and institutional affiliations.

1. Introduction

Tracked vehicles have been popularized because of more contact area with the ground, which leads to better float and traction than wheeled vehicles, making them suitable for rough and relatively saturated terrain. Tracked vehicles are used in various fields, such as mining, forestry, agriculture, planetary exploration, army, and construction [1]. Tractive performance of tracked vehicles including propulsion and resistance is very important for terrain trafficability and is influenced by vehicle and soil factors [2]. Grousers are devices intended to increase the traction of the continuously tracked vehicle on soil or snow; this is done by increasing contact with the ground with teeth equipped on crawler tracks, similar to conventional tire treads; usually, grousers are made up of hardened forged steel. The soil-track interaction tool includes two aspects: forces arising at the interface between the soil and the tool, such as thrust, lateral force, and vertical force and displacement of soil particles also known as soil disturbance [3,4]. The proper design and selection of soil-track interaction tools depends largely on the mechanical behavior of the soils [5].

Soil conditions greatly affect the traction performance of off-road or tracked/wheeled vehicles. Soil properties are important to generate traction for the soil track system [2]. The key to off-road vehicle performance prediction lies in the proper evaluation of the mechanical properties of the field [6]. Traction performance of the vehicle significantly influenced by the mechanical properties of the soil and it has been reported that the variation of soil water contents (MC) also affects the traction ability of a tracked vehicle [7,8].

The tractive performance of off-road vehicle prediction depends upon the appropriate assessment of mechanical properties (soil cohesion, soil adhesion, and angle of internal and external friction, etc.) of the soil. Mobility states a relationship between soil and the vehicle. To assess the mobility and traction of vehicles, it is important to determine the mechanical properties of soil, which are believed to be related to the mobility of vehicles. Terrain topology and soil parameters also influence vehicle performance, besides to the intrinsic vehicle characteristics. To correctly determine the mechanical properties in terms of mobility of off-road vehicles, it is necessary to carry out measurements under load conditions similar to those imposed by vehicles. The vertical load applied by the off-road vehicle to a soil causes sinkage, while the horizontal load created by the track/wheels on the soil surface causes the development of shear strength and the associated slip [2,9,10]. Determining the mobility and traction of off-road vehicles requires a thorough understanding of soil mechanics since it is necessary to use a track/wheel to determine power transfer between the vehicle and the soil. Traction as a force derived from the interaction between track and soil. It could be affected by two types of factors: soil conditions and shoe dimensions [11]. The effects of the soil on tractive performance were obtained on soft conditions over 50% [12].

The methods developed for the study of the traction of a track–soil interaction is empirical, semi-empirical, and computer methods [13]. The empirical model was originally developed by the U.S. Army Waterway Experimental Station (WES) to meet military needs in predicting the performance of land-based off-road vehicles, including concepts of "go/no go", mobility index, etc. Robert G applied a pilot model of traction performance for rubber tracks on agricultural soils [14]. A semi-empirical approach named bevameter technique for the prediction of traction of a tracked vehicle was developed by Bekker [2]. The tractive performance of the single-track shoe (grouser) could be predicted by knowing the normal stress distribution and shear at the front of the soil path and 3D geometry of the contact surface. Hiroshi made a computer approach by using "The Discrete Element Method (DEM)" for the specific ability to assess total traction potential based on interaction studies of computing reactions and soil behaviors, regardless of the type of method chosen for predicting traction, as soil parameters should be known at first [15].

The sinkage of soil occurs when a known load is applied on the surface of the soil, the area under the load sinks in the soil at a certain depth until the resistance of the soil is balanced with the applied force. Soil pressure resistance uses two parameters for characterization: soil cohesion, internal friction angle [16]. In the past several decades, a set of studies has been conducted to estimate traction performance for a variety of tracked components [17–20] but still, the interaction study among grouse height, moisture contents, and soil mechanical properties on the traction performance of the single-track shoe is needed.

The present research was based on the semi-empirical method developed by Bekker to assess the impacts of grouser height on the traction performance of a grouser of the single-track shoe in certain soil conditions under sensor-based soil bin.

2. Materials and Methods

2.1. Expeerimental Setup

The experimental setup was consisted of penetration test and direct shear test (Figure 1).

The tractive performance parameters were determined by deriving the Equations (1)–(16), the amounts of sinkage were less than the height of the grouser when the vertical load on the single grouser shoe was applied, and the entire track shoe in the field had been shown to have more sinkage. Between the grouse and the plate, the vertical load is shared.

The proposed model and schematic depiction of forces acting on a single-track shoe are shown in Figure 2.

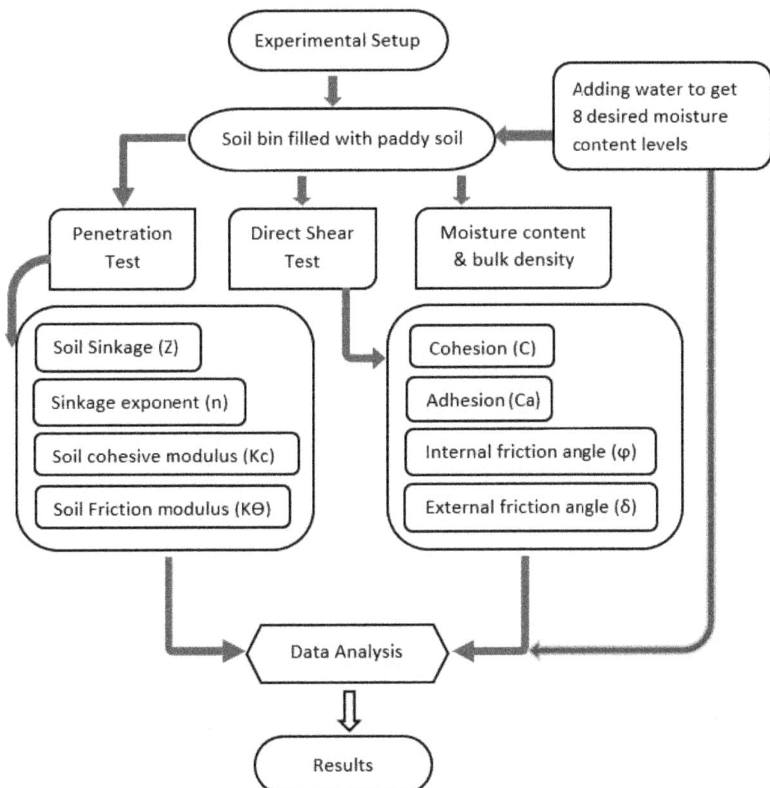

Figure 1. Flow diagram of the experimental procedure.

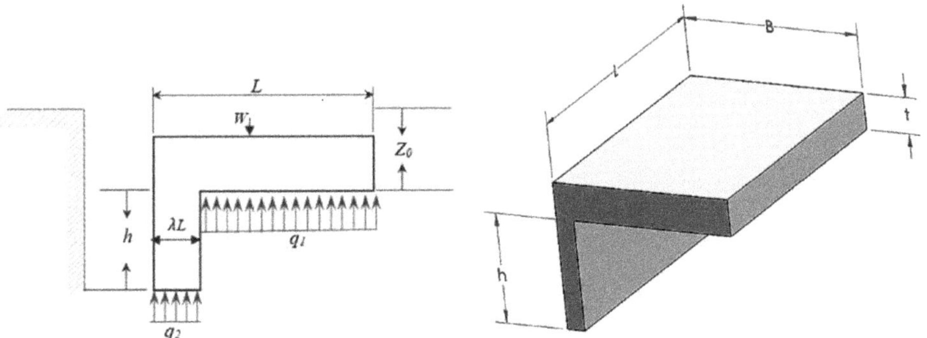

Figure 2. Forces acting on single track shoe.

The vertical load was measured by the following Equation:

$$W = q_1 \lambda LB + q_2(1 - \lambda)LB \tag{1}$$

where W is vertical load, L is track shoe pitch, B is the width of single track shoe, and λ is the ratio of grouser plate thickness to the length of single track shoe. Whereas q_1 is the contact pressure of the top surface of the grouser and q_2 is the pressure at the bottom surface of single track shoe, was expressed as follows:

$$q_1 = K(h + Z_0)^n \tag{2}$$

$$q_2 = KZ_0{}^n \tag{3}$$

where K is the modulus of soil deformation, n is soil deformation index or sinkage exponent, h is grouser height and Z is soil sinkage. Furthermore, K was determined by Equation (4) as:

$$K = \frac{k_c}{B} + k_\varnothing \tag{4}$$

where k_c is soil cohesion within the deformation modulus, $k\phi$ friction modulus of soil, and B is the width of the single-track shoe. So, the load Equation could make the following relationship:

$$W = \left(\frac{k_c}{B} + k_\varnothing\right) LB\{(h + Z_0)^n \lambda + Z_0^n (1 - \lambda)\} \tag{5}$$

The pressure and soil sinkage were computed according to Bekker's pressure-sinkage relationship as:

$$p = (\frac{k_c}{B} + k_\theta)z^n \tag{6}$$

Soil sinkage is accountable for the increase in running resistance. The running resistance of a single-track shoe was found by Equation (7):

$$R = \frac{k_c + Bk_\varnothing}{n+1}\{(h + Z_0)^{(n+1)}\lambda + Z_0^{(n+1)}(1 - \lambda)\} \tag{7}$$

Soil thrust is the utmost vital parameter of traction of the single-track shoe. the thrust is the resultant of the shearing force on the tip surface of grouser (F_1), the shearing force of the grouser shoe's lateral sides (F_2), and the shearing force on the bottom surface of soil beneath the spacing surface (F_3). The F_1 was measured by Equation (8):

$$F_1 = \lambda LB(C_a + q_1 tan\delta) \tag{8}$$

where C_a is soil adhesion, δ is the external friction angle of the soil. Furthermore, the shearing force of grouser's lateral sides is divided into 3 parts as F_{sg1}, F_{sg2}, and F_{ss}. Moreover, they could be shown by the following relationships:

$$F_2 = 2(F_{sg1} + F_{sg2} + F_{ss}) \tag{9}$$

$$F_{sg1} = \lambda hL[C_a + q_1 tan\delta tan\varphi\left(45 - \frac{\varphi}{2}\right)\left\{\frac{\gamma_t(2Z_0 + h)}{2} tan\left(45 - \frac{\varphi}{2}\right) - 2C\right\}] \tag{10}$$

where F_{sg1} is the shearing force at grouser's lateral side, φ is the internal friction angle of soil, γ_t is the soil density and C is the soil cohesion. Whereas the shearing force of spacing's lateral side (F_{sg2}) was measured by Equation (11).
If $Z < t$ then:

$$F_{sg2} = Z_0L[C_a + q_1 tan\delta tan\varphi\left(45 - \frac{\varphi}{2}\right)\left\{\frac{\gamma_t Z_0}{2} tan\left(45 - \frac{\varphi}{2}\right) - 2C\right\}] \tag{11}$$

If $Z > t$ then:

$$F_{sg2} = tL[C_a + q_1 tan\delta tan\varphi\left(45 - \frac{\varphi}{2}\right)\left\{\frac{\gamma_t(2Z_0 - t)}{2} tan\left(45 - \frac{\varphi}{2}\right) - 2C\right\}] \tag{12}$$

where t is the thickness of shoe spacing. Finally, the shearing force of the lateral side of the soil below the spacing surface (F_{ss}) was measured by the following Equation (13):

$$F_{ss} = (1 - \lambda)hL[C + \{q_2 + \frac{\gamma_t(h + 2Z_0)}{2}\}\, tan^2\left(45 - \frac{\varphi}{2}\right)\, tan\varphi - 2Ctan\left(45 - \frac{\varphi}{2}\right)tan\varphi] \tag{13}$$

Furthermore, the shearing force on the bottom surface of soil beneath the spacing surface (F_3) was calculated as:

$$F_3 = (1 - \lambda)LB(C + q_3 tan\varphi) \tag{14}$$

The total soil thrust is the addition of all thrusts including F_1, F_2 and F_3 as following:

$$F = F_1 + F_2 + F_3 \tag{15}$$

The subtraction of total soil thrust from running resistance is the traction (T), it can be expressed as Equation (16):

$$T = F - R \tag{16}$$

2.2. Experimental Work

The experiment was performed at the key laboratory of modern agricultural equipment and technology, school of agricultural equipment engineering, Jiangsu University, Zhenjiang, P.R China. A 500 mm wide, 1500 mm long, and 500 mm deep soil bin was used for the mechanical properties of soil. A factorial design experiment was conducted considering soil moisture content as first factor with eight number of levels (7.5, 12, 16.7, 21.5, 26.2, 30.7, 35.8, and 38%) and grouse height with three number of levels (60, 55, 45 mm). Figure 2 shown the flow diagram of the experimental work.

2.3. Dimensions of Single-Track Shoe

The details of the dimension of the single track-shoe model used for the research-based study are shown in Table 1.

Table 1. Dimension details of single-track shoe model.

Parameters	Dimensions (mm)
Length, L	100
Width, B	150
Height of grouser, h	60, 55, 45
Grouser thickness ratio, λL	6
The thickness of the shoe plate, t	40

2.4. Soil Preparation

The soil was taken from the experimental site of the school of agricultural equipment engineering and sun-dried. The dried soil was crushed, and a sieve analysis test was performed [1]. The soil was filled in the soil bin layer by layer, and at the 100 mm height, a little wooden roller was utilized to move it to and fro twice. The way toward putting the soil was rehashed until the elevation of the soil bin arrived at 400 mm. To get uniformity and higher moisture content, the calculated amount of water was added, mixed thoroughly, and left for 24 h [16]. To minimize moisture loss by evaporation, the soil bin was lined with a plastic sheet (polyethylene). Soil samples have been taken from the soil bin at three different random places. Averaging the moisture content of the three samples reported for each test. Besides which, an oven-dried method was used to measure soil moisture content [21]. The experimental work was divided into two parts, soil physical properties (soil texture and dry bulk density), direct shear test, and penetration test.

2.5. Direct Shear Test

The strength characteristics of soil were tested with a strain-controlled direct shear test equipment (ZJ Nanjing Soil Instrument Factory Co., Ltd. Nanjing, China Figure 3). The most important part of this device is the lower and upper blocks. The sample of soil is always put into the upper box; the lower box includes a circular soil or steel plate to check cohesion/adhesion and friction angle. A horizontal shear force is applied to the shear, and the shear stress during the failure of the soil sample was obtained under pressure. Furthermore, the shear strength (τ), internal friction angle (ϕ), and cohesion (C) of the soil were determined by Coulomb's law (Equation (17)).

$$\tau = C + \sigma tan\phi \tag{17}$$

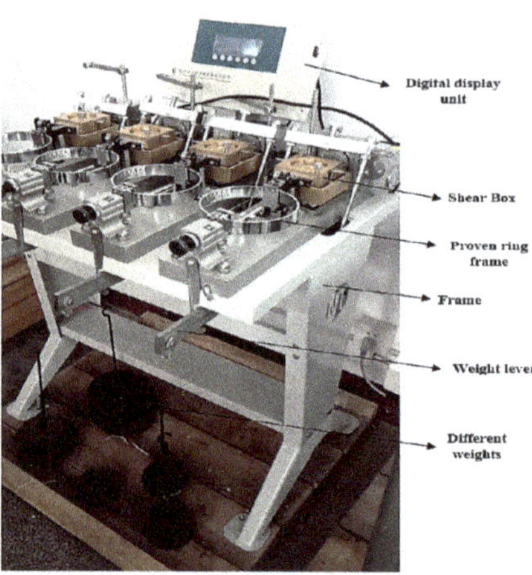

Figure 3. Strain controlled direct shear test apparatus ZJ type.

2.6. Penetration Test

The penetration test of soil was determined by the newly designed device (Figures 4 and 5). The penetration test device consisted of a steel frame, wheels, soil bin (500 × 500 × 1500 mm), and penetration system. An AC gear motor with spur rack and speed control system (ASLONG-5F4, 120 W) was used to push the penetration plates into the soil. Penetration plates used for the experiment were 10 × 30 × 40 and 10 × 25 × 40 mm. The penetration test was repeated thrice for each moisture content. The design concept was in line with Tiwari et al. [22] in which a testing facility was designed and developed to determine the traction of the tire.

2.7. Data Acquisition and Analysis

The data acquisition system comprised the load cell (ATO-LCS-DYLY-106), linear displacement transducer (ATO-LDSR, 400 mm), DAQ device (Ni-6009), and LabVIEW software. The penetration force and sinkage were measured by a load cell and linear displacement transducer, respectively. Data acquisition was made by the DAQ device with help of LabVIEW. The Origin (version 2018) software was used for data analysis and the gauss-newton method was used for non-linear curve fitting.

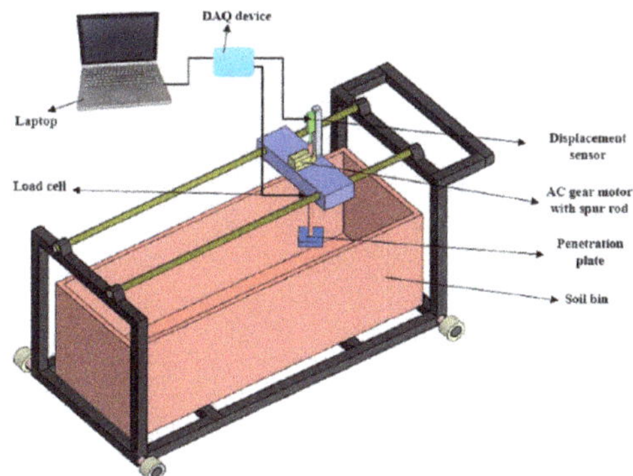

Figure 4. Designed device for the Penetration test.

Figure 5. Original picture of the designed penetration device

3. Results and Discussion

3.1. Soil Reaction Force and Soil Sinkage

The assumption for the soil reaction force/pressure sinkage test was similar to traditional bevameter. The test results of soil reaction force and sinkage were obtained by penetration test with two different sized plates for eight levels of moisture contents (7.5, 12, 16.7, 21.5, 26.2, 30.7, 35.8, and 38%, respectively) (Figures 6–9). Obtained results showed the comparative trend over both penetration plates at all the moisture contents. The average maximum sinkage for plates 1 and 2 was 202.405 and 204.178 mm at 38% and the minimum was 44.942 and 53.103 mm at 7.5% soil moisture content, respectively. Similarly, the greater force soil reaction force was 195.126 and 195.465 N at 38% and smaller was

128.7 and 117.782 N at 7.5% moisture content for plates 1 and 2, respectively. Consequently, the soil sinkage increased with an increase in soil moisture content, the reason being that the soil becomes softer with the addition of water, thus allowing the plate to penetrate more quickly in soil and resulting more sinkage. The increase trend was observed over moisture contents for both plates. The overall results of sinkage and soil reaction force/pressure were found significant ($p < 0.05$). The main effect of soil sinkage on moisture content was observed by 38%. The same findings have been recorded previously that plate sinkage with high moisture content is increased [23]. At 35% moisture content the plate sinkage was high [24]. The maximum sinkage varied linearly with the pressure applied in moist soil [25]. When the soil reaction force increases the soil sinkage increases [6,26–30].

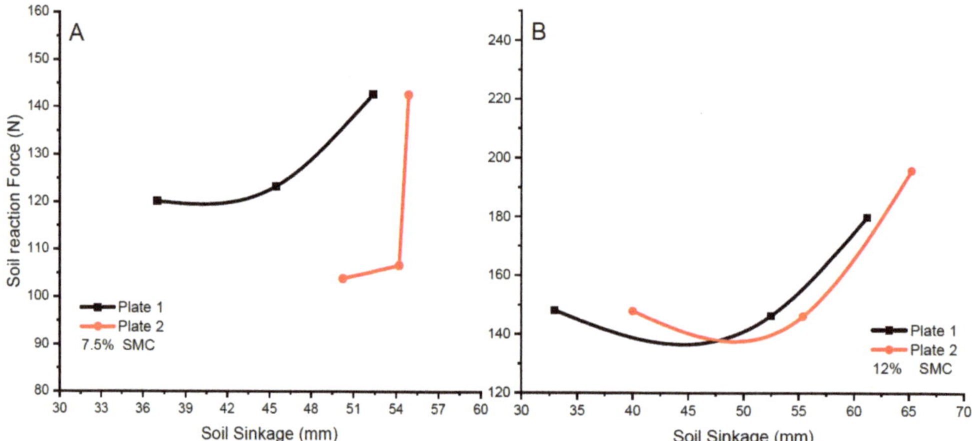

Figure 6. Soil sinkage and reaction force for plate 1 and 2 at 7.5% (**A**) and 12% (**B**).

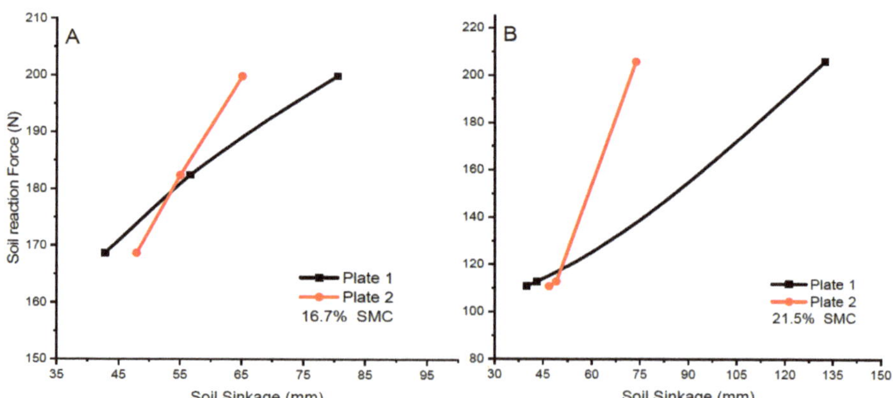

Figure 7. Soil sinkage and reaction force for plate 1 and 2 at 16.7% (**A**) and 21.5% (**B**).

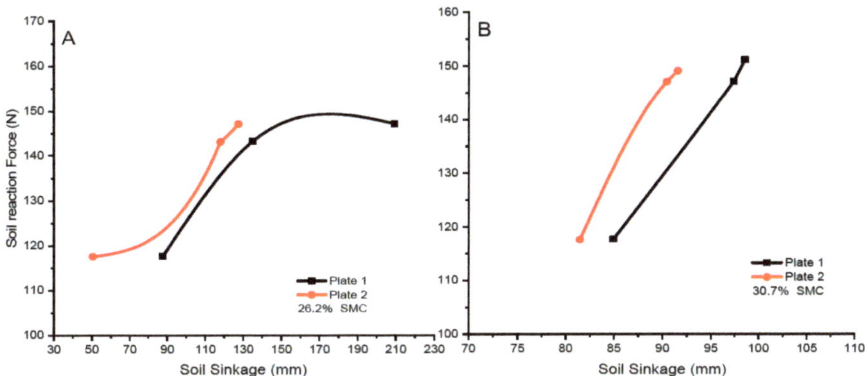

Figure 8. Soil sinkage and reaction force for plate 1 and 2 at 26.2% (**A**) and 30.7% (**B**).

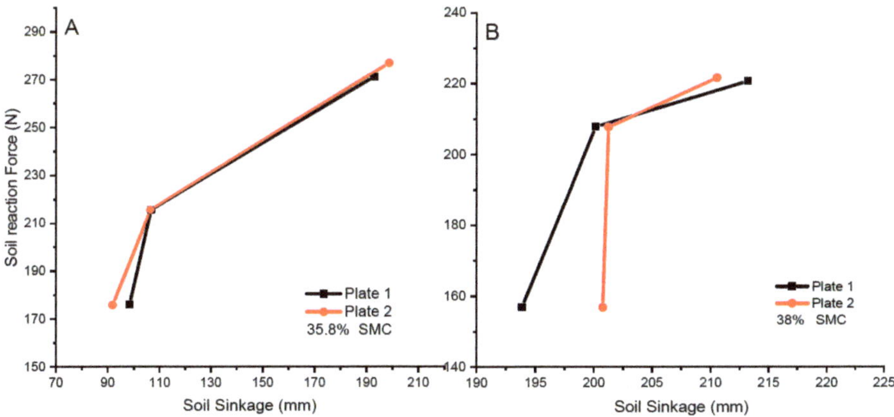

Figure 9. Soil sinkage and reaction force for plates 1 and 2 at 35.8% (**A**) and 38% (**B**).

3.2. Soil Cohesion Modulus (K_c), Friction Modulus ($K\phi$), and Sinkage Exponent (n) at Moisture Contents

The Bekker's pressure sinkage model parameters, including soil cohesion modulus (K_c), soil friction modulus ($K\phi$), and sinkage exponent (n) were derived after penetration test for eight moisture content. The relationship between soil cohesion and friction modulus is shown in Figure 10. The values of both parameters were initially increased with an increase in moisture content to 30.7% and then began decreasing until the final moisture content 38%. The findings of the tests suggest that the effect of the water content on the soil cohesive modulus is not apparent. Forces between soil particles may influence the cohesive properties of soil; furthermore, the interaction between soil moisture content and various forces may differ. For the friction modulus of soil, initially, a rapid increase followed by a gradual increase with the increase in moisture content was observed.

The maximum soil cohesion and friction modulus were 0.096 mN/m^{n+1} and 0.242 mN/m^{n+2} recorded at 30.7% soil moisture content. Similarly, the minimum Kc was 0.042 mN/m^{n+1} at 38% and Kϕ were 0.110 mN/m^{n+2} observed at 7.5% moisture content. The sinkage exponent (Figure 11) of soil was increased when the moisture content increased to 12%, then gradually decreased with an increase in moisture content to 38%. The effect of pore pressure build-up may be the internal cause of this phenomenon. The rise of the sinkage exponent indicates the increase in sinkage of the vehicle under the same pressure. The adjusted R square of sinkage exponent was 0.94309 found by non-linear curve fitting using the Gauss model. The present results are in close agreement with previously reported that the soil cohesive and friction modulus increased with the increase in the moisture content to a certain level, and after that, it decreased with the increase in the moisture [24,31–34].

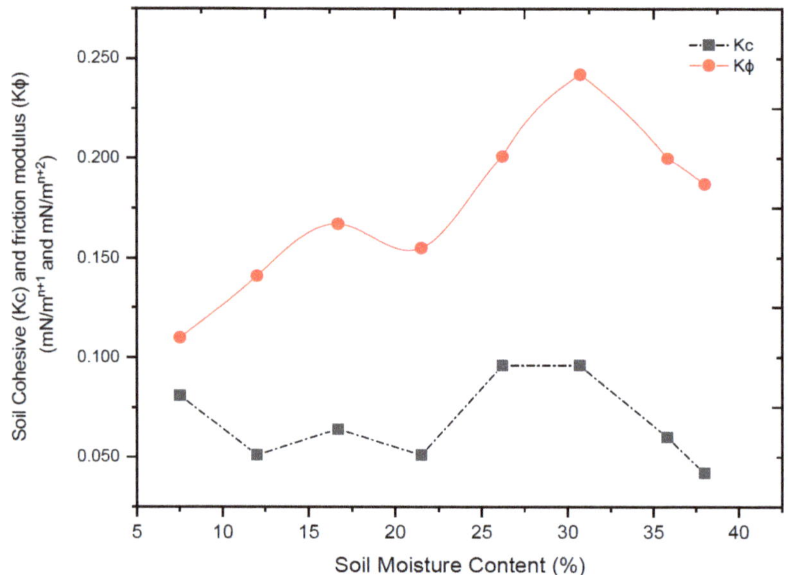

Figure 10. Soil cohesive (K_c) and friction modulus ($K\phi$) at moisture contents (7.5–38%).

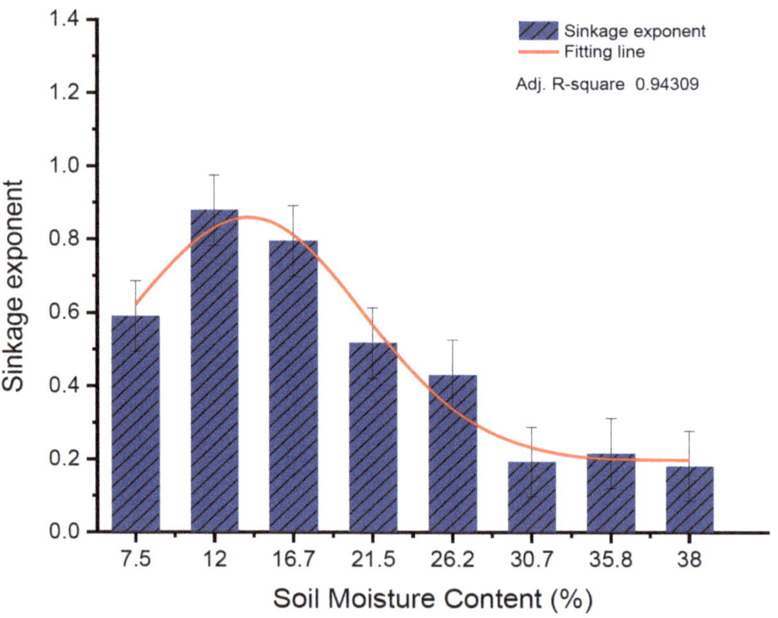

Figure 11. Soil sinkage exponent (n) at moisture contents (7.5–38%).

3.3. The Cohesion of Soil (C) at Different Moisture Contents

The experimental results of soil cohesion were obtained by direct shear test for eight moisture contents (Figure 12). The plot showed the rapid rise in cohesion with moisture content until 12%, then it was gradually dropped with further increase in soil moisture content. The mean maximum cohesion was 0.84 kPa at 12% while a minimum 0.112 kPa at 35.8% moisture content. The non-linear curve fitting was performed by Gauss model. the adj. R-square of soil cohesion was 0.98504. The present results are in line with Jun et.al, who found the increase in cohesion with moisture content until certain values then begin decrease in cohesion [35–37]. Other studies reported the significant effect of soil mechanical properties on moisture content [38,39].

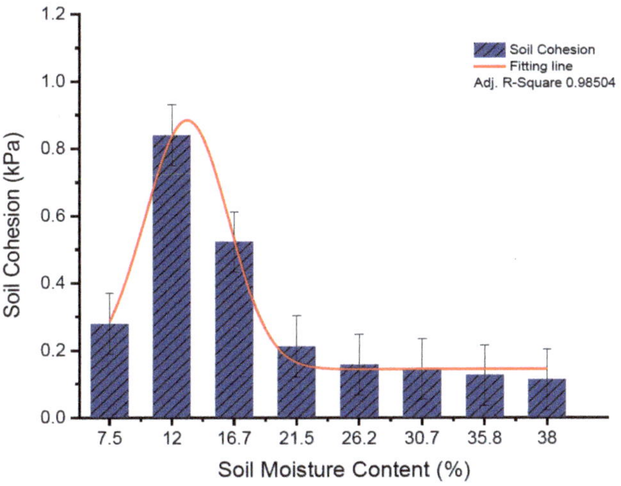

Figure 12. Soil cohesion at moisture contents (7.5–38%).

3.4. Soil Adhesion (C_a) at Different Moisture Contents

The relationship between soil adhesion and soil moisture content is shown in Figure 13. The plot discloses the parabolic curve of results. It has been found that soil adhesion is higher when the soil is wet. Initially, adhesion increased with increasing moisture content, it was higher at 26.2% moisture content, and as the moisture content further increased, soil adhesion decreased. The average greater adhesion was 0.976 kPa at 26.2%, similarly, the smaller was 0.108 kPa at 7.5% moisture content. Statistically, the results were significant ($p < 0.05$). The Gauss model was used to fit the data and to get best fitting line, the adjusted R^2 value was 0.94648. The previous results reported that the soil moisture content significantly affected soil cohesion and adhesion [7,37,40,41]. Present results are in line with those reported results.

3.5. Internal and External Friction Angle

The experimental results of soil internal and external friction angles were obtained by the direct shear test and shown in Figure 14. The comparative trend between both angles was observed in results, initially, both angles slowly decreased with an increase in moisture content to 21.5%, but when moisture content further increased the internal friction angle initiated to increase and the external friction angle remains decreased. The maximum internal and external friction was 25.07 and 20.82 observed at 26.2 and 30.7% moisture content, respectively. Moreover, the minimum 21.02, and 19.43 were recorded at 16.7 and 38%. The gauss model was used to perform non-linear curve fitting. The adjusted coefficient of determinations was 0.84497 and 0.8655 calculated for internal and external friction angle, respectively. Previous research reported that a decrease in external friction angle was observed with increasing soil moisture content [42,43]. Additionally, another previous report indicated that soil moisture increases with a decrease in the angle of internal and external friction [44,45].

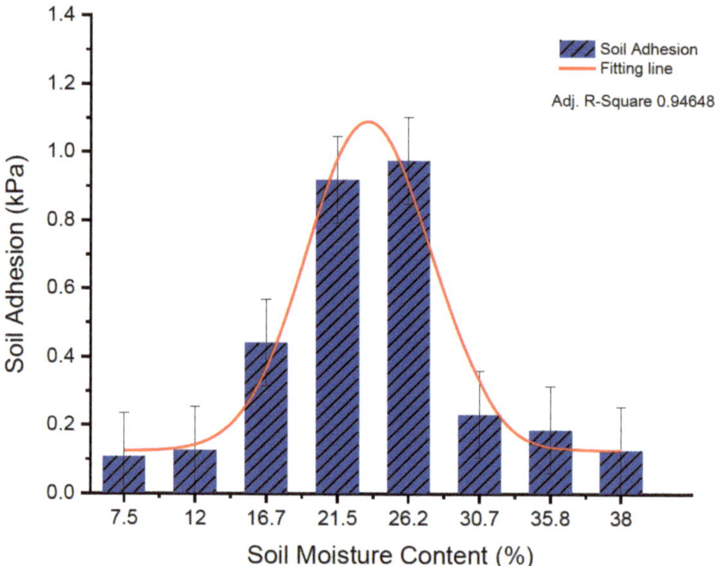

Figure 13. Soil Adhesion at moisture contents (7.5–38%).

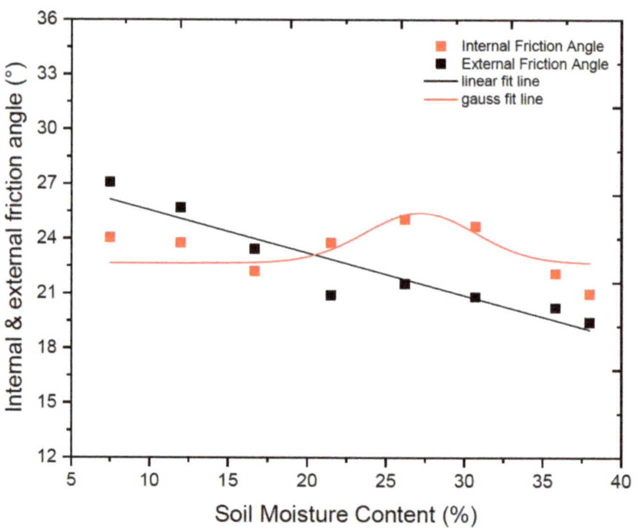

Figure 14. Internal and external friction angle of soil at moisture contents (7.5–38%).

3.6. Soil Bulk Density

Bulk density offers the most basic details on the solid, water and air proportions and seems to be important to any study. The results of soil bulk density at eight moisture contents (7.2, 12, 16.7, 21.5, 26.2, 30.7, 35.8, and 38%) were shown in Figure 15. As one can see from the figure that the soil bulk density linearly increased with an increase in soil moisture content. The major increase in soil bulk density was found at 38%, while the lower was at 7.5% moisture content. The reason for the change in soil bulk density is when water is applied to the soil to reach optimum moisture levels, the soil form varies from semisolid to liquid because of capillary force. The influence of soil bulk density on moisture content was significant ($p < 0.05$), the R^2 was 0.9609 calculated by linear regression. The present results are in line with the previously reported studies found the maximum change in soil dry bulk density at 30% moisture content [37,46,47].

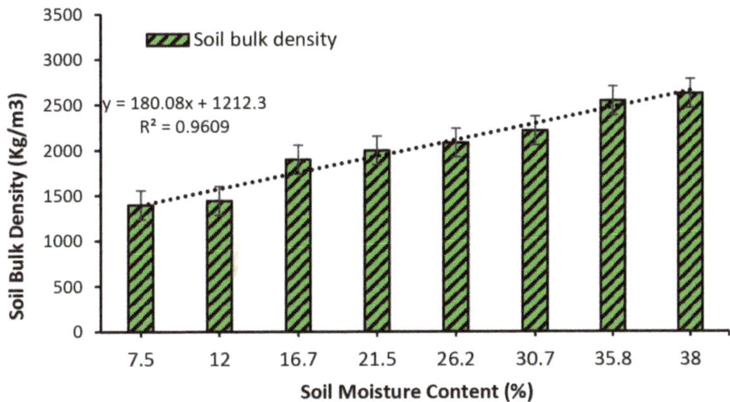

Figure 15. Soil bulk density at moisture contents (7.5–38%).

3.7. Tractive Performance Parameters of Single-Track Shoe

3.7.1. The Thrust Generated at Lateral Sides and Bottom of the Grouser

The total soil thrust is the addition of thrust generated at the tip surface of grouser (F_1), lateral sides of grouser and spacing, the surface of spacing (F_2), and the bottom surface of grouser (F_3). Figure 16 shows the results of thrust at the tip of grouser (FF_1) with different heights, results reveal that all grouser heights had the same trend over moisture content, but the high value (2.732 KN) was recorded 21.5% with 45 mm grouser height and it was steadily decreased with the rise in moisture content until the end of the experiment. F_2 results were shown in Figure 17, the results indicated that the thrust was increased with a rise in moisture content until 16.7%, and then slowly decreased with further increase in moisture content, the maximum thrust of F2 was (8.867 KN) observed for 60 mm height at 16.7%. Similarly, the F3 has also the same pattern (Figure 18), the more value of thrust was (6.821 KN) recorded with 45 mm height at 16.7% moisture content. The gauss model was used for non-linear curve fitting to obtain best fit line and coefficient of determination.

The total thrust generated at the single-track shoe was presented in Figure 19, the thrust with 45, 55, 60 mm height grew before 16.7% moisture content, while moisture content was increased all results were decreased until the end of 38%. The maximum thrust 18.293 KN was calculated as 45 mm grouser height at 16.7%, while a minimum 9.38 KN was observed with 60 mm height at 16.5% moisture content. The coefficient of determination for total thrust was 0.92673, 0.86866 and 0.79471 observed for 45, 55 and 60 mm grouser height, respectively. The results evaluated that the major increase in thrust was observed for 45 mm grouser height as compared to other heights; it was seen that grouser with small height had the best performance. Ge et al. reported that soil thrust decreased with an increase in moisture content [34]. Other studies reported that an increase in soil moisture content causes a decrease in the peak values of the thrust [48,49].

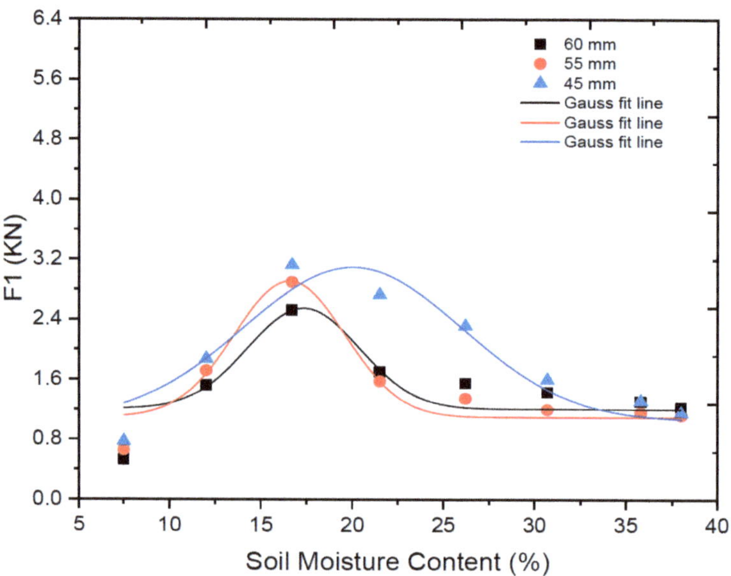

Figure 16. Soil thrust at grouser tip surface with three grouser heights at moisture contents (7.5–38%).

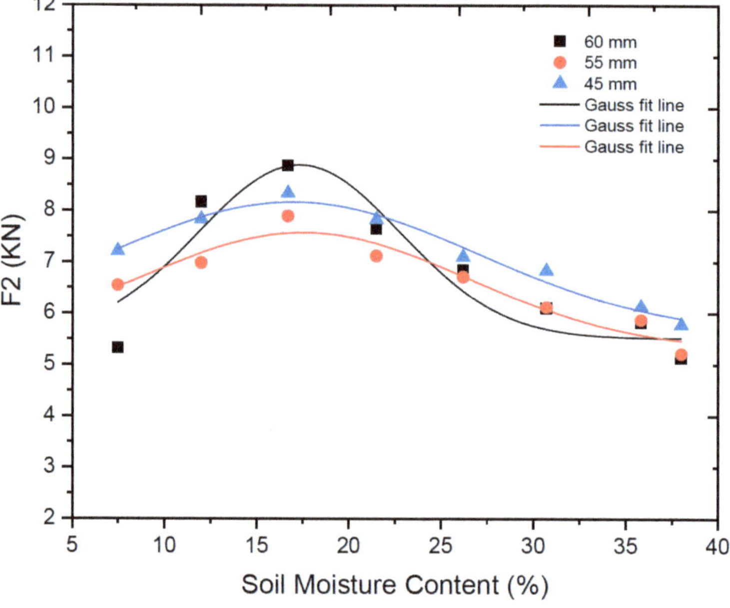

Figure 17. Soil thrust at grouser and spacing lateral sides and spacing surface with three grouser heights at moisture contents (7.5–38%).

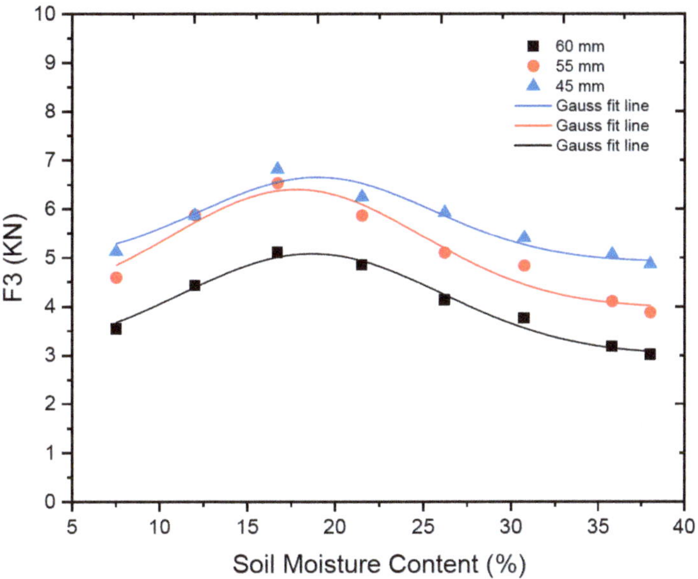

Figure 18. Soil thrust at the bottom surface of grouser with three grouser heights at moisture contents (7.5–38%).

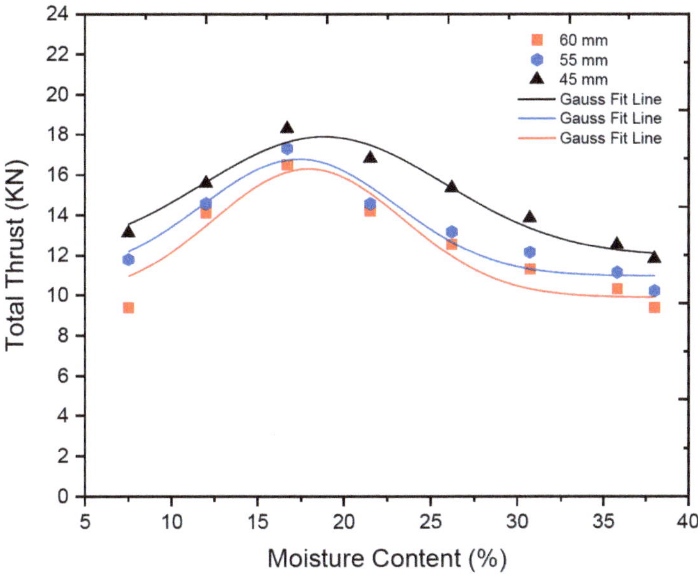

Figure 19. Total soil thrust generated at single track shoe with three grouser heights at moisture contents (7.5–38%).

3.7.2. Motion (Running) Resistance of the Single-Track Shoe

The Running resistance of single-track shoe with three grouser heights at eight moisture contents is shown in Figure 20. The result running resistance showed the variation from 1.07 to 4.03, 0.89 to 3.78, and 0.87 to 3.42 KN for 45, 55 and 60-mm grouser height, respectively. The linear increasing trend was found in results because moisture content directly proportionated the running resistance. The major rise in running resistance was recorded with 45 mm height at 38% moisture content as compared to 55 and 60 mm height. It is clear from the results that, when the contact area was increased the running resistance decreases because of less sinkage, it also leads to increased net traction. The linear fitting was performed to get best fit, the coefficient of determination for running resistance was 0.99612, 0.98938 and 0.99232 calculated for 45, 55 and 60 mm grouser height, respectively Present results are in agreement with the reported results that increase in running resistance was determined with the increase in moisture content [37,50,51]. Another study reported that the softer, wetter, and more slippery the soil, the smaller the traction and greater the running resistance [52]. Actually, the dynamic transformation of the soil flow produced by the grouser can be found, as the grouser height is adequately high [19].

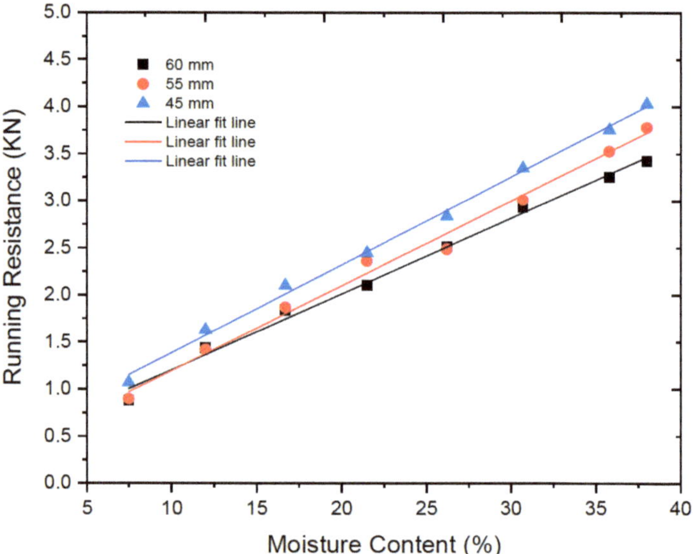

Figure 20. Grouser running resistance of single-track shoe with three grouser heights at moisture contents (7.5–38%).

3.7.3. Tractive Effort (Traction) of Single-Track Shoe

Figure 21 shows the results of the tractive effort of a single-track shoe with three grouser heights for eight moisture contents. The variation in the result was found for each height at all moisture content, initially, the traction of single-track shoe increased with an increased in moisture content until 16.7%, then it decreased steadily until the end of the experiment 38%. The mean maximum traction 16.189, 15.452, and 14.66 KN, similarly minimum was 7.78, 6.43, and 5.97 KN was determined for 45, 55 and 60 mm grouser height at 16.7% and 38% moisture content, respectively. Results indicated the more traction with smaller grouser height (45 mm) as compared to larger heights, and when soil is wetter the tractive effort was less. The gauss model was used to perform non-linear curve fitting and coefficient of determination, adjusted R square was 0.92468, 0.91853 and 0.97225 for 45, 55 and 60 mm height, respectively. Previous studies reported that soft soil can drastically reduce the traction performance [53,54], traction efficiency of two-wheel agricultural tractor reduced in soft soil [37,55], the moisture content between 15% and 20% is ideal for the net traction, greater the moisture content lower the traction [56]. Presented results are in close agreement with these results. Spacing between the grouse also influences on the sinkage, and gross traction [20].

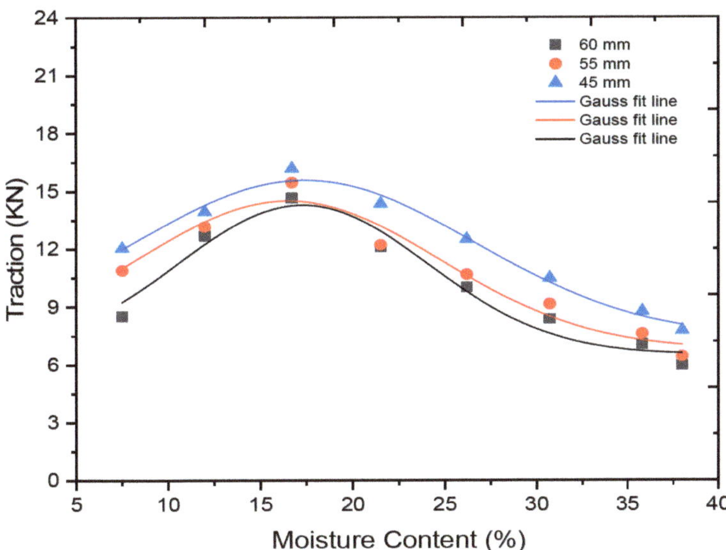

Figure 21. Traction of the single-track shoe with three grouser heights at moisture contents (7.5–38%).

4. Conclusions

The effect of grouser height on the tractive performance of single grouser shoe at eight moisture content was studied in this research. The semi-empirical method for tractive performance of grouser was used, which is based on Bekker's bevameter method. Soil cohesive and friction modulus were increased with moisture content while the sinkage exponent was decreased. Soil sinkage increased with an increase in moisture content, soil cohesion decreased, and adhesion was initially increased to 21.5%, then decreased until the end level 38%. Soil dry bulk density varied from 1394 to 2621 Kg/m^3. Maximum soil thrust was observed at 4.5 cm grouser height at 16.5% moisture content. The running resistance was decreased with a rise in moisture content, the major decrease was in 4.5 cm grouser height at 38% moisture content. Traction is the difference of soil thrust to running resistance, the grouser height 4.5 cm height showed the best results, followed by 5.5 and 6 cm grouser height at moisture contents (7.5–38%). It could be concluded that an off-road tracked vehicle (crawler combine harvester) with 4.5 cm grouser height of single-track shoe could be operated towards moderate moisture content (16.7–21.5%) under paddy soil for achieving better traction.

Author Contributions: Conceptualization, L.Y. and S.A.S.; methodology, L.Y., S.A.S., M.Z. and F.A.C.; investigation, L.Y., S.A.S. and M.Z.; data collection, S.A.S., formal analysis, F.A., F.A.C. and I.A.; writing-review and editing, L.Y., S.A.S., M.H.T. and F.A.; supervision, L.Y. All authors have read and agreed to the published version of the manuscript.

Funding: This research was funded by the National Natural Science Foundation, 51975256, a project funded by Priority Academic Program of the Development of Jiangsu Higher Education Institutions (PAPD).

Institutional Review Board Statement: Not applicable.

Informed Consent Statement: Not applicable.

Data Availability Statement: The data presented in this study are available in the main body of this manuscript.

Conflicts of Interest: The authors declare no conflict of interest.

References

1. Wong, J.Y. *Theory of Ground Vehicles*, 3rd ed.; John Wiley & Sons: New York, NY, USA, 2001.
2. Bekker, M.G. *Introduction of Terrain Vehicle Systems*; University of Michigan Press: Ann Arbor, MI, USA, 1969.
3. Conte, O.; Levien, R.; Debiasi, H.; Stürmer, S.L.K.; Mazurana, M.; Müller, J. Soil disturbance index as an indicator of seed drill efficiency in no-tillage agrosystems. *Soil Tillage Res.* **2011**, *114*, 37–42. [CrossRef]
4. Chen, Y.; Munkholm, L.J.; Nyord, T. A discrete element model for soil–sweep interaction in three different soils. *Soil Tillage Res.* **2013**, *126*, 34–41. [CrossRef]
5. Rajaram, G.; Erbach, D.C. Drying stress effect on mechanical behaviour of a clay-loam soil. *Soil Tillage Res.* **1998**, *49*, 147–158. [CrossRef]
6. Abubakar, M.S.; Ahmad, D.; Othman, J.; Sulaiman, S. Mechanical properties of paddy soil in relation to high clearance vehicle mobility. *Aust. J. Basic Appl. Sci.* **2010**, *4*, 906–913.
7. Lyasko, M. Multi-pass effect on off-road vehicle tractive performance. *J. Terramech.* **2010**, *47*, 275–294. [CrossRef]
8. Wong, J.Y. Methods for evaluating tracked vehicle performance. In *Terramechanics and Off-Road Vehicle Engineering*; Elsevier: Amsterdam, The Netherlands, 2010; pp. 155–176; ISBN 978-0-7506-8561-0.
9. Chang, B.-S.; Baker, W.J. Soil parameters to predict the performance of off-road vehicles. *J. Terramech.* **1973**, *9*, 13–31. [CrossRef]
10. Soltynski, A. The mobility problem in agriculture. *J. Terramech.* **1979**, *16*, 139–149. [CrossRef]
11. Bodin, A. Development of a tracked vehicle to study the influence of vehicle parameters on tractive performance in soft terrain. *J. Terramech.* **1999**, *36*, 167–181. [CrossRef]
12. Zoz, F.M. Predicting tractor field performance. *Trans. ASAE* **1972**, *15*, 0249–0255. [CrossRef]
13. Tiwari, V.K.; Pandey, K.P.; Pranav, P.K. A review on traction prediction equations. *J. Terramech.* **2010**, *47*, 191–199. [CrossRef]
14. Grisso, R.; Perumpral, J.; Zoz, F. An empirical model for tractive performance of rubber-tracks in agricultural soils. *J. Terramech.* **2006**, *43*, 225–236. [CrossRef]
15. Nakashima, H.; Yoshida, T.; Wang, X.L.; Shimizu, H.; Miyasaka, J.; Ohdoi, K. Comparison of gross tractive effort of a single grouser in two-dimensional DEM and experiment. *J. Terramech.* **2015**, *62*, 41–50. [CrossRef]
16. Abou-Zeid, A.S. Distributed Soil Displacement and Pressure Associated with Surface Loading. Ph.D. Thesis, University of Saskatchewan, Saskatoon, SK, Canada, 2004.
17. Nagaoka, K.; Sawada, K.; Yoshida, K. Shape effects of wheel grousers on traction performance on sandy terrain. *J. Terramech.* **2020**, *90*, 23–30. [CrossRef]
18. Li, J.; Liu, S.; Dai, Y. Effect of grouser height on tractive performance of tracked mining vehicle. *J. Braz. Soc. Mech. Sci. Eng.* **2017**, *39*, 2459–2466. [CrossRef]
19. Yamanoa, Y.; Nagaokaa, K.; Yoshidaa, K. PIV analysis of soil deformation beneath a grouser wheel. In Proceedings of the 10th Asia-Pacific Conference ISTVS, Kyoto, Japan, 11–13 July 2018.
20. Yokoyama, A.; Nakashima, H.; Shimizu, H.; Miyasaka, J.; Ohdoi, K. Effect of open spaces between grousers on the gross traction of a track shoe for lightweight vehicles analyzed using 2D DEM. *J. Terramech.* **2020**, *90*, 31–40. [CrossRef]
21. Reeb, J.E.; Milota, M.R.; Association, W.D.K. Moisture content by the oven-dry method for industrial testing. *Engineering* **1999**. Available online: ir.library.oregonstate.edu (accessed on 10 August 2020).
22. Tiwari, V.K.; Pandey, K.P.; Sharma, A.K. Development of a tyre traction testing facility. *J. Terramech.* **2009**, *46*, 293–298. [CrossRef]
23. Massah, J.; Noorolahi, S. Effects of tillage operation on soil properties from Pakdasht, Iran. *Int. Agrophys.* **2008**, *22*, 143–146.
24. Bahnasy, A.M.F. Utilization of different soil sinkage plates to predict tire inflation pressure and its sinkage under different soil conditions. *Misr. J. Agric. Eng.* **2012**, *29*, 531–552. [CrossRef]
25. Upadhyaya, S.K.; Wulfsohn, D.; Mehlschau, J. An instrumented device to obtain traction related parameters. *J. Terramech.* **1993**, *30*, 1–20. [CrossRef]
26. Gotteland, P.; Benoit, O. Sinkage tests for mobility study, modelling and experimental validation. *J. Terramech.* **2006**, *43*, 451–467. [CrossRef]
27. Liu, K.; Ayers, P.; Howard, H.; Anderson, A. Lateral slide sinkage tests for a tire and a track shoe. *J. Terramech.* **2010**, *47*, 407–414. [CrossRef]
28. Rashidi, M.; Fakhri, M.; Sheikhi, M.; Azadeh, S.; Razavi, S. Evaluation of Bekker model in predicting soil pressure-sinkage behaviour under field conditions. *Middle East J. Sci. Res.* **2012**, *12*, 1364–1369.
29. Malý, V.; Kučera, M. Determination of mechanical properties of soil under laboratory conditions. *Res. Agric. Eng.* **2014**, *60*, S66–S69. [CrossRef]
30. Baek, S.-H.; Shin, G.-B.; Lee, S.-H.; Yoo, M.; Chung, C.-K. Evaluation of the slip sinkage and its effect on the compaction resistance of an off-road tracked vehicle. *Appl. Sci.* **2020**, *10*, 3175. [CrossRef]
31. Zhang, J.X.; Sang, Z.Z.; Gao, L.R. Adhesion and friction between soils and solids. *Nung Yeh Chi Hsieh Hsueh Pao Trans. Chin. Soc. Agric. Mach.* **1986**, *17*, 32–40.
32. Zhao, J.; Wang, W.; Sun, Z.; Su, X. Improvement and verification of pressure-sinkage model in homogeneous soil. *Trans. Chin. Soc. Agric. Eng.* **2016**, *32*, 60–66.
33. Yang, C.; Yang, G.; Liu, Z.; Chen, H.; Zhao, Y. A method for deducing pressure–sinkage of tracked vehicle in rough terrain considering moisture and sinkage speed. *J. Terramech.* **2018**, *79*, 99–113. [CrossRef]

34. Ge, J.; Zhang, D.; Wang, X.; Cao, C.; Fang, L.; Duan, Y. Tractive performances of single grouser shoe affected by different soils with varied moisture contents. *Adv. Mech. Eng.* **2019**, *11*. [CrossRef]
35. Huang, K.; Wan, J.W.; Chen, G.; Zeng, Y. Testing study of relationship between water content and shear strength of unsaturated soils. *Rock Soil Mech.* **2012**, *33*, 2600–2604.
36. Zydroń, T.; Dąbrowska, J. The influence of moisture content on shear strength of cohesive soils from the landslide area around Gorlice. *AGH J. Min. Geoengin* **2012**, *36*, 309–317.
37. Jun, G.; Xiulun, W.; Kito, K. Comparing tractive performance of steel and rubber single grouser shoe under different soil moisture contents. *Int. J. Agric. Biol. Eng.* **2016**, *9*, 11–20.
38. Mouazen, A.M.; Ramon, H.; De Baerdemaeker, J. SW—Soil and water: Effects of bulk density and moisture content on selected mechanical properties of sandy loam soil. *Biosyst. Eng.* **2002**, *83*, 217–224. [CrossRef]
39. Bravo, E.L.; Suárez, M.H.; Cueto, O.G.; Tijskens, E.; Ramon, H. Determination of basics mechanical properties in a tropical clay soil as a function of dry bulk density and moisture. *Revista Ciencias Técnicas Agropecuarias* **2012**, *21*, 5–11.
40. Adeniran, K.A.; Babatunde, O.O. Investigation of wetland soil properties affecting optimum soil cultivation. *J. Eng. Sci. Technol.* **2010**, *3*, 23–26. [CrossRef]
41. Abbaspour-Gilandeh, Y.; Fazeli, M.; Roshanianfard, A.; Hernández-Hernández, J.L.; Fuentes Penna, A.; Herrera-Miranda, I. Effect of different working and tool parameters on performance of several types of cultivators. *Agriculture* **2020**, *10*, 145. [CrossRef]
42. Komandi, G. On the mechanical properties of soil as they Affect traction. *J. Terramech.* **1992**, *29*, 373–380. [CrossRef]
43. Han, Z.; Li, J.; Gao, P.; Huang, B.; Ni, J.; Wei, C. Determining the shear strength and permeability of soils for engineering of new paddy field construction in a hilly mountainous region of Southwestern China. *Int. J. Environ. Res. Public Health* **2020**, *17*, 1555. [CrossRef]
44. McKyes, E.; Nyamugafata, P.; Nyamapfene, K.W. Characterization of cohesion, friction and sensitivity of two hardsetting soils from Zimbabwe. *Soil Tillage Res.* **1994**, *29*, 357–366. [CrossRef]
45. Rajaram, G.; Erbach, D.C. Effect of wetting and drying on soil physical properties. *J. Terramech.* **1999**, *36*, 39–49. [CrossRef]
46. Mari, I.A.; Changying, J.; Leghari, N.; Chandio, F.A.; Arslan, C.; Hassan, M. Impact of tillage operation on soil physical, mechanical and rhelogical properties of paddy soil. *Bulg. J. Agric. Sci.* **2015**, *21*, 940–946.
47. Shaikh, S.A.; Yaoming, L.; Chandio, F.A.; Tunio, M.H.; Ahmad, F.; Mari, I.A.; Solangi, K.A. Effect of wheat residue incorporation with tillage management on physico-chemical properties of soil and sustainability of maize production. *Fresenius Environ. Bull.* **2020**, *29*, 10.
48. Hermawan, W.; Oida, A.; Yamazaki, M. The characteristics of soil reaction forces on a single movable lug. *J. Terramech.* **1997**, *34*, 23–35. [CrossRef]
49. Xie, Z.; Wang, X.L.; Ge, J. Effect of soil moisture content on tractive performance of single grouser shoe in tracked vehicle. *Int. Inf. Inst. Tokyo Inf. Koganei* **2019**, *22*, 229–240.
50. Wang, X.L.; Ito, N.; Kito, K. Studies on grouser shoe dimension for optimum tractive performance (part 2) effect on thrust, rolling resistance and traction. *Agric. Mach. J. Jpn.* **2002**, *64*, 55–60.
51. Keen, A.; Hall, N.; Soni, P.; Gholkar, M.D.; Cooper, S.; Ferdous, J. A review of the tractive performance of wheeled tractors and soil management in lowland intensive rice production. *J. Terramech.* **2013**, *50*, 45–62. [CrossRef]
52. Abubakar, M.S.; Ahmad, D.; Othman, J.; Suleiman, S. Present state of research on development of a high clearance vehicle for paddy fields. *Res. J. Agric. Biol. Sci.* **2009**, *5*, 489–497.
53. Senatore, C. Prediction of Mobility, Handling, and Tractive Efficiency of Wheeled Off-Road Vehicles. Ph.D. Thesis, Virginia Tech, Blacksburg, VA, USA, 2010.
54. Rasool, S.; Raheman, H. Suitability of rubber track as traction device for power tillers. *J. Terramech.* **2016**, *66*, 41–47. [CrossRef]
55. McKyes, E. *Soil Cutting and Tillage*; Elsevier: New York, NY, USA, 1985.
56. Schreiber, M.; Kutzbach, H.D. Influence of soil and tire parameters on traction. *Res. Agric. Eng.* **2008**, *54*, 43–49. [CrossRef]

 *sustainability*

Article

Effect of K and Zn Application on Biometric and Physiological Parameters of Different Maize Genotypes

Hafiz Muhammad Ali Raza [1], Muhammad Amjad Bashir [1,2], Abdur Rehim [1,2,*], Qurat-Ul-Ain Raza [1], Kashif Ali Khan [2], Muhammad Aon [1], Muhammad Ijaz [2], Shafeeq Ur Rahman [3], Fiaz Ahmad [4] and Yucong Geng [5,*]

1   Department of Soil Science, Faculty of Agricultural Sciences and Technology, Bahauddin Zakariya University, Multan 60800, Pakistan; maraza1524@gmail.com (H.M.A.R.); 2017y90100004@caas.cn (M.A.B.); quratulain1111@yahoo.com (Q.-U.-A.R.); aonses2009@yahoo.com (M.A.)
2   College of Agriculture Bahadur Sub-Campus Layyah, Bahauddin Zakariya University, Multan 60800, Pakistan; rana13tda@yahoo.com (K.A.K.); muhammad.ijaz@bzu.edu.pk (M.I.)
3   Farmland Irrigation Research Institute, Chinese Academy of Agricultural Sciences, Xinxiang 453003, China; malikshafeeq1559@gmail.com
4   Department of Agricultural Engineering, Bahauddin Zakariya University, Multan 60800, Pakistan; fiazahmad@bzu.edu.pk
5   Key Laboratory of Nonpoint Source Pollution Control, Ministry of Agriculture, Institute of Agricultural Resources and Regional Planning, Chinese Academy of Agricultural Sciences, Beijing 100081, China
*   Correspondence: abdur.rehim@bzu.edu.pk (A.R.); gengyucong@caas.cn (Y.G.)

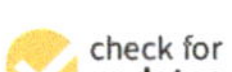
check for updates

Citation: Ali Raza, H.M.; Bashir, M.A.; Rehim, A.; Raza, Q.-U.-A.; Khan, K.A.; Aon, M.; Ijaz, M.; Ur Rahman, S.; Ahmad, F.; Geng, Y. Effect of K and Zn Application on Biometric and Physiological Parameters of Different Maize Genotypes. *Sustainability* **2021**, *13*, 13440. https://doi.org/10.3390/su132313440

Academic Editors: Emanuele Radicetti and Roberto Mancinelli

Received: 23 October 2021
Accepted: 30 November 2021
Published: 4 December 2021

Publisher's Note: MDPI stays neutral with regard to jurisdictional claims in published maps and institutional affiliations.

Abstract: Potassium (K) and zinc (Zn) are mineral nutrients required for adequate plant growth, enzyme activation, water retention and photosynthetic activities. However, Pakistani soils are alkaline and have serious problems regarding Zn deficiency. The current study aims at finding the nutrient–nutrient interaction of K and Zn to affect maize plants' (i) physiological processes and (ii) productivity. For this purpose, a pot experiment was conducted at the research area of the Department of Soil Science, Faculty of Agricultural Science and Technology, Bahauddin Zakariya University, Multan. Two maize genotypes, DK-6142 (hybrid) and Neelam (non-hybrid), were used with three K fertilizer doses, i.e., 0, 60 and 100 kg ha^{-1} in all possible combinations with three Zn fertilizer doses, i.e., 0, 16 and 24 kg ha^{-1}. The treatments were replicated under a completely randomized block design. The results elucidated that the combined application of K and Zn with K60 + Zn16 treatment significantly increased agronomic, productive, and physiological attributes. It has improved fresh biomass (89%), dry biomass (94%), membrane stability index (142%), relative water content (200%) and chlorophyll contents (191%) as compared to the control. Moreover, the mineral uptake of K and Zn was significantly improved with their maximum fertilization rate in hybrid genotype compared to non-hybrid and CK.

Keywords: potassium; zinc; maize; physiological attributes

1. Introduction

Maize (*Zea mays* L.) is a C$_4$ plant belonging to the family Poaceae. In Pakistan, it is considered the third most important cereal crop, after wheat and rice. Moreover, various human communities consume it as staple food and animal feed in different parts of the world, including Africa, America and Asia [1,2]. It is also certified as a rich source of protein content (up to 15%). In 2019–2020, the total maize cultivation area was 1404.2 thousand hectares, while maize production was 7.88 million tonnes in Pakistan, contributing about 0.6% in Gross Domestic Production (GDP) [3]. Because of its accumulative significance, progress in the agronomic physiognomies of maize has received great reverence in Pakistan [4].

Potassium (K) is one of the primary macronutrients required by plants to complete their life span. Its fertilization stimulates root growth significantly, enhancing water uptake from soil to the plant body [5,6]. In the same way, it decreases the transpiration rate and helps in water conservation during drought conditions [7]. In addition, it has a significant contribution in activating many enzymes, maintains the turgor pressure, the electric potential gradient and the balance between cations and anions [8,9]. According to [10], K content present in plant tissues is capable of the appropriate functioning of numerous processes directly involved in crop growth, development, and yield. Consequently, the regular application of a balanced dose of K is necessary to maintain soil fertility, optimum crop growth and yield for a longer run.

Zinc (Zn) plays a crucial role in the metabolic activities of major proteins among plants and humans and is hence considered as an essential nutrient for all organisms [5,11]. According to [12], Zn is indirectly involved in carbohydrate metabolism, it is a structural constituent of many enzymes involved in different processes such as photosynthesis, the metabolism of different proteins, pollen germination and maintaining membrane integrity also provide resistance against various pathogens. Pakistani soils have low organic matter contents (0.4–0.7%) and are abundant in calcium carbonate ($CaCO_3$), mainly maintaining a high pH (7.5–8.4). Moreover, the cation exchange capacity (CEC) is formally controlled by Calcium (Ca) and such factors are key reasons to limit the plant-available Zn and to reduce the crop yield in absence of Zn fertilizer [5]. However, Zn application in these soils helps to improve the nutritional status of crop and enhance grain production [13]. Maize is a source of macro- and micronutrients in the human diet, which helps in a balanced diet. The deficiency of Zn in maize could induce Zn deficiency in humans using maize as staple food [14,15]. According to Krebs [16], the problems produced due to Zn deficiency are very common in children. According to the previous studies, maize genotype plays a vital role in grain Zn uptake [17–19]. The concentration of Zn present in maize grain is about 20 mg kg^{-1} on an average basis [20,21], so its bioavailability in humans/animals is insufficient.

Limited knowledge is available in the literature about the interaction of these two mineral nutrients (K and Zn) and their effectiveness on maize growth and yield. So, it is promising that the soil with a marginal deficiency of these nutrients have no capacity to ensure optimum yield of maize without balanced fertilization of K and Zn. This gives rise to the need for the current study to investigate the efficiency of different Zn and K rates applied to maize genotypes (hybrid and non-hybrid) for efficient mineral uptake. The major aims of current study were to (i) identify the nutrient–nutrient interaction between K and Zn in the agroecosystem, (ii) elucidate the effectiveness of the combined application of K and Zn on maize physiology and (iii) define the impacts of K and Zn fertilization on yield in different genotypes.

2. Material and Methods

2.1. Site Description

To investigate the above-mentioned aims, a pot experiment was conducted at the research area of the Department of Soil Science (30.258° N, 71.515° E), Faculty of Agricultural Sciences and Technology, Bahauddin Zakariya University, Multan, Pakistan. Multan city is located in the arid and subtropical region of Pakistan [22]. The maximum recorded temperature is approximately 54 °C, and the lowest recorded temperature is approximately 4.5 °C. The average rainfall is roughly 186 mm. Dust storms are of frequent occurrence within the city. The soil used in the experiment was alkaline calcareous and deficient in plant-available Zn (0.54 mg kg^{-1}) and K (75.0 mg kg^{-1}). The physicochemical properties of soil used for pot experiment are as given in Table 1.

2.2. Experiment Details

The study was conducted in a wire-house under ambient conditions. The moisture contents were visually checked every day, and the pots were irrigated frequently (3–4 times a week) to maintain the water contents at field capacity level. The soil used in the experiment was sieved through 2 mm mesh. Soil samples were used to measure the soil textural class [23,24], organic matter [25] and extractable K [26]. Electrical conductivity (inoLab® Cond 7110, Xylem Analytics, Weilheim, Germany) and pH (METTLER TOLEDO, Jenway, UK) of soil samples were measured through soil extract and saturated soil paste, respectively. Soil sample was taken in a conical flask with AB-DTPA extraction solution and shaken for 15 min at 180 cycles/minute, then filtrated. Atomic absorption spectrophotometry was used to determine the AB-DTPA extractable Zn in 1:2 soil solution (Thermo Scientific 3000 Series, Waltham, MA, USA; [27,28]). A three-factorial, completely randomized design was followed where maize genotypes, i.e., hybrid Monsanto 'DK-6142' and non-hybrid 'Neelam', were used in all possible combination with three levels of K (0, 60 and 100 kg ha^{-1}) and three levels of Zn (0, 16 and 24 kg ha^{-1}). Approximately, 20 kg soil was added to each pot having 38 cm height, 27 cm diameter, 0.0572 m^2 surface area and 0.02175 m^3 volume. The pots were maintained at an adequate distance from each other, they were at (25 cm × 75 cm) row × plant distance to support easy movement during cultivation of plants. The experiment was conducted for one growth season, and the treatments were replicated three times.

Table 1. Physicochemical properties of soil used in the experiment.

Character	Unit	Value
Texture		Loam
Sand	%	47.6
Silt	%	34.2
Clay	%	18.2
PHs	—	8.43
EC	dS m^{-1}	0.72
CaCO$_3$	%	4.67
Bulk density	Mg m^{-3}	1.30
Organic matter	%	0.57
AB-DTPA Extractable Zn	mg kg^{-1}	0.54
Extractable K	mg kg^{-1}	75.0

Five seeds of each genotype were sown in each pot. After one week of germination, thinning was conducted up to three plants per pot. Recommended dose of N (1.3 g per pot) and P$_2$O$_5$ fertilizer (1 g per pot) were used in the form of urea (contains 46% N; manufactured by Fauji Fertilizer Company Limited (FFC), Rawalpind, Pakistan) and diammonium phosphate (DAP contains 46% P, 18% N, manufactured by FFC, Pakistan), respectively. Urea was applied in two equal splits (i.e., at the time of sowing and 25 days after germination), while full recommended dose of DAP was added just before sowing. Potassium chloride (KCl) fertilizer also termed as Muriate of potash (MOP; 60% K$_2$O; manufactured by FFC, Pakistan) and zinc sulfate monohydrate (ZnSO$_4$·H$_2$O; 27% Zn; manufactured by FFC, Pakistan) were used as sources of K and Zn, respectively. As per treatment plan, K- and Zn-containing fertilizers were applied in soil as substrate just before sowing. All the fertilizers were thoroughly mixed with 20 kg soil.

2.3. Plant Harvest

Shoots and cobs were harvested at crop maturity (120 days after germination). After air-drying, each plant sample was placed in an oven at 65 °C for 48 h until constant weight was achieved. After oven drying, samples were ground and digested using a diacid mixture with HClO$_4$:HNO$_3$ in 1:2 ratio (72% and 69% pure) [29]. For digestion, ground plant samples and HNO$_3$ were added to the flasks and stood overnight. Afterward, heated at 125° for 1 h using hot plate. Then, the samples were cooled down to avoid drying out, and HClO$_4$ was added until colorless solution was obtained. Digested plant samples were used for further analysis. All the chemicals used in the experiment were bought from Eastern Scientific Corporation Private limited and were of analytical grade manufactured by Merck.

2.4. Plant Analysis

2.4.1. Morphological and Yield Parameters

Morphological and yield parameters, including plant height (cm), fresh weight (g), dry weight (g), cob length (cm) and 1000 grain weight (g) were measured after harvesting.

2.4.2. Physiological Parameters

The plant physiological parameters of all the treatments were recorded in the morning (between 09:30 to 10:30 am). Fully expanded second leaf of each plant was selected for the measurements of gaseous exchange parameters and chlorophyll content at principal growth stage 5: Inflorescence emergence (BBCH-51). Gaseous exchange parameters such as photosynthetic rate (A), transpiration rate (E), stomatal conductance (Gs) and internal CO_2 concentration (Ci) were recorded using infrared gas analyzer (IRGA: type 225 mk3; Analytical Development Company, Hoddesdon, UK). Chlorophyll contents (SPAD value) were measured using SPAD-meter (SPAD-502, Minolta, Osaka, Japan). Relative water contents (RWC) were estimated using the following formula [5]:

$$RWC = (\text{Fresh weight} - \text{Dry weight}) / (\text{Turgid weight} - \text{Dry weight}) \times 100$$

Membrane stability index (MSI) was calculated by the formula [30] as given below:

$$MSI = \{I - (C1/C2)\} \times 100$$

2.4.3. Ionic Parameters

At maturity, the straw and cobs samples were collected, and oven dried until constant weight was achieved. Afterward, the grains were removed and both straw and grain samples were ground. Following wet digestion method, the straw and grain samples were placed in diacid mixture (HNO_3: $HClO_4$, 2:1 ratio) separately in conical flasks overnight, then digested separately on hot plate (at 150 °C for 1 h, afterward at 235 °C for 30 min). The concentration of K was determined by flame photometer (FP 6410, Shanghai Jingke, China) [23], and Zn was determined using atomic absorption spectrophotometer (Thermo Scientific 3000 Series, Waltham, MA, USA).

2.5. Statistical Analysis

A completely randomized design (CRD) with three-factorial arrangement was analyzed. Statistix 9® (Analytical Software, Tallahassee, FL, USA) was used for analysis of variance (ANOVA) and Microsoft Excel 2013® (Microsoft Corporation, Redmond, WA, USA) were used for data computations. Least significant difference (LSD) test based on Duncan multiple comparisons was used to identify the significant differences among the treatment means [31]. The treatments were tested in three replicates and the data presented with standard deviation in the graphs.

3. Results

3.1. Effect of K and Zn Fertilization on Plant Height

All the treatments showed a significant difference ($p \leq 0.05$) in the plant heights of both genotypes (hybrid and non-hybrid) with K and/or Zn application as compared to the control (Figure 1). Plant height increased significantly in the hybrid cultivar (DK-6142) with the application of K60, Zn16 (56.84%) followed by K100, Zn16 (53.42%) and K100, Zn24 (50.71%) in comparison with CK (K0, Zn0).

3.2. Effect of Potassium and Zinc Application on Plant Fresh and Dry Biomass (g Pot^{-1})

Biomass production is considered as an important factor for fodder production for maize. The application of mineral nutrients such as K and Zn increased biomass production. A significant increase ($p \leq 0.05$) was observed in the fresh and dry biomass of both maize genotypes (hybrid and non-hybrid) through the combined application of Zn and K ($ZnSO_4$

+ MOP) in comparison with CK (Figure 2). Plant fresh biomass was significantly higher in the hybrid cultivar with the application of K60, Zn16 (89.73%) followed by K100, Zn16 (81.46%) and the inbred cultivar with the application of K60, Zn16 (74.03%) as compared to CK. Moreover, the plant dry biomass was increased with the application of K60, Zn16 (94.94%) followed by K100, Zn16 (85.08%) and K100, Zn24 (75.42%) in DK-6142 maize hybrid with respect to CK.

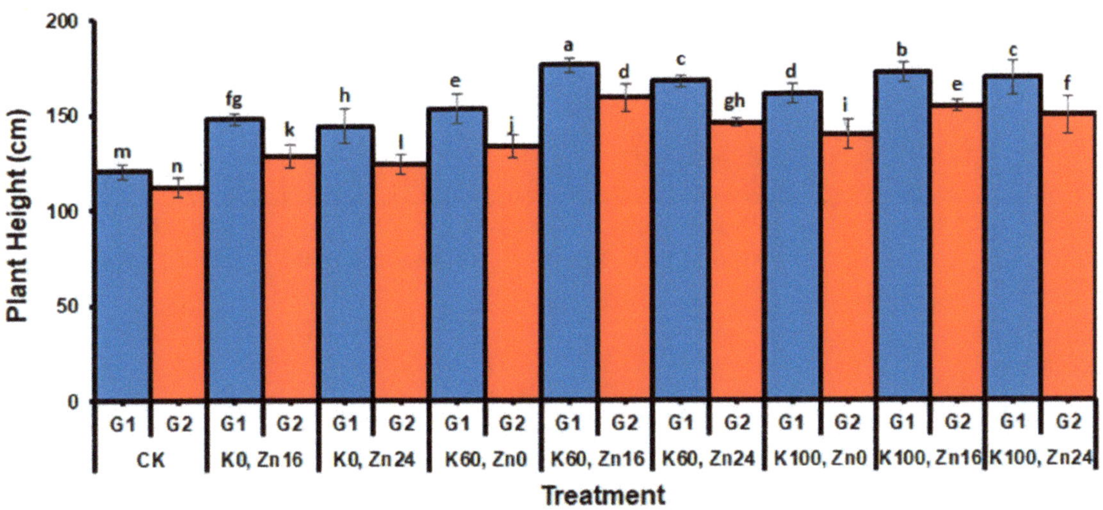

Figure 1. Effect of various levels of K and Zn on plant height of hybrid (G1) and non-hybrid (G2) genotypes of a maize plant; K0, K60 and K100 represent 0, 60 and 100 kg K ha^{-1}, respectively; Z0, Z16 and Z24 represent 0, 16 and 24 kg Zn ha^{-1}, respectively. The presented data are the means ± SD ($n = 3$). The lowercase letters indicate statistically significant differences according to the LSD test.

3.3. Effect of Potassium and Zinc on Membrane Stability Index (MSI) and Relative Water Content (RWC)

The application of mineral nutrients such as K and Zn increased the physiological parameters, i.e., the MSI and RWC, in the maize plants. A significant increase ($p \leq 0.05$) was observed in the MSI (%) and RWC (%) of both maize genotypes through the combined application of Zn and K (ZnSO$_4$ + MOP) compared to maize under control conditions where no Zn and K was applied (Figure 3). The maximum MSI was observed in the hybrid variety with the application of K60, Zn16 (2.42 fold) followed by K100, Zn16 (2.35 fold) and K100, Zn24 (2.23 fold) as compared to CK. Meanwhile, the RWC was significantly enhanced in the DK-6142 hybrid with the application of K60, Zn16 (3.0 fold) followed by K100, Zn16 (2.73 fold) and the inbred 'Neelam' genotype with the application of K60, Zn16 (2.70 fold) as compared to CK.

3.4. Effect of Potassium and Zinc on Chlorophyll Contents (SPAD Value) and Cob Length (cm)

Comparatively, DK-6142 performed better than Neelam under various fertilization treatments. The chlorophyll content (SPAD value) was significantly increased ($p \leq 0.05$) in the DK-6142 maize hybrid with the application of K60, Zn16 (2.91 fold) followed by K100, Zn16 (2.67 fold) and the Neelam inbred genotype with the application of K60, Zn16 (2.65 fold) as compared to CK (Figure 4). Moreover, cob length was increased with the application of K60, Zn16 in maize hybrid (68.47%) and inbred cultivar (60.71%) followed by maize hybrid with the application of K100, Zn16 (60.16%).

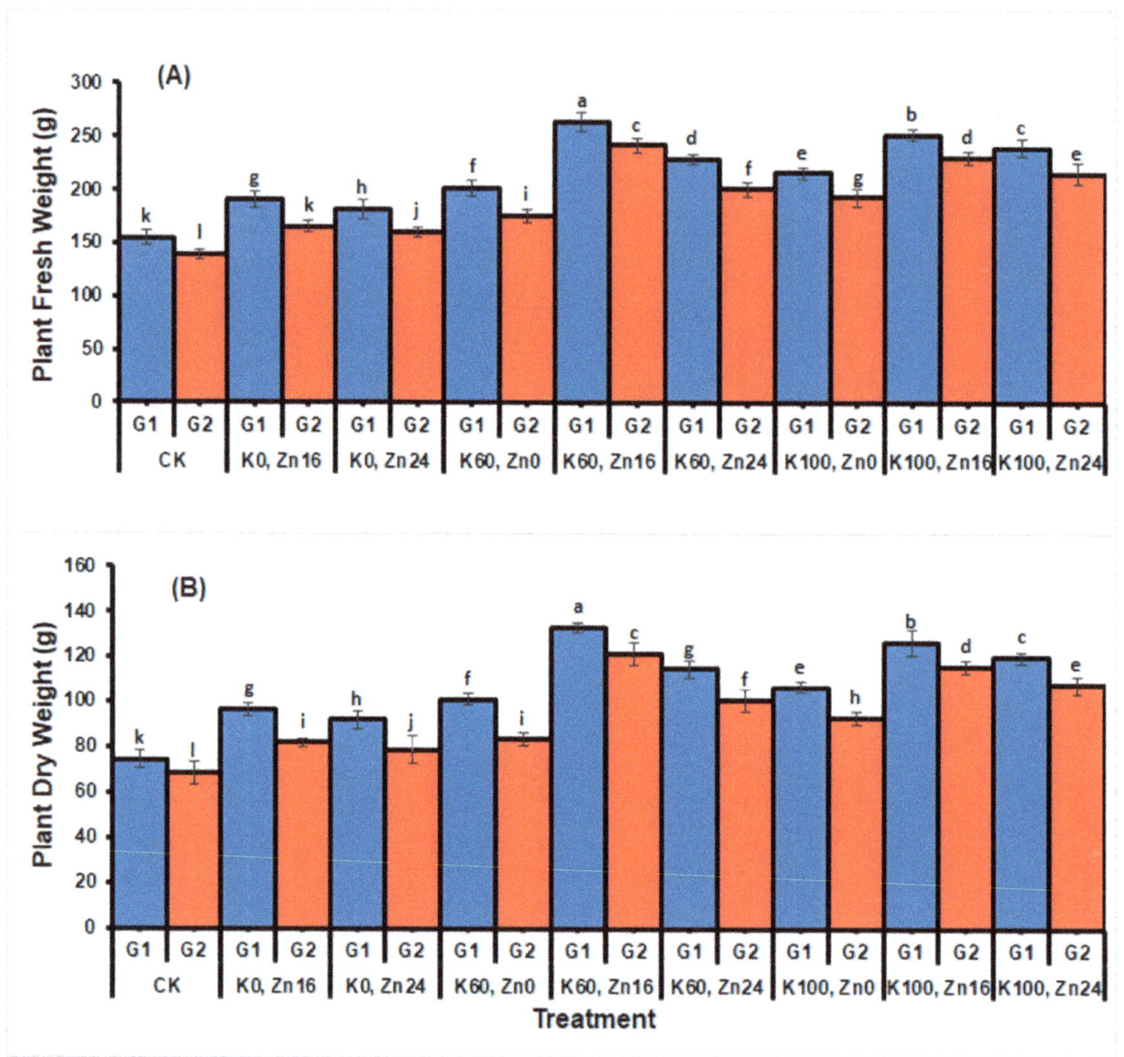

Figure 2. Effect of MOP and ZnSO$_4$ on (**A**) fresh weight and (**B**) dry weight of hybrid (G1) and non-hybrid (G2) genotypes of a maize plant; K0, K60 and K100 represent 0, 60 and 100 kg K ha^{-1}, respectively; Z0, Z16 and Z24 represent 0, 16 and 24 kg Zn ha^{-1}, respectively. The presented data are the means $\pm$ SD (n = 3). The lowercase letters indicate statistically significant differences according to the LSD test.

3.5. Effect of Potassium and Zinc on Gas Exchange Parameters

Gas exchange parameters such as photosynthetic rate (A), transpiration rate (E), stomatal conductance (Gs), and internal CO$_2$ concentration (Ci) are important and the main components of plant growth and development. A significant increase ($p \leq 0.05$) was observed in gas exchange parameters i.e., A and E (Figures 5 and 6). Photosynthetic rate significantly improved with the application of K60, Zn16 in maize hybrid (3.54 fold) and inbred genotype (3.15 fold) followed by hybrid DK-6142 with the application of K100, Zn16 (3.05 fold), respectively.

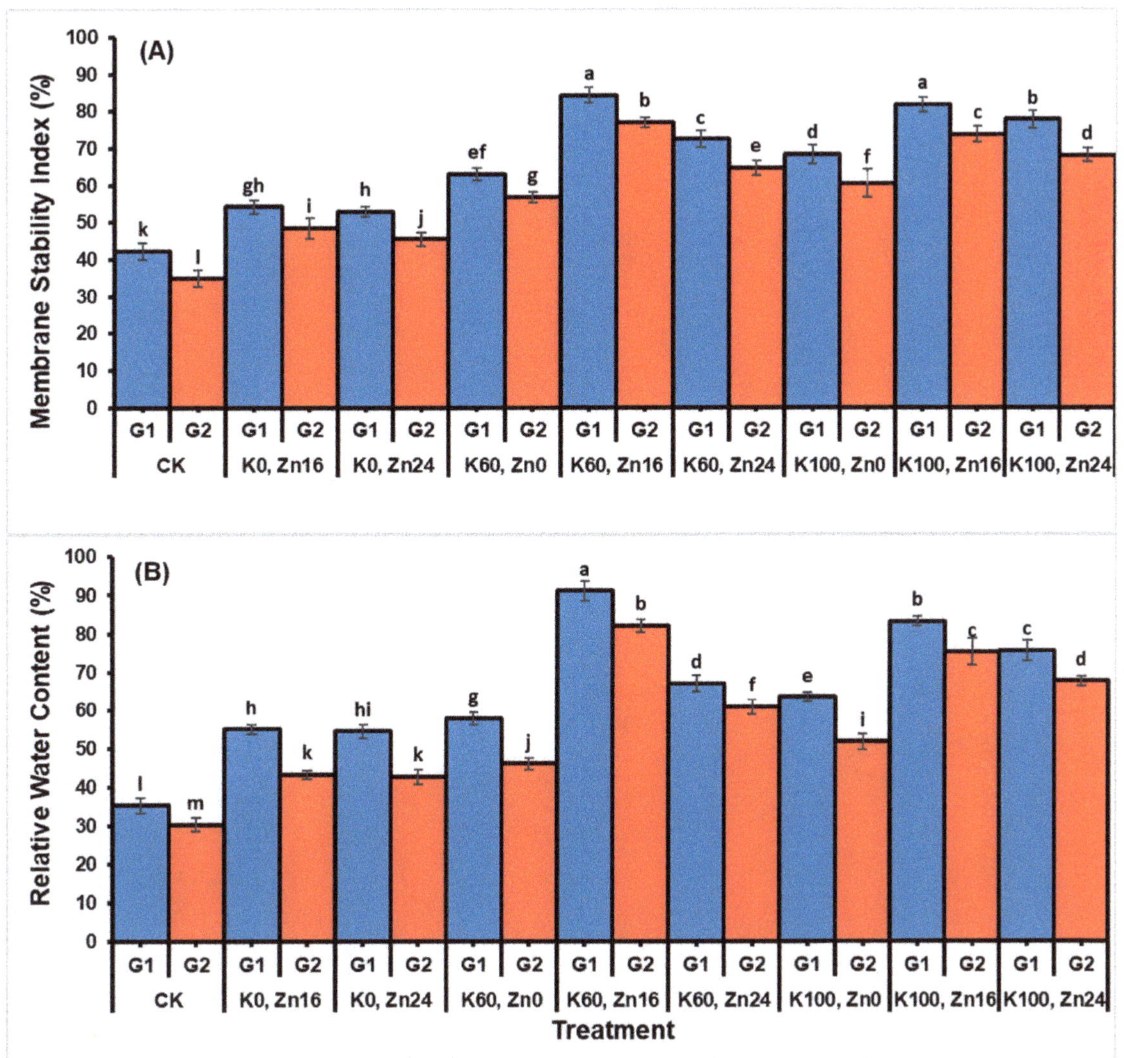

Figure 3. Effect of MOP and ZnSO$_4$ on (**A**) Membrane stability index and (**B**) Relative water content of hybrid (G1) and non-hybrid (G2) genotypes of a maize plant; K0, K60 and K100 represent 0, 60 and 100 kg K ha^{-1}, respectively; Z0, Z16 and Z24 represent 0, 16 and 24 kg Zn ha^{-1}, respectively. The presented data are the means ± SD (n = 3). The lowercase letters indicate statistically significant differences according to the LSD test.

Transpiration rate also enhanced with the application of K60, Zn16 in DK-6142 maize hybrid (4.38 fold) and Neelam genotype (4.08 fold) followed by DK-6142 hybrid with K100, Zn16 (3.94 fold) in comparison with CK. Moreover, stomatal conductance increased in DK-6142 with the application of K60, Zn16 and in Neelam with K60, Zn16 followed by K100, Zn16 in hybrid cultivar showed similar results (3.58 fold) in relation to CK.

Internal CO$_2$ concentration were also increased with the application of K60, Zn16 in hybrid (2.20 fold) and inbred (2.05 fold) cultivars followed by DK-6142 with the application of K100, Zn16 (103.01%) in comparison with CK.

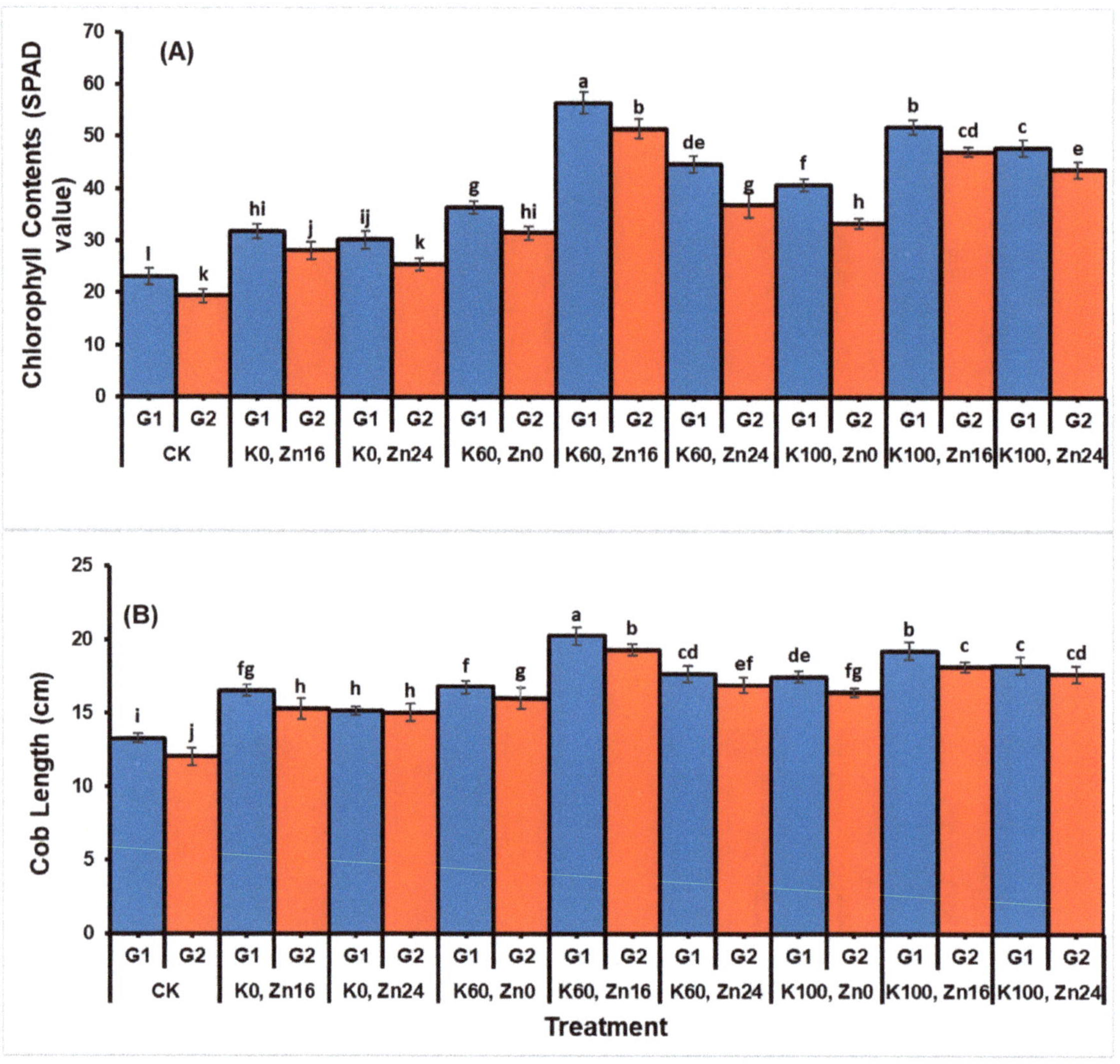

Figure 4. Effect of MOP and ZnSO^4 on (**A**) chlorophyll contents and (**B**) cob length of hybrid (G1) and non-hybrid (G2) genotypes of a maize plant; K0, K60 and K100 represent 0, 60 and 100 kg K ha^{-1}, respectively; Z0, Z16 and Z24 represent 0, 16 and 24 kg Zn ha^{-1}, respectively. The presented data are the means ± SD (*n* = 3). The lowercase letters indicate statistically significant differences according to the LSD test.

3.6. Effect of Potassium and Zinc on 1000 Grain Weight and Grain Yield of a Maize

A significant increase ($p \leq 0.05$) was observed in the 1000 grain weight (g) and grain yield (g pot^{-1}) of maize in all treatments compared to the control (Figure 7). The results showed that the 1000 grain weight was increased with the application of K60, Zn16 in the DK-6142 maize hybrid (30.99%) and in the inbred Neelam (26.33%) genotypes, followed by DK-6142 at K100, Zn16 (25.68%). So, it was clear from the data that the hybrid genotype performed better than the non-hybrid genotype. Moreover, the grain yield was also improved with the application of K60, Zn16 in the hybrid (39.74%) and inbred genotypes (35.27%) followed by K100, Zn16 (32.70%) in hybrid cultivar as compared to CK.

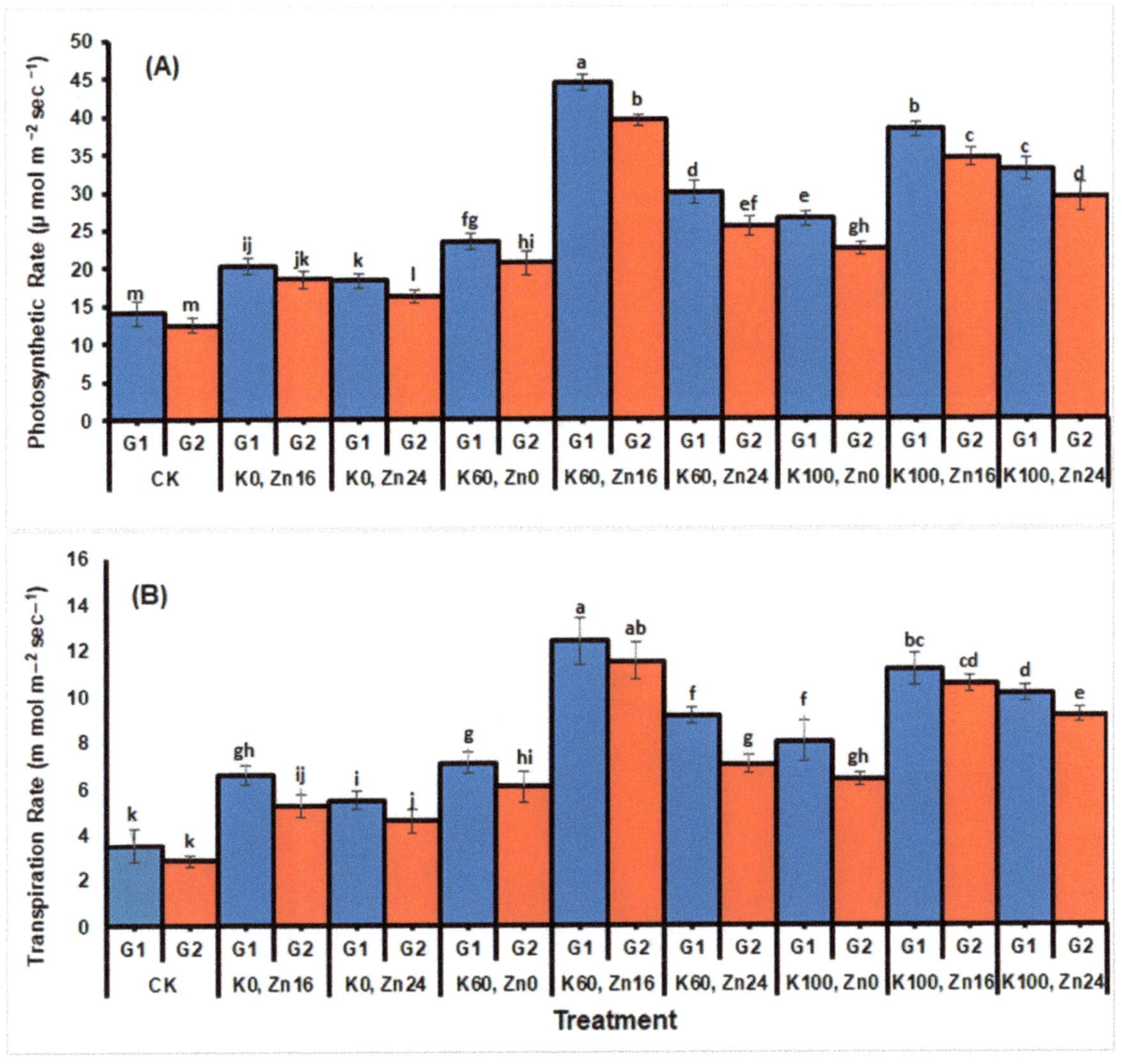

Figure 5. Effect of MOP and $ZnSO_4$ on (**A**) photosynthetic rate and (**B**) transpiration rate of hybrid (G1) and non- hybrid (G2) genotypes of a maize plant; K0, K60 and K100 represent 0, 60 and 100 kg K ha^{-1}, respectively; Z0, Z16 and Z24 represent 0, 16 and 24 kg Zn ha^{-1}, respectively. The presented data are the means ± SD (*n* = 3). The lower case letters indicate statistically significant differences according to the LSD test.

3.7. Effect of Potassium and Zinc on K Concentration in Straw and Grain of a Maize

A significant increase ($p \leq 0.05$) was observed in the K concentration in the straw and grain of the maize through the combined application of K and Zn compared to under control conditions (Figure 8). The straw K concentration improved in the hybrid cultivar with the application of K100, Zn24 (94.06%), followed by K100, Zn0 (93.28%) and K100, Zn16 (92.25%). It clarifies that the hybrid cultivar performed better than the inbred plants.

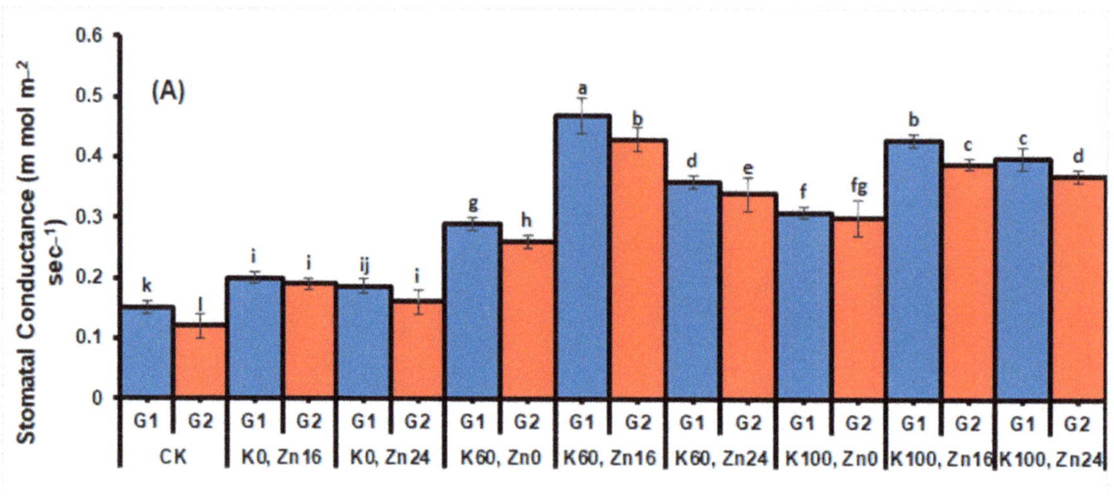

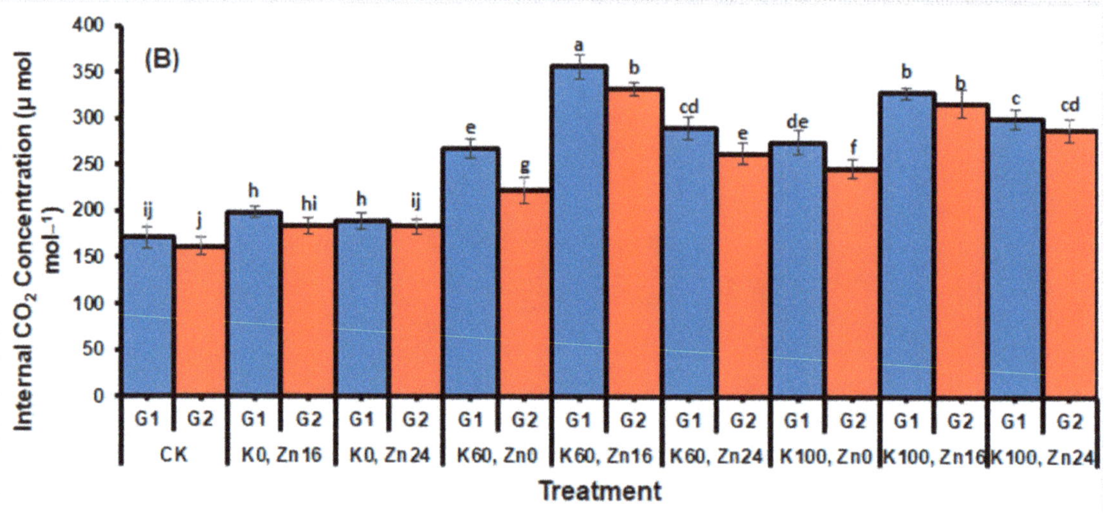

Figure 6. Effect of MOP and $ZnSO_4$ on (**A**) stomatal conductance and (**B**) internal CO_2 concentration of hybrid (G1) and non- hybrid (G2) genotypes of a maize plant; K0, K60 and K100 represent 0, 60 and 100 kg K ha^{-1}, respectively; Z0, Z16 and Z24 represent 0, 16 and 24 kg Zn ha^{-1}, respectively. The presented data are the means $\pm$ SD ($n = 3$). The lower case letters indicate statistically significant differences according to the LSD test.

On the other hand, a significant increase ($p \leq 0.05$) was observed in the case of K concentration in maize grain, through the various combined application of $ZnSO_4$ and MOP than the control (Figure 8). The grain K concentration increased in the hybrid variety with the application of K100, Zn24 (43.48%), followed by K100, Zn16 (42.39%) and K100, Zn0 (40.58%).

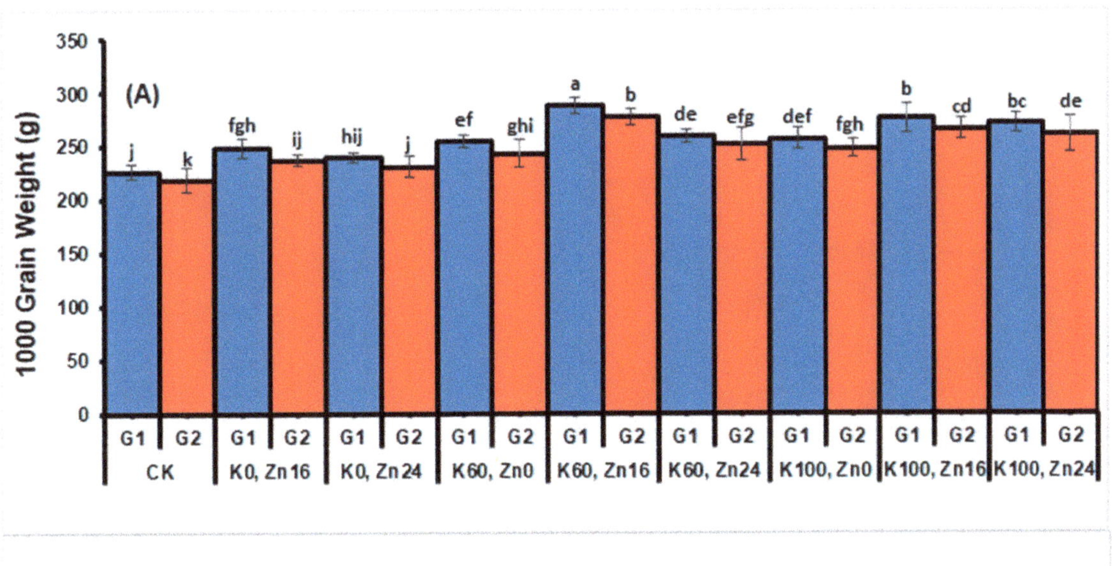

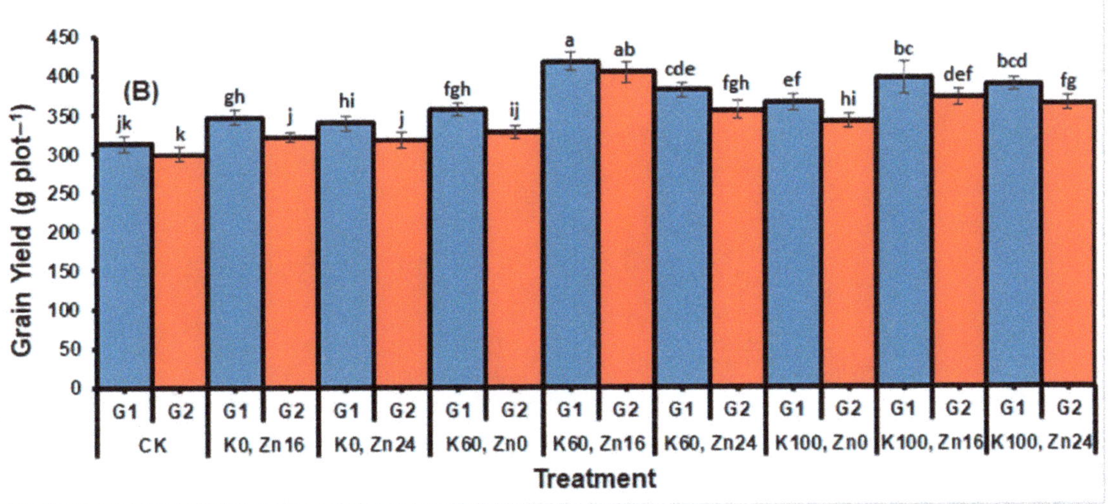

Figure 7. Effect of MOP and $ZnSO_4$ on (**A**) 1000 grain weight and (**B**) grain yield (g pot^{-1}) of hybrid (G1) and non-hybrid (G2) genotypes of a maize plant; K0, K60 and K100 represent 0, 60 and 100 kg K ha^{-1}, respectively; Z0, Z16 and Z24 represent 0, 16 and 24 kg Zn ha^{-1}, respectively. The presented data are the means ± SD (n = 3). The lowercase letters indicate statistically significant differences according to the LSD test.

3.8. Effect of Potassium and Zinc on Zn Concentration in Straw and Grain of a Maize

Due to various combined applications of $ZnSO_4$ + MOP, a significant increase ($p \leq 0.05$) was observed in the Zn concentration in the straw (mg kg^{-1}) and grain (mg kg^{-1}) of maize compared to the control (Figure 9). The zinc concentration in straw improved in G1 significantly with the application of K0, Zn24 (3.40 fold), followed by K100, Zn24 (3.35 fold) and K60, Zn24 (3.21 fold). In the case of the Zn concentration in grain (mg kg^{-1}), G1 showed better results with the application of K0, Zn24 (2.63 fold). followed by K100, Zn24 (2.58 fold) and K60, Zn24 (2.46 fold) compared with CK.

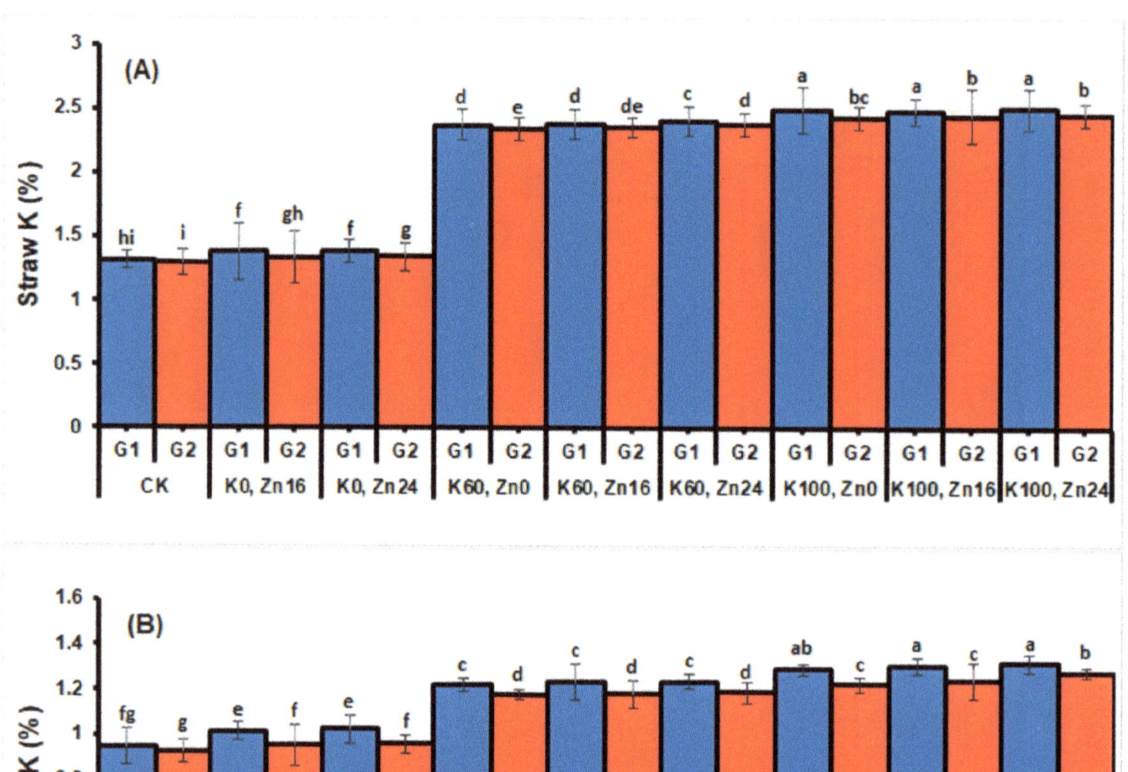

Figure 8. Effect of MOP and $ZnSO_4$ on (**A**) straw K and (**B**) grain K of hybrid (G1) and non-hybrid (G2) genotypes of a maize plant; K0, K60 and K100 represent 0, 60 and 100 kg K ha^{-1}, respectively; Z0, Z16 and Z24 represent 0, 16 and 24 kg Zn ha^{-1}, respectively. The presented data are the means ± SD (*n* = 3). The lowercase letters indicate statistically significant differences according to the LSD test.

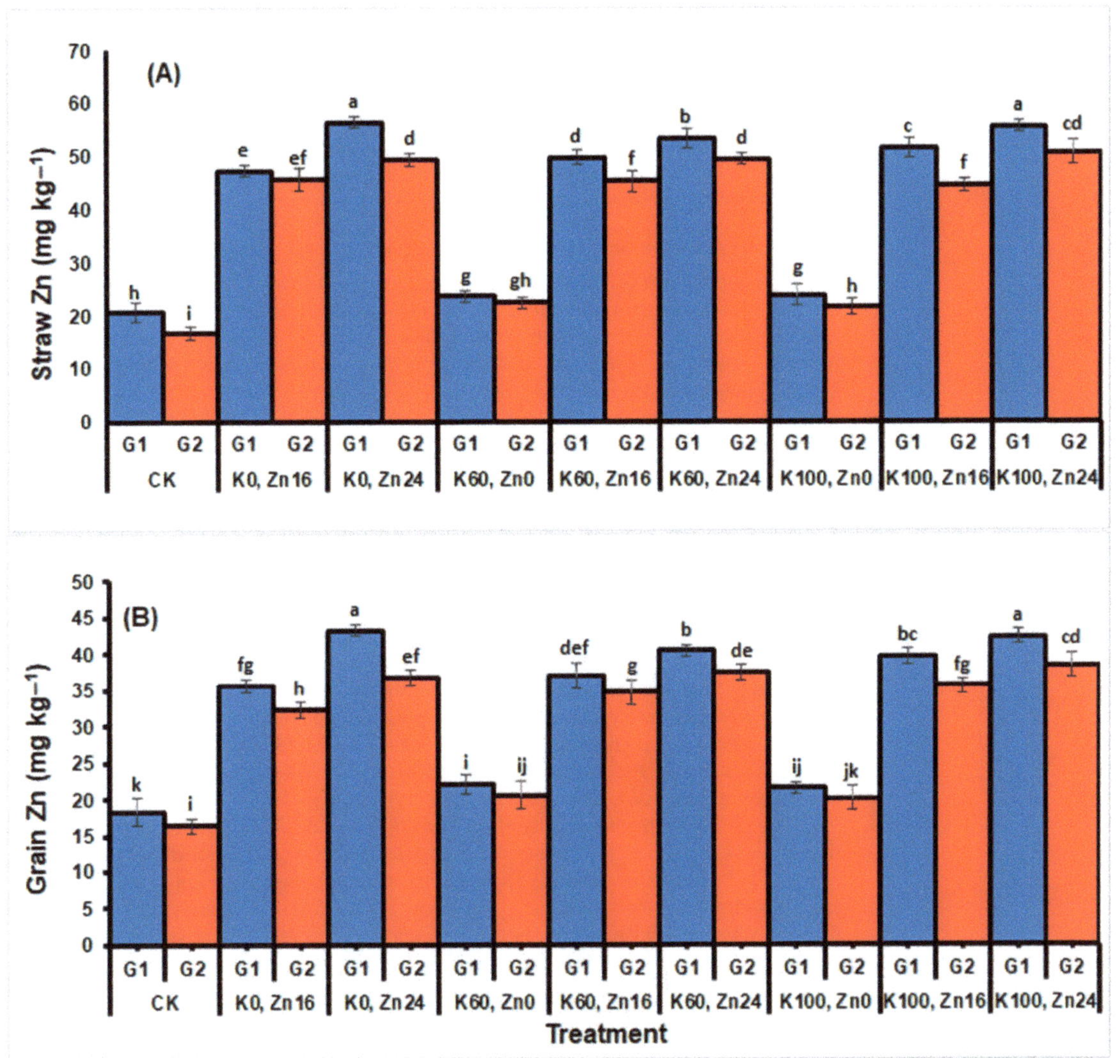

Figure 9. Effect of MOP and ZnSO$_4$ on (**A**) straw Zn and (**B**) grain Zn of hybrid (G1) and non-hybrid (G2) genotypes of a maize plant; K0, K60 and K100 represent 0, 60 and 100 kg K ha^{-1}, respectively; Z0, Z16 and Z24 represent 0, 16 and 24 kg Zn ha^{-1}, respectively. The presented data are the means ± SD (n = 3). The lowercase letters indicate statistically significant differences according to the LSD test.

4. Discussion

4.1. Effect of Potassium and Zinc Application on Growth Attributes of Maize Genotypes

Zinc deficiency is common in Pakistani soils, which are alkaline calcareous in natural conditions. Other important reasons for the K and Zn deficiency is the frequent use of tubewell water and less availability of canal water [32]. So, under such conditions, a combined fertilization approach plays an important role in improving crop growth and yield. In the present study, the combined application of MOP and ZnSO$_4$ caused a significant increase in the plant height along with the fresh and dry biomass of maize because the combined application of K and Zn improves the root growth which helps in increasing the uptake of K and also increasing the crop growth by improving the photosynthetic rate [33]. This increase is due to the involvement of Zn in many enzymatic reactions, the synthesis of chlorophyll, stomatal regulation, protein synthesis and carbohydrates transformation [5,34]. On the other hand, K also plays an important role in enhancing crop growth [7,35,36]. Potassium is also involved in increasing plants' dry biomass [6] because it increases the photosynthetic activity, which results in increasing the number of carbohydrates. According to Marschner [9], K increases the dry biomass of a plant under stress condition by enhancing the carboxylation efficiency.

4.2. Effect of Potassium and Zinc Application on Gas Exchange Parameters of Maize Genotypes

The application of K and Zn improved the gas exchange parameters (A, E, Ci and Gs) of both maize genotypes. Actually, both these nutrients have a key role in enzymatic activities such as CO_2 fixation. In the present study, the gas exchange parameters of maize hybrid DK-6142 increased by K and Zn application. Similar findings were also observed by [35,37], as they also reported the beneficial impact of K and Zn for the gas exchange parameters. According to [38], the photosynthetic rate is controlled by the application of K because it is directly involved in the opening and closing of stomata. Moreover, Zn maintains the water contents at tissue level at high rates [9]. Both the stomatal and non-stomatal factors are involved in controlling the direct relationship of photosynthesis with the seed yield and dry matter production of crop plants [37].

4.3. Effect of Potassium and Zinc Application on Physiological Attributes of Maize Genotypes

The physiological attributes such as the membrane stability index, the relative water content and the chlorophyll contents are the key factors for the estimation of plant growth parameters. The application of K and Zn improved all the mentioned physiological attributes of the maize genotypes. These results are also in agreement with findings of [39], who reported that the RWC and the osmotic potential were significantly improved through the application of Zn. These results can be related to the findings of some previous studies in which it was concluded that increase in leaf Zn due to supplemental Zn increased the leaf turgor and the RWC of soybean [40] and wheat [41].

The photosynthesis rate is limited in response to dehydration due to the closure of the stomata and defects in metabolic processes, and the total amount of chlorophyll is reduced [42]. Potassium and Zn application improve conditions to increase chlorophyll concentration and photosynthesis. The chlorophyll content in living plants is one of the important factors for preserving photosynthetic capacity [43]. The researchers concluded that possibly the deficiency in micro-nutrients could prevent the activity of a number of antioxidant enzymes, resulting in oxidative damage to chlorophyll [13,44]. Increasing the chlorophyll content is attributed to increasing the nutrients availability, especially Zn and K, and increasing the availability of other elements. Zinc application at different stages of plant growth may result in less nutrients loss and consequently increases nutrients' availability for the plants which in turn increase the chlorophyll content. Zinc application does not directly affect the formation of chlorophyll, but it can affect the concentration of elements that are part of the chlorophyll molecule, such as iron and magnesium [45]. Various studies have observed similar findings [5,46].

4.4. Effect of Potassium and Zinc Application on Yield Parameters of Maize Genotypes

The data clarified that the application of K and Zn significantly increased the cob length, 1000 grain weight and grain yield of maize plants. Potassium plays a vital role in crop growth and yield such as it is involved in water use efficiency (WUE), the division of cells, stomatal regulation, protein synthesis as well as hydrocarbon formation and their transfer towards cereal grains [9]. According to Broadley [11], various physiochemical processes directly or indirectly depend on Zn fertilization, which play an important role in enhancing dry matter production as well as the grain yield of cereals crops [35,47]. The combined application of K and Zn has a positive impact on the 1000 grain weight of a maize plants because the enzymes involved in carbohydrate synthesis are Zn dependent, and Zn plays a role in stomatal regulation and in the transformation of photosynthetase from source to sink [9]. So, K improves the plant's growth and yield due to significant enhancements in yield attributes [48], and also from its different role in different plant metabolic procedures [49].

4.5. Effect of Potassium and Zinc Application on Ionic Contents of Maize Genotypes

The results revealed that the combined application of K100, Zn24 significantly increased straw K (94.06%) and grain K (43.48%), whereas K0, Zn24 increased Zn concentration in straw (240.58%) and grain (163.84%). There are different essential nutrients which are present in soil with different concentrations. Some of these nutrients interact with each other; these interactions may be antagonistic or synergistic in nature. According to Maleki [50], the root surface is directly related to the absorption of nutrients, which is due to the positive interaction of K and Zn that enhance the lateral as well as fibrous root system of plant. This beneficial interaction also increases the concentration of N, P and K in soil solution, which affects the root system that uptakes more nutrients and increases the cop growth and development [33]. The concentration of K increases with the increase in K application [5,6], which may be due to the increase in the concentration of K in the soil solution as well as in the exchange site that is evident from the greater K concentration in the crop plant [51].

According to Liu [5,52], the positive effect of Zn application on the K content and its uptake by wheat and rice can be determined by the Zn-induced increase in the root system, which is due to the formation and polar transportation of indole acetic acid (IAA) that could affect more absorption of K. According to the work of Harris [2], two important methods that could be used to increase the plant Zn concentration are seed priming as well as soil zinc fertilization. Other important things for better Zn uptake are the soil bioavailable Zn and the surface area of plant roots [47].

5. Conclusions

The current study revealed that a combined fertilization (Zn + K) approach significantly enhanced the agronomic, physiological, growth and yield parameters of two maize genotypes compared to those under control conditions. The combined fertilization of Zn + K increases the K concentration in the grain and straw of maize, while the Zn concentration increases in the grain and straw of maize genotypes under solely Zn application. It was also concluded that the hybrid maize genotype (DK-1642) performed better than the non-hybrid (Neelam) genotype.

Author Contributions: Conceptualization, H.M.A.R., M.A.B., Q.-U.-A.R. and A.R.; methodology, H.M.A.R., M.A.B., Q.-U.-A.R., F.A. and A.R.; software, H.M.A.R., Q.-U.-A.R. and M.A.; validation, M.I., S.U.R., M.A.B., Y.G. and A.R.; formal analysis, M.A.B., Q.-U.-A.R., K.A.K. and S.U.R.; investigation, H.M.A.R., Q.-U.-A.R. and M.A.B.; resources, A.R., F.A. and Y.G.; writing—original draft preparation, H.M.A.R. and Q.-U.-A.R.; writing—review and editing, M.A.B., A.R., M.I., K.A.K., M.A. and Y.G.; supervision, A.R. and M.A.B.; project administration, Y.G. and A.R.; funding acquisition, Y.G., A.R. All authors have read and agreed to the published version of the manuscript.

Funding: This study has not received any funding.

Institutional Review Board Statement: Not applicable.

Informed Consent Statement: Not applicable.

Conflicts of Interest: The authors declare no conflict of interest.

References

1. Menkir, A. Genetic variation for grain mineral content in tropical-adapted maize inbred lines. *Food Chem.* **2008**, *110*, 454–464. [CrossRef]
2. Harris, D.; Rashid, A.; Miraj, G.; Arif, M.; Shah, H. "On-farm" seed priming with zinc sulphate solution-A cost-effective way to increase the maize yields of resource-poor farmers. *Field Crop. Res.* **2007**, *102*, 119–127. [CrossRef]

3. Pakistan Bureau of Statistics. *Agricultural Statistic of Pakistan*; Minestry of Food, Agriculture and Livestock: Islamabad, Pakistan, 2019.
4. Mehdi, S.S.; Ahsan, M. Coefficient of Variation, Inter-relationship and Heritability Estimates for Some Seedling Traits in Maize in C1 Recurrent Selection Cycle. *Pak. J. Biol. Sci.* **1999**, *3*, 181–182. [CrossRef]
5. Raza, H.M.A.; Bashir, M.A.; Rehim, A.; Jan, M.; Raza, Q.-U.-A.; Berlyn, G.P. Potassium and zinc co-fertilization provide new insights to improve maize (*Zea mays* L.) physiology and productivity. *Pak. J. Bot.* **2021**, *53*, 2059–2065. [CrossRef]
6. Rehim, A.; Saleem, J.; Bashir, M.A.; Imran, M.; Naveed, S.; Sial, M.U.; Ahmed, F. Potassium and Boron Fertilization Approaches to Increase Yield and Nutritional Attributes in Maize Crop. *Technol. Dev.* **2018**, *37*, 69–77. [CrossRef]
7. Umar, S. Moinuddin Genotypic differences in yield and quality of groundnut as affected by potassium nutrition under erratic rainfall conditions. *J. Plant Nutr.* **2002**, *25*, 1549–1562. [CrossRef]
8. Marschner, H. *Marschner's Mineral Nutrition of Higher Plants*; Elsevier: Amsterdam, The Netherlands, 2012; Volume 89.
9. Marschner, H. Mineral Nutrition of Higher Plants. *Miner. Nutr. High. Plants* **1995**, 537–595. [CrossRef]
10. Singh, J.; Sharma, H.L.; Singh, C.M. Effect of levels and phases of potassium application on growth parameters of Rice and Wheat. *Inti. Symp. A Decad. Potassium Res.* **2000**, *16*, 35–40.
11. Broadley, M.R.; White, P.J.; Hammond, J.P.; Zelko, I.; Lux, A. Zinc in plants: Tansley review. *New Phytol.* **2007**, *173*, 677–702. [CrossRef]
12. Singh, K. Response of Zinc Fertilization to Wheat on Yield, Quality, Nutrients Uptake and Soil Fertility Grown In a Zinc Deficient Soil. *Eur. J. Acad. Essays* **2014**, *1*, 22–26.
13. Cakmak, I. Enrichment of cereal grains with zinc: Agronomic or genetic biofortification? *Plant Soil* **2008**, *302*, 1–17. [CrossRef]
14. Chomba, E.; Westcott, C.M.; Westcott, J.E.; Mpabalwani, E.M.; Krebs, N.F.; Patinkin, Z.W.; Palacios, N.; Hambidge, K.M. Zinc Absorption from Biofortified Maize Meets the Requirements of Young Rural Zambian Children. *J. Nutr.* **2015**, *145*, 514–519. [CrossRef]
15. Nuss, E.T.; Tanumihardjo, S.A. Maize: A paramount staple crop in the context of global nutrition. *Compr. Rev. Food Sci. Food Saf.* **2010**, *9*, 417–436. [CrossRef]
16. Krebs, N.F. Update on zinc deficiency and excess in clinical pediatric practice. *Ann. Nutr. Metab.* **2013**, *62*, 19–29. [CrossRef] [PubMed]
17. Bänziger, M.; Long, J. The potential for increasing the iron and zinc density of maize through plant-breeding. *Food Nutr. Bull.* **2000**, *21*, 397–400. [CrossRef]
18. Maziya-Dixon, B.; Kling, J.G.; Menkir, A.; Dixon, A. Genetic variation in total carotene, iron, and zinc contents of maize and cassava genotypes. *Food Nutr. Bull.* **2000**, *21*, 419–422. [CrossRef]
19. Fahad, S.; Hussain, S.; Saud, S.; Hassan, S.; Shan, D.; Chen, Y.; Deng, N.; Khan, F.; Wu, C.; Wu, W.; et al. Grain Cadmium and Zinc Concentrations in Maize Influenced by Genotypic Variations and Zinc Fertilization. *Clean-Soil Air Water* **2015**, *43*, 1433–1440. [CrossRef]
20. Ortiz-Monasterio, J.I.; Palacios-Rojas, N.; Meng, E.; Pixley, K.; Trethowan, R.; Peña, R.J. Enhancing the mineral and vitamin content of wheat and maize through plant breeding. *J. Cereal Sci.* **2007**, *46*, 293–307. [CrossRef]
21. Reeves, D.R. Metal-accumulating plants. In *Phytoremediation Toxic Metals Using Plants to Clean Up Environment*; John Wiley & Sons: Hoboken, NJ, USA, 2000.
22. Rehim, A.; Khan, M.; Imran, M.; Bashi, M.A.; Ul-Allah, S.; Khan, M.N.; Hussain, M. Integrated use of farm manure and synthetic nitrogen fertilizer improves nitrogen use efficiency, yield and grain quality in wheat. *Ital. J. Agron.* **2020**, *15*, 29–34. [CrossRef]
23. Bashir, M.A.; Rehim, A.; Liu, J.; Imran, M.; Liu, H.; Suleman, M.; Naveed, S. Soil survey techniques determine nutrient status in soil profile and metal retention by calcium carbonate. *Catena* **2019**, *173*, 141–149. [CrossRef]
24. Barth, H.G.; Sun, S.T. Particle Size Analysis. *Anal. Chem.* **1989**, *61*, 143R–152R. [CrossRef]
25. Walkley, A. A critical examination of a rapid method for determining organic carbon in soils—effect of variations in digestion conditions and of inorganic soil constituents. *Soil Sci.* **1947**, *63*, 251–264. [CrossRef]
26. Richards, L.A. Diagnosis and improving of saline and alkaline soils. *Soil Sci. Soc. Am. J.* **1954**, *18*, 348.
27. Bashir, M.A.; Wang, H.; Pan, J.; Khoshnevisan, B.; Sun, W.; Zhai, L.; Zhang, X.; Wang, N.; Rehim, A.; Liu, H. Variations in soil nutrient dynamics and their composition in rice under integrated rice-crab co-culture system. *J. Clean. Prod.* **2021**, *281*, 125222. [CrossRef]
28. Sahrawat, K.L. Macro- and micronutrients removed by upland and lowland rice cultivars in West Africa. *Commun. Soil Sci. Plant Anal.* **2000**, *31*, 717–723. [CrossRef]
29. Westerman, R.L.; Jones, J.B.; Case, V.W. Sampling, Handling, and Analyzing Plant Tissue Samples. 1990. Available online: https://doi.org/10.2136/sssabookser3.3ed.c15 (accessed on 29 November 2021).
30. Sairam, R.K. Effect of Moisture Stress on Physiological Activities of Two Contrasting Wheat Genotypes. *Indian J. Exp. Biol.* **1994**, *32*, 584–593.
31. Steel, R.; Torrie, J.; Dickey, D. *Principles and Procedures of Statistics: A Biometrical Approach*, 3rd ed.; McGraw Hill Book Co.: New York, NY, USA, 1997.
32. Maqsood, M.A.; Hussain, S.; Aziz, T.; Ahmad, M.; Naeem, M.A.; Ahmad, H.R.; Kanwal, S.; Hussain, M. Zinc indexing in wheat grains and associated soils of southern Punjab. *Pak. J. Agric. Sci.* **2015**, *52*, 431–438.

33. Jat, G.; Majumdar, S.P.; Jat, N.K.; Mazumdar, S.P. Potassium and zinc fertilization of wheat (Triticum aestivum) in Western arid zone of India. *Indian J. Agron.* **2013**, *58*, 67–71.
34. Stepien, A.; Wojtkowiak, K. Effect of foliar application of Cu, Zn, and Mn on yield and quality indicators of winter wheat grain. *Chil. J. Agric. Res.* **2016**, *76*, 220–227. [CrossRef]
35. Sharma, P.N.; Kumar, N.; Bisht, S.S. Effect of zinc deficiency on chlorophyll content, photosynthesis and water relations of cauliflower plants. *Photosynthetica* **1994**, *30*, 353–359.
36. Bahadur, L.; Tiwari, D.D.; Mishra, J.; Gupta, B.R. Nutrient Management in Rice-Wheat Sequence under Sodic Soil. *J. Indian Soc. Soil Sci.* **2013**, *61*, 341–346.
37. Pessarakli, M. *Handbook of Photosynthesis*; Taylor & Francis: Boca Raton, FL, USA, 2005; ISBN 9781420027877.
38. Shabala, S.; Schimanski, L.J.; Koutoulis, A. Heterogeneity in bean leaf mesophyll tissue and ion flux profiles: Leaf electrophysiological characteristics correlate with the anatomical structure. *Ann. Bot.* **2002**, *89*, 221–226. [CrossRef] [PubMed]
39. Subbarao, G.V.; Wheeler, R.M.; Stutte, G.W.; Levine, L.H. Low potassium enhances sodium uptake in red-beet under moderate saline conditions. *J. Plant Nutr.* **2000**, *23*, 1449–1470. [CrossRef] [PubMed]
40. Weisany, W.; Sohrabi, Y.; Heidari, G.; Siosemardeh, A.; Ghassemi-Golezani, K. Changes in antioxidant enzymes activity and plant performance by salinity stress and zinc application in soybean (*Glycine max* L.). *Plant Omics* **2012**, *5*, 60–67.
41. Sharma, N.; Gupta, N.K.; Gupta, S.; Hasegawa, H. Effect of NaCl salinity on photosynthetic rate, transpiration rate, and oxidative stress tolerance in contrasting wheat genotypes. *Photosynthetica* **2005**, *43*, 609–613. [CrossRef]
42. Mafakheri, A.; Siosemardeh, A.; Bahramnejad, B.; Struik, P.C.; Sohrabi, Y. Effect of drought stress and subsequent recovery on protein, carbohydrate contents, catalase and peroxidase activities in three chickpea (cicer arietinum) cultivars. *Aust. J. Crop Sci.* **2011**, *5*, 1255–1260.
43. Jiang, Y.; Huang, B. Drought and heat stress injury to two cool-season turfgrasses in relation to antioxidant metabolism and lipid peroxidation. *Crop Sci.* **2001**, *41*, 436–442. [CrossRef]
44. Cakmak, I. Plant nutrition research: Priorities to meet human needs for food in sustainable ways. *Plant Soil* **2002**, *247*, 3–24. [CrossRef]
45. Kaya, C.; Higgs, D. Growth enhancement by supplementary phosphorus and iron in tomato cultivars grown hydroponically at high zinc. *J. Plant Nutr.* **2001**, *24*, 1861–1870. [CrossRef]
46. Iqbal, Z.; Abbas, F.; Ibrahim, M.; Ayyaz, M.M.; Ali, S.; Mahmood, A. Surveillance of heavy metals in maize grown with wastewater and their impacts on animal health in peri-urban areas of multan, Pakistan. *Pak. J. Agric. Sci.* **2019**, *56*, 321–328. [CrossRef]
47. Hussain, A.; Ali, S.; Rizwan, M.; Zia ur Rehman, M.; Hameed, A.; Hafeez, F.; Alamri, S.A.; Alyemeni, M.N.; Wijaya, L. Role of Zinc–Lysine on Growth and Chromium Uptake in Rice Plants under Cr Stress. *J. Plant Growth Regul.* **2018**, *37*, 1413–1422. [CrossRef]
48. Zulkarnain, W.M.; Ismail, M.R.; Ashrafuzzaman, M.; Saud, H.M.; Haroun, I.C. Rice growth and yield under rain shelter house as influenced by different water regimes. *Int. J. Agric. Biol.* **2009**, *11*, 566–570.
49. Mohd Zain, N.A.; Ismail, M.R. Effects of potassium rates and types on growth, leaf gas exchange and biochemical changes in rice (Oryza sativa) planted under cyclic water stress. *Agric. Water Manag.* **2016**, *164*, 83–90. [CrossRef]
50. Maleki, A.; Fazel, S.; Naseri, R.; Rezaei, K.; Heydari, M. The effect of potassium and zinc sulfate application on grain yield of maize under drought stress conditions. *Adv. Environ. Biol.* **2014**, *8*, 890–893.
51. Bukhsh, M.A.A.H.A.; Ahmad, R.; Iqbal, J.; Mudassar Maqbool, M.; Ali, A.; Ishaque, M.; Hussain, S. Nutritional and physiological significance of potassium application in maize hybrid crop production. *Pak. J. Nutr.* **2012**, *11*, 187–202.
52. Naveed, S.; Rehim, A.; Imran, M.; Bashir, M.A.; Anwar, M.F.; Ahmad, F. Organic manures: An efficient move towards maize grain biofortification. *Int. J. Recycl. Org. Waste Agric.* **2018**, *7*, 189–197. [CrossRef]

 *sustainability*

Review

Mapping Paddy Rice with Satellite Remote Sensing: A Review

Rongkun Zhao [1], Yuechen Li [1,2,*] and Mingguo Ma [1,2]

1   Chongqing Jinfo Mountain Karst Ecosystem National Observation and Research Station, School of Geographical Sciences, Southwest University, Chongqing 400715, China; zrk1998@email.swu.edu.cn (R.Z.); mmg@swu.edu.cn (M.M.)
2   Chongqing Engineering Research Center for Remote Sensing Big Data Application, School of Geographical Sciences, Southwest University, Chongqing 400715, China
*   Correspondence: liyuechen@swu.edu.cn; Tel.: +86-23-682-53912

Abstract: Paddy rice is a staple food of three billion people in the world. Timely and accurate estimation of the paddy rice planting area and paddy rice yield can provide valuable information for the government, planners and decision makers to formulate policies. This article reviews the existing paddy rice mapping methods presented in the literature since 2010, classifies these methods, and analyzes and summarizes the basic principles, advantages and disadvantages of these methods. According to the data sources used, the methods are divided into three categories: (I) Optical mapping methods based on remote sensing; (II) Mapping methods based on microwave remote sensing; and (III) Mapping methods based on the integration of optical and microwave remote sensing. We found that the optical remote sensing data sources are mainly MODIS, Landsat, and Sentinel-2, and the emergence of Sentinel-1 data has promoted research on radar mapping methods for paddy rice. Multisource data integration further enhances the accuracy of paddy rice mapping. The best methods are phenology algorithms, paddy rice mapping combined with machine learning, and multisource data integration. Innovative methods include the time series similarity method, threshold method combined with mathematical models, and object-oriented image classification. With the development of computer technology and the establishment of cloud computing platforms, opportunities are provided for obtaining large-scale high-resolution rice maps. Multisource data integration, paddy rice mapping under different planting systems and the connection with global changes are the focus of future development priorities.

Keywords: optical remote sensing; microwave remote sensing; phenology-based method

Citation: Zhao, R.; Li, Y.; Ma, M. Mapping Paddy Rice with Satellite Remote Sensing: A Review. *Sustainability* **2021**, *13*, 503. https://doi.org/10.3390/su13020503

Received: 3 December 2020
Accepted: 31 December 2020
Published: 7 January 2021

Publisher's Note: MDPI stays neutral with regard to jurisdictional claims in published maps and institutional affiliations.

1. Introduction

Paddy rice, as a major staple food, feeds almost half the world's population [1]. As the population grows, the demand for food grows. In terms of water use, about one-quarter to one-third of the world's freshwater resources are used for paddy rice irrigation [2]. Paddy rice fields are a major source of methane (CH_4) emissions [3]. Globally, methane (CH_4) emissions from paddy rice account for more than 10% of the total amount of CH_4 in the atmosphere [4]. Methane is the second most abundant greenhouse gas after carbon dioxide [5]. Paddy rice fields serve as habitats for birds, ducks, and other species, which are the origin of highly pathogenic avian influenza [6]. Therefore, the development of paddy rice distribution maps is of great significance for understanding and assessing the environmental conditions of food security, climate change, disease transmission and water use at regional, national and global levels [7].

An in-depth understanding of paddy rice cultivation and physiology is the premise of paddy rice mapping. The general physical characteristics of different crops are different, and the characteristics of paddy rice at different growth stages are also different. The paddy rice growth period can be divided into four stages [7]: (1) from sowing to transplanting in the nursery stage (~1 month), (2) from the transplanting to the heading stage (1.5 to 3 months), (3) from the heading to the reproductive stage with flowering (~1 month, including start, heading and flowering, stem elongation and panicle development), and (4) from flowering to mature stages at full maturity (~1 month, including milk stage, dough, and ripe grains). The morphology of paddy rice at the main growth stages is shown in Figure 1. Paddy rice is the only crop that needs extensive water during the growing phase and is the only staple that needs transplanting. Therefore, paddy rice can be identified by studying the sensitive spectral bands or indices during the period of water, soil, and vegetation mixing. Temporal variation in water–soil–vegetation composition is a key factor in paddy rice identification.

Figure 1. The example of paddy rice growth stages.

In previous research, some scholars summarized and analyzed the content related to rice mapping. Dong et al. [7] discussed the evolution of rice mapping methods from the 1980s to 2015 and summarized the methods used to characterize each stage and future development trends. Claudia et al. [1] mainly discussed the basic work of rice mapping. Based on a large number of studies, they summarized the characteristics of rice mapping (such as sensors, vegetation index, biomass) and summarized the application fields of different satellite sensors. Mostafa et al. [8] discussed the applicability of remote sensing images to rice area mapping and yield prediction. Methods and limitations of mapping and yield prediction using different remote sensing sensors are briefly described. Niel et al. [9] mainly discussed the current status of the application of remote sensing technology in rice planting areas in Australia, including crop identification, area measurement, and yield prediction. The research of these scholars is of great significance to the understanding of rice mapping methods. However, there are also some deficiencies: (I) The systematic induction of rice mapping methods is not comprehensive enough. (II) Detailed introductions of the method principles are limited. (III) There is a lack of comparison and evaluation of different methods. Based on the above analysis, this paper conducts a systematic review of rice mapping methods under recent remote sensing technology to correctly select rice mapping methods suitable for specific research purposes.

Paddy rice mapping algorithms are diverse. These methods include supervised classification and unsupervised classification methods, phenology algorithms, and object-oriented image classification. Data sources include optical remote sensing and microwave remote sensing. To facilitate the subsequent development of new methods, this study reviews paddy rice mapping methods in the literature since 2010. The content mainly include three aspects: (I) Methods based on optical remote sensing data and their advantages and disadvantages; (II) Methods based on microwave remote sensing data and their advantages and disadvantages; (III) Methods based on the integration of optical remote sensing data and microwave remote sensing data and their advantages and disadvantages. Finally, we summarize the development trend of paddy rice mapping methods, as well as the challenges and future direction of development.

2. Key Feature Statistics in Remote Sensing

We used Web of Science to collect rice-related papers published in major international remote sensing journals and some agricultural journals between 2010 and 2020. The search keywords were rice, remote sensing, and mapping. Afterwards, further screening was carried out by reading the abstract, and highly relevant papers were selected for review and analysis.

The main data sources are shown in Table 1. The most commonly used optical remote sensing satellites are Landsat, MODIS and HJ-1A/B, with spatial resolutions of 30 m and 500 m, respectively. In recent research on radar data, Sentinel-1 data are often used, with a spatial resolution of 10 m and temporal resolution less than 10 days. In addition, we made statistics on the relevant characteristics of the data sources in the integration method. It is worth mentioning that the integration method is mostly based on Landsat data set in optical images, extracting all high-quality images in a period of time for research, and the time resolution is 8 days~16 days (Table 2). These satellite sensors have the potential to obtain multitemporal and multispectral reflectance data on farmland. These data can be used to derive the time series of vegetation indices (VIs), which are calculated as a function of the red, green, blue, and infrared spectral bands (see the major VIs in Table 3).

Table 1. Main satellite data sources used for paddy rice mapping.

Satellite	Sensor	Spatial Resolution	Temporal Resolution	Free or Charge	Literature Number
Landsat	MSS+TM (Landsat-5) ETM+ (Landsat-7) OLI (Landsat-8)	30 m	16 days	Free	16
Terra/Aqua	MODIS	250–1000 m	1–2 days	Free	22
HJ-1A/B	CCD1/2	30 m	2–4 days	Free	3
SPOT	HRV (SPOT1~3) VGT (SPOT-4) HRG/HRS/VGT (SPOT-5)	1 km	1 day	Charge	2
Sentinel-2	MSI	10–20 m	5 day	Free	7
Sentinel-1	SAR	5–40 m	12 days	Free	14
COSMO-SkyMed	SAR	3–15 m	16 days	Charge	1
TerraSAR-X	SAR	3–10 m	11 days	Charge	1
ENVISAT	ASAR	20–500 m	35 days	Free	2
RADARSAT-1	SAR	10–100 m	24 days	Charge	1
RADARSAT-2	SAR	3–100 m	24 days	Charge	2
ALOS-2	PALSAR-2	25 m	14 days	Charge	3

Note: MSS—Multispectral Scanner; TM—Thematic Mapper; ETM+—Enhanced Thematic Mapper Plus; OLI—Operational Land Imager; CCD—Charge Coupled Device; HRV—High Resolution Visible; VGT—VEGETATION; HRG—High Resolution Geometric Imaging Instrument; HRS—High Resolution Stereoscopic Imaging Instrument; MSI—Multi-Spectral Instrument; SAR—Synthetic Aperture Radar; ASAR—Advanced Synthetic Aperture Radar; Literature Number—The number of literature using this data in the literatures included in this study.

Table 2. Introduction to Integration Method Data Sources.

Integrated Data Sources	Integrated Spatial Resolution	Integrated Time Resolution	Ref.
Landsat ETM+\OLI	30 m	8 days	[10]
Landsat 8 OLI MODIS	30 m	16 days	[11]
Landsat TM\ETM+\OLI	30 m	<16 days	[12]
Landsat TM\ETM+	30 m	$\leq$16 days	[13,14]
Landsat ETM+\OLI	30 m	16 days	[15]
Sentinel-2 MODIS	10 m	16 days	[16]

Table 3. Indices used in paddy rice mapping methods.

Index	Abbreviation	Formula	Literature Number
Normalized Difference Vegetation Index	NDVI	$\dfrac{\rho_{nir}-\rho_{red}}{\rho_{nir}+\rho_{red}}$	27
Enhanced Vegetation Index	EVI	$2.5 \times \dfrac{\rho_{nir}-\rho_{red}}{\rho_{nir}+6\times\rho_{red}-7.5\times\rho_{blue}+1}$	14
Two-band Enhanced Vegetation Index	EVI2	$2.5 \times \dfrac{\rho_{nir}-\rho_{red}}{\rho_{nir}+2.4\times\rho_{red}+1}$	1
Land Surface Water Index	LSWI	$\dfrac{\rho_{nir}-\rho_{swir}}{\rho_{nir}+\rho_{swir}}$	16
Normalized Difference Snow Index	NDSI	$\dfrac{\rho_{green}-\rho_{swir}}{\rho_{green}+\rho_{swir}}$	3
Normalized Difference Water Index	NDWI	$\dfrac{\rho_{green}-\rho_{nir}}{\rho_{green}+\rho_{nir}}$	1
Normalized Difference Flood Index	NDFI	$\dfrac{\rho_{swir}-\rho_{red}}{\rho_{swir}+\rho_{red}}$	1

Note: Surface reflectance values from the blue (ρ_{blue}), green (ρ_{green}), red (ρ_{red}), Near Infrared (NIR) (ρ_{nir}) and Shortwave Infrared(SWIR) (ρ_{swir}) bands.

3. Taxonomy

In the past 10 years, scholars have carried out many studies on the paddy rice mapping method and further improved the method's precision based on its predecessors. The methods can be divided into the following three categories, based on the difference in data sources: optical remote sensing mapping method, microwave remote sensing mapping method, and integrated method (Table 4). According to the method principle, the optical remote sensing mapping method is further divided into four categories. Generally, the microwave remote sensing mapping method extracts the variation in the backscattering coefficient of paddy rice during the growing period, which can be divided into two categories according to the principle of the method. The methods of optical and microwave integration are divided into two categories according to the principle of the method: complementary method and comparative method. Among the above methods, time series analysis methods such as phenology-based methods are relatively common, and object-oriented, deep learning, and data integration methods have been relatively innovative in the past 10 years.

Table 4. Categories of existing paddy rice mapping methods.

Main Classification	Specific Methods	Refs.
Optical Remote Sensing-Based Mapping Methods	Machine learning	[13,17–23]
	Time series similarity method	[24,25]
	Vegetation index feature-based method	[10–12,14,26–41]
	Object-based image analysis	[42–44]
Microwave Remote Sensing-Based Mapping Methods	Empirical model	[45–48]
	Machine learning	[49–58]
Integration of Optical and Microwave Remote Sensing-Based Mapping Methods	Complementary method	[16,59–65]
	Comparison class method	[66–69]

4. Evolution of Paddy Rice Mapping Methods

Remote sensing platforms can repeatedly observe the Earth's surface and collect a variety of data, so several remotely based methods have been developed to map paddy rice areas around the world. There are three types of methods based on different data sources. These methods are described in the following sections.

4.1. Optical Remote Sensing-Based Mapping Methods

Optical remote sensing sensors have been used extensively for mapping paddy rice areas around the world. The earliest method of paddy rice monitoring was to extract paddy rice by using remote sensing images and classification methods. Later, with the emergence of the vegetation index, phenological algorithm, cloud computing, and machine learning, the precision of paddy rice mapping based on optical remote sensing was constantly improved.

4.1.1. Machine Learning

Machine learning methods are commonly used methods of rice mapping, including traditional machine learning and deep learning. Traditional machine learning includes supervised and unsupervised classification, such as ISODATA, decision tree (DT), random forest (RF), support vector machine (SVM). The principle of this type of method is to first collect images and sample training data and determine the decision rules to extract rice based on characteristic parameters.

Supervised classification is based on the samples provided by the known training area to obtain feature parameters to establish decision rules. Unsupervised classification obtains feature parameters through computer agglomeration statistical analysis of images to establish decision rules. The latter is an image classification method without a priori classification criteria. The input into the classifier is mainly preprocessed spectral images [13,21]. In recent years, normalized difference vegetation index (NDVI) temporal curves have also been used as the characteristic parameter for classification [18]. Manjunath et al. [17] used multitemporal SPOT VGT NDVI data for analysis. The ISODATA clustering method is used to distinguish between paddy rice areas and non-paddy rice areas. Then, the auxiliary data set is used to further subdivide the areas. Similarly, Okamoto et al. [13] used this method to extract paddy rice fields in Heilongjiang. The difference is that this study used Landsat TM/ETM+ as the data source. Gumma et al. used MODIS data products combined with the k-means clustering algorithm to map the paddy rice area. In 2011 and 2015, they used the same method to map paddy rice in different regions [18,19]. The results were relatively good, with a correlation of more than 90% with local statistics. Paddy rice mapping is also carried out using supervised classification methods, such as SVM [20] and RF [21]. The advantage of this method is its strong operability. The basic principles are easy to understand. The difficulty of the supervised method may be the collection of training samples. However, Google Earth's high-resolution images and the global geo-referenced field photo library (http://eomf.ou.edu/photos/) provide convenience. The disadvantage of this method is that the validity of the image will affect the accuracy of the results.

For example, cloudy and foggy areas, broken terrain areas, and mixed pixel problems will affect the results. In addition, the threshold settings in supervised classification and unsupervised classification methods will change according to the study area.

Deep learning performs well in image recognition and signal processing. In optical remote sensing, the CNN method is mainly used. Convolutional neural network (CNN) is well applied in the field of image analysis. In terms of scene classification, the CNN algorithm has higher classification accuracy than traditional algorithms. CNN is composed of several layers with different functions: input layer, convolution layer, pool layer, fully connected layer and output layer. The input layer is used to import training data, and the convolutional layer is used to extract features. The steps of applying this method to rice mapping are as follows: (1) Use the University of California Merced land-use data set, land-use/land-cover (LULC) Map, Google Earth high-resolution images, and field survey data to pretrain the model. (2) Input the original image into the model and output the result. In this step, in addition to the spectral data, NDVI, Land Surface Temperature (LST), and related phenological information can also be imported into the model [22]. Common training outputs classification results. Zhao et al. [23] combined CNN classification results with the results of NDVI under the DT to achieve further classification and output the final classification results.

The accuracy of this method is generally high. The overall accuracy is greater than 93%. The advantage is improved classification of complex surfaces and the broken landscapes. The disadvantage is that complex models require a lot of data for training. If the tagged data are not enough to support the entire training process, the deep learning model will have poor results. Therefore, the correct amount of training data guarantees the reliability and rationality of the training model.

4.1.2. Time Series Similarity Method

A new method that appeared in recent years is the time series similarity method of dynamic time warping (DTW) distance [24,25]. Time series similarity measures are used to describe the characteristics of data changes over time. DTW distance was initially applied to text data matching, speech processing, and visual pattern recognition. The research shows that algorithms based on the nonlinear bending technique can obtain high recognition and matching accuracy. The steps in this method are as follows: first, establish the standard NDVI sequence curve of the paddy rice growth cycle through field sample data, and then determine the threshold based on the DTW distance between the NDVI time series of standard paddy rice growth and the NDVI time series of the pixels to extract the paddy rice field. The principle of the time sequence similarity method based on the DTW distance is as follows:

Suppose two time series, i.e., $S_1(t) = \{s_1^1, s_2^1, \cdots, s_m^1\}$, $S_2(t) = \{s_1^2, s_2^2, \cdots, s_n^2\}$, with respective lengths of m and n. Construct an $m \times n$ matrix $A_{m \times n}$ and define the distance between each element as $a_{ij} = d\left(s_i^1, s_j^2\right) = \sqrt{\left(s_i^1, s_j^2\right)^2}$. In the matrix $A_{m \times n}$, a winding path is set by a group of adjacent matrix elements, and notes for W = {w_1,w_2,$\cdots$,w_k} and the k_{th} element in W are defined as $w_k = \left(a_{ij}\right)_k$; this path meets the following conditions:

Monotonicity constraint: $w_k = a_{ij}, w_{k+1} = a_{i'j'}, i' \geq i,\ j' \geq j$,
Continuity constraint: $w_k = a_{ij}, w_{k+1} = a_{i'j'},\ i' \leq i+1,\ j' \leq j+1$,,
Endpoint constraint: $w_1 = a_{11}, w_k = a_{mn}$.

This element satisfies the condition $0 \leq i - i', 0 \leq j - j' \leq 1$, and thus, $DTW(S_1, S_2) = min\frac{1}{K}\sum_{i=1}^{k} W_i$. The DTW algorithm can be summarized by applying ideal dynamic programming to find the best (i.e., least bending) cost path, as shown in Formula (1):

$$\begin{cases} D(1,1) = a_{11} \\ D(i,j) = a_{ij} + min\{D(i-1,j-1), D(i,j-1), D(i-1,j)\} \end{cases} \tag{1}$$

where i = 2,3 ... m, j =2,3 ... n, $D\,(m, n)$ is the minimum cumulative value of the winding paths.

The DTW distance can reflect the similarity and difference between the standard paddy rice growth NDVI time series and the NDVI time series of a pixel. In the DTW algorithm, when the DTW distance is short, the curve of the NDVI time series shows high similarity. We performed correlation analysis on the NDVI time series and ground truth data to determine the DTW distance threshold for identifying single- and multi-cropping paddy rice. Assuming that the DTW distance of the pixel is greater than the threshold shown, the pixel is unlikely to be paddy rice.

In 2014, Guan et al. [24] extracted rice areas from Southeast Asia and initially explored the applicability of this method in cloudy and rainy areas with good results. In 2018, the same team used this method to extract rice areas in Vietnam, and the results correlated well with statistical data ($R^2 = 0.809$). This result showed once again the potential of this method for rice mapping in monsoon regions and multiple cropping systems with diverse cultivation processes [25].

The accuracy of this method is good, and the overall accuracy is 83%. The advantage of this method is that it is suitable for cloudy and rainy areas, and the similarity analysis based on DTW distance can solve the overall curve deviation caused by the flexibility of paddy rice planting. This method has good application potential in different crops and different cropping systems. The disadvantages are the determination of the empirical model threshold and the determination of the NDVI standard curve. Affected by the spatial resolution of satellite data, the accuracy of the national scale is high and that of the provincial scale is low.

4.1.3. Vegetation Index Feature-Based Method

The third method is the vegetation index feature-based method. This method can be divided into two categories. One is the features are obtained through mathematical analysis. The threshold formula is established by mathematical analysis of the vegetation index time series curve. The other is the phenology algorithm. The principle is to extract paddy rice, which is grown on flooded soils, based on the unique physical characteristics. NDVI < Land Surface Water Index (LSWI) or Enhanced Vegetation Index (EVI) < LSWI during the flooding period of paddy rice, but the EVI value of other vegetation (non-flooded) is usually greater than the LSWI value.

Mathematical methods include correlation analysis, analysis of variance, and normal distribution. The principle of the correlation analysis method is to extract 100 sample pixels to generate the NDVI time profile curve and calculate the average [26]. Then, the correlation coefficient of 100 pixels is calculated to set the threshold for rice extraction. Then, the symbol test method is used to evaluate the difference between each pair of data from two related samples to compare the significance of the two samples. The variance analysis method uses multitemporal image data to calculate the time series curves of the vegetation index and calculate the standard deviation and variance of the vegetation index in each pixel, and then determines the threshold range by Formula (2). If the pixel value falls within the threshold range, it is determined as a paddy rice pixel [27]. The normal distribution method has the following assumption: the probability distribution function (PDF) of the land cover type follows a normal distribution [28]. We use the mean and standard deviation of each land cover type to define its normal distribution function, and two parameters are obtained from the training data set. The key to correctly distinguishing one specific land cover type is to minimize the overlaps between the target and the neighboring ordinary PDFs. For two land cover types L1 and L2, assuming L1~$N(\mu_1, \sigma_1{}^2)$ and L2~$N(\mu_2, \sigma_2{}^2)$, then the intersection between L1 and L2 is calculated by Formula (3).

$$V_{mean} - (nS) < x_1 < V_{mean} + (nS), \tag{2}$$

$$x_2 = \frac{\sigma_1\mu_2 + \sigma_2\mu_1}{\sigma_1 + \sigma_2} \tag{3}$$

where V_{mean}, n, S, μ, σ, x_1, and x_2 are the average of the variance of the paddy rice field, the maximum distance from the standard deviation, the standard deviation of the variance of the paddy rice field, and the average variance of the image to be classified, the mean of each land cover type, the standard deviation of each land cover type, the intersection between two land cover types. Generally, the two land cover types can be thought separable if x_2 is outside of $[\mu - \sigma, \mu + \sigma]$.

Chen used statistical methods to classify double-cropping paddy rice in Taiwan [26]. In addition, this paper also compared the accuracy of different smoothing methods with different NDVI time curves. Studies have shown that classification methods based on empirical mode decomposition (EMD) filtered data produce better classification results than wavelet transform. Nuarsa et al. [27] used the method of variance analysis and MODIS images to extract paddy rice from Bali, Indonesia. The results were good, and the kappa coefficient reached 0.8371. Wang et al. [28] used a normal distribution to process the threshold value of the vegetation index curve for paddy rice extraction in the eastern plains of China. This method was mainly applied to single-season rice. This method is only applied to the key phenological phase images of paddy rice growth. In addition, some studies have used the difference in NDVI during the critical phenology period to define the threshold for paddy rice mapping [12]. Liu et al. [29] proposed a subpixel method that used the relationship between the coefficient of variation (CV) of the LSWI and the planting fraction to estimate the planting fraction of paddy rice. The new method calculated the scale of paddy rice area based on the CV of the LSWI determined for pure water bodies and upland pixels, which can be automatically obtained from the MCD12Q1 land cover product. The overall accuracy was 88%.

The overall accuracy of this method is greater than 85%, and the kappa coefficient is greater than 0.7. The method has the advantages of simple principles and easy operation. The disadvantage is that the applicability of cloudy areas needs to be investigated. Mixed pixels and boundary effects will reduce the classification accuracy. Furthermore, it remains to be studied whether the accuracy of the method will be improved under the conditions of improved image spatial resolution, extended time series, and large-scale research areas.

The use of the phenology algorithm began in approximately 2000. Xiao et al. discovered the characteristics of the vegetation index and conducted paddy rice extraction studies in large areas such as South Asia and central and southern regions [30,31]. The results were good and showed the effectiveness of the phenology algorithm in paddy rice mapping. The previous method has some drawbacks. For example, the critical time window for paddy rice growth is obtained based on a large amount of agricultural phenology data. Incomplete agricultural phenology data in some areas will hinder the implementation of this method.

In recent years, paddy rice mapping methods have been continuously improved. The improvement is reflected in the use of high-resolution data sources, the increase in the complexity of the study area, the study of long-term sequences, and the increase in auxiliary materials (phenology information, other land cover masks, etc.). First, we will discuss high-resolution data sources. Previously, the MOD09A1 MODIS product was mostly used, but it has a spatial resolution of 500 m. For precision agriculture, there will still be mistakes. Subsequent studies used Landsat images and HJ-1A/B with a spatial resolution of 30 m, and Sentinel-2 with a spatial resolution of 10 m [10,11,14,28,33,40]. The accuracy has been further improved. Other studies have considered the issue of temporal resolution. MODIS and Landsat data have been integrated, and these data were then combined with a phenology algorithm for paddy rice mapping [10,32,41]. Second, the complexity of the study area also has an impact. Early studies were mostly concentrated in South Asia and other regions, and summer rainfall was mostly taken into consideration. With the expansion of paddy rice in Northeast Asia, the research area moved northward [14,33,34]. Compared with South Asia, the impact of early spring snowmelt should be considered due to the climate of the northeast region. Some scholars have studied the changes in the area of paddy rice in high temperature disaster areas [35]. Initially, research focused on paddy rice extraction in a specific area in a certain year to verify the accuracy of the algorithm. Subsequent related studies focused on long-term sequence studies to study the expansion of paddy rice fields and changes in the planting area [14,33,36]. Finally, the increase in auxiliary information should also be considered. Some recent studies have attempted to use surface temperature or air temperature to define the time window that defines the temperature that should be reached during the key growing period of paddy rice, effectively excluding the effects of summer rainfall and early spring snow melt on monitoring [10,11,14,33–38]. Other relevant mask data also include cloud cover, snow cover, seasonal water cover, evergreen vegetation, and DEM. The algorithm flow chart is shown in Figure 2. The statistical data brought by the state's advocacy for refined agriculture have greatly facilitated the extraction of paddy rice. In addition, some studies have used the phenology algorithm to extract the spatial distribution of paddy rice with different planting intensities, which showed the potential of the phenology algorithm in describing two- and three-season paddy rice [39]. Some studies have added the results of field spectrometer measurements on the basis of previous optical remote sensing data to verify the changes in the rice vegetation index curve [32].

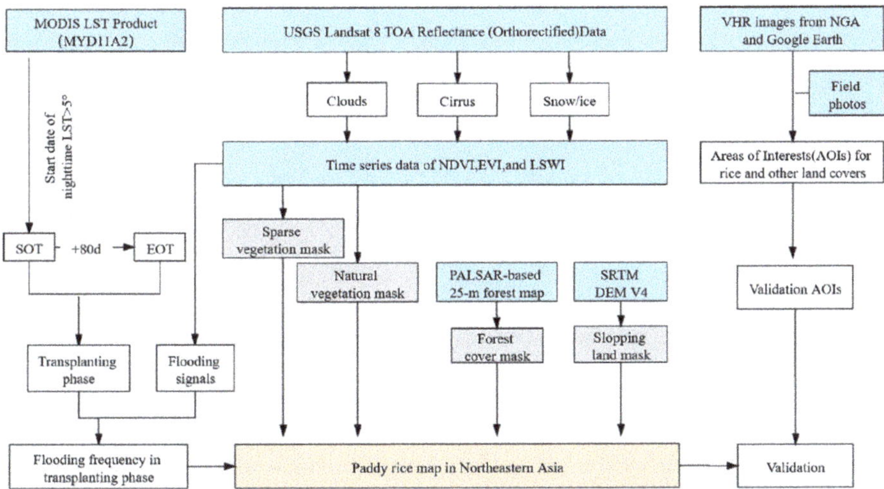

Figure 2. The workflow for phenology and pixel-based paddy rice mapping, major modules include time window determination of the rice transplanting phase (starting point: SOT, ending point: EOT), Landsat data preprocessing, phenology- and pixel-based mapping for non-cropland masks and paddy rice flooding, validation based on the areas of interest (AOIs) from very high-resolution (VHR) images and field photos [33].

The accuracy of rice mapping methods based on phenology is usually high, exceeding 80%. The advantage of this method is that it is suitable for long-term sequence dynamic analysis and large-scale observations. Based on phenological observations, the rice growth period can be accurately identified, reducing the need for data processing work. The principle of the method is simple and operable. The disadvantages of this method include errors in the cloud coverage area, mixed pixel problems, and limited observations over scattered landscapes. The recognition accuracy of clouds is high, but the recognition accuracy of cloud shadows is usually low. Because cloud shadow pixels usually meet the threshold of LSWI–EVI > 0, they may affect paddy rice field mapping. In addition, the inundation of the surface caused by extreme precipitation events can also affect paddy rice mapping [10].

4.1.4. Object-Based Image Analysis

There are three key steps of the object-based image analysis method: (1) segmentation of generated image object; (2) determination of features based on feature extraction of objects; and (3) classification (multiple classification methods). Su [42] focused on using phenology to classify paddy rice under the object-based image analysis framework. The main purpose of this framework is to study the applicability of phenology in the localization of paddy rice based on object-based image analysis. The image segmentation is performed using the multiresolution segmentation algorithm in eCognition software. Then it is classified based on the neural network classification method. Singha et al. [43], in order to improve the segmentation quality, improved the fusion criterion on the basis of the commonly used fractal network evolution method, and a new segmentation algorithm was proposed. An unsupervised scale selection method was proposed to determine the optimal scale parameters for image segmentation, and to automate the process of determining scale parameters. After segmentation, geometric, spectral and texture features were extracted and input into the subsequent classification process. Paddy fields and non-paddy fields were classified by a random forest classifier. Zhang et al. [44] also performed image segmentation by using the multiresolution segmentation algorithm in eCognition 9.0 software. The prototype objects were classified by using the random tree (RT) classifier.

The accuracy of this method is generally better than that of other methods. The overall accuracy is over 90%, and the kappa >0.82. The advantages of this method are that geometric information, texture information and spectral information can be used simultaneously to improve the extraction accuracy, and the method analyzes objects by integrating neighborhood information rather than pixels, which will reduce the "salt and pepper" effect when rendering heterogeneous landscapes to classify paddy rice fields more accurately. Object-based image analysis shows advantages in identifying broken paddy rice fields. The disadvantage of this method is that the accuracy of the method is related to the accuracy of data, cloud pollution, spatial resolution, and processing of mixed pixels. In addition, image segmentation is still a challenging problem. Improving the quality of image segmentation is a key factor.

4.2. Microwave Remote Sensing-Based Mapping Methods

The use of a microwave source is a second type of mapping method for paddy rice. The first spaceborne synthetic aperture radar (SAR) sensor for paddy rice mapping used data from the European Remote Sensing Satellite 1 (ERS-1), which showed good results [45]. These groundbreaking studies were often limited to small-scale studies due to a lack of high-quality ground truth images, single polarization, or large data volumes. Subsequent research began to focus on using multiple SAR sensors to improve rice mapping over large land areas, and ERS-1, ERS-2, and RADARSAT were used to test various algorithms. Recent research included RADARSAT-2 data, combined optical and SAR data, object-oriented crop mapping, and Sentinel-1 C-band SAR data. Sentinel-1 satellite data can be obtained freely and openly all over the world, further promoting large-scale rice monitoring operations using radar data.

The main advantage of microwave remote sensing is theoretically the ability to acquire images under any weather conditions, such as cloud cover, rain, snow, and solar irradiance. In most cases, paddy rice cultivation is carried out during the rainy season when overcast and rainy weather prevails. Therefore, the radar image collected by the microwave sensor is an excellent image source for mapping paddy rice areas. In the growth process of paddy rice, the time series change in the radar backscatter coefficient is the key factor to distinguish paddy rice areas. The characteristic of the backscattering coefficient in the growth stage of paddy rice is that in the nutrition and reproduction stage, the backscattering increases continuously until it reaches the maximum at the heading stage. With the development of paddy rice phenology, stems elongate and leaf area, plant water and biomass increase. These changes increase the area available for radar wave reflection, leading to an increase in measured backscatter. After the heading stage, due to plant water, leaf area and biomass begin to decrease, the aforementioned scattering effect is reduced, resulting in a decrease in SAR backscatter. This time backscattering behavior is illustrated in Figure 3, which is based on multiyear advanced synthetic aperture radar (ASAR) wide swath mode (WSM) time series data and shows the SAR backscattering behavior with triple-cropped rice growing stages.

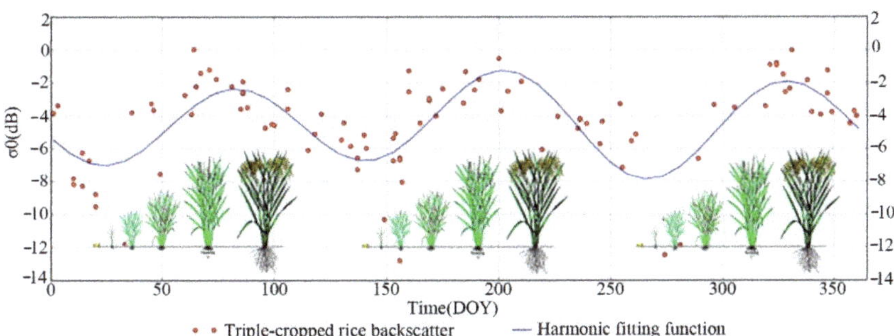

Figure 3. SAR backscatter behavior with triple-cropped rice growing stages based on a multiyear ASAR WSM time series [49].

4.2.1. Empirical Model

The earliest method of applying radar data to paddy rice mapping was to observe the changes in the backscattering coefficient during the paddy rice growth cycle to establish an empirical model. The principle of this method is to establish a mathematical formula based on the change in the backscattering coefficient during the paddy rice growth cycle, determine the threshold, coefficient and other parameters, and extract and map the paddy rice according to the parameters. In 2001, Shao et al. [46] investigated the backscattering behavior of paddy rice throughout the growth cycle, and paddy rice monitoring and extraction were carried out according to its characteristics. An empirical model of the paddy rice growth cycle and backscattering coefficient was established with an accuracy of 91%. However, the disadvantage of this method is that it has a single channel and a fixed angle of incidence. It is difficult to estimate multiple parameters for a target, and the target recognition ability needs to be strengthened.

In the past few years, with the advancement of algorithms and the diversity of data, empirical models have also been developed. In 2011, Bouvet used multitrack wide-swath data sets combined with former methods, using temporal backscatter changes as a classification feature for mapping. Compared with the previously used single-track narrow-swath data sets, this method can significantly increase the observation frequency and the size of the mapping areas [47]. The disadvantage is that the establishment of the empirical model must use the existing detailed land cover data to establish the equation and determine the classification threshold. When no ground information is available, the values in the previous literature are used, and there will be errors.

Radar data contain band information of different frequencies, and most previous methods have used C-band information. In 2018, Hoa et al. [48] used COSMO-SkyMed X-band SAR data to analyze the changes in the SAR intensity over time for short- and long-period paddy rice varieties and field seeding periods in the Anjiang region of the Mekong Delta. First, based on the survey data, a comprehensive analysis of the characteristics and cultivation techniques of paddy rice crops in the region was carried out. Then they analyzed the differences of backscattering intensity between paddy rice and other land cover types in this area under vertical transmission/vertical receive (VV), horizontal transmission/horizontal receive (HH) and HH/VV, and obtained indicators closely related to paddy rice mapping. Paddy rice fields were distinguished from other LULCs, and indicators derived from HH polarization could be used to map other LULCs (water, forests, and built-up areas). These maps can be used as auxiliary data to improve the accuracy of the results. The results showed that, due to the vertical structure of the paddy rice plants, this ratio was a good indicator for paddy rice field mapping. Vertically polarized waves are more attenuated than horizontally polarized waves, so the ratio of the backscattered intensity of HH and VV is higher over paddy rice than most other land cover types. The accuracy of the paddy rice planting area has been found to be as high as 92%. However, for provincial and national surveying and mapping, the coverage of satellite data sources is a limitation. In this case, this method is more suitable for large coverage data with frequent repetition cycles.

The advantage of this method is that the idea is relatively simple, and the threshold can be developed for analysis and extraction after determining the threshold according to the data extraction features of the long-term sequence. The disadvantage is that this method depends on long-term observation data and is limited by data availability. The temporal resolution must meet the needs of the paddy rice growth cycle. On the other hand, there must be accurate prior knowledge in the study area to facilitate the establishment of equations and verification of results. The universality of the method is also limited. The backscattering coefficients of paddy rice will show different characteristics in different regions, and the parameters will change.

4.2.2. Machine Learning

In recent years, the machine learning method has mostly been used for paddy rice mapping based on radar data, which extracts the eigenvalues of the backscatter coefficient and inputs these values into the classifier for paddy rice mapping. Classification models mainly include traditional machine learning models (DT, SVM, RF) and deep learning models such as CNN and recurrent neural network (RNN). There are similar methods based on optical remote sensing data. The principles of these two methods are similar. The difference is the input of the training sample. The former method's input data include optical images and vegetation index curves. The latter inputs the backscatter coefficient value extracted after radar image processing. Both methods will consider the input of phenology information and texture feature information to improve the accuracy of the results.

In 2015, Nguyen et al. [49] normalized the data collected over many years and multitrack SAR with a statistical method and then classified it through a knowledge-based DT method. This study obtained an overall accuracy of 85.3%, kappa coefficient of 0.74. He et al. [50] used the backscattering coefficient and its combination with phenological information as inputs to the DT classifier for classification, and HH/VV, VV/VH, and HH/VH ratios were found to have the greatest potential for phenology monitoring. The overall accuracy level of 86.2% was obtained in this study. In March 2017, the Sentinel-2 satellite was launched. The following radar data research mostly used Sentinel-2 data as the data source. In 2019, the temporal behavior of the SAR backscattering coefficient over 832 plots containing different crop types was analyzed. Using the derived metrics, paddy rice plots were mapped through two different methods of DT and RF. The overall accuracy is high; the former has an overall accuracy of 96.3%, and the latter is 96.6% [51]. In addition, researchers have done further research on the combination of SAR and deep learning. Wang et al. [55] used crowdsourced data, Sentinel-2 and DigitalGlobe images, and CNN to map crop types with an overall accuracy of 74%. Secondly, we consider the RNN model. The commonly used method in the RNN model is the long short-term memory (LSTM) model and its improvements, such as bidirectional LSTM (Bi-LSTM). Researchers use the model and backscatter coefficient time series data to achieve a paddy rice map. Compared with traditional machine learning models [56–58], the research results show that the overall accuracy of RNN model results is 95%, and the accuracy of deep learning models is better than traditional machine learning models. One point to mention is that different radar polarization methods have different results. In 2016, Hoang and others used SAR to map paddy rice crops in the Mekong Delta [52]. This study used two methods of single polarization, dual polarization and full polarization to map paddy rice, and the classification accuracy increased with the complexity of the polarization method. In 2018, Lasko et al. used a random forest algorithm and Sentinel-1 radar time series images to draw a double-season and single-season paddy rice map of Hanoi, Vietnam, with resolutions of 10 and 20 m, respectively, using VV and VH polarization methods [53].

The overall accuracy of the 10-meter VV and VH polarization was the highest (93.5%). Subsequent research can focus on the comparison of multipolarized SAR data with different frequencies (C, X, L) to obtain the optimal combination.

Moreover, in 2017, Clauss et al. [54] proposed a method of drawing paddy rice planted area maps using Sentinel-1 time series using superpixel segmentation and phenology-based DTs. Superpixel segmentation is the establishment of a spatially averaged backscattering time series, which has the characteristics of robustness to speckles and reduces the amount of data to be processed. However, the classifier depends on the phenology-based empirical thresholds of the research site. If this method is applied to other regions, it is recommended to adapt the threshold parameters.

The advantage of this method is that paddy rice mapping is carried out by means of machine learning, feature extraction is performed using a large amount of data, and the overall accuracy is improved. However, this method relies on the input of training data to determine the parameters, and different regions will result in different parameters. The completeness and diversity of the training data determine the accuracy of this method.

In general, the accuracy of extraction algorithms based on optical remote sensing improves with the improvement of the method and the improvement of data quality. Most of the time series studies focus on annual series changes. The study areas are relatively large, covering the national scale, and these data generally have high spatial resolution. From the original spatial resolution of 500 m to the current spatial resolution of 30 m, it has been continuously improved, and the characteristics of the data are mainly large-scale. The research mainly focuses on the dynamic changes in the paddy rice area and the changes in the centroid of the paddy rice planting in the region. Extraction algorithms based on microwave remote sensing and rice monitoring based on the backscattering coefficient generally have high accuracy, approximately 90%, and the time series are mainly concentrated on the monthly scale. The study area is mostly within the province and city, with a resolution of 10 m, and its largest advantage is the tropics, where cloudy and rainy conditions dominate.

4.3. Integration of Optical and Microwave Remote Sensing-Based Mapping Methods

Optical remote sensing images and microwave data have their respective advantages. To improve the data accuracy, integrated analysis using both methods is essential. The integration methods are mainly the following, and the accuracy of the results is higher than that of a single data source.

4.3.1. Complementary Method

The main principle of this method is to first obtain the rice extraction layer with optical remote sensing or radar data and then supplement the layered data from another data research institute or use these two data sources as the input layer for the classifier for a comprehensive analysis. This method mainly includes the following complementary methods: (I) The phenological information is determined based on the optical data. Radar images are collected based on phenological information for rice mapping. (II) The optical features of rice and the radar features are input into the classifier together for rice mapping. (III) The results are output separately based on the two data sources. The intersection of the two results is treated as the final result.

Using the first type of method, Asilo et al. [59] extracted paddy rice planting information based on MODIS and SAR images, and the results indicated that MODIS can be used to guide SAR image acquisition and planning to a large extent. Torbick et al. [60] conducted a large-scale paddy rice extraction experiment in Myanmar. In this study, Landsat 8 and other data were used to generate a large-scale land cover map, and then the radar image backscatter coefficient was used to create a detailed range of paddy rice masks. Using the second type of method, Mansaray et al. [61] focused on rice extraction in Shanghai, China. By combining the backscatter coefficient of the radar image with the vegetation index, the decision-making classification method was used to extract rice. Tian et al. [62] used the characteristics of the backscattering coefficient and NDVI to enhance image information and combined this information with k-means unsupervised classification to determine the rice area of Poyang Lake in China. Fiorillo et al. [64] used Sentinel1 and Sentinel-2 data to extract rice spectra and backscatter coefficient features in degraded areas, and input them to the RF classifier together. The combination of Sentinel-1 and Sentinel-2 dense time series provided reliable predictors for RF classification, and the results were good. The overall accuracy is greater than 80%. Chen et al. [65] applied this method in a multi-cloud area and used the Google Earth Engine (GEE) platform. Overall accuracy is 66%. In the third type of method, Guo et al. [63] proposed an optical SAR collaborative paddy rice extraction method. The characteristics of rice growth were collected and analyzed under optical images and SAR for classification. Based on the rule that pixels with one of the classification results as rice are classified as rice, a collaborative fusion method was developed. In one area of Australia, the overall accuracy rate reached 94.7%. Ramadhani et al. [16] first extracted rice using Sentinel-1 and -2 and MODIS data, respectively combined with the SVM classification method, and then fused the two results to generate a multitemporal rice map. The advantage of this method is that it combines the advantages of two data sources. This method also effectively avoids the defects of a single data source. To a certain extent, the accuracy of the results has been improved. However, shortcomings still exist. For example, the spectral similarity of different crops is one shortcoming. Both data sources suffer from this problem. Whether data fusion effectively avoids this problem remains to be studied.

4.3.2. Comparison Class Method

The principle of this method is mainly based on different data combination methods, different classification methods, and the results of different regions to obtain the optimal combination of methods for paddy rice extraction. For example, the results of the same data input to different classifiers can be compared, and the results of radar data in different polarization modes combined with the same optical index can also be compared. Comparisons between pixel-based classification and area-based classification have also been conducted.

Onojeghuo et al. [66] took the Sanjiang Plain in northeast China as the research area, utilized NDVI images and dual-polarization (VH/VV) SAR as input data, and used RF and SVM machine learning classification algorithms to perform paddy rice mapping. The results showed that the RF algorithm applied to multitemporal VH polarization and NDVI data produced the highest classification accuracy (96.7%). Zhang et al. [67] first performed image preprocessing on Google Earth Engine (GEE) and combined the pixel-based classification results with object-based segmentation results to output a paddy rice area map. The combination of the two methods eliminated the noise that is common in medium- and high-resolution pixel classification and brought the rice planting area closer to official statistics. As a result, rice maps with a resolution of 10 m were established in Heilongjiang, Hunan and Guangxi provinces of China, with a total accuracy of approximately 90%. In the same year, Yang et al. [68] combined the characteristics of multiple watershed and mountainous areas in Wuhua County, South China, and used region-based and pixel-based methods to map the paddy rice planting area. The results showed that the accuracy of the area-based method was 1.18% higher than that of the pixel-based method (91.38%). The area-based method mainly eliminates the influence of speckle noise. Park et al. [69] classified paddy rice based on different data input combinations (original image, vegetation index, backscatter coefficient) combined with RF and SVM. The results showed that the fusion optics and SAR data had the highest accuracy. In this study, the Paddy Rice Mapping Index (PMI) was established based on the spectral and phenological characteristics of paddy rice, which could be used to extract paddy rice over a large area.

In fact, this kind of method is complementary to the first method. Here, we focus on the comparison between different methods. Researchers can choose the appropriate method according to their own research needs. For the advantages and disadvantages, please refer to the advantages and disadvantages of the first method. There is limited literature on data fusion, and such studies have only appeared in recent years. These studies catered to the development trend of multisource data. Therefore, the problem of how to achieve the best fusion effect will be a focus of future work.

5. Discussion

5.1. Method Evolution Trend

From 2010 to 2020, paddy rice mapping methods were continuously innovated, following the development of science and technology (Figure 4). Previously, paddy rice mapping methods mainly consisted of images combined with simple classification methods, vegetation indices, etc. In 2010, the main method was still the phenology algorithm. Studies performed algorithm verification with different data sources in different regions. In 2011, radar data began to enter the field of paddy rice mapping. The emergence of radar data brought opportunities for paddy rice mapping in cloudy and rainy areas. The further development of the method improved the mapping accuracy of paddy rice by combining the method with computer technology. These methods include object-oriented, cloud computing, deep learning, and machine learning. GEE is a platform for online visualization computing analysis and processing to use Google's abundant computing resources for large-scale geospatial data processing [15]. The collocation of remote sensing methods, data, and processing infrastructure will help create high-resolution remote sensing products that cover a large scale. With the continuous launch of satellites, the functionality of satellites has been continuously enhanced, and the resolution of satellite data has also been continuously improved. Researchers have begun to focus on multisource data fusion, large-scale paddy rice extraction, and surface temperature data to improve phenology algorithms. The future development direction is actually very clear. The first is the use of Cubsat, GF series satellites, and satellite fusion data sets (harmonized Landsat and Sentinel-2(HLS) data set [70]). High-resolution satellite data provide a reliable foundation for mapping methods. The use of data sets can reduce the calculation procedures of researchers and further promote the development of mapping methods. Secondly, the combination of computer technology for image processing greatly improves the data processing efficiency and accuracy to a certain extent. In addition, the development of the unmanned aerial vehicle (UAV) has also provided a direction for paddy rice mapping and is suitable for the development of fine agriculture.

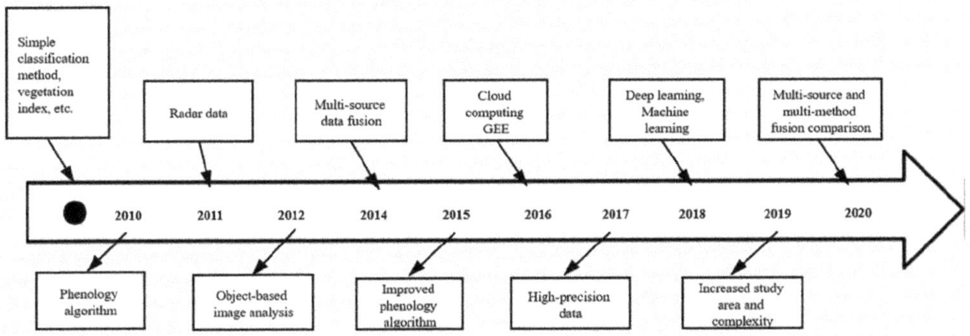

Figure 4. Evolution of methods.

5.2. Research Challenges

Although the methods have been continuously improved in recent years and the accuracy of paddy rice mapping has been continuously improved, challenges remain.

The discussion of previous methods mentioned that cloud cover is a major problem in paddy rice mapping. However, with the continuous improvement of methods and data sources, the impact of cloud cover has been reduced. The fusion of multisource data, which involves the combination of MODIS and Landsat data, has filled the gaps in the time series of the area covered by clouds [10,32]. The emergence of free SAR data and its cloud-penetrating characteristics also effectively solved the problem of cloud cover. Some studies have established time series models to eliminate the effects of cloud cover [36].

The problem of data verification remains. As previously mentioned, the improvement of statistical data in various regions and the emergence of high-resolution images have improved the quality of data verification. However, verifying the data remains a challenge because, based on artificial statistics, you cannot be sure of their accuracy. However, this challenge may gradually be overcome. With the continuous development of paddy rice mapping methods, a large number of research results have emerged, and there are overlapping areas in the study area; you can refer to the literature for data and method comparisons.

Therefore, the greatest problem is the versatility of the method. In different research areas, the threshold settings used by the methods are different, and different research areas may have different results; thus, a universal method is lacking. Most of the methods mentioned in this paper rely on the vegetation index and the time-varying curve of the backscattering coefficient for research. Machine learning and deep learning rely on training samples to improve the classification accuracy. In different regions and under different climatic conditions, the growth cycle of paddy rice will change, and the characteristic curve will change, which will change the number of features. Training samples also need to be re-extracted. There are also studies on using pixel segmentation algorithms based on different regions to jointly take eigenvalues to verify the generalizability of the method. However, the representative area in the study remains still to be studied [54].

6. Conclusions

Through a review and analysis of paddy rice remote sensing mapping methods applied over the past 10 years, the paddy rice mapping methods are divided into three categories according to the data source and then subdivided according to the principle of the method. Overall, there are many studies on paddy rice mapping using optical images, with MODIS, Landsat, and Sentinel-2 as the main data sources. With the emergence of Sentinel-1 data, research on the extraction of paddy rice based on radar data has gradually increased, effectively eliminating the problem of clouds and fog in optical images. The emergence of the concept of multisource data fusion has also brought good news to rice mapping, greatly improving the accuracy of rice extraction. Among these methods, there are classic methods and innovative methods. The best method in optical remote sensing is paddy rice mapping based on paddy rice phenology. Many studies have confirmed the applicability of this method in different regions, indicating the high overall accuracy. Innovative methods include the spectral matching method and threshold method using a combination of mathematical principles. Both of these methods rely on a time series graph of the vegetation index. The effects in cloudy and foggy areas need to be considered, and the effect may be improved by using a combination of the spatiotemporal data fusion models. With the development of computer technology in recent years, object-oriented and machine learning methods have emerged in the field of paddy rice mapping. In microwave remote sensing, the method combined with machine learning has high overall accuracy. When training the model, paddy rice phenology information and texture information can also be combined to improve model accuracy. The combination of optical remote sensing data and microwave data is a development direction for paddy rice extraction in the future. The advantages of the two types of data complement each other. Optical images can provide guidance for radar data, and radar data can provide assistance for optical data.

Combined with the above analysis, the following insight is obtained. One feature is multisource data fusion, which realizes rice mapping from a system perspective. The development of technology and the emergence of more accurate data sources with increased spatial and temporal resolution provide new opportunities for rice monitoring. With the continuous innovation of algorithms and the continuous improvement of computing power, methods such as cloud platforms, GEE, and machine learning have emerged. Radar remote sensing images and optical remote sensing images can be effectively combined to better realize rice identification and monitoring. Integrated systems are the focus of future research. The second feature is the extraction of rice area under different planting systems. Most of the previous research focused on the accuracy of the algorithm. Most previous studies did not mention the rice area under different planting systems. Therefore, when paying attention to the extraction of rice and non-rice regions, it is necessary to focus on the analysis of the difference in yield caused by the difference in the paddy rice internal planting system. These results can provide powerful help for global food security. The third feature is to attach importance to issues such as global change and the ecological environment. Globalization is currently a major trend. On the basis of accurate paddy rice mapping, we must integrate global changes. Studies should pay attention to the environmental problems caused by rice growth, which will provide a better understanding of the response and adaptation of agricultural systems to global climate change.

Author Contributions: Conceptualization, Y.L., M.M. and R.Z.; methodology, Y.L. and R.Z.; formal analysis, R.Z.; writing—original draft preparation, R.Z.; writing—review and editing, Y.L. and M.M. All authors have read and agreed to the published version of the manuscript.

Funding: This research was funded by the National Natural Science Foundation of China, grant number: 41571419, 41830648, 41771453.

Institutional Review Board Statement: Not applicable.

Informed Consent Statement: Not applicable.

Data Availability Statement: No new data were created or analyzed in this study. Data sharing is not applicable to this article.

Acknowledgments: In this study, the papers were downloaded from the Web of Science and China National Knowledge Infrastructure, thanks to the retrieval and download services of the above websites.

Conflicts of Interest: The authors declare no conflict of interest.

References

1. Kuenzer, C.; Knauer, K. Remote sensing of rice crop areas. *Int. J. Remote Sens.* **2013**, *34*, 2101–2139. [CrossRef]
2. Bouman, B. How much water does rice use. *Rice Today* **2009**, *8*, 28–29.
3. Sass, R.L.; Fisher, F.M.; Ding, A.; Huang, Y. Exchange of methane from ricefields: National, regional, and global budgets. *J. Geophys. Res. Atmos.* **1999**, *104*, 26943–26951. [CrossRef]
4. Sass, R.L.; Cicerone, R.J. Photosynthate allocations in rice plants: Food production or atmospheric methane? *Proc. Natl. Acad. Sci. USA* **2002**, *99*, 11993–11995. [CrossRef]
5. Yan, X.; Akiyama, H.; Yagi, K.; Akimoto, H. Global estimations of the inventory and mitigation potential of methane emissions from rice cultivation conducted using the 2006 Intergovernmental Panel on Climate Change Guidelines. *Glob. Biogeochem. Cycle* **2009**, *23*, GB2002. [CrossRef]
6. Gilbert, M.; Golding, N.; Zhou, H.; Wint, G.R.; Robinson, T.P.; Tatem, A.J.; Lai, S.; Zhou, S.; Jiang, H.; Guo, D.; et al. Predicting the risk of avian influenza A H7N9 infection in live-poultry markets across Asia. *Nat. Commun.* **2014**, *5*, 4116. [CrossRef] [PubMed]
7. Dong, J.; Xiao, X. Evolution of regional to global paddy rice mapping methods: A review. *ISPRS J. Photogramm. Remote Sens.* **2016**, *119*, 214–227. [CrossRef]
8. Mosleh, M.; Hassan, Q.; Chowdhury, E. Application of remote sensors in mapping rice area and forecasting its production: A review. *Sensors* **2015**, *15*, 769–791. [CrossRef]
9. Niel, T.G.V.; Mcvicar, T.R. Current and potential uses of optical remote sensing in rice-based irrigation systems: A review. *Aust. J. Agric. Res.* **2004**, *55*, 155–185. [CrossRef]
10. Qin, Y.; Xiao, X.; Dong, J.; Zhou, Y.; Zhu, Z.; Zhang, G.; Du, G.; Jin, C.; Kou, W.; Wang, J.; et al. Mapping paddy rice planting area in cold temperate climate region through analysis of time series Landsat 8 (OLI), Landsat 7 (ETM+) and MODIS imagery. *ISPRS J. Photogramm. Remote Sens.* **2015**, *105*, 220–233. [CrossRef]
11. Zhou, Y.; Xiao, X.; Qin, Y.; Dong, J.; Zhang, G.; Kou, W.; Jin, C.; Wang, J.; Li, X. Mapping paddy rice planting area in rice-wetland coexistent areas through analysis of Landsat 8 OLI and MODIS images. *Int. J. Appl. Earth Obs. Geoinf.* **2016**, *46*, 1–12. [CrossRef] [PubMed]

12. Jiang, M.; Xin, L.; Li, X.; Tan, M.; Wang, R. Decreasing rice cropping intensity in Southern China from 1990 to 2015. *Remote Sens.* **2018**, *11*, 35. [CrossRef]
13. Okamoto, K.; Kawashima, H. Estimating total area of paddy fields in Heilongjiang, China, around 2000 using Landsat thematic mapper/enhanced thematic mapper plus data. *Remote Sens. Lett.* **2016**, *7*, 533–540. [CrossRef]
14. Dong, J.; Xiao, X.; Kou, W.; Qin, Y.; Zhang, G.; Li, L.; Jin, C.; Zhou, Y.; Wang, J.; Biradar, C.; et al. Tracking the dynamics of paddy rice planting area in 1986–2010 through time series Landsat images and phenology-based algorithms. *Remote Sens. Environ.* **2015**, *160*, 99–113. [CrossRef]
15. Shew, A.M.; Ghosh, A. Identifying dry-season rice-planting patterns in bangladesh using the Landsat archive. *Remote Sens.* **2019**, *11*, 1235. [CrossRef]
16. Ramadhani, F.; Pullanagari, R.; Kereszturi, G.; Procter, J. Automatic mapping of rice growth stages using the integration of SENTINEL-2, MOD13Q1, and SENTINEL-1. *Remote Sens.* **2020**, *12*, 3613. [CrossRef]
17. Manjunath, K.R.; More, R.S.; Jain, N.K.; Panigrahy, S.; Parihar, J.S. Mapping of rice-cropping pattern and cultural type using remote-sensing and ancillary data: A case study for South and Southeast Asian countries. *Int. J. Remote Sens.* **2015**, *36*, 6008–6030. [CrossRef]
18. Gumma, M.K. Mapping rice areas of South Asia using MODIS multitemporal data. *J. Appl. Remote Sens.* **2011**, *5*, 053547. [CrossRef]
19. Gumma, M.K.; Mohanty, S.; Nelson, A.; Arnel, R.; Mohammed, I.A.; Das, S.R. Remote sensing based change analysis of rice environments in Odisha, India. *J. Environ. Manag.* **2015**, *148*, 31–41. [CrossRef]
20. Clauss, K.; Yan, H.; Kuenzer, C. Mapping rice in China in 2002, 2005, 2010 and 2014 with MODIS time series. *Remote Sens.* **2016**, *8*, 434. [CrossRef]
21. Yin, Q.; Liu, M.; Cheng, J.; Ke, Y.; Chen, X. Mapping rice planting area in Northeastern China using spatiotemporal data fusion and phenology-based method. *Remote Sens.* **2019**, *11*, 1699. [CrossRef]
22. Zhang, M.; Lin, H.; Wang, G.; Sun, H.; Fu, J. Mapping paddy rice using a convolutional neural network (CNN) with Landsat 8 datasets in the Dongting Lake area, China. *Remote Sens.* **2018**, *10*, 1840. [CrossRef]
23. Zhao, S.; Liu, X.; Ding, C.; Liu, S.; Wu, C.; Wu, L. Mapping rice paddies in complex landscapes with convolutional neural networks and phenological metrics. *GISci. Remote Sens.* **2019**, *57*, 37–48. [CrossRef]
24. Guan, X.; Huang, C.; Liu, G.; Xu, Z.; Liu, Q. Extraction of paddy rice area using a DTW distance based similarity measure. *Resour. Sci.* **2014**, *36*, 267–272.
25. Guan, X.; Huang, C.; Liu, G.; Meng, X.; Liu, Q. Mapping rice cropping systems in Vietnam using an NDVI-based time-series similarity measurement based on DTW distance. *Remote Sens.* **2016**, *8*, 19. [CrossRef]
26. Chen, C.-F.; Huang, S.-W.; Son, N.-T.; Chang, L.-Y. Mapping double-cropped irrigated rice fields in Taiwan using time-series satellite pour l'observation de la Terre data. *J. Appl. Remote Sens.* **2011**, *5*, 053528. [CrossRef]
27. Nuarsa, I.W.; Nishio, F.; Hongo, C.; Mahardika, I.G. Using variance analysis of multitemporal MODIS images for rice field mapping in Bali Province, Indonesia. *Int. J. Remote Sens.* **2012**, *33*, 5402–5417. [CrossRef]
28. Wang, J.; Huang, J.; Zhang, K.; Li, X.; She, B.; Wei, C.; Gao, J.; Song, X. Rice fields mapping in fragmented area using multi-temporal HJ-1A/B CCD images. *Remote Sens.* **2015**, *7*, 3467–3488. [CrossRef]
29. Liu, W.; Dong, J.; Xiang, K.; Wang, S.; Han, W.; Yuan, W. A sub-pixel method for estimating planting fraction of paddy rice in Northeast China. *Remote Sens. Environ.* **2018**, *205*, 305–314. [CrossRef]
30. Xiao, X.; Boles, S.; Liu, J.; Zhuang, D.; Frolking, S.; Li, C.; Salas, W.; Moore, B. Mapping paddy rice agriculture in Southern China using multi-temporal MODIS images. *Remote Sens. Environ.* **2005**, *95*, 480–492. [CrossRef]
31. Xiao, X.; Boles, S.; Frolking, S.; Li, C.; Babu, J.Y.; Salas, W.; Moore, B. Mapping paddy rice agriculture in South and Southeast Asia using multi-temporal MODIS images. *Remote Sens Environ.* **2006**, *100*, 95–113. [CrossRef]
32. Teluguntla, P.; Ryu, D.; George, B.; Walker, J.; Malano, H. Mapping flooded rice paddies using time series of MODIS imagery in the Krishna River Basin, India. *Remote Sens.* **2015**, *7*, 8858–8882. [CrossRef]
33. Dong, J.; Xiao, X.; Menarguez, M.A.; Zhang, G.; Qin, Y.; Thau, D.; Biradar, C.; Moore, B. Mapping paddy rice planting area in Northeastern Asia with Landsat 8 images, phenology-based algorithm and Google Earth Engine. *Remote Sens. Environ.* **2016**, *185*, 142–154. [CrossRef]
34. Dong, J.; Xiao, X.; Zhang, G.; Menarguez, M.A.; Choi, C.Y.; Qin, Y.; Luo, P.; Zhang, Y.; Moore, B. Northward expansion of paddy rice in Northeastern Asia during 2000–2014. *Geophys. Res. Lett.* **2016**, *43*, 3754–3761. [CrossRef]
35. Dou, Y.; Huang, R.; Mansaray, L.R.; Huang, J. Mapping high temperature damaged area of paddy rice along the Yangtze River using moderate resolution imaging spectroradiometer data. *Int. J. Remote Sens.* **2019**, *41*, 471–486. [CrossRef]
36. Zhang, G.; Xiao, X.; Biradar, C.M.; Dong, J.; Qin, Y.; Menarguez, M.A.; Zhou, Y.; Zhang, Y.; Jin, C.; Doughty, R.B.; et al. Spatiotemporal patterns of paddy rice croplands in China and India from 2000 to 2015. *Sci. Total Environ.* **2017**, *579*, 82–92. [CrossRef]
37. Zhang, G.; Xiao, X.; Dong, J.; Kou, W.; Jin, C.; Qin, Y.; Zhou, Y.; Wang, J.; Menarguez, M.A.; Biradar, C. Mapping paddy rice planting areas through time series analysis of MODIS land surface temperature and vegetation index data. *ISPRS J. Photogramm. Remote Sens.* **2015**, *106*, 157–171. [CrossRef]
38. Sousa, D.; Small, C. Mapping and monitoring rice agriculture with multisensor temporal mixture models. *Remote Sens.* **2019**, *11*, 181. [CrossRef]

39. Biradar, C.M.; Xiao, X. Quantifying the area and spatial distribution of double- and triple-cropping croplands in India with multi-temporal MODIS imagery in 2005. *Int. J. Remote Sens.* **2011**, *32*, 367–386. [CrossRef]

40. Son, N.-T.; Chen, C.-F.; Chen, C.-R.; Guo, H.-Y. Classification of multitemporal Sentinel-2 data for field-level monitoring of rice cropping practices in Taiwan. *Adv. Space Res.* **2020**, *65*, 1910–1921. [CrossRef]

41. Ding, M.; Guan, Q.; Li, L.; Zhang, H.; Liu, C.; Zhang, L. Phenology-based rice paddy mapping using multi-source satellite imagery and a fusion algorithm applied to the Poyang Lake Plain, Southern China. *Remote Sens.* **2020**, *12*, 1022. [CrossRef]

42. Su, T. Efficient paddy field mapping using Landsat-8 imagery and object-based image analysis based on advanced fractel net evolution approach. *GISci. Remote Sens.* **2016**, *54*, 354–380. [CrossRef]

43. Singha, M.; Wu, B.; Zhang, M. An object-based paddy rice classification using multi-spectral data and crop phenology in Assam, Northeast India. *Remote Sens.* **2016**, *8*, 479. [CrossRef]

44. Zhang, M.; Lin, H. Object-based rice mapping using time-series and phenological data. *Adv. Space Res.* **2019**, *63*, 190–202. [CrossRef]

45. Ribbes, F.; Le Toan, T. Use of ERS-1 SAR data for ricefield mapping and rice crop parameters retrieval. *IGARSS* **1996**, *4*, 1983–1985. [CrossRef]

46. Shao, Y.; Fan, X.; Liu, H.; Xiao, J.; Ross, S.; Brisco, B.; Brown, R.; Staples, G. Rice monitoring and production estimation using multitemporal RADARSAT. *Remote Sens. Environ.* **2001**, *76*, 310–325. [CrossRef]

47. Bouvet, A.; Le Toan, T. Use of ENVISAT/ASAR wide-swath data for timely rice fields mapping in the Mekong River Delta. *Remote Sens. Environ.* **2011**, *115*, 1090–1101. [CrossRef]

48. Phan, H.; Le Toan, T.; Bouvet, A.; Nguyen, L.; Pham Duy, T.; Zribi, M. Mapping of rice varieties and sowing date using X-band SAR data. *Sensors* **2018**, *18*, 316. [CrossRef]

49. Nguyen, D.; Clauss, K.; Cao, S.; Naeimi, V.; Kuenzer, C.; Wagner, W. Mapping rice seasonality in the Mekong Delta with multi-year Envisat ASAR WSM data. *Remote Sens.* **2015**, *7*, 15868–15893. [CrossRef]

50. He, Z.; Li, S.; Wang, Y.; Dai, L.; Lin, S. Monitoring rice phenology based on backscattering characteristics of multi-temporal RADARSAT-2 datasets. *Remote Sens.* **2018**, *10*, 340. [CrossRef]

51. Bazzi, H.; Baghdadi, N.; El Hajj, M.; Zribi, M.; Minh, D.H.T.; Ndikumana, E.; Courault, D.; Belhouchette, H. Mapping paddy rice using Sentinel-1 SAR time series in Camargue, France. *Remote Sens.* **2019**, *11*, 887. [CrossRef]

52. Hoang, H.K.; Bernier, M.; Duchesne, S.; Tran, Y.M. Rice mapping using RADARSAT-2 dual- and quad-pol data in a complex land-use watershed: Cau River Basin (Vietnam). *IEEE JSTARS* **2016**, *9*, 3082–3096. [CrossRef]

53. Lasko, K.; Vadrevu, K.P.; Tran, V.T.; Justice, C. Mapping double and single crop paddy rice with Sentinel-1A at varying spatial scales and polarizations in Hanoi, Vietnam. *IEEE JSTARS* **2018**, *11*, 498–512. [CrossRef] [PubMed]

54. Clauss, K.; Ottinger, M.; Kuenzer, C. Mapping rice areas with Sentinel-1 time series and superpixel segmentation. *Int. J. Remote Sens.* **2017**, *39*, 1399–1420. [CrossRef]

55. Wang, S.; di Tommaso, S.; Faulkner, J.; Friedel, T.; Kennepohl, A.; Strey, R.; Lobell, D.B. Mapping crop types in Southeast India with smartphone crowdsourcing and deep learning. *Remote Sens.* **2020**, *12*, 2957. [CrossRef]

56. Crisóstomo de Castro Filho, H.; Abílio de Carvalho Júnior, O.; Ferreira de Carvalho, O.L.; Pozzobon de Bem, P.; dos Santos de Moura, R.; Olino de Albuquerque, A.; Rosa Silva, C.; Guimaraes Ferreira, P.H.; Fontes Guimaraes, R.; Trancoso Gomes, R.A. Rice crop detection using LSTM, Bi-LSTM, and machine learning models from sentinel-1 time series. *Remote Sens.* **2020**, *12*, 2655. [CrossRef]

57. Zhao, H.; Chen, Z.; Jiang, H.; Jing, W.; Sun, L.; Feng, M. Evaluation of three deep learning models for early crop classification using sentinel-1A imagery time series—A case study in Zhanjiang, China. *Remote Sens.* **2019**, *11*, 2673. [CrossRef]

58. Ndikumana, E.; Ho Tong Minh, D.; Baghdadi, N.; Courault, D.; Hossard, L. Deep recurrent neural network for agricultural classification using multitemporal SAR Sentinel-1 for Camargue, France. *Remote Sens.* **2018**, *10*, 1217. [CrossRef]

59. Asilo, S.; de Bie, K.; Skidmore, A.; Nelson, A.; Barbieri, M.; Maunahan, A. Complementarity of two rice mapping approaches, characterizing strata mapped by hypertemporal MODIS and rice paddy identification using multitemporal SAR. *Remote Sens.* **2014**, *6*, 12789–12814. [CrossRef]

60. Torbick, N.; Chowdhury, D.; Salas, W.; Qi, J. Monitoring rice agriculture across myanmar using time series Sentinel-1 assisted by Landsat-8 and PALSAR-2. *Remote Sens.* **2017**, *9*, 119. [CrossRef]

61. Mansaray, L.; Huang, W.; Zhang, D.; Huang, J.; Li, J. Mapping rice fields in urban Shanghai, Southeast China, using Sentinel-1A and Landsat 8 datasets. *Remote Sens.* **2017**, *9*, 257. [CrossRef]

62. Tian, H.; Wu, M.; Wang, L.; Niu, Z. Mapping early, middle and late rice extent using Sentinel-1A and Landsat-8 data in the Poyang Lake Plain, China. *Sensors* **2018**, *18*, 185. [CrossRef]

63. Guo, Y.; Jia, X.; Paull, D.; Benediktsson, J.A. Nomination-favoured opinion pool for optical-SAR-synergistic rice mapping in face of weakened flooding signals. *ISPRS J. Photogramm. Remote Sens.* **2019**, *155*, 187–205. [CrossRef]

64. Fiorillo, E.; di Giuseppe, E.; Fontanelli, G.; Maselli, F. Lowland rice mapping in Sédhiou Region (Senegal) using sentinel 1 and sentinel 2 data and random forest. *Remote Sens.* **2020**, *12*, 3403. [CrossRef]

65. Chen, N.; Yu, L.; Zhang, X.; Shen, Y.; Zeng, L.; Hu, Q.; Niyogi, D. Mapping paddy rice fields by combining multi-temporal vegetation index and synthetic aperture radar remote sensing data using google earth engine machine learning platform. *Remote Sens.* **2020**, *12*, 2992. [CrossRef]

66. Onojeghuo, A.O.; Blackburn, G.A.; Wang, Q.; Atkinson, P.M.; Kindred, D.; Miao, Y. Rice crop phenology mapping at high spatial and temporal resolution using downscaled MODIS time-series. *GISci. Remote Sens.* **2018**, *55*, 659–677. [CrossRef]
67. Zhang, X.; Wu, B.; Ponce-Campos, G.; Zhang, M.; Chang, S.; Tian, F. Mapping up-to-date paddy rice extent at 10 m resolution in china through the integration of optical and synthetic aperture radar images. *Remote Sens.* **2018**, *10*, 1200. [CrossRef]
68. Yang, H.; Pan, B.; Wu, W.; Tai, J. Field-based rice classification in Wuhua county through integration of multi-temporal Sentinel-1A and Landsat-8 OLI data. *Int. J. Appl. Earth Obs. Geoinf.* **2018**, *69*, 226–236. [CrossRef]
69. Park, S.; Im, J.; Park, S.; Yoo, C.; Han, H.; Rhee, J. Classification and mapping of paddy rice by combining Landsat and SAR time series data. *Remote Sens.* **2018**, *10*, 447. [CrossRef]
70. Claverie, M.; Ju, J.; Masek, J.G.; Dungan, J.L.; Vermote, E.F.; Roger, J.-C.; Skakun, S.V.; Justice, C. The harmonized landsat and sentinel-2 surface reflectance data set. *Remote Sens. Environ.* **2018**, *219*, 145–161. [CrossRef]

sustainability

Article

Response of Different Band Combinations in Gaofen-6 WFV for Estimating of Regional Maize Straw Resources Based on Random Forest Classification

Huawei Mou [1,2,3], Huan Li [1,2,4], Yuguang Zhou [1,5,6,7,*] and Renjie Dong [1,2,4,5,8]

[1] Bioenergy and Environment Science & Technology Laboratory, College of Engineering, China Agricultural University, Beijing 100083, China; mouhuawei@cau.edu.cn (H.M.); huanli828@cau.edu.cn (H.L.); rjdong@cau.edu.cn (R.D.)

[2] Key Laboratory of Clean Production and Utilization of Renewable Energy, Ministry of Agriculture and Rural Affairs, Beijing 100083, China

[3] Department of Agricultural Systems Management, University of Missouri, Columbia, MO 65211, USA

[4] National Center for International Research of BioEnergy Science and Technology, Ministry of Science and Technology, Beijing 100083, China

[5] State R&D Center for Efficient Production and Comprehensive Utilization of Biobased Gaseous Fuels, National Energy Administration, Beijing 100083, China

[6] National Energy R&D Center for Biomass, China Agricultural University, Beijing 100193, China

[7] Prataculture Machinery and Equipment Research Center, College of Engineering, China Agricultural University, Beijing 100083, China

[8] Yantai Institute, China Agricultural University, Yantai 264670, China

* Correspondence: zhouyg@cau.edu.cn; Tel.: +86-10-6273-7858

Citation: Mou, H.; Li, H.; Zhou, Y.; Dong, R. Response of Different Band Combinations in Gaofen-6 WFV for Estimating of Regional Maize Straw Resources Based on Random Forest Classification. *Sustainability* **2021**, *13*, 4603. https://doi.org/10.3390/su13094603

Academic Editor: C. Ronald Carroll

Received: 10 March 2021
Accepted: 13 April 2021
Published: 21 April 2021

Publisher's Note: MDPI stays neutral with regard to jurisdictional claims in published maps and institutional affiliations.

Abstract: Maize straw is a valuable renewable energy source. The rapid and accurate determination of its yield and spatial distribution can promote improved utilization. At present, traditional straw estimation methods primarily rely on statistical analysis that may be inaccurate. In this study, the Gaofen 6 (GF-6) satellite, which combines high resolution and wide field of view (WFV) imaging characteristics, was used as the information source, and the quantity of maize straw resources and spatial distribution characteristics in Qihe County were analyzed. According to the phenological characteristics of the study area, seven classification classes were determined, including maize, buildings, woodlands, wastelands, water, roads, and other crops, to explore the influence of sample separation and test the responsiveness to different land cover types with different waveband combinations. Two supervised classification methods, support vector machine (SVM) and random forest (RF), were used to classify the study area, and the influence of the newly added band of GF-6 WFV on the classification accuracy of the study area was analyzed. Furthermore, combined with field surveys and agricultural census data, a method for estimating the quantity of maize straw and analyzing the spatial distribution based on a single-temporal remote sensing image and random forests was proposed. Finally, the accuracy of the measurement results is evaluated at the county level. The results showed that the RF model made better use of the newly added bands of GF-6 WFV and improved the accuracy of classification, compared with the SVM model; the two red-edge bands improved the accuracy of crop classification and recognition; the purple and yellow bands identified non-vegetation more effectively than vegetation, thus minimizing the "salt-and-pepper noise" of classification results. However, the changes to total classification accuracy were not obvious; the theoretical quantity of maize straw in Qihe County in 2018 was 586.08 kt, which reflects an error of only 2.42% compared to the statistical result. Hence, the RF model based on single-temporal GF-6 WFV can effectively estimate regional maize straw yield and spatial distribution, which lays a theoretical foundation for straw recycling.

Keywords: GF-6; maize; straw; support vector machine; random forest; red-edge wavelength

1. Introduction

China is a large country known for its abundant agricultural resources, with agricultural production comprising a significant proportion of its national economy [1]. Crop straw, as a by-product of agricultural production [2], is an indispensable production material in vast rural areas [3]. In recent years, as the rural energy structure has shifted, fewer people have used straw as an energy resource for rural life because of its large volume and scattered distribution, as well as the low degree of industrialization [4]. Furthermore, owing to the regional, seasonal, and structural surplus of straw becoming increasingly prominent, a large amount of straw is still not fully utilized, which severely restricts the development of circular agriculture in China [5,6]. Although the government advocates for the return of straw to the field [7], straw is often discarded or burned in large amounts to allow for timely sowing at the start of the growing season, leading to the serious waste of resources and environmental pollution [8]. Straw is beneficial if used, but it is harmful if it is discarded [9]. With the development of renewable energy technology, biotechnology, circular agriculture, and environmental science, the value of crop straw as a renewable energy source has gradually been increasing and become widely accepted [10], which can be used as bio-fertilizer, feed, raw materials, fuels, and base materials. Therefore, studying the quantity and spatial distribution of straw resources in China and promoting the comprehensive utilization of straw resources are necessary for promoting rural building and sustainable agricultural development in China [11].

Although the current straw quantity was estimated in many studies, there are several limitations to extant methods and findings [12]. Firstly, the low resolution of statistical data is generally used for the analysis of straw at the county level or above, which limits the value of the data for detailed spatial analyses [13]. Secondly, the quantity of straw resources cannot be calculated in time, because the agricultural census can only be completed in the next year at the earliest. Moreover, if we aim to realize the comprehensive utilization of straw, we must not only estimate the straw yield but also fully consider the spatial distribution of regional straw [14]. The effective recycling and utilization of crop straw resources can be realized more efficiently by combining the relationship between the supply and demand of regional straw resources and optimizing straw recycling and comprehensive utilization [15].

With the in-depth application of remote sensing technology in crop area extraction, growth monitoring [16], and yield estimation [17], the use of remote sensing technology to analyze the yield and spatial distribution of straw has become a major development direction for straw resource investigation [18]. The spatial characteristics of crop planting in China exhibit complex structures and fragmentations. Therefore, in the estimation of large-scale straw quantity, the data to be processed are very large when the high-resolution remote sensing image is used, while the low-resolution remote sensing image will lead to a rapid decline in the measurement accuracy [19]. Classification using single-temporal remote sensing images of the "key phenological period" combined with multi-characteristic parameters and sensitive bands has become an important method for current crop type identification [20]. The response characteristics of different wavebands to different crops can be used to optimize the combination of wavebands, so that the spectral difference and Class Separability between different crop types are significantly improved, and finally, the accurate investigation and analysis of different crop straw resources can be realized [21].

Maize, which accounts for approximately one-fifth of grain crops in China, is the third-largest grain crop after rice and wheat [22]. Therefore, the quantity and spatial distribution of maize straw in the region are of great significance to the collection, storage, and transportation as well as comprehensive utilization of straw.

The Gaofen-6 (GF-6) satellite, planned in China's high-resolution major special series satellites, adds four bands with central wavelengths of 710 nm, 750 nm, 425 nm, and 610 nm, which can provide richer spectral information for agricultural research [23]. This technological advancement is important for improving the spectral information characteristics of China's medium and high-resolution satellites [24].

In order to improve the comprehensive utilization and accelerate the development of the scale, industrialization, and commercialization of straw, the quantity and spatial distribution of straw need to be studied, to plan the recycling network and the site selection of the utilization factory of straw. In this study, the effects of different wavebands on the classification of different land cover types were analyzed based on GF-6 satellite imagery, and the quantity and spatial distribution characteristics of maize straw in Qihe County were estimated to provide data support for recycling and effective utilization of regional straw. The research contents included: (1) exploring and analyzing the impact of different band combinations on samples separability, (2) analyzing the classification accuracy of the support vector machine (SVM) and random forest (RF) classification models under different band combinations for different land cover types, and (3) proposing a method for estimating the quantity and spatial distribution of maize straw based on planting area.

2. Materials and Methods

2.1. Research Area

The study area is located in Qihe County, which is in the southernmost part of Dezhou City, Shandong Province, China, at latitude range 36°24′37″–37°1′44″ N and longitude range 116°23′28″–116°57′35″ E (Figure 1). The annual average temperature is 15 °C throughout the year, which indicates a warm temperate and sub-humid monsoon climate zone, with four distinct seasons and mild weather patterns. The land area of the study area is approximately 1411 km², of which arable land comprises 840 km². It is flat, with an average elevation of 26 m (mean sea level), and is an alluvial plain in the lower reaches of the Yellow River. It is also an important food production area in Shandong Province. The main food crops are winter wheat and maize; peanuts, soybeans, and cotton are also planted.

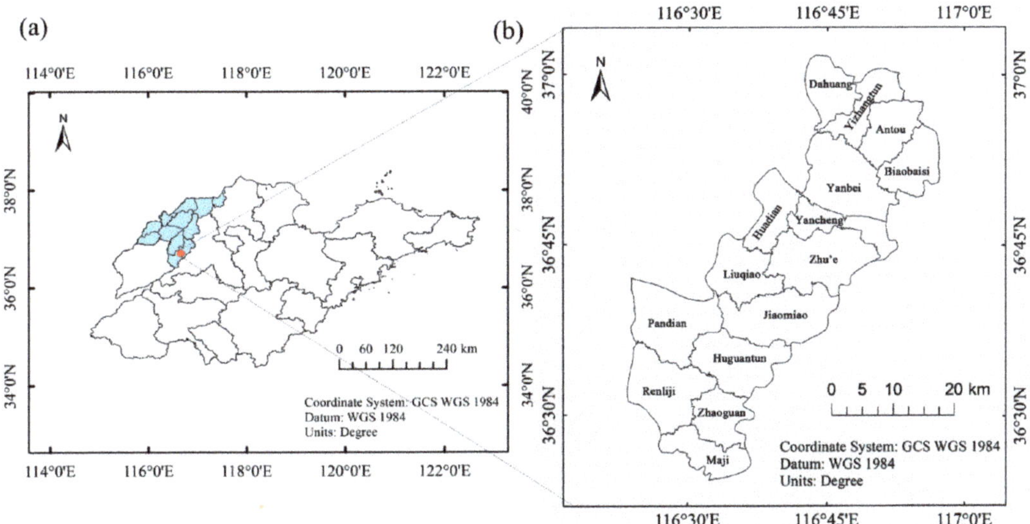

Figure 1. Research area: (**a**) Shandong Province; (**b**) Qihe County.

2.2. Data Source and Preprocessing

The GF-6 was successfully launched on 2 June 2018, and is mainly used in precision agriculture observation and forestry resource investigation. An 8-band complementary metal-oxide-semiconductor detector was employed in China, equipped with a 2 m panchromatic/8 m multi-spectral high-resolution camera and a 16 m multi-spectral medium-resolution wide field of view (WFV) camera. The details of GF6-WFV are shown in Table 1. For the first time in China, the "red-edge" band, which can effectively reflect the unique spectral characteristics of crops, was added, which has greatly improved the monitoring of agriculture, forestry, grassland, and other resources.

Table 1. Parameters of Gaofen 6 (GF-6) wide field of view (WFV).

	Parameters	Spectral Range (nm)	Spatial Resolution (m)	Swath Width (km)
WFV	B1 (Blue)	450–520	16	800
	B2 (Green)	520–590		
	B3 (Red)	630–690		
	B4 (Near Infrared)	770–890		
	B5 (Red-edge 1)	690–730		
	B6 (Red-edge 2)	730–770		
	B7 (Purple)	400–450		
	B8 (Yellow)	590–630		

The image of the research area taken on 9 September 2018, was selected for analysis, as this date corresponds to the maize filling period. The 1A-level image downloaded from the China Centre for Resources Satellite Data and Application (CCRSDA) (http://www.cresda.com/CN/index.shtml (accessed on 7 February 2020)) must be preprocessed by radiometric calibration, atmospheric correction, and orthorectification [25], and all preprocessing performed in ENVI (Version 5.3, Research System Inc., Boulder, CO, USA). Atmospheric correction was performed using the fast line-of-sight atmospheric analysis of the spectral hypercubes model [26], and the spectral response function was provided by the CCRSDA. The rational polynomial coefficients model based on rational functions was used to further orthorectify without control points. The 2019 agricultural census data including planting area and yields of maize were obtained from the local government website (http://dztj.dezhou.gov.cn/n3100530/n3100065/index.html (accessed on 27 April 2020)), and the administrative boundary vector data of the study area were downloaded from Resource and Environmental Science and Data Center (http://www.resdc.cn/data.aspx?DATAID=202 (accessed on 7 February 2020)). SuperMap (iDEesktop 8C, SuperMap Software Co., Ltd., Beijing, China) was used to process these data and transform the coordinate system. All spatial data were converted into the universal transverse Mercator (WGS84 UTM 45N) projection.

Two types of samples were used in this study, namely training and verification samples, most of which were obtained through ground surveys using OvitalMap (V8.7.1, Beijing Ovital Software Co., Ltd., Beijing, China) in June 2018. In addition, with the support of higher spatial resolution image data, historical data, and expert knowledge, we also acquired a portion of training samples through manual visual interpretation. A total of 689 samples were acquired in this study, including maize, buildings, woodlands, wastelands, water, roads, and other crops. According to the proportions of different land cover types, 250 samples were randomly selected as verification samples and the rest were used for training samples. The training samples were used to classify land cover types in the research area. The supervised classification method was used to obtain the planting area of maize, from which the yield and distribution of maize straw were estimated. The verification sample was used to evaluate the classification accuracy of different land cover types. All samples were randomly collected to cover the entire study area as much as possible and they were quadrats of single crops to better avoid noise and ensure classification accuracy.

When the maize was being harvested in October 2018, three quadrats of 5 m × 10 m were selected in the southern, central, and northern regions of Qihe County, respectively, to count the number of the maize planted and the weight of straw (15% moisture content). These data will be used for the estimation of maize straw yield.

2.3. Classification of Land Cover Types Using Different Bands Combinations

The maize in the research area of the acquired image was in the grain-filling stage, and the main land cover types were determined by ground investigation as maize, buildings, wasteland, water, woodland, roads, vegetables, cotton, and soybean. Soybean and cotton were planted less than the other crops, and their spectral characteristics were similar to those of vegetables. Hence, vegetables, cotton, soybeans, and a very small number of other crops were classified as other crop types, and the identification and statistics of maize were the focus of classification in this study. In summary, we divided the study area into seven final land cover types, including maize, buildings, woodland, wasteland, water, roads, and other crops. Firstly, layer stacking was performed on the preprocessed image, and five schemes (Table 2) were designed for the newly added bands for experimentation. Second, two types of machine learning—SVM [27] and RF [28]—were used to classify the research area [29]. Finally, the classification results of the two classification methods were analyzed on the influence of the red-edge, purple, and blue bands on the recognition of various land cover types to verify the improvement of the classification accuracy of the newly added band of GF-6 WFV compared to GF1/WFV.

Table 2. Classification schemes with different band combinations.

Schemes	Operating Bands
S1	B1, B2, B3, B4
S2	B1, B2, B3, B4, B5
S3	B1, B2, B3, B4, B6
S4	B1, B2, B3, B4, B5, B6
S5	B1, B2, B3, B4, B5, B6, B7, B8

SVM is based on a statistical learning theory, trying to find an optimal hyperplane as a decision function in high-dimensional space. The number of free parameters used in the SVM does not depend on the number of input features, and the reduction in the number of features is not required to avoid overfitting. SVM provides a generic mechanism to fit the surface of the hyperplane to the data through the use of a kernel function, such as linear, polynomial, or sigmoid curve. RF is a combination of tree predictors which exhibits superior performance in cases with noise and weak discrimination data and is insensitive to the initialization of parameters [30]. Compared to SVM, the number of user-defined parameters in RF is less than the number required for SVMs and easier to define. In this paper, the training of SVM with a linear kernel was performed. ENMAP-BOX [31] was used for RF classification.

2.4. Classes Separability Assessment

Class separability, which is a measure of similarity between classes, can be determined from these values. There are four widely used quantitative measures for class separability: divergence, transformed divergence (TD), Bhattacharyya distance, and Jeffries–Matusita distance (JM) [32]. Divergence is one of the most popular separability measures used in remote sensing, which can be calculated by the mean and variance-covariance matrices of the data representing feature classes. The TD is the standardized form of divergence, which can minimize the effect of several well-separated classes that may increase the average divergence value and make the divergence measure misleading. The Bhattacharyya distance and the JM can be used to estimating the probability of correct classification, and the JM can suppress high separability values by transforming the Bhattacharyya distance values to a specific range.

2.5. Maize Straw Estimation

The goal of county-level straw estimation was to determine the type and quantity of straw resources. The total theoretical quantity of straw was considered the maximum quantity that can be produced in a certain area each year [33]. This value was estimated by taking the crop planting area and straw resource density, and the equation used is as follows:

$$P_T = \sum_{i=1}^{n} D_i \cdot A_i \tag{1}$$

where: P_T is the theoretical total quantity of straw (t); i is the number of different crop straws, and the maize straw was counted in this study, thus, $i = 1$; D_i indicates the straw resource density of the ith crop (t/km^2), and A_i is the plantation area of the ith crop (km^2).

$$D_i = 1000 \times \left(\sum_{j=1}^{n} \frac{C_{ij}}{S_{ij}} \right) / j \tag{2}$$

where: j is a different sampling region; C_{ij} is the theoretical total quantity of straw resources of the ith crop in area j (kg); S_{ij} is the plantation area of ith crop in area j (m^2).

2.6. Accuracy Verification

A confusion matrix [34] was used to evaluate the accuracy of the classification results based on the verification samples of the ground survey. The evaluation indicators include overall accuracy (OA), user accuracy (UA), production accuracy (PA), and kappa coefficient (KC). The OA and KC reflect the overall classification effect, while the PA and UA represent omission and misclassification errors, respectively. Since accuracy is not necessarily normally distributed, the non-parametric Wilcoxon test for paired samples was conducted to evaluate the changes in OA and PA of each land cover type between different bands combination. Besides, the planting area of maize can also be verified by statistical census data.

According to the yield of maize and the straw-grain ratio, the straw yield of maize could be calculated; that is, the theoretical total quantity of maize straw used as validation data was the product of maize yield and straw-grain ratio, and the maize yield was obtained using annual census data [35].

3. Results and Discussion

3.1. Band Reflectivity Analysis

All training samples (179, 68, 50, 19, 28, 49, and 46 samples of maize, buildings, woodlands, wastelands, water, roads, and other crops) were used to perform pixel information statistics for different feature types, and reflectivity curves of different land cover types were drawn, as shown in Figure 2. Vegetation and non-vegetation have significant differences in spectral characteristics. We found that buildings, roads, and wasteland exhibit higher reflectivity in visible wavelengths (B1–B3 and B7–B8), and buildings and other land cover types were strongly separated. Roads and wasteland were significantly different in B4–B6, which can be easily distinguished. Water exhibits strong NIR waves absorption, so the reflectance of water in B4 is lower than that of all other land cover types [36]. Maize, woodland, and other plants exhibit low reflectivity of visible wavelengths due to the absorption of chlorophyll, and their spectral characteristics are similar [37]. Thus, it is difficult to distinguish each plan type, although the reflectivity is higher in B4 (NIR) and B6 (Red-edge 2). In B4, the reflectivity of different land cover types was as follows: other plants > maize > wasteland > woodland > roads > building > water. The reflectivity of wasteland and woodland were similar, and the difference between roads and buildings was also small, but both were easily distinguished in B1–B3 and B5, as they have good separation.

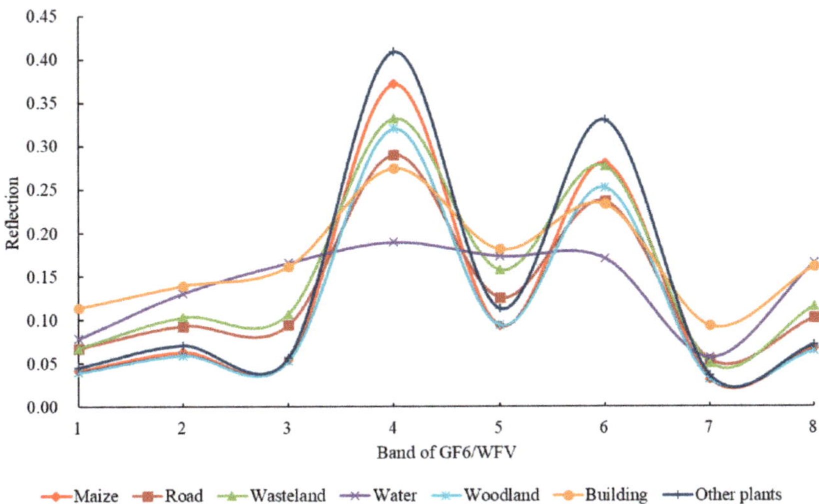

Figure 2. Spectral curves of different land cover types.

3.2. Class Separability

In this study, JM and TD were used to measure the separability between maize and other land cover types in the research area. The range of JM and TD are within [0, 2]. As a general rule, values in the range of [0.0, 1.0) indicate a very poor class separability; values in the range of [1.0, 1.9) indicate a poor separability; and values in the range of [1.9, 2.0] indicate a relatively good separability.

The class separability of the samples in Table 3 was analyzed, and we found that the separability of different land cover types was significantly different, in whether there was the participation of the new band of GF-6 satellite. Compared with S1, the JM and TD between maize and other plants in S2 increased from 1.32 and 1.43 to 1.80 and 1.97, respectively, and the values between maize and woodland also increased from 1.43 and 1.81 to 1.67 and 1.95, which indicated that B5 can significantly enhance the separability of maize from other plants and wood-lands, but the spectra between them still exhibit a large overlap. By comparing the JM between maize and other land cover types in S1 and S2, we found that B5 also contributes partly to the separability of maize from wasteland and roads. The JM between maize and all land cover types in S3 was unchanged or decreased than that in S2, except between maize and woodland, which indicated that the contribution of B6 to the separability between maize and other land cover types was less than that of B5, but the contribution to the separability between maize and woodland was greater than that of B5. To sum up, the superposition of B5 and B6 can further increase the separability of maize and other land cover types, specifically for the distinction between maize and woodland. In S2, the JM between maize and all other land cover types was greater than 1.8, and the TD was more than 1.9, indicating that when the red-edge wavelength was involved in the calculation, the separability of maize and other land cover types was very high. The purple and yellow bands added in S4 and S5, respectively, can increase the separability of maize and other land cover types in a certain range because they still slightly increase JD, but their effect on improving TD was not obvious.

Table 3. Separability between maize and other land cover types under different schemes (S1–S5).

	S1		S2		S3		S4		S5	
	JM	TD	JM	TD	JM	TD	JM	TD	JM	TD
Building	2.00	2.00	2.00	2.00	2.00	2.00	2.00	2.00	2.00	2.00
Woodland	1.43	1.81	1.67	1.95	1.77	1.95	1.83	1.98	1.86	1.99
Other plants	1.32	1.43	1.80	1.97	1.71	1.97	1.87	1.99	1.89	2.00
Water	2.00	2.00	2.00	2.00	2.00	2.00	2.00	2.00	2.00	2.00
Wasteland	1.87	2.00	1.99	2.00	1.97	2.00	1.99	2.00	2.00	2.00
Road	1.85	2.00	1.91	2.00	1.88	2.00	1.92	2.00	1.95	2.00

3.3. Maize Identification and Classification Results

Using the same training samples, SVM and RF classification models were used to classify the remote sensing image under the S1–S5 schemes. The classification results are shown in Table 4.

Table 4. Classification results based on support vector machine (SVM) and random forest (RF) models in Qihe County.

		S1	S2	S3	S4	S5
SVM	OA	84.05%	90.02%	89.80%	91.10%	91.53%
	KC	0.80	0.87	0.87	0.88	0.89
RF	OA	85.57%	93.24%	87.38%	93.15%	94.18%
	KC	0.82	0.91	0.84	0.91	0.92

The OA of the SVM and RF models under the S1 scheme was quite low at 84.05% and 85.57%, respectively, and the KC was 0.80 and 0.82, respectively. Under the S2 scheme, the classification accuracy of the two classification methods was significantly improved, reaching more than 90%; specifically, the OA of the RF model was 93.24%, and the KC was 0.91. This indicates that Red-edge 1 can effectively improve the classification accuracy of SVM and RF models in the research area. The OA of SVM in the S3 scheme was lower than that in the S2 scheme but remained 5.75% higher than that in the S1 scheme. The total accuracy of RF in the S3 scheme exhibited a significant decline and was far lower than that of the SVM model, but it remained 1.81% higher than that of the S1 scheme. These results indicate that Red-edge 2 can effectively improve the recognition ability of the SVM model on land cover types, but the improvement is lower than that of Red-edge 1. This may be because Red-edge 2 is significantly correlated with NIR (B4; $R^2 = 0.991$; Figure 3a), resulting in feature redundancy. The classification accuracy of the SVM model did not change significantly in the S4 and S5 schemes, and the OA was maintained at approximately 91%, with a KC of 0.89. The OA of the RF model in the S4 scheme increased rapidly compared with the S3 scheme and exceeded the classification result accuracy of the SVM model. The OA was 93.15%, and the KC was 0.91, which was almost the same as the result of the S2 scheme. The OA of the RF model in the S5 scheme was the highest, but it did not change significantly compared with the S4 scheme. The results showed that purple and yellow bands had a limited influence on the classification results. We speculate that one of the reasons may be that B7 was significantly correlated with B1 ($R^2 = 0.986$; Figure 3b), and B8 was significantly correlated with B2 and B3 ($R^2 = 0.982$ and $R^2 = 0.981$; Figure 3c,d).

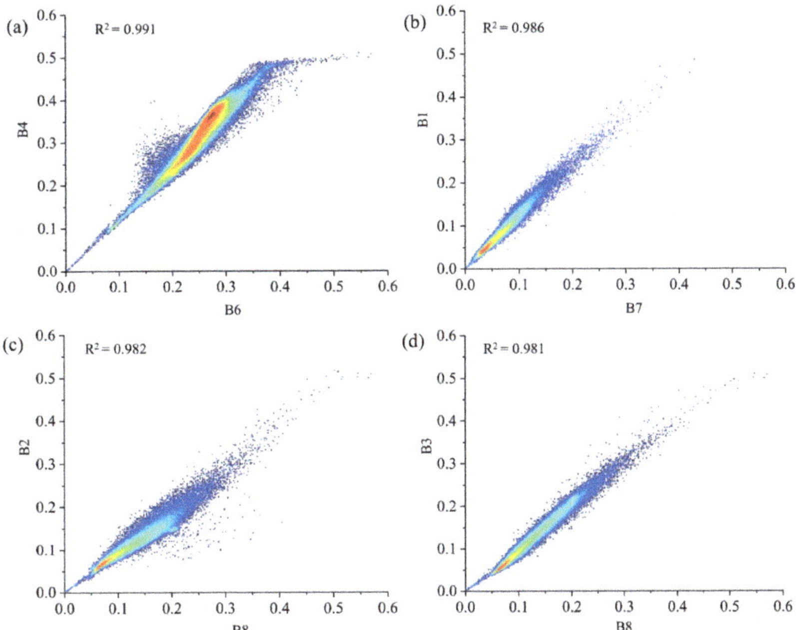

Figure 3. Correlation analysis between various bands: (**a**) B4 and B6; (**b**) B1 and B7; (**c**) B2 and B8; (**d**) B3 and B8.

The impact of the newly added bands on the classification results of various land cover types was specifically analyzed, and a confusion matrix was established for the RF model results under the S1–S5 scheme through verification samples (Table 5). Difference analysis results for each land cover type between different band combinations are shown in Table 6.

The generated confusion matrix indicates that the PA of buildings and water under the S1 scheme was higher, reaching 97.50% and 97.49%, respectively. However, a misclassification between other crops and maize was observed, and the PA of maize is 82.36%. Furthermore, many wastelands and roads were mistakenly divided into buildings. From Figure 4b,h, we intuitively found that there were a large number of road and woodland spots in the RF classification results, as well as a large number of misclassified wastelands.

Under the S2 scheme, with the addition of Red-edge 1, the classification results were significantly improved, especially for woodlands and other plants, and the PA increased from 84.45% and 59.31% to 90.29% and 72.07%, respectively. In particular, the smoothness of the classification results of woodland on both sides of the road was improved (Figure 4c). The classification accuracy of maize reached 97.57%, which was an increase of 15.21% compared to the S1 scheme. However, the classification effect of roads was not significantly improved, and a large number of misclassified discontinuous roads were mixed in the maize fields. It showed that the Red-edge 1 band significantly contributed to vegetation classification, but its role in the recognition of non-vegetation land cover types was limited.

The classification results of S3 with the addition of Red-edge 2 were similar to those of S1 without a significant difference. Although the PA of maize, woodland, and other crops was improved, the PA of water bodies and roads decreased. These results indicated that Red-edge 2 had no obvious effect on improving the classification of land cover types in the research area, particularly for non-vegetation. We intuitively determined that the addition of Red-edge 2 significantly improved the recognition of continuous roads (Figure 4d) and the removal of wasteland patches (Figure 4f). However, the fragmentation of the roads in the field was not reduced. This may be because the resolution of the remote sensing image used was too low to identify a narrow road in the field.

The classification results of the S4 scheme with two red-edge bands were similar to those of the S2 scheme, but the misclassification and omission of almost all vegetation were significantly reduced—specifically, the ability to distinguish maize from other plants was very strong. The classification results exhibited better completeness, and the continuity and smoothness of the boundary of the spots were also improved (Figure 4e). However, Figure 4k shows that discontinuous road spots remained in the classification results. This indicates that when two red-edge bands exist simultaneously, Red-edge 1 plays a major role in classifying land cover types, and Red-edge 2 can be superimposed to improve the classification accuracy of buildings and wasteland.

Table 5. Confusion matrix for verifying classification accuracy of RF model (unit: %).

		Building	Woodland	Other Plants	Water	Wasteland	Road	Maize
	Building	97.50	0	0	0.95	16.91	12.17	0
	Woodland	0.16	84.45	0.31	0.08	3.76	0.58	3.91
	Other plants	0	0	59.31	0.01	0	0	12.27
S1	Water	0	7.88	0.09	97.49	1.36	0	0
	Wasteland	1.61	1.17	21.58	0.08	64.65	2.63	0.17
	Road	0.73	4.46	0.09	1.20	11.78	83.78	1.29
	Maize	0	2.04	18.62	0.18	1.53	0.85	82.36
	Building	97.69	0	0	0.83	15.28	12.48	0
	Woodland	0.18	90.29	1.50	0.07	3.22	0	1.47
	Other plants	0	0	72.07	0.01	0	0	0.08
S2	Water	0	0.73	0	97.46	1.53	0	0
	Wasteland	1.55	1.27	18.18	0.07	71.63	2.05	0.01
	Road	0.58	6.17	1.77	1.37	6.11	84.09	0.86
	Maize	0	1.53	6.49	0.18	2.24	1.38	97.57
	Building	97.81	0	0	1.05	20.73	12.48	0
	Woodland	0.05	84.59	0.44	0.13	2.62	0	2.02
	Other plants	0	0.05	59.8	0	0	0	9.8
S3	Water	0	8.14	0	97.44	0.93	0	0
	Wasteland	1.55	1.06	21.76	0.07	65.19	2.63	0.12
	Road	0.6	4.74	0.31	1.14	9.00	83.42	1.17
	Maize	0	1.41	17.7	0.17	1.53	1.47	86.89
	Building	97.97	0	0	1.03	20.68	12.39	0
	Woodland	0.05	88.41	0.71	0.07	2.89	0.04	1.33
	Other plants	0	0.02	77.14	0.01	0	0	0.29
S4	Water	0	2.86	0.04	97.46	1.96	0	0
	Wasteland	1.55	1.31	18.09	0.07	68.63	2.32	0.02
	Road	0.42	5.94	0.44	1.17	3.76	83.91	0.81
	Maize	0.02	1.46	3.57	0.18	2.07	1.34	97.55
	Building	97.42	0	0	0	14.72	2.81	0
	Woodland	0.07	87.00	0.18	0	2.08	0	1.00
	Other plants	0	0.02	70.93	0	0	0	0.20
S5	Water	0	10.34	0.18	100	3.12	0	0
	Wasteland	1.47	1.17	19.35	0	75.97	2.08	0
	Road	1.04	0.67	0.57	0	1.59	94.89	0.52
	Maize	0	0.81	8.80	0	2.52	0.23	98.28

Table 6. Difference analysis results for each land cover type between different band combinations.

No.	Contrast of Bands Different Combination	*p*-Value
1	S2 to S1	0.0156
2	S3 to S1	0.0781
3	S4 to S1	0.0156
4	S5 to S1	0.0156
5	S3 to S2	0.0234
6	S4 to S2	0.6875
7	S5 to S2	0.3828
8	S4 to S3	0.0078
9	S5 to S3	0.0156
10	S5 to S4	0.3828

Significance levels: 0.05.

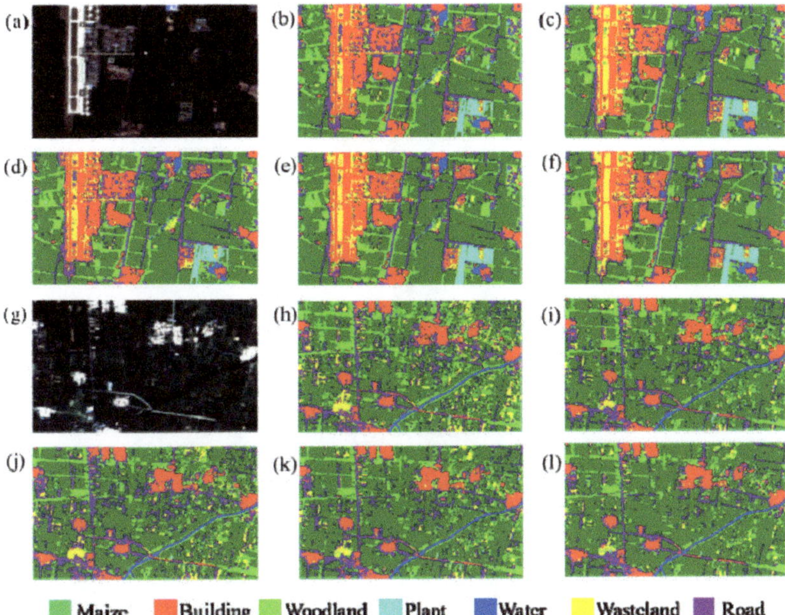

Figure 4. Classification results of RF models under different band combination schemes: (**a**) Original image of area A; (**b**–**f**) Classification results of S1–S5 in area A; (**g**) Original image of area B; (**h**–**l**) Classification results of S1–S5 in area B.

The S5 scheme with the purple and yellow bands added to the S4 scheme is a remote sensing image that includes all bands of GF-6 WFV and the classification results were not significantly improved. But compared with S4, the PA of roads, wastelands, and water was significantly improved from 83.91%, 68.63%, and 97.46% to 94.89%, 75.97%, and 100%, respectively, indicating that the purple and yellow bands were more likely to respond to non-vegetation and increase their classification accuracy, but the PA of woodland and other vegetation decreased. From the perspective of the entire research area, the purple and yellow bands effectively reduced the "salt-and-pepper phenomenon" in the classification results (Figure 4f,l), thereby improving the accuracy of maize producers to 98.28%.

To sum up, we found that the overall classification accuracy was the best when the RF model was used in the S5 scheme, and the land cover types of Qihe County were extracted as shown in Figure 5. The planting area of maize was estimated with the class statistical tool in ENVI, which was 774.18 km^2. According to agricultural statistics, the maize planting area in Qihe County in 2018 was 757.73 km^2, which falls within a 2.17% error from our calculated value. As shown in Figure 6, the maize plantation area was the largest, accounting for 54.87% in Qihe County in 2018. These results indicate that maize is a primary food crop in Qihe County; this facilitates a notably high quantity of maize straw being produced in the region, which exhibits great potential for recycling and utilization.

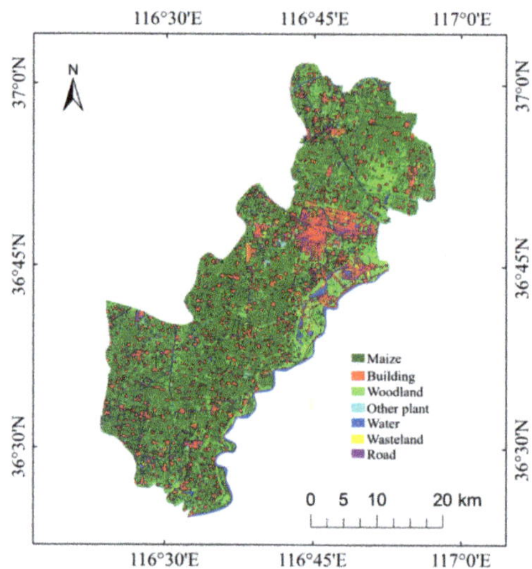

Figure 5. Land cover types in Qihe County based on the RF model in 2018.

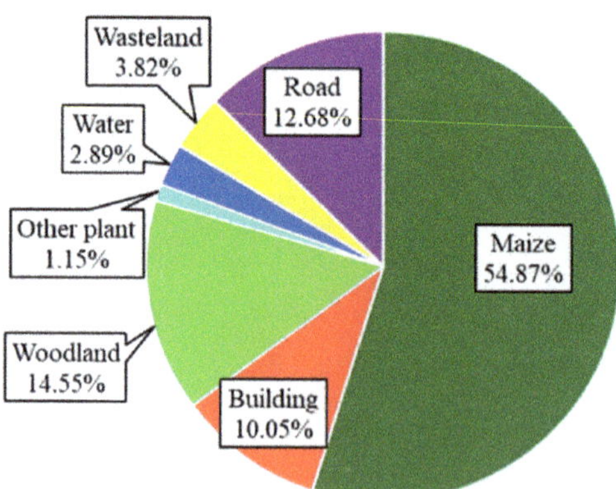

Figure 6. Land cover type proportions based on the RF model in 2018.

3.4. Spatial Distribution of Maize Straw

Averaging the field survey data in Table 7, the density of maize straw was calculated. According to Equation (1), the spatial distribution of maize straw in 2018 (Figure 7) was obtained by multiplying the planting area (each pixel is 16 m × 16 m) with the density of straw. The distribution of maize straw was the widest in southern and northeastern Qihe County, and the average distribution density in the northernmost and southernmost regions was slightly lower. The central and northern areas exhibit the lowest quantities and average densities because these areas are characterized by numerous towns.

Table 7. The density of maize straw in the sampling area.

No.	Area (m × m)	No. of Plants	Weight of Straw (kg)	Density of Straw (t/km^2)
1	10 × 5	336	40.24	804.8
2	10 × 5	308	35.63	712.6
3	10 × 5	322	37.46	749.2
		Average		755.53

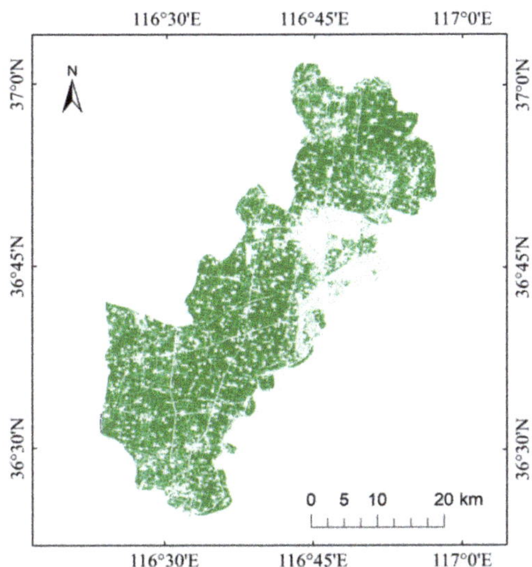

Figure 7. Spatial distribution of maize straw in Qihe County in 2018.

The total quantity and average density of maize straw in each township were further evaluated, as shown in Figure 8. Although the straw quantities of Liuqiao and An'tou were 47.12 kt and 39.07 kt, respectively, because of the small area of cultivated land, the average maize straw density was 543.46 t/km^2 and 525.58 t/km^2, respectively, so these two towns were very suitable for maize straw recycling and utilization. Larger townships in the central and southern regions, including Pandian, Renliji, Jiaomiao, and Huguantun, had relatively large maize straw yields and relatively high average density, so the potential for the utilization of maize straw resources was also notable. In northern towns such as Dahuang, Yizhangtun, Biaobaisi, and Huadian and the southern towns such as Zhaoguan and Maji, the total theoretical maize straw quantity was 25–32 kt, with the average density of straw being 380–470 t/km^2, which indicates that these towns were also suitable for maize straw recycling and utilization. Yancheng had a small proportion of cultivated land and the lowest average density and yield of maize straw. Due to the development of towns and secondary industries, the average density of maize straw in Zhu'A and Yanbei was also low, indicating that the utilization potential of straw resources is relatively lesser.

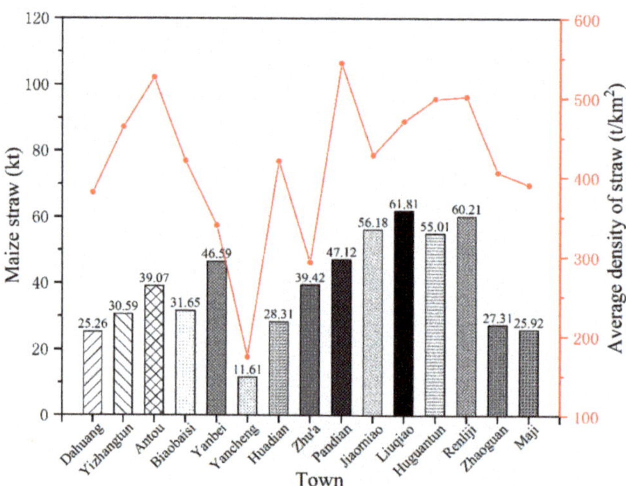

Figure 8. Statistics of maize straw in each township of Qihe County in 2018.

4. Conclusions

The results of this study show that it is feasible to use SVM and RF models to estimate the yield and spatial distribution of maize straw by combining GF6/WFV, field survey data, and agricultural census. The land cover types in Qihe County were more accurately identified by the RF model, which consequently improved the estimation accuracy of the yield and spatial distribution of the maize straw. The main conclusions are as follows:

1. Both SVM and RF models can effectively identify and aid in the classification of land cover types in the research area. The RF model exhibits improved classification accuracy compared to that of SVM when the newly added band of GF-6 WFV was used;

2. The addition of two red-edge bands increased the separability of land cover types with large differences in red-side spectral characteristics and generally significantly improved the overall classification accuracy and reduced the misclassification and omission of crops. Red-edge 1 can improve the recognition accuracy of land cover types more than Red-edge 2 in Qihe County. In this study, the classification accuracy and KC of the RF model increased from 85.57% and 0.82 to 93.15% and 0.91, respectively, after adding two red-edge bands;

3. The response of purple and yellow bands to non-vegetation was more obvious than that to vegetation, which increased the classification accuracy of non-vegetation and slightly reduced the "salt-and-pepper noise" in the classification results. However, the effects of the two bands on the classification accuracy of vegetation and the total classification accuracy were not obvious;

4. The theoretical total quantity of maize straw in Qihe County was 586.08 kt in 2018, which reflected only a 2.42% error from the statistical result. Maize straw in Qihe County was planted, excluding in central and northern urban areas. Among them, the southern and northeastern regions exhibited the widest distribution areas and highest average densities, followed by the northernmost and southernmost regions. The central and northern urban areas exhibited the lowest average distribution densities.

5. Future Work

The research method of this study still exhibits some limitations that should be addressed. After inspection, it was found that the misjudged pixels were primarily those located at the junctions of various land cover types. Many mixed pixels that were difficult to distinguish, even by visual interpretation, were formed in the image because of the superimposed spectral characteristics of different land cover types at the junctions [38]. Concurrently, the spatial resolution of the remote sensing image also limited the accuracy of extraction to a large extent [39]. Therefore, the focus of further research is to fuse higher resolution spatial information, and a discriminant model based on spatial relationship knowledge and mixed pixel decomposition will also be considered to improve extraction accuracy.

Author Contributions: The manuscript was written through the contributions of all authors. Conceptualization, Y.Z. and R.D.; Data curation, H.M.; Formal analysis, H.M. and H.L.; Funding acquisition, Y.Z.; Investigation, H.L.; Methodology, H.M.; Supervision, R.D.; Writing—original draft, H.M.; Writing—review and editing, H.L. and Y.Z. All authors have read and agreed to the published version of the manuscript.

Funding: This study was supported by the National Natural Science Foundation of China (Grant No. U20A2086; 51806242), the Special Project on Innovation Methodology, Ministry of Science and Technology of China (No. 2020IM020900); and the Yantai Educational-Local Synthetic Development Project (Grant No. 2019XDRHXMXK25 and No. 2019XDRHXMQT36).

Institutional Review Board Statement: Not applicable.

Informed Consent Statement: Not applicable.

Data Availability Statement: The data that supported the findings of this study can be available from the corresponding author upon reasonable request.

Acknowledgments: We appreciate the supports from the Key Laboratory of Clean Production and Utilization of Renewable Energy, Ministry of Agriculture and Rural Affairs, at China Agricultural University; the National Joint R&D Center for International Research of BioEnergy Science and Technology, Ministry of Science and Technology, at China Agricultural University; the National Energy R&D Center for Biomass, National Energy Administration of China, at China Agricultural University; and Beijing Municipal Key Discipline of Biomass Engineering.

Conflicts of Interest: The authors declare no competing financial and non-financial interests.

References

1. Jiao, X.; Mongol, N.; Zhang, F. The transformation of agriculture in China: Looking back and looking forward. *J. Integr. Agric.* **2018**, *17*, 755–764. [CrossRef]
2. El-Dewany, C.; Awad, F.; Zaghloul, A.M. Utilization of rice straw as a low-cost natural by-product in agriculture. *Int. J. Environ. Pollut. Environ. Model.* **2018**, *1*, 91–102.
3. Nie, P.; Sousa-Poza, A.; Xue, J. Fuel for life: Domestic cooking fuels and women's health in rural China. *Int. J. Environ. Res. Public Health* **2016**, *13*, 810. [CrossRef]
4. Cui, M.; Zhao, L.; Tian, Y.; Meng, H.; Sun, L.; Zhang, Y.; Wang, F.; Li, B. Analysis and evaluation on energy utilization of main crop straw resources in China. *Trans. Chin. Soc. Agric. Eng.* **2008**, *24*, 291–296. (In Chinese)
5. Wang, Y.; Bi, Y.; Gao, C. The assessment and utilization of straw resources in China. *Agric. Sci. China* **2010**, *9*, 1807–1815. [CrossRef]
6. Ren, J.; Yu, P.; Xu, X. Straw utilization in China—Status and recommendations. *Sustainability* **2019**, *11*, 1762. [CrossRef]
7. Hong, J.; Ren, L.; Hong, J.; Xu, C. Environmental impact assessment of corn straw utilization in China. *J. Clean. Prod.* **2016**, *112*, 1700–1708. [CrossRef]
8. Wang, B.; Shen, X.; Chen, S.; Bai, Y.; Yang, G.; Zhu, J.; Shu, J.; Xue, Z. Distribution characteristics, resource utilization and popularizing demonstration of crop straw in southwest China: A comprehensive evaluation. *Ecol. Indic.* **2018**, *93*, 998–1004. [CrossRef]
9. Abdel-Mohdy, F.A.; Abdel-Halim, E.S.; Abu-Ayana, Y.M.; El-Sawy, S.M. Rice straw as a new resource for some beneficial uses. *Carbohyd. Polym.* **2009**, *75*, 44–51. [CrossRef]
10. Venturini, G.; Pizarro-Alonso, A.; Münster, M. How to maximise the value of residual biomass resources: The case of straw in Denmark. *Appl. Energy* **2019**, *250*, 369–388. [CrossRef]
11. Bi, Y.Y.; Wang, Y.J.; Gao, C.Y. Straw Resource quantity and its regional distribution in China. *J. Agric. Mech. Res.* **2010**, *3*, 1–7.
12. Long, H.; Li, X.; Wang, H.; Jia, J. Biomass resources and their bioenergy potential estimation: A review. *Renew. Sust. Energy Rev.* **2013**, *26*, 344–352. [CrossRef]
13. Moxey, A.; Mcclean, C.J.; Allanson, P. Transforming the spatial basis of agricultural census cover data. *Soil Use Manag.* **2010**, *11*, 21–25. [CrossRef]

14. Galanopoulos, C.; Odierna, A.; Barletta, D.; Zondervan, E. Design of a wheat straw supply chain network in Lower Saxony, Germany through optimization. *Comput. Aided Chem. Eng.* **2017**, *40*, 871–876.

15. Zhou, L.; Gu, W.; Zhang, Q. Logistics mode and network planning for recycle of crop straw resources. *Asian Agric. Res.* **2013**, *5*, 87.

16. Ballesteros, R.; Ortega, J.F.; Hernandez, D.; Del Campo, A.; Moreno, M.A. Combined use of agro-climatic and very high-resolution remote sensing information for crop monitoring. *Int. J. Appl. Earth Obs. Geoinf.* **2018**, *72*, 66–75. [CrossRef]

17. Kern, A.; Barcza, Z.; Marjanović, H.; Árendás, T.; Fodor, N.; Bónis, P.; Bognár, P.; Lichtenberger, J. Statistical modelling of crop yield in Central Europe using climate data and remote sensing vegetation indices. *Agric. For. Meteorol.* **2018**, *260*, 300–320. [CrossRef]

18. Atzberger, C. Advances in remote sensing of agriculture: Context description, existing operational monitoring systems and major information needs. *Remote Sens.* **2013**, *5*, 949–981. [CrossRef]

19. Duveiller, G.; Defourny, P. A conceptual framework to define the spatial resolution requirements for agricultural monitoring using remote sensing. *Remote Sens. Environ.* **2010**, *114*, 2637–2650. [CrossRef]

20. Vrieling, A.; Skidmore, A.K.; Wang, T.; Meroni, M.; Ens, B.J.; Oosterbeek, K.; O'Connor, B.; Darvishzadeh, R.; Heurich, M.; Shepherd, A. Spatially detailed retrievals of spring phenology from single-season high-resolution image time series. *Int. J. Appl. Earth Obs. Geoinf.* **2017**, *59*, 19–30. [CrossRef]

21. Wang, L.; Liu, J.; Yang, F.; Yao, B.; Shao, J.; Yang, L. Rice recognition ability basing on GF-1 multi-temporal phases combined with near infrared data. *Trans. Chin. Soc. Agric. Eng.* **2017**, *33*, 196–202. (In Chinese)

22. Li, S.; Zhao, J.; Dong, S.; Zhao, M.; Li, C.; Cui, Y.; Liu, Y.; Gao, J.; Xue, J.; Wang, L. Advances and prospects of maize cultivation in China. *Sci. Agric. Sin.* **2017**, *50*, 1941–1959.

23. Xu, W. A LM-2D Launches GF-6 Satellite. *Aerosp. China* **2019**, *19*, 60.

24. Tong, X.; Zhao, W.; Xing, J.; Fu, W. In Status and development of china high-resolution earth observation system and application. In Proceedings of the 2016 IEEE International Geoscience and Remote Sensing Symposium (IGARSS), Beijing, China, 10–15 July 2016; pp. 3738–3741.

25. Young, N.E.; Anderson, R.S.; Chignell, S.M.; Vorster, A.G.; Lawrence, R.; Evangelista, P.H. A survival guide to Landsat preprocessing. *Ecology* **2017**, *98*, 920–932. [CrossRef]

26. Lantzanakis, G.; Mitraka, Z.; Chrysoulakis, N. Comparison of physically and image based atmospheric correction methods for Sentinel-2 satellite imagery. In *Perspectives on Atmospheric Sciences*; Bais, A., Nastos, P., Eds.; Springer International Publishing: Cham, Switzerland, 2017; pp. 255–261.

27. Mountrakis, G.; Im, J.; Ogole, C. Support vector machines in remote sensing: A review. *ISPRS J. Photogramm. Remote Sens.* **2011**, *66*, 247–259. [CrossRef]

28. Breiman, L. Random forests. *Mach. Learn.* **2001**, *45*, 5–32. [CrossRef]

29. Thanh Noi, P.; Kappas, M. Comparison of random forest, k-nearest neighbor, and support vector machine classifiers for land cover classification using Sentinel-2 imagery. *Sensors* **2018**, *18*, 18. [CrossRef]

30. Du, P.; Xia, J.; Chanussot, J.; He, X. Hyperspectral remote sensing image classification based on the integration of support vector machine and random forest. In Proceedings of the 2012 IEEE International Geoscience and Remote Sensing Symposium, Munich, Germany, 22–27 July 2012; pp. 174–177.

31. Van der Linden, S.; Rabe, A.; Held, M.; Jakimow, B.; Leitão, P.J.; Okujeni, A.; Schwieder, M.; Suess, S.; Hostert, P. The EnMAP-Box—A Toolbox and Application Programming Interface for EnMAP Data Processing. *Remote Sens.* **2015**, *7*, 11249–11266. [CrossRef]

32. Gambarova, Y.M.; Gambarov, A.Y.; Rustamov, R.B.; Zeynalova, M.H. Remote sensing and GIS as an advance space technologies for rare vegetation monitoring in Gobustan State National Park, Azerbaijan. *J. Geogr. Inf. Syst.* **2010**, *2*, 93. [CrossRef]

33. Weiser, C.; Zeller, V.; Reinicke, F.; Wagner, B.; Majer, S.; Vetter, A.; Thraen, D. Integrated assessment of sustainable cereal straw potential and different straw-based energy applications in Germany. *Appl. Energy* **2014**, *114*, 749–762. [CrossRef]

34. Tharwat, A. Classification assessment methods. *Appl. Comput. Inform.* **2020**, *17*, 168–192. [CrossRef]

35. Ai, B.; Sheng, Z.; Zheng, L.; Shang, W. Collectable amounts of straw resources and their distribution in China. In *Advances in Engineering Research, Proceedings of the International Conference on Advances in Energy, Environment and Chemical Engineering, Changsha, China, 26–27 September 2015*; Atlantis Press: Paris, France, 2015; pp. 441–444.

36. Jawak, S.D.; Luis, A.J. Improved land cover mapping using high resolution multiangle 8-band WorldView-2 satellite remote sensing data. *J. Appl. Remote Sens.* **2013**, *7*, 73573. [CrossRef]

37. Martinez, L.J.; Ramos, A. Estimation of chlorophyll concentration in maize using spectral reflectance. *Int. Arch. Photogramm. Remote Sens. Spat. Inf. Sci.* **2015**, *40*, 65. [CrossRef]

38. Löw, F.; Duveiller, G. Defining the spatial resolution requirements for crop identification using optical remote sensing. *Remote Sens.* **2014**, *6*, 9034–9063. [CrossRef]

39. Kempeneers, P.; Sedano, F.; Seebach, L.; Strobl, P.; San-Miguel-Ayanz, J. Data fusion of different spatial resolution remote sensing images applied to forest-type mapping. *IEEE Trans. Geosci. Remote Sens.* **2011**, *49*, 4977–4986. [CrossRef]

Review

Functionalized Carbon Nanotubes (CNTs) for Water and Wastewater Treatment: Preparation to Application

Mian Muhammad-Ahson Aslam [1], Hsion-Wen Kuo [1,*], Walter Den [2,*], Muhammad Usman [3], Muhammad Sultan [4,*] and Hadeed Ashraf [4]

[1] Department of Environmental Science and Engineering, Tunghai University, No. 1727, Section 4, Taiwan Boulevard, Xitun District, Taichung City 407, Taiwan; ahson17@gmail.com
[2] Department of Science and Mathematics, Texas A&M University—San Antonio, One University Way, San Antonio, TX 78224, USA
[3] Institute for Water Resources and Water Supply, Hamburg University of Technology, Am Schwarzenberg—Campus 3, 20173 Hamburg, Germany; muhammad.usman@tuhh.de
[4] Department of Agricultural Engineering, Bahauddin Zakariya University, Multan 60800, Pakistan; hadeedashraf15@gmail.com
* Correspondence: hwkuo@thu.edu.tw (H.-W.K.); walter.den@tamusa.edu (W.D.); muhammadsultan@bzu.edu.pk (M.S.); Tel.: +886-(4)2359-0121 (ext. 3363) (H.-W.K.); +1-(210)784-2815 (W.D.); +92-333-610-8888 (M.S.)

Citation: Aslam, M.M.-A.; Kuo, H.-W.; Den, W.; Usman, M.; Sultan, M.; Ashraf, H. Functionalized Carbon Nanotubes (CNTs) for Water and Wastewater Treatment: Preparation to Application. *Sustainability* **2021**, *13*, 5717. https://doi.org/10.3390/su13105717

Academic Editors: Muhammad Sultan, Yuguang Zhou, Redmond R. Shamshiri and Aitazaz A. Farooque

Received: 24 April 2021
Accepted: 10 May 2021
Published: 19 May 2021

Abstract: As the world human population and industrialization keep growing, the water availability issue has forced scientists, engineers, and legislators of water supply industries to better manage water resources. Pollutant removals from wastewaters are crucial to ensure qualities of available water resources (including natural water bodies or reclaimed waters). Diverse techniques have been developed to deal with water quality concerns. Carbon based nanomaterials, especially carbon nanotubes (CNTs) with their high specific surface area and associated adsorption sites, have drawn a special focus in environmental applications, especially water and wastewater treatment. This critical review summarizes recent developments and adsorption behaviors of CNTs used to remove organics or heavy metal ions from contaminated waters via adsorption and inactivation of biological species associated with CNTs. Foci include CNTs synthesis, purification, and surface modifications or functionalization, followed by their characterization methods and the effect of water chemistry on adsorption capacities and removal mechanisms. Functionalized CNTs have been proven to be promising nanomaterials for the decontamination of waters due to their high adsorption capacity. However, most of the functional CNT applications are limited to lab-scale experiments only. Feasibility of their large-scale/industrial applications with cost-effective ways of synthesis and assessments of their toxicity with better simulating adsorption mechanisms still need to be studied.

Keywords: carbon nanotubes; surface modification; heavy metals; adsorption; water and wastewater treatment

1. Introduction

Rapid urbanization and industrialization has significantly increased the clean water demands in the domestic, industrial, and agricultural sectors [1–3]. Meanwhile, large quantities of pollutants including organic, inorganic, and biological contaminants are being released into the water bodies from these sectors [4–6]. Eccentric waters such as brackish, storm, and wastewater are being used depending upon the purposes [7,8]. Increasingly, use of these waters has also increased the urgent concern about the burden of negative impacts on the surrounding environment; one of the tremendous challenges confronting mankind is the exploration of green and sustainable methods to overcome these shortcomings [9–13]. Keeping in mind the current situation of water and wastewater treatment status, the technologies are not sustainable to meet healthy requirements for surrounding environment and community health [14,15].

Historically, numerous techniques and methods have been investigated for advanced treatment of water and wastewater [16]. The most common, adsorption, was proven to be the improved technique to remove a variety of pollutants including organic and inorganic contaminants present in water and wastewater [17,18]. Limited treatment efficiency was reported by using conventional adsorbents due their small surface area, limited number of active sites, deficiency in selectivity, and low adsorption kinetics [19]. These shortcomings of conventional adsorbents have been addressed in recent advancements of nano-adsorbents owing to their high surface area coupled with a higher number of active sites, tunable pore size, fast kinetics, and improved surface chemistry [10,20–24].

The nanomaterials can be used for treatment of water and desalination as well and also reveal properties including electron affinity, mechanical strength, and flexibility during functionalization [25–27]. Carbon nanomaterials (CNs) such as carbon nanotubes (CNTs) are supposed to be a promising material to break down the tradeoff concerning selectivity and adsorption, resulting in an increase of the economics of adsorption technology [25]. As a result, CNTs, as an adsorbent for treatment of water, have attained the focus of countless scholars over the previous few decades who are projected to carry on the exploration and developments in the field of CNs [28].

Numerous significant articles have been published on nanomaterials applied for the treatment of water and wastewater in previous few years [29–37]. Despite rapid developments, innovations, and applications of CNT-based nanomaterials, there is an increasing need for an across-the-board review of the synthesis of CNTs, functionalization of surface modifications, and finally their application to remove aqueous contaminants and to identify potential directions. This is the main motivation of the current review article.

This review attempts to address a brief history of CNTs, synthesis, purification, and functionalization, followed by the application of these nanomaterials for eliminating organics, inorganics, and microorganisms present in water and wastewater samples.

1.1. Historical Background

The discovery of CNTs was reviewed in 2006 by Monthioux and Kuznetsov, showing that the science has seemed to remain controversial [38]. Most literature mentions that nanotubes were discovered by Sumio Iijima [38]. However, Radushkevich and Lukyanovich explained the synthesis process of CNTs with 50 nm diameter [39]. Oberlin et al. explained the vapor phase growth technique for the synthesis of carbon fibers; the synthesized tubes consisted of turbostratic stacks of carbon layers (i.e., describing a crystal structure in which basal planes have slipped out of alignment) [40]. In addition, Abrahamson et al. [41] described the arc discharge method for carbon fiber synthesis using carbon anodes. Later on, scientists described the thermal catalytic disproportionation of CO for the synthesis of CNs. Transmission electron microscopy (TEM) and X-ray diffraction pattern (XRD) were used to characterize the synthesized CNs, and as a result they believed that CNTs can be formed by a graphene layer turning into a tubular shape. They also concluded that two types of promising arrangements, such as a helix-shaped spiral and circular arrangements in the form of a graphene hexagonal network, can result by turning the graphene layer into a tabular shape [42]. Later, a US patent was issued in 1987 on carbon nanofibers synthesis, the diameter ranging from 3.5 to 70 nm and five times greater in length than the diameter [43].

Back to the dates in 1950s, after the disclosure of CNTs by Iijima, projection of surprising properties of single-walled carbon nanotubes (SWCNTs) made by Dunlap and colleagues also attracted the attention of researchers around the world. At this time, after the discoveries and exploration of SWCNTs by Bethune and Iijima independently at IBM (Shiba, Minato) and Nippon Electric Co., Ltd. (Tokyo, Japan), respectively, the research on CNs and their specific methods of production was extended [44,45].

The above findings seem to be the extension of Fullerence's discovery. Arc discharge technology had previously been applied for the production of laboratory-scale Buckminster fullerenes [46,47]. CNTs are being studied after the report published in 1991 by Iijima [45] was fundamental, because it put CNTs in the limelight [38].

1.2. Types of CNTs and Structure

CNTs are composed of carbon atoms organized in a progression of fused benzene rings, which are pleated into a cylindrical shape. This new sort of man-made nano-material has a place with fullerene family and is treated as carbon's third allotrope as well as sp^2 and sp^3 forms of graphite and diamond, respectively [42,48,49].

Generally, there are two types of CNTs [50] on the bases of number of layers shown in Figure 1:

1. Single-walled carbon nanotubes (SWCNTs)
2. Multi-walled carbon nanotubes (MWCNTs).

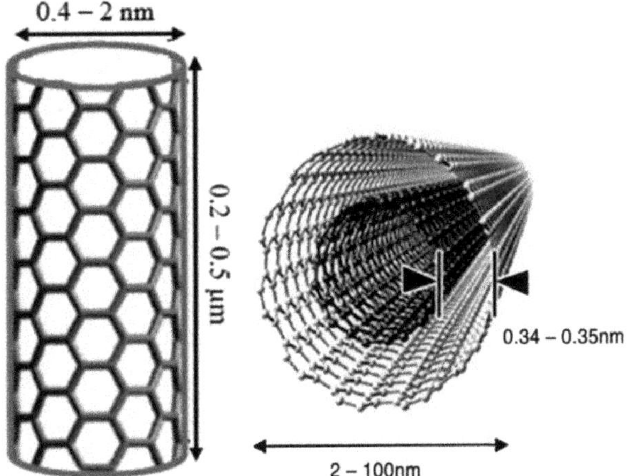

Figure 1. Types of typical dimensions of CNTs, SWCNTs (**left**) and MWCNTs (**right**).

CNTs are made by a sheet of graphene when rolled into a cylindrical shape, which may have a capped or open end, usually in a hexagonal form close packed with a diameter at a small scale of 1 nm, and a few microns long. SWCNTs (Figure 1) with a diameter as small as 0.4 to 2 nm are made by the single sheet of graphene rolled into a cylindrical shape, while MWCNTs (Figure 1) with an outer and inner diameter ranging from 2–100 nm and 1–3 nm, respectively, and a several microns in length are made up of two or more sheets of graphene incasing a hollow core in the same way as in SWCNTs [49,51,52].

Based on the chemistry, there are two zones of CNTs: the sidewall and tip. A significant aspect in controlling these distinctive properties emanates after the change in the tube-like structure due to entrapment of graphene layers into a cylindrical shape. Figure 2 shows different structures of rolled SWCNTs based on graphene sheets. Depending on alignment of the cylinder axis relative to the hexagonal matrix, the CNTs structure can be stipulated by chiral carrier in three ways, armchair, chiral, and zigzag, illustrated by their chirality index (n,m). Geometric arrangement of carbon atoms present at the layer of nanotubes is responsible for the foundation of zigzag (m = 0) and armchair (n = m) CNTs, whereas the structure of the nanotube with the two enantiomorphs on the right side is chiral (n ≠ m) [53,54].

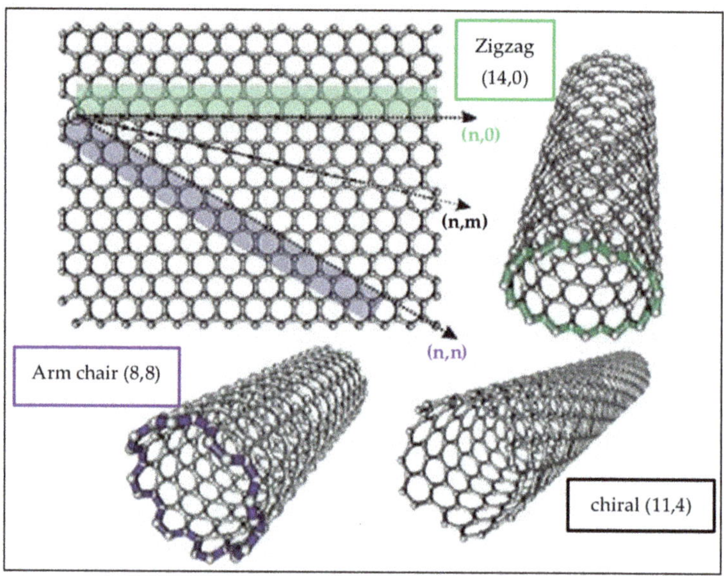

Figure 2. Roll-up of graphene sheet into different types of CNTs [54,55].

Recent reviews are good to find detailed elucidations of the structure of CNTs [48,49,56–61]. Here, Table 1 summarizes a comparison of the properties of SWCNTs and MWCNTs.

Table 1. Comparison between properties of SWCNTs and MWCNTs [49,62,63].

Properties	SWNTs	MWCNTs
Layer type	Single graphene layer	Multiple graphene layer
Catalyst requirement	Essential during synthesis	No need during synthesis
Bulk or massive production	Difficult	Easy
Purity level	Low	Large
Defect's level	High	Low
Characterization	Easy	Difficult
Manage	Easily twisted	Cannot be twisted easily
Specific gravity	About 0.8 g/cm^3	Less than 1.8 g/cm^3
Elastic modulus	About 1.4 TPa	Ranging from 0.3 to 1 TPa
Strength	Ranging from 50 to 500 GPa	Ranging from 10 to 60 GPa
Electrical conductivity	Ranging from 102 to 106 S/cm	Ranging from 103 to 105 S/cm
Electron mobility	About 105 cm^2/(V s)	Ranging from 104 to 105 cm^2/(V s)
Thermal conductivity	About 6000 W/(m K)	About 2000 W/(m K)
Coefficient of thermal expansion	Greater than 1.1×10^{-3} K^{-1}	About -1.37×10^{-3} K^{-1}
Thermal stability in air	Ranging from 600 to 800 °C	Ranging from 600 to 800 °C
Resistivity	Ranging from 50 to 500 μΩ cm	Ranging from 50 to 500 μΩ cm
Specific Surface Area	Ranging from 400 to 900 m^2/g	Ranging from 200 to 400 m^2/g

2. Synthesis of CNTs

Typically, there are three extensive methods for the synthesis of CNTs as given bellow:

(a) Arc discharge
(b) Laser ablation
(c) Chemical vapor deposition (CVD).

CNTs are produced by using energy and carbon sources in all the synthesis methods. A carbon electrode or a gas and an electric current or heat is used as a carbon and energy source, while using arc discharge or CVD methods, respectively, for the synthesis of CNTs, whereas a laser beam is used as an energy source during the laser ablation method. Table 2 presents a detailed summary of the efficiencies of the CNT synthesis methods. These methods are based on the formation of a single or a consortium of carbon atoms that sack to recombine into CNTs.

CNT synthesis mechanisms have been debated in detail by Cassell et al. [64] and Sinnott et al. [65]. It is believed that by using the metallic catalyst in the CVD method for CNT synthesis, the cylindrical shaped graphene tube is formed by initial deposition of carbon atoms on the used catalyst surface [66]. It was also concluded that the size of particles of the used catalyst also play an important role in the CNTs diameter, as the catalyst particles in a smaller size produce SWCNTs with a diameter of a few nanometers, while the larger particles tend to produce MWCNTs [67].

The arc discharge method between graphite electrodes is the first method of producing CNTs [68]. Briefly, in this method, direct current of 50 to 100 A and about 20 volts of potential difference is established between a graphite electrode pair in the presence of one of the inert gases containing helium or argon with a pressure of 500 Torr [69,70]. The carbon electrode surface evaporates and forms a cylindrical-shaped tube structure because of the high temperature generated due to the discharge of electric current in low pressure, inert gas, and catalyst [44,70,71]. MWCNTs can be synthesized via an arc discharge method without a metallic catalyst; on the other hand, mixed-metal catalysts, for example iron, cobalt, and nickel, are necessary for SWCNTs fabrication [72]. In general, higher levels of structural precision are noted in CNTs produced via the arc discharge method [73]; however, different variables such as chamber temperature, concentration and type of catalyst, hydrogen presence, etc., may affect the structure and size of synthesized CNTs [74]. Recently, nickel-filled CNTs were synthesized via a local arc discharge method in liquid ethanol [75], nitrogen-doped CNTs via vaporization of boron nitride [76], low-cost SWCNTs via an arc discharge method in open air [77], and SWCNTs and MWCNTs via a hot plasma arc discharge method [78].

The use of laser ablation to synthesize CNTs was first reported by Guo and colleagues in 1995 [79,80]. Briefly, a graphite object is targeted by a laser beam with high energy in the presence of argon at 800–1200 °C temperature and 500 Torr pressure [69,70,81,82]. In this method, a laser pulse provides an energy source, and a graphite object serves as a source of carbon. Soot deposition of carbon can be avoided by uniform evaporation of the target resulting because of continuous applications of laser pulses. The larger size particles after the first laser beam are broken down into smaller ones by the successive beams. Later, the smaller size particles are produced into the CNT structure. Commonly, transition metals are used as catalysts in this method. Rope-shaped CNTs can be found by using a laser ablation method with the diameter ranging from 10 to 20 nm and about 100 mm long [83].

CVD is a well-liked method for bulk fabrication of CNTs around the globe. Typically, in this method, carbon monoxide or hydrocarbons gases are used as a source of carbon, while 500 to 1100 °C of temperature is used as the source of energy. The carbon atoms are deposited by the decomposition of the used carbon source and shaped into CNTs [84,85]. Briefly, the decomposition of gas (carbon source) occurs because of high temperature when transferred into reaction chamber together with the carrier gas and generates a substrate of carbon atoms on the surface of catalyst to form CNTs [86]. When compared to other methods of synthesis of CNTs, the CVD is the most common route for relatively large-scale production of CNTs as it is simple in operation, higher in yield, and economic and has a high rate of deposition and good control over the morphology of tubes during the synthesis process [87–89] Cassell et al. [64] studied that CNTs in bulk can be produced, especially SWCNTs via the CVD method, by using acetylene as a source of carbon deposition in the presence of iron and cobalt and zeolite or silica as a carrier support material. They also concluded that SWCNTs can be produced on a largescale when a mixture of H_2 and CH_4 is deposited on the catalyst (Co or Ni), and MgO as carrier support material is used.

Table 2. A summary of CNTs synthesis strategies and their efficiencies.

Parameters		Arc Discharge Method	Chemical Vapor Deposition	Laser Ablation (Vaporization)	Ref.
Method	Source of energy	Direct current	Temperature (ignition)	High intensity laser beam	[90,91]
	Source of carbon	Carbon or graphite electrodes	Hydrocarbon gases or carbon monoxide (CH$_4$, CO, or acetylene)	Graphite object	
Temperature (°C)		3000 to 4000	500 to 1100	About 3000	[84,85]
Cost per unit synthesis		Costly	Economic	Costly	[83,92]
CNTs selectivity		Less	High	Less	[93]
Availability of carbon source		Complex	Easy	Difficult	[94]
Purification level		More	Less	More	[95]
Nature of synthesis process		Batch	Continuous	Batch	[64,96]
Control on synthesis parameters		Difficult	Easy	Difficult	[97,98]
Energy requirement		High	Low	High	[99]
Design of reactor		Hard	Simple and easy	Hard	[100]
Nanotube graphitization		High	Moderate	High	[101–103]
Typical yield		30 to 90%	20 to 100%	Up to 70%	[84,91,95,104–106]
Typical Diameter	SWCNTs	0.6 to 1.4 nm	0.6 to 4 nm	1 to 2 nm	[87,91,107,108]
	MWCNTs	Inner: 1 to 3 nm Outer: ~10 nm	0.1 to several nanometers	10 to 20 nm	[79,83,91,104,109]

Table 2. *Cont.*

Parameters	Arc Discharge Method	Chemical Vapor Deposition	Laser Ablation (Vaporization)	Ref.
Advantages	1. Synthesis of both SWCNTs and MWCNTs is easy 2. MWCNTs can be produced without any catalyst 3. Costly process but less than laser ablation method 4. Synthesis of CNTs is possible in open air 5. High degree of structural perfection	1. Bulk production is easy 2. More extensive length CNTs than other methods 3. Simple and easy process 4. Quite pure 5. Alignment of produced CNTs is good 6. Diameter and number of layers can be controlled	1. Primarily for SWCNTs 2. Diameter of CNTs can be controlled 3. Lower numbers of defects 4. High degree of structural perfection 5. Tubes' length can vary from 5 to 20 mm	[98,110–113]
Disadvantages	1. Received with some structural defects 2. Short and randomly distributed in length and direction 3. Lot of structural purification is needed 4. Contains carbon impurities	1. Only used to produce MWCNTs 2. Higher structural defect density	1. Costly technique due to expensive lasers beams 3. Power needs are high 4. Low yield	[65,86,98,109]
Figures				[114]

3. CNT Purifications

Some of the impurities include an amorphous phase of carbon, particles of particular metals, or any other carrier material associated with CNTs that will have an effect on their execution performance [115]. Some typical purification technologies and their characteristics are discussed in Table 3. On average, the CNTs synthesized via the CVD method showed a purity level ranging from 5% to 10% [116]. Therefore, a broad purification of CNTs is necessary before being used for different applications. The detection and identification of different impurities associated with CNTs using different techniques have been discussed in Table 4. It is believed that the CNT structure may be affected to some extent when removing impurities, so there is always need for a compromise with the final structure after purification process [116]. The common CNTs purification methods are discussed below:

1.   Oxidation
2.   Acid treatment
3.   Surfactant based sonication.

Oxidation is a decent manner to remove carbon [117–122] and metal [117,121,123–126] impurities associated with CNTs. One of the main shortcomings that occur using this process of purification is the oxidation of CNTs themselves along with the impurities, but fortunately, the loss of CNTs is smaller than the impurities [114]. The reason to oxidize these impurities is more defects or open structures associated with them. This is another reason that the attachment of these impurities is often observed with a used metallic catalyst, and this metal catalyst may also play a role in oxidation [117,118,123,124,127]. There are some factors, such as type of oxidant, time of oxidation, temperature, metal contents, and environment, which can affect the oxidation efficiency and final yield.

Typically, the method of acid treatment is used to eliminate the metallic impurities associated with CNTs. First of all, by oxidizing or sonication of the CNTs, the surface of associated impurities (metals) is made apparent to acid until the solvation and finally CNTs collect in suspension. It has been observed in a number of studies that by using HNO_3 for the purpose of acid treatment, it only affects the metallic impurities rather than the CNTs or other carbon containing impurities [117,118,122,125]. By using HCl for this purpose, the impacts on CNTs and carbon impurities are also observed to a small extent [117,123,126]. Acid treatment for purification of CNTs in diluted form (4M HCl) can show same results as by the HNO_3, but the metal surface must be apparent to the applied acid to make the solvation [128].

Although purified CNTs are produced relatively by acid reflux, the nanotubes amalgamate, and the impurities that they capture are very hard or sometime impossible to remove by filtration [129]. Therefore, a surfactant-based sonication process is implemented generally by dissolving sodium dodecyl benzene sulfate (SDBS) in ethyl or methyl alcohol solution for this purpose. Since after sonication CNTs took longer to settle down, ultrafiltration is required and then annealed at a high temperature (about 1000 °C) in the presence of N_2 for 4 h. Annealing of CNTs is performed to optimize their structure. Surfactant-based sonication has been presented to be an effective method for removing tangled impurities associated with amalgamated CNTs [116].

Table 3. Typical purification technologies and their characteristics [130,131].

Technologies	Methods		Characteristics	Limitations
			Advantages	
Physical method	Filtration		1. Non-destructive 2. Retains the inherency and intrinsic structure necessary to elucidate the properties of CNTs 3. More suitable as an auxiliary step in combination with chemical purification 4. Improve crystallinity 5. High selectivity to metal 6. CNTs can be separated on the bases of difference in length and conductivity	1. Not very effective 2. CNT samples need to be extremely dispersed 3. Purification of samples can be done in a limited quantity at a time
	Centrifugation			
	Solubilization with functional groups			
	High temperature annealing			
	Chromatography, electrophoresis			
Chemical method	Gas phase	Air, Cl_2, H_2O, HCl, H_2O, Ar, O_2, $C_2H_2F_4$, SF_6	1. Opens the lid of the CNTs without affecting sidewalls or associated functional groups 2. Eliminates polyhedral and amorphous carbon and metallic impurities at the cost of substantial amounts of CNTs or damage to the CNT structure 3. Leads to functional groups 4. Does not disrupt or affect the alignments of CNTs	1. Low yield 2. Produces more defects on sidewalls, breaks into different shorter length, and also the alignment and structure are affected greatly, thus limiting the final applications of CNTs
	Liquid phase	HNO_3, H_2O_2, HCl, Mixture of acid or $KMnO_4$, Microwave in inorganic acid		
	Electrochemical	Alkali or acid solution		
Multi step method	Oxidation, sonication, centrifugation, filtration, wet grinding, and HIDE		1. High-purity with respect to metal 2. Metal free, improving crystallinity 3. Effectively removes carbonaceous and metallic impurities 4. Better purification yield due to the early removal of metallic impurities that can oxidize CNTs	1. Low yield
	Filtration/magnetic filtration, oxidation, annealing			
	Filtration, sonication in HNO_3, HF, H_2O_2, or SDS			
	Annealing at high temperature, extraction			

Table 4. Impurities associated with CNTs and their detection techniques.

Technique	Residual Material	Assessment Techniques	Advantages	Limitations
Thermo-gravimetric analysis (TGA)	Carbonaceous impurities Metal impurities	After oxidation of material, the residual metallic impurities are calculated by weighing ash and the carbonaceous impurities by area ratio of DTG	Accurate measurement of impurities	Completely oxidize/destroy the CNTs
Raman spectrometry	Carbonaceous impurities Structure defects Conductivity characteristics	The pure CNTs are associated with G-band by RBM as well as no D-band	Conductivity features and quality of CNTs can be measure along with their diameter	Difficult or even unacceptable for MWCNTs and metallic contents
Electron microscopy (SEM, TEM)	Defects in CNTs Amorphous carbon	Directly observes and qualitatively evaluates the adhesion defects on the CNT wall, the amount of amorphous carbon, fullerene	Absolute scrutiny can be undertaken	Can analyze the sample in a very small amount
UV–vis-NIR	Carbonaceous impurities conductivity characteristics	Absorption spectroscopy or reflectance spectroscopy in the ultraviolet-visible spectral region	Conductivity features and contents of CNTs can be analyzed exactly	A standard sample is needed with 100% purity
X-ray photoelectron spectroscopy (XPS)	Support material/functional groups (fine alumina, magnesium oxide, silica, zeolite, etc.)	Quantitatively characterizes the type and contents of functional groups or support materials	Analysis of functional groups on CNTs can be undertaken exactly	Unacceptable for purity
Energy-dispersive X-ray spectroscopy (EDS)	Metal impurities	Analytical technique used for chemical and/or elemental analysis of a sample	Contents and traces of different elements can be analyzed	Evaluation of the contents of CNTs is invalid

4. Functionalization of CNTs

The non-polar nature of graphene layers makes the CNTs naturally hydrophobic. The hydrophobic property of CNTs is indispensable for the adsorption of aromatic contaminants like benzene and anthracene. A very strong complexion is formed due to π electrons present on the graphene layer making CNTs, between aromatic contaminants and the tube surface [132]. The surface affinity of CNTs can be modulated to a variety of contaminants in water and wastewater after the purification and surface functionalization. Higher adsorption of benzene was obtained by using CNTs as compared to activated carbon (AC) because of strong interaction between benzene rings and the surface of CNTs due its hydrophobicity [133]. Figure 3 shows the different routes and schemes of CNT functionalization to increase their affinity for water and wastewater contaminates, which can be subsequently captured on the surface of CNTs used. Moreover, the functionalization of CNTs can be divided into three categories, shown in Figure 4 [134]:

1. With π conjugated network of CNTs through covalent bonds;
2. Attachment of different chemical groups via non-covalent bonds by using hydrophobicity of CNTs such as hydrogen bonds, π–π interactions, or ionic bonds;
3. Inline filling (endohedral) of hollow tubes of CNTs. The two methods are more common for CNTs functionalization and variously used by the researchers.

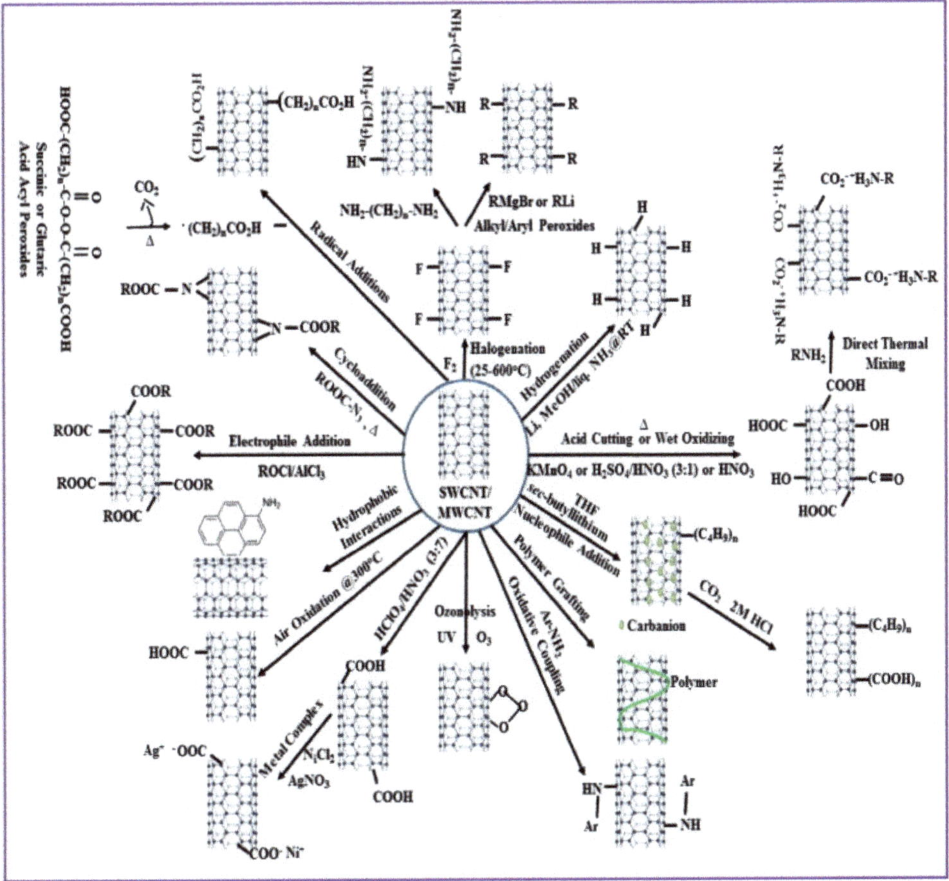

Figure 3. Functionalization routes of CNTs and associated functional groups [132].

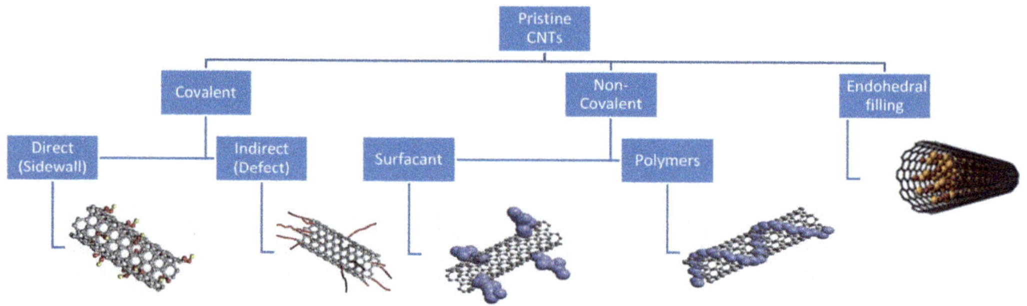

Figure 4. Methods of functionalization of carbon nanotubes.

CNTs are unique because of their distinctive properties such as adsorption capability, permeability, morphology, and physicochemical properties. There are several disadvantages that are also associated with raw CNTs, such as their low dispersion in solutions and low adsorption capacity for bulk fabrication of CNTs with organic and inorganic composites. In fact, aggregation is the main problem for low contaminant adsorption efficiencies by the original CNT samples [33,53,135–137] and also obscures the process of membrane preparation [138]. The π–π interactions and van der Waals force between CNTs are responsible for the less dispersion, which results in tight fit bundles and aggregation of CNTs [139]. As the number of graphite layers of CNTs decreases from MWCNTs to SWCNTs, the tendency to bunch increases [140]. Therefore, to overcome these drawbacks, the chemical reactivity and contaminant adsorption capacity of CNTs must be improved by increasing their dispersion rate, and this is done by functionalizing the nanotube [53,135–137,139,141]. Furthermore, solubility of CNTs can be increased by their functionalization, which causes them to repel each other [50]. Table 5 provides a compression between adoption capacities and the corresponding surface area of pristine and oxidized CNTs treated with different acids [142].

Table 5. Adsorption capacity of CNTs and corresponding surface area [142].

CNTs	Adsorption Capacity (mg/g)	Surface Area (m^2/g)
Pristine	1.1	82.2
H_2O_2 oxidized	2.6	130.0
HNO_3 oxidized	5.1	84.3
$KMnO_4$ oxidized	11.0	128.
NaOCl oxidized	47.4	94.9

Different physical or chemical processes like oxidation, impregnation, or grafting (Figure 3) are used for the functionalization of CNTs [143,144]. During the process of functionalization, the covalent or non-covalent bonds of particular functional groups result on the end or sidewall of CNTs. Functionalization is preferred over covalent bonding, because non-covalent bonding does not influence the structure and surface area of CNTs [135]. Advantages and the limitation of covalent and non-covalent functionalization of CNTs are discussed in Table 6. Functionalization is generally done to increase the dispersibility as well as contaminant removal efficiency of CNTs, thereby improving the water or wastewater treatment application capability of the CNTs or promoting membrane fabrication [145]. Higher dispersibility in polar solvent (water) was found by covalent functionalization of CNTs with the phenolic group by 1,3 dypolar cyclo addition; covalent functionalization of CNTs with phenol groups by 1, 3-dipolar cyclo-addition was found, which facilitated

CNTs' amalgamation into the polymer matrix [139]. Oyetade et al. [141] found that MWC-NTs efficiently absorbed Pb^{2+} and Zn^{2+} from aqueous samples after functionalizing with nitrogen. The increase in adsorption was due to higher surface area and more adsorption sites linked to CNTs.

Table 6. Benefits and limitations of covalent and non-covalent functionalization [134].

Methods	Benefit(s)	Limitation(s)
Covalent functionalization	Highly stable bonds are formed	Intrinsic characteristics are damaged Structural defects CNTs Aggregation of CNTs
Non-covalent functionalization	Simple and easy procedure CNTs structure is maintained with minimum defects Electronic characteristics of CNTs are not affected	Stability of bonds is weak

Oxidation of raw CNTs with HNO_3, H_2SO_4, HCl, H_2O_2, $KMnO_4$, and NaOCl, or sometimes a combination of some of these (Figure 5), has often been exercised to introduce oxidized functional groups [146,147]. Generally, oxidation improves the dispersibility and enhances the ability to adsorb certain harmful contaminants in water and wastewater at the expense of fractional damage to the surface of CNTs, as described earlier. Furthermore, the surface of CNTs can also be modified with the addition of oxygen containing functional groups by performing oxygen–plasma action. In addition to oxidation and plasma action for surface modification, the CNTs can also be successfully modified with the addition of metal oxides like Al_2O_3 [148], MnO_2 [149], and Fe_3O_4 [150], which deliver another way to coat the surface of CNTs, thereby increasing the contaminant removal efficiencies of CNTs [53,135–137]. The potential surface modification schemes of CNTs used for targeted contaminants are shown in Figure 5.

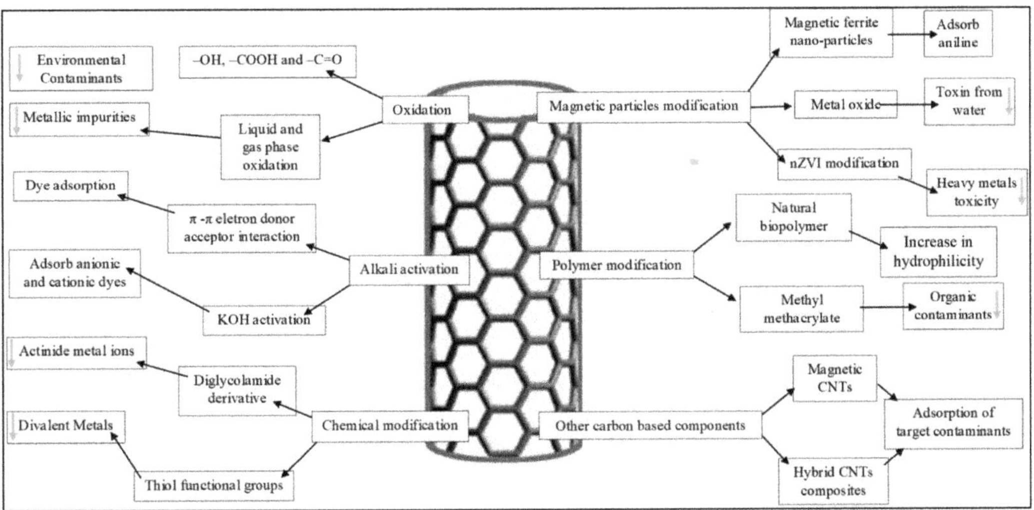

Figure 5. Schematic representation of surface modification schemes of CNTs used for targeted contaminants (green arrows refer to decrease in final effluent concentration).

5. Characterization of CNTs

Intrinsic properties of CNTs make them fascinating and desirable candidates for diverse remediation. Characterization of CNTs plays an important role due to their distinctive properties. Numerous comments and debates have been published in past decades on different techniques and strategies used to evaluate CNTs [151–154]. Techniques used for the characterization of CNTs are divided into four groups: microscopy and diffraction, thermal, spectroscopic, and separation techniques [155]. Sometimes research includes one more group, the magnetic measurement technique. Table 7 shows different characterization techniques used for the evaluation of CNTs. It must be noted more than one technique is prominent for the characterization of CNTs; techniques employed alone are not fully characterized, nor they are absolutely quantitative. Even though qualitative analysis of CNTs can be done by electron microscopy, scanning electron microscopy (SEM) evaluates the nanostructure of the tubes, and transmission electron microscopy (TEM) is used for further precise inspection, generally detecting the defects in CNTs [129,156].

Table 7. Different analytical techniques used for the characterization of carbon nanotubes [155].

Characterization Techniques	Used for Studying
Microscopy and diffraction techniques	[157–159]
SEM	Morphological analysis (diameter and length), aggregation state
TEM/HR-TEM	Morphological analysis of internal structure (diameter, number of layers and distance between them)
AFM	Morphological analysis of internal structure (diameter, number of layers and distance between them)
Scanning tunneling microscopy	Morphological analysis of internal structure (diameter, number of layers and distance between them)
Neutron diffraction	Morphological analysis of bulk samples
XRD	Morphological analysis of bulk samples
Spectroscopic techniques	[139,160,161]
Raman spectroscopy	Purity and presence of by-products, diameter distribution, (n, m) chirality
IR and FT-IR	Purity, functionalization by attaching functional groups to the sidewall
UV–vis and NIR	Dispersion efficiency, diameter and length distribution, purity
Fluorescence spectroscopy	Size, dispersion efficiency, (n, m) chirality
XPS and EDS	Elemental composition, functionalization (covalent and non-covalent)
Thermal techniques	[162]
TGA	Purity and presence of by-products, quality control of synthesis and manufacture processes
Separation techniques	
Size exclusion chromatography	Purification, separation by size (length)
Capillary electrophoresis	Purification, separation by size (length, diameter, and cross-section)
Field flow fractionation	Fractionation by size (length)
Ultracentrifugation	Separation by chirality, electronic type, length, and enantiomeric identity
Magnetic techniques	[158,163,164]
Vibrating sample magnetometry	Magnetic properties
Alternating gradient magnetometry	Magnetic properties
Superconducting quantum interference device	Magnetic properties

The ultrastructure of different types of species including organic, inorganic, and biological species can be evaluated by using very popular techniques known as SEM and TEM. The scanning of the sample in SEM generates an image when the targeted electron ray interacts with sample of CNTs. Generally, the technique is used to analysis the morphology (length and diameter) of nanomaterials [123,155,165] to assess the quality of prepared CNTs. For example, Figure 6a shows a SEM image of the as-prepared CNTs [148], and Figure 6b shows alignment of CNTs synthesized by using a horizontal quartz tube housed in muffle furnace. Average length of the tubes was 70 μm measured by using SEM [166]. Sometimes, it is also used to validate the surface modifications in terms of functional reactions that occur on the surface of CNTs [167]. In the case where the required measurement exceeds 1 to 20 nm resolution while using SEM, then TEM is used [156,167]. Small dimensions in CNTs such as interlayer distance, diameter, and number of graphene sheets can be easily examined (Figure 6c,d) by targeting the sample with a high energy electron beam of up to 300 keV [168,169]. It should be noted that the functional groups (organic and inorganic) that modify the surface of CNTs can also be evaluated by using TEM [161,170,171]. Moreover, structural integrity, surface functionalization, and defect son the surface of CNTs caused by the oxidation (acidic or basic) to introduce oxygen containing functional groups like hydroxyl, carbonyl, and carboxylic acid groups have also been studied by using both SEM and TEM techniques [155,172].

The image of the atomic structure and crystal structure information can be obtained by using high resolution transmission electron microscopy (HR-TEM) [173]. A high phase contrast image as small as the crystal unit can be obtained. This technology is widely used for advanced characterization of materials, allowing access to information on just-in-time defects, stacking faults, deposits, and grain boundaries. In addition to the morphology of the MWCNTs in the HR-TEM image, the direct measurements can be made on the MWCNT walls. For example, the number of walls constituting the nanotubes can be directly counted and recorded as control parameters in subsequent experiments in case the number of such walls needs to be changed. In addition, the interplanar distance between the walls can be accurately measured and compared with the crystal structure data table of the carbon structure and its respective diffraction pattern.

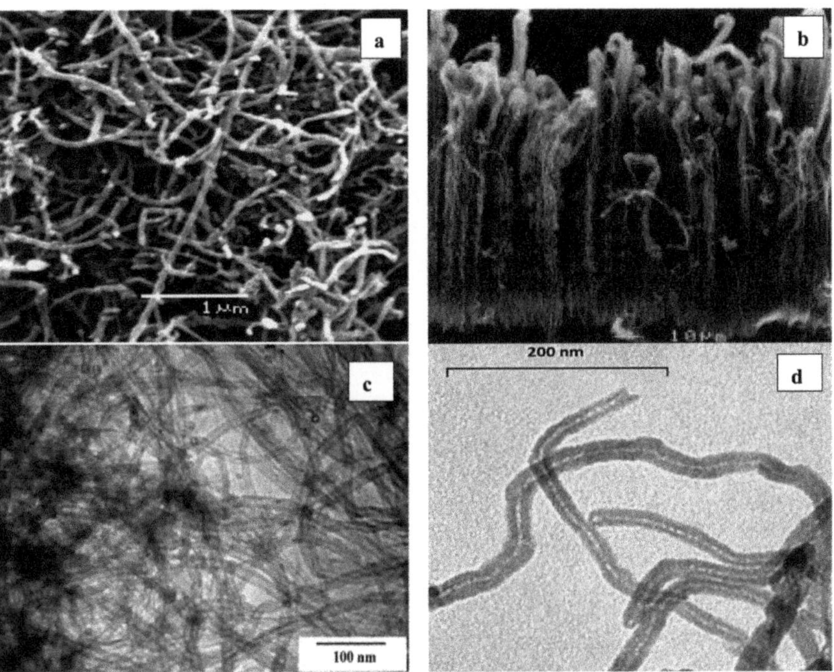

Figure 6. (**a**) SEM image of a bulk sample of multi-walled carbon nanotubes [155], (**b**) SEM image showing vertical aligned CNTs [174], (**c**) high-magnified TEM images of CNTs grown on unreduced catalyst [168] and, (**d**) TEM image of a bulk sample of multi-walled carbon nanotubes [155].

The chemical state and structure of CNTs can be obtained by using an X-ray photoelectron spectroscopy (XPS) technique [114]. However, the data obtained from this technique are used to examine the structural modification of CNTs before and after the chemical interactions with organics, inorganics, or gaseous adsorption. The investigation of CNTs by using XPS is done after the incorporation of nitrogen into the nanotube [175]. Due to the polar nature of the carbon–nitrogen bond, the peak shift before and after the modification is evidence of nitrogen incorporation [176,177]. Furthermore, the technology demonstrates that carbon nanofibers are more similar to carbon oxides than various graphites [178]. Fluorinated functionalization of SWCNTs was also studied by using XPS; the results concluded that three peaks of sp^2, sp^3, and carboxyl groups (284.3, 285, and 288.5 eV, respectively) were associated with C1s of un-doped nanotubes [114]. The observed carboxyl group (288.5 eV) was similar to in nanofibers [179].

Crystallinity of CNTs can be obtained by using an X-ray diffraction (XRD) technique [90,176]. X-ray diffraction patterns of graphite and CNTs are very close to each other because of their inherent properties. The XRD pattern of CVD-synthesized MWCNTs is shown in Figure 7, illustrating a peak similar to graphite (002), and the measured layer spacing can be obtained from Bragg's law, while the other peaks (family of (hk 0) peaks)) can be obtained because the mono-graphene layer makes the honeycomb matrix [114]. Therefore, the curve obtained by XRD does not distinguish between the microstructure information of CNTs and graphite; nevertheless, it is helpful for purity analysis of the sample [102,103]. The XRD pattern revealed that straight CNTs with a good alignment on the surface of the substrate did not show the peak, i.e., 022 [114]. For CNTs aligned vertically to their substrate surface, the XRD pattern is not collected because of the scattering of the beams inside the sample. Therefore, the 002 peak is always lowered for better-aligned CNTs [180]. In addition, several other types of parameters such as bundle size, mean diameter, and diameter dispersion can also be studied by using the XRD technique. The peak 10 is greatly influenced by these parameters in terms of its location and thickness [181,182].

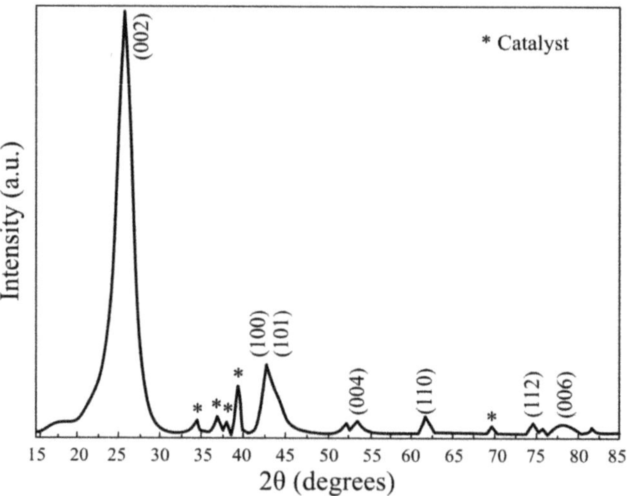

Figure 7. XRD pattern of MWCNTs synthesized by CVD [114].

Qualitative evaluation for nanotubes' surface can be done by using Fourier transform infrared spectroscopy (FT-IR) [183]. The sample is characterized by passing the infrared radiations through it, and the part of radiations absorbed by the sample at specific energy is determined. Each functional group is identified by a particular range of frequency with associated absorption peaks. The infrared spectrum of the original CNTs (Figure 8) shows a characteristic band of about 1600 cm^{-1} associated with aromatic rings (C=C bond) of rolled graphene layers. Sometimes, the peaks of 3800 to 3200 cm^{-1} (O-H stretch) and 1700 particles that absorb moisture into the atmosphere or due to certain purification processes [184]. In addition, a band of 2910 to 2940 cm^{-1} was also detected for CNTs, which is related to the vibration of C-H stretching methylene (CH_2) [185]. The FTIR spectra can also suggest surface modification of CNTs. For example, in Figure 8b, a new peak near 1450 cm^{-1} appeared that was assigned to asymmetric CH_2 bending. This peak is typically interpreted as evidence of defects in the structure of CNTs. In Figure 8c, a new peak near 1735 cm^{-1} suggested a carbonyl stretch of the carboxylic acid group. In addition, a double peak at approximately 2900 cm^{-1} was attributed to the loss of aromaticity due to the oxygen functional groups. The thermal stability and proportion of volatile compounds of nanotubes can be analyzed by an analytical technique called thermogravimetric analysis (TGA). The analysis is made by heating the sample directly in the air or inert gases (He/Ar) while recording the change in its weight with respect to elevated temperature [176,186,187]. In some cases, the analysis is made in the presence of N_2 or He with poor oxygen atmosphere (1% to 5% O_2) to slow oxidation [188]. During the TGA analysis of the CNT sample in the air atmosphere, the weight loss of the sample (Figure 9) is usually due to carbon oxidation to CO_2, while the solid oxides after the oxidation of metallic catalyst are responsible for the superposition of the sample [165,189,190]. Generally, the percent yields of carbon deposits are determined by using TGA. Usually, during the oxidation of the sample the weight occurs in the temperatures ranging from 200 to 680 °C [190]. The contents of carbon can be calculated by obtaining the percentage of $(m_1 - m_2)/m_1$, where m_1 is the weight of the sample before oxidation and m_2 is the weight of the sample after the oxidation [11].

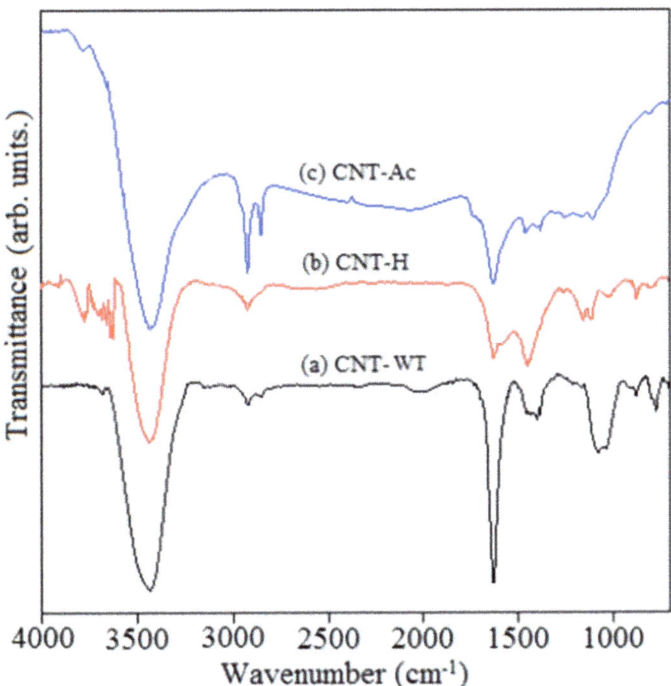

Figure 8. FT-IR spectra of CNTs: (**a**) pristine CNTs (CNTs-WT), (**b**) CNTs treated with HCl (CNT-H), and (**c**) CNTs treated with mixture of H_3SO_4 and HNO_3 (CNTs-AC) [191].

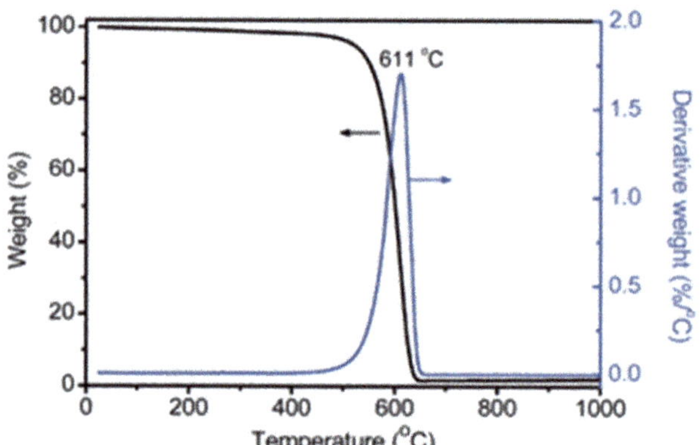

Figure 9. TGA analysis of the CNT sample [190].

The structural features of CNTs can be defined by their crystalline structure, chiral carrier, and single- or multi-walled features [192]. Crystalline arrangement of nanotubes can be characterized by the ID/IG ratio determined from the RAMAN spectroscopy. ID/IG indicates the ratio between the organized and unstructured carbon in the CNTs and uses the intensity of defective and graphitic carbon (D and G bands) at the high wavenumbers in the RAMAN absorption spectrum (Figure 10) [193]. ID/IG is a good quality indicator for CNTs, and the low ID/IG ratio is characteristic of highly graphitized structures; the laboratory reported mass of MWCNT is 0.65, and industrial grade MWCNT is 2.04 [90,194].

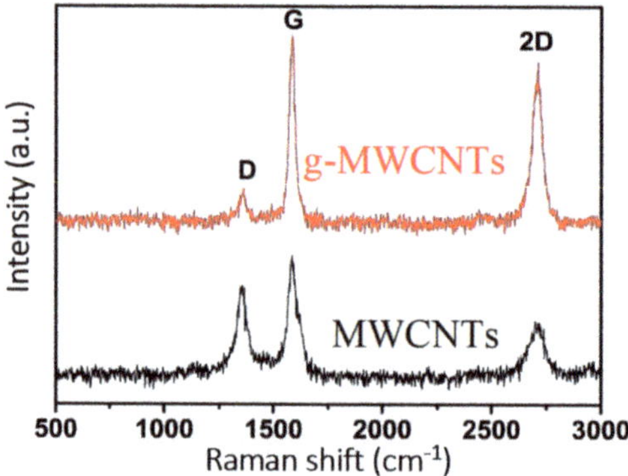

Figure 10. Raman spectra of pristine (MWCNTs) and graphitized MWCNTs (g-MWCNTs) [193].

6. Applications of CNTs

6.1. Removal of Heavy Metals

Removal of heavy metals from water and wastewater by using surface modified CNTs has been studied extensively [143,195–197]. Contaminant adsorption mainly occurs at four possible types of sites that are available on the CNTs such as outer and inner grooves and interstitial channels mas shown in Figure 11, but the inner region of the nanotubes is less adsorbed [53,135–137,198]. Bahgat et al. [143] used functionalized MWCNTs for the adsorption of heavy metals and concluded that adsorption of the metals occurs because of a number of adsorption spots available on the tubes' surfaces. In another study, higher adsorption of Zn^{2+} ions was observed by the plasma-treated CNTs due to the availability of more oxygen-containing functional groups. The mechanism of surface complexation was responsible for the adsorption of cationic ions, as the adsorption sites increased due to deprotonating functional groups [197].

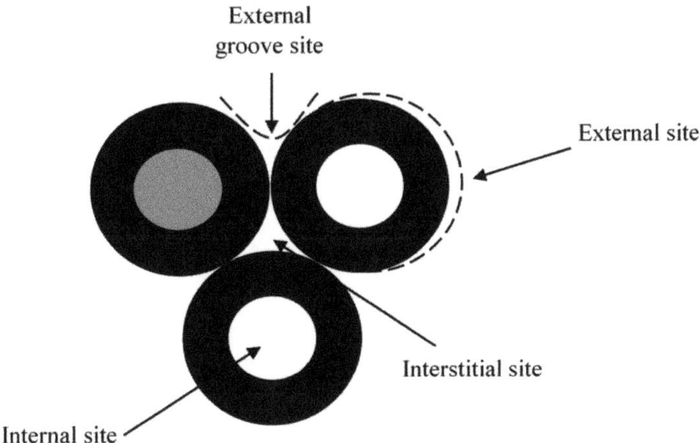

Figure 11. Different adsorption sites on CNTs [198].

Surface modification of CNTs improved their affinity for heavy metal ions and metalloid adsorption [199,200] by incorporating iron oxide and oxidation [16,201–203], coating with oxides of nonmagnetic metals [204], introducing a thiol functional group [82], and functionalizing with sulfur [205]. According to Addo Ntim and Mitra [203], the different oxides of iron such as magnetite (Fe_3O_4), maghemite (γ-Fe_2O_3), hematite (α-Fe_2O_3), and goethite (α-FeO) are very useful for the removal of trace heavy metals present in water and wastewater [170].

Generally, the process of surface modification of functionalization of CNTs is responsible for the adsorption mechanisms of heavy metals [53,191,206]. Table 8 shows an explanation of previously used functionalized CNTs and their adsorption capacities and removal mechanisms. Surface area, porous structure, functional groups, and interaction mechanisms between the absorbate and absorbents are the main attributes of CNTs for their heavy metal adsorption capability [53,136,137,143]. In addition, these properties enhance the heavy metal removal efficiencies onto the polymer film incorporated by functional CNTs [135].

Physical and chemisorption mechanisms of adsorption have been observed between the CNTs modified with metal oxides and heavy metal ions [203]. In addition, the contaminant with particular hydration energy, ionic radius, and potential of hydrolysis also affects the adsorption capacities of CNTs. A study conducted by Hu et al. [207] confirms the above statement, when higher removal of Pb^{+2} was observed than Cu^{+2} by using MWCNTs modified with iron oxide under the same experimental conditions. The adsorption performance of CNTs for heavy metals is also affected by the presence of other organics. For example, enhanced removal of Cd^{+2} ion was observed by using oxidized MWCNTs in the presence of 1–naphthol, while there was no effect recorded during the presence of Cd^{+2} ion on the removal of 1–naphthol in similar conditions [208]. These types of results mainly occur because of the distinct interaction mechanisms that are responsible for the adsorption of different types of pollutants [170].

Solution pH also plays an important role in the adsorption capacity and mechanisms of CNTs. For example, higher pH is favorable for endospheric interactions, while the lower pH facilitates extracellular interaction and/or ion exchange of targeted metal ions and surface functional groups of CNTs [208]. Moreover, as the pH increases (alkaline conditions), the charge density of functionalized CNTs moves towards more negativity, which is efficient for the adsorption of cationics, while, at lower pH (acidic conditions), because of protonating functional groups, the positive charge density increases, which repels the cationic metals, resulting in the lower efficiency of CNTs [53,136,137]. In a study, oxidized and ethylenediamine-doped MWCNTs were used for the adsorption of

Cd^{+2} ions from aqueous samples. The results concluded that both types of absorbents removed Cd^{+2} ions strongly depending on pH ranging from 8 to 9 [209]. According to Rao and coworkers [146], the best adsorption capacity of nanotubes was observed in the pH ranging from 7 to 10. In addition to this particular range of pH, the ionization and competition between different species may also happen [210–213]. For example, an effective adsorption of Pb^{+2} ions on functionalized CNTs was reported during the co-existence of sodium dodecylbenzenesulfonate (DBS), while the adsorption of lead was significantly reduced in the presence of benzalkonium chloride [214]. The higher adsorption of lead ions might be due to anionic surfactant formation between Pb^{+2} and DBS, which are very complex compounds. In addition, the charge density (negative or positive) on the surface of CNTs makes the different interaction mechanisms of metal ions. For example, chemical interaction occurred between the N-dopped magnetic CNTs and Cr(III), while an electrostatic interaction was observed between the acid oxidized CNTs and Cr(III) [215].

Regardless of the costs, the CNTs are more efficient in terms of their adsorption and desorption phases than the AC, as reported in many studies [146,216]. Lu and coworkers [216] reported a slight decrease in adsorption and desorption of Ni^{+2} while using CNTs, and on the other hand, a sharp decrease was observed in the case of AC [216].

The adsorption of heavy metals by the functional CNTs depends on the interactions between available functional groups on the surface of CNTs and particular contaminants rather than the size of nanotubes. For example, higher adsorption of As(III) and As(V) was reported by Addo Ntim and Mitra [203] by using zirconia-modified CNTs of the same diameter (20 to 40 nm) than by using the CNTs coated with iron oxide. These results demonstrate higher arsenic adsorption by zirconia nanocomposites than the CNTs modified with iron oxide with similar range of diameters. Therefore, based on these results, further investigations on the adsorption capacities of functionalized CNTs are needed in terms of their surface area rather than the size CNTs.

Table 8. Functional CNT based nanomaterials used for the removal of heavy metal ions from aqueous samples.

Adsorbate	Adsorbent	Surface Area (m^2/g)	Diameter (nm)	Q_e/RE	Experimental Conditions			Removal Mechanism	Model	Comments	Ref.
					pH	IC	AL				
As(III)	MWCNTs	9.1	10–40	91%	6	40 µg/L	2.0 g/L	Liquid film diffusion, ion exchange	Tempkin, Dubinin-Radushkevic, Langmuir, Freundlich	In column operation, the removal As(III) was up to 13.5 µg/L	[217]
	Zero-valent iron-doped MWCNT	-	-	200 mg/g	4	10 mg/L	0.2–4.0 g/L	Ion exchange, surface complexation	Langmuir	Maximum As(III) removal efficiency was 98.5%	[218]
	Floating catalyst CNTs (FCNT)	74	18.6	1.22 mg/g	6.5	0.1–10 mg/L	1 g/L	Electrostatic attraction, surface complexation	Langmuir	Potential adsorbent for removal to total arsenic	[219]
	Oxidized-FCNT	129	10.7	1.90 mg/g							
	Heat-treated oxidized CNTs (FCNT-HOX)	168	7	5.99 mg/g							
	Zero-valent iron immobilized on MWCNTs	78.78		111.1 mg/g	7	0.1–1 mg/L	0.05 g	Surface complexation	Langmuir	Reusability of adsorbent was up to 5 cycles	[219]
	MWCNT-ZrO$_2$	152	20–40	2 mg/g	6	100 µg/L	100 mg/10 mL	Chemisorption/physisorption	Langmuir	The adsorption capacity of AS (III) is not associated with pH value	[203]
	Iron-oxide-coated multi-walled carbon nanotubes	153	20–40	1.723 mg/g	4	100 µg/L	10 mg/10 mL	Electrostatic interaction, surface complexation	Langmuir	Suggesting that modifying MWCNTs with other groups can develop potential adsorbents for water treatment	[203]

Table 8. *Cont.*

Adsorbate	Adsorbent	Surface Area (m²/g)	Diameter (nm)	Q_e/RE	Experimental Conditions			Removal Mechanism	Model	Comments	Ref.
					pH	IC	AL				
As(V)	Iron-oxide-coated SWCNTs	-	-	49.65 mg/g	4	5–50 mg/L	-	Surface complexation	Freundlich	Adsorption was very fast for low concentration of As(V)	[220]
	MWCNTs	9.1	10–40	92%	6	40 µg/L	2.0 g/L	Liquid film diffusion, ion exchange	Tempkin, Dubinin-Radushkevic, Langmuir, Freundlich	In column operation, the removal As(III) was up to 14.0 µg/L	[217]
	Zero valent iron doped MWCNTs	-	-	250 mg/g	4	10 mg/L	0.2–4.0 g/L	Ion exchange, surface complexation	Langmuir	Maximum As(V) removal efficiency was 98.5%	[218]
	Floating catalyst CNTs (FCNT)	74	18.6	0.88 mg/g	6.5	0.1–10 mg/L	1 g/L	Electrostatic attraction, surface complexation	Langmuir	Potential adsorbent for removal to total Arsenic	[219]
	Oxidized-FCNT	129	10.7	2.51 mg/g							
	Heat-treated oxidized CNTs (FCNT-HOX)	168	7	6.37 mg/g							
	3-(2-aminoethylamino) propy-ltrimethoxysilane modified MWCNTs	-	-	8.01 mg/g	2.2	1.0 mg/L	40 mg	-	-	Cr(IV) was selectively adsorbed in the micro-column packed with adsorbent	[221]
	Zero-valent iron immobilized on MWCNTs	78.78		167 mg/g	7	0.1–1 mg/L	0.05 g	Surface complexation	Langmuir	Successfully applied to ground water with high pH	[219]

292

Table 8. *Cont.*

Adsorbate	Adsorbent	Surface Area (m^2/g)	Diameter (nm)	Q_e/RE	Experimental Conditions			Removal Mechanism	Model	Comments	Ref.
					pH	IC	AL				
	Iron(III)-oxide-coated ethylenediamine functionalized MWCNTs	198.5	5–10	23.5 mg/g	4	100 µg/L	50 mg/10 mL	Ion exchange	Langmuir	Greater efficiency to remove As(V) due to enormous adsorbing sites	[222]
	MWCNT–zirconia nanohybrid	152	20–40	5.0 mg/g	6	100 µg/L	100 mg/10 mL	Chemisorption/ physisorption	Langmuir	The adsorption capacity of As(V) is not associated with pH value	[223]
	Iron-oxide-coated MWCNTs	153	20–40	0.189 mg/g	4	100 µg/L	10 mg/10 mL	Electrostatic interaction, surface complexation	Langmuir	Modifying MWCNTs with other groups can develop potential adsorbents for water treatment	[203]
	Iron oxide/carbon nanotubes/chitosan magnetic composite film	64.4		66.25 mg/g	2–10	100 mg/L	0.3 mg/g	Electrostatic	Langmuir	Decrease in efficiency was 12% after reusing the adsorbent for ten cycles	[224]
Cr(III)	Nitrogen-doped magnetic carbon nanoparticles	-	-	83.7 mmol/g	8	200 mg/L	10 mg/500 mL	Chemical adsorption	Langmuir	10-fold greater removals than activated carbon due to large SSA	[225]
	Acid modified MWCNTs	-	23	0.5 mg/g	7	1 mg/L	120 mg/500 mL	Electrostatic interaction	Pseudo-second order	Increasing removal of Cr with increasing the dose of CNTs	[215]

293

Table 8. *Cont.*

Adsorbate	Adsorbent	Surface Area (m^2/g)	Diameter (nm)	Q$_e$/RE	Experimental Conditions			Removal Mechanism	Model	Comments	Ref.
					pH	IC	AL				
Cr(IV)	Iron oxide/carbon nan-otubes/chitosan magnetic composite film	64.4	-	449.3 mg/g	10-Feb	100 mg/L	0.3 mg/g	Electrostatic	Langmuir	Decrease in efficiency was 6% after reusing the adsorbent for ten cycles	[224]
	Nitrogen-doped magnetic CNTs	116.4	-	970.9 mg/g	1	40–1000 mg/L	0.5–3.5 g/L	Surface com-plexation	Langmuir	Recycled adsorbent was successfully used for excellent electrochemical reduction of CO$_2$	[163]
	Chitosan-modified MWCNTs	-	30–50	164.0 mg/g	2	50 mg/L	50 mg	Electrostatic	Langmuir	Adsorbent can be recycled up to 4 times	[226]
	Magnetic iron oxide MWCNTs	-	~50	42.0 mg/g	2	5 mg/L	0.4–1.0 g/L	Electrostatic	Langmuir	Absorbent highly showed durability, selectivity, easy regeneration ability	[227]
	Chitin magnetite MWCNTs	69.1	-	100%	2	50 mg/L	-	Physical	-	Removal of Cr(IV) was enhanced after mixing MWCNTs with chitin	[228]

Table 8. *Cont.*

Adsorbate	Adsorbent	Surface Area (m²/g)	Diameter (nm)	Q_e/RE	Experimental Conditions			Removal Mechanism	Model	Comments	Ref.
					pH	IC	AL				
	Magnetic MWCNTs	200	20–40	16.23 mg/g	3	25 mg/L	-	-	Langmuir	The adsorption capacity of adsorbent increases with initial concentration of Cr(VI) and contact time, but decreases with the increase of adsorbent dosage	[229]
295	3-(2-aminoethylamino)propyltrimethoxysilane-modified MWCNTs	-	-	9.79 mg/g	2.2	1.0 mg/L	40 mg	-	-	Cr(IV) was selectively adsorbed in the micro-column packed with adsorbent	[221]
	Activated-carbon-coated CNTs	-	10–20	9.0 mg/g	2	0.2–0.5 mg/L	2 mg/50 mL	-	Langmuir	The f-CNT can be used largely for the removal of Cr ions	[215]
	Ceria-supported CNTs nanoparticles	-	20–80	31.55 mg/g	7	35.3 mg/L	100 mg/100 mL	Ion exchange	Langmuir	Suggesting that CeO_2/ACNTs has high potential for heavy metal removals	[230]

Table 8. *Cont.*

Adsorbate	Adsorbent	Surface Area (m^2/g)	Diameter (nm)	Q_e/RE	Experimental Conditions			Removal Mechanism	Model	Comments	Ref.
					pH	IC	AL				
Pb(II)	Thiol-functionalizedM-WCNTs/Fe_3O_4	97.367	-	65.4 mg/g	6.5	50 mg/L	100 mg/100 mL	Lewis acid–base interactions	Langmuir	The adsorbent removed heavy metal ions effectively at various pH values	[231]
	Magnetic MWCNTs	295.4	-	N/A	6	100 mg/L	1000 mg	-	Experimental	High removal efficiency due to intrinsic properties, large SSA, and porous structure	[232]
	MWCNTs/Fe_3O_4	108.37	10–20	22.04 mg/g	5.3	30 mg/L	500 mg/1000 mL	Electrostatic, hydrophobic, and π–π interactions	Langmuir	Easily regenerate the adsorbent by external magnetic field after several cycles	[233]
	MWCNTs/Fe_3O_4 modified with 3-aminopropyltrie-thoxysilane	90.68	10–20	75.02 mg/g	5.3	30 mg/L	500 mg/1000 mL	Electrostatic, hydrophobic, and π–π interactions	Langmuir	Easily regenerate the adsorbent by external magnetic field after several cycles	[233]
	MWCNTs grafted/PAAM membrane	-	-	98%	-	10 mg/L	1000 mg/1000 mL	Electrostatic interaction	-	The f-CNT membrane potentially enhances the water flux and removal of heavy metals	[234]

Table 8. Cont.

Adsorbate	Adsorbent	Surface Area (m^2/g)	Diameter (nm)	Q_e/RE	Experimental Conditions			Removal Mechanism	Model	Comments	Ref.
					pH	IC	AL				
	Oxidized CNT sheets	-	-	117.65 mg/g	7	1200 mg/L	50 mg/25 mL	Chemical interaction	Langmuir	Considering the oxidize CNT sheets promising nanomaterial for adsorption	[235]
	MWCNTs grafted with 2-Vinylpyridine	-	-	37.0 mg/g	6	10 mg/L	640 mg/1000 mL	Ion exchange, electrostatic interaction	Langmuir	Showed high suitability for preconcentration and immobilization of heavy metal ions from water	[236]
297	Oxidized MWCNTs	142.29	10–30	0.021 mmol/g	4.1	0.83 mmol/L	0.75 g/L	Chemical, electrostatic, hydrophobic, and π–π interactions	Langmuir	High removal efficiency toward heavy metal ions in wastewater	[214]
	Alumina-coated MWCNTs	-	-	99%	Different	-	10 mg/25 mL	N/A	-	The composite can be used largely to remove lead from industrial wastewater. Adsorption efficiency increased with the pH (3 to 7)	[236]
	Nitrogen-doped magnetic carbon nanoparticles		-	6.74 mmol/g	8	200 mg/L	10 mg/500 mL	Chemical adsorption	Experimental	High removal efficiency toward Pb compared to Cr	[225]

Table 8. *Cont.*

Adsorbate	Adsorbent	Surface Area (m^2/g)	Diameter (nm)	Q_e/RE	Experimental Conditions			Removal Mechanism	Model	Comments	Ref.
					pH	IC	AL				
	Titanium Dioxide /MWCNT composites	-	-	137.0 mg/g	6	10 mg/L	20 mg/10 mL	-	Langmuir	Important adsorption ability to remove large amount of Pb(II) in short period	[204]
Pb(II)	Oxidize MWCNTs	-	20–30	-	-	10 mg/L	3000 mg/1000 mL	-	-	The sorption of Pb largely depends on foreign ions and ionic strength	[237]
	Manganese oxide-coated CNTs	275	2.60	78.74 mg/g	5	30 mg/L	50 mg/100 mL	Electrostatic interaction, surface complexation	Langmuir	300% greater adsorption capacity than raw CNTs	[238]
	Acidified MWCNTs	237.3	29.0	85 mg/g	5	50 mg/L	25 mg/50 mL	Physical adsorption	Langmuir	The regeneration of Pb increasing with decreasing pH and can be used for several cycles	[239]
Cd(II)	Alumina-decorated MWCNTs	109.8	10–20	27.21 mg/g	7	1 mg/L	50 mg/L	Electrostatic interaction, physical adsorption, surface precipitation	Langmuir	Capable of removing both metallic and organic Contaminants	[240]

Table 8. *Cont.*

Adsorbate	Adsorbent	Surface Area (m²/g)	Diameter (nm)	Q_e/RE	Experimental Conditions			Removal Mechanism	Model	Comments	Ref.
					pH	IC	AL				
	Oxidized MWCNTs	78.5	16.09	24.15 mg/g	-	5 mg/L	1 mg/10 mL	Chemisorption	Langmuir	The sorption capacity is strongly dependent on pH due to surface charge and showed best performance in the pH ranging from 6 to 10	[209]
	Ethylenediamine-functionalized MWCNTs	101.2	21.25	25.7 mg/g	-	5 mg/L	1 mg/10 mL	Chemisorption	Langmuir	The sorption capacity is strongly dependent on pH due to surface charge and showed best performance in the pH ranging from 6 to 10	[209]
	Oxidized CNT sheets	-	-	92.59 mg/g	7	1200 mg/L	50 mg/25 mL	Chemical interaction	Langmuir	Excellent removal of heavy metal ions	[235]
	Acid-modified CNTs	170	10–20	4.35 mg/g	7	-	50 mg	Electrostatic interaction	Langmuir	Potential material for water purification	[241]
	MWCNTs modified with Chitosan	-	60–100	-	-	-	2000 mg	Electrostatic interaction	-	The removal efficiency increases with increase of mass of both MWCNTs and chitosan	[242]

Table 8. *Cont.*

Adsorbate	Adsorbent	Surface Area (m^2/g)	Diameter (nm)	Q_e/RE	Experimental Conditions			Removal Mechanism	Model	Comments	Ref.
					pH	IC	AL				
	MnO_2-coated CNTs	110.4	30–50	58.82 mg/g	5–7	10 mg/L	200 mg/20 mL	Electrostatic interaction	Langmuir	Higher adsorption affinity to other heavy metals rather than Hg	[149]
	Thiol-derivatized SWCNTs	-	-	131.58 mg/g	5	40 mg/L	0.25 mg/mL	Electrostatic interaction	Langmuir	Easily des-orb/regenerate Hg after treatment of water	[243]
Hg(II)	Amino and thiolated functionalized-MWCNTs	-	5–10	84.66 mg/g	6	100 mg/L	60 mg	Physisorption	Langmuir	Highly efficient removal from real wastewater and further research is necessary to commercialize	[244]
	Iodide-incorporated MWCNT (CNT-I)	153	10–20	123.45 mg/g	6	100–500 mg/L	2500 mg/1000 mL	Ion exchange	Langmuir	Successfully used for the adsorption and desorption of Hg(II)	[205]
	Sulphur-containing CNTs	-	-	72.8 µg/g	12.15	0.1mg/L	100 mg/20 mL	Chemisorption	Freundlich	Greater treatment ability for industrial wastewater containing Hg and other anions and cations	[245]
Hg(II)	Thiol-functionalized-MWCNTs/Fe_3O_4	97.2	-	65.52 mg/g	6.5	50 mg/L	1000 mg/100 mL	Lewis acid–base interactions	Langmuir	Better removal of heavy metals in different pH concentration	[231]

Table 8. *Cont.*

Adsorbate	Adsorbent	Surface Area (m^2/g)	Diameter (nm)	Q_e/RE	Experimental Conditions			Removal Mechanism	Model	Comments	Ref.
					pH	IC	AL				
	Oxidized MWCNTs	-	-	3.83 mg/g	7	10–100 µg/L	25 mg/50 mL	Electrostatic interaction	Langmuir	Small diameter of CNTs removing greater amount of Hg(II) from aqueous solution	[246]
	Functionalized MWCNTs	250	10–25	2.42 mg/g	10	1.1 mg/L	0.09 g	Electrostatic interaction	Langmuir	Excellent potential for the removal of heavy metal ions	[247]
Zn(II)	Oxidized CNTs	-	-	74.63 mg/g	7	1200 mg/L	50 mg/25 mL	Chemical interaction	Langmuir	Economically feasible material with excellent heavy metal ion removal efficiency without any CNTs leakage	[235]
	Chitosan-MWCNTs	-	60–100	N/A	7	-	200 mg	Electrostatic interaction	N/A	The removal efficiency increases with increase of mass of both MWCNTs and Chitosan	[242]
	Nitrogen-doped magnetic carbon nanoparticles	-	-	9.31 mmol/g	8	12.82 mg/L	10 mg/500 mL	Chemical adsorption	Langmuir	Higher specific surface area and nitrogen make the nanomaterial an excellent adsorbent	[225]
Zn(II)	Oxidized MWCNTS	-	14	0.27 mmol/g	6.5–6.8	15 mg/L	5 mg/5 mL	Electrostatic interaction	Langmuir	Further research is necessary to understand the full mechanism	[199]

Table 8. *Cont.*

Adsorbate	Adsorbent	Surface Area (m^2/g)	Diameter (nm)	Q_e/RE	Experimental Conditions			Removal Mechanism	Model	Comments	Ref.
					pH	IC	AL				
Cu(II)	Sodium-hypochlorite-treated MWCNTs	-	<10	34.36 mg/g	-	60 mg/L	50 mg/100 mL	Electrostatic interaction	Langmuir	Zinc ion could be easily regenerated, and the adsorbent can be used for many cycles	[211]
	Sulfonated MWCNTs	28.7	-	43.16 mg/g	5	20 mg/L	25 mg/50 mL	Electrostatic interaction, surface complexation	D–R model	Enabling CNTs for wastewater treatment and composite formation or physical blending	[248]
	Magnetic MWCNTs	-	10–20	38.91 mg/g		30 mg/L	200 mg/1000 mL	Electrostatic interaction, physical interaction	-	Easily regenerate the Cu after removal from polluted water	[232]
	Oxidized CNT sheets	-	-	64.93 mg/g	7	200 mg/L	50 mg/25 mL	Chemical interaction	Langmuir	Considering the oxidize CNT sheets promising nanomaterial for heavy metal adsorption	[235]
	Chitosan/poly(vinyl) functionalized MWCNTs	-	5–20	11.1 mg/g	5.5	30 mg/L	0.5–2 wt%	Ion exchange	Langmuir	No loss in the adsorption capacity after four regeneration cycles	[249]
	MWCNTs modified with Chitosan	-	60–100	>95%	-	-	2000 mg	Electrostatic interaction	-	The removal efficiency increases with increase of mass of both MWCNTs and chitosan	[242]

Table 8. *Cont.*

Adsorbate	Adsorbent	Surface Area (m^2/g)	Diameter (nm)	Q_e/RE	Experimental Conditions			Removal Mechanism	Model	Comments	Ref.
					pH	IC	AL				
	Chitosan-grafted MWCNTs	-	-	24.0 mg/g	-	10mg/L	1000 mg/1000 mL	N/A	-	Effective preconcentration and solidification of heavy metals in aqueous samples	[250]
Co(II)	Poly(acrylic acid)-grafted MWCNTs	-	-	1.66×10^{-4} mol/g	6.8	1.69×10^{-4} mol/L	1.0 g/L	Surface complexation	Langmuir	Promising ability to use in water purification	[251]
	MWCNT$_S$/ iron oxide	-	-	0.18 mmol/g	6.4	4.2 mg/L	0.5 g/L	Ion exchange, surface complexation	Langmuir	Highlights the interaction between heavy metals and organic substances in wastewater	[252]
	Oxidized CNT sheets	-	-	85.74 mg/g	7	1200 mg/L	50 mg/25 mL	Chemical interaction	Langmuir	Considering the oxidized CNT sheets promising material for the removal of heavy metal ions	[235]
Ni(II)	HNO$_3$-treated MWCNTs	102	10–20	17.86 mg/g	6.5	20 mg/L	0.8 g/L	Ion exchange	Langmuir	Better removal efficiency toward heavy metal ions	[253]
	MWCNTs modified with Chitosan	-	60–100	90%	-	-	2000 mg	Electrostatic interaction	-	The removal efficiency increases with an increase of mass of both MWCNTs and Chitosan	[242]
	Nitrogen-doped magnetic carbon nanoparticles	-	-	8.06 mmol/g	8	12.82 mg/L	10 mg/500 mL	Chemical adsorption	Langmuir	The removal efficiency was not very good for Ni compared to Cr	[225]

Table 8. *Cont.*

Adsorbate	Adsorbent	Surface Area (m^2/g)	Diameter (nm)	Q_e/RE	Experimental Conditions			Removal Mechanism	Model	Comments	Ref.
					pH	IC	AL				
	Poly(acrylic acid) (PAA)-oxidized MWCNTs	197	-	6.615×10^{-6} mol/g	5.4	5 mg/L	0.8 g/L	Electrostatic interaction, π–π interaction	Langmuir	Effective preconcentration and solidification of Ni(II) in liquid samples	[60]
	NaClO-modified SWCNTs	-	380	47.86 mg/g	7	10–80 mg/L	50 mg/100 mL	Electrostatic interaction	Langmuir	High removal affinity to heavy metals and can be used for water treatment	[216]
	MWCNTs/ Iron oxide	-	-	9.18 mg/g	-	6 mg/L	0.75 g/L	Ion exchange	Langmuir	Promising candidate for the solidification and preconcentration of heavy metal ions as well as for radionuclides from water	[201]
	Oxidized MWCNTs	-	5.5–14	49.26 mg/g	-	10–200 mg/L	20 mg/50 mL	Electrostatic interaction	-	Greater adsorption ability than raw MWCNTs in water	[254]
	Oxidized MWCNTs	197	10–30	>80%	8	6–20 mg/L	50 mg/200 mL	Electrostatic interaction	-	Excellent material for the adsorption of metal ions.	[255]
Ni(II)	MWCNTs	40–600	40–60	6.09 mg/g	7	25 mg/L	5 g/L	Ion exchange, surface complexation, chemical interaction	-	Excellent sorption of Ni^{+2} ions with smaller equilibrium time	[256]

Table 8. *Cont.*

Adsorbate	Adsorbent	Surface Area (m²/g)	Diameter (nm)	Q_e/RE	Experimental Conditions			Removal Mechanism	Model	Comments	Ref.
					pH	IC	AL				
U(II)	Diglycolamide-functionalized MWCNT (DGA-MWCNTs)	300–600	-	133.74 mg/g	7	-	1–10 mg	-	Langmuir	Adsorption efficiency increased with the increasing dose of adsorbent and temperature	[257]
Sr(II)	Oxidized-MWCNTs	-	-	36%	2-11	-	3 g/L	-	Diffuse layer model	Adsorption efficiency increased with increasing pH but decreased with the ionic strength	[258]
Eu (III)	Oxidized-MWCNTs	-	-	96%	2-11	-	3 g/L	-	Diffuse layer model	Higher adsorption efficiency for Eu(III) than Sr(II)	[258]

Q_e = Max adsorption capacity (mg/g); RE = Removal efficiency; IC = Initial concentration; AL = Adsorbent loading; - = Data not available

6.2. Removal of Organics

Organics due to human actions [259], animals, or plant deterioration [260] are present in water and wastewater, usually in the form of dissolved and/or suspended particulate matters [135]. Zare et al. [261] reported that the adsorption efficiencies of CNTs for organic dyes can be enhanced after the functionalization. Methylene blue and orange were effectively removed from the water matrix by using oxidized MWCNTs [262,263] as compared to other types of modified adsorbents. Duman et al. [264] successfully used a novel nanocomposite, MWCNTs/carrageenan/Fe_3O_4, for effective removal of crystal violet. New nanocomposites can also be used for CNT modification to increase the adsorption capacity of cationic dyes. Sadegh and coworkers [265] reported that the adsorption of amide black can be significantly enhanced by using MWCNT-COOH-cysteamine. The results showed a maximum adsorption capacity of 131 mg g^{-1} for MWCNT-COOH-cysteamine, while it was 90 mg g^{-1} for MWCNT-COOH [265].

The adsorption capacities of CNT composites can be reduced effectively because of competition that occurred between several types of organic contaminants present in water and wastewater at the same time [266], similarly to as described for heavy metal adsorption. However, as reported by Ali et al. [135], the initial concentration did not have any significant effect on the removal capacities of CNTs for the removal of inorganic contaminants. As a result, customization and modifications of the surface of CNTs make them significant for selective adsorption of organic contaminants. Moreover, Wang et al. showed about 95% efficiency of the CNT composite for pharmaceutical and personal care product (PPCP) removal by increasing the specific surface area and aromatic ring of CNTs [267,268]. Jahangiri-Rad et al. [269] showed an adsorption capacity of 496 mg g^{-1} by using oxidized SWCNTs with large specific surface for the removal of blue 29 dye.

Adsorption capacities and mechanisms responsible for the interaction between organic contaminants and functionalized CNTs at different experimental conditions are described in Table 9. The increase in oxygen-containing functional groups on the surface of nanotubes results in a decrease in natural organic maters (NOM), because the π–π interaction decreases due to more electrostatic repulsion [270]. The same type of results were shown in the case of higher pH [270]. Therefore, π–π interactions are responsible between the large specific surface area of functionalized CNTs and NOM for the adsorption area of CNT [271,272]. Yang et al. [208] also reported similar results, as organics may be absorbed by π–π interactions occurring between the surface of CNTs and aromatic rings of 1–naphthol.

CNTs have also been effectively used for the removal of pesticides [170]. Oxidized and as prepared MWCNTs showed excellent removal efficiency for diuron at pH $\geq$ 7.0 [273]. Oxidation of CNTs leads to the increase of their surface area, and pore size results in the higher adsorption of diuron [170]. Hamdi et al. [274] reported a reduction in uptake of chlordane and p,p'-dichlorodiphenyldichloroethylene from 78% to 23% in the roots of a lettuce crop by the addition of CNTs functionalized with amino group. The pesticides (1-pyrenebutyric, 2,4-dichlorophenoxyacetic, and diquat dibromide) were absorbed more (up to 70.6%) on semiconductor CNTs than metallic CNTs [275]. The lower density of electrons on the surface of semiconductor CNTs were found to facilitate higher adsorption [275]. On the other hand, during the batch modes, the pesticide removal was found to be limited [276]. SWCNTs and MWCNTs were used for the removal of diquat dibromide in the fixed bed system. The results were not higher than those of the batch system, but the pesticide absorbed completely in the fixed bed system by increasing the time [276].

The CNTs are also being used for nano-filtration of contaminants from aqueous solutions [162,277,278]. For adsorbents, the selectivity of nano-filters can be controlled by attaching different functional groups to the surface of CNTs [279]. Beside the hydrophobicity of CNTs, molecular dynamics simulations showed a weak interaction between water and nanotubes [280]. Hummer [281] explained that friction-free water flow was caused by the nano-scale pore size, which makes interaction energy smaller and finally lowers the interactions with water [281].

Table 9. Functionalized CNTs used for organic pollutants removal.

Adsorbent	Dye Pollutants	Surface area (m^2/g)	Q (mg/g)	Removal percentage (%)	Optimum conditions	Remarks	Ref.
Oxidized SWCNT	Basic red 46 (BR 46)	400	49.45	-	pH 9, IC = 150 mg/L, AL = 0.05 g, Contact time = 100 min, 298 K,	Exothermic process favored at lower temperature range Orderly adsorption of dye due to negative entropy	[282]
HNO3-oxidized MWCNTs	Bromothymol blue (BTB)	96.8	55	97	pH 1, IC = 30 mg/L, AL = 0.02 g, T = 293.15 K,	Endothermic process of adsorption significantly affected by pH, initial concentration, sorbent dosage, and contact time	[262]
Functionalized CNT/Mg(Al)O	Congo red	148	1250	94	pH 7, AL = 30 mg, contact time = 75 min	Strong electrostatic interactions between dye particles and functional groups associated with the surface of nanomaterial	[283]
Magnetic MWCNTs-Fe3C nanocomposite	Direct red 23	38.7	85.5	-	pH 3.7, IC = 54 mg/L, AL = 0.04 g, T = 333 K,	Spontaneous endothermic adsorption process due to positive enthalpy	[284]
SWCNT–COOH	Malachite Green	400	22.33	-	pH 7, IC = 10 mg/L, 300 K,	Adsorption significantly affected by ionic strength, initial concentration, sorbent mass, contact time, and temperature More active functional groups on SWCNT-NH2 adsorbed more dye than SWCNT-COOH	[285]
SWCNT-NH2	Malachite Green	400	29.36	-	pH 7, IC = 10 mg/L, T = 300 K,		
SWCNT–COOH	Methyl orange	400	25	-	pH 7, IC = 10 mg/L, T = 300 K,		
SWCNT-NH2	Methyl orange	400	27.15	-	pH 7, IC = 10 mg/L, T = 300 K,		
Oxidized MWCNTs	Methyl orange	165	10	-	AL = 20 mg/L, T = 313 K, stirring speed = 500 rpm	Initially, rapid adsorption was observed, but it slowed down with the time As the mixture temperature, agitation speed, and initial concentration increased, the adsorption efficiency also increased	[286]

Table 9. *Cont.*

Adsorbent		Q (mg/g)	Surface area (m2/g)	Removal percentage (%)	Optimum conditions	Remarks	Ref.
Functionalized-CNTs loaded TiO_2	Methyl orange	-	42.85	100	pH 6.5, IC = 5 mg/L, contact time = 30 min, T = 298 K	Highly active hydroxyl and amine functional groups made TiO_2-CNT composite an effective adsorbent	[287]
Thiol-functionalized MWCNT (MWCNT-SH)	Methylene blue	400	166.67	-	pH 6, IC = 10 mg/L, AL = 20 mg, T = 298 K, Contact time = 60 min,	As the temperature and initial concentration increased, the adsorption efficiency also increased	[288]
Adsorbent	**Phenol and its derivatives pollutants**	**Q (mg/g)**	**Surface area (m2/g)**	**Removal percentage (%)**	**Optimum conditions**	**Remarks**	**Ref.**
KOH-modified MWCNTs	Bisphenol-A	0.20 mmol/g	494.48	-	pH 6, IC = 40 mg/L, contact time = 5 min, T = 298 K,	As the pH increased, the adsorption capacity decreased because of deprotonating; both the negatively charged function groups and adsorbates repel each other	[289]
HNO_3-modified MWCNTs	Bisphenol-A	0.59 mmol/g	153.79	-	pH 6, IC = 40 mg/L, contact time = 30 min, T = 298 K,		
$SOCl_2/NH_4OH$-modified CNT	Bisphenol A	69.93	94.8	-	pH 6.5, IC = 10 mg/L, AL = 0.125 g/L, T = 280 K,	Adsorption efficiency increased with the initial concentrations	[290]
NH_3-treated MWCNTs	Chlorophenols (CP)	110.3	195	-	pH 3.8, T = 298 K,	The adsorption capacity increased due to higher pores size, π–π interactions, and hydrophobicity of nanocomposite Effective nanomaterial with smaller equilibrium time	[291]
HNO_3 and $KMnO_4$-Functionalized MWCNTs	Phenol	76.92	-	88	IC = 500 mg/L, Agitation speed = 200 rpm, T = 298 K	Adsorption capacity can be greatly affected by pH and adsorbent mass	[292]
Oxidized SWNTs	*p*-Nitrophenol (PNP)	206	-	97.9	IC = 0.01 mg/mL, agitation time = 30 min T = 293 K	Open ends of nanotubes, higher surface area, and the functional groups (hydroxyl, carbonyl, and carboxyl) were responsible for higher adsorption	[293]

Table 9. *Cont.*

Adsorbent	Contaminant	Q			Conditions	Remarks	Ref
Nitrogen-doped carbon nanotubes (CNx)	Phenol	0.16 mmol/g	102	-	pH 7, AL = 0.6mmol/L, 298 K,	π–π interaction occurred between the functional groups and phenol; more oxidized CNTs adsorbed less phenol	[294]
MWCNT-COOH	Phenol	0.15 mmol/g	-	-	IC = 0.417 mg/L, AL = 10 mg, T = 293 K	Higher adsorption of CP than phenol resulted because of the different solubility of these contaminants	[295]
	3-Chlorophenol (CP)	0.37 mmol/g	-	95	IC = 1.25 mg/L, AL = 10 mg, T = 293 K		
Acid-functionalized MWCNT (MWCNT-COOH)	2-Nitrophenol	256.41	197.83	-	pH 5.5, IC = 45 mg/L, T = 298 K,	Excellent adsorbent due to strong interactions between 2–Nitrophenol and surface functional groups	[296]

Q = adsorption capacity, IC = Initial concentration, AL = Adsorbent loading and T = Temperature.

6.3. Removal of Microorganisms

Bacteriological contaminants, deteriorating the assimilative capacity of water bodies leading to diverse impacts on surrounding environment, are a major challenge [297]. Bacteriological contaminants are often found in surface waters, wastewaters, and respective treatment plants [145,298]. CNTs, with their diverse range of surface and functional characteristics, have a high-affinity interaction with biological contaminants. CNTs have been proven for their higher adsorption capacities, inactivation efficiencies of viral or bacterial spores, and greater antimicrobial potential than conventional sorbents because of their greater surface area [170,299,300].

Previous studies that have reported that CNTs can inactivate or remove a variety of microorganisms, including bacteria, are shown in Table 10. The inactivation of *E. coli* was observed by using SWCNTs because of its penetration into the bacterial cell wall [301]. In addition, surface-modified CNTs with different chemical groups destroy the cell wall of microorganisms more strongly than the raw and polymer-grafting CNT membranes [135]. Furthermore, the straight interaction of microbes (i.e., *E. coli*) and functionalized CNTs can cause adverse effects on the metabolisms and morphological structure of the cell wall of bacteria [135]. The penetration of CNTs into the cell wall of microorganisms is the main reason for their higher inactivation affinity [301].

CNTs functionalized with silver nanoparticles showed excellent ability to inactivate microorganisms. For example, Ihsanullah and his coworkers [302] synthesized and successfully used silver-doped CNTs for the inactivation of *E. coli* bacteria. The results proved that 100% of the bacteria were killed due to toxicity of synthesized silver doped nanomaterial [302]. In addition, many researchers reported that the diameter of nanotubes can be an important factor of concern for inactivation pathogenic microorganisms (Table 10) in water and wastewater.

The single kinetics of CNTs also reveal the CNTs' ability to eliminate pathogens in water and wastewater treatment, and microbes remain on the CNT surface based on deep filtration mechanisms [303,304]. Brady and coworkers [299] used a poly vinylidene fluoride-based SWCNT filter for excellent removal of *E. coli* bacteria at low pressure. The results show that the cells were completely captured by the filter [299]. In addition to filtrations, an excellent removal efficiency of MS2 virus was observed by filtering the sample through a controlled nano porous CNT based filter at a pressure of about 8 to 11 bar [304]. Beside the filtrations, CNTs are widely used in water and wastewater treatment as an antimicrobial agent, as described earlier. This behavior makes them a replacement for chemical disinfectants as a new method for controlling pathogens [301,305–308]. The applications of CNTs for disinfection treatment in water and wastewater avoid the materialization of unsafe disinfectant by-products like trihalomethanes, aldehydes, and haloacetic acids due to their low oxidation state and solubility in water. Therefore, it is necessary to promote the dispersibility of these compounds; a surfactant or a polymer such as sodium dodecylbenzenesulfonate, polyvinylpyrrolidone is used [309]. Due to the excellent mechanical properties on CNTs, they act as scaffolds for antimicrobial agents, such as silver nanoparticles [310,311] and antibacterial lysozyme [307,309].

Table 10. Functionalized CNTs used for disinfection.

Contaminants	Adsorbents		AL	IC	RE	Removal Mechanism	Comments	Ref
	Types							
Escherichia coli (E. coli)	Silver-doped CNT membrane		-	1×10^6 CFU/mL	100%	-	All the bacteria were inactivated by membrane with 10% silver loadings in 60 min only	[312]
	Silver-nanoparticle-loaded CNTs		2.5 µg/mL	10^6 CFU/mL	89%	-	Effectively inactivate the pathogen from wastewater effluents, resistance toward bacterial adhesion	[313]
	Chitosan/CNT nanocomposites		2 wt%	1.5×10^8 to 5.0×10^8 CFU/mL	2.89 log reduction	Physical interaction and surface complexation	Higher antimicrobial activity at the low contact time (10 min) and low concentration (1%)	[314]
	Acidic-conditioned MWCNTs		200 µg/mL	10^6 to 10^9 CFU/mL	-	Steric obstruction	Inactivation of pathogens was due to both MWCNT functionalization and nutrition level	[315]
	1-octadecanol-functionalized MWCNTS		0.2 g/100 mL	3.5×10^7 CFU/mL	100%	Polarization	The interaction of microwaves with f-CNTs is an innovative approach that has the potential to be employed for water disinfection	[316]
Staphylococcusaureus	CNT–Ag nanohybrid		2.5 µg/mL	10^6 CFU/mL	100%	-	Effectively inactivate the pathogen from wastewater effluents, resistance toward bacterial adhesion	[313]
	Chitosan/CNTs nanocomposites		2 wt%	1.5×10^8 to 5.0×10^8 CFU/mL	4.9 log reduction	Physical interaction and surface complexation	Higher antimicrobial activity at the low contact time (10 min) and low concentration (1%)	[314]
Aspergillus flavus	Chitosan/CNTs nanocomposites		2 wt%	1.5×10^8 to 5.0×10^8 CFU/mL	5.5 log reduction	Physical interaction and surface complexation	Higher antimicrobial activity at the low contact time (10 min) and low concentration (1%)	[314]

RE = removal efficiency (%)/log reduction; AL = adsorbent loading; IC = initial concentration, mg/L; CFU/mL = colony-forming unit per milliliter.

311

7. Conclusions

Purifying water from assorted contaminants is challenging, and carbon nanotube-based nanocomposites can provide simple as well as effective water decontamination/disinfection. This review shows that functionalized CNTs are a new generation of pollution management materials. These materials have excellent adsorption capacities and work effectively in removing organics, inorganics, and biological species. Various sorption mechanisms include physical adsorption, electrostatic interaction, surface complexation, and chemical interactions between surface functional groups and metal ions. The effects of pH, CNT dosage, time, ionic strength, temperature, and surface charge on the adsorption of heavy metal ions on carbon nanotube surface were also discussed. Even functionalized CNTs also has antibacterial efficacy against Gram-positive and Gram-negative bacteria.

Almost all the studies show effective removal of contaminants and have been performed using deionized water, but the potential of functionalized CNT nanomaterials needs to be verified under real water conditions. In wastewater, carbonates, phosphates, and silicates can successfully compete with target metals/organics for adsorption sites in nanostructures. Other important components of source water and wastewater are natural organic materials, such as fulvic acid and humic acid, which will occupy the surface of CNTs and thus affect the adsorption of contaminants on the nanostructures. The effectiveness of CNT-based nanotechnology should be evaluated under real water conditions.

Many researchers have also focused on evaluating the adsorption–desorption capacity to make it a cost-effective adsorbent for use in wastewater treatment. However, more studies are encouraged to check the feasibility of reuse.

Author Contributions: Conceptualization, W.D. and M.M.-A.A.; resources, W.D. and H.-W.K.; writing—original draft preparation, M.M.-A.A.; writing—review and editing, H.-W.K., W.D., M.S., M.U., M.M.-A.A.; visualization, H.A.; supervision, H.-W.K. All authors have read and agreed to the published version of the manuscript.

Funding: This research received no external funding.

Institutional Review Board Statement: Not applicable.

Informed Consent Statement: Not applicable.

Data Availability Statement: Data are contained within the article.

Acknowledgments: This study was done at the Department of Environmental Science and Engineering, Tunghai University, Taiwan. The authors gratefully acknowledge the funding support by the Research Council of Texas A&M University San Antonio, and the Ministry of Science and Technology of R.O.C. (Taiwan), 107-2221-E-029-001-MY2. The authors would also like to thank Yu-Ling Wei, Chiung-Fen Chang, and Meng-Hau Sung for providing financial and analytical support to M.M.A.A. We acknowledge support of the Hamburg University of Technology (TUHH) by enabling open access publishing through funding programme Open Access Publishing.

Conflicts of Interest: The authors declare no conflict of interest.

References

1. Ali, I.; Li, J.; Peng, C.; Qasim, M.; Khan, Z.M.; Naz, I.; Sultan, M.; Rauf, M.; Iqbal, W.; Sharif, H.M.A. 3-Dimensional membrane capsules: Synthesis modulations for the remediation of environmental pollutants—A critical review. *Crit. Rev. Environ. Sci. Technol.* **2020**, 1–62. [CrossRef]
2. Usman, M.; Waseem, M.; Mani, N.; Andiego, N. Optimization of soil aquifer treatment by chemical oxidation with hydrogen peroxide addition. *Pollution* **2018**, *4*, 369–379. [CrossRef]
3. Usman, M. New Applications of Fine-Grained Iron Oxyhydroxides as Cost-Effective Arsenic Adsorbents in Water Treatment. Ph.D. Thesis, Technische Universität Hamburg, Hamburg, Germany, 2020. [CrossRef]
4. Islam, T.; Peng, C.; Ali, I.; Li, J.; Khan, Z.M.; Sultan, M.; Naz, I. Synthesis of rice husk-derived magnetic biochar through Liquefaction to adsorb anionic and cationic dyes from aqueous solutions. *Arab. J. Sci. Eng.* **2021**, *46*, 233–246. [CrossRef]
5. Khan, S.U.; Farooqi, I.H.; Usman, M.; Basheer, F. Energy efficient rapid removal of arsenic in an electrocoagulation reactor with hybrid Fe/Al electrodes: Process optimization using CCD and kinetic modeling. *Water* **2020**, *12*, 2876. [CrossRef]

6. Usman, M.; Katsoyiannis, I.; Rodrigues, J.H.; Ernst, M. Arsenate removal from drinking water using by-products from conventional iron oxyhydroxides production as adsorbents coupled with submerged microfiltration unit. *Environ. Sci. Pollut. Res.* **2020**. [CrossRef] [PubMed]
7. Amjed, M.A.; Peng, C.; Dai, M.; Chang, Q.; Ali, I.; Sultan, M.; Naz, I.; Farooq, M.Z.; Kashif, M. Recent updates on the solar-assisted biochar production and potential usage for water treatment. *Fresenius Environ. Bull.* **2020**, *29*, 5616–5632.
8. Cheng, N.; Wang, B.; Wu, P.; Lee, X.; Xing, Y.; Chen, M.; Gao, B. Adsorption of emerging contaminants from water and wastewater by modified biochar: A review. *Environ. Pollut.* **2021**, *273*, 116448. [CrossRef] [PubMed]
9. Aslam, M.M.A.; Khan, Z.M.; Sultan, M.; Niaz, Y.; Mahmood, M.H.; Shoaib, M.; Shakoor, A.; Ahmad, M. Performance evaluation of trickling filter-based wastewater treatment system utilizing cotton sticks as filter media. *Polish J. Environ. Stud.* **2017**, *26*, 1955–1962. [CrossRef]
10. Khan, Z.M.; Kanwar, R.M.A.; Farid, H.U.; Sultan, M.; Arsalan, M.; Ahmad, M.; Shakoor, A.; Aslam, M.M.A. Wastewater evaluation for multan, pakistan: Characterization and agricultural reuse. *Polish J. Environ. Stud.* **2019**, *28*, 2159–2174. [CrossRef]
11. Aslam, M.M.A.; Den, W.; Kuo, H.W. Encapsulated chitosan-modified magnetic carbon nanotubes for aqueous-phase CrVI uptake. *J. Water Process Eng.* **2021**, *40*, 101793. [CrossRef]
12. Den, W.; Wang, C.J. Removal of silica from brackish water by electrocoagulation pretreatment to prevent fouling of reverse osmosis membranes. *Sep. Purif. Technol.* **2008**, *59*, 318–325. [CrossRef]
13. Su, Y.N.; Lin, W.S.; Hou, C.H.; Den, W. Performance of integrated membrane filtration and electrodialysis processes for copper recovery from wafer polishing wastewater. *J. Water Process Eng.* **2014**, *4*, 149–158. [CrossRef]
14. Ali, I.; Peng, C.; Khan, Z.M.; Naz, I.; Sultan, M.; Ali, M.; Abbasi, I.A.; Islam, T.; Ye, T. Overview of microbes based fabricated biogenic nanoparticles for water and wastewater treatment. *J. Environ. Manag.* **2019**, *230*, 128–150. [CrossRef]
15. Ali, I.; Peng, C.; Khan, Z.M.; Naz, I.; Sultan, M. An overview of heavy metal removal from wastewater using magnetotactic bacteria. *J. Chem. Technol. Biotechnol.* **2018**, *93*, 2817–2832. [CrossRef]
16. Gupta, V.K.; Agarwal, S.; Saleh, T.A. Chromium removal by combining the magnetic properties of iron oxide with adsorption properties of carbon nanotubes. *Water Res.* **2011**, *45*, 2207–2212. [CrossRef]
17. Usman, M.; Belkasmi, A.I.; Katsoyiannis, I.A.; Ernst, M. Pre-deposited dynamic membrane adsorber formed of microscale conventional iron oxide-based adsorbents to remove arsenic from water: Application study and mathematical modeling. *J. Chem. Technol. Biotechnol.* **2021**. [CrossRef]
18. Chai, W.S.; Cheun, J.Y.; Kumar, P.S.; Mubashir, M.; Majeed, Z.; Banat, F.; Ho, S.-H.; Show, P.L. A review on conventional and novel materials towards heavy metal adsorption in wastewater treatment application. *J. Clean. Prod.* **2021**, *296*, 126589. [CrossRef]
19. Usman, M.; Zarebanadkouki, M.; Waseem, M.; Katsoyiannis, I.A.; Ernst, M. Mathematical modeling of arsenic(V) adsorption onto iron oxyhydroxides in an adsorption-submerged membrane hybrid system. *J. Hazard. Mater.* **2020**, *400*, 123221. [CrossRef] [PubMed]
20. Kyzas, G.Z.; Matis, K.A. Nanoadsorbents for pollutants removal: A review. *J. Mol. Liq.* **2015**, *203*, 159–168. [CrossRef]
21. Trujillo-Reyes, J.; Peralta-Videa, J.R.; Gardea-Torresdey, J.L. Supported and unsupported nanomaterials for water and soil remediation: Are they a useful solution for worldwide pollution? *J. Hazard. Mater.* **2014**, *280*, 487–503. [CrossRef]
22. Sharma, V.K.; McDonald, T.J.; Kim, H.; Garg, V.K. Magnetic graphene–carbon nanotube iron nanocomposites as adsorbents and antibacterial agents for water purification. *Adv. Colloid Interface Sci.* **2015**, *225*, 229–240. [CrossRef] [PubMed]
23. Qu, X.; Alvarez, P.J.J.; Li, Q. Applications of nanotechnology in water and wastewater treatment. *Water Res.* **2013**, *47*, 3931–3946. [CrossRef] [PubMed]
24. Usman, M.; Katsoyiannis, I.; Mitrakas, M.; Zouboulis, A.; Ernst, M. Performance evaluation of small sized powdered ferric hydroxide as arsenic adsorbent. *Water* **2018**, *10*, 957. [CrossRef]
25. Goh, K.; Karahan, H.E.; Wei, L.; Bae, T.-H.; Fane, A.G.; Wang, R.; Chen, Y. Carbon nanomaterials for advancing separation membranes: A strategic perspective. *Carbon N. Y.* **2016**, *109*, 694–710. [CrossRef]
26. Goh, P.S.; Ismail, A.F.; Hilal, N. Nano-enabled membranes technology: Sustainable and revolutionary solutions for membrane desalination? *Desalination* **2016**, *380*, 100–104. [CrossRef]
27. Goh, P.S.; Matsuura, T.; Ismail, A.F.; Hilal, N. Recent trends in membranes and membrane processes for desalination. *Desalination* **2016**, *391*, 43–60. [CrossRef]
28. Gupta, V.K.; Agarwal, S.; Saleh, T.A. Synthesis and characterization of alumina-coated carbon nanotubes and their application for lead removal. *J. Hazard. Mater.* **2011**, *185*, 17–23. [CrossRef]
29. Santhosh, C.; Velmurugan, V.; Jacob, G.; Jeong, S.K.; Grace, A.N.; Bhatnagar, A. Role of nanomaterials in water treatment applications: A review. *Chem. Eng. J.* **2016**, *306*, 1116–1137. [CrossRef]
30. Xu, P.; Zeng, G.M.; Huang, D.L.; Feng, C.L.; Hu, S.; Zhao, M.H.; Lai, C.; Wei, Z.; Huang, C.; Xie, G.X.; et al. Use of iron oxide nanomaterials in wastewater treatment: A review. *Sci. Total Environ.* **2012**, *424*, 1–10. [CrossRef]
31. Pendergast, M.M.; Hoek, E.M.V. A review of water treatment membrane nanotechnologies. *Energy Environ. Sci.* **2011**, *4*, 1946. [CrossRef]
32. Bethi, B.; Sonawane, S.H.; Bhanvase, B.A.; Gumfekar, S.P. Nanomaterials-based advanced oxidation processes for wastewater treatment: A review. *Chem. Eng. Process. Process Intensif.* **2016**, *109*, 178–189. [CrossRef]
33. Holmes, A.B.; Gu, F.X. Emerging nanomaterials for the application of selenium removal for wastewater treatment. *Environ. Sci. Nano* **2016**, *3*, 982–996. [CrossRef]

34. Lee, J.; Jeong, S.; Liu, Z. Progress and challenges of carbon nanotube membrane in water treatment. *Crit. Rev. Environ. Sci. Technol.* **2016**, *46*, 999–1046. [CrossRef]
35. Olivera, S.; Muralidhara, H.B.; Venkatesh, K.; Guna, V.K.; Gopalakrishna, K.; Kumar, K.Y. Potential applications of cellulose and chitosan nanoparticles/composites in wastewater treatment: A review. *Carbohydr. Polym.* **2016**, *153*, 600–618. [CrossRef]
36. Ong, C.S.; Goh, P.S.; Lau, W.J.; Misdan, N.; Ismail, A.F. Nanomaterials for biofouling and scaling mitigation of thin film composite membrane: A review. *Desalination* **2016**, *393*, 2–15. [CrossRef]
37. Stefaniuk, M.; Oleszczuk, P.; Ok, Y.S. Review on nano zerovalent iron (nZVI): From synthesis to environmental applications. *Chem. Eng. J.* **2016**, *287*, 618–632. [CrossRef]
38. Monthioux, M.; Kuznetsov, V.L. Who should be given the credit for the discovery of carbon nanotubes? *Carbon N. Y.* **2006**, *44*, 1621–1623. [CrossRef]
39. Radushkevich, L.V.; Lukyanovich, V.M. About the structure of carbon formed by thermal decomposition of carbon monoxide on iron substrate. *J. Phys. Chem.* **1952**, *26*, 88–95.
40. Oberlin, A.; Endo, M.; Koyama, T. Filamentous growth of carbon through benzene decomposition. *J. Cryst. Growth* **1976**, *32*, 335–349. [CrossRef]
41. Abrahamson, J.; Wiles, P.G.; Rhoades, B.L. Structure of carbon fibres found on carbon arc anodes. *Carbon N. Y.* **1999**, *37*, 1873–1874. [CrossRef]
42. Hirlekar, R.; Yamagar, M.; Garse, H.; Vij, M.; Kadam, V.; Vidyapeeth, B. Carbon nanotubes and its applications: A review. *Asian J. Pharm. Clin. Res.* **2009**, *2*, 17–27.
43. Tennent, H.G.; Barber, J.J.; Hoch, R. Carbon Fibrils, Method for Producing Same and Compositions Containing Same. U.S. Patent 4,663,230, 5 May 1987.
44. Bethune, D.S.; Kiang, C.H.; de Vries, M.S.; Gorman, G.; Savoy, R.; Vazquez, J.; Beyers, R. Cobalt-catalysed growth of carbon nanotubes with single-atomic-layer walls. *Nature* **1993**, *363*, 605–607. [CrossRef]
45. Iijima, S.; Ichihashi, T. Single-shell carbon nanotubes of 1-nm diameter. *Nature* **1993**, *363*, 603–605. [CrossRef]
46. Krätschmer, W.; Lamb, L.D.; Fostiropoulos, K.; Huffman, D.R. Solid C60: A new form of carbon. *Nature* **1990**, *347*, 354–358. [CrossRef]
47. Kroto, H.W.; Heath, J.R.; O'Brien, S.C.; Curl, R.F.; Smalley, R.E. C60: Buckminsterfullerene. *Nature* **1985**, *318*, 162–163. [CrossRef]
48. Liu, Z.; Sun, X.; Nakayama-Ratchford, N.; Dai, H. Supramolecular Chemistry on Water-Soluble Carbon Nanotubes for Drug Loading and Delivery. *ACS Nano* **2007**, *1*, 50–56. [CrossRef]
49. Singh, B.G.P.; Baburao, C.; Pispati, V.; Pathipati, H.; Muthy, N.; Prassana, S.R.V.; Rathode, B.G. Carbon nanotubes. A novel drug delivery system. *Int. J. Res. Pharm. Chem.* **2012**, *2*, 523–532.
50. Lam, C.; James, J.T.; McCluskey, R.; Arepalli, S.; Hunter, R.L. A review of carbon nanotube toxicity and assessment of potential occupational and environmental health risks. *Crit. Rev. Toxicol.* **2006**, *36*, 189–217. [CrossRef] [PubMed]
51. Bekyarova, E.; Ni, Y.; Malarkey, E.B.; Montana, V.; McWilliams, J.L.; Haddon, R.C.; Parpura, V. Applications of Carbon Nanotubes in Biotechnology and Biomedicine. *J. Biomed. Nanotechnol.* **2005**, *1*, 3–17. [CrossRef]
52. He, H.; Pham-Huy, L.A.; Dramou, P.; Xiao, D.; Zuo, P.; Pham-Huy, C. Carbon Nanotubes: Applications in Pharmacy and Medicine. *Biomed. Res. Int.* **2013**, *2013*, 1–12. [CrossRef]
53. Reilly, R.M. Carbon Nanotubes: Potential Benefits and Risks of Nanotechnology in Nuclear Medicine. *J. Nucl. Med.* **2007**, *48*, 1039–1042. [CrossRef] [PubMed]
54. Xie, X.; Mai, Y.; Zhou, X. Dispersion and alignment of carbon nanotubes in polymer matrix: A review. *Mater. Sci. Eng. R Rep.* **2005**, *49*, 89–112. [CrossRef]
55. Ihsanullah, A.A.; Al-Amer, A.M.; Laoui, T.; Al-Marri, M.J.; Nasser, M.S.; Khraisheh, M.; Atieh, M.A. Heavy metal removal from aqueous solution by advanced carbon nanotubes: Critical review of adsorption applications. *Sep. Purif. Technol.* **2016**, *157*, 141–161. [CrossRef]
56. Balasubramanian, K.; Burghard, M. Chemically Functionalized Carbon Nanotubes. *Small* **2005**, *1*, 180–192. [CrossRef] [PubMed]
57. Monea, B.F.; Ionete, E.I.; Spiridon, S.I.; Ion-Ebrasu, D.; Petre, E. Carbon Nanotubes and Carbon Nanotube Structures Used for Temperature Measurement. *Sensors* **2019**, *19*, 2464. [CrossRef]
58. Digge, M.; Moon, R.; Gattani, S. Application of Carbon Nanotubes in Drug Delivery: A Review. *Int. J. PharmTech Res.* **2011**, *4*, 839–847.
59. Kateb, B.; Yamamoto, V.; Alizadeh, D.; Zhang, L.; Manohara, H.M.; Bronikowski, M.J.; Badie, B. Multi-walled Carbon Nanotube (MWCNT) Synthesis, Preparation, Labeling, and Functionalization. In *Immunotherapy of Cancer*; Humana Press: Totowa, NJ, USA, 2010; pp. 307–317.
60. Liao, H.; Paratala, B.; Sitharaman, B.; Wang, Y. Applications of Carbon Nanotubes in Biomedical Studies. In *Biomedical Nanotechnology*; Humana Press: Totowa, NJ, USA, 2011; pp. 223–241.
61. Usui, Y.; Haniu, H.; Tsuruoka, S.; Saito, N. Carbon nanotubes innovate on medical technology. *Med. Chem* **2012**, *2*, 1–6. [CrossRef]
62. Yang, D.; Yang, F.; Hu, J.; Long, J.; Wang, C.; Fu, D.; Ni, Q. Hydrophilic multi-walled carbon nanotubes decorated with magnetite nanoparticles as lymphatic targeted drug delivery vehicles. *Chem. Commun.* **2009**, 4447. [CrossRef]
63. Zhang, Y.; Bai, Y.; Yan, B. Functionalized carbon nanotubes for potential medicinal applications. *Drug Discov. Today* **2010**, *15*, 428–435. [CrossRef]

64. Cassell, A.M.; Raymakers, J.A.; Kong, J.; Dai, H. Large Scale CVD Synthesis of Single-Walled Carbon Nanotubes. *J. Phys. Chem. B* **1999**, *103*, 6484–6492. [CrossRef]
65. Sinnott, S.B.; Andrews, R.; Qian, D.; Rao, A.M.; Mao, Z.; Dickey, E.C.; Derbyshire, F. Model of carbon nanotube growth through chemical vapor deposition. *Chem. Phys. Lett.* **1999**, *315*, 25–30. [CrossRef]
66. Vinciguerra, V.; Buonocore, F.; Panzera, G.; Occhipinti, L. Growth mechanisms in chemical vapour deposited carbon nanotubes. *Nanotechnology* **2003**, *14*, 655. [CrossRef]
67. Helveg, S.; López-Cartes, C.; Sehested, J.; Hansen, P.L.; Clausen, B.S.; Rostrup-Nielsen, J.R.; Abild-Pedersen, F.; Nørskov, J.K. Atomic-scale imaging of carbon nanofibre growth. *Nature* **2004**, *427*, 426–429. [CrossRef] [PubMed]
68. Maser, W.K.; Benito, A.M.; Martınez, M.T. Production of carbon nanotubes: The light approach. *Carbon N. Y.* **2002**, *40*, 1685–1695. [CrossRef]
69. Kingston, C.T.; Simard, B. Fabrication of Carbon Nanotubes. *Anal. Lett.* **2003**, *36*, 3119–3145. [CrossRef]
70. Hutchison, J.L.; Kiselev, N.A.; Krinichnaya, E.P.; Krestinin, A.V.; Loutfy, R.O.; Morawsky, A.P.; Muradyan, V.E.; Obraztsova, E.D.; Sloan, J.; Terekhov, S.V.; et al. Double-walled carbon nanotubes fabricated by a hydrogen arc discharge method. *Carbon N. Y.* **2001**, *39*, 761–770. [CrossRef]
71. Shi, Z.; Lian, Y.; Zhou, X.; Gu, Z.; Zhang, Y.; Iijima, S.; Zhou, L.; Yue, K.T.; Zhang, S. Mass-production of single-wall carbon nanotubes by arc discharge method11This work was supported by the National Natural Science Foundation of China, No. 29671030. *Carbon N. Y.* **1999**, *37*, 1449–1453. [CrossRef]
72. Sano, N.; Wang, H.; Chhowalla, M.; Alexandrou, I.; Amaratunga, G.A.J. Synthesis of carbon "onions" in water. *Nature* **2001**, *414*, 506–507. [CrossRef]
73. Li, H.; Guan, L.; Shi, Z.; Gu, Z. Direct Synthesis of High Purity Single-Walled Carbon Nanotube Fibers by Arc Discharge. *J. Phys. Chem. B* **2004**, *108*, 4573–4575. [CrossRef]
74. Imasaka, K.; Kanatake, Y.; Ohshiro, Y.; Suehiro, J.; Hara, M. Production of carbon nanoonions and nanotubes using an intermittent arc discharge in water. *Thin Solid Films* **2006**, *506–507*, 250–254. [CrossRef]
75. Sagara, T.; Kurumi, S.; Suzuki, K. Growth of linear Ni-filled carbon nanotubes by local arc discharge in liquid ethanol. *Appl. Surf. Sci.* **2014**, *292*, 39–43. [CrossRef]
76. Ben Belgacem, A.; Hinkov, I.; Yahia, S.B.; Brinza, O.; Farhat, S. Arc discharge boron nitrogen doping of carbon nanotubes. *Mater. Today Commun.* **2016**, *8*, 183–195. [CrossRef]
77. Berkmans, J.; Jagannatham, M.; Reddy, R.; Haridoss, P. Synthesis of thin bundled single walled carbon nanotubes and nanohorn hybrids by arc discharge technique in open air atmosphere. *Diam. Relat. Mater.* **2015**, *55*, 12–15. [CrossRef]
78. Su, Y.; Zhang, Y. Carbon nanomaterials synthesized by arc discharge hot plasma. *Carbon N. Y.* **2015**, *83*, 90–99. [CrossRef]
79. Guo, T.; Nikolaev, P.; Thess, A.; Colbert, D.T.; Smalley, R.E. Catalytic growth of single-walled manotubes by laser vaporization. *Chem. Phys. Lett.* **1995**, *243*, 49–54. [CrossRef]
80. Guo, T.; Nikolaev, P.; Rinzler, A.G.; Tomanek, D.; Colbert, D.T.; Smalley, R.E. Self-assembly of tubular fullerenes. *J. Phys. Chem.* **1995**, *99*, 10694–10697. [CrossRef]
81. Nagy, J.B.; Bister, G.; Fonseca, A.; Méhn, D.; Kónya, Z.; Kiricsi, I.; Horváth, Z.E.; Biró, L.P. On the Growth Mechanism of Single-Walled Carbon Nanotubes by Catalytic Carbon Vapor Deposition on Supported Metal Catalysts. *J. Nanosci. Nanotechnol.* **2004**, *4*, 326–345. [CrossRef]
82. Bandaru, P.R. Electrical Properties and Applications of Carbon Nanotube Structures. *J. Nanosci. Nanotechnol.* **2007**, *7*, 1239–1267. [CrossRef]
83. Thess, A.; Lee, R.; Nikolaev, P.; Dai, H.; Petit, P.; Robert, J.; Xu, C.; Lee, Y.H.; Kim, S.G.; Rinzler, A.G.; et al. Crystalline Ropes of Metallic Carbon Nanotubes. *Science* **1996**, *273*, 483–487. [CrossRef]
84. Journet, C.; Bernier, P. Production of carbon nanotubes. *Appl. Phys. A Mater. Sci. Process.* **1998**, *67*, 1–9. [CrossRef]
85. Ebbesen, T.W.; Hiura, H.; Fujita, J.; Ochiai, Y.; Matsui, S.; Tanigaki, K. Patterns in the bulk growth of carbon nanotubes. *Chem. Phys. Lett.* **1993**, *209*, 83–90. [CrossRef]
86. José-Yacamán, M.; Miki-Yoshida, M.; Rendón, L.; Santiesteban, J.G. Catalytic growth of carbon microtubules with fullerene structure. *Appl. Phys. Lett.* **1993**, *62*, 202–204. [CrossRef]
87. Ren, Z.F. Synthesis of Large Arrays of Well-Aligned Carbon Nanotubes on Glass. *Science* **1998**, *282*, 1105–1107. [CrossRef]
88. Oliver, J. *Global Markets and Technologies for Carbon Nanotubes: NAN024F BCC Research*; BCC Publishing: Wellesley, MA, USA, 2015.
89. Kumar, S.; Rani, R.; Dilbaghi, N.; Tankeshwar, K.; Kim, K.-H. Carbon nanotubes: A novel material for multifaceted applications in human healthcare. *Chem. Soc. Rev.* **2017**, *46*, 158–196. [CrossRef] [PubMed]
90. Ferreira, F.V.; Franceschi, W.; Menezes, B.R.C.; Biagioni, A.F.; Coutinho, A.R.; Cividanes, L.S. Synthesis, Characterization, and Applications of Carbon Nanotubes. In *Carbon-Based Nanofillers and Their Rubber Nanocomposites*; Elsevier: Amsterdam, The Netherlands, 2019; pp. 1–45.
91. Donaldson, K.; Aitken, R.; Tran, L.; Stone, V.; Duffin, R.; Forrest, G.; Alexander, A. Carbon Nanotubes: A Review of Their Properties in Relation to Pulmonary Toxicology and Workplace Safety. *Toxicol. Sci.* **2006**, *92*, 5–22. [CrossRef]
92. Ando, Y.; Zhao, X.; Sugai, T.; Kumar, M. Growing carbon nanotubes. *Mater. Today* **2004**, *7*, 22–29. [CrossRef]

93. Muataz, A.A.; Fakhrul-Razi, A.; Dayang, B.A.R.; El-Sadig, M.; Chuah, G.T.; Maan, F.A.; Sunny, I.; Faizah, Y.; Abdul Hamid, M.; Halim, M. Production of vapor growth carbon fiber (vgcf) by using cvd. In Proceedings of the 17th Symposium of Malaysian Chemical Engineers (SOMChE 2003) "Role of Chemical Engineers for Sustainability of Small Medium Industries (SMI)", Penang, Malaysia, 29–30 December 2003; pp. 596–602.

94. Cao, Z.; Sun, Z.; Guo, P.; Chen, Y. Effect of acetylene flow rate on morphology and structure of carbon nanotube thick films grown by thermal chemical vapor deposition. *Front. Mater. Sci. China* **2007**, *1*, 92–96. [CrossRef]

95. Fonseca, A.; Hernadi, K.; Piedigrosso, P.; Colomer, J.-F.; Mukhopadhyay, K.; Doome, R.; Lazarescu, S.; Biro, L.P.; Lambin, P.; Thiry, P.A.; et al. Synthesis of single- and multi-wall carbon nanotubes over supported catalysts. *Appl. Phys. A Mater. Sci. Process.* **1998**, *67*, 11–22. [CrossRef]

96. Zheng, L.X.; O'Connell, M.J.; Doorn, S.K.; Liao, X.Z.; Zhao, Y.H.; Akhadov, E.A.; Hoffbauer, M.A.; Roop, B.J.; Jia, Q.X.; Dye, R.C.; et al. Ultralong single-wall carbon nanotubes. *Nat. Mater.* **2004**, *3*, 673–676. [CrossRef]

97. Kim, N.S.; Lee, Y.T.; Park, J.; Han, J.B.; Choi, Y.S.; Choi, S.Y.; Choo, J.; Lee, G.H. Vertically Aligned Carbon Nanotubes Grown by Pyrolysis of Iron, Cobalt, and Nickel Phthalocyanines. *J. Phys. Chem. B* **2003**, *107*, 9249–9255. [CrossRef]

98. Bustero, I.; Ainara, G.; Isabel, O.; Roberto, M.; Inés, R.; Amaya, A. Control of the Properties of Carbon Nanotubes Synthesized by CVD for Application in Electrochemical Biosensors. *Microchim. Acta* **2006**, *152*, 239–247. [CrossRef]

99. Mohamed, M.M.; Ghanem, M.A.; Khairy, M.; Naguib, E.; Alotaibi, N.H. Zinc oxide incorporated carbon nanotubes or graphene oxide nanohybrids for enhanced sonophotocatalytic degradation of methylene blue dye. *Appl. Surf. Sci.* **2019**, *487*, 539–549. [CrossRef]

100. Kong, J.; Cassell, A.M.; Dai, H. Chemical vapor deposition of methane for single-walled carbon nanotubes. *Chem. Phys. Lett.* **1998**, *292*, 567–574. [CrossRef]

101. Pérez-Cabero, M.; Monzón, A.; Rodrıguez-Ramos, I.; Guerrero-Ruız, A. Syntheses of CNTs over several iron-supported catalysts: Influence of the metallic precursors. *Catal. Today* **2004**, *93–95*, 681–687. [CrossRef]

102. Zhu, S.; Su, C.-H.; Lehoczky, S.L.; Muntele, I.; Ila, D. Carbon nanotube growth on carbon fibers. *Diam. Relat. Mater.* **2003**, *12*, 1825–1828. [CrossRef]

103. Zhu, W.Z.; Miser, D.E.; Chan, W.G.; Hajaligol, M.R. Characterization of multiwalled carbon nanotubes prepared by carbon arc cathode deposit. *Mater. Chem. Phys.* **2003**, *82*, 638–647. [CrossRef]

104. Collins, P.G.; Avouris, P. Nanotubes for electronics. *Sci. Am.* **2000**, *283*, 62–69. [CrossRef]

105. Colomer, J.-F.; Stephan, C.; Lefrant, S.; Van Tendeloo, G.; Willems, I.; Kónya, Z.; Fonseca, A.; Laurent, C.; Nagy, J. Large-scale synthesis of single-wall carbon nanotubes by catalytic chemical vapor deposition (CCVD) method. *Chem. Phys. Lett.* **2000**, *317*, 83–89. [CrossRef]

106. Ebbesen, T.W.; Ajayan, P.M. Large-scale synthesis of carbon nanotubes. *Nature* **1992**, *358*, 220–222. [CrossRef]

107. Ren, Z.F.; Huang, Z.P.; Wang, D.Z.; Wen, J.G.; Xu, J.W.; Wang, J.H.; Calvet, L.E.; Chen, J.; Klemic, J.F.; Reed, M.A. Growth of a single freestanding multiwall carbon nanotube on each nanonickel dot. *Appl. Phys. Lett.* **1999**, *75*, 1086–1088. [CrossRef]

108. Yudasaka, M.; Kikuchi, R.; Matsui, T.; Ohki, Y.; Yoshimura, S.; Ota, E. Specific conditions for Ni catalyzed carbon nanotube growth by chemical vapor deposition. *Appl. Phys. Lett.* **1995**, *67*, 2477–2479. [CrossRef]

109. Yudasaka, M.; Kikuchi, R.; Ohki, Y.; Ota, E.; Yoshimura, S. Behavior of Ni in carbon nanotube nucleation. *Appl. Phys. Lett.* **1997**, *70*, 1817–1818. [CrossRef]

110. Eklund, P.C.; Pradhan, B.K.; Kim, U.J.; Xiong, Q.; Fischer, J.E.; Friedman, A.D.; Holloway, B.C.; Jordan, K.; Smith, M.W. Large-Scale Production of Single-Walled Carbon Nanotubes Using Ultrafast Pulses from a Free Electron Laser. *Nano Lett.* **2002**, *2*, 561–566. [CrossRef]

111. Maser, W.K.; Muñoz, E.; Benito, A.M.; Martınez, M.T.; de la Fuente, G.F.; Maniette, Y.; Anglaret, E.; Sauvajol, J.-L. Production of high-density single-walled nanotube material by a simple laser-ablation method. *Chem. Phys. Lett.* **1998**, *292*, 587–593. [CrossRef]

112. Bolshakov, A.P.; Uglov, S.A.; Saveliev, A.V.; Konov, V.I.; Gorbunov, A.A.; Pompe, W.; Graff, A. A novel CW laser–powder method of carbon single-wall nanotubes production. *Diam. Relat. Mater.* **2002**, *11*, 927–930. [CrossRef]

113. Scott, C.D.; Arepalli, S.; Nikolaev, P.; Smalley, R.E. Growth mechanisms for single-wall carbon nanotubes in a laser-ablation process. *Appl. Phys. A Mater. Sci. Process.* **2001**, *72*, 573–580. [CrossRef]

114. Aqel, A.; El-Nour, K.M.M.A.; Ammar, R.A.A.; Al-Warthan, A. Carbon nanotubes, science and technology part (I) structure, synthesis and characterisation. *Arab. J. Chem.* **2012**, *5*, 1–23. [CrossRef]

115. Gong, K.; Ci, L. Process for purification of carbon nanotubes. *Carbon* **2008**, *46*, 2003–2025.

116. Hou, P.; Bai, S.; Yang, Q.; Liu, C.; Cheng, H. Multi-step purification of carbon nanotubes. *Carbon N. Y.* **2002**, *40*, 81–85. [CrossRef]

117. Hajime, G.; Terumi, F.; Yoshiya, F.; Toshiyuki, O. *Method of Purifying Single Wall Carbon Nanotubes from Metal Catalyst Impurities*; JP Patent Filed & Issued; Honda Giken Kogyo Kabushiki Kaisha: Tokyo, Japan, 2002.

118. Borowiak-Palen, E.; Pichler, T.; Liu, X.; Knupfer, M.; Graff, A.; Jost, O.; Pompe, W.; Kalenczuk, R.; Fink, J. Reduced diameter distribution of single-wall carbon nanotubes by selective oxidation. *Chem. Phys. Lett.* **2002**, *363*, 567–572. [CrossRef]

119. Huang, S.; Dai, L. Plasma Etching for Purification and Controlled Opening of Aligned Carbon Nanotubes. *J. Phys. Chem. B* **2002**, *106*, 3543–3545. [CrossRef]

120. Chiang, Y.-C.; Chen, C.-H.; Chiang, Y.-C.; Chen, S.-L. Circulating inclined fluidized beds with application for desiccant dehumidification systems. *Appl. Energy* **2016**, *175*, 199–211. [CrossRef]

121. Harutyunyan, A.R.; Pradhan, B.K.; Chang, J.; Chen, G.; Eklund, P.C. Purification of Single-Wall Carbon Nanotubes by Selective Microwave Heating of Catalyst Particles. *J. Phys. Chem. B* **2002**, *106*, 8671–8675. [CrossRef]
122. Farkas, E.; Elizabeth Anderson, M.; Chen, Z.; Rinzler, A.G. Length sorting cut single wall carbon nanotubes by high performance liquid chromatography. *Chem. Phys. Lett.* **2002**, *363*, 111–116. [CrossRef]
123. Chiang, I.W.; Brinson, B.E.; Huang, A.Y.; Willis, P.A.; Bronikowski, M.J.; Margrave, J.L.; Smalley, R.E.; Hauge, R.H. Purification and Characterization of Single-Wall Carbon Nanotubes (SWNTs) Obtained from the Gas-Phase Decomposition of CO (HiPco Process). *J. Phys. Chem. B* **2001**, *105*, 8297–8301. [CrossRef]
124. Chiang, I.W.; Brinson, B.E.; Smalley, R.E.; Margrave, J.L.; Hauge, R.H. Purification and Characterization of Single-Wall Carbon Nanotubes. *J. Phys. Chem. B* **2001**, *105*, 1157–1161. [CrossRef]
125. Kajiura, H.; Tsutsui, S.; Huang, H.; Murakami, Y. High-quality single-walled carbon nanotubes from arc-produced soot. *Chem. Phys. Lett.* **2002**, *364*, 586–592. [CrossRef]
126. Moon, J.-M.; An, K.H.; Lee, Y.H.; Park, Y.S.; Bae, D.J.; Park, G.-S. High-Yield Purification Process of Singlewalled Carbon Nanotubes. *J. Phys. Chem. B* **2001**, *105*, 5677–5681. [CrossRef]
127. Doi, M.; Ikuga, Y.; Kimura, T.; Mitsuzuka, H. Laminated Sheet and Diaper and Article for Sanitary Use. JP Patent JPH10713A, Application JP8116796A, 10 May 1996.
128. Bandow, S.; Rao, A.M.; Williams, K.A.; Thess, A.; Smalley, R.E.; Eklund, P.C. Purification of Single-Wall Carbon Nanotubes by Microfiltration. *J. Phys. Chem. B* **1997**, *101*, 8839–8842. [CrossRef]
129. Sajid, M.I.; Jamshaid, U.; Jamshaid, T.; Zafar, N.; Fessi, H.; Elaissari, A. Carbon nanotubes from synthesis to in vivo biomedical applications. *Int. J. Pharm.* **2016**, *501*, 278–299. [CrossRef]
130. Hou, P.-X.; Liu, C.; Cheng, H.-M. Purification of carbon nanotubes. *Carbon N. Y.* **2008**, *46*, 2003–2025. [CrossRef]
131. Sato, Y.; Ogawa, T.; Motomiya, K.; Shinoda, K.; Jeyadevan, B.; Tohji, K.; Kasuya, A.; Nishina, Y. Purification of MWNTs Combining Wet Grinding, Hydrothermal Treatment, and Oxidation. *J. Phys. Chem. B* **2001**, *105*, 3387–3392. [CrossRef]
132. Das, R.; Abd Hamid, S.B.; Ali, M.E.; Ismail, A.F.; Annuar, M.S.M.; Ramakrishna, S. Multifunctional carbon nanotubes in water treatment: The present, past and future. *Desalination* **2014**, *354*, 160–179. [CrossRef]
133. Long, R.Q.; Yang, R.T. Carbon Nanotubes as Superior Sorbent for Dioxin Removal. *J. Am. Chem. Soc.* **2001**, *123*, 2058–2059. [CrossRef]
134. Jun, L.Y.; Mubarak, N.M.; Yee, M.J.; Yon, L.S.; Bing, C.H.; Khalid, M.; Abdullah, E.C. An overview of functionalised carbon nanomaterial for organic pollutant removal. *J. Ind. Eng. Chem.* **2018**, *67*, 175–186. [CrossRef]
135. Ali, S.; Rehman, S.A.U.; Luan, H.-Y.; Farid, M.U.; Huang, H. Challenges and opportunities in functional carbon nanotubes for membrane-based water treatment and desalination. *Sci. Total Environ.* **2019**, *646*, 1126–1139. [CrossRef]
136. Ihsanullah; Al Amer, A.M.; Laoui, T.; Abbas, A.; Al-Aqeeli, N.; Patel, F.; Khraisheh, M.; Atieh, M.A.; Hilal, N. Fabrication and antifouling behaviour of a carbon nanotube membrane. *Mater. Des.* **2016**, *89*, 549–558. [CrossRef]
137. Ihsanullah; Al-Khaldi, F.A.; Abu-Sharkh, B.; Abulkibash, A.M.; Qureshi, M.I.; Laoui, T.; Atieh, M.A. Effect of acid modification on adsorption of hexavalent chromium (Cr(VI)) from aqueous solution by activated carbon and carbon nanotubes. *Desalin. Water Treat.* **2016**, *57*, 7232–7244. [CrossRef]
138. Garzia Trulli, M.; Sardella, E.; Palumbo, F.; Palazzo, G.; Giannossa, L.C.; Mangone, A.; Comparelli, R.; Musso, S.; Favia, P. Towards highly stable aqueous dispersions of multi-walled carbon nanotubes: The effect of oxygen plasma functionalization. *J. Colloid Interface Sci.* **2017**, *491*, 255–264. [CrossRef] [PubMed]
139. Georgakilas, V.; Bourlinos, A.; Gournis, D.; Tsoufis, T.; Trapalis, C.; Mateo-Alonso, A.; Prato, M. Multipurpose Organically Modified Carbon Nanotubes: From Functionalization to Nanotube Composites. *J. Am. Chem. Soc.* **2008**, *130*, 8733–8740. [CrossRef] [PubMed]
140. Bounos, G.; Andrikopoulos, K.S.; Moschopoulou, H.; Lainioti, G.C.; Roilo, D.; Checchetto, R.; Ioannides, T.; Kallitsis, J.K.; Voyiatzis, G.A. Enhancing water vapor permeability in mixed matrix polypropylene membranes through carbon nanotubes dispersion. *J. Memb. Sci.* **2017**, *524*, 576–584. [CrossRef]
141. Oyetade, O.A.; Skelton, A.A.; Nyamori, V.O.; Jonnalagadda, S.B.; Martincigh, B.S. Experimental and DFT studies on the selective adsorption of Pb^{2+} and Zn^{2+} from aqueous solution by nitrogen-functionalized multiwalled carbon nanotubes. *Sep. Purif. Technol.* **2017**, *188*, 174–187. [CrossRef]
142. Ali, I. New Generation Adsorbents for Water Treatment. *Chem. Rev.* **2012**, *112*, 5073–5091. [CrossRef]
143. Bahgat, M.; Farghali, A.A.; El Rouby, W.M.A.; Khedr, M.H. Synthesis and modification of multi-walled carbon nano-tubes (MWCNTs) for water treatment applications. *J. Anal. Appl. Pyrolysis* **2011**, *92*, 307–313. [CrossRef]
144. Zhang, Y.; Wu, B.; Xu, H.; Liu, H.; Wang, M.; He, Y.; Pan, B. Nanomaterials-enabled water and wastewater treatment. *Nanoimpact* **2016**, *3–4*, 22–39. [CrossRef]
145. Upadhyayula, V.K.K.; Deng, S.; Mitchell, M.C.; Smith, G.B. Application of carbon nanotube technology for removal of contaminants in drinking water: A review. *Sci. Total Environ.* **2009**, *408*, 1–13. [CrossRef]
146. Rao, G.; Lu, C.; Su, F. Sorption of divalent metal ions from aqueous solution by carbon nanotubes: A review. *Sep. Purif. Technol.* **2007**, *58*, 224–231. [CrossRef]
147. Ren, X.; Chen, C.; Nagatsu, M.; Wang, X. Carbon nanotubes as adsorbents in environmental pollution management: A review. *Chem. Eng. J.* **2011**, *170*, 395–410. [CrossRef]

148. Liang, J.; Li, L.; Chen, D.; Hajagos, T.; Ren, Z.; Chou, S.-Y.; Hu, W.; Pei, Q. Intrinsically stretchable and transparent thin-film transistors based on printable silver nanowires, carbon nanotubes and an elastomeric dielectric. *Nat. Commun.* **2015**, *6*, 7647. [CrossRef]
149. Moghaddam, H.K.; Pakizeh, M. Experimental study on mercury ions removal from aqueous solution by MnO_2/CNTs nanocomposite adsorbent. *J. Ind. Eng. Chem.* **2015**, *21*, 221–229. [CrossRef]
150. Tang, W.-W.; Zeng, G.-M.; Gong, J.-L.; Liu, Y.; Wang, X.-Y.; Liu, Y.-Y.; Liu, Z.-F.; Chen, L.; Zhang, X.-R.; Tu, D.-Z. Simultaneous adsorption of atrazine and Cu (II) from wastewater by magnetic multi-walled carbon nanotube. *Chem. Eng. J.* **2012**, *211–212*, 470–478. [CrossRef]
151. Lehman, J.H.; Terrones, M.; Mansfield, E.; Hurst, K.E.; Meunier, V. Evaluating the characteristics of multiwall carbon nanotubes. *Carbon N. Y.* **2011**, *49*, 2581–2602. [CrossRef]
152. Hassellöv, M.; Readman, J.W.; Ranville, J.F.; Tiede, K. Nanoparticle analysis and characterization methodologies in environmental risk assessment of engineered nanoparticles. *Ecotoxicology* **2008**, *17*, 344–361. [CrossRef]
153. Belin, T.; Epron, F. Characterization methods of carbon nanotubes: A review. *Mater. Sci. Eng. B* **2005**, *119*, 105–118. [CrossRef]
154. Thostenson, E.T.; Ren, Z.; Chou, T.-W. Advances in the science and technology of carbon nanotubes and their composites: A review. *Compos. Sci. Technol.* **2001**, *61*, 1899–1912. [CrossRef]
155. Herrero-Latorre, C.; Álvarez-Méndez, J.; Barciela-García, J.; García-Martín, S.; Peña-Crecente, R.M. Characterization of carbon nanotubes and analytical methods for their determination in environmental and biological samples: A review. *Anal. Chim. Acta* **2015**, *853*, 77–94. [CrossRef]
156. Täschner, C.; Pácal, F.; Leonhardt, A.; Spatenka, P.; Bartsch, K.; Graff, A.; Kaltofen, R. Synthesis of aligned carbon nanotubes by DC plasma-enhanced hot filament CVD. *Surf. Coat. Technol.* **2003**, *174–175*, 81–87. [CrossRef]
157. Xiao, L.; Ha, J.W.; Wei, L.; Wang, G.; Fang, N. Determining the Full Three-Dimensional Orientation of Single Anisotropic Nanoparticles by Differential Interference Contrast Microscopy. *Angew. Chem.* **2012**, *124*, 7854–7858. [CrossRef]
158. Korneva, G.; Ye, H.; Gogotsi, Y.; Halverson, D.; Friedman, G.; Bradley, J.-C.; Kornev, K.G. Carbon Nanotubes Loaded with Magnetic Particles. *Nano Lett.* **2005**, *5*, 879–884. [CrossRef]
159. Jia, C.L.; Mi, S.B.; Faley, M.; Poppe, U.; Schubert, J.; Urban, K. Oxygen octahedron reconstruction in the $SrTiO_3$/$LaAlO_3$ heterointerfaces investigated using aberration-corrected ultrahigh-resolution transmission electron microscopy. *Phys. Rev. B* **2009**, *79*, 081405. [CrossRef]
160. Gao, C.; Stading, M.; Wellner, N.; Parker, M.L.; Noel, T.R.; Mills, E.N.C.; Belton, P.S. Plasticization of a Protein-Based Film by Glycerol: A Spectroscopic, Mechanical, and Thermal Study. *J. Agric. Food Chem.* **2006**, *54*, 4611–4616. [CrossRef] [PubMed]
161. Sun, Y.-P.; Fu, K.; Lin, Y.; Huang, W. Functionalized Carbon Nanotubes: Properties and Applications. *Acc. Chem. Res.* **2002**, *35*, 1096–1104. [CrossRef] [PubMed]
162. Jin, S.; Fallgren, P.H.; Morris, J.M.; Chen, Q. Removal of bacteria and viruses from waters using layered double hydroxide nanocomposites. *Sci. Technol. Adv. Mater.* **2007**, *8*, 67–70. [CrossRef]
163. Huang, J.; Cao, Y.; Qin, B.; Zhong, G.; Zhang, J.; Yu, H.; Wang, H.; Peng, F. Highly efficient and acid-corrosion resistant nitrogen doped magnetic carbon nanotubes for the hexavalent chromium removal with subsequent reutilization. *Chem. Eng. J.* **2019**, *361*, 547–558. [CrossRef]
164. Correa-Duarte, M.A.; Grzelczak, M.; Salgueiriño-Maceira, V.; Giersig, M.; Liz-Marzán, L.M.; Farle, M.; Sierazdki, K.; Diaz, R. Alignment of Carbon Nanotubes under Low Magnetic Fields through Attachment of Magnetic Nanoparticles. *J. Phys. Chem. B* **2005**, *109*, 19060–19063. [CrossRef]
165. Rinzler, A.G.; Liu, J.; Dai, H.; Nikolaev, P.; Huffman, C.B.; Rodríguez-Macías, F.J.; Boul, P.J.; Lu, A.H.; Heymann, D.; Colbert, D.T.; et al. Large-scale purification of single-wall carbon nanotubes: Process, product, and characterization. *Appl. Phys. A Mater. Sci. Process.* **1998**, *67*, 29–37. [CrossRef]
166. Gommes, C.; Blacher, S.; Masenelli-Varlot, K.; Bossuot, C.; McRae, E.; Fonseca, A.; Nagy, J.-B.; Pirard, J.-P. Image analysis characterization of multi-walled carbon nanotubes. *Carbon N. Y.* **2003**, *41*, 2561–2572. [CrossRef]
167. Jimeno, A.; Goyanes, S.; Eceiza, A.; Kortaberria, G.; Mondragon, I.; Corcuera, M.A. Effects of Amine Molecular Structure on Carbon Nanotubes Functionalization. *J. Nanosci. Nanotechnol.* **2009**, *9*, 6222–6227. [CrossRef]
168. Chai, S.-P.; Zein, S.H.S.; Mohamed, A.R. The effect of reduction temperature on Co-Mo/Al2O3 catalysts for carbon nanotubes formation. *Appl. Catal. A Gen.* **2007**, *326*, 173–179. [CrossRef]
169. Kiang, C.-H.; Endo, M.; Ajayan, P.M.; Dresselhaus, G.; Dresselhaus, M.S. Size Effects in Carbon Nanotubes. *Phys. Rev. Lett.* **1998**, *81*, 1869–1872. [CrossRef]
170. Sarkar, C.; Chowdhuri, A.R.; Kumar, A.; Laha, D.; Garai, S.; Chakraborty, J.; Sahu, S.K. One pot synthesis of carbon dots decorated carboxymethyl cellulose- hydroxyapatite nanocomposite for drug delivery, tissue engineering and Fe^{3+} ion sensing. *Carbohydr. Polym.* **2018**, *181*, 710–718. [CrossRef]
171. Wepasnick, K.A.; Smith, B.A.; Bitter, J.L.; Howard Fairbrother, D. Chemical and structural characterization of carbon nanotube surfaces. *Anal. Bioanal. Chem.* **2010**, *396*, 1003–1014. [CrossRef]
172. Datsyuk, V.; Kalyva, M.; Papagelis, K.; Parthenios, J.; Tasis, D.; Siokou, A.; Kallitsis, I.; Galiotis, C. Chemical oxidation of multiwalled carbon nanotubes. *Carbon N. Y.* **2008**, *46*, 833–840. [CrossRef]
173. Liu, M.; Cowley, J.M. Structures of carbon nanotubes studied by HRTEM and nanodiffraction. *Ultramicroscopy* **1994**, *53*, 333–342. [CrossRef]

174. Cao, A.; Zhang, X.; Xu, C.; Wei, B.; Wu, D. Tandem structure of aligned carbon nanotubes on Au and its solar thermal absorption. *Sol. Energy Mater. Sol. Cells* **2002**, *70*, 481–486. [CrossRef]
175. Droppa, R.; Hammer, P.; Carvalho, A.C.; dos Santos, M.; Alvarez, F. Incorporation of nitrogen in carbon nanotubes. *J. Non-Cryst. Solids* **2002**, *299–302*, 874–879. [CrossRef]
176. Chong, C.T.; Tan, W.H.; Lee, S.L.; Chong, W.W.F.; Lam, S.S.; Valera-Medina, A. Morphology and growth of carbon nanotubes catalytically synthesised by premixed hydrocarbon-rich flames. *Mater. Chem. Phys.* **2017**, *197*, 246–255. [CrossRef]
177. Hammer, P.; Victoria, N.M.; Alvarez, F. Effects of increasing nitrogen concentration on the structure of carbon nitride films deposited by ion beam assisted deposition. *J. Vac. Sci. Technol. A Vac. Surfaces Film.* **2000**, *18*, 2277. [CrossRef]
178. Pham-Huu, C.; Keller, N.; Roddatis, V.V.; Mestl, G.; Schlögl, R.; Ledoux, M.J. Large scale synthesis of carbon nanofibers by catalytic decomposition of ethane on nickel nanoclusters decorating carbon nanotubes. *Phys. Chem. Chem. Phys.* **2002**, *4*, 514–521. [CrossRef]
179. Lee, Y.S.; Cho, T.H.; Lee, B.K.; Rho, J.S.; An, K.H.; Lee, Y.H. Surface properties of fluorinated single-walled carbon nanotubes. *J. Fluor. Chem.* **2003**, *120*, 99–104. [CrossRef]
180. Cao, A.; Xu, C.; Liang, J.; Wu, D.; Wei, B. X-ray diffraction characterization on the alignment degree of carbon nanotubes. *Chem. Phys. Lett.* **2001**, *344*, 13–17. [CrossRef]
181. Rols, S.; Almairac, R.; Henrard, L.; Anglaret, E.; Sauvajol, J.-L. Diffraction by finite-size crystalline bundles of single wall nanotubes. *Eur. Phys. J. B* **1999**, *10*, 263–270. [CrossRef]
182. Kuzmany, H.; Plank, W.; Hulman, M.; Kramberger, C.; Grüneis, A.; Pichler, T.; Peterlik, H.; Kataura, H.; Achiba, Y. Determination of SWCNT diameters from the Raman response of the radial breathing mode. *Eur. Phys. J. B* **2001**, *22*, 307–320. [CrossRef]
183. Santos, T.C.; dos Gates, R.S.; de Tinôco, I.F.F.; Zolnier, S.; da Baêta, F.C. Behavior of Japanese quail in different air velocities and air temperatures. *Pesqui. Agropecuária Bras.* **2017**, *52*, 344–354. [CrossRef]
184. Cividanes, L.S.; Brunelli, D.D.; Antunes, E.F.; Corat, E.J.; Sakane, K.K.; Thim, G.P. Cure study of epoxy resin reinforced with multiwalled carbon nanotubes by Raman and luminescence spectroscopy. *J. Appl. Polym. Sci.* **2013**, *127*, 544–553. [CrossRef]
185. Puangjan, A.; Chaiyasith, S.; Taweeporngitgul, W.; Keawtep, J. Application of functionalized multi-walled carbon nanotubes supporting cuprous oxide and silver oxide composite catalyst on copper substrate for simultaneous detection of vitamin B2, vitamin B6 and ascorbic acid. *Mater. Sci. Eng. C* **2017**, *76*, 383–397. [CrossRef]
186. Ferreira, F.V.; Francisco, W.; Menezes, B.R.C.; Brito, F.S.; Coutinho, A.S.; Cividanes, L.S.; Coutinho, A.R.; Thim, G.P. Correlation of surface treatment, dispersion and mechanical properties of HDPE/CNT nanocomposites. *Appl. Surf. Sci.* **2016**, *389*, 921–929. [CrossRef]
187. Silambarasan, D.; Surya, V.J.; Iyakutti, K.; Asokan, K.; Vasu, V.; Kawazoe, Y. Gamma (γ)-ray irradiated multi-walled carbon nanotubes (MWCNTs) for hydrogen storage. *Appl. Surf. Sci.* **2017**, *418*, 49–55. [CrossRef]
188. Arepalli, S.; Nikolaev, P.; Gorelik, O.; Hadjiev, V.G.; Holmes, W.; Files, B.; Yowell, L. Protocol for the characterization of single-wall carbon nanotube material quality. *Carbon N. Y.* **2004**, *42*, 1783–1791. [CrossRef]
189. Alvarez, W.E.; Kitiyanan, B.; Borgna, A.; Resasco, D.E. Synergism of Co and Mo in the catalytic production of single-wall carbon nanotubes by decomposition of CO. *Carbon N. Y.* **2001**, *39*, 547–558. [CrossRef]
190. Bahr, J.L.; Yang, J.; Kosynkin, D.V.; Bronikowski, M.J.; Smalley, R.E.; Tour, J.M. Functionalization of Carbon Nanotubes by Electrochemical Reduction of Aryl Diazonium Salts: A Bucky Paper Electrode. *J. Am. Chem. Soc.* **2001**, *123*, 6536–6542. [CrossRef]
191. Liu, W.-W.; Aziz, A.; Chai, S.-P.; Mohamed, A.R.; Tye, C.-T. Preparation of iron oxide nanoparticles supported on magnesium oxide for producing high-quality single-walled carbon nanotubes. *New Carbon Mater.* **2011**, *26*, 255–261. [CrossRef]
192. Anoshkin, I.V.; Nefedova, I.I.; Lioubtchenko, D.V.; Nefedov, I.S.; Räisänen, A.V. Single walled carbon nanotube quantification method employing the Raman signal intensity. *Carbon N. Y.* **2017**, *116*, 547–552. [CrossRef]
193. Xue, Y.; Zheng, S.; Sun, Z.; Zhang, Y.; Jin, W. Alkaline electrochemical advanced oxidation process for chromium oxidation at graphitized multi-walled carbon nanotubes. *Chemosphere* **2017**, *183*, 156–163. [CrossRef]
194. Faraji, S.; Yildiz, O.; Rost, C.; Stano, K.; Farahbakhsh, N.; Zhu, Y.; Bradford, P.D. Radial growth of multi-walled carbon nanotubes in aligned sheets through cyclic carbon deposition and graphitization. *Carbon N. Y.* **2017**, *111*, 411–418. [CrossRef]
195. Pillay, K.; Cukrowska, E.M.; Coville, N.J. Multi-walled carbon nanotubes as adsorbents for the removal of parts per billion levels of hexavalent chromium from aqueous solution. *J. Hazard. Mater.* **2009**, *166*, 1067–1075. [CrossRef]
196. Yu, F.; Ma, J.; Wang, J.; Zhang, M.; Zheng, J. Magnetic iron oxide nanoparticles functionalized multi-walled carbon nanotubes for toluene, ethylbenzene and xylene removal from aqueous solution. *Chemosphere* **2016**, *146*, 162–172. [CrossRef]
197. Usman Farid, M.; Luan, H.-Y.; Wang, Y.; Huang, H.; An, A.K.; Jalil Khan, R. Increased adsorption of aqueous zinc species by Ar/O 2 plasma-treated carbon nanotubes immobilized in hollow-fiber ultrafiltration membrane. *Chem. Eng. J.* **2017**, *325*, 239–248. [CrossRef]
198. Agnihotri, S.; Mota, J.P.B.; Rostam-Abadi, M.; Rood, M.J. Theoretical and Experimental Investigation of Morphology and Temperature Effects on Adsorption of Organic Vapors in Single-Walled Carbon Nanotubes. *J. Phys. Chem. B* **2006**, *110*, 7640–7647. [CrossRef]
199. Cho, H.-H.; Wepasnick, K.; Smith, B.A.; Bangash, F.K.; Fairbrother, D.H.; Ball, W.P. Sorption of Aqueous Zn[II] and Cd[II] by Multiwall Carbon Nanotubes: The Relative Roles of Oxygen-Containing Functional Groups and Graphenic Carbon. *Langmuir* **2010**, *26*, 967–981. [CrossRef]

200. Yu, X.-Y.; Luo, T.; Zhang, Y.-X.; Jia, Y.; Zhu, B.-J.; Fu, X.-C.; Liu, J.-H.; Huang, X.-J. Adsorption of Lead(II) on O_2-Plasma-Oxidized Multiwalled Carbon Nanotubes: Thermodynamics, Kinetics, and Desorption. *ACS Appl. Mater. Interfaces* **2011**, *3*, 2585–2593. [CrossRef]
201. Addo Ntim, S.; Mitra, S. Removal of Trace Arsenic To Meet Drinking Water Standards Using Iron Oxide Coated Multiwall Carbon Nanotubes. *J. Chem. Eng. Data* **2011**, *56*, 2077–2083. [CrossRef] [PubMed]
202. Chen, C.; Hu, J.; Shao, D.; Li, J.; Wang, X. Adsorption behavior of multiwall carbon nanotube/iron oxide magnetic composites for Ni(II) and Sr(II). *J. Hazard. Mater.* **2009**, *164*, 923–928. [CrossRef] [PubMed]
203. Daneshvar Tarigh, G.; Shemirani, F. Magnetic multi-wall carbon nanotube nanocomposite as an adsorbent for preconcentration and determination of lead (II) and manganese (II) in various matrices. *Talanta* **2013**, *115*, 744–750. [CrossRef]
204. Zhao, X.; Jia, Q.; Song, N.; Zhou, W.; Li, Y. Adsorption of Pb(II) from an Aqueous Solution by Titanium Dioxide/Carbon Nanotube Nanocomposites: Kinetics, Thermodynamics, and Isotherms. *J. Chem. Eng. Data* **2010**, *55*, 4428–4433. [CrossRef]
205. Gupta, A.; Vidyarthi, S.R.; Sankararamakrishnan, N. Enhanced sorption of mercury from compact fluorescent bulbs and contaminated water streams using functionalized multiwalled carbon nanotubes. *J. Hazard. Mater.* **2014**, *274*, 132–144. [CrossRef]
206. Gupta, V.K.; Moradi, O.; Tyagi, I.; Agarwal, S.; Sadegh, H.; Shahryari-Ghoshekandi, R.; Makhlouf, A.S.H.; Goodarzi, M.; Garshasbi, A. Study on the removal of heavy metal ions from industry waste by carbon nanotubes: Effect of the surface modification: A review. *Crit. Rev. Environ. Sci. Technol.* **2016**, *46*, 93–118. [CrossRef]
207. Hu, J.; Zhao, D.; Wang, X. Removal of Pb(II) and Cu(II) from aqueous solution using multiwalled carbon nanotubes/iron oxide magnetic composites. *Water Sci. Technol.* **2011**, *63*, 917–923. [CrossRef]
208. Yang, S.; Guo, Z.; Sheng, G.; Wang, X. Investigation of the sequestration mechanisms of Cd(II) and 1-naphthol on discharged multi-walled carbon nanotubes in aqueous environment. *Sci. Total Environ.* **2012**, *420*, 214–221. [CrossRef]
209. Vuković, G.D.; Marinković, A.D.; Čolić, M.; Ristić, M.Đ.; Aleksić, R.; Perić-Grujić, A.A.; Uskoković, P.S. Removal of cadmium from aqueous solutions by oxidized and ethylenediamine-functionalized multi-walled carbon nanotubes. *Chem. Eng. J.* **2010**, *157*, 238–248. [CrossRef]
210. Stafiej, A.; Pyrzynska, K. Solid phase extraction of metal ions using carbon nanotubes. *Microchem. J.* **2008**, *89*, 29–33. [CrossRef]
211. Lu, C.; Chiu, H. Adsorption of zinc(II) from water with purified carbon nanotubes. *Chem. Eng. Sci.* **2006**, *61*, 1138–1145. [CrossRef]
212. Weng, C.-H.; Huang, C.P. Adsorption characteristics of Zn(II) from dilute aqueous solution by fly ash. *Colloids Surf. A Physicochem. Eng. Asp.* **2004**, *247*, 137–143. [CrossRef]
213. Boehm, H. Surface oxides on carbon and their analysis: A critical assessment. *Carbon N. Y.* **2002**, *40*, 145–149. [CrossRef]
214. Li, J.; Chen, S.; Sheng, G.; Hu, J.; Tan, X.; Wang, X. Effect of surfactants on Pb(II) adsorption from aqueous solutions using oxidized multiwall carbon nanotubes. *Chem. Eng. J.* **2011**, *166*, 551–558. [CrossRef]
215. Atieh, M.A.; Bakather, O.Y.; Tawabini, B.S.; Bukhari, A.A.; Khaled, M.; Alharthi, M.; Fettouhi, M.; Abuilaiwi, F.A. Removal of Chromium (III) from Water by Using Modified and Nonmodified Carbon Nanotubes. *J. Nanomater.* **2010**, *2010*, 1–9. [CrossRef]
216. Lu, C.; Liu, C.; Rao, G.P. Comparisons of sorbent cost for the removal of Ni2+ from aqueous solution by carbon nanotubes and granular activated carbon. *J. Hazard. Mater.* **2008**, *151*, 239–246. [CrossRef]
217. Ali, I. Microwave assisted economic synthesis of multi walled carbon nanotubes for arsenic species removal in water: Batch and column operations. *J. Mol. Liq.* **2018**, *271*, 677–685. [CrossRef]
218. Alijani, H.; Shariatinia, Z. Effective aqueous arsenic removal using zero valent iron doped MWCNT synthesized by in situ CVD method using natural α-Fe_2O_3 as a precursor. *Chemosphere* **2017**, *171*, 502–511. [CrossRef]
219. Sankararamakrishnan, N.; Chauhan, D.; Dwivedi, J. Synthesis of functionalized carbon nanotubes by floating catalytic chemical vapor deposition method and their sorption behavior toward arsenic. *Chem. Eng. J.* **2016**, *284*, 599–608. [CrossRef]
220. Ma, M.-D.; Wu, H.; Deng, Z.-Y.; Zhao, X. Arsenic removal from water by nanometer iron oxide coated single-wall carbon nanotubes. *J. Mol. Liq.* **2018**, *259*, 369–375. [CrossRef]
221. Peng, H.; Zhang, N.; He, M.; Chen, B.; Hu, B. Simultaneous speciation analysis of inorganic arsenic, chromium and selenium in environmental waters by 3-(2-aminoethylamino) propyltrimethoxysilane modified multi-wall carbon nanotubes packed microcolumn solid phase extraction and ICP-MS. *Talanta* **2015**, *131*, 266–272. [CrossRef]
222. Veličković, Z.; Vuković, G.D.; Marinković, A.D.; Moldovan, M.-S.; Perić-Grujić, A.A.; Uskoković, P.S.; Ristić, M.Đ. Adsorption of arsenate on iron(III) oxide coated ethylenediamine functionalized multiwall carbon nanotubes. *Chem. Eng. J.* **2012**, *181–182*, 174–181. [CrossRef]
223. Addo Ntim, S.; Mitra, S. Adsorption of arsenic on multiwall carbon nanotube–zirconia nanohybrid for potential drinking water purification. *J. Colloid Interface Sci.* **2012**, *375*, 154–159. [CrossRef]
224. De Marques Neto, J.O.; Bellato, C.R.; de Silva, D.C. Iron oxide/carbon nanotubes/chitosan magnetic composite film for chromium species removal. *Chemosphere* **2019**, *218*, 391–401. [CrossRef]
225. Shin, K.-Y.; Hong, J.-Y.; Jang, J. Heavy metal ion adsorption behavior in nitrogen-doped magnetic carbon nanoparticles: Isotherms and kinetic study. *J. Hazard. Mater.* **2011**, *190*, 36–44. [CrossRef]
226. Huang, Y.; Lee, X.; Macazo, F.C.; Grattieri, M.; Cai, R.; Minteer, S.D. Fast and efficient removal of chromium (VI) anionic species by a reusable chitosan-modified multi-walled carbon nanotube composite. *Chem. Eng. J.* **2018**, *339*, 259–267. [CrossRef]
227. Lu, W.; Li, J.; Sheng, Y.; Zhang, X.; You, J.; Chen, L. One-pot synthesis of magnetic iron oxide nanoparticle-multiwalled carbon nanotube composites for enhanced removal of Cr(VI) from aqueous solution. *J. Colloid Interface Sci.* **2017**, *505*, 1134–1146. [CrossRef]

228. Salam, M.A. Preparation and characterization of chitin/magnetite/multiwalled carbon nanotubes magnetic nanocomposite for toxic hexavalent chromium removal from solution. *J. Mol. Liq.* **2017**, *233*, 197–202. [CrossRef]

229. Huang, Z.; Wang, X.; Yang, D. Adsorption of Cr(VI) in wastewater using magnetic multi-wall carbon nanotubes. *Water Sci. Eng.* **2015**, *8*, 226–232. [CrossRef]

230. Di, Z.-C.; Ding, J.; Peng, X.-J.; Li, Y.-H.; Luan, Z.-K.; Liang, J. Chromium adsorption by aligned carbon nanotubes supported ceria nanoparticles. *Chemosphere* **2006**, *62*, 861–865. [CrossRef] [PubMed]

231. Zhang, C.; Sui, J.; Li, J.; Tang, Y.; Cai, W. Efficient removal of heavy metal ions by thiol-functionalized superparamagnetic carbon nanotubes. *Chem. Eng. J.* **2012**, *210*, 45–52. [CrossRef]

232. Wang, G.; Gao, Z.; Tang, S.; Chen, C.; Duan, F.; Zhao, S.; Lin, S.; Feng, Y.; Zhou, L.; Qin, Y. Microwave Absorption Properties of Carbon Nanocoils Coated with Highly Controlled Magnetic Materials by Atomic Layer Deposition. *ACS Nano* **2012**, *6*, 11009–11017. [CrossRef] [PubMed]

233. Ji, L.; Zhou, L.; Bai, X.; Shao, Y.; Zhao, G.; Qu, Y.; Wang, C.; Li, Y. Facile synthesis of multiwall carbon nanotubes/iron oxides for removal of tetrabromobisphenol A and Pb(ii). *J. Mater. Chem.* **2012**, *22*, 15853. [CrossRef]

234. Yang, L.; Jiang, S.; Zhao, Y.; Zhu, L.; Chen, S.; Wang, X.; Wu, Q.; Ma, J.; Ma, Y.; Hu, Z. Boron-Doped Carbon Nanotubes as Metal-Free Electrocatalysts for the Oxygen Reduction Reaction. *Angew. Chem. Int. Ed.* **2011**, *50*, 7132–7135. [CrossRef] [PubMed]

235. Tofighy, M.A.; Mohammadi, T. Adsorption of divalent heavy metal ions from water using carbon nanotube sheets. *J. Hazard. Mater.* **2011**, *185*, 140–147. [CrossRef]

236. Ren, X.; Shao, D.; Zhao, G.; Sheng, G.; Hu, J.; Yang, S.; Wang, X. Plasma Induced Multiwalled Carbon Nanotube Grafted with 2-Vinylpyridine for Preconcentration of Pb(II) from Aqueous Solutions. *Plasma Process. Polym.* **2011**, *8*, 589–598. [CrossRef]

237. Xu, D.; Wang, Z. Role of multi-wall carbon nanotube network in composites to crystallization of isotactic polypropylene matrix. *Polymer* **2008**, *49*, 330–338. [CrossRef]

238. Wang, S.; Gong, W.; Liu, X.; Yao, Y.; Gao, B.; Yue, Q. Removal of lead(II) from aqueous solution by adsorption onto manganese oxide-coated carbon nanotubes. *Sep. Purif. Technol.* **2007**, *58*, 17–23. [CrossRef]

239. Wang, H.J.; Zhou, A.L.; Peng, F.; Yu, H.; Chen, L.F. Adsorption characteristic of acidified carbon nanotubes for heavy metal Pb(II) in aqueous solution. *Mater. Sci. Eng. A* **2007**, *466*, 201–206. [CrossRef]

240. Liang, J.; Liu, J.; Yuan, X.; Dong, H.; Zeng, G.; Wu, H.; Wang, H.; Liu, J.; Hua, S.; Zhang, S.; et al. Facile synthesis of alumina-decorated multi-walled carbon nanotubes for simultaneous adsorption of cadmium ion and trichloroethylene. *Chem. Eng. J.* **2015**, *273*, 101–110. [CrossRef]

241. Al-Khaldi, F.A.; Abusharkh, B.; Khaled, M.; Atieh, M.A.; Nasser, M.S.; Saleh, T.A.; Agarwal, S.; Tyagi, I.; Gupta, V.K. Adsorptive removal of cadmium(II) ions from liquid phase using acid modified carbon-based adsorbents. *J. Mol. Liq.* **2015**, *204*, 255–263. [CrossRef]

242. Salam, M.A.; Makki, M.S.I.; Abdelaal, M.Y.A. Preparation and characterization of multi-walled carbon nanotubes/chitosan nanocomposite and its application for the removal of heavy metals from aqueous solution. *J. Alloys Compd.* **2011**, *509*, 2582–2587. [CrossRef]

243. Bandaru, N.M.; Reta, N.; Dalal, H.; Ellis, A.V.; Shapter, J.; Voelcker, N.H. Enhanced adsorption of mercury ions on thiol derivatized single wall carbon nanotubes. *J. Hazard. Mater.* **2013**, *261*, 534–541. [CrossRef]

244. Hadavifar, M.; Bahramifar, N.; Younesi, H.; Li, Q. Adsorption of mercury ions from synthetic and real wastewater aqueous solution by functionalized multi-walled carbon nanotube with both amino and thiolated groups. *Chem. Eng. J.* **2014**, *237*, 217–228. [CrossRef]

245. Pillay, K.; Cukrowska, E.M.; Coville, N.J. Improved uptake of mercury by sulphur-containing carbon nanotubes. *Microchem. J.* **2013**, *108*, 124–130. [CrossRef]

246. El-Sheikh, A.H.; Al-Degs, Y.S.; Al-As'ad, R.M.; Sweileh, J.A. Effect of oxidation and geometrical dimensions of carbon nanotubes on Hg(II) sorption and preconcentration from real waters. *Desalination* **2011**, *270*, 214–220. [CrossRef]

247. Mubarak, N.M.; Alicia, R.F.; Abdullah, E.C.; Sahu, J.N.; Haslija, A.B.A.; Tan, J. Statistical optimization and kinetic studies on removal of Zn2+ using functionalized carbon nanotubes and magnetic biochar. *J. Environ. Chem. Eng.* **2013**, *1*, 486–495. [CrossRef]

248. Ge, Y.; Li, Z.; Xiao, D.; Xiong, P.; Ye, N. Sulfonated multi-walled carbon nanotubes for the removal of copper (II) from aqueous solutions. *J. Ind. Eng. Chem.* **2014**, *20*, 1765–1771. [CrossRef]

249. Salehi, E.; Madaeni, S.S.; Rajabi, L.; Vatanpour, V.; Derakhshan, A.A.; Zinadini, S.; Ghorabi, S.; Ahmadi Monfared, H. Novel chitosan/poly(vinyl) alcohol thin adsorptive membranes modified with amino functionalized multi-walled carbon nanotubes for Cu(II) removal from water: Preparation, characterization, adsorption kinetics and thermodynamics. *Sep. Purif. Technol.* **2012**, *89*, 309–319. [CrossRef]

250. Shao, D.; Hu, J.; Chen, C.; Sheng, G.; Ren, X.; Wang, X. Polyaniline Multiwalled Carbon Nanotube Magnetic Composite Prepared by Plasma-Induced Graft Technique and Its Application for Removal of Aniline and Phenol. *J. Phys. Chem. C* **2010**, *114*, 21524–21530. [CrossRef]

251. Chen, H.; Li, J.; Shao, D.; Ren, X.; Wang, X. Poly(acrylic acid) grafted multiwall carbon nanotubes by plasma techniques for Co(II) removal from aqueous solution. *Chem. Eng. J.* **2012**, *210*, 475–481. [CrossRef]

252. Wang, Q.; Li, J.; Chen, C.; Ren, X.; Hu, J.; Wang, X. Removal of cobalt from aqueous solution by magnetic multiwalled carbon nanotube/iron oxide composites. *Chem. Eng. J.* **2011**, *174*, 126–133. [CrossRef]

253. Mobasherpour, I.; Salahi, E.; Ebrahimi, M. Removal of divalent nickel cations from aqueous solution by multi-walled carbon nano tubes: Equilibrium and kinetic processes. *Res. Chem. Intermed.* **2012**, *38*, 2205–2222. [CrossRef]

254. Kandah, M.I.; Meunier, J.-L. Removal of nickel ions from water by multi-walled carbon nanotubes. *J. Hazard. Mater.* **2007**, *146*, 283–288. [CrossRef]

255. Chen, C.; Wang, X. Adsorption of Ni(II) from Aqueous Solution Using Oxidized Multiwall Carbon Nanotubes. *Ind. Eng. Chem. Res.* **2006**, *45*, 9144–9149. [CrossRef]

256. Abdel-Ghani, N.T.; El-Chaghaby, G.A.; Helal, F.S. Individual and competitive adsorption of phenol and nickel onto multiwalled carbon nanotubes. *J. Adv. Res.* **2015**, *6*, 405–415. [CrossRef]

257. Deb, A.K.S.; Ilaiyaraja, P.; Ponraju, D.; Venkatraman, B. Diglycolamide functionalized multi-walled carbon nanotubes for removal of uranium from aqueous solution by adsorption. *J. Radioanal. Nucl. Chem.* **2012**, *291*, 877–883. [CrossRef]

258. Chen, C.; Hu, J.; Xu, D.; Tan, X.; Meng, Y.; Wang, X. Surface complexation modeling of Sr(II) and Eu(III) adsorption onto oxidized multiwall carbon nanotubes. *J. Colloid Interface Sci.* **2008**, *323*, 33–41. [CrossRef]

259. Tousova, Z.; Oswald, P.; Slobodnik, J.; Blaha, L.; Muz, M.; Hu, M.; Brack, W.; Krauss, M.; Di Paolo, C.; Tarcai, Z.; et al. European demonstration program on the effect-based and chemical identification and monitoring of organic pollutants in European surface waters. *Sci. Total Environ.* **2017**, *601–602*, 1849–1868. [CrossRef]

260. Sillanpää, M. *Natural Organic Matter in Water*, 1st ed.; Elsevier: Amsterdam, The Netherlands, 2015; ISBN 9780128015032.

261. Zare, K.; Gupta, V.K.; Moradi, O.; Makhlouf, A.S.H.; Sillanpää, M.; Nadagouda, M.N.; Sadegh, H.; Shahryari-ghoshekandi, R.; Pal, A.; Wang, Z.; et al. A comparative study on the basis of adsorption capacity between CNTs and activated carbon as adsorbents for removal of noxious synthetic dyes: A review. *J. Nanostruct. Chem.* **2015**, *5*, 227–236. [CrossRef]

262. Ghaedi, M.; Khajehsharifi, H.; Yadkuri, A.H.; Roosta, M.; Asghari, A. Oxidized multiwalled carbon nanotubes as efficient adsorbent for bromothymol blue. *Toxicol. Environ. Chem.* **2012**, *94*, 873–883. [CrossRef]

263. Mahmoodian, H.; Moradi, O.; Shariatzadeha, B.; Salehf, T.A.; Tyagi, I.; Maity, A.; Asif, M.; Gupta, V.K. Enhanced removal of methyl orange from aqueous solutions by poly HEMA–chitosan-MWCNT nano-composite. *J. Mol. Liq.* **2015**, *202*, 189–198. [CrossRef]

264. Duman, O.; Tunç, S.; Polat, T.G.; Bozoğlan, B.K. Synthesis of magnetic oxidized multiwalled carbon nanotube-κ-carrageenan-Fe 3 O 4 nanocomposite adsorbent and its application in cationic Methylene Blue dye adsorption. *Carbohydr. Polym.* **2016**, *147*, 79–88. [CrossRef]

265. Sadegh, H.; Zare, K.; Maazinejad, B.; Shahryari-ghoshekandi, R.; Tyagi, I.; Agarwal, S.; Gupta, V.K. Synthesis of MWCNT-COOH-Cysteamine composite and its application for dye removal. *J. Mol. Liq.* **2016**, *215*, 221–228. [CrossRef]

266. Wang, Y.; Huang, H.; Wei, X. Influence of wastewater precoagulation on adsorptive filtration of pharmaceutical and personal care products by carbon nanotube membranes. *Chem. Eng. J.* **2018**, *333*, 66–75. [CrossRef]

267. Wang, S.; Liang, S.; Liang, P.; Zhang, X.; Sun, J.; Wu, S.; Huang, X. In-situ combined dual-layer CNT/PVDF membrane for electrically-enhanced fouling resistance. *J. Memb. Sci.* **2015**, *491*, 37–44. [CrossRef]

268. Wang, Y.; Zhu, J.; Huang, H.; Cho, H.-H. Carbon nanotube composite membranes for microfiltration of pharmaceuticals and personal care products: Capabilities and potential mechanisms. *J. Memb. Sci.* **2015**, *479*, 165–174. [CrossRef]

269. Jahangiri-Rad, M.; Nadafi, K.; Mesdaghinia, A.; Nabizadeh, R.; Younesian, M.; Rafiee, M. Sequential study on reactive blue 29 dye removal from aqueous solution by peroxy acid and single wall carbon nanotubes: Experiment and theory. *Iran. J. Environ. Health Sci. Eng.* **2013**, *10*, 5. [CrossRef] [PubMed]

270. Engel, M.; Chefetz, B. Adsorption and desorption of dissolved organic matter by carbon nanotubes: Effects of solution chemistry. *Environ. Pollut.* **2016**, *213*, 90–98. [CrossRef]

271. Ajmani, G.S.; Goodwin, D.; Marsh, K.; Fairbrother, D.H.; Schwab, K.J.; Jacangelo, J.G.; Huang, H. Modification of low pressure membranes with carbon nanotube layers for fouling control. *Water Res.* **2012**, *46*, 5645–5654. [CrossRef] [PubMed]

272. Yang, X.; Lee, J.; Yuan, L.; Chae, S.-R.; Peterson, V.K.; Minett, A.I.; Yin, Y.; Harris, A.T. Removal of natural organic matter in water using functionalised carbon nanotube buckypaper. *Carbon N. Y.* **2013**, *59*, 160–166. [CrossRef]

273. Deng, J.; Shao, Y.; Gao, N.; Deng, Y.; Tan, C.; Zhou, S.; Hu, X. Multiwalled carbon nanotubes as adsorbents for removal of herbicide diuron from aqueous solution. *Chem. Eng. J.* **2012**, *193–194*, 339–347. [CrossRef]

274. Hamdi, H.; De La Torre-Roche, R.; Hawthorne, J.; White, J.C. Impact of non-functionalized and amino-functionalized multiwall carbon nanotubes on pesticide uptake by lettuce (*Lactuca sativa* L.). *Nanotoxicology* **2015**, *9*, 172–180. [CrossRef] [PubMed]

275. Rocha, J.-D.R.; Rogers, R.E.; Dichiara, A.B.; Capasse, R.C. Emerging investigators series: Highly effective adsorption of organic aromatic molecules from aqueous environments by electronically sorted single-walled carbon nanotubes. *Environ. Sci. Water Res. Technol.* **2017**, *3*, 203–212. [CrossRef]

276. Dichiara, A.B.; Harlander, S.F.; Rogers, R.E. Fixed bed adsorption of diquat dibromide from aqueous solution using carbon nanotubes. *RSC Adv.* **2015**, *5*, 61508–61512. [CrossRef]

277. Srivastava, A.; Srivastava, O.N.; Talapatra, S.; Vajtai, R.; Ajayan, P.M. Carbon nanotube filters. *Nat. Mater.* **2004**, *3*, 610–614. [CrossRef]

278. Tahaikt, M.; El Habbani, R.; Ait Haddou, A.; Achary, I.; Amor, Z.; Taky, M.; Alami, A.; Boughriba, A.; Hafsi, M.; Elmidaoui, A. Fluoride removal from groundwater by nanofiltration. *Desalination* **2007**, *212*, 46–53. [CrossRef]

279. Fornasiero, F.; Park, H.G.; Holt, J.K.; Stadermann, M.; Grigoropoulos, C.P.; Noy, A.; Bakajin, O. Ion exclusion by sub-2-nm carbon nanotube pores. *Proc. Natl. Acad. Sci. USA* **2008**, *105*, 17250–17255. [CrossRef]

280. Noy, A.; Park, H.G.; Fornasiero, F.; Holt, J.K.; Grigoropoulos, C.P.; Bakajin, O. Nanofluidics in carbon nanotubes. *Nano Today* **2007**, *2*, 22–29. [CrossRef]

281. Hummer, G.; Rasaiah, J.C.; Noworyta, J.P. Water conduction through the hydrophobic channel of a carbon nanotube. *Nature* **2001**, *414*, 188–190. [CrossRef] [PubMed]

282. Moradi, O. Adsorption Behavior of Basic Red 46 by Single-Walled Carbon Nanotubes Surfaces. *Fuller. Nanotub. Carbon Nanostructures* **2013**, *21*, 286–301. [CrossRef]

283. Yang, S.; Wang, L.; Zhang, X.; Yang, W.; Song, G. Enhanced adsorption of Congo red dye by functionalized carbon nanotube/mixed metal oxides nanocomposites derived from layered double hydroxide precursor. *Chem. Eng. J.* **2015**, *275*, 315–321. [CrossRef]

284. Konicki, W.; Pełech, I.; Mijowska, E.; Jasińska, I. Adsorption of anionic dye Direct Red 23 onto magnetic multi-walled carbon nanotubes-Fe3C nanocomposite: Kinetics, equilibrium and thermodynamics. *Chem. Eng. J.* **2012**, *210*, 87–95. [CrossRef]

285. Setareh Derakhshan, M.; Moradi, O. The study of thermodynamics and kinetics methyl orange and malachite green by SWCNTs, SWCNT-COOH and SWCNT-NH2 as adsorbents from aqueous solution. *J. Ind. Eng. Chem.* **2014**, *20*, 3186–3194. [CrossRef]

286. Zhao, D.; Zhang, W.; Chen, C.; Wang, X. Adsorption of Methyl Orange Dye Onto Multiwalled Carbon Nanotubes. *Procedia Environ. Sci.* **2013**, *18*, 890–895. [CrossRef]

287. Ahmad, A.; Razali, M.H.; Mamat, M.; Mehamod, F.S.B.; Anuar Mat Amin, K. Adsorption of methyl orange by synthesized and functionalized-CNTs with 3-aminopropyltriethoxysilane loaded TiO2 nanocomposites. *Chemosphere* **2017**, *168*, 474–482. [CrossRef]

288. Robati, D.; Mirza, B.; Ghazisaeidi, R.; Rajabi, M.; Moradi, O.; Tyagi, I.; Agarwal, S.; Gupta, V.K. Adsorption behavior of methylene blue dye on nanocomposite multi-walled carbon nanotube functionalized thiol (MWCNT-SH) as new adsorbent. *J. Mol. Liq.* **2016**, *216*, 830–835. [CrossRef]

289. Shih, M.-W.; Chin, C.-J.M.; Yu, Y.-L. The role of oxygen-containing groups on the adsorption of bisphenol-A on multi-walled carbon nanotube modified by HNO3 and KOH. *Process Saf. Environ. Prot.* **2017**, *112*, 308–314. [CrossRef]

290. Kuo, C.-Y. Comparison with as-grown and microwave modified carbon nanotubes to removal aqueous bisphenol A. *Desalination* **2009**, *249*, 976–982. [CrossRef]

291. Liao, Q.; Sun, J.; Gao, L. Adsorption of chlorophenols by multi-walled carbon nanotubes treated with HNO3 and NH3. *Carbon N. Y.* **2008**, *46*, 553–555. [CrossRef]

292. Mubarak, N.M.; Sazila, N.; Nizamuddin, S.; Abdullah, E.C.; Sahu, J.N. Adsorptive removal of phenol from aqueous solution by using carbon nanotubes and magnetic biochar. *Nano World J.* **2017**, *3*, 32–37. [CrossRef]

293. Yao, Y.-X.; Li, H.-B.; Liu, J.-Y.; Tan, X.-L.; Yu, J.-G.; Peng, Z.-G. Removal and Adsorption of p-Nitrophenol from Aqueous Solutions Using Carbon Nanotubes and Their Composites. *J. Nanomater.* **2014**, *2014*, 1–9. [CrossRef]

294. Diaz-Flores, P.E.; López-Urías, F.; Terrones, M.; Rangel-Mendez, J.R. Simultaneous adsorption of Cd^{2+} and phenol on modified N-doped carbon nanotubes: Experimental and DFT studies. *J. Colloid Interface Sci.* **2009**, *334*, 124–131. [CrossRef]

295. Tóth, A.; Törőcsik, A.; Tombácz, E.; László, K. Competitive adsorption of phenol and 3-chlorophenol on purified MWCNTs. *J. Colloid Interface Sci.* **2012**, *387*, 244–249. [CrossRef]

296. Arasteh, R.; Masoumi, M.; Rashidi, A.M.; Moradi, L.; Samimi, V.; Mostafavi, S.T. Adsorption of 2-nitrophenol by multi-wall carbon nanotubes from aqueous solutions. *Appl. Surf. Sci.* **2010**, *256*, 4447–4455. [CrossRef]

297. Kassem, A.; Ayoub, G.M.; Malaeb, L. Antibacterial activity of chitosan nano-composites and carbon nanotubes: A review. *Sci. Total Environ.* **2019**, *668*, 566–576. [CrossRef]

298. Smith, S.C.; Rodrigues, D.F. Carbon-based nanomaterials for removal of chemical and biological contaminants from water: A review of mechanisms and applications. *Carbon N. Y.* **2015**, *91*, 122–143. [CrossRef]

299. Brady-Estévez, A.S.; Kang, S.; Elimelech, M. A Single-Walled-Carbon-Nanotube Filter for Removal of Viral and Bacterial Pathogens. *Small* **2008**, *4*, 481–484. [CrossRef]

300. Lu, C.; Su, F. Adsorption of natural organic matter by carbon nanotubes. *Sep. Purif. Technol.* **2007**, *58*, 113–121. [CrossRef]

301. Kang, S.; Pinault, M.; Pfefferle, L.D.; Elimelech, M. Single-Walled Carbon Nanotubes Exhibit Strong Antimicrobial Activity. *Langmuir* **2007**, *23*, 8670–8673. [CrossRef]

302. Ihsanullah; Asmaly, H.A.; Saleh, T.A.; Laoui, T.; Gupta, V.K.; Atieh, M.A. Enhanced adsorption of phenols from liquids by aluminum oxide/carbon nanotubes: Comprehensive study from synthesis to surface properties. *J. Mol. Liq.* **2015**, *206*, 176–182. [CrossRef]

303. Bohonak, D.; Zydney, A. Compaction and permeability effects with virus filtration membranes. *J. Memb. Sci.* **2005**, *254*, 71–79. [CrossRef]

304. Mostafavi, S.T.; Mehrnia, M.R.; Rashidi, A.M. Preparation of nanofilter from carbon nanotubes for application in virus removal from water. *Desalination* **2009**, *238*, 271–280. [CrossRef]

305. Savage, N.; Diallo, M.S. Nanomaterials and Water Purification: Opportunities and Challenges. *J. Nanopart. Res.* **2005**, *7*, 331–342. [CrossRef]

306. Li, Q.; Mahendra, S.; Lyon, D.Y.; Brunet, L.; Liga, M.V.; Li, D.; Alvarez, P.J.J. Antimicrobial nanomaterials for water disinfection and microbial control: Potential applications and implications. *Water Res.* **2008**, *42*, 4591–4602. [CrossRef]

307. Nepal, D.; Balasubramanian, S.; Simonian, A.L.; Davis, V.A. Strong Antimicrobial Coatings: Single-Walled Carbon Nanotubes Armored with Biopolymers. *Nano Lett.* **2008**, *8*, 1896–1901. [CrossRef]

308. Cortes, P.; Deng, S.; Smith, G.B. The Adsorption Properties of Bacillus atrophaeus Spores on Single-Wall Carbon Nanotubes. *J. Sens.* **2009**, *2009*, 1–6. [CrossRef]
309. Ong, Y.T.; Ahmad, A.L.; Zein, S.H.S.; Tan, S.H. A review on carbon nanotubes in an environmental protection and green engineering perspective. *Braz. J. Chem. Eng.* **2010**, *27*, 227–242. [CrossRef]
310. Yuan, W.; Jiang, G.; Che, J.; Qi, X.; Xu, R.; Chang, M.W.; Chen, Y.; Lim, S.Y.; Dai, J.; Chan-Park, M.B. Deposition of Silver Nanoparticles on Multiwalled Carbon Nanotubes Grafted with Hyperbranched Poly(amidoamine) and Their Antimicrobial Effects. *J. Phys. Chem. C* **2008**, *112*, 18754–18759. [CrossRef]
311. Morones, J.R.; Elechiguerra, J.L.; Camacho, A.; Holt, K.; Kouri, J.B.; Ramírez, J.T.; Yacaman, M.J. The bactericidal effect of silver nanoparticles. *Nanotechnology* **2005**, *16*, 2346–2353. [CrossRef]
312. Ihsanullah; Laoui, T.; Al-Amer, A.M.; Khalil, A.B.; Abbas, A.; Khraisheh, M.; Atieh, M.A. Novel anti-microbial membrane for desalination pretreatment: A silver nanoparticle-doped carbon nanotube membrane. *Desalination* **2015**, *376*, 82–93. [CrossRef]
313. Nie, C.; Yang, Y.; Cheng, C.; Ma, L.; Deng, J.; Wang, L.; Zhao, C. Bioinspired and biocompatible carbon nanotube-Ag nanohybrid coatings for robust antibacterial applications. *Acta Biomater.* **2017**, *51*, 479–494. [CrossRef]
314. Morsi, R.E.; Alsabagh, A.M.; Nasr, S.A.; Zaki, M.M. Multifunctional nanocomposites of chitosan, silver nanoparticles, copper nanoparticles and carbon nanotubes for water treatment: Antimicrobial characteristics. *Int. J. Biol. Macromol.* **2017**, *97*, 264–269. [CrossRef]
315. Chi, M.-F.; Wu, W.-L.; Du, Y.; Chin, C.-J.M.; Lin, C.-C. Inactivation of Escherichia coli planktonic cells by multi-walled carbon nanotubes in suspensions: Effect of surface functionalization coupled with medium nutrition level. *J. Hazard. Mater.* **2016**, *318*, 507–514. [CrossRef]
316. Al-Hakami, S.M.; Khalil, A.B.; Laoui, T.; Atieh, M.A. Fast Disinfection of Escherichia coli Bacteria Using Carbon Nanotubes Interaction with Microwave Radiation. *Bioinorg. Chem. Appl.* **2013**, *2013*, 1–9. [CrossRef]

sustainability

Article

Compost Inoculated with Fungi from a Mangrove Habitat Improved the Growth and Disease Defense of Vegetable Plants

Fuad Ameen * and Ali A. Al-Homaidan

Department of Botany and Microbiology, College of Science, King Saud University, Riyadh 11451, Saudi Arabia; homaidan@KSU.EDU.SA
* Correspondence: fuadameen@ksu.edu.sa

Abstract: Municipal organic wastes could be exploited as fertilizers, having been given the ability to suppress plant diseases by the inoculation of the waste with certain fungi in the composting process. Our aim was to develop a novel fertilizer using composting in combination with fungi associated with mangrove forests. Nine fungal species were isolated from a mangrove forest habitat and screened for their activity against five phytopathogenic fungi, their plant-growth promotion ability, and their phosphate solubilization ability. Two fungal isolates, *Penicillium vinaceum* and *Eupenicillium hirayama*, were inoculated into organic waste before the composting experiment. After 90 days, the physico-chemical properties of the compost (color, moisture, pH, C:N ratio and cation exchange capacity (CEC)) indicated the maturity of the compost. The C:N ratio decreased and the CEC value increased most in the compost with the inoculum of both mangrove fungi. The vegetable plants grown in the mangrove fungi-inoculated composts had a higher vigor index than those grown in the control compost. The seeds collected from the plants grown in the fungi-inoculated composts had higher disease defense ability than the seeds collected from the control compost. The results indicated that the properties of the fungi shown *in vitro* (antagonistic and plant-growth promotion) remained in the mature compost. The seeds of the plants acquired disease defense ability, which is a remarkable observation that is useful in sustainable agriculture.

Keywords: municipal waste; biological fertilizer; phytopathogens; sustainable agriculture; organic waste treatment

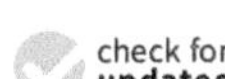
check for
updates

Citation: Ameen, F.; Al-Homaidan, A.A. Compost Inoculated with Fungi from a Mangrove Habitat Improved the Growth and Disease Defense of Vegetable Plants. *Sustainability* **2021**, *13*, 124. https://dx.doi.org/10.3390/su13010124

Received: 20 November 2020
Accepted: 19 December 2020
Published: 24 December 2020

Publisher's Note: MDPI stays neutral with regard to jurisdictional claims in published maps and institutional affiliations.

1. Introduction

Organic wastes from communities are still largely dumped or burned worldwide. In landfills, organic waste is degraded to methane and carbon dioxide, two of the many gases that are responsible for climate change. Burning instead produces harmful gaseous emissions, and the nutrients the waste contains are lost. In order to be able to return the valuable nutrients to agricultural soils, novel practices without huge investments in waste management are needed. When organic wastes are sorted, treated, and used in agriculture locally, many benefits are reached, especially in the areas of rapid population growth. One area where almost all municipal solid wastes are dumped in landfills is Saudi Arabia [1]. About 40% of the waste dumped is organic, and composting as a waste management practice is limited to small experiments [2].

The composting of organic wastes diminishes the environmental problems of waste disposal, and it has several other benefits. The addition of compost to farmland improves the fertility of the soil by increasing the amount of nutrients available to plants and organic matter. Organic matter, in turn, improves the water-holding capacity of soil. Farming in sandy desert areas, such as in Saudi Arabia, is totally based on irrigation [3] and, therefore, the development of organic waste recycling in these farmlands is of crucial importance.

The plant-growth promotion after the addition of compost is well known in agricultural fields [3–5]. Furthermore, the disease suppression ability of compost has been recognized recently [6,7]. Microbes not only transform the waste into fertile compost but also, when certain microbes are used in the composting, the microbes themselves might provide benefits. This has been observed previously. Certain microbial inoculums have been observed to enhance the microbial activity and speed the composting process, as well as being observed to enhance the soil suppressiveness against pathogens [8–11]. However, more information about potential beneficial microbes is needed. One habitat where such useful active microbes might be common is mangrove forests. Mangrove fungi, such as *Trichoderma viride* and *Steptomyces* sp., have been shown to promote plants' growth [12,13]. Moreover, the inhibition of phytopathogens has been observed for mangrove-associated fungi [14]. Phytopathogens such as *Pythium aphanidermatum*, *Fusarium solani* and *Rhizoctonia solani* were inhibited by fungi isolated from a mangrove habitat [14–16]. Therefore, we chose the mangrove forest habitat to search for beneficial fungi in composting.

Compost must be mature in order to be non-toxic to plants [17]. This can be assessed by its color, and by measuring its pH, C:N ratio, and cation exchange capacity [18]. In searching for beneficial fungi in composting, the enzyme activities, activities against phytopathogens, and plant-growth promotion abilities are useful to screen. The measurement of the activity against phytopathogens can be carried out with a traditional dual culture method. The biological disease control potential can be assessed as hydrogen cyanide production [19]. The plant growth promotion can be assessed as the production of plant hormones, such as indole acetic acid and gibberellic acid [20,21]. Phosphate solubilization ability of the microbes is useful to assess because it would be beneficial in sustainable agriculture [22].

Our aim was to develop a fertilizer that promotes vegetable plant growth and suppresses plant diseases by modifying the composting process. We inoculated waste with mangrove-associated fungi and assessed the final product for its ability to improve plant (pepper, aubergine and tomato) growth and suppress phytopathogens. Moreover, we took a novel approach in which we assessed the ability of the plants grown in the compost to produce seeds that defend themselves against phytopathogens.

2. Materials and Methods

2.1. Preliminary Screening for Mangrove-Associated Fungi

We isolated fungi from a mangrove habitat and screened their enzyme activities, activity against phytopathogens, and plant-growth promotion ability. Floating debris and sediment were separately collected 40 km south of Jeddah and 45 km south of Jazan on the Saudi Red Sea coast (Figure 1) in May 2018. The dominant mangrove species was *Avicennia marina* [23]. The samples were collected in sterile bottles and transported in an icebox to the laboratory.

The homogenized fresh sediment (1 g) was mixed with 1 mL of distilled water and 100 µL Tween-20, and inoculated in a serial dilution on MF1 medium [10] with dextrose (20 g L^{-1}) as a carbon source and streptomycin (500 mg L^{-1}) to avoid bacterial growth.

The plant material of the debris samples was first surface-sterilized with 2.5% hypochlorite for 2 min. After washing with sterile distilled water, small pieces of debris were placed on MF1 medium. All of the plates were incubated at 30 ± 2 °C for 5–7 days and the pure cultures were collected for further studies. All of the chemicals and reagents used in the present study were of analytical grade, and were procured from Sigma Aldrich (St. Louis, MO, USA).

Figure 1. Two sampling sites along the Red Sea Coast.

2.1.1. Activity against Phytopathogens

The ability of mangrove fungi to inhibit phytopathogens was measured using the dual-culture technique. The plant pathogens *Fusarium oxysporum, F. solani, Aspergillus flavus, A. niger,* and *Bipolaris oryzae*—stored King Saud University, Department of Botany and Microbiology (Saudi Arabia)—and nine mangrove fungal isolates (*Aspergillus nomius; A. terreus; Alternaria alternata; Cladosporium sphaerospermum; E. hirayamae; Geosmithia pallida; P. vinaceum; Paecilomyces variotii* and *Phialophora alba*) were inoculated on potato dextrose agar (PDA). All of the combinations of five pathogens and nine mangrove fungi were prepared by placing mycelia plugs (5 mm) on petri dishes (90 mm) with 6 cm distance from each other. After a five days of incubation, the plates on which the phytopathogens did not grow or grew slowly were chosen for further study.

2.1.2. Plant-Growth Promotion

The indole acetic acid (IAA) production method was adopted from Khan et al. [24]. The indole acetic acid (IAA) production was confirmed with High Performance Liquid Chromatography (HPLC). The agar plugs of mangrove fungi were transferred aseptically to an Erlenmeyer flask (1 L) containing 500 mL PD broth containing 0.2% tryptophan, and were incubated at room temperature with shaking at $110 \times g$ for 10 days. The culture broth was centrifuged at $10,000 \times g$ for 10 min and the supernatant was collected. The culture filtrate was extracted twice with an equal volume of ethyl acetate, and the organic phase was concentrated by the use of a rotary evaporator below 40 °C in order to obtain the crude extract powder [25]. The resultant crude extract was collected and re-suspended in methanol. The HPLC (Merck KGaA, Darmstadt, Germany) analysis was performed using a mixture of methanol and water containing 0.5% acetic acid in the ratio of 60:40 with a flow rate of 1 mL/min at 280 nm. For the gibberellic acid confirmation, the fungi were incubated in Czapek dox broth, as above. The HPLC analysis of the extract prepared as above was performed using acetonitrile and 0.01% H_3PO_4 in water (60:40) with a flow rate of 0.6 mL/min at 206 nm [26].

The hydrogen cyanide production was measured by inoculating the fungal cultures onto PDA containing 4.4 g/L glycine [27]. A Whatman filter paper no.1 soaked in 0.5% picric acid in 2% Na_2CO_3 was placed on the plate. The plates were sealed and incubated at 30 °C for 4 days. The color change of the filter paper from yellow to red was considered to be a positive result. Un-inoculated growth medium was used as negative control.

The 1 aminocyclopropane-1-carboxylic acid (ACC) deaminase production of the isolates was screened using Dworkin-Foster's (DF) salt minimal medium (4 g L^{-1} KH_2PO_4, 6 g L^{-1} Na_2HPO_4, 0.2 g L^{-1} $MgSO_4 \cdot 7H_2O$, 1 mg L^{-1} $FeSO_4 \cdot H_2O$, 10 µg L^{-1} H_3BO_3, 10 µg L^{-1} $MnSO_4$, 70 µg L^{-1} $ZnSO_4$, 50 µgL^{-1} $CuSO_4$, 10 µg L^{-1} MoO_3, 2 g L^{-1} glucose, 2 g L^{-1} gluconic acid, 2 g L^{-1} citric acid, 12 g L^{-1} agar) amended with 0.2% (w/v) (NH$_4$)2SO$_4$ [28]. The growth of this medium after 4 days of incubation was considered to be a positive result.

The phosphate solubilization was analyzed using Pikovskaya medium (HiMedia Laboratories, Bengaluru, India) containing 2.4 mg/mL bromophenol blue [29]. The media inoculated with the isolates were incubated for 48 h and observed for the formation of the yellow zone around the colony indicating phosphate solubilization.

2.1.3. Extracellular Enzyme Activity

The extracellular enzyme (cellulase, laccase, lipase, protease, amylase and chitinase) activities of the isolates were screened using the methods based on culturing the microbes on different media, as described by Maria et al. [30].

2.1.4. Molecular Identification of Fungi

The isolates were incubated in 50 mL potato dextrose broth (PDB) at room temperature for 5 days in an orbital shaker. The fungal mycelia (500 mg) were weighed for the total genomic DNA analysis, as described by White et al. [31]. ITS1 (5′-TCC gTA ggT gAA CCT gCg g-3′) and ITS4 (5′-TCC TCC gCT TAT TgA TAT gC-3′) primers were used. PCR was performed with an initial denaturation at 95 °C for 5 min, followed by 35 cycles of denaturation at 94 °C for 1 min, annealing at 55.5 °C for 2 min, extension at 72 °C for 2 min, and a final extension at 72 °C for 10 min. The amplified product was checked in 1.2% (w/v) agarose gel, purified using a cleanup kit (Mo-Bio Ultra Clean) and used for the DNA sequencing (Genetic analyzer 3130-Applied Bio-systems). The sequence data were analyzed with BLAST software.

2.2. Composting Experiment

We chose two isolates of mangrove fungi and carried out an experiment in order to produce compost. We assessed the ability of the compost to improve the soil fertility and suppress phytopathogens, as well as the ability of the plants grown in the compost to produce seeds that defend themselves against phytopathogens.

Raw materials and organic wastes—namely hay, leaves, fruit wastes, newspaper, vegetable wastes, rice husks and wheat straw—were collected and stored in the laboratory. The wastes (400 g each) were sterilized in a conical flask (1000 mL^{-1}). Three replicates of the treatments with two mangrove fungal isolates (*P. vinaceum* and *E. hirayamae*), their combination, and a control (Sterilized compost) were prepared. In addition, an unsterilized control treatment was prepared. The wastes were inoculated with the spore suspensions of the mangrove fungi, at 5% v/v for the single fungi treatments and 2.5% v/v each for the combined treatment. All of the flasks were incubated at 30 ± 2 °C for three months.

2.2.1. Physio-Chemical Properties

The color of the compost was assessed visually. The dry weight of the compost was determined at 105 °C for 4 h, and the loss on ignition was determined at 800 °C overnight. The pH was measured in distilled water (1:10; sample: water v/v). The carbon and nitrogen percentage was measured according to Iqbal et al. [32] using the digestive mixture of Na_2SO_4 and $CuSO_4·5H_2O$ (10:1).

The cation exchange capacity was measured using the method described by [18]. Compost (2 g) and 0.1 M HCL (100 mL^{-1}) were shaken for 2 h and precipitated with 1% silver nitrate. The solution was mixed in a flask with 0.25 M barium acetate, shaken and titrated with 0.1 M NaOH. The alkaline cations were calculated from the amount of sodium hydroxide solution used.

2.2.2. Plant-Growth Promoting Activity

The seeds of *Capsicum annuum* (pepper), *Solanum melongena* (aubergine) and *S. lycopersicum* (tomato) were first surface-sterilized in 2% NaOCl (bleach) for 1 min, washed several times with sterile distilled water, immersed in 70% ethanol, and finally rinsed three times with sterile distilled water in order to remove epiphytes. The efficiency of the surface sterilization was confirmed by spreading 100 µL of the last rinse water onto PDA, which was then incubated for 14 days. The surface-sterilized seeds were placed on sterile filter paper to dry for 20 min.

The seeds were placed on tissue paper on petri dishes and saturated with sterile distilled water for germination. After 5 days, the germinated seeds were transferred into plastic black pots (10 replicates) with compost. At the ripening stage after 60 days, the root length and shoot length were measured manually, and the leaf number, number of fruits, and number of seeds were counted. The dry mass and wet mass were monitored. For the vigor index, the mean root length and the mean shoot length were summed and multiplied by the percentage of the seeds germinated.

2.2.3. Disease Defense Ability of Seeds

The seeds from the fruits of the plants grown (as explained above) in the mature composts were collected. In total, 10 seeds of *S. lycopersicum*, 20 seeds of *S. melongena* and 50 seeds of *C. annuum* were placed as three replicates on petri dishes and exposed to the spore suspensions of (5% v/v) of four plant pathogens *F. solani*, *F. oxysporum*, *A. flavus*, and *B. oryzae*, separately. The germinated seeds were calculated after 48 h, which indicated the disease defense ability of the seeds.

2.3. Statistical Analysis

A two-way analysis of variance (ANOVA) (*P. vinaceum* and *E. hirayamae* and their interaction) was carried out in order to find out the effect of the mangrove fungi on the quality of the compost. In addition, a *t*-test was carried out in order to compare the sterilized control and the unsterilized control. $p < 0.05$ was considered to be a significant effect. RStudio Version (3.5.1) was used [33]. The data were log transformed when needed (tested using Shapiro–Wilk test) in order to fulfil the assumptions.

3. Results

3.1. Preliminary Screening of the Fungal Isolates

Two isolates, *P. vinaceum* (Accession No. KP033201) and *E. hirayamae* (Accession No. KM979606), showed the highest antifungal, enzymatic and plant-growth promotion activities *in vitro*. *Penicillium vinaceum* had relatively high antifungal activity against the plant pathogens *F. oxysporum*, *A. flavus*, *B. oryzae*, and *F. solani* (Table 1). *Eupenicillium hirayamae* showed activity against *A. niger*.

Table 1. *In vitro* activity of nine fungi isolated from a mangrove habitat against five fungal phytopathogens, measured as the zone of inhibition (mm). - indicates no inhibition.

Mangrove Fungi	*Fusarium oxysporum*	*Aspergillus niger*	*Aspergillus flavus*	*Bipolaris oryzae*	*Fusarium solani*
Positive control	12 ± 0.18	13 ± 0.11	10 ± 0.14	11 ± 0.13	14 ± 0.19
E. hirayamae	16 ± 0.10	18 ± 0.18	18.5 ± 0.10	14.2 ± 0.12	17 ± 0.11
C. sphaerospermum	-	-	-	-	10 ± 0.21
A. terreus	8 ± 0.23	4 ± 0.21	-	-	-
P. vinaceum	17 ± 0.12	14 ± 0.13	19 ± 0.21	15 ± 0.12	19 ± 0.13
P. alba	-	-	6 ± 0.23	-	8 ± 0.21
G. pallida	1.5 ± 0.24	-	-	-	4 ± 0.25
A. nomius	5 ± 0.14	-	-	3 ± 0.15	-
A. alternata	-	3 ± 0.18	-	-	-
P. variotii	2.5 ± 0.13	-	-	7.5 ± 0.17	-

Penicillium vinaceum and *E. hirayamae* showed strong enzyme activities, including: amylase, protease, chitinase, cellulose, and laccase (*P. vinaceum* only) (Table 2). Medium activity was detected for lipase (*P. vinaceum and E. hirayamae*) and laccase (*E. hirayamae* only). All of the other isolates had weak enzyme activities.

Table 2. Enzyme activities of nine fungi isolated from a mangrove habitat. +++ indicates strong activity, ++ indicates medium activity, and + indicates weak activity.

Mangrove Fungi	Amylase	Protease	Chitinase	Cellulase	Laccase	Lipase
E. hirayamae	+++	+++	+++	+++	++	++
C. sphaerospermum				+	+	
A. terreus		+	++			+
P. vinaceum	+++	+++	+++	+++	+++	++
P. alba	+					
G. pallida	+	+		+		+
A. nomius	+					
A. alternata		+				
P. variotii						+

Penicillium vinaceum and *E. hirayamae* showed potential plant-growth promotion abilities. The IAA and gibberellic acid production of both *P. vinaceum* and *E. hirayamae* were observed (Table 3) and confirmed by HPLC analysis. Moreover, these isolates also showed phosphate solubilization activity and hydrogen cyanide production.

Table 3. Variables indicating the plant-growth promotion activity of fungi isolated from a mangrove habitat. +++ indicates strong activity, ++ indicates medium activity, + indicates weak activity, and - indicates no activity.

Mangrove Fungi	Acc Deaminase	IAA Production	Phosphate Solubilization	HCN Production
E. hirayamae	+++	+++	+++	+++
C. sphaerospermum	+	+	-	-
A. terreus	+	-	-	-
P. vinaceum	+++	+++	+++	++
P. alba	-	+	-	-
G. pallida	+	-	-	-
A. nomius	+	+	-	-
A. alternata	-	+	-	-
P. variotii	-	+	-	-

3.2. Composting Experiment

After 12 days of incubation, the volume of the compost decreased and decoloration appeared. After 90 days of incubation, the color of the compost was changed to a blackish-brown in the fungal treatments; in the sterilized control, the color was a light brown; and in the unsterilized control, it was brown. For each physico-chemical variable (Table 4), the two-way ANOVA indicated that both of the fungal treatments (*P. vinaceum* and *E. hirayamae*) had significant effects. The interaction term was not significant, indicating that the combination treatment with the two fungi was not more effective than the single-fungus treatments. The moisture percentage of the mangrove fungi-treated samples was 20% higher than that of sterilized control (Table 4). The pH of the fungi-treated samples was 7.4–7.5, and whereas in the sterilized control it was 7. The total N_2% was more than doubled in fungal treatments (5.0–5.4%) compared to the sterilized control (2.4%), and conversely, the C:N ratio was more than doubled in sterilized control (17.9). The cation exchange capacity value was almost doubled in the fungal treatments (65 cmolc kg^{-1}) compared to the sterilized control (34 cmolc kg^{-1}). The results of the unsterilized control were set between the sterilized control and the fungal treatments. The *t*-test showed significant differences for moisture, N, C:N ratio and CEC results, which were all, however, nearer to the sterilized control than the fungal treatments.

Table 4. Physico-chemical properties of the composts produced by inoculating two individual mangrove fungi (*E. hirayamae* and *P. vinaceum*) and their combination in comparison to a sterilized waste control. The unsterilized control refers to the treatment of raw waste. * indicates a significant effect ($p < 0.05$) of the fungal species, or a significant interaction (combined effect) in a two-way ANOVA. An 'a' indicates a significant difference between the two control treatments (*t*-test).

Treatment	*E. hirayamae*	*P. vinaceum*	*E. hirayamae* *P. vinaceum*	Sterilized Control	Unsterilized Control
Compost property					
Color	Blackish brown	Blackish brown	Blackish brown	Light brown	Brown
Moisture (%)	38 *	40 *	40	12	20 a
pH	7.5 *	7.4 *	7.4	7.0	7.2
OM (%)	33 *	36 *	34	43	40
N (%)	5.1 *	5.0 *	5.4	2.4	3.0 a
C:N ratio	6.4 *	7.2 *	6.2	17.9	13.3 a
CEC (cmol(+) kg^{-1})	64 *	65 *	64	34	39 a

All of the plant-growth promotion measures were higher in the fungal treatments than in the sterilized control, as seen from the values relative to the sterilized control (Figure 2). Most values were also higher in unsterilized than unsterilized control. For the vigor index, the two-way ANOVA showed the significant effects of *P. vinaceum* and *E. hirayamae*, as well as their significant interaction for the three plant species (Figure 3). The highest vigor index was observed for *S. melongena* (5200) in the combined treatment of the two fungal species. The single fungal treatments gave about 25% lower indices. For comparison, in the sterilized control, the vigor index was less than half that of the combined treatment (2000). The *t*-test showed no difference in the vigor index between the sterilized control and the unsterilized control.

The disease incidence of the plants grown in the composts was 60–70% in the unsterilized control and 50–70% in the sterilized control (Table 5). For each plant species, the difference was significant (*t*-test). However, the difference was not consistent within the plant species. For *S. melongena*, the disease incidence was higher, whereas for the other two plant species it was lower in the sterilized control than in the unsterilized control. The two-way ANOVA indicated significant effects for both *P. vinaceum* and *E. hirayamae*, as well as a significant interaction. In the combined treatment, the disease incidence was 0% for each plant species. For other fungi and plant species, the disease incidence varied between 10% and 20%.

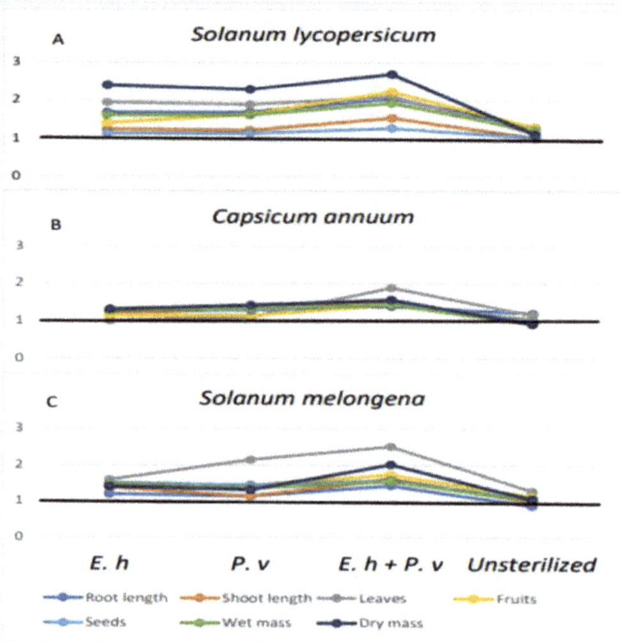

Figure 2. Relative plant growth variables (mean, $n = 3$, $0.01 < SD < 0.4$) in the different composting treatments. Sterilized control = 1. The treatments refer to *E. hirayamae*, *P. vinaceum*, *E. hirayamae* + *P. vinaceum*, and the unsterilized control.

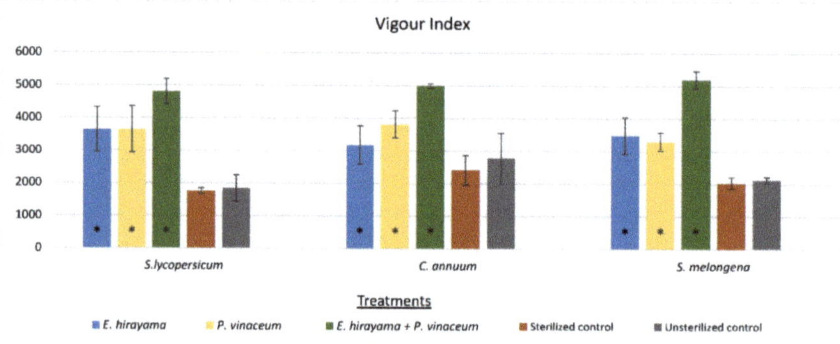

Figure 3. Vigor index of three plants (*S. lycopersicum*, *C. annuum*, *S. melongena*) grown in the composts inoculated with two mangrove fungi (*P. vinaceum* and *E. hirayamae*) and their combination compared to sterilized waste control. Unsterilized control refers to the treatment of raw waste. * indicates a significant effect ($p < 0.05$) of the fungal species or a significant interaction (combined effect) in two-way ANOVA (no significant effects was observed). a indicates a significant difference between the sterilized and unsterilized control treatments (*t*-test, $p < 0.05$).

The disease defense ability of the seeds produced by the plants grown in the composts was assessed with a seed germination test after exposure to plant pathogens. The germination test gave similar results for each plant species and all of the pathogens (Figure 4). The germination was, in general, clearly higher in the fungal treatments than in the control treatments, with the combined fungal treatment being the highest. The only non-significant effects of *P. vinaceum* and *E. hirayamae* were for the combination of the plant *S. lycopersicum* and the pathogen *F. oxysporum*. The interaction term was significant except for the combination of the plant *S. melongena* and the pathogen *F. oxysporum*. Only in three cases did the unsterilized control and the sterilized control differ significantly in the *t*-test, with the unsterilized control showing a higher germination percentage. The germination varied between 87% and 90% for the combined fungal treatment, and 17% and 40% in the control treatments for the plant *S. lycopersicum*. The respective values for the second plant *C. annuum* were 93–96% and 6–15%, and for the third plant *S. melongena*, they were 87–93% and 13–25%.

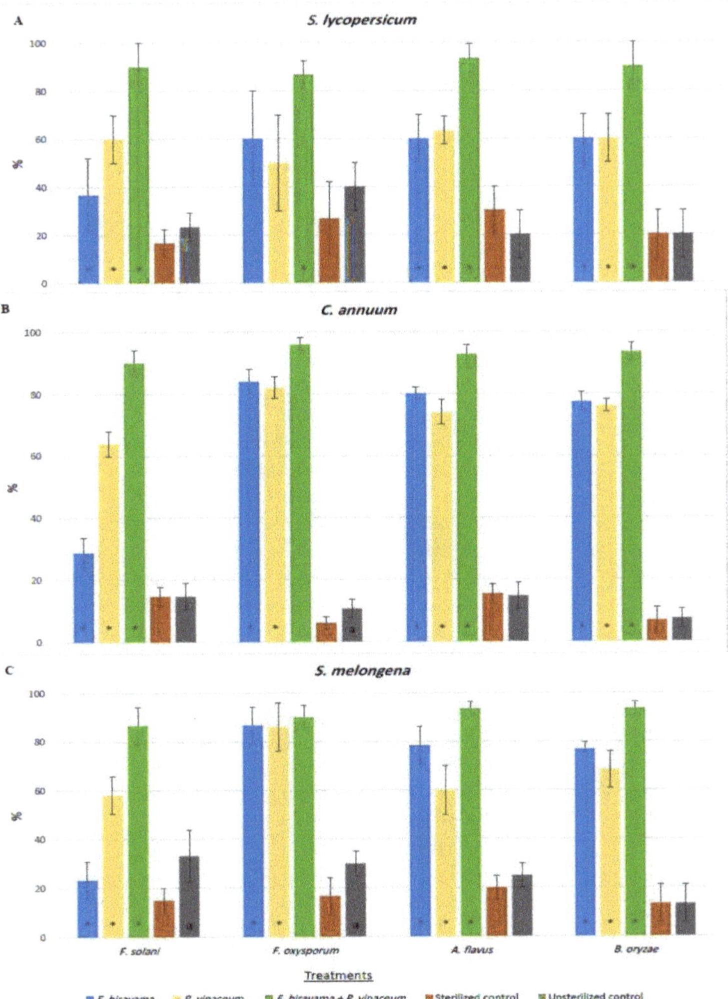

Figure 4. Germination of the seeds of three plants (*S. lycopersicum*, *C. annuum*, *S. melongena*) after exposure to four plant pathogens (*F. solani*, *F. oxysporum*, *A. flavus*, *B. oryzae*) and grown in the composts inoculated with two mangrove fungi (*P. vinaceum* and *E. hirayamae*) and their combination compared to sterilized waste control. Unsterilized control refers to the treatment of raw waste. * indicates a significant effect ($p < 0.05$) of the fungal species or a significant interaction (combined effect) in two-way ANOVA. a indicates a significant difference between the sterilized and unsterilized control treatments (*t*-test, $p < 0.05$).

Table 5. Disease incidence of the three plant species grown in the composts inoculated with two mangrove fungi (*E. hirayamae* and *P. vinaceum*) and their combination, compared to a sterilized waste control. The unsterilized control refers to the treatment of raw waste. * indicates a significant effect ($p < 0.05$) of the fungal species and a significant interaction (combined effect) in the two-way ANOVA. An 'a' indicates a significant difference between the control treatments (*t*-test, $p < 0.05$).

Treatment	Sterilized Control	*E. hirayamae*	*P. vinaceum*	*E. hirayamae P. vinaceum*	Unsterilized Control
Plant species					
S. lycopersicum	50 ± 5	20 ± 2 *	10 ± 2 *	0 ± 0 *	60 ± 7
C. annuum	50 ± 1	10 ± 3 *	10 ± 1 *	0 ± 0 *	70 ± 5 a
S. melongena	70 ± 3	10 ± 2 *	10 ± 1 *	0 ± 0 *	60 ± 4 a

4. Discussion

Municipal waste composting, especially small-scale composting, provides benefits from agriculture to the circular economy [34,35]. The use of compost reduces the need for fertilizers in farming; compost addition has been shown to increase yields more than nitrogen, phosphorus, and potassium (NPK) chemical fertilization alone [20]. Moreover, compost has been shown to reduce many plant diseases [7,35]. We focused on further ameliorating the quality of compost by inoculating fungi into the composting process. We showed that the fungi in the compost not only increased vegetable growth remarkably but also suppressed phytopathogens. It seems that when the fungi secreted plant hormones *in vitro*, they improved the quality of the compost and the plants grew better. More importantly, when the fungi showed activity against phytopathogens *in vitro*, the seeds collected from the plants grown in the compost were able to defend themselves against the phytopathogens.

The addition of compost has been shown to improve the fertility, water holding capacity, and suppressiveness of soil against pathogens. All of these properties can be enhanced by inoculating certain microbes into the composting process. In choosing the microbes, one important aspect is the tolerance of the microbes against the harmful pollutants the waste contains, which is an aspect that has not often been considered previously. Municipal organic waste, even if it is sorted, contains small amounts of harmful materials such as plastic, heavy metals, and oil products [36]. Therefore, fungi that are not only able to tolerate pollutants but also to degrade recalcitrant materials might be useful additions to compost. Heavy-metal–tolerant organisms have been found in mangrove habitats [37,38]. In addition, recalcitrant substrates such as oil products have been shown to be degraded by mangrove habitat fungi. Nine out of 45 fungal isolates originating from a mangrove habitat were able to degrade engine oil *in vitro* [39]. Petroleum and cellulosic materials were degraded most efficiently by the fungal species used in this experiment, *E. hirayamae* [39–42]. *Eupenicillium hirayamae* did not efficiently degrade polyethylene, which was degraded most with a combination of fungal species [41]. In general, *E. hirayamae* in combination with the other fungal species, among them the one in our experiment, *P. vinaceum*, degraded the materials most efficiently [39–42]. *Alternaria alternata* appeared to degrade polyethene the most efficiently in a previous experiment [42]. However, in our preliminary screening of the fungi, *A. alternata* did not indicate any plant-growth promoting or antifungal activities *in vitro*.

We chose the two species showing the highest activities *in vitro*, as well as the ability to degrade recalcitrant materials in our previous studies. The latter property is especially important in Saudi Arabia, where plastic and other oil-based materials are common pollutants in municipal waste [43]. We inoculated waste with two fungi that were isolated from a mangrove habitat. The fungi *P. vinaceum* and *E. hirayamae* appeared to produce compost that doubled the yield of tomatoes, aubergines and peppers, and increased the vegetative growth of the plants by more than double. The highest increase was observed in the amount of seeds in the fruits.

Another important aspect in the assessment of the quality of a compost product is how it suppresses pathogens. Compost produced without any microbial inoculation has been shown to suppress many soil-borne pathogens. *Fusarium* sp.-caused diseases have been observed to be suppressed by different kinds of compost additions [44]. *Fusarium oxysporum* was suppressed by different types of compost [45]. Compost leachate was observed to suppress *A. niger* [46]. *Fusarium oxysporum* and *A. niger* were included in our experiment. Other pathogens—such as *Pythium ultimum, Rhizoctonia solani, Phytophthora* sp., *Sclerotinia minor,* and *Verticillium dahlia*—have been suppressed by the addition of different types of composts [47].

Disease suppression might be improved by the inoculation of waste with microbes. The bacterium *Streptomyces* sp., when used in composting, was observed to suppress several pathogens [48]. Filamentous fungi (*Trichoderma, Fusarium, Gliocladium, Aspergillus* and *Penicillium*) have also been inoculated and suppressed pathogens successfully, as reviewed recently [35]. The review did not include our mangrove-associated fungal species *P. vinaceum* and *E. hirayamae*, which we, therefore, can add to the list. In our experiment, the vegetables grew best in the compost in which the combination of two mangrove fungi had been inoculated to originally-sterilized wastes. Furthermore, the seeds of the most vigorous plants were able to suppress phytopathogens remarkably. To the best of our knowledge, this continuum from the plants to the seeds was reported here for the first time.

The physico-chemical properties of the composts indicated that both fungi individually and in combination produced mature compost. The lack of significant interaction between the species indicated that all of the fungal treatments produced similar compost concerning the physico-chemical properties measured. However, the microbial community in the compost has been shown to be of importance, as reviewed recently by [35]. This was observed in our study as well. Contrary to the physico-chemical variables, the biological variables gave significant interactions between the fungal species; the combined effects were showed some improvement over the others. We observed this phenomenon first for the disease incidence of the plants grown in the composts, and second for the improved germination of the seeds produced by the plants grown in the composts. This indicates that both of the microbes had a remarkable role in improving the quality of the compost. This is an important observation from a biological control point of view.

5. Conclusions

Composting is a far-too-seldom used technique in waste treatment, although many of its benefits have been shown. The addition of compost improves soil fertility and suppresses plant diseases, which allows the reduction of the use of chemical fertilizers and disease control agents. The properties of compost can be further ameliorated by inoculating certain microbes into the composting process, as we showed in our composting experiment. The inoculation of two mangrove fungal species improved the properties of the compost produced. We showed that the growth of vegetables increased and the phytopathogens were suppressed in the compost produced. The remarkable observation was that the ability of plants to defend themselves against phytopathogens was observed in the vegetable seeds grown in the compost. This continuum is an important observation for sustainable agriculture. Many environmental and economic benefits encourage us to take up composting as a waste treatment practice. It would be especially beneficial in areas where poor waste management facilities exist.

Author Contributions: Both authors (F.A. and A.A.A.-H.) contributed to the data analysis and the drafting of the manuscript. Both authors (F.A. and A.A.A.-H.) approved the final manuscript and are accountable for all aspects of the work. All authors have read and agreed to the published version of the manuscript.

Funding: This project was funded by the National Plan for Science, Technology and Innovation(MAARIFAH), King Abdulaziz City for Science and Technology, Kingdom of Saudi Arabia, Award Number (2-17-01-001-0051).

Conflicts of Interest: The authors declare no conflict of interest.

References

1. Ouda, O.K.M.; Cekirge, H.M.; Raza, S.A.R. An assessment of the potential contribution from waste-to-energy facilities to electricity demand in Saudi Arabia. *Energy Convers. Manag.* **2013**, *75*, 402–406. [CrossRef]
2. Ayilara, M.S.; Olanrewaju, O.S.; Babalola, O.O.; Odeyemi, O. Waste Management through Composting: Challenges and Potentials. *Sustainability* **2020**, *12*, 4456. [CrossRef]
3. Chen, T.; Zhang, S.; Yuan, Z. Adoption of solid organic waste composting products: A critical review. *J. Clean. Prod.* **2020**, *272*, 122712. [CrossRef]
4. Adugna, G. A review on impact of compost on soil properties, water use and crop productivity. *Acad. Res. J. Agric. Sci. Res.* **2016**, *4*, 93–104.
5. Amlinger, F.; Götz, B.; Dreher, P.; Geszti, J.; Weissteiner, C. Nitrogen in biowaste and yard waste compost: Dynamics of mobilisation and availability—A review. *Eur. J. Soil Biol.* **2003**, *39*, 107–116. [CrossRef]
6. De Corato, U. Disease-suppressive compost enhances natural soil suppressiveness against soil-borne plant pathogens: A critical review. *Rhizosphere* **2020**, *13*, 100192. [CrossRef]
7. Noble, R.; Coventry, E. Suppression of soil-borne plant diseases with composts: A review. *Biocontrol Sci. Technol.* **2005**, *15*, 3–20. [CrossRef]
8. Postma, J.; Montanari, M.; van den Boogert, P.H.J.F. Microbial enrichment to enhance the disease suppressive activity of compost. *Eur. J. Soil Biol.* **2003**, *39*, 157–163. [CrossRef]
9. Sharma, A.; Sharma, R.; Arora, A.; Shah, R.; Singh, A.; Pranaw, K.; Nain, L. Insights into rapid composting of paddy straw augmented with efficient microorganism consortium. *Int. J. Recycl. Org. Waste Agric.* **2014**, *3*, 54. [CrossRef]
10. Song, C.; Zhang, Y.; Xia, X.; Qi, H.; Li, M.; Pan, H.; Xi, B. Effect of inoculation with a microbial consortium that degrades organic acids on the composting efficiency of food waste. *Microb. Biotechnol.* **2018**, *11*, 1124–1136. [CrossRef] [PubMed]
11. Sharma, A.; Saha, T.N.; Arora, A.; Shah, R.; Nain, L. Efficient Microorganism compost benefits plant growth and improves soil health in calendula and marigold. *Hortic. Plant J.* **2017**, *3*, 67–72. [CrossRef]
12. Kumar, A.; Brar, N.S.; Pal, S.; Singh, P. Available soil macro and micronutrients under rice wheat cropping system in District Tarn Taran of Punjab. *Ecol. Enviton. Conserv.* **2017**, *23*, 202–207.
13. Damare, V.S.; Kajawadekar, K.G. A preliminary study on L-asparaginase from mangrove detritus-derived fungi and its application in plant growth promotion. *MycoAsia J.* **2020**, *4*.
14. Sureshkumar, P.; Kavitha, S. Bioprospecting potential of mangrove fungus from vellar estuary, southeast coast of india for biocontrol of damping off on mustard. *Res. J. Biotechnol.* **2019**, *14*, 72–78.
15. Hamzah, T.N.T.; Lee, S.Y.; Hidayat, A.; Terhem, R.; Faridah-Hanum, I.; Mohamed, R. Diversity and characterization of endophytic fungi isolated from the tropical mangrove species, *Rhizophora mucronata*, and identification of potential antagonists against the soil-borne fungus, *Fusarium solani*. *Front. Microbiol.* **2018**, *9*, 1707. [CrossRef]
16. Al-Shibli, H.; Dobretsov, S.; Al-Nabhani, A.; Maharachchikumbura, S.S.N.; Rethinasamy, V.; Al-Sadi, A.M. *Aspergillus terreus* obtained from mangrove exhibits antagonistic activities against *Pythium aphanidermatum*-induced damping-off of cucumber. *PeerJ* **2019**, *7*, e7884. [CrossRef]
17. He, X.-T.; Traina, S.J.; Logan, T.J. Chemical properties of municipal solid waste composts. *J. Environ. Qual.* **1992**, *21*, 318–329. [CrossRef]
18. Harada, Y.; Inoko, A. The measurement of the cation-exchange capacity of composts for the estimation of the degree of maturity. *Soil Sci. Plant Nutr.* **1980**, *26*, 127–134. [CrossRef]
19. Nandi, M.; Selin, C.; Brawerman, G.; Fernando, W.G.D.; de Kievit, T. Hydrogen cyanide, which contributes to *Pseudomonas chlororaphis* strain PA23 biocontrol, is upregulated in the presence of glycine. *Biol. Control.* **2017**, *108*, 47–54. [CrossRef]
20. Ozimek, E.; Jaroszuk-Ściseł, J.; Bohacz, J.; Korniłłowicz-Kowalska, T.; Tyśkiewicz, R.; Słomka, A.; Nowak, A.; Hanaka, A. Synthesis of indoleacetic acid, gibberellic acid and ACC-deaminase by *Mortierella* strains promote winter wheat seedlings growth under different conditions. *Int. J. Mol. Sci.* **2018**, *19*, 3218. [CrossRef]
21. Malhotra, M.; Srivastava, S. Stress-responsive indole-3-acetic acid biosynthesis by *Azospirillum brasilense* SM and its ability to modulate plant growth. *Eur. J. Soil Biol.* **2009**, *45*, 73–80. [CrossRef]
22. Alori, E.T.; Glick, B.R.; Babalola, O.O. Microbial phosphorus solubilization and its potential for use in sustainable agriculture. *Front. Microbiol.* **2017**, *8*, 971. [CrossRef] [PubMed]
23. Ameen, F.; Moslem, M.; Hadi, S. *Biodegradation of Urban Waste by Mangrove Fungi*; Lap Lambert Academic Publishing GmbH & Company KG: Saarbrücken, Germany, 2015; p. 236.
24. Khan, A.L.; Hamayun, M.; Kang, S.-M.; Kim, Y.-H.; Jung, H.-Y.; Lee, J.-H.; Lee, I.-J. Endophytic fungal association via gibberellins and indole acetic acid can improve plant growth under abiotic stress: An example of *Paecilomyces formosus* LHL10. *BMC Microbiol.* **2012**, *12*, 3. [CrossRef] [PubMed]
25. Mehmood, A.; Khan, N.; Irshad, M.; Hamayun, M.; Husna, I.; Javed, A.; Hussain, A. IAA producing endopytic fungus *Fusariun oxysporum* wlw colonize maize roots and promoted maize growth under hydroponic condition. *Eur. Exp. Biol.* **2018**, *8*, 24. [CrossRef]
26. Waqas, M.; Khan, A.L.; Kamran, M.; Hamayun, M.; Kang, S.-M.; Kim, Y.-H.; Lee, I.-J. Endophytic fungi produce gibberellins and indoleacetic acid and promotes host-plant growth during stress. *Molecules* **2012**, *17*, 10754–10773. [CrossRef]

27. Harman, G.E.; Howell, C.R.; Viterbo, A.; Chet, I.; Lorito, M. *Trichoderma* species—Opportunistic, avirulent plant symbionts. *Nat. Rev. Microbiol.* **2004**, *2*, 43–56. [CrossRef]

28. Sambrook, H.C. *Molecular Cloning: A Laboratory Manual*; Cold Spring Harbor: New York, NY, USA, 1989.

29. Ameen, F.; AlYahya, S.A.; AlNadhari, S.; Alasmari, H.; Alhoshani, F.; Wainwright, M. Phosphate solubilizing bacteria and fungi in desert soils: Species, limitations and mechanisms. *Arch. Agron. Soil Sci.* **2019**, *65*, 1446–1459. [CrossRef]

30. Maria, G.L.; Sridhar, K.R.; Raviraja, N.S. Antimicrobial and enzyme activity of mangrove endophytic fungi of southwest coast of India. *J. Agric. Technol.* **2005**, *1*, 67–80.

31. White, T.J.; Bruns, T.; Lee, S.; Taylor, J.W. Amplification and direct sequencing of fungal ribosomal RNA genes for phylogenetics. *Pcr. Protoc. A Guide Methods Appl.* **1990**, *18*, 315–322.

32. Iqbal, M.K.; Nadeem, A.; Sherazi, F.; Khan, R.A. Optimization of process parameters for kitchen waste composting by response surface methodology. *Int. J. Environ. Sci. Technol.* **2015**, *12*, 1759–1768. [CrossRef]

33. RStudio Team. *Integrated Development for R*; RStudio, Inc.: Boston, MA, USA, 2015.

34. Mu, D.; Horowitz, N.; Casey, M.; Jones, K. Environmental and economic analysis of an in-vessel food waste composting system at Kean University in the US. *Waste Manag.* **2017**, *59*, 476–486. [CrossRef] [PubMed]

35. De Corato, U. Agricultural waste recycling in horticultural intensive farming systems by on-farm composting and compost-based tea application improves soil quality and plant health: A review under the perspective of a circular economy. *Sci. Total Environ.* **2020**, *738*, 139840. [CrossRef] [PubMed]

36. Alessi, A.; Lopes, A.d.C.P.; Müller, W.; Gerke, F.; Robra, S.; Bockreis, A. Mechanical separation of impurities in biowaste: Comparison of four different pretreatment systems. *Waste Manag.* **2020**, *106*, 12–20. [CrossRef] [PubMed]

37. Almahasheer, H. High levels of heavy metals in Western Arabian Gulf mangrove soils. *Mol. Biol. Rep.* **2019**, *46*, 1585–1592. [CrossRef] [PubMed]

38. Al-Homaidan, A.A.; Al-Ghanayem, A.A.; Al-Qahtani, H.S.; Al-Abbad, A.F.; Alabdullatif, J.A.; Alwakeel, S.S.; Ameen, F. Effect of sampling time on the heavy metal concentrations of brown algae: A bioindicator study on the Arabian Gulf coast. *Chemosphere* **2020**, *263*, 127998. [CrossRef]

39. Ameen, F.; Hadi, S.; Moslem, M.; Al-Sabri, A.; Yassin, M.A. Biodegradation of engine oil by fungi from mangrove habitat. *J. Gen. Appl. Microbiol.* **2015**, *61*, 185–192. [CrossRef]

40. Ameen, F.; Moslem, M.; Hadi, S.; Al-Sabri, A.E. Biodegradation of diesel fuel hydrocarbons by mangrove fungi from Red Sea Coast of Saudi Arabia. *Saudi J. Biol. Sci.* **2016**, *23*, 211–218. [CrossRef]

41. Ameen, F.; Moslem, M.A.; Hadi, S.; Al-Sabri, A. Biodegradation of cellulosic materials by marine fungi isolated from South Corniche of Jeddah, Saudi Arabia. *J. Pure Appl. Microbiol.* **2014**, *8*, 3617–3626.

42. Ameen, F.; Moslem, M.; Hadi, S.; Al-Sabri, A.E. Biodegradation of Low Density Polyethylene (LDPE) by Mangrove Fungi From the Red Sea Coast. *Prog. Rubber Plast. Recycl. Technol.* **2015**, *31*, 125. [CrossRef]

43. Anjum, M.; Miandad, R.; Waqas, M.; Ahmad, I.; Alafif, Z.O.A.; Aburiazaiza, A.S.; Barakat, M.A.; Akhtar, T. Solid waste management in Saudi Arabia: A review. *J. Appl. Agric. Biotechnol.* **2016**, *1*, 13–26.

44. Moosa, A.; Sahi, S.T.; Haq, I.-U.; Farzand, A.; Khan, S.A.; Javaid, K. Antagonistic potential of Trichoderma isolates and manures against Fusarium wilt of tomato. *Int. J. Veg. Sci.* **2017**, *23*, 207–218. [CrossRef]

45. Suárez-Estrella, F.; Vargas-Garcia, C.; Lopez, M.J.; Capel, C.; Moreno, J. Antagonistic activity of bacteria and fungi from horticultural compost against *Fusarium oxysporum* f. sp. melonis. *Crop. Prot.* **2007**, *26*, 46–53. [CrossRef]

46. Özer, N.; Köycü, N.D. The ability of plant compost leachates to control black mold (*Aspergillus niger*) and to induce the accumulation of antifungal compounds in onion following seed treatment. *BioControl* **2006**, *51*, 229–243. [CrossRef]

47. Postma, J.; Nijhuis, E.H. *Pseudomonas chlororaphis* and organic amendments controlling *Pythium* infection in tomato. *Eur. J. Plant Pathol.* **2019**, *154*, 91–107. [CrossRef]

48. Al-Dhabi, N.A.; Esmail, G.A.; Mohammed Ghilan, A.-K.; Valan Arasu, M. Composting of Vegetable Waste Using Microbial Consortium and Biocontrol Efficacy of *Streptomyces* Sp. Al-Dhabi 30 Isolated from the Saudi Arabian Environment for Sustainable Agriculture. *Sustainability* **2019**, *11*, 6845. [CrossRef]

 *sustainability*

Article

Effects of the COVID-19 Pandemic on Food Security and Agriculture in Iran: A Survey

Abdullah Kaviani Rad [1], **Redmond R. Shamshiri** [2,*], **Hassan Azarm** [3], **Siva K. Balasundram** [4] and **Muhammad Sultan** [5]

1   Department of Soil Science, School of Agriculture, Shiraz University, Shiraz 71946-85111, Iran; arad@adaptiveagrotech.com
2   Leibniz Institute for Agricultural Engineering and Bioeconomy, 14469 Potsdam-Bornim, Germany
3   Department of Agricultural Economics, School of Agriculture, Shiraz University, Shiraz 71946-85111, Iran; hazarm@shirazu.ac.ir
4   Department of Agriculture Technology, Faculty of Agriculture, University Putra Malaysia, Serdang 43400, Selangor, Malaysia; siva@upm.edu.my
5   Department of Agricultural Engineering, Bahauddin Zakariya University, Bosan Road, Multan 60800, Pakistan; muhammadsultan@bzu.edu.pk
*   Correspondence: rshamshiri@atb-potsdam.de; Tel.: +49-(0)331-569-9410

Abstract: The consequences of COVID-19 on the economy and agriculture have raised many concerns about global food security, especially in developing countries. Given that food security is a critical component that is affected by global crises, beside the limited studies carried out on the macro-impacts of COVID-19 on food security in Iran, this paper is an attempt to address the dynamic impacts of COVID-19 on food security along with economic and environmental challenges in Iran. For this purpose, a survey was conducted with the hypothesis that COVID-19 has not affected food security in Iran. To address this fundamental hypothesis, we applied the systematic review method to obtain the evidence. Various evidences, including indices and statistics, were collected from national databases, scientific reports, field observations, and interviews. Preliminary results revealed that COVID-19 exerts its effects on the economy, agriculture, and food security of Iran through six major mechanisms, corresponding to a 30% decrease in the purchasing power parity in 2020 beside a significant increase in food prices compared to 2019. On the other hand, the expanding environmental constraints in Iran reduce the capacity of the agricultural sector to play a crucial role in the economy and ensure food security, and in this regard, COVID-19 forces the national programs and budget to combat rising ecological limitations. Accordingly, our study rejects the hypothesis that COVID-19 has not affected food security in Iran.

Keywords: COVID-19; Iran; food security; economic crisis; agriculture; food supply chain

 check for **updates**

Citation: Rad, A.K.; Shamshiri, R.R.; Azarm, H.; Balasundram, S.K.; Sultan, M. Effects of the COVID-19 Pandemic on Food Security and Agriculture in Iran: A Survey. *Sustainability* **2021**, *13*, 10103. https://doi.org/10.3390/su131810103

Academic Editor: Riccardo Testa

Received: 8 August 2021
Accepted: 7 September 2021
Published: 9 September 2021

Publisher's Note: MDPI stays neutral with regard to jurisdictional claims in published maps and institutional affiliations.

1. Introduction

Global economic growth was projected to reach 3.2% in 2019 and 3.5% in 2020, and emerging economies were expected to grow by 4.1% and 4.7%, as well as 6.5% growth for developing Asia in 2019–2020 [1]. However, the outbreak of a novel SARS-CoV-2 from Wuhan province (China) known as "COVID-19" [2] quickly changed all the predictions about the future of the global and regional economy. Although the global lockdowns assisted to control the disease outbreak, this pandemic damaged many economic sectors such as industries [3], tourism [4], trade and business activities [5], and agriculture [6]. Thus, investors have removed $83 billion from emerging markets since the beginning of the crisis, and the most significant capital outflow ever recorded [7]. While there is no system to determine the actual economic damage from the pandemic [8], it is expected to impose the most influential threat to the future of trade [9], and global economic growth was forecasted to reach −3.2%, and the most extensive global recession was created during the COVID-19 pandemic, as well as declining the GDP by 5.2% in 2020 [10–12] that pushed 34 million people into severe poverty [13]. Due to the sharp economic losses, the concerns about the food security situation during the COVID-19 outbreak period aroused researchers' attention.

Food security, characterized as access to safe and sufficient food, is affected by economic crises. Pandemics can cause damage to the global economy [14], and the COVID-19 pandemic has short and long-term effects on food security and human health worldwide [15]. At this time, COVID-19 has infected more than 218 million and led to the death of 4.5 million people [16]. It appears that the economic consequences of the COVID-19 on food security are more significant in developing countries [17]. The $64 million loss caused by lockdown each day in Bangladesh, the inability of 16% of Indian urban households to access government food aid, and the 14% reduction in food security for Mexican families are examples of food insecurity caused by the pandemic [18–20].

Iran, as a developing country with a population of 82 million people, has also been affected by the various consequences of COVID-19. In February 2020, some countries were battling the pandemic, and after China, Iran appeared as one of the earliest COVID-19 epidemic countries [21]. Now, COVID-19 has infected more than 4.9 million people with more than 107,000 victims in Iran [22], leading to a unique economic crisis. Enhancement of health sector costs (+28%) [23] and financial losses, caused by lockdown restrictions, led to a remarkable decline in revenue, growth of unemployment, and interruptions in transportation, industry, services, oil [24], tourism [25], and agriculture [26,27]. Restrictions on imports-exports generated hurdles for food transportation; meanwhile, consumers and producers faced difficulties that ultimately led to declining farmers' income and striking damage to agriculture [28]. However, the agriculture sector is challenged by environmental limitations such as climate change and water crisis.

After considering the mentioned concerns, these questions were created; (i) are the merged economic consequences of the COVID-19 and ecological constraints synergistic? (ii) What is the connection between agricultural constraints and the consequences of the COVID-19 on food security? In other words, (iii) what are the dynamic impacts of the pandemic on food security along with economic and environmental challenges in Iran? We began with the hypothesis that COVID-19 has not been linked to agricultural constraints and has no synergistic impressions. Since the shocks of COVID-19 on food security in Iran and the environmental approach in agricultural economics, examinations of the pandemic have not been considered; hence, the present study aims to investigate the various effects of COVID-19 on food security status in Iran with a new specific approach.

2. Materials and Methods

Since previous statistical studies on food security in Iran were primarily on a local scale, the present study proposed to draw a comprehensive schematic of the impacts of the COVID-19 crisis on food security in Iran. Hence, a systematic reviewing was applied as an appropriate method to collect evidence for hypothesis investigation to address the research questions. For this purpose, keywords food security, COVID-19, global lockdowns, trade disruption, agriculture, food supply chain, food insecurity, economic crisis, climate change, water crisis, natural disasters, environmental sustainability, and ecological constraints were searched in Google Scholar, Scopus, and other international and national databases. Global and Iranian databases, including statistics of governmental organizations, were adopted to explore the consequences of the pandemic on food security in Iran. The literature review of this study was conducted in three scopes: COVID-19 and food security, economic results of the COVID-19, and agricultural sustainability of Iran. Obtained evidence were associated to World Bank, International Monetary Fund (IMF), FAO, United Nations, WHO, OECD, World Trade Organization (WTO), and Iranian national organizations such as Central Bank (CBI) and Statistical Center of Iran (SCI), Department of Environment (DOE), and Iran Parliament Research Center (IPRC). After assembling the relevant information, the results of the study were presented in a classified style.

3. Results and Discussion

3.1. COVID-19 and New Economic Challenges

Iran was challenged by significant economic problems such as high inflation, rising exchange rate, and declining industrial activities prior to the COVID-19 outbreak, and the economic growth and unemployment rates were -7.6% and 10.6%, as well as economic growth without oil 0%, agriculture 3.2%, oil -37%, and industry -2.3% in 2019 [29]. Hence, high inflation has been one of the significant economic concerns in Iran due to the destructive consequences on the Iranian monetary value [30]. In this regard, Sultan-Tavih et al. [31] reported that rising inflation has significant adverse impacts on Iran's economic growth, and its long-term effects are more severe than short-term. The high inflation rate has led to a notable rise in the final price of products for consumers in recent years [32]. Accordingly, Iran has the sixth-highest inflation rate globally [33], and its per capita income was reduced by 34% between 2011 and 2019, while the number of countries that provided 80% of Iran's export revenue has decreased from 23 to 9 in 2001–2018. More so, statistics revealed that Iran's oil exports increased by two million barrels per day after the Joint Comprehensive Plan of Action (JCPOA) agreement [34]; nevertheless, with the return of sanctions in May 2018, oil exports were estimated to be less than 500,000 barrels per day [35], which reduced foreign exchange earnings and a worsening government budget deficit.

Therefore, Iran's economy has faced COVID-19 after two difficult years between 2018 and 2020 because the high inflation rate of 2018–2019 damaged many industries and business activities, and Iran's exports decreased by 30% in March 2020 compared to 2019. Additionally, Iran's per capita income has been split during the pandemic outbreak in comparison to 2019, due to a significant decline in the oil price (falling oil price to $40 per barrel). The point inflation rate also reached 41.4% in September 2020 due to the COVID-19 economic consequences; thus, Iranian households have spent 41.4% more than September 2019 for buying the same products. Meanwhile, the point inflation rate for foods and beverages, as well as services, grew $9\text{–}40.5\%$ and $7\text{–}41.9\%$, respectively [36]. The COVID-19 pandemic has rigorously changed employment and salary in numerous labor-intensive activities, specifically in the informal sector. Thus, the employment opportunities have declined by more than one million jobs. Although poverty had reached 14% before COVID-19, it was estimated that a decline in incomes and increasing household costs, due to inflation, will drive poverty by 20% in 2020 [37]. During the lockdowns period, growth in the value of financial transactions was 41% lower than a month before the pandemic [38].

Consequently, COVID-19 had the most significant impact on livelihood and employment in Iran because 25% of Iran's employees are in the service sector [39]. The governmental revenue level was decreased due to severe financial consequences of COVID-19, such as the growing exchange rate. Rising exchange rates have affected inflation and led to higher prices for vital goods [40]. It seems that COVID-19 accelerated the growth of the currency exchange rate due to the reduction in exports by disruption of international trade in 2020 [41]. The average exchange rate has risen seven times from 2017 to September 2020. In other words, the value of Iran's national currency has decreased by more than 590% in the mentioned period. The dollar currency value was about 130 thousand Rial in 2019, but it reached 280 thousand Rial in September 2020 ($+111\%$) (1 US dollar = 27,000 Rial) [42]. Figure 1 shows that the exchange rate has been significantly increased since 2011.

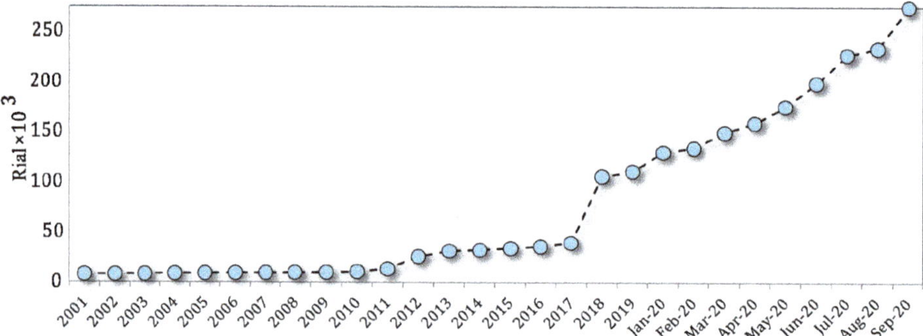

Figure 1. Rising exchange rate in Iran from 2001 to 2020; data source: [43].

Pakravan-Charvadeh et al. [44] demonstrated that Iranian families decreased consumption of some groups of foods during the pandemic outbreak due to new financial problems; therefore, COVID-19 can negatively impact food security by decreasing people's income [45]. Consequently, inflation led to food insecurity through poverty spread and reduction of income [46]. Additionally, controlling the goods price and monopoly pricing are severe problems of Iran's economy [47]. Figure 2 shows the impact of inflation on the increasing price index from 2001 to 2020. In addition, the growth trend of the exchange rate has led to an increase in production cost; hence, a significant rise in the final product price has been observed.

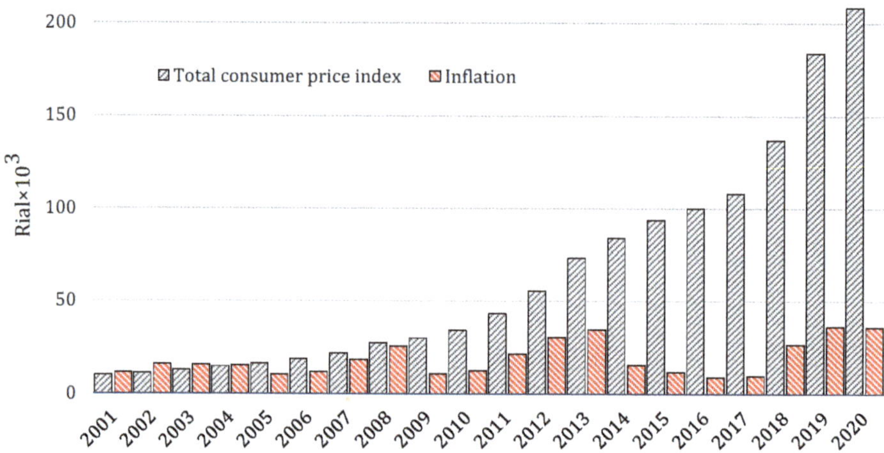

Figure 2. Demonstration of the increasing average inflation rate and total consumer price index from 2001 to 2020; data source: [48,49].

3.2. COVID-19, Food System and Supply Chain

By cutting the connections between farmers and food markets, COVID-19 shocked the food supply chain and reduced food accessibility. Shocks to the food supply chain by the pandemic can be double in developing and low-income countries [50–52] and endanger the lives of millions of people and smallholders [53]. COVID-19 affects the total food demand and supply system in Iran by creating imbalance due to the following factors: (i) reduction of household demand due to an unwillingness to use the services sector for preventing the COVID-19 outbreak, (ii) reduction of consumer demand due to declining incomes for households who lost their jobs in lockdown restriction, (iii) reduction of household demand for durable goods and increase of money-savings, and (iv) reduction of exports due to border closure. Motevali et al. [54] reported that rice acts as an imperative product in Iran and Iranian families need 3,200,000 tons of rice per year, and 50 million people will be without food if the government does not import this product; the ban on rice exports in India, Pakistan, and Thailand in 2020 increased price of rice in Iran and reduced its imports by 50%. Between January and August 2020, approximately 450,000 tons of rice were imported by the government, while in the same period of the previous year, 950,000 tons were imported [55]; thus, rice price in Iran has been increased by 13% [56]. Therefore, the food supply chain imbalance by COVID-19 in Iran has imposed a significant threat to food security in poor households. Arouna et al. [57] confirmed the same result for West Africa, and similar evidence demonstrated that fifteen countries observed more than a 10% increase in food prices, and fourteen countries have announced export bans for twenty various products in the first quarter of 2020, according to World Food Program Market Monitoring [58]. Rice prices in exporter countries, such as Russia and Vietnam, grew up, and they limited wheat and rice exports due to domestic demand enhancement. Philippines and Saudi Arabia have bought surplus rice and wheat, and the price of rice in Myanmar increased due to simultaneously rising demand and declining seasonal harvests [59,60]. In addition, Iran's government reduced subsidies for chemical fertilizers, which led to a significant fertilizer price enhancement in 2020 [61]. This excessive increase in prices has reduced farmers' demand for chemical fertilizers, and low crop yield and reduction of farmers' income were expected in 2020–2021. Moreover, the highest wages in the agricultural sector were observed in fruit picking (58.1%), disking by tractor (44.6%), and weeding (43.1%) [62]; hence, growing the price of agricultural inputs and workers' wages will reduce farmers' willingness to produce in the next cultivation seasons. From the global perspective, the low availability of pesticides and their high price has threatened crop protection in 2020 [63]. A similar global economic crisis in 2008–2009 raised prices of agricultural inputs and products, such as fertilizers and cereal grains, simultaneously and threatened food security in many countries (Figure 3) [64]. It can be concluded that COVID-19, similar to the economic crisis of 2008–2009, has increased the production costs and price of crop products, and a similar disturbance was recognized in Iran.

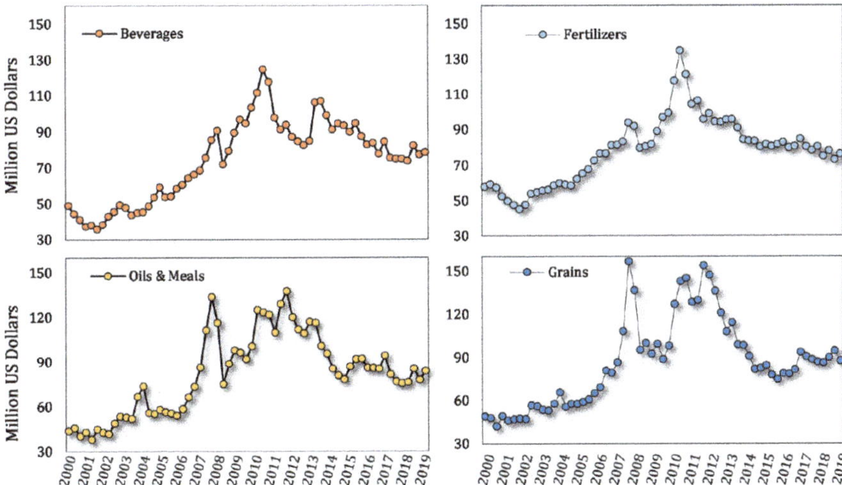

Figure 3. The global growth in monthly prices of agricultural products and inputs from 2000 to 2019, data source: [63].

Since the COVID-19 applied a general mechanism to affect the food supply chain and interrupt food trade between countries, the consequences for the food supply chain in Iran were similar to other countries. Furthermore, the results of other studies in many countries can be correlated to Iran. COVID-19 has effectively changed food consumption patterns and disturbed the food supply chain in rural sub-Saharan Africa [53,65]. Restrictions on access to agricultural inputs and labor in Senegal changed the production of four critical grains such as rice, maize, sorghum, and millet [66]. By evaluating the shocks of the pandemic toward food security and livelihoods of Senegalese families, Middendorf et al. [67] found that 82.5% of households had difficulty accessing sufficient food. These studies demonstrated the vulnerabilities of the food supply chain, in developing countries, to COVID-19 [68]. Figure 4 shows the dynamic mechanisms of COVID-19 to affect the food supply chain and has symbolized that rising food prices ultimately increase poverty and food insecurity.

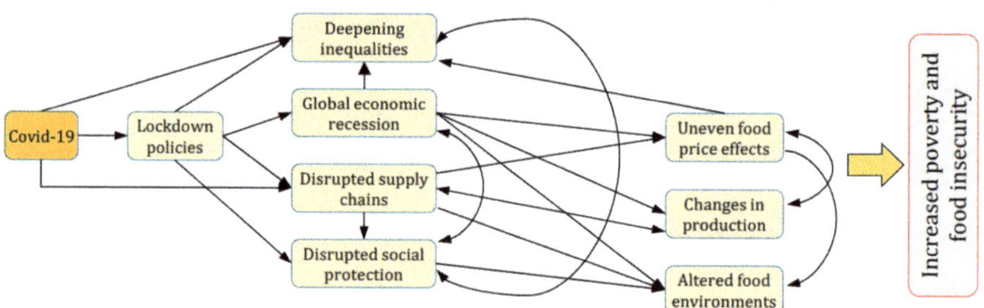

Figure 4. The dynamic mechanisms of COVID-19 that endanger food security; adapted from FAO [69].

After the extensive pandemic outbreak in Iran, the government has decided to pay a subsidy to low-income households in order to prevent food insecurity; however, it was not an efficient strategy to improve food security [70]. By considering the relative increase in subsidy and workers' wages in 2020–2021, it is observed that the purchasing power of every Iranian reduced by 30% in comparison to previous years [71]; furthermore, the government's ineffective policies to combat the consequences of the COVID-19 economic crisis have led to significant growth in the price of foods and a decline in people's revenue simultaneously, which leads to malnutrition in low-income households. Figure 5 shows the significant increase in different foods price from 2017 to 2020.

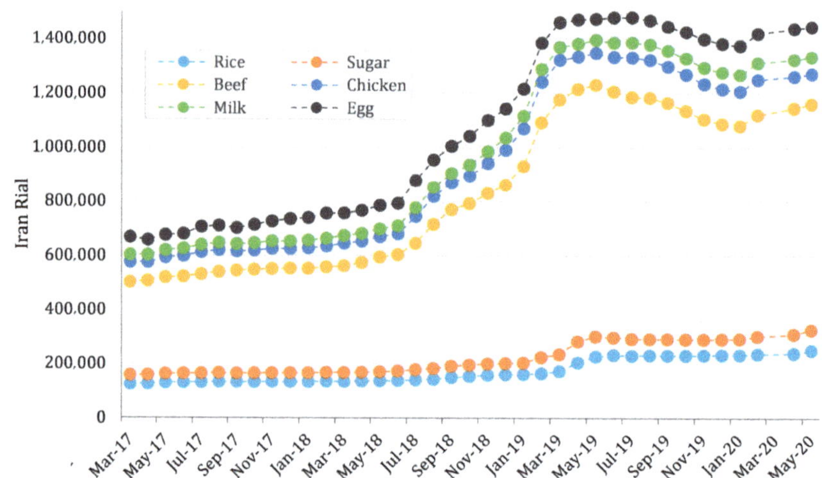

Figure 5. Rising prices of foods and beverages in Iran from 2017 to 2020 (Rial per kg^{-1} for rice, beef, sugar, and chicken; Rial per liter for milk; Rial per piece for egg), data source: [71].

The highest price enhancement in the food-beverages class is related to beef, milk, egg, and chicken [72]. Food prices have been significantly increased during the COVID-19 outbreak period compared to 2019, and the highest price in 2020 was related to wheat, lentils, watermelon, and apricots by 44.7%, 106.7%, 112.7%, and 46.4%, respectively. The red meat in Iran is higher-priced than in other markets worldwide; consequently, many low-income households have reduced their meat consumption due to its high price [73]. Therefore, the new economic crisis can shift the demand for nutritious foods such as vegetable oils and meat to cereal grains and poor-quality foods [74] through reducing purchasing power. Layani et al., [75] found that after the enhancement in the price of eight food groups, including meat, cooking oil, cereals, sugar, coffee, tea, dairy products, vegetables, and fruits, the abundance of vulnerable families raised by 10.63%. Despite the crucial role of a healthy diet in combating COVID-19, nutritional support is essential to prevent infection of individuals [76]. Hence, the people will be malnourished and more vulnerable to COVID-19 if they cannot consume nutritious foods such as vegetables, meat, dairy, grains, and fruits due to poverty and the high price of foods. COVID-19 has made people search for a nutritious diet, including foods that boost the immune system [77,78]; accordingly, food price monitoring by the government is necessary during the COVID-19 outbreak [79]. Consequently, COVID-19 also has indirect effects on infecting people in addition to direct impacts, and Figure 6 shows the indirect and direct consequences of the pandemic on human health.

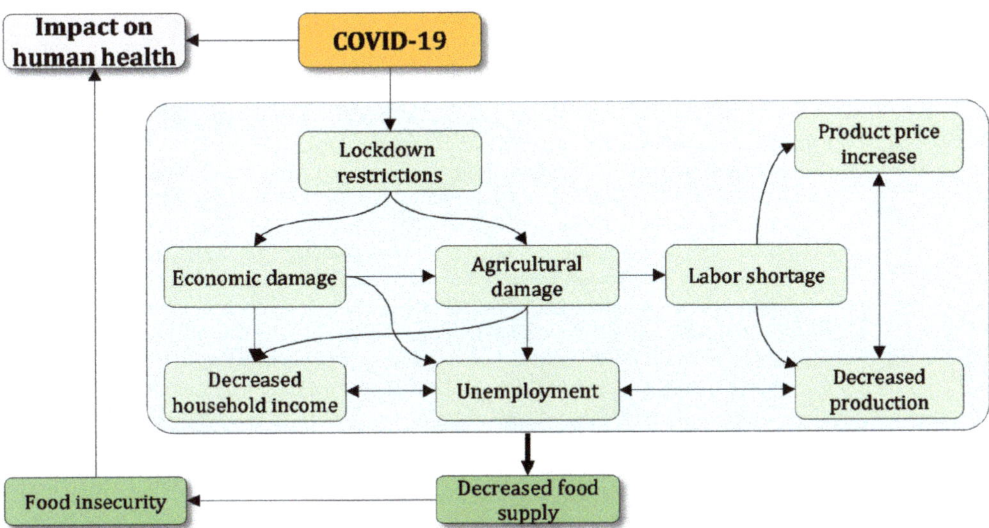

Figure 6. Various indirect and direct impacts of COVID-19 on human health.

According to obtained evidence of the COVID-19 results on food security, there are similar mechanisms that the pandemic influences food security through, which include a healthy diet [80], food supply chain [81], income, employment [82], national economic situation, and agriculture sustainability [26]. Therefore, the hypothesis that COVID-19 does not influence food safety was rejected, owing to its adverse consequences on food security through above six general mechanisms, which are obviously in Iran and other countries. Figure 7 shows that food security depends on the economy, household income, employment, food supply chain, and healthy diet, and the impact mechanism of the lockdown restrictions and economic crisis on food security in Iran has also been indicated.

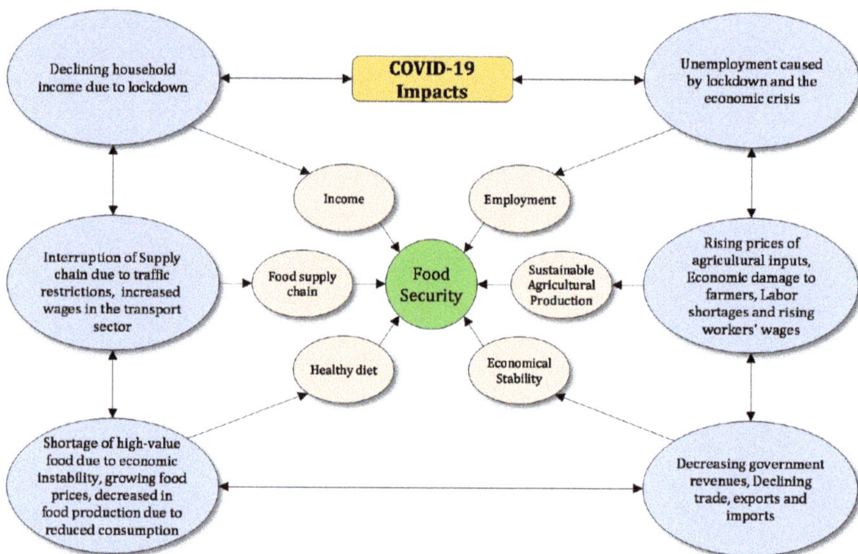

Figure 7. The profound effects of COVID-19 on food security by using six general mechanisms.

3.3. COVID-19, Challenges toward Sustainable Agriculture

As an important economic sector, agriculture has an intense connection with the economy (8% of Iran's GDP), supports food security [83–85], and activates the food system, which including connections between farmers, food factories, and the food supply chain [86]. COVID-19 cut these relationships and limited people's access to nutritious foods through lockdown restrictions [87]. Numerous evidence indicated that general consequences of the pandemic in the agricultural sector, including labor shortages, reduction of planting and harvesting, and reduction of agricultural trade, created an imbalance in the food system of many countries [88–90]. Although ecological constraints, such as climate change, water crisis, and land degradation, could significantly damage the prospects for economic growth and lead to food insecurity [91]. Growing agricultural limitations such as water shortage, desertification, and soil degradation are serious threats, and according to United Nations [92], 6.6% and 10.6% of the world's lands are arid and semi-arid, threatening the revenue of more than 1 billion people in 100 countries. Globally, 24% of the lands are being destroyed, and the portion of farmlands and rangelands is approximately 20–25%. Approximately 1.5 billion people are dependent on these degraded lands directly. The global degraded lands could produce 20 million tons of grain per year; thus, land degradation leads to a financial loss of $42 billion in revenue annually [92]. However, previous studies have only investigated the final consequences of COVID-19 toward agriculture and food security from an economic approach. Climate change imperils human prosperity [93], and according to a study by Fuentes et al. [94], climate change and COVID-19 potentially create a global tremendous economic loss; furthermore, most of their outcomes are observed in vulnerable communities [95]. Hence, the damages caused by COVID-19 have a synergistic influence with the financial loss of ecological constraints.

Agriculture can support a dynamic and sustainable economy and maintains food security if ecological constraints are reduced. The consequences of the pandemic outbreak in Iran can be divided into stable and new challenges. Environmental constraints besides the disturbance of the food system can continuously threaten food security in Iran. Climate change, water crisis, soil erosion, and salinity are critical agents that reduce crop productivity [96–98]. Tehrani et al. [97] evaluated more than 20,000 soil samples and found that more than 50% of Iran's soils are deficient in nutrients. Thus, despite the simultaneous COVID-

19 crisis and ecological limitations [99], farmers are experiencing a massive challenge for producing crops in Iran. FAO has also reported that reducing available water resources, land degradation, and climate change will have long-term adverse effects on Iran's food security, although maintaining food security in global crises is a significant challenge [100] that requires attention to environmental health [101]. If the government fails to combat extensive environmental constraints due to the financial loss created by the pandemic, the synergistic effects of both crises will threaten food security in the future. A schematic of the stable and new challenges caused by the COVID-19 has been shown in Figure 8, and each ecological limitation is discussed in the next sections.

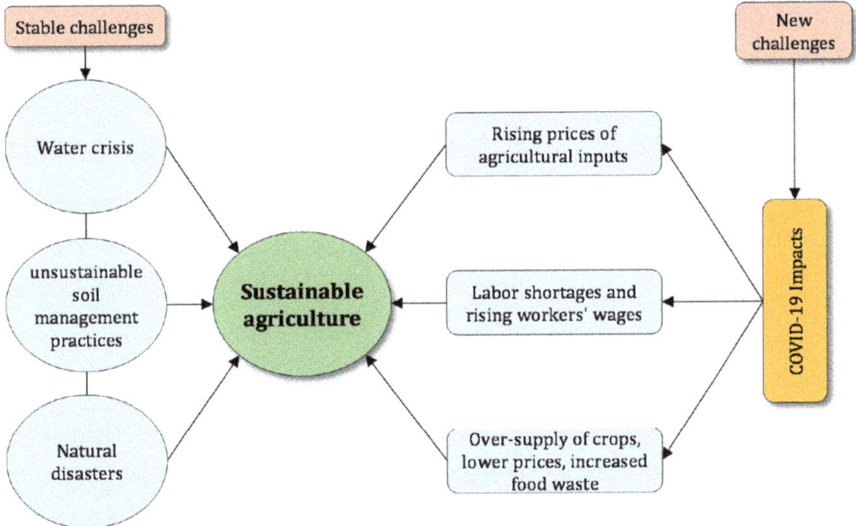

Figure 8. Impact of new challenges created by COVID-19 and continuous challenges on agricultural sustainability.

3.3.1. Water Crisis

Every person needs at least 2000 m^3 of water per year, but people that are living in arid lands have access to 1300 m^3 only; hence water shortages affect between one and two billion people. Furthermore, 50% of the population of the world will live in countries under extreme water deficit by 2030, according to the climate change scenario. Water plays a crucial role in the agriculture, and the farming sector is responsible for 70% of total freshwater consumption [102]. According to a study by Zarei [103], most Middle Eastern countries face water-energy-food insecurity due to ineffective water management strategies. The general climate of Iran is arid and semi-arid, and drought is a prominent trait of Iran's climate [104]. Approximately 90% of water consumption is associated with the agricultural sector, which is consuming five billion cubic meters of groundwater resources per year in Iran [105]; consequently, many regions of Iran have experienced significant insufficient water resources for farming due to the over-exploitation [106], and the crop production has been threatened by drought. Figures 9 and 10 show that the water crisis scenario will threaten areas under cultivation in Iran by 2030. In this regard, Soltani et al. [107] showed that self-sufficiency in crop production would be reduced from 83 to 39% in Iran if water consumption for agriculture will be limited by 2030.

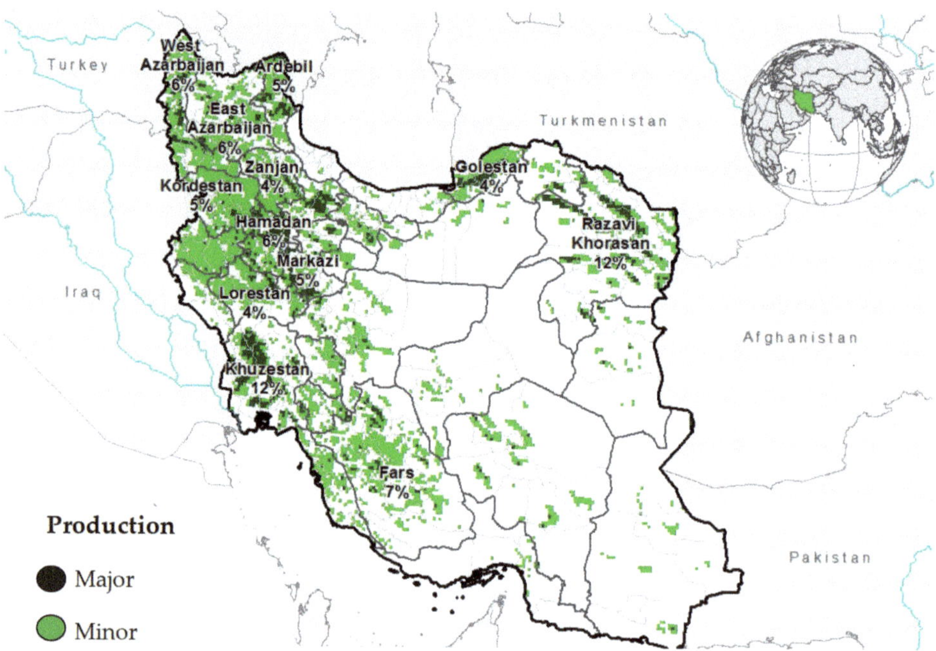

Figure 9. Map of farmlands under cultivation (wheat) in Iran in 2021; source: [108].

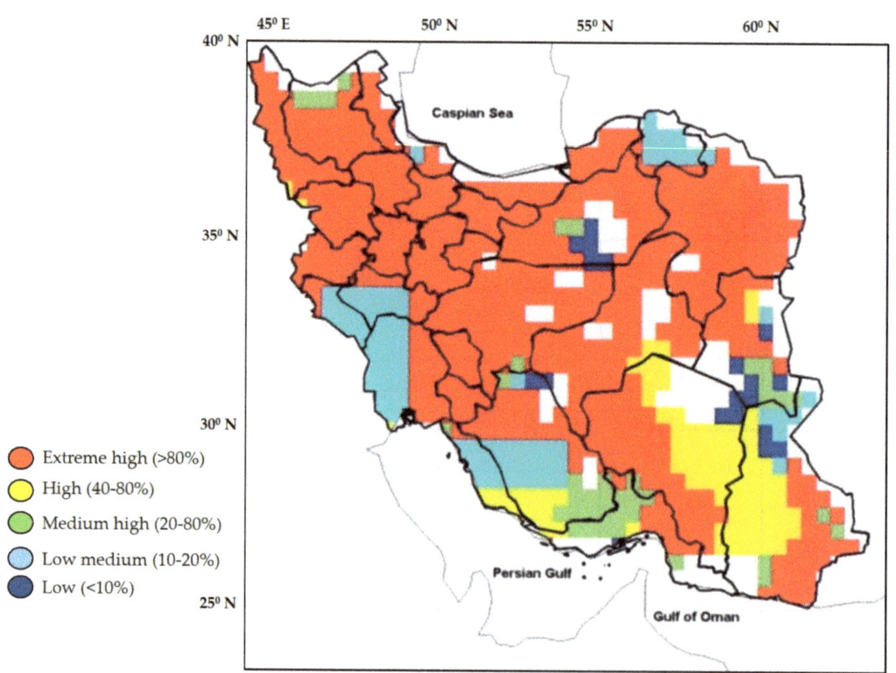

Figure 10. Map of areas under water stress in Iran according to the water crisis scenario in 2030; source: [109].

3.3.2. Soil Salinity and Erosion

Through using the UNESCO World Soil Map (1970–1980), FAO estimated that global total saline lands are 397 million hectares, and 38.7 (2%) and 195.1 (6.3%) million hectares have been located in Africa and Asia, respectively [110]. As a rising concern that limits sustainable agricultural production, soil salinity encompasses the large areas of the Central Plateau, southern coastal plains, and Khuzestan in Iran (Figure 11) [111]. From 6.8 million hectares of saline agricultural land, 4.3 million hectares are saline, and approximately 2.5 million hectares have other restrictions such as soil erosion and groundwater scarcity.

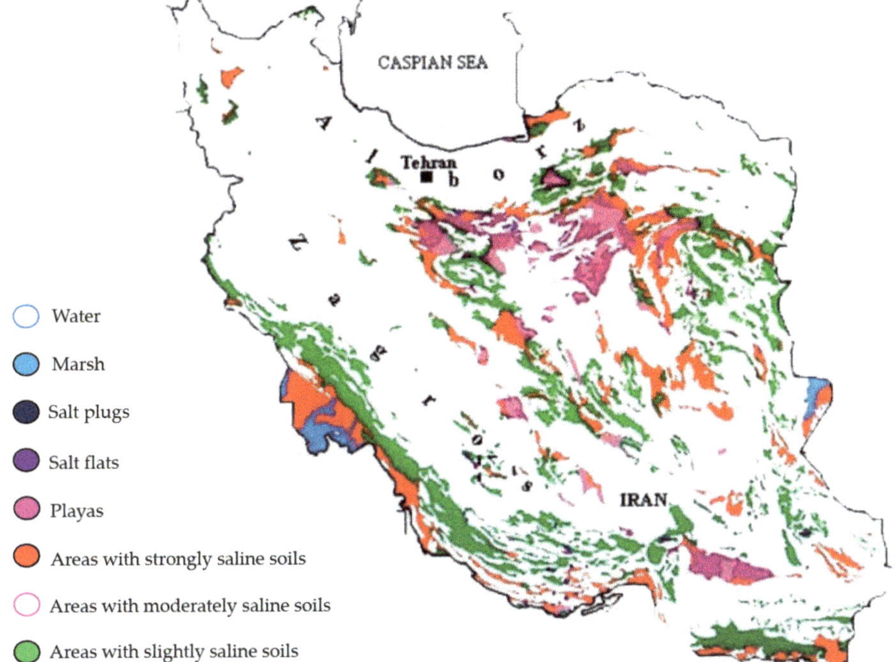

Figure 11. Map of saline soils in Iran; source [112].

Anthropogenic activities and land-use change also lead to soil erosion, which degrades 20–30 billion tons of soil per year and severely decreases soil productivity; hence, extreme soil erosion has been predicted for developing regions such as Southeast Asia, South America, and sub-Saharan Africa [113]. In a study by Mosaffaie and Talebi [114], water erosion using the EPM model was estimated at 975 million tons per year, and the total volume of erosion sediment was also estimated at 129 million tons in Iran. Hence, limited soil productivity of farmlands in Iran limits crops production, and Figure 12 shows that extensive areas of lands of this country cannot perform as arable lands, due to soil erosion or specific topography traits.

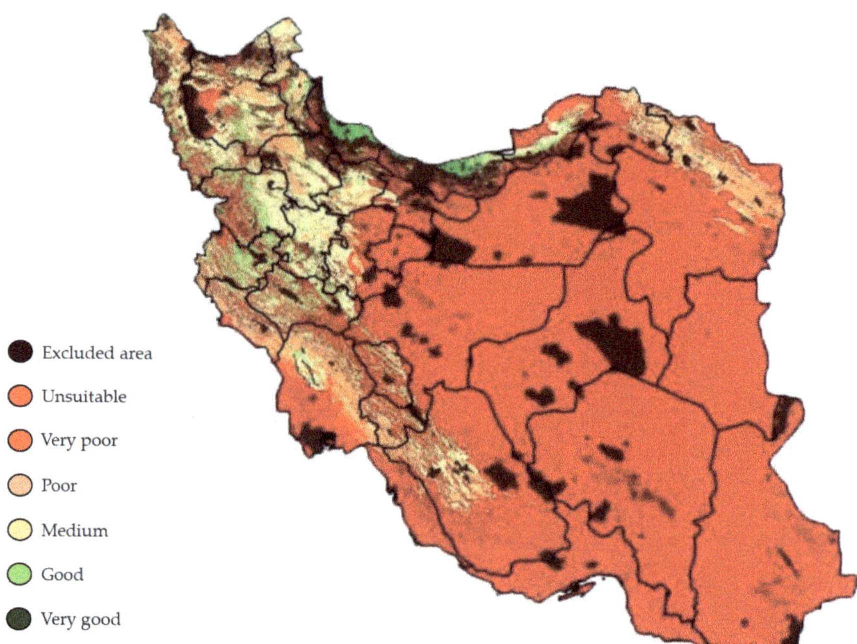

Figure 12. Soil suitability map of lands of Iran for cultivation; source: [115].

According to the calculations of Kohneshahri and Sadeghi [116], economic losses of soil erosion were 31% of the agriculture sector value-added in 2000. Hence, annual loss by soil erosion has been estimated at $56 billion in Iran [117]. Water erosion caused by flooding led to a 100-billion-dollar loss in 2019 because the floods in the three areas of Dez, Karun, and Karkheh were about 12 thousand billion cubic meters, which led to the erosion of 3.5 billion tons of soil [118]. Water erosion also removes 23–42 and 14.6–26.4 million tons of nitrogen and phosphorus from farmlands, and their costs are $1.45 and $5.26 per kilogram; accordingly, the financial losses are $33–60 and $77–140 billion for wasted nitrogen and phosphorus per year [119]. A lack of a strategic plan and insufficient knowledge about the consequences of soil salinity and erosion are the critical challenges in Iran [120]; consequently, soil erosion and salinity threaten sustainable agricultural production by reducing land fertility and crop productivity.

3.3.3. Deforestation and Land-Use Change

According to a study by Mirakhor-Lou and Akhavan [121], the deforestation rate is 0.74% per year in northern Iran, and the area of Hyrcanian forests in the three provinces of Gilan, Golestan, and Mazandaran was 1,811,788 hectares in 2004, and in 2016, it was estimated at 1,650,498 hectares. Forests in the northern regions of Iran (Gilan, Mazandaran, and Golestan provinces) have been decreased by 9% between 2004 and 2016 due to land-use change. Higher than 86 million hectares (52%) of Iran's lands are pastures, which contain more than 7000 plant species, and the livelihood of about 916,000 rural and nomadic households depends on the use of rangelands. The situation of Iran's rangelands is also unfavorable, and a decreasing trend has been observed in Figure 13; accordingly, SCI shows that the total area of Iran's rangelands has been reduced between 2002 to 2012 [122], and the reduction in rangelands vegetation will be extreme if the overuse of rangelands for livestock grazing increases due to the high price of livestock fodder.

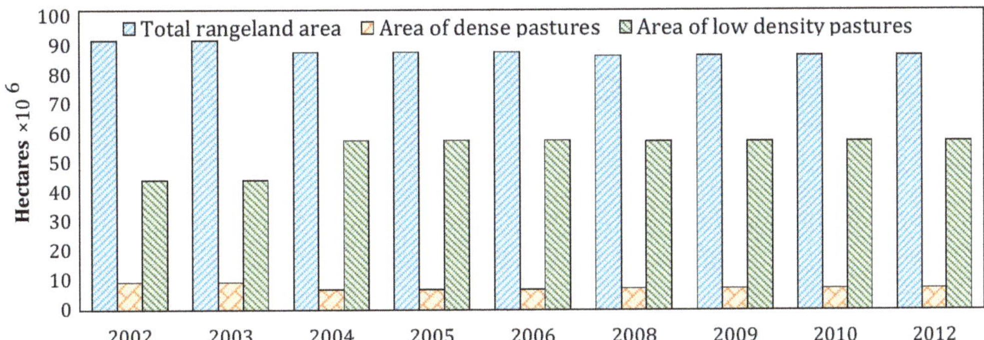

Figure 13. Reducing rangelands (dense, semi-dense, and low) in Iran from 2002 to 2012, data source: [121].

3.3.4. Natural Disasters

Sudden and irregular rainfall is a significant climatic train that creates big floods in Iran, and the difference between the average rainfall in the water year 2018–2019 and 2017–2018 was 46–181%, which led to severe floods in the provinces of Golestan, Lorestan, and Khuzestan in 2019 [123]. In addition to floods, locust attacks also cause significant damage to the agricultural sector. Desert locusts (*Schistocerca gregaria*) are found in the Middle East, Asia, and Africa and have destructive effects on crop yields. According to the Iran Agriculture Commission of the Parliament, this crisis caused a ten billion Rial (370 million US dollar) financial loss, in the southern provinces of Iran, in 2020 [124]. A similar locust attack in April 2012 affected the plantation in southern Indian Ocean Island, and FAO estimated that four million people were at risk of food insecurity in rural Madagascar in June-July 2013 [125]. The locusts have caused enormous damage in Ethiopia and Somalia because a small group of 80 million locusts can destroy the share food of 35,000 people per day and seriously threaten food security. This crisis has shocked the food supply chain of 23 countries in 2020 [126]. The cost of control for locusts increased from $1 million to $100 million in West Africa in 2003. A joint assessment by the Ethiopian government and FAO shows an outbreak of locusts destroyed 356,286 metric grains, destroying 197,163 hectares of crops and 1.3 million hectares of pastures. Then, COVID-19 exacerbated the locust crisis by disrupting the supply chain of pesticides [127], and Northern Somalia and asymptotic regions of eastern Ethiopia and borders of Djibouti, Pakistan, India, Sudan, and Oman were at a high risk of locust outbreaks [128].

4. Conclusions and Recommendations

Along with the appearance of the pandemic in 2020, the economic recession led to food insecurity in many countries due to financial loss and food prices enhancement. Among the developing countries, Iran has been affected by macroeconomic problems such as inflation and high currency exchange rate. Iran's per capita income significantly decreased, and rising inflation and the dollar exchange rate increased prices of various foods and beverages due to the rise in production costs in recent years and during the pandemic. Food shortages and their price enhancement endangered food security directly, and oversupply reduced some product prices, leading to financial losses for producers and farmers, ultimately threatening food security. Besides the complex economic effects of the pandemic, ecological constraints such as a water crisis, salinity, extreme soil erosion, deforestation, and natural disasters have severe adverse consequences on food security in Iran. Hence, COVID-19 causes direct and indirect damages and interruptions to the agriculture sector by limiting the production of some crops and disregarding environmental protection policies, due to economic losses, because stable ecological problems such as salinity and soil erosion can threaten sustainable agricultural production in the future. In response to the main hypotheses of this study, it can be demonstrated that consequences of the COVID-19 and ecological constraints have synergistic impacts. The enormous financial loss can pause support for small-scale farming and agricultural research projects. Therefore, governments must improve food security and environmental health through promoting sustainable agriculture measures and eco-friendly policies. It is necessary to implement sustainable agriculture policies during the COVID-19 outbreak period in Iran, and other countries, by considering ecological constraints. It should be highlighted that, while sustainable agricultural development can provide sufficient nutritious food, protect environmental health, and presumably support mitigate the financial losses in the post-COVID-19 crisis, it is still difficult to improve the economic consequences of pandemics. Therefore, future studies may involve investigating strategies that support the agricultural sector to endure the long-term economic consequences in the post-pandemic and maintaining food security by means of developing agroecology training courses, accelerating agricultural projects via sustainable and eco-friendly approaches, and assisting small-scale farming projects by providing site-specific solutions, automated instruments, and new breeds of seeds and plants to compensate for the drought and salinity problems.

Author Contributions: Conceptualization, A.K.R. and R.R.S.; methodology, A.K.R. and R.R.S.; software, R.R.S.; validation, R.R.S., S.K.B. and M.S.; formal analysis, R.R.S. and S.K.B.; investigation, A.K.R. and R.R.S.; resources, R.R.S., S.K.B. and M.S.; data curation, A.K.R. and H.A.; writing—original draft preparation, A.K.R. and R.R.S.; writing—review and editing, A.K.R., R.R.S., H.A., S.K.B. and M.S.; visualization, A.K.R. and R.R.S.; supervision, R.R.S. and H.A.; project administration, R.R.S., H.A.; funding acquisition, S.K.B. All authors have read and agreed to the published version of the manuscript.

Funding: This research received no external funding.

Data Availability Statement: Data is contained within the article.

Acknowledgments: The authors would like to acknowledge the editorial support from Adaptive AgroTech Consultancy Network, and technical assistant from Benjamin Mahns of the Leibniz Institute for Agricultural Engineering and Bioeconomy (ATB).

Conflicts of Interest: The authors declare no conflict of interest.

References

1. IMF. World Economic Outlook, July 2019. Available online: https://www.imf.org/en/Publications/WEO/Issues/2019/07/18/WEOupdateJuly2019 (accessed on 23 July 2019).
2. Kumar, A.; Singh, R.; Kaur, J.; Pandey, S.; Sharma, V.; Thakur, L.; Kumar, N. Wuhan to World: The COVID-19 Pandemic. *Front. Cell. Infect. Microbiol.* **2021**, *11*, 596201. [CrossRef]
3. Debata, B.; Patnaik, P.; Mishra, A. COVID-19 pandemic! It's impact on people, economy, and environment. *J. Public Aff.* **2020**, *20*, e2372. [CrossRef]
4. IMF. Tourism-Dependent Economies Are among Those Harmed the Most by the Pandemic. Available online: https://www.imf.org/external/pubs/ft/fandd/2020/12/impact-of-the-pandemic-on-tourism-behsudi.htm (accessed on 1 December 2020).
5. WTO. COVID-19 and World Trade. Available online: https://www.wto.org/english/tratop_e/covid19_e/covid19_e.htm (accessed on 7 May 2021).
6. Meuwissen, M.P.M.; Feindt, P.H.; Slijper, T.; Spiegel, A.; Finger, R.; de Mey, Y.; Reidsma, P. Impact of COVID-19 on farming systems in Europe through the lens of resilience thinking. *Agric. Syst.* **2021**, *191*, 103152. [CrossRef]
7. IMF. The Great Lockdown: Worst Economic Downturn since the Great Depression. Available online: https://www.imf.org/en/News/Articles/2020/03/23/pr2098-imf-managing-director-statement-following-a-g20-ministerial-call-on-the-coronavirus-emergency (accessed on 23 March 2020).
8. Statista. Impact of the Coronavirus Pandemic on the Global Economy—Statistics & Facts. Available online: https://www.statista.com/topics/6139/COVID-19-impact-on-the-global-economy/ (accessed on 11 June 2021).
9. WTO. Frequently Asked Questions: The WTO and COVID-19. Available online: https://www.wto.org/english/tratop_e/covid19_e/faqcovid19_e.htm (accessed on 1 July 2021).
10. Josephson, A.; Kilic, T.; Michler, J.D. Socioeconomic impacts of COVID-19 in low-income countries. *Nat. Hum. Behav.* **2020**, *5*, 557–565. [CrossRef] [PubMed]
11. Diffenbaugh, N.S.; Field, C.B.; Appel, E.A.; Azevedo, I.L.; Baldocchi, D.D.; Burke, M.; Wong-Parodi, G. The COVID-19 lockdowns: A window into the Earth System. *Nat. Rev. Earth Environ.* **2020**, *1*, 470–481. [CrossRef]

12. World Bank. The Global Economic Outlook during the COVID-19 Pandemic: A Changed World. Available online: https://www.worldbank.org/en/news/feature/2020/06/08/the-global-economic-outlook-during-the-COVID-19-pandemic-a-changed-world (accessed on 8 June 2020).

13. United Nations. COVID-19 to Slash Global Economic Output by \$8.5 Trillion over Next Two Years. Available online: https://www.un.org/en/desa/COVID-19-slash-global-economic-output-85-trillion-over-next-two-years (accessed on 1 July 2021).

14. Han, S.; Roy, P.K.; Hossain, M.I.; Byun, K.H.; Choi, C.; Ha, S.D. COVID-19 pandemic crisis and food safety: Implications and inactivation strategies. *Trends Food Sci. Technol.* **2021**, *109*, 25–36. [CrossRef] [PubMed]

15. Rivington, M.; King, R.; Duckett, D.; Iannetta, P.; Benton, T.G.; Burgess, P.J.; Keay, C. UK food and nutrition security during and after the COVID-19 pandemic. *Nutr. Bull.* **2021**, *46*, 88–97. [CrossRef] [PubMed]

16. Worldometers. Coronavirus Cases, Updated: 1 September 2021. Available online: https://www.worldometers.info/coronavirus/ (accessed on 1 September 2021).

17. Erokhin, V.; Gao, T. Impacts of COVID-19 on Trade and Economic Aspects of Food Security: Evidence from 45 Developing Countries. *Int. J. Env. Res. Public Health* **2020**, *17*, 5775. [CrossRef]

18. Kesar, S.; Abraham, R.; Lahoti, R.; Nath, P.; Basole, A. Pandemic, informality, and vulnerability: Impact of COVID-19 on livelihoods in India. *Can. J. Dev. Stud./Rev. Can. D'études Du Dév.* **2021**, *42*, 1–20. [CrossRef]

19. Mottaleb, K.A.; Mainuddin, M.; Sonobe, T. COVID-19 induced economic loss and ensuring food security for vulnerable groups: Policy implications from Bangladesh. *PLoS ONE* **2020**, *15*, e0240709. [CrossRef] [PubMed]

20. Gaitan-Rossi, P.; Vilar-Compte, M.; Teruel, G.; Perez-Escamilla, R. Food insecurity measurement and prevalence estimates during the COVID-19 pandemic in a repeated cross-sectional survey in Mexico. *Public Health Nutr.* **2021**, *24*, 412–421. [CrossRef] [PubMed]

21. WHO. In Middle East COVID-19 Hotspot Iran, WHO Walks the Talk. Available online: https://www.who.int/news-room/feature-stories/detail/in-middle-east-COVID-19-hotspot-iran-who-walks-the-talk (accessed on 23 December 2020).

22. Iran's Health Ministry. Coronavirus Cases in Iran, Updated: 1 September 2021. Available online: https://behdasht.gov.ir (accessed on 1 September 2021). (In Persian)

23. World Bank. Islamic Republic of Iran. Available online: https://www.worldbank.org/en/country/iran/overview#1 (accessed on 30 March 2021).

24. OECD. The Impact of Coronavirus (COVID-19) and the Global Oil Price Shock on the Fiscal Position of Oil-Exporting Developing Countries. Available online: https://www.oecd.org/coronavirus/policy-responses/the-impact-of-coronavirus-COVID-19-and-the-global-oil-price-shock-on-the-fiscal-position-of-oil-exporting-developing-countries-8bafbd95/ (accessed on 30 September 2020).

25. United Nations. Global Economy Could Lose over \$4 Trillion Due to COVID-19 Impact on Tourism. Available online: https://unctad.org/news/global-economy-could-lose-over-4-trillion-due-COVID-19-impact-tourism (accessed on 30 June 2021).

26. OECD. The Impact of COVID-19 on Agricultural Markets and GHG Emissions. Available online: https://www.oecd.org/coronavirus/policy-responses/the-impact-of-COVID-19-on-agricultural-markets-and-ghg-emissions-57e5eb53/ (accessed on 8 December 2020).

27. Pak, A.; Adegboye, O.A.; Adekunle, A.I.; Rahman, K.M.; McBryde, E.S.; Eisen, D.P. Economic Consequences of the COVID-19 Outbreak: The Need for Epidemic Preparedness. *Front. Public Health* **2020**, *8*, 241. [CrossRef]

28. FAO. FAO in the Islamic Republic of Iran. Available online: http://www.fao.org/iran/news/detail-events/zh/c/1271192/ (accessed on 15 June 2020).

29. SCI. Economic Growth Rate Spring 2020. Available online: https://www.amar.org.ir/news/ID/13081 (accessed on 1 September 2020). (In Persian)

30. Arman, S.A.; Ghorbannejad, M.; Kafili, V. A Look at Inflation in Iran: The VARX Approach. *Iran. J. Appl. Econ. Stud.* **2017**, *6*, 99–121. (In Persian) [CrossRef]

31. Sultan-Tavih, M.; Akbari, M.A.; Rassaian, A. Investigating the relationship between inflation and economic growth in Iran using a rolling linear regression model. *J. Monet. Bank. Res.* **2013**, *6*. Available online: http://jmbr.mbri.ac.ir/browse.php?a_code=A-10-24-81&slc_lang=fa&sid=1 (accessed on 19 September 2020). (In Persian)

32. SCI. Consumer Price Index. Available online: https://www.amar.org.ir/news/ID/13201 (accessed on 9 November 2020).

33. Trading Economics. Inflation Rate. Available online: https://tradingeconomics.com/country-list/inflation-rate (accessed on 14 July 2021).

34. OPEC. World Oil Outlook. 2018. Available online: https://www.opec.org/opec_web/en/publications/340.htm (accessed on 15 July 2020).

35. Reuters. Iran's Oil Exports Jump in September Defying Sanctions: Tanker Trackers. Available online: https://www.reuters.com/article/us-iran-oil-exports-idUSKCN26G1VA (accessed on 25 September 2020).

36. IPRC. Assessing the Macroeconomic Impacts of Coronavirus Outbreak in Iran (First Edition). Available online: https://rc.majlis.ir/fa/report/show/1510373 (accessed on 27 April 2020). (In Persian)

37. Takeuchi, R.; Kiku, L.; Atamanov, A. *Welfare and Distributional Impacts of Inflation and the COVID-19 Outbreak in the Islamic Republic of Iran (English)*; Policy Research Working Paper; No. WPS 9558; COVID-19 (Coronavirus); World Bank Group: Washington, DC, USA, 2021; Available online: http://documents.worldbank.org/curated/en/288661614696489389/Welfare-and-Distributional-Impacts-of-Inflation-and-the-COVID-19-Outbreak-in-the-Islamic-Republic-of-Iran (accessed on 2 March 2021).

38. Hoseini, M.; Valizadeh, A. The effect of COVID-19 lockdown and the subsequent reopening on consumption in Iran. *Rev. Econ. Househ.* **2021**, *19*, 373–397. [CrossRef] [PubMed]
39. SCI. Data on Expenses and Income of Iranian Households. Available online: https://www.amar.org.ir/ (accessed on 14 July 2020). (In Persian)
40. IPRC. Changes in Price Index and Inflation in 2020. Available online: https://rc.majlis.ir/fa/news/show/1620540 (accessed on 10 October 2020). (In Persian)
41. TCCIM. Total Import and Export Statistics of Iran. Available online: https://www.tccim.ir/ImpExpStats_TarrifCustomCountry.aspx?slcImpExp=Export&slcCountry=&sYear=1397&mode=doit (accessed on 15 July 2021). (In Persian)
42. IPRC. Investigating Monetary Developments and Instability in Commodity Prices. Available online: http://rc.majlis.ir/fa/report/show/1620606 (accessed on 11 October 2020). (In Persian)
43. CBI. Economic Time Series Database. 2020. Available online: https://www.cbi.ir (accessed on 15 July 2021).
44. Pakravan-Charvadeh, M.R.; Mohammadi-Nasrabadi, F.; Gholamrezai, S.; Vatanparast, H.; Flora, C.; Nabavi-Pelesaraei, A. The short-term effects of COVID-19 outbreak on dietary diversity and food security status of Iranian households (A case study in Tehran province). *J. Clean Prod.* **2021**, *281*, 124537. [CrossRef] [PubMed]
45. World Bank. Food Security and COVID-19. Available online: https://www.worldbank.org/en/topic/agriculture/brief/food-security-and-COVID-19 (accessed on 14 September 2020).
46. Pourkazemi, M.H.; Beiranvand, A.; Delfan, M. Designing a Warning System for Hyperinflation for Iran's Economy. *Q. J. Econ. Res. Policies.* **2016**, *23*, 145–166. Available online: http://qjerp.ir/article-1-961-fa.html (accessed on 19 September 2020). (In Persian)
47. Shakeri, A.; Mohammadi, T.; Rajabi, F. The effect of pricing power on inflation in the Iran's economy. *Econ. Res. J.* **2015**, *15*, 37–60. Available online: http://joer.atu.ac.ir/article_1822.html (accessed on 19 September 2020). (In Persian)
48. SCI. Consumer Price and Production Index Data. Available online: https://www.amar.org.ir/ (accessed on 19 September 2020). (In Persian)
49. CBI. Inflation Rate and the Price Index of Consumer Goods and Services. 2017. Available online: https://www.cbi.ir/Inflation/Inflation_fa.aspx (accessed on 15 July 2021).
50. World Bank. Latest Commodity Prices Published. Available online: https://www.worldbank.org/en/research/commodity-markets (accessed on 2 October 2020).
51. Paslakis, G.; Dimitropoulos, G.; Katzman, D.K. A call to action to address COVID-19-induced global food insecurity to prevent hunger, malnutrition, and eating pathology. *Nutr. Rev.* **2021**, *79*, 114–116. [CrossRef] [PubMed]
52. Zurayk, R. Pandemic and Food Security: A View from the Global South. *J. Agric. Food Syst. Community Dev.* **2020**, 1–5. [CrossRef]
53. Huss, M.; Brander, M.; Kassie, M.; Ehlert, U.; Bernauer, T. Improved storage mitigates vulnerability to food-supply shocks in smallholder agriculture during the COVID-19 pandemic. *Glob. Food Secur.* **2021**, *28*. [CrossRef]
54. Motevali, A.; Hashemi, S.J.; Tabatabaeekoloor, R. Environmental footprint study of white rice production chain-case study: Northern of Iran. *J. Environ. Manag.* **2019**, *241*, 305–318. [CrossRef]
55. Donyae Eqtesad Newspaper. Taste Expensive on the Table of 50 Million Iranians. Available online: https://www.donya-e-eqtesad.com/fa/tiny/news-3679607 (accessed on 5 August 2020). (In Persian)
56. Donyae Eqtesad Newspaper. Rice Price According to Statistics. Available online: https://www.donya-e-eqtesad.com/fa/tiny/news-3684480 (accessed on 22 August 2020). (In Persian)
57. Arouna, A.; Soullier, G.; Villar, P.M.; Demont, M. Policy options for mitigating impacts of COVID-19 on domestic rice value chains and food security in West Africa. *Glob. Food Secur.* **2020**, *26*, 100405. [CrossRef] [PubMed]
58. CSIS (Center for Strategic International Studies). COVID-19 and Food Security, Updated: 24 April 2020. Available online: https://www.csis.org/programs/global-food-security-program/COVID-19-and-food-security (accessed on 26 January 2021).
59. FAO. Regional Roundups (Southern Africa), Prices of Maize Remained Generally Firm. Available online: http://www.fao.org/giews/food-prices/regional-roundups/detail/en/c/1313842/ (accessed on 13 October 2020).
60. Falkendal, T.; Otto, C.; Schewe, J.; Jägermeyr, J.; Konar, M.; Kummu, M.; Puma, M.J. Grain export restrictions during COVID-19 risk food insecurity in many low- and middle-income countries. *Nat. Food* **2021**, *2*, 11–14. [CrossRef]
61. Khorasan Newspaper. An Increase of 5 Times the Rate of Pesticides and Fertilizers on the Verge of Growing Season. 2020. Available online: http://www.khorasannews.com/newspaper/page/20486/1/707966/0 (accessed on 28 September 2020). (In Persian)
62. SCI. The Average Price of Products and Cost of Agricultural Services in Spring 2020, Updated: 22 August 2020. Available online: https://www.amar.org.ir/ (accessed on 22 August 2020). (In Persian)
63. World Bank. Commodity Markets Outlook (Implications of COVID-19 for Commodities). 2020. Available online: https://openknowledge.worldbank.org/bitstream/handle/10986/33624/CMO-April-2020.pdf (accessed on 17 April 2020).
64. United Nations. Food. Available online: https://www.un.org/en/global-issues/food (accessed on 1 October 2020).
65. Nchanji, E.B.; Lutomia, C.K.; Chirwa, R.; Templer, N.; Rubyogo, J.C.; Onyango, P. Immediate impacts of COVID-19 pandemic on bean value chain in selected countries in sub-Saharan Africa. *Agric. Syst.* **2021**, *188*, 103034. [CrossRef]
66. Jha, P.K.; Araya, A.; Stewart, Z.P.; Faye, A.; Traore, H.; Middendorf, B.J.; Prasad, P.V.V. Projecting potential impact of COVID-19 on major cereal crops in Senegal and Burkina Faso using crop simulation models. *Agric. Syst.* **2021**, *190*, 103107. [CrossRef] [PubMed]

67. Middendorf, B.J.; Faye, A.; Middendorf, G.; Stewart, Z.P.; Jha, P.K.; Prasad, P.V.V. Smallholder farmer perceptions about the impact of COVID-19 on agriculture and livelihoods in Senegal. *Agric. Syst.* **2021**, *190*, 103108. [CrossRef]
68. Fan, S.; Teng, P.; Chew, P.; Smith, G.; Copeland, L. Food System Resilience and COVID-19—Lessons from the Asian Experience. *Glob. Food Secur.* **2021**, *28*. [CrossRef]
69. FAO. Impacts of COVID-19 on Food Security and Nutrition: Developing Effective Policy Responses to Address the Hunger and Malnutrition Pandemic. 2020. Available online: http://www.fao.org/agroecology/database/detail/en/c/1310872/ (accessed on 15 July 2021).
70. Pakravan-Charvadeh, M.R.; Khan, H.A.; Flora, C. Spatial analysis of food security in Iran: Associated factors and governmental support policies. *J. Public Health Policy* **2020**, *41*, 351–374. [CrossRef] [PubMed]
71. SCI. Average Prices of Selected Food Items. Available online: https://www.amar.org.ir/ (accessed on 22 July 2020). (In Persian)
72. SCI. Report of Consumer Price Index. Available online: https://www.amar.org.ir/Portals/0/News/1399/shg99.52.pdf (accessed on 22 August 2020). (In Persian)
73. Cheraghi, D.; Gholipour, S. An overview of the significant challenges of red meat in Iran. *Bus. Rev.* **2010**, *8*, 89–110. Available online: https://www.sid.ir/fa/journal/ViewPaper.aspx?id=112549 (accessed on 19 September 2020). (In Persian)
74. Bene, C. Resilience of local food systems and links to food security-A review of some important concepts in the context of COVID-19 and other shocks. *Food Secur.* **2020**, 1–18. [CrossRef]
75. Layani, G.; Bakhshoodeh, M.; Aghabeygi, M.; Kurstal, Y.; Viaggi, D. The impact of food price shocks on poverty and vulnerability of urban households in Iran. *Bio-Based Appl. Econ.* **2020**, *9*, 109–125. [CrossRef]
76. Heydari, K.; Zarvar, P. Estimating the effects of bread price reform and energy carriers on the household budget. *Strategy Mag.* **2011**, *19*, 181–195. Available online: http://rahbord.csr.ir/WebUsers/rahbord/UploadFiles/OK/1394091161533525000483-F.pdf (accessed on 19 September 2020). (In Persian)
77. Mayasari, N.R.; Ho DK, N.; Lundy, D.J.; Skalny, A.V.; Tinkov, A.A.; Teng, I.C.; Chang, J.S. Impacts of the COVID-19 Pandemic on Food Security and Diet-Related Lifestyle Behaviors: An Analytical Study of Google Trends-Based Query Volumes. *Nutrients* **2020**, *12*, 3103. [CrossRef]
78. Savary, S.; Akter, S.; Almekinders, C.; Harris, J.; Korsten, L.; Rotter, R.; Watson, D. Mapping disruption and resilience mechanisms in food systems. *Food Secur.* **2020**, *12*, 695–717. [CrossRef]
79. Gustafson, S. Food Crises Increase Around the World: 2020 Global Report on Food Crises Released. Available online: http://www.foodsecurityportal.org/food-crises-increase-around-world-2020-global-report-food-crises-released (accessed on 22 April 2020).
80. FAO. Eating Healthy before, during and after COVID-19. Available online: http://www.fao.org/fao-stories/article/en/c/1392499/ (accessed on 7 April 2021).
81. Barman, A.; Das, R.; De, P.K. Impact of COVID-19 in food supply chain: Disruptions and recovery strategy. *Curr. Res. Behav. Sci.* **2021**, *2*. [CrossRef]
82. Tandon, A.; Roubal, T.; McDonald, L.; Cowley, P.; Palu, T.; de Oliveira Cruz, V.; Eozenou, P.; Cain, J.; Teo, H.S.; Schmidt, M.; et al. *Economic Impact of COVID-19: I Plications for Health Financing in Asia and Pacific*; Health, Nutrition and Population Discussion Paper; World Bank: Washington, DC, USA, 2020; Available online: https://openknowledge.worldbank.org/handle/10986/34572 (accessed on 28 September 2020).
83. Pawlak, K.; Kołodziejczak, M. The Role of Agriculture in Ensuring Food Security in Developing Countries: Considerations in the Context of the Problem of Sustainable Food Production. *Sustainability* **2020**, *12*, 5488. [CrossRef]
84. Poudel, P.B.; Poudel, M.R.; Gautam, A.; Phuyal, S.; Tiwari, C.K.; Bashyal, N.; Bashyal, S. COVID-19 and its Global Impact on Food and Agriculture. *J. Biol. Today's World* **2020**, *9*, 221. [CrossRef]
85. Zarei, M.; Rad, A. COVID-19, Challenges and Recommendations in Agriculture. *J. Bot. Res.* **2020**, *2*. [CrossRef]
86. Cable, J.; Jaykus, L.A.; Hoelzer, K.; Newton, J.; Torero, M. The impact of COVID-19 on food systems, safety, and security-a symposium report. *Ann. N. Y. Acad. Sci.* **2021**, *1484*, 3–8. [CrossRef] [PubMed]
87. FAO. COVID-19 and Its Impact on Agri-Food Systems, Food Security and Nutrition: Implications and Priorities for the Africa Region. Available online: www.fao.org (accessed on 27 October 2020).
88. Mardones, F.O.; Rich, K.M.; Boden, L.A.; Moreno-Switt, A.I.; Caipo, M.L.; Zimin-Veselkoff, N.; Baltenweck, I. The COVID-19 Pandemic and Global Food Security. *Front. Vet. Sci.* **2020**, *7*, 578508. [CrossRef] [PubMed]
89. Glaros, A.; Alexander, C.; Koberinski, J.; Scott, S.; Quilley, S.; Si, Z. A systems approach to navigating food security during COVID-19: Gaps, opportunities, and policy supports. *J. Agric. Food Syst. Community Dev.* **2021**, *10*, 1–13. [CrossRef]
90. Chiwona-Karltun, L.; Amuakwa-Mensah, F.; Wamala-Larsson, C.; Amuakwa-Mensah, S.; Abu Hatab, A.; Made, N.; Bizoza, A.R. COVID-19: From health crises to food security anxiety and policy implications. *Ambio* **2021**, *50*, 794–811. [CrossRef]
91. United Nations. World Economic Situation and Prospects 2019. Available online: https://www.un.org/development/desa/dpad/publication/world-economic-situation-and-prospects-2019/ (accessed on 21 January 2019).
92. United Nations. Deserts and Fight against Desertification. Available online: https://www.un.org/en/events/desertification_decade/whynow.shtml (accessed on 1 October 2020).
93. Markard, J.; Rosenbloom, D. A tale of two crises: COVID-19 and climate. *Sustain. Sci. Pract. Policy.* **2020**, *16*, 53–60. [CrossRef]
94. Fuentes, R.; Galeotti, M.; Lanza, A.; Manzano, B. COVID-19 and Climate Change: A Tale of Two Global Problems. *Sustainability* **2020**, *12*, 8560. [CrossRef]

95. Botzen, W.; Duijndam, S.; van Beukering, P. Lessons for climate policy from behavioral biases towards COVID-19 and climate change risks. *World Dev.* **2021**, *137*, 105214. [CrossRef]
96. FAO. The Islamic Republic of Iran at a Glance. 2019. Available online: http://www.fao.org/iran/fao-in-iran/iran-at-a-glance/en/ (accessed on 29 November 2019).
97. Tehrani, M.M.; Belali, M.R.; Moshiri, F.; Daryashenas, A. Fertilizer Recommendation and Estimation in Iran: Challenges and Solutions. *J. Soil Res.* **2012**, *26*, 123–144. Available online: https://www.sid.ir/fa/journal/ViewPaper.aspx?id=163754 (accessed on 15 July 2021). (In Persian)
98. Barani, N.; Karami, A.; Borazjani, M.A. The impact of climatic changes on total horticultural production and food security in agro-ecological zones of Iran. *J. Water Clim. Chang.* **2020**, *11*, 1712–1723. [CrossRef]
99. Rasul, G. Twin challenges of COVID-19 pandemic and climate change for agriculture and food security in South Asia. *Environ. Chall.* **2021**, *2*, 100027. [CrossRef]
100. Farcas, A.C.; Galanakis, C.M.; Socaciu, C.; Pop, O.L.; Tibulca, D.; Paucean, A.; Socaci, S.A. Food Security during the Pandemic and the Importance of the Bioeconomy in the New Era. *Sustainability* **2021**, *13*, 150. [CrossRef]
101. Cole, M.B.; Augustin, M.A.; Robertson, M.J.; Manners, J.M. The science of food security. *NPJ Sci. Food* **2018**, *2*, 14. [CrossRef] [PubMed]
102. United Nations. Water and Food Security, Updated: 23 October 2014. Available online: https://www.un.org/waterforlifedecade/food_security.shtml (accessed on 15 July 2021).
103. Zarei, M. The Water-Energy-Food Nexus: A Holistic Approach for Resource Security in Iran, Iraq, and Turkey. *Water-Energy Nexus* **2020**, *3*, 81–94. [CrossRef]
104. Osoly, N.; Hosseini, M. Water Scarcity Is a Challenge for Agricultural Development, 3rd Congress of Agricultural Extension and Education Sciences. 2009. Available online: https://civilica.com/doc/131942/ (accessed on 26 January 2012). (In Persian)
105. IPRC. Water Crisis Management (Comparative Study). Available online: https://rc.majlis.ir/fa/mrc_report/show/1077820 (accessed on 20 October 2018).
106. Javan, J.; Fal Soleiman, M. Water Crisis and the Importance of Water Productivity in Agriculture in Dry Area of Iran (Case Study: Birjand Plain). *J. Geogr. Dev.* **2008**, *6*, 115–138. (In Persian) [CrossRef]
107. Soltani, A.; Alimagham, S.M.; Nehbandani, A.; Torabi, B.; Zeinali, E.; Zand, E.; van Ittersum, M.K. Future food self-sufficiency in Iran: A model-based analysis. *Glob. Food Secur.* **2020**, *24*, 100351. [CrossRef]
108. USDA. Iran Crop Production Data, Updated: August 2021. Available online: https://ipad.fas.usda.gov/countrysummary/default.aspx?id=IR (accessed on 1 September 2021).
109. Caldera, U.; Bogdanov, D.; Fasihi, M.; Aghahosseini, A.; Breyer, C. Securing future water supply for Iran through 100% renewable energy powered desalination. *Int. J. Sustain. Energy Plan. Manag.* **2019**, *23*, 39–54. [CrossRef]
110. FAO. Soils Portal, Salt-Affected Soils. 2019. Available online: http://www.fao.org/soils-portal/soil-management/management-of-some-problem-soils/salt-affected-soils/more-information-on-salt-affected-soils/en/ (accessed on 19 September 2020).
111. Momeni, A. Geographical distribution and salinity levels of Iranian soil resources. *J. Soil Res.* **2010**, *24*, 203–215. Available online: https://www.sid.ir/fa/journal/ViewPaper.aspx?id=133124 (accessed on 19 September 2020). (In Persian)
112. Fathi, M.; Rezaei, M. Soil Salinity in the Central Arid Region of Iran: Esfahan Province. In *Developments in Soil Salinity Assessment and Reclamation*; Shahid, S., Abdelfattah, M., Taha, F., Eds.; Springer: Dordrecht, The Netherlands, 2013. [CrossRef]
113. Borrelli, P.; Robinson, D.A.; Fleischer, L.R.; Lugato, E.; Ballabio, C.; Alewell, C.; Meusburger, K.; Modugno, S.; Schütt, B.; Ferro, V.; et al. An assessment of the global impact of 21st-century land-use change on soil erosion. *Nat. Commun.* **2013**, *8*, 1–3. [CrossRef]
114. Mosaffaie, J.; Talebi, A. A statistical look at the situation of water erosion in Iran. *J. Ext. Dev. Watershed Manag.* **2014**, *2*. Available online: http://wmji.ir/en/ManuscriptDetail?mid=126 (accessed on 19 September 2020). (In Persian)
115. Mesgaran, M.B.; Madani, K.; Hashemi, H.; Azadi, P. Iran's Land Suitability for Agriculture. *Sci. Rep.* **2017**, *7*, 1–12. [CrossRef]
116. Kohneshahri, L.A.; Sadeghi, H. Estimating the economic effects of soil erosion in Iran. *Sustain. Growth Dev. Res.* **2005**, *5*, 87–100. Available online: https://www.sid.ir/fa/journal/ViewPaper.aspx?id=55522 (accessed on 3 September 2020). (In Persian)
117. DOE. $ 56 Billion in Soil Damage Per year in Iran, Code: 7c88c30. Available online: https://doe.ir/portal/home/?news/196210/963034/1090302 (accessed on 6 December 2019). (In Persian)
118. DOE. One Hundred Billion Dollars in Soil Erosion in the Spring 2019 Flood in Iran, Code: d0f5e78. Available online: https://doe.ir/portal/home/?news/196210/963034/982819/ (accessed on 8 May 2019). (In Persian)
119. FAO. Status of the World's Soil Resources. 2015. Available online: http://www.fao.org/3/a-i5199e.pdf (accessed on 4 December 2015).
120. Ranjbar, G.; Anosheh, H.P. A look at salinity research in Iran with emphasis on improving crop production. *Iran. J. Crop Sci.* **2015**, *17*, 165–178. Available online: http://agrobreedjournal.ir/article-1-525-fa.html (accessed on 19 September 2020). (In Persian)
121. Mirakhor-Lou, K.; Akhavan, R. Area changes of Hyrcanian Forests from 2004 to 2016. *Iran. Nat. J.* **2017**, *2*, 40–45. (In Persian) [CrossRef]
122. SCI. Area of Forests, Pastures, and Desert Phenomena in Iran. Available online: https://www.amar.org.ir/ (accessed on 17 July 2017). (In Persian)
123. IPRC. Investigation and Analysis of Flood Events in April 2019 (Rainfall Status and Reservoirs of the Country's Dams). Available online: https://rc.majlis.ir/fa/report/show/1142662 (accessed on 19 May 2019). (In Persian)

124. IPRC. The Credit of 300 Billion Toman to Deal with Locusts. Available online: https://rc.majlis.ir/fa/news/show/1507167 (accessed on 23 April 2020). (In Persian)
125. FAO. Locust Plague Campaign Gets Results in Madagascar. 2014. Available online: http://www.fao.org/news/story/en/item/210810/icode/ (accessed on 15 July 2021).
126. Manning, L. Safeguard global supply chains during a pandemic. *Nat. Food* **2021**, *2*, 10. [CrossRef]
127. World Bank. The Locust Crisis: The World Bank's Response. Available online: https://www.worldbank.org/en/news/factsheet/2020/04/27/the-locust-crisis-the-world-banks-response (accessed on 1 July 2020).
128. FAO. Desert Locust Situation, Updated: 5 October 2020. Available online: http://www.fao.org/ag/locusts/en/info/info/index.html (accessed on 15 July 2021).

Communication

Will Farmers Accept Lower Gross Margins for the Sustainable Cultivation Method of Mixed Cropping? First Insights from Germany

Vanessa Bonke [1,2,*], Marius Michels [2] and Oliver Musshoff [2]

1. Centre of Biodiversity and Sustainable Land Use, Georg-August-University, Goettingen, Buesgenweg 1, 37077 Goettingen, Germany
2. Department of Agricultural Economics and Rural Development, Georg-August-University Goettingen, Platz der Goettinger Sieben 5, 37073 Goettingen, Germany; marius.michels@agr.uni-goettingen.de (M.M.); oliver.musshoff@agr.uni-goettingen.de (O.M.)
* Correspondence: vanessa.bonke@agr.uni-goettingen.de

Abstract: A decline in the legume cultivation has contributed to the biodiversity loss within the agricultural production across Europe. One possibility to include legumes into the production and promote sustainability is mixed cropping with legumes and non-legumes. However, the adoption of mixed cropping is challenging for farmers and information about the profitability is scarce. If mixed cropping should become a widely established production method, it is essential to gain an understanding of famers' evaluation of the profitability mixed cropping needs to reach. Therefore, this article provides first empirical insights into farmers stated willingness to accept gross margin changes compared to current production possibilities. Based on a survey with results from 134 German non-adopters conducted in 2018 we can distinguish conventional farmers with a positive, neutral and negative willingness to accept reductions in gross margins as the trade-off for ecological benefits. Using an ordered logistic model we find that risk attitude, risk perception, the number of measures performed for ecological focus areas, the farmer's age and being located in the south of Germany influence their willingness to accept gross margin changes compared to currently produced cereals.

Keywords: willingness to accept; gross margin; mixed cropping; ordered logistic regression

Citation: Bonke, V.; Michels, M.; Musshoff, O. Will Farmers Accept Lower Gross Margins for the Sustainable Cultivation Method of Mixed Cropping? First Insights from Germany. *Sustainability* **2021**, *13*, 1631. https://doi.org/10.3390/su13041631

Academic Editor: Aitazaz A. Farooque
Received: 12 January 2021
Accepted: 1 February 2021
Published: 3 February 2021

1. Introduction

In accordance with the Sustainable Development Goals of the United Nations, restoring biodiversity and promoting sustainable production patterns are among some of the goals sustainable agriculture should achieve [1] Promoting changes towards a more ecologically beneficial production and reducing negative environmental externalities of current production patterns are therefore core challenges the agricultural sector has to address [2]. A substantial decline in the cultivation of legumes has reduced the provision of ecosystem services and contributed to the biodiversity loss within agricultural production patterns in the European Union (EU) [3,4]. Legumes are able to fixate atmospheric nitrogen (N) through symbiosis with rhizobia in their root system. They can thus reduce the need for mineral N fertilizers which in turn can decrease nitrate leaching and potentially ground water pollution [5]. However, conventional crop rotations in the EU are largely dominated by cereals nowadays. In 2019, around 121 Mio hectare (ha) of cereals were cultivated within the EU, whereas the cultivation of grain legumes amounted to approximately 5 Mio ha [6]. Enhancing legume production has therefore been a political objective. The EU's Common Agricultural Policy (CAP) regulations currently include legumes as part of the greening restrictions, in particular for the provision of Ecological Focus Areas (EFA), in order to encourage adoption by farmers. Nevertheless, the implementation of a pesticide ban with the last adaptation of regulations is assumed to be the reason for a decline from 40% to 24% of legumes in EFA between 2017 and 2018 on the EU level [7].

The plant production in Germany is likewise characterized by a large share of cereals, with winter wheat being the most dominantly cultivated cereal by far [8]. Due to the current agronomic practices and favorable climatic conditions winter varieties in particular are highly productive and profitable [9]. This has led to Germany being the second largest producer of wheat within the EU [10] but it is also part of the reason why Germany is the second largest consumer of mineral N fertilizers [11]. The cereal production is heavily reliant on the use of fertilizers and chemical crop protection products. In contrast, the cultivation of grain legumes has steadily declined over the past decades due to the low economic competitiveness [12]. Even though a slight increase in the cultivated area has been observed in recent years [12], the area cultivated with grain legumes still amounts to only approximately 0.2 Mio ha versus over 6 Mio ha of cereals [8]. As a result there has been a national political focus on enhancing the legume production [12].

One possibility to (re)introduce more legumes into crop rotations and promote sustainable intensification is mixed cropping [13,14]. Mixed cropping, also known as the cultivation of mixed stands or intercropping, is by definition the simultaneous production of two or more coexisting crops on the same area of land [15]. The simultaneous cultivation of legumes and non-legumes in particular can contribute to the provision of ecosystem services while maintaining productivity by utilizing basic ecological concepts and the legume's ability to fixate atmospheric nitrogen [16–19]. Combining non-legumes with legumes reduces the intraspecific competition between the non-legumes with respect to the uptake of mineral N. Simultaneously the legumes in mixed stands fixate atmospheric N which is further promoted by the cereals uptake of soil mineral N [16]. These effects inter alia facilitate a sustainable cultivation. However, the adoption of mixed cropping with main crops (e.g., wheat and faba bean, oat and pea) is especially associated with a number of challenges for European farmers that have been recognized and discussed [14,17,18,20]. In a recent review, Mamine and Farés [14] provide a detailed overview of barriers and levers associated with the mixed cropping of wheat and peas in Europe. Their review highlights the challenges along the value chain for such a specialized production method. For instance, the options for chemical crop protection in mixed stands are limited. This is due to the fact that a treatment which is beneficial for one crop species can be detrimental for the other species in the mixed stand. Hence, the production and yield risk is higher for mixed cropping.

Socio-economic research related to the adoption of mixed cropping with grain legumes in Europe, and Germany in particular, is scarce. Lemken et al. [20] were the first who presented empirical results with respect to early adopters of mixed cropping in Germany. Bonke and Musshoff [21] provided deeper insights into the motivation of farmers to adopt mixed cropping and empirically evaluated the barriers that hinder the widespread adoption. To the best of our knowledge no study to date focuses on the profitability dimension of mixed cropping from a farmer's point of view. Still, there is consensus in the literature that adoption across Europe could be facilitated by the implementation of an environmental scheme [14,17,21] and that a lack of research related to the economic efficiency in high-input agricultural systems persists [13,22]. If a shift within the conventional agriculture sector towards the extensive and more sustainable cultivation of mixed stands should be achieved, the assessment of farmers' willingness to accept (WTA) profitability changes is imperative. Especially considering the many associated challenges that have been identified, like technical barriers and difficulties in crop protection [14,18], the adoption of mixed cropping is also associated with a higher risk from the farmers' perspective. Assessing farmers' WTA profitability changes is consequently the first step to evaluate to what extent financial incentives are necessary to facilitate widespread adoption.

While providing additional ecological benefits compared to sole cropping, mixed cropping also maintains the productivity to generate agricultural output from the land [23]. Moreover, some of the ecological benefits associated with mixed cropping can affect the farm directly. For instance by improving soil quality and reducing the need for synthetic N-fertilizers [24]. Especially the possibility to reduce synthetic N-fertilizers is one aspect that has direct economic effects, for example by reducing production costs. Other ecological benefits, like the increased biodiversity and reduced nitrate leaching [18], are ecosystem services that have positive effects beyond the scope of the farm. In this respect, mixed cropping is broadly comparable to conservation agriculture, where production is aimed to be more environmentally friendly while agricultural output is maintained. Nonetheless, from a farmer's point of view switching from sole to mixed cropping with main crops is a far greater conversion than from, e.g., conservational tillage to no-tillage.

Chouinard et al. [25] derived an expanded utility framework for farmers' selection of conservation practices that incorporates self and social interest with respect to the environment as components of a farmer's utility. They hypothesized that some farmers would be willing to forgo parts of profits for eco-friendly agricultural practices, and found empirical support that some farmers are indeed willing to choose less profitable production methods. Their results further show that farmers are heterogeneous with respect to their willingness to pay for an increase in environmental benefits. Several other studies have also shown that farmers are heterogeneous with respect to their WTA payments for ecosystem services and that farmers are not behaving in a strictly profit maximizing way, e.g., [26–28]. If the objective is to establish mixed cropping as a cultivation method within the production portfolio of farms in the long-run, it is hence also essential to get an understanding whether farmers evaluation of the profitability levels mixed cropping needs to reach are heterogeneous. Against this background, we formulated the research questions: (1) Are farmers willing to forgo profits for the sustainable cultivation method of mixed cropping? (2) Do farmers' risk attitude and their perception of the risk associated with mixed cropping influence the willingness to accept profitability changes?

We aim to distinguish groups of farmers that are heterogeneous in their WTA to provide impulses on which non-adopters to target first to facilitate mixed cropping adoption. Non-adopters who have a positive WTA reduction in profitability compared to current dominant production possibilities will be the ones who will demand the lowest financial incentives for adoption. This makes them of particular interest for researchers and policy makers.

2. Materials and Methods

2.1. Willingness to Accept Reduced Gross Margins

To elicit whether farmers are willing to accept reductions in gross margins (€/ha), i.e., forego profits, and are heterogeneous in their willingness to accept we chose a three-step approach. Changing the production towards mixed stands implies that farmers have to reduce their production of pure stands. Thus, with respect to property rights and the economic consequences in line with the contingent valuation literature, the appropriate concept is the evaluation of the WTA [29]. Since cereals vastly dominate the agricultural production in Germany, the gross margin of the cereal was chosen as the reference point. This was done in order to provide a realistic and familiar scenario for the farmers as the status quo comparison. To account for the forgone gross margin of the cereal production, and thus capturing the majority of opportunity costs for adopting mixed cropping, in the first step farmers were asked, whether they would be willing to cultivate mixed stands with main crops if the gross margin is equal to that of the cereal in the pure stand. Farmers who gave a positive response to the first question were subsequently asked if they would be willing to accept lower gross margins in the mixed stands when considering the associated ecological benefits. Only farmers who indicated that they were willing to accept a lower gross margin were presented with a third open-ended question that asked how much of the gross margin they would be willing to give up for the cultivation of mixed stands considering the additional ecological benefits (in %) (Figure 1).

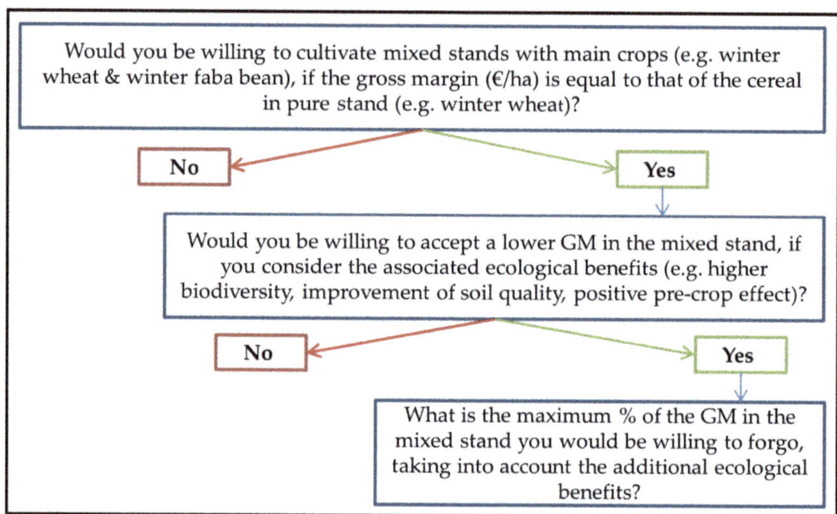

Figure 1. Three Step Questioning Approach (GM—Gross Margin).

This three-step approach allows us to distinguish three groups of farmers that are heterogeneous in their WTA foregone profits in exchange for the ecologically advantageous production method of mixed cropping (Table 1). However, an equal gross margin between the pure stand and the mixed stand does not account for all cost components associated with a change towards mixed cropping. Using the gross margins abstracts from the possibility that additional investment might be necessary [14], that learning costs arise [30], that variable labor costs could be higher or that income risks increase. On the other side, the positive pre-crop effect of mixed cropping, which inter alia can reduce costs for N fertilizer in the following crop [17], is also not included in the gross margins. Nevertheless, since the gross margin is one of the most common profitability criteria that farmers are familiar with and base their production decisions on, using the gross margins allows for the most realistic baseline in our case.

Table 1. Classification of Groups by Willingness to Accept (WTA).

Group	Description	Implication
"negative" WTA	Not willing to accept equal gross margin for mixed cropping	$GM_{Mixed} > GM_{Pure}$
"neutral" WTA	Willing to accept equal gross margin for mixed cropping	$GM_{Mixed} = GM_{Pure}$
"positive" WTA	Willing to accept lower gross margin for mixed cropping	$GM_{Mixed} < GM_{Pure}$

GM_{Mixed}—Gross Margin Mixed Stand; GM_{Pure}—Gross Margin Pure Stand.

Furthermore, we explicitly decided to frame the questions without implying any relation to a political scheme or subsidy. For one, implying that an equal gross margin would be achieved by politically subsidized payments increases the likelihood of a bias that farmers state the demand of higher payments [29]. Second, there are several possibilities how an equal gross margin could be achieved, for instance by enhancements in the yield through optimizing plant breeding and agronomy. Moreover, reducing the WTA question to changes in the gross margin and abstracting from potential subsidy payments and regulatory implications, allows for a more direct insight in the WTA as a trade-off between economic and ecological benefits.

2.2. Risk Attitude and Risk Perception

As briefly outlined in the introduction, the adoption of mixed cropping is related to many challenges from a farmer's point of view [14,20]. Since mixed cropping is a new production method for famers addressed in this study, its implementation is associated with higher risks, not at least due to the fact that farmers do not have experience in the cultivation of mixed stands. Adopting mixed cropping hence also implies a change in the income risk for the farmer. Dörschner and Musshoff [31] demonstrated based on a normative model that considering changes in income risk as well as the risk attitude of the farmer can considerably influence WTA compensation payments for agri-environmental related measures. Moreover, empirical studies have shown that the farmers' risk attitude and their perception of risks associated with a measure can substantially influence their willingness to adopt agri-environmental related measures and accept payments [32–34].

Neglecting changes in income risk and the risk attitude can therefore cause major changes in the magnitude of farmers' WTA. Our evaluation of the WTA based on the gross margins cannot explicitly account for changes in income risk. Nevertheless, we hypothesize that the farmers' perception of risk associated with the cultivation of mixed stands does influence their WTA and can serve as an indicator. Likewise, we assume that farmers' subjective risk attitude influences their WTA. Therefore, we used the 11-point scale proposed by Dohmen et al. [35] to assess the farmer's attitude toward risk (How do you personally rate yourself: Are you generally a person who is willing to take risks or do you try to avoid risks? 0 = "not at all willing to take risks", ... , 10 = "very willing to take risks").

Furthermore, we included an indicator for the farmer's risk perception of mixed cropping which was measured on a 5-point Likert scale ("The cultivation of mixed stands is associated with a higher risk" 1 = "totally disagree", ... , 5 = "totally agree"). If the risk perception and risk attitude have an influence on the farmer's WTA, this has implications for the design of subsidy-schemes and contractual agreements for the production of mixed cropping, and in a broader context other sustainable practices as well.

2.3. Ordered Logistic Regression

The sequential questioning allows us to distinguish three separate groups with respect to the WTA. These groups can be ranked on an ordinal scale by design: negative WTA, neutral, and positive WTA reductions in gross margins (see Table 1). For such an ordinal limited dependent variable, in our case with three distinct categories, the ordered logistic regression can be used [36]. In accordance with our conceptual framework we estimate a model that includes the farmers' risk attitude and risk perception. Moreover, socio-demographic and farm variables are included as explanatory variables (Table 2) to assess if these can explain the heterogeneity in the WTA. Therefore, the following model specification is estimated:

$$\begin{aligned} \text{WTA Group}_i = \quad & \beta_0 + \beta_1 \text{ Risk attitude} + \beta_2 \text{ Risk perception} + \beta_3 \text{ Farm size} \\ & + \beta_4 \text{Full time farm} + \beta_5 \text{Legumes} + \beta_6 \text{ No. EFA measures} \\ & + \beta_7 \text{ Region} + \beta_8 \text{ Rented land} + \beta_9 \text{ Training enterprise} \\ & + \beta_{10} \text{ Age} + \beta_{11} \text{College degree} + \varepsilon_i \end{aligned} \tag{1}$$

where i represents the individual respondent and ε_i is assumed to be an error term with a logistic distribution. The included farm and farmer related variables were chosen since these are objectively measurable characteristics among the features that have been found to statistically significantly influence farmers' adoption behavior of conservation and agri-environmental practices, e.g., [20,34,37]. Including objectively measurable variables supports an easier distinction between potential target groups for voluntary agri-environmental schemes.

Statistical analysis was conducted with STATA 15. Regression results are displayed in form of odds ratios. An odds ratio larger than one thereby implies a positive effect of the independent variable on the dependent variable WTA. This means an increase in the independent variable increases the likelihood of being in a higher WTA group. An important assumption for the ordered logit model is the parallel regression assumption. This indicates that the coefficients are equal across all ordinal stages of the dependent variable. If this holds true, there is only one set of coefficients to be estimated, because the relationship between each pair of stages is the same. The Brant test is applied to validate this critical model assumption [38].

2.4. Sample and Descriptive Statistics

The data used to analyze the presented research question were gathered as part of a survey that focused on German farmers' motivations and the perceived barriers for the adoption of mixed cropping. The survey was conducted online between September and November 2018 [21] . The online questionnaire was designed in the way that farmers who are non-adopters of mixed cropping were subsequently presented with the WTA questions. Thus, after removal of incomplete surveys, the sample used in this analysis comprised of 134 conventional German farmers who are non-adopters of mixed cropping. Focusing on non-adopters has two reasons: First, it can be assumed that these farmers all have the same level of real experience with mixed cropping, namely none, which implies the hypothetical bias is the same for all of them. Providing an explanation about mixed cropping with main crops at the beginning of the survey and giving examples of crop combinations (with pictures) that fulfill this specification ensured that all participants had the same specification in mind when answering the questions.

Second, it is the non-adopters who will need additional financial incentives compared to the real world status quo to include mixed cropping in their production, making them especially relevant for our research question. Farmers with a positive WTA lower gross margins in exchange for the environmentally friendly production of mixed stands are the ones that will be of particular interest for the implementation of voluntary environmental-schemes, as those farmers will participate with lower financial incentives.

The descriptive statistics (Table 2) show that our sample comprises of highly educated young farmers, who work on large farms. Our sample is not representative for the German farms [39]. However, especially considering the long-term sustainable development of agricultural production, focusing on younger farmers delivers more relevant results since these are the farmers who will make the production decisions for years to come. Likewise, it is the production decisions on larger farms that will influence the sustainability of production to a greater extent.

Based on our three-step approach, we can classify 23.13% of our sample as having a "negative" WTA, 23.13% as having a "neutral" WTA (=0) and 53.73% of having a "positive" WTA reductions in gross margins for the adoption of mixed cropping (Table 1). With respect to our first research question, this implies that farmers with a "positive" WTA are willing to forgo profits for mixed cropping, farmers with a "neutral" WTA are not willing to forgo profits, and farmers with a "negative" WTA would even demand higher profits for mixed cropping. This result confirms that farmers are heterogeneous in their willingness to forgo profits as a trade-off for the more sustainable cultivation method of mixed cropping. Detailed sociodemographic characteristics for the separate groups as well as the indicators for the risk attitude and the risk perception of mixed cropping are also depicted in Table 2. Though our share of famers' with a "positive" WTA might seem high, it parallels the results of Chouinard et al. [25] that a share of farmers is willing to forego profits for environmental benefits. Since parts of the ecological benefits associated with mixed cropping will also directly positively influence the farm itself, this share seems plausible.

Table 2. Descriptive Results for Identified WTA Groups.

Variable		Full Sample N = 134 (100%) Mean (SD|Median)	"Negative" WTA N = 31 (23.13%) Mean (SD|Median)	"Neutral" WTA N = 31 (23.13%) Mean (SD|Median)	"Positive" WTA N = 72 (53.73%) Mean (SD|Median)
Risk attitude (of the farmer) [a]		6.31 (1.73|6)	5.77 (1.87|6)	6.32 (1.87|6)	6.54 (1.57|7)
Risk perception (of mixed cropping) [b]		3.61 (1.02|4)	4.00 (1.06|4)	3.52 (1.06|4)	3.48 (0.95|4)
Farm size (ha of arable land)		329.80 (488|130)	209.61 (317|130)	428.16 (545|200)	339.19 (517|118)
Full time farm		0.93	0.96	0.94	0.92
Legumes (produced as main crop)		0.27	0.26	0.23	0.29
No. EFA measures [c]		2.99 (1.35|3)	2.52 (1.43|2)	2.90 (1.40|3)	3.22 (1.25|3)
Region (% of farms in region)	North	0.44	0.41	0.29	0.51
	East	0.13	0.10	0.16	0.14
	West	0.21	0.13	0.29	0.21
	South	0.21	0.35	0.26	0.14
Rented land (% rented of total land)		0.54	0.53	0.55	0.53
Training enterprise (educate junior farmers)		0.75	0.80	0.84	0.68
Age (in years)		35.80 (11.37|33)	39.71 (11.31|37)	38.87 (11.03|37)	32.79 (10.78|29)
College degree (at least B.Sc.)		0.63	0.55	0.58	0.68

[a] Measured on the 11-point scale proposed by Dohmen et al. [35]: How do you personally rate yourself: Are you generally a person who is willing to take risks or do you try to avoid risks? 0 = "not at all willing to take risks", …, 10 = "very willing to take risks" [b] Measured on 5-point Likert scale: "The cultivation of mixed stands is associated with a higher risk" 1 = "totally disagree", … , 5 = "totally agree" [c] Under the current CAP farmers have to maintain 5% of their land as Ecological Focus Area (EFA), several measures for the provision of this EFA are recognized and farmers can decide freely how many different measures they implement [40] .

Farmers who indicated that they would be willing to accept a lower gross margin for the cultivation of mixed stands were presented with the last question of how much of the gross margin they would be willing to forego. They could type in their response or use a slide bar to settle on a percentage value between 0% and 100%. The mean value of the accepted gross margin reduction is 13.08% (SD 5.96) and the median is 10% for the N = 72 farmers that answered the question (Figure 2). This means on average these farmers would cultivate mixed stands if the mixed stands reach 86.92% of the profitability of currently dominantly produced cereals.

Since the WTA question is hypothetical and did not have any real financial consequences, the stated values can be biased towards a higher positive WTA. The depicted values thus have to be interpreted with caution and have to be validated in future studies. Nevertheless, the stated values can provide a first impression about farmers WTA gross margin reductions and that farmers are willing to accept lower profits has also been shown in other studies. For example, Buckley et al. [26] had a share of 16% in their sample of Irish farmers that were willing to participate in a scheme to provide buffer strips to reduce water pollution and had a WTA payments of 0€/ha. In other words, those farmers were willing to reduce the agricultural output without any compensation, thus also forego profits. Our reference scenario was an equal profitability to the current production; thus, it is plausible that a comparatively large share will accept reductions in the gross margin since the ecological benefits will also have a direct positive influence on their farms.

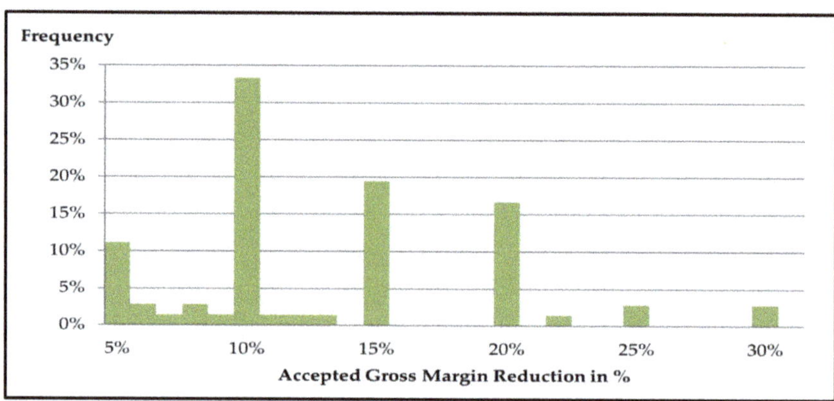

Figure 2. Distribution of responses to the question: "What is the maximum % of the gross margin in the mixed stand you would be willing to forgo, taking into account the additional ecological benefits?" (N = 72, Group "positive" WTA).

3. Regression Results and Discussion

The results of the ordered logistic regression for the dependent variable WTA group are depicted in Table 3. Due to the comparatively small sample size we applied a backward selection procedure to reduce the set of explanatory variables based on their statistical significance. The Akaike Information Criterion (AIC), the Bayesian Information Criterion (BIC) and a likelihood ratio test between the reduced and the full model (LR $\chi^2(8)$ = 3.74; p = 0.8796), imply a better fit of the reduced model. Therefore, the reduced model is depicted below in Table 3, but the full model is included in Appendix A. The likelihood ratio test for the model is statistically significant, rejecting the null-hypothesis that all coefficients are zero (Log-likelihood = -119.78, LR $\chi^2(5)$ = 31.40; p < 0.001). The Brant test is not statistically significant indicating that the assumption of proportional odds is not violated and the model is suitable ($\chi^2(5)$ = 2.676; p = 0.750). This implies that the coefficients are equal across all ordinal stages of the dependent variable WTA group. To test for multicollinearity issues the variance inflation factors (VIF) were calculated, which should not surpass the value of 5 [41]. With a maximum value of 1.04 for our model, we conclude that multicollinearity is not problematic.

For the variable Risk attitude, we find a statistically significant effect. The odds ratio of 1.1962 implies that with a one unit increase in the risk attitude score (towards risk seeking), the odds of a positive WTA versus a negative- and neutral WTA are 1.1962 times higher ceteris paribus. This result is in line with previous research [33,42,43]. For instance, Serra et al. [32] show that risk aversion negatively influences the adoption of organic farming practices.

For the variable Risk perception, we find an odds ratio smaller than 1 which is statistically significant at the 10% level. This indicates that holding all other values constant, a one unit increase in the risk perception score (towards higher risk associated with mixed cropping), the odds of being in the group of positive WTA is 0.7083 times lower than being in the negative or neutral WTA group. Risk perception can be interpreted as a proxy for changes in production risk. To illustrate, an example are changes in input risks due to the fact that seed varieties have not been selected and bred for the cultivation in mixed stands [44]. Since varieties are not well adapted for mixed cropping, yield risks increase compared to sole cropping [14]. In line with this, Hannus and Sauer [33] show that a high risk perception of a sustainability standard decreases the likelihood of German farmers participation.

Table 3. Ordinal Regression Results for the dependent variable WTA Group [a] after Backward Selection Procedure (N = 134).

Variable	Odds Ratio	SE	*p*-Value [b]	[95% CI]
Risk attitude (of the farmer)	1.1962	0.1231	0.082 *	[0.9777;1.4635]
Risk perception (of mixed cropping)	0.7083	0.1278	0.056 *	[0.4973;1.0087]
No. EFA measures	1.3432	0.1821	0.029 **	[1.0299;1.7520]
Region South	0.3432	0.1361	0.007 ***	[0.1371;0.7359]
Age	0.9472	0.0150	0.001 ***	[0.9182;0.9771]
Likelihood ratio $\chi^2(5)$	31.40 ($p < 0.001$)			
McFadden pseudo R^2	0.116			
Nagelkerke pseudo R^2	0.241			
Brant test $\chi^2(5)$	2.676 ($p = 0.750$)			
VIF	mean = 1.02 max = 1.04			
AIC	253.57			
BIC	273.85			

SE—Standard Error; CI—Confidence Interval; VIF—Variance Inflation Factor; AIC—Akaike Information Criterion, BIC—Bayesian Information Criterion. [a] Includes the ordinal categories: negative, neutral, and positive WTA. [b] Asterisks indicate different levels of significance (*** $p < 0.01$, ** $p < 0.05$, * $p < 0.10$).

Combined, the effect of risk attitude and perception indicate that a negative WTA is partially caused by the demand of a risk premium for the adoption of this innovative cultivation method. Reaching a similar profitability in mixed cropping compared to the current cereal production will not be enough for the farmers that are categorized in the "negative" WTA group. They will demand a higher profitability in exchange for the increased risks associated with mixed cropping, which does not necessarily mean that those farmers do not value the environmental benefits. It is more likely to imply that the increased risk outweighs the ecological benefits for them. The demand of a risk premium is in line with expected utility theory and has been demonstrated to increase payments necessary for the adoption of more environmentally friendly production measures [31]. For mixed cropping this implies that reducing the risks associated with the cultivation can be a lever to facilitate adoption and reduce requirements for the profitability. In non-monetary terms, this could for instance be facilitated by enhancing the information availability for farmers, which also calls for further research in plant sciences [14,21].

We also find a statistically significant effect for the number of measures a farmer provides for the Ecological Focus Areas (No. EFA measures). The odds ratio larger than one indicates that an increasing number of different measures implemented on the farm, increases the likelihood of being in higher ranked WTA group. Within the CAP regulations, most conventional farmers have to dedicate a share of 5% of their arable land for the provision of the EFA [40] How many different measures they perform is their decision. It has been argued that German farmers predominantly choose measures that they already perform or that fit well within their established production patterns, like catch crops [45,46]. Hence, performing a higher number of different measures for the EFA can be interpreted as a measurable indicator for farmers whose arable production already includes more ecologically beneficial traits. These farmers thus might have a higher preference for environmentally friendly production methods.

For the variable Region South, we see a statistically significant negative effect for having a positive WTA. The agricultural production in the south of Germany is characterized by comparatively small farms with the highest share of grassland across Germany [39,47] The high share of grassland compared to arable land can be a reason why these farmers display a lower WTA to accept reduced gross margins in their arable production.

With respect to the Age of the farmers, we find a negative effect that is statistically significant at the 1% level. A one-year increase in age, ceteris paribus, decreases the odds of having a positive WTA. While the results of Lemken et al. [20] also show a negative effect of the age on farmers mixed cropping adoption tendencies, the effect was not statistically significant. There is no clear consensus with respect to age of the farmer in studies related to agri-environmental measures, both positive and negative effects have been observed [48].

4. Implications, Limitations, and Future Directions

The presented results provide important first insights about the extent of the financial incentives that would be necessary to facilitate mixed cropping in German agriculture and potentially other European countries where the technological lock-in around dominant cereal crops is prevalent. We deliver a starting point into further research with respect to farmers' willingness to accept payments for the adoption of mixed cropping. Our results support that heterogeneity between farmers with respect to the trade-off between economic and ecological benefits exists for the case of mixed cropping. A group of farmers is willing to forgo profits for the adoption of mixed cropping compared to the current dominant cereal production. Nevertheless, another group of farmers would demand an even higher profitability of mixed stands compared to the cereal production. The results of the ordered logistic regression imply that the farmers' risk attitude and their perception of risks associated with mixed cropping statistically significantly influence their WTA. This indicates that the trade-off between risk and profitability is at least partially responsible for demanding a higher profitability in mixed stands. Our results also show that the farmers' age and the number of measures they perform for the provision of EFA positively influence their WTA. Thus, younger farmers who perform a larger number of EFA measures, are not located in the south of Germany, have a lower risk aversion, and a lower perception of the risks associated with mixed cropping are the farmers who will demand the lowest financial incentives for adoption.

One lever that will facilitate the acceptance of mixed cropping and reduce profitability requirements is reducing the risk associated with the cultivation. Assessing possibilities to reduce risk, for instance by decoupling incentives from the produced marketable output in mixed stands in contractual agreements, poses a potential future research agenda. However, implementing restrictions on the cultivation is likely to increase the requirement for payments.

Our results are based on a comparatively small sample that is not representative for Germany; this has to be considered for the external validity of the results. The small sample limits the generalizability of the results for the population of German farmers. To validate the results, further research is needed which should preferably be based on a representative sample. Nevertheless, our results indicate that there is a share of German farmers that will forgo profits for the adoption of mixed cropping. Our results can give valuable implications for the design of voluntary agri-environmental schemes with respect to mixed cropping, as the adoption of voluntary measures itself is subject to a self-selection bias. The question design with respect to the WTA was hypothetical in nature. This could have led to the farmers overstating their WTA, which is a common criticism of stated preference approaches. It will therefore be necessary to further investigate farmers true WTA with respect to mixed cropping. Nonetheless, our results can serve as a starting point for future research that could for example use incentivized choice experiments to elicit farmers WTA. Furthermore, we cannot distinguish between farmers with a true "negative" WTA and protest responses. Protest responses can also bias the results. Since we explicitly did not frame mixed cropping in a political context, we believe this bias to be negligible.

Against the background that mixed cropping faces many challenges that have to be overcome to be widely adopted in the EU and further research is needed along different steps of the value chain, our results are a first impulse to what extent financial incentives will be necessary to facilitate adoption. How financial incentives could be provided effectively is a further research question that can be addressed in the future.

Author Contributions: Conceptualization, V.B. and O.M.; Methodology, V.B. and M.M.; Formal analysis, V.B. and M.M.; Investigation, V.B.; Writing—Original draft preparation, V.B.; Writing—review and editing, V.B., M.M. and O.M.; Visualization, V.B.; Funding Acquisition, O.M.; Supervision, O.M. All authors have read and agreed to the published version of the manuscript.

Funding: This research was funded by the German Federal Ministry of Education and Research, grant number FKZ 031A351A, within the framework of the IMPAC^3 project. IMPAC^3 is a project of the Centre of Biodiversity and Sustainable Land Use at the Georg-August-University Goettingen. We acknowledge support by the Open Access Publication Funds of the Goettingen University for the APC.

Data Availability Statement: The datasets analyzed for the present article are available from the corresponding author on reasonable request.

Acknowledgments: We gratefully acknowledge the valuable comments and suggestions of four anonymous referees and the editors of Sustainability.

Conflicts of Interest: The authors declare no conflict of interest. The funders had no role in the design of the study; in the collection, analyses, or interpretation of data; in the writing of the manuscript, or in the decision to publish the results.

Appendix A

Table A1. Ordinal Regression for dependent variable WTA Group [a] with Full Set of Explanatory Variables (N = 134).

Variable	Odds Ratio	SE	*p*-Value [b]	[95% CI]
Risk attitude (of the farmer)	1.2058	0.1274	0.076 *	[0.9803;1.4831]
Risk perception (of mixed cropping)	0.6739	0.1277	0.037 **	[0.4649;0.9769]
Farm size	1.0001	0.0004	0.759	[0.9993;1.0010]
Full time	0.6066	0.5394	0.574	[0.1061;3.4660]
Legumes [c]	1.1863	0.5179	0.696	[0.5041;2.7913]
No. EFA measures	1.3918	0.2018	0.023 **	[1.0476;1.8492]
Region East	0.6765	0.4313	0.540	[0.1939;2.3599]
Region West	0.7303	0.3497	0.511	[0.2858;1.8661]
Region South	0.2773	0.1362	0.009 ***	[0.1058;0.7264]
Rented land	1.0052	0.0079	0.506	[0.9899;1.0209]
Training enterprise	0.5392	0.2952	0.259	[0.1844;1.5767]
Age	0.9534	0.0168	0.007 ***	[0.9209;0.9870]
College degree	1.1028	0.4456	0.809	[0.4995;2.4346]
Likelihood ratio $\chi^2(13)$	35.14 ($p < 0.001$)			
McFadden pseudo R^2	0.130			
Nagelkerke pseudo R^2	0.266			
Brant test $\chi^2(13)$	10.77 ($p = 0.630$)			
VIF	mean = 1.28 max = 1.62			
AIC	265.83			
BIC	309.29			

SE-Standard Error; CI-Confidence Interval; VIF-Variance Inflation Factor; AIC-Akaike Information Criterion, BIC-Bayesian Information Criterion. [a] Includes the ordinal categories: negative, neutral, and positive WTA. [b] Asterisks indicate different levels of significance (*** $p < 0.01$, ** $p < 0.05$, * $p < 0.10$). [c] Producing Legumes as a main crop can be understood as a proxy for farmers' having an improved understanding of the ecological benefits and maybe the risks associated with the cultivation of mixed cropping. The VIF values suggest however that the correlation between the variables Legumes and Risk perception is low.

References

1. United Nations. Resolution Adopted by the General Assembly on 25 September 2015; Transforming Our World: The 2030 Agenda for Sustainable Development. Available online: https://www.un.org/en/development/desa/population/migration/generalassembly/docs/globalcompact/A_RES_70_1_E.pdf (accessed on 22 January 2021).

2. Melchior, I.C.; Newig, J. Governing Transitions towards Sustainable Agriculture—Taking Stock of an Emerging Field of Research. *Sustainability* **2021**, *13*, 528. [CrossRef]

3. Zander, P.; Amjath-Babu, T.S.; Preissel, S.; Reckling, M.; Bues, A.; Schläfke, N.; Kuhlman, T.; Bachinger, J.; Uthes, S.; Stoddard, F.; et al. Grain legume decline and potential recovery in European agriculture: A review. *Agron. Sustain. Dev.* **2016**, *36*, 1–20. [CrossRef]

4. Meynard, J.-M.; Charrier, F.; Fares, M.'h.; Le Bail, M.; Magrini, M.-B.; Charlier, A.; Messéan, A. Socio-technical lock-in hinders crop diversification in France. *Agron. Sustain. Dev.* **2018**, *38*, 116. [CrossRef]

5. Peoples, M.B.; Brockwell, J.; Herridge, D.F.; Rochester, I.J.; Alves, B.J.R.; Urquiaga, S.; Boddey, R.M.; Dakora, F.D.; Bhattarai, S.; Maskey, S.L.; et al. The contributions of nitrogen-fixing crop legumes to the productivity of agricultural systems. *Symbiosis* **2009**, *48*, 1–17. [CrossRef]

6. FAOSTAT. Food and Agriculture Organization of the United Nations. *Statistics Database*, Production. Crops. Available online: www.fao.org/faostat/en/#data/QC (accessed on 14 October 2020).

7. Alliance Environnement. *Evaluation of the impact of the CAP on Habitats, Landscapes, Biodiversity. Final Report*; Publications Office of the European Union: Luxembourg, 2020. [CrossRef]

8. Federal Statistical Office. Land- und Forstwirtschaft, Fischerei. Landwirtschaftliche Bodennutzung Anbau auf dem Ackerland. 2019. Available online: https://www.statistischebibliothek.de/mir/receive/DEHeft_mods_00130381. (accessed on 23 January 2021).

9. Lüttringhaus, S.; Gornott, C.; Wittkop, B.; Noleppa, S.; Lotze-Campen, H. The Economic Impact of Exchanging Breeding Material: Assessing Winter Wheat Production in Germany. *Front. Plant Sci.* **2020**, *11*. [CrossRef] [PubMed]

10. Statistical Office of the European Union (EUROSTAT). Agricultral Production-Crops. Available online: https://ec.europa.eu/eurostat/statistics-explained/index.php?title=Agricultural_production_-_crops#Cereals (accessed on 25 January 2021).

11. Statistical Office of the European Union (EUROSTAT). Agri-Environmental Indicator-Mineral Fertiliser Consumption. Available online: https://ec.europa.eu/eurostat/statistics-explained/index.php?title=Agri-environmental_indicator_-_mineral_fertiliser_consumption (accessed on 25 January 2021).

12. Federal Ministry of Food and Agriculture. Ackerbohne, Erbse & Co. Die Eweißpflanzenstrategie des Bundesministeriums für Ernährung und Landwirtschaft zur Förderung des Leguminosenanbaus in Deutschland. 2020. Available online: https://www.bmel.de/SharedDocs/Downloads/DE/Broschueren/EiweisspflanzenstrategieBMEL.pdf?__blob=publicationFile&v=4 (accessed on 22 January 2021).

13. Rosa-Schleich, J.; Loos, J.; Mußhoff, O.; Tscharntke, T. Ecological-economic trade-offs of Diversified Farming Systems—A review. *Ecol. Econ.* **2019**, *160*, 251–263. [CrossRef]

14. Mamine, F.; Farès, M. Barriers and Levers to Developing Wheat–Pea Intercropping in Europe: A Review. *Sustainability* **2020**, *12*, 6962. [CrossRef]

15. Andrews, D.J.; Kassam, A.H. The Importance of Multiple Cropping in Increasing World Food Supplies. *Mult. Crop.* **1976**, *27*, 1–10. [CrossRef]

16. Hauggaard-Nielsen, H.; Jørnsgaard, B.; Kinane, J.; Jensen, E.S. Grain legume–cereal intercropping: The practical application of diversity, competition and facilitation in arable and organic cropping systems. *Renew. Agric. Food Syst.* **2008**, *23*, 3–12. [CrossRef]

17. Bedoussac, L.; Journet, E.-P.; Hauggaard-Nielsen, H.; Naudin, C.; Corre-Hellou, G.; Jensen, E.S.; Prieur, L.; Justes, E. Ecological principles underlying the increase of productivity achieved by cereal-grain legume intercrops in organic farming. A review. *Agron. Sustain. Dev.* **2015**, *35*, 911–935. [CrossRef]

18. Jensen, E.S.; Carlsson, G.; Hauggaard-Nielsen, H. Intercropping of grain legumes and cereals improves the use of soil N resources and reduces the requirement for synthetic fertilizer N: A global-scale analysis. *Agron. Sustain. Dev.* **2020**, *40*, 1–9. [CrossRef]

19. Wezel, A.; Casagrande, M.; Celette, F.; Vian, J.-F.; Ferrer, A.; Peigné, J. Agroecological practices for sustainable agriculture. A review. *Agron. Sustain. Dev.* **2014**, *34*, 1–20. [CrossRef]

20. Lemken, D.; Spiller, A.; von Meyer-Höfer, M. The Case of Legume-Cereal Crop Mixtures in Modern Agriculture and the Transtheoretical Model of Gradual Adoption. *Ecol. Econ.* **2017**, *137*, 20–28. [CrossRef]

21. Bonke, V.; Musshoff, O. Understanding German farmer's intention to adopt mixed cropping using the theory of planned behavior. *Agron. Sustain. Dev.* **2020**, *40*, 1–14. [CrossRef]

22. Martin-Guay, M.-O.; Paquette, A.; Dupras, J.; Rivest, D. The new Green Revolution: Sustainable intensification of agriculture by intercropping. *Sci. Total Environ.* **2018**, *615*, 767–772. [CrossRef]

23. Gaba, S.; Lescourret, F.; Boudsocq, S.; Enjalbert, J.; Hinsinger, P.; Journet, E.-P.; Navas, M.L.; Wery, J.; Louarn, G.; Malezieux, E.; et al. Multiple cropping systems as drivers for providing multiple ecosystem services: From concepts to design. *Agron. Sustain. Dev.* **2015**, *35*, 607–623. [CrossRef]

24. Pelzer, E.; Bazot, M.; Makowski, D.; Corre-Hellou, G.; Naudin, C.; Rifai, M.A.; Baranger, E.; Bedoussac, L.; Biarnes, V.; Boucheny, P.; et al. Pea–wheat intercrops in low-input conditions combine high economic performances and low environmental impacts. *Euro. J. Agron.* **2012**, *40*, 39–53. [CrossRef]

25. Chouinard, H.H.; Paterson, T.; Wandschneider, P.R.; Ohler, A.M. Will Farmers Trade Profits for Stewardship? Heterogeneous Motivations for Farm Practice Selection. *Land Econ.* **2008**, *84*, 66–82. [CrossRef]

26. Buckley, C.; Hynes, S.; Mechan, S. Supply of an ecosystem service—Farmers' willingness to adopt riparian buffer zones in agricultural catchments. *Environ. Sci. Policy* **2012**, *24*, 101–109. [CrossRef]

27. Marr, E.J.; Howley, P. The accidental environmentalists: Factors affecting farmers' adoption of pro-environmental activities in England and Ontario. *J. Rur. Stud.* **2019**, *68*, 100–111. [CrossRef]
28. Da Motta, R.S.; Ortiz, R.A. Costs and Perceptions Conditioning Willingness to Accept Payments for Ecosystem Services in a Brazilian Case. *Ecol. Econ.* **2018**, *147*, 333–342. [CrossRef]
29. Whittington, D.; Adamowicz, W.; Lloyd-Smith, P. Asking Willingness-to-Accept Questions in Stated Preference Surveys: A Review and Research Agenda. *Annu. Rev. Resour. Econ.* **2017**, *9*, 317–336. [CrossRef]
30. Lemken, D.; Knigge, M.; Meyerding, S.; Spiller, A. The Value of Environmental and Health Claims on New Legume Products: A Non-Hypothetical Online Auction. *Sustainability* **2017**, *9*, 1340. [CrossRef]
31. Dörschner, T.; Musshoff, O. Cost-oriented evaluation of ecosystem services under consideration of income risks and risk attitudes of farmers. *J. Environ. Manage.* **2013**, *127*, 249–254. [CrossRef] [PubMed]
32. Serra, T.; Zilberman, D.; Gil, J.M. Differential uncertainties and risk attitudes between conventional and organic producers: The case of Spanish arable crop farmers. *Agric. Econ.* **2008**, *39*, 219–229. [CrossRef]
33. Hannus, V.; Sauer, J. Are Farmers as Risk-averse as They Think They Are? *Proc. Food Sys. Dyn.* **2020**, *0*, 165–173. [CrossRef]
34. Bartkowski, B.; Bartke, S. Leverage Points for Governing Agricultural Soils: A Review of Empirical Studies of European Farmers' Decision-Making. *Sustainability* **2018**, *10*, 3179. [CrossRef]
35. Dohmen, T.; Falk, A.; Huffman, D.; Sunde, U.; Schupp, J.; Wagner, G.G. Individual Risk Attitudes: Measurement, Determinants, and Behavioral Consequences. *J. Euro. Econ. Assoc.* **2011**, *9*, 522–550. [CrossRef]
36. Verbeek, M. *A Guide to Modern Econometrics*, 3rd ed.; John Wiley & Sons, Inc: Hoboken, NJ, USA, 2008.
37. Knowler, D.; Bradshaw, B. Farmers' adoption of conservation agriculture: A review and synthesis of recent research. *Food Policy* **2007**, *32*, 25–48. [CrossRef]
38. Williams, R. Understanding and interpreting generalized ordered logit models. *J. Math. Socio.* **2016**, *40*, 7–20. [CrossRef]
39. The German Farmers' Association. Situationsbericht 2019/20. 2020. Available online: https://www.bauernverband.de/situationsbericht (accessed on 5 September 2020).
40. Federal Ministry of Food and Agriculture. Umsetzung der EU-Agrarreform in Deutschland–Ausgabe 2015. 2015. Available online: www.bmel.de/SharedDocs/Downloads/Broschueren/UmsetzungGAPinD.pdf?__blob=publicationFile#page=44 (accessed on 5 September 2020).
41. Curto, J.D.; Pinto, J.C. The corrected VIF (CVIF). *J. Appl. Stat.* **2011**, *38*, 1499–1507. [CrossRef]
42. Wang, J.; Yang, C.; Ma, W.; Tang, J. Risk preference, trust, and willingness-to-accept subsidies for pro-environmental production: An investigation of hog farmers in China. *Environ. Econ. Policy Stud.* **2020**, *22*, 405–431. [CrossRef]
43. Greiner, R.; Patterson, L.; Miller, O. Motivations, risk perceptions and adoption of conservation practices by farmers. *Agric. Sys.* **2009**, *99*, 86–104. [CrossRef]
44. Siebrecht-Schöll, D. Züchterische Analyse von acht Winterackerbohnengenotypen für den Gemengeanbau mit Winterweizen. Doctoral Dissertation, Georg-August-Universität, Göttingen, Germany, 2019.
45. Zinngrebe, Y.; Pe'er, G.; Schueler, S.; Schmitt, J.; Schmidt, J.; Lakner, S. The EU's ecological focus areas–How experts explain farmers' choices in Germany. *Land Use Policy* **2017**, *65*, 93–108. [CrossRef]
46. Weller von Ahlefeld, P.J.; Michels, M. Die Reform der Gemeinsamen Agrarpolitik 2013–Ein Literaturüberblick zur Umsetzung und Effektivität der Greening-Maßnahmen. *Ber. Landwirtsch.* **2019**, *97*, 1–26. [CrossRef]
47. Federal Statistical Office. Statistisches Jahrbuch 2019. Kapitel 19 Land- und Forstwirtschaft. 2019. Available online: www.destatis.de/DE/Themen/Querschnitt/Jahrbuch/_inhalt.html (accessed on 12 October 2020).
48. Brown, C.; Kovács, E.; Herzon, I.; Villamyor-Tomas, S.; Albizua, A.; Galanki, A.; Grammatikopolu, I.; McCracken, D.; Alkan, J.O.; Zinngrebe, Y. Simplistic understandings of farmer motivations could undermine the environmental potential of the common agricultural policy. *Land Use Policy* **2021**, *101*, 105136. [CrossRef]

 *sustainability*

Article

Are Farms Located in Less-Favoured Areas Financially Sustainable? Empirical Evidence from Polish Farm Households

Radosław Pastusiak [1,*], Michał Soliwoda [1], Magdalena Jasiniak [1], Joanna Stawska [2] and Joanna Pawłowska-Tyszko [3]

1. Corporate Finance Department, Faculty of Economics and Sociology, University of Lodz, 90-213 Lodz, Poland; michal.soliwoda@uni.lodz.pl (M.S.); magdalena.jasiniak@uni.lodz.pl (M.J.)
2. Central Banking and Financial Intermediation Department, Faculty of Economics and Sociology, University of Lodz, 90-213 Lodz, Poland; joanna.stawska@uni.lodz.pl
3. Department of Farm Accountancy, Institute of Agricultural and Food Economics—NRI, 00-002 Warsaw, Poland; Joanna.Pawlowska-Tyszko@ierigz.waw.pl
* Correspondence: radoslaw.pastusiak@uni.lodz.pl

Abstract: The topic of farms that deal with environmental constraints is an ongoing agricultural policy issue, including within the Common Agricultural Policy. We propose empirical evidence based on a sample of Farm Accountancy Data Network (FADN) farm households, evaluate the influence of chosen factors on financially sustainable farm development and verify less-favoured area (LFA) farms' growth compared with non-LFA households. To specify farm households, we use the Sustainable Growth Challenge (SGC) model and DuPont decomposition based on financial measures and indicators that were adopted from corporate finance. It is concluded that the differences in SGC and revenue growth values between LFA and non-LFA farms mainly results from the system of subsidising LFA farms that receive compensation for farming in areas with adverse environmental conditions. Generally, the impact of agricultural policies on LFA and non-LFA farms is significant and may weaken the effect on LFA. With the exception of education, other sociodemographic factors do not highly influence farm efficiency. Along with improvements in the quality of human capital (e.g., higher education level), awareness of subsidies, and debt and innovative solutions increases. The interest in precision agriculture and agriculture 4.0 is also growing, which directly translates into better technological and financial efficiency of farms.

Keywords: less-favoured areas; sustainable agriculture; agricultural policy; farm profitability

Citation: Pastusiak, R.; Soliwoda, M.; Jasiniak, M.; Stawska, J.; Pawłowska-Tyszko, J. Are Farms Located in Less-Favoured Areas Financially Sustainable? Empirical Evidence from Polish Farm Households. *Sustainability* **2021**, *13*, 1092. https://doi.org/10.3390/su13031092

Academic Editor: Aitazaz A. Farooque

Received: 7 December 2020
Accepted: 8 January 2021
Published: 21 January 2021

Publisher's Note: MDPI stays neutral with regard to jurisdictional claims in published maps and institutional affiliations.

1. Introduction

EU Member States are required to provide a special subsidy to certain farmers to compensate for the disadvantages associated with the management of less-favoured areas (LFAs). This is designed to prevent the depopulation of rural areas and the loss of their agricultural character. In Poland, 58.7% of the utilised agricultural area (UAA) has been classified as restricted. Of this, 1.7% of the UAA is included in the first group and comprises mountain/highland areas with shorter vegetation periods due to the altitude and/or slope; the second group—47% of the UAA—includes areas other than mountain areas facing significant natural constraints, such as inadequate climatic conditions, low soil productivity or steep slopes (outside areas considered mountainous); the third group, which makes up 10% of the UAA, includes areas affected by specific constraints where land management is carried out and includes the protection or improvement of the environment and provision of appropriate landscape. In Poland, biophysical criteria, which relate directly to the properties of soils and slopes, are the most important; climatic factors are not very significant. Poland is characterised by a large share of agricultural areas with restrictions. In most cases, these restrictions result from the properties of soils [1] (p. 27).

The traditional concept of "sustainability" in agriculture is based on three pillars that include economic, social and environmental dimensions. This is highly related to the "triple bottom line" approach, which links economic, environmental and social aspects of sustainability. On the one hand, the perspective of farm sustainability and its sustainable development is based on the balance between farming goals, its operators and the environment. Therefore, a detailed analysis of the economic and financial condition of farm households should refer to the concept of "three pillars." On the other hand, the managerial approach in financial management underlines the concept of balance between farm growth and access to financing. The concepts of financial sustainability (including sustainable growth) at the microscale level (e.g., for enterprises—Higgins [2]; for farms—Escalante et al. [3,4], Mishra et al. [5,6]) integrates equity growth rate, financial leverage and growing sales revenues on agricultural products. This is very important in agriculture, which is considered one of the riskiest businesses. A high number of farm households, the price risk level and sensitivity to weather events and climate changes results in high income variability. This partially justifies public policies that comprise targeted support instruments (e.g., direct payments to agricultural sector operators), such as the Common Agricultural Policy (CAP) of the European Union (EU). The modern concept of sustainability includes the stakeholder interests and long-term growth objectives (e.g., the concept of optimal growth).

Although environmental constraints may negatively affect the economic and financial condition of farms located on LFAs, the Rural Development Programme (RDP) with LFA subsidies can weaken external business conditions. Monitoring the in-depth financial behaviour of farms located on LFAs may shed light on the financial sustainability of farm households with some peculiarities (compared with non-LFA farms). Furthermore, the nexus between sociodemographic characteristics of farms operators and financial sustainability of farm households is interesting from the perspective of rural polices that seek to improve human and social capital in rural areas (e.g., through courses for active farmers).

The objective of the paper is to verify whether LFA farms are financially sustainable and evaluate the influence of selected factors on financial behaviour. In addition, the paper aims to provide a comparison between the financial situation of LFA and non-LFA farms.

The following hypotheses are formulated:

Hypotheses 1 (H1). *The financial behaviour of LFA farms (analysed by the Sustainable Growth Challenge model) is significantly more sustainable.*

We treat the sustainable growth challenge (SGC) level as a proxy for farms' financial balance. The farm may be described as financially sustainable when its SGC level reaches 0. We assume that LFA farms face more environmental constraints (e.g., hills, lower soil quality) and limitations in production management that highly influence their total output, and consequently, their income and profitability.

Hypotheses 2 (H2). *The DuPont decomposition of LFA farms significantly differs from non-LFA farms; thus, there are different drivers of farm efficiency in LFA and non-LFA areas.*

The DuPont decomposition is expected to show the main drivers of financial efficiency. Given that LFA farms benefit from the CAP subsidies (in particular from the RDP, inter alia, LFA subsidies) they may achieve good financial results; however, their financial growth will depend not only on sales growth, as in the case of non-LFA farms, but also on other factors.

Hypotheses 3 (H3). *Sociodemographic characteristics of farm operators differentiate the SGC value.*

We assume that the SGC values of LFA farms are not only affected by their higher subsidy rate. Other factors may be significant to improving LFA farms' efficiency. We chose age, gender and the level of education of the farm manager as basic sociodemographic factors. This relates to previous findings, such as the impact of sociodemographic features on the financial situation of farms (mainly, profitability). The development of human capital in rural areas may improve the quality of financial management of farm households.

This paper significantly contributes to the empirical literature on the economics of farm households. The issue of farms that deal with environmental constraints is still a lively debate in agricultural policies, including the CAP. We propose empirical evidence based on a sample of Farm Accountancy Data Network (FADN) farm households. Our article may extend the scope of financial analysis of LFA farms, which play a significant role in Polish agriculture. Furthermore, at the sectoral level, identifying success factors for farm households may be important to designing development paths.

Section 1 of this article briefly presents the literature review focused on the issue of financial sustainability in agriculture. Section 2 describes FADN data and the methodology in our paper. Then, Section 3 presents our empirical findings and discusses the results. Our article concludes with final remarks and recommendations.

2. Financial Sustainability in Agriculture: Literature Review

2.1. Financial Sustainability

Sustainability is considered a highly interdisciplinary concept. For example, it may be understood in the context of economic activity, such as the idea of sustainable business development. In this context, some terminological problems may arise from the need to emphasise the processes for managing risk in the financial, social and environmental categories—that is, the concentration of activities on profit, people and planet [7]. Another terminological problem related to the definition of sustainability is the so-called "elasticities" of economic entities in dynamic categories, which means that economic entities can survive crises because they are associated with "healthy" economic, financial, social and environmental systems [8]. The sustainability of business entities is associated with criteria such as economic efficiency in the areas of innovation, well-being or productivity; human rights and social equity, such as sensitivity to poverty, local communities or respect for human rights; concern for the environment, including around climate change, land use and biodiversity [8].

In the context of agriculture, sustainability can be considered at various levels, ranging from a specific field, crop or other agricultural activity, and farms at local, regional, national, as well as continental and global levels. A set of sustainable development indicators has been developed with regards to sociological criteria (related to a farm), economic criteria (based on net farm income) and social criteria [9].

Escalante et al. [4] have emphasised that the sustainable growth paradigm (SGP) introduced in the late 70s by Higgins [2] plays an important role in linking production volumes (as a consequence of sales revenues) with farmers' financial decisions. In the theory of corporate finance, there is an apparatus and tools for quantifying this type of sustainability (in financial models). The concept of the sustainable growth rate (SGR), which indicates what an economic operator can afford without increasing leverage, is of key importance.

The SGR is useful when it comes to determining the sustainability rate. This is related to the principle that retained earnings should be the main source of new equity, and both the value of sales revenues and assets cannot grow faster than retained earnings plus additional debt [10] (p. 139), [11].

Based on research concerning agricultural sector operators in the USA, Mishra [6] has come to very important conclusions: the type of production, contraction and specialisation are important determinants of the capital structure—more precisely, of the ratio of assets to equity.

Escalante et al. [4] have noted that the planned growth in agriculture is mainly based on long-term forecasts. Current development is affected by the volatility of agri-food prices and the level of yields. The situation is unfavourable if the planned growth rate exceeds the sustainable growth rate because it becomes necessary to acquire external financing, such as loans and credits. In the opposite situation where planned growth rate is below the sustainable growth rate, assets are not fully used and resources are retained, usually in an unproductive way. US economists argue that farm income higher than expected is a source of risk because it leads to increased cash flow and higher working capital demand.

2.2. Less-Favoured Areas as a Sensitive Issue in Agricultural Policy

Areas strongly affected by natural handicaps, such as difficult climatic conditions, steep slopes in mountainous areas or low soil productivity are characterised as less-favoured areas. Farming in LFA is considered high risk.

The LFA schemes were included in the Rural Development Policies of 2007–2013 and 2014–2020. According to the Polish Ministry of Agriculture and Rural Development [12], "Payments for areas with natural or other specific restrictions (so-called LFA support) are a measure of the Rural Development Programme intended to facilitate farmers' continued agricultural use of the land and to enable them to maintain the landscape values of rural areas, as well maintain and promote sustainable systems of agricultural activity in areas with substandard natural conditions. As a result, this support is intended to increase the vitality of rural areas and help maintain biological diversity. LFA support takes the form of an area payment and parcels located in communes or districts designated according to strictly defined rules in EU regulations are eligible for aid."

According to Polish Ministry of Agriculture and Rural Areas [12], in 2018, new rules for delimitation of less-favoured areas (LFA) were announced. In the Polish context, the most important conditions are the biophysical criteria relating to soil properties. Under these new rules for Member States, when designating LFAs with natural constraints, it became necessary to exclude from support areas where natural constraints occur but have been overcome by intensification of production or production practices—the procedure of narrowing or "fine-tuning" the areas. After eliminating areas that have overcome natural constraints, the area of LFAs with natural constraints is 46.0% of agricultural land. In connection with the loss of the status of LFA with natural constraints by some areas and taking into account the spatial concentration of the effects of the new delimitation, measures were proposed to mitigate the effects of the new delimitation of the LFA: (i) the delimitation of a new category of a LFA specific type based on national criteria (unfavourable conditions of natural and touristic value); (ii) transitional payments to farmers who, as a result of the new delimitation, would lose support under the LFA lowland type; (iii) targeting part of the measures under RDP 2014–2020 on measures supporting administrative units (or regions) where there would be the greatest loss in LFA areas.

In 2019, new rules of delimitation of LFA areas were adopted, according to criteria established by the European Commission. The changes mainly concerned areas with natural constraints, i.e., the so-called lowland type LFAs (zone I and zone II). As a result, new lowland LFA areas were designated in our country: LFA zones with natural constraints I and II (representing, respectively, 28.5% and 18.5% of utilised agricultural areas, UAA in Poland) and additionally LFA type zone I—characterised by high natural value (7.0% of UAA in Poland). It should also be added that rules of delimitation of LFA areas, in particular zone II (related to piedmont and mountain areas (3.0% of the UAA in Poland) and zone II covering mainly mountain areas (1.7% of the UAA in Poland) were updated. Important criteria, including socioeconomic ones, were added (average farm size less than 7.5 ha; occurrence of soils threatened by water erosion; agricultural activity discontinued in at least 25% of the total number of farms; the share of permanent grassland in the agricultural land structure higher than 40%) [13].

The LFA transitional payment has been applied in Poland since 2019. It is a support payment for areas which, as a result of new delimitation criteria, have lost their lowland type LFA status (I or II). For such areas, there is a possibility of applying transitional support in 2019–2020 in the form of degressive payments for beneficiaries in areas that have lost their LFA status due to the new delimitation. In 2019, the support was the level of no more than 80% of the average LFA payment in RDP 2007–2013; in 2020 it was 25 EUR/ha. This support applies only to areas that have so far qualified as lowland type LFA I or II—not specific or mountain type LFA.

Regulation (EU) No 130: in 2019, new rules of delimitation of LFA areas were adopted, according to criteria established by the European Commission. The changes mainly concerned areas with natural constraints, i.e., the so-called lowland LFAs (Zone I and Zone II). As a result, new LFA areas in lowland areas were designated in our country: LFA zone with natural constraints I and II (representing, respectively, 28.5% and 18.5% of the UAA in Poland) and additionally LFA zone specific type zone I characterised by high natural value (7.0% of agricultural land, UAA in Poland). It should also be added that as part of work on a new delimitation of LFA areas, the LFA zone specific type zone II covering mainly submontane areas (3.0% of the UAA in Poland) and LFA zone specific type zone II covering mainly mountain areas (1.7% of the UAA in Poland) were updated. Very important criteria, including socioeconomic ones, should be added (average farm size is less than 7.5 ha; occurrence of soils threatened by water erosion; agricultural activity has been discontinued in at least 25% of the total number of farms; the share of permanent grassland in the agricultural land structure is higher than 40%). The 5/2013 of the European Parliament and of the Council on support for rural development by the European Agricultural Fund for Rural Development [14,15] will be in force temporarily until 2022. On 1 December 2020 [16], the Parliament's Agriculture and Rural Development Committee endorsed the agreement with the Council, including "the two-year duration of the transitional period, ending on 31 December 2022, and the extension of the multiannual rural development projects focused on environment and climate measures, and on organic farming".

Thus, environmental policies strongly highlight cofunding farming in LFAs, as well as promoting proper land management and protecting environment natural sources. As many studies show (e.g., Bigman [17], Fan et al. [18], Ruben et al. [19]), many categories are affected by environmental policies implemented on LFAs, including price and market policies, public service and investment, institutions and governance. To enhance the competitiveness of these areas and improve market access, main actions concentrate on investments in infrastructure, human capital and technology [20].

However, apart from a proper understanding of rural development issues, it is also essential to identify effective and profitable activities.

Environmental constraints affect farming in two main ways: they can increase costs and/or decrease profitability, significantly reducing the opportunity to intensify production both in situ and via land-use change. This also influences credit risk evaluation and household opportunities to develop [21].

However, it is also argued that despite the production disadvantages of LFAs in comparison with favoured areas, they may have also a comparative advantage in some types of agricultural production or non-farm activities. Alternative use of the labour force in these areas can make the production profitable. The varied situation in LFAs can allow them to use their different comparative advantages provided that the necessary investments in infrastructure and institutions are made. There is growing evidence to suggest that investments in LFAs can contribute to relatively high rates of return and to reduce poverty in some countries [22].

2.3. Measurement of Sustainability in Agriculture (Sectoral and Farm Household Level)

The category of growth, especially with regards to financial development in the case of family farms, is particularly ambiguous as a result of various measures to assess the size of these entities. It should be emphasised that in Poland, a farm's capital is important because of the dominant share (on average 80%) in its capital structure. The predominance of equity in the financing of the agricultural sector, as well as at the level of an individual farm, has the following implications: greater security of the agricultural sector functioning, due to lower risk of "own funds," but at the expense of a lower ability to create equity [23]. While there is a consensus around "(propogating) the idea of sustainable development," the numerous attempts to concretise it with various indicators of its measurement point to a considerable complexity in methodological foundations [24]. The emphasis is also on the paradigm of sustainable growth of farms (referring to the Higgins model from corporate finance) and the notion that the growth rate of sales of these farms should not change the ratio between equity and debt [4].

The sustainable growth rate (SGR) proposed by Higgins [2] determines the maximum rate of growth in company sales to avoid exhaustion of financial resources. His approach was related to the growth phase of companies when financial needs are more pressing. According to Higgins [2], when companies sell no new equity, have a target divided policy and want to maintain their capital structure, retained earnings create additional equity and enterprises can borrow sufficient financial sources to maintain their capital structure.

The concept of sustainable development combines the paradigm of sustainability with the theory of capital structure, both in the areas of corporate finance (see Modigliani and Miller [25], Myers [26], DeAngelo and Masulis [27], Jensen and Meckling [28]) and agriculture finance (see Barry et al. [29], Lagerkvist et al. [30]). Disappointment in the current neoclassical approach in finance, particularly following the global financial crisis beginning in 2007, as indicated by Stiglitz and Eggertsson's articles [31,32], led to an increased interest in the paradigm of sustainability in finance (see Rezende [33]). The SGC concept is used to understand the economic conditions and business decisions made by farmers [4].

When discussing issues related to the sustainable growth of farms, it is worth mentioning the risk associated with the financial management of farms. In the view of Escalante et al. [4] and Wauters et al. [34], the "risk balancing" (RB) hypothesis combines operational, financial and investment decisions of the farmer. It refers to the situation in which he aspires to an optimal level of total risk (TR), balancing economic risk (business risk, BR)—an inherent risk related to the market environment and natural factors—as well as financial risk (FR) as an additional risk resulting from debt financing. FR is linked to the level of BR by the leveraging effect and includes, for example, credit risk. RB behaviour means that policies lowering the level of BR may turn out to be ineffective, reducing the level of farm TR by increasing the leverage effect. The behaviour of RB also assumes that the farmer is characterised by risk aversion:

$$\alpha \leq TR = BR + FR = \frac{\sigma_{NOI}}{\mu_{NOI}} + \frac{\sigma_{NOI}}{\mu_{NOI}} \frac{I}{\mu_{NOI} - I} \leq \beta$$

where:

α is a minimum level of total risk; BR, a level of economic (business) risk; FR, the financial risk level; NOI, net operating income; I, liabilities arising from debt service; σ, μ, standard deviation (SD) and mean of variables, respectively; β, risk constraint.

Depending on the relationship between TR and β, one can distinguish several farm behavioural strategies against risk. In simple terms, RB behaviour occurs when β is constant, although the BR and FR levels are changed (in opposite directions).

The issue of growth, including the increase of one's own capital in agriculture, is related to the theory of the enterprise. While research on this subject is being developed abroad (see work by Viira [35]) there are significantly fewer studies regarding the small and medium-sized enterprises (SME sector) or family farming.

Balezentis and Novickyte [36] studied Lithuanian family farms' profitability and growth from 2005 to 2015 using aggregate data from the FADN database for different farming types and Lithuanian regions. They presented a sustainable growth ratio (SGR) and the SGC ratio to verify whether Lithuanian family farms' growth had been sustainable. Balezentis and Novickyte [36] found that Lithuanian family farms were characterised by negative profitability growth. Their analyses showed that farm operators should "exploit all internal resources, use cost control and improve the scale of operations." Particular attention should be paid to specialist cereal, oilseeds and protein crops, general field cropping and horticultural family farms, which are characterised by relatively higher profitability and growth compared with other farm types.

Viira et al. [35] investigated the impact of social and economic (including financial) factors on the probabilities of farm growth, decline and exit relative to the previous farm size. They based their findings on survey data and agricultural registers and employed multinomial logit estimation to build econometric models. The farm growth probability was highest in the 40–49 year age group. The farm operator's level of education increased farm growth sustainability. The availability of successors significantly reduced farm exit probability, and the level of education of the farm operator increased the farm growth probability. They found that the impact of off-farm work on farm growth was negative. The size of the farm was a significant determinant of remaining in business.

Figure 1 presents various determinants of farm growth based on a critical literature overview.

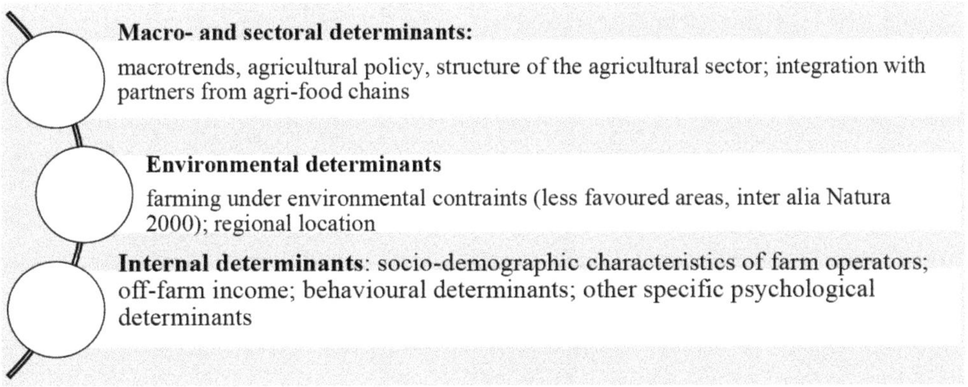

Macro- and sectoral determinants:

macrotrends, agricultural policy, structure of the agricultural sector; integration with partners from agri-food chains

Environmental determinants

farming under environmental contraints (less favoured areas, inter alia Natura 2000); regional location

Internal determinants: socio-demographic characteristics of farm operators; off-farm income; behavioural determinants; other specific psychological determinants

Figure 1. Determinants of farm growth—a conceptual synthesis. Source: Authors' own literature studies.

It is projected that the agricultural sector will have structurally lower yield growth rates, will experience substantial direct and indirect damages from climate change and may face additional costs from stringent economy-wide mitigation policies. Due to these multiple challenges, many governments are exploring policies that can stimulate sustainable agricultural productivity growth by exploiting synergies with mitigation and adaptation climate objectives [37].

3. Materials and Methods

3.1. Materials

The farm-level data for our analyses were collected by the Polish FADN:

- The variables/margins used are fully consistent with FADN Standard Results published annually by the Directorate-General for Agriculture and Rural Development of the European Commission.
- The FADN field of observation covers commercial holdings. In practice, the FADN field of observation covers farms producing at least 90% of the standard output value generated by all the farms in a given country (so-called commercial holdings).
- "Polish FADN farms sample is representative according to three grouping criteria: FADN region, economic size and type of farming. Currently, more than 12,000 farms deliver data for the Polish FADN survey" [38].

On average, during the period of 2010–2017, there were 11,699 individual farms in the FADN sample (Appendix A—Table A1). The number of farms was a statistically representative sample for the observation field of the Polish FADN, which averaged 733,000 commodity farms in Poland during that period. The number of farms in particular years was relatively stable, allowing for obtaining reliable and comparable results for long-term analyses. In the years 2010–2017 in Poland, 56% of the farms represented by FADN were located on LFA, including 55% in lowland areas and 1% in highland/mountainous areas. In the analysed period, the area of agricultural land of farms located in LFAs was comparable to the area used by farms located in favourable conditions. The much smaller area was used by farms located in less-favoured mountain areas. This area was 46% smaller than that of other farms. Furthermore, the area of land used for agriculture was comparable in all analysed groups of farms. The value of total production of LFA farms was about 8.4% and 59.4% higher than that obtained under unfavourable conditions in lowland and mountain farms, respectively. However, the indicator of total production value per 1 ha of UAA was slightly more favourable and there were no indications of disproportions between less favourable farms and farms with favourable farming conditions. The differences were 10.5% in relation to lowland farms and 32.2% in relation to the rest of LFA farms. In the analysed years, the value of production in favourable and disadvantaged areas in the years 2010–2012 increased; after 2012, it decreased until 2016. The situation was similar for the total output per 1 ha of UAA. The aforementioned results confirm that more difficult conditions reduce the income of both lowland and mountain areas. The economic impact of farm conditions is cumulative. Lowland and highland LFA farms received lower income per farm than units farming in favourable areas. This difference may result from the higher cost of fertilisers and plant protection products incurred by less-favoured farms, especially in highland areas, which are lower than in favourable farms, allowing for relatively high profitability.

3.2. Methods

We employed financial analysis methods for farm households: the SGC model and the DuPont decomposition base on financial measures and indicators that were adapted from corporate finance to the specificity of farm households.

The SGC model may be operationalised by financial indicators. SGC is regarded as the difference between the growth in sales (in our article: sustainable revenues (SRev)) and the sustainable growth rate (SGR) [2]. The sustainable growth relationship presents how increases in sales via increased productivity or marketing activities has to be managed. Balanced growth occurs when the percentage change in sales from one period to the next is equal to the SGR. If this happens, the value of SGC indicators is equal to 0, indicating that managers do not have to change the profit margin, asset turnover or leverage (3). There are two opposite situations:

- A positive SGC (targeted revenues increase faster than the SGR), which indicates that financial adjustments need to be made.
- A negative SGC (the SGR increases faster than the targeted revenues), which indicates that the utilisation of existing resources should be improved.

The *SGC* model [2,39] is also used to measure the disparity between real and sustainable growth rates, which is represented by the difference between sales or revenue growth and sustainable growth rates [3,4]:

$$SGC = \ln \frac{Revenue_t}{Revenue_{t-1}} - g$$

where *Revenue* represents the gross farm income (SE410) variable from FADN.

The growth model can be written as follows, considering accounting identity:

$$g = ROE \frac{Equity_{end}}{Equity_{beginning}}$$

where g is the sales growth rate; ROE; $Equity_{end}$, equity at the end of the period; $Equity_{beginning}$, equity at the beginning of the period; $Equity_{end}$ and $Equity_{beginning}$ represent the net worth (SE501) variable from FADN at both the end and beginning of the year.

The DuPont analysis is regarded as a useful financial technique used to decompose the different drivers of the return on equity (ROE)—a proxy for the financial efficiency of enterprises. The DuPont model has roots in corporate finance and is a useful tool for assessing the financial position of enterprises. This underlines the nexus between operating and financial performance. The DuPont model provides "the roadmap for business and managerial decision making" and information for farm businesses to analyse and make decisions The DuPont decomposition has several advantages at the farm-level management, including prediction of debt level. Mishra et al. (p. 325) [6] "used a financial approach based on the DuPont expansion to investigate the impact of demographics, specialisation, tenure, vertical integration, farm type, and regional location on the three levers of performance (ROE)—namely, net profit margins, asset turnover ratio, and asset-to-equity ratio." Furthermore, Tigner [40] regarded the DuPont system (analysis based on the DuPont model) as "a useful tool for farm/ranch managers analysing financial performance".

There is a plethora of empirical articles related to the application of the DuPont model that is based on decomposition of the ROE as a relatively good proxy for financial efficiency of farms (Melvin et al., [41]; Mishra et al. [6]; Nehring et al. [42]; Grashuis [43]). For example, Grashuis [43] decomposed ROE into five ratios related to efficiency, productivity and leverage.

We followed the methodological approach that was proposed by Balezentis, Novickyte and Namiotko [44] who presented the DuPont decomposition as:

- Profit margin (PM) = (Farm Net Income (SE 420)—Family Remuneration (PL FADN))/ Gross Farm Income (SE410)
- Asset Turnover (AT) = Gross Farm Income (SE410)/Total Assets (SE436)
- Equity Multiplier (EM) = Total Assets (SE436)/Net Worth (SE501)

To verify each individual hypothesis, we used statistical methods:

H1-H2: the Mann–Whitney U test to check whether two independent samples were drawn from the same population with the same distributions

H3: the Kruskal–Wallis test (H3), to see whether medians of more than two populations are different and the abovementioned Mann–Whitney U test.

4. Results

To verify farms' sustainable growth, we employed the SGC indicator. On average, farms located on non-LFA areas had a more favourable level of SGC during the period analysed than farms located on LFA (2.06 versus 4.36, respectively). Table 1 presents the analysis of the medians for SGC in the LFA and non-LFA areas; we note that the medians for SGC (closer to 0 indicating a more favourable situation) were recorded in 2011–2012 and 2016–2017 and the total in 2010–2017 in non-LFA. Medians for SGC were closer to 0 in LFAs only in 2010 and 2013 (Table 2). Moreover, in the years 2014–2015, the results are statistically insignificant. However, it should be emphasised that the median calculated in total for the years 2010–2017 turned out to be statistically significant and farms located in non-LFA areas show a more favourable SGC level, which confirms the H1 hypothesis. When verifying H1 for individual years, we consider that H1 is partially confirmed because in 2011–2012 and 2016–2017, non-LFA farms were characterised by levels of SGC that were more desirable in terms of financial sustainability, while in 2010 and 2013, farms in LFAs were more balanced in financial terms.

Thus, we can only partially confirm hypothesis H1: The financial behaviour of LFA farms (analysed by the SGC model) is significantly more sustainable. This results from the higher subsidy rate of LFA farms (based on RDP LFA payments), oriented as a public compensation for lower productivity in areas with environmental restrictions. Payment rates for farming in Polish LFA are described in Appendix A—Table A7. Subsidy rates (measured as the sum of subsidies to operating activities/crop production and animal output) in the research period were 18.8% and 26.6% for non-LFA and LFA farms, respectively. The amount of this sum of subsidies was higher than for LFA farms (1400,17 PLN per 1 ha of UAAs). RDP subsidies accounted for over 18.5% of operating subsidies (of which as much as 82% were LFA payments) in the case of LFA farms. Farms without environmental restrictions benefited less from RDP subsidies, representing only 5% of their sum of operating subsidies. All differences in subsidy rates, the amount of the subsidies to operating activities per hectare of UAAs and the share of RDP subsidies were statistically significant (p-values < 0.001). Additionally, the level of vertical and horizontal integration of the LFA farms was much lower compared to the other farms. This hinders the development of market relations within food chains.

Table 1. Medians of Sustainable Growth Challenge (SGC) for farms in Less Favoured Areas (LFA) and non—Less Favoured Areas (non-LFA).

Year	Group	N	Median	Year	Group	N	Median
2010	Non-LFA	4584	24.72	2015	Non-LFA	4854	−6.65
	LFA	5918	22.00		LFA	6111	−6.65
	Total	10,502	23.22		Total	10,965	−6.65
2011	Non-LFA	4325	8.27	2016	Non-LFA	4690	3.12
	LFA	5626	11.36		LFA	6073	8.69
	Total	9951	10.17		Total	10,763	6.32
2012	Non-LFA	4331	1.19	2017	Non-LFA	4533	12.54
	LFA	5591	−2.11		LFA	6097	15.41
	Total	9922	−0.93		Total	10,630	14.02
2013	Non-LFA	4456	−9.59	2010–2017	Non-LFA	36,690	3.78
	LFA	5744	−2.69		LFA	47,345	4.95
	Total	10,200	−5.49		Total	84,035	4.42
2014	Non-LFA	4917	−2.86				
	LFA	6185	−4.44				

Source: own computation based on Farm Accountancy Data Network (FADN) data [38].

Table 2. Sustainable Growth Challenge (SGC) of farms located in LFAs and non-LFAs—Mann–Whitney U test.

Specification	Mann–Whitney U-Test	
	z	Prob > $\vert z \vert$
2010	−3.046	0.0023
2011	−3.068	0.0022
2012	4.677	0.0000
2013	−6.791	0.0000
2014	1.407	0.1596
2015	−0.057	0.9544
2016	−6.668	0.0000
2017	−3.205	0.0014
2010–2017	−4.375	0.0000

Note: SGC was calculated only for the 2010–2017 period, which resulted from the specifics of this indicator; p-values < 0.05 were shaded. Source: own computation based on FADN data [38].

Analysing the results of the Mann–Whitney U test (Table 2), we note that in the 2010–2014 and 2016–2017 or the total in 2010–2017, farms located in LFAs and farms located in other areas (non-LFA) had differentiated SGC levels. These results were confirmed by the median test, wherein the period under study the years for which the test results are statistically significant to prevail, except for 2014–2015.

Using the following relationship: SGC = SRev − SGR, we notice in Figure 2 that over the period considered, SGC overlaps to the greatest extent with SRev; SGC fluctuates similarly to an increase in sales revenues (SGC and its components for LFA and non-LFA farm households are described in Appendix A—Table A8). SGR as a variable related to the ROE is more stable and subject to smaller fluctuations. It also seems that SGC for all farms is more similar to SGC for LFA farms. In the years when SRev assumed negative values, the SGC level also decreased with slight changes in SGR (i.e., changes in equity). This may mean that the agricultural sector in these periods of declines could have been experiencing falling prices of agricultural commodities and thus lower revenues [4].

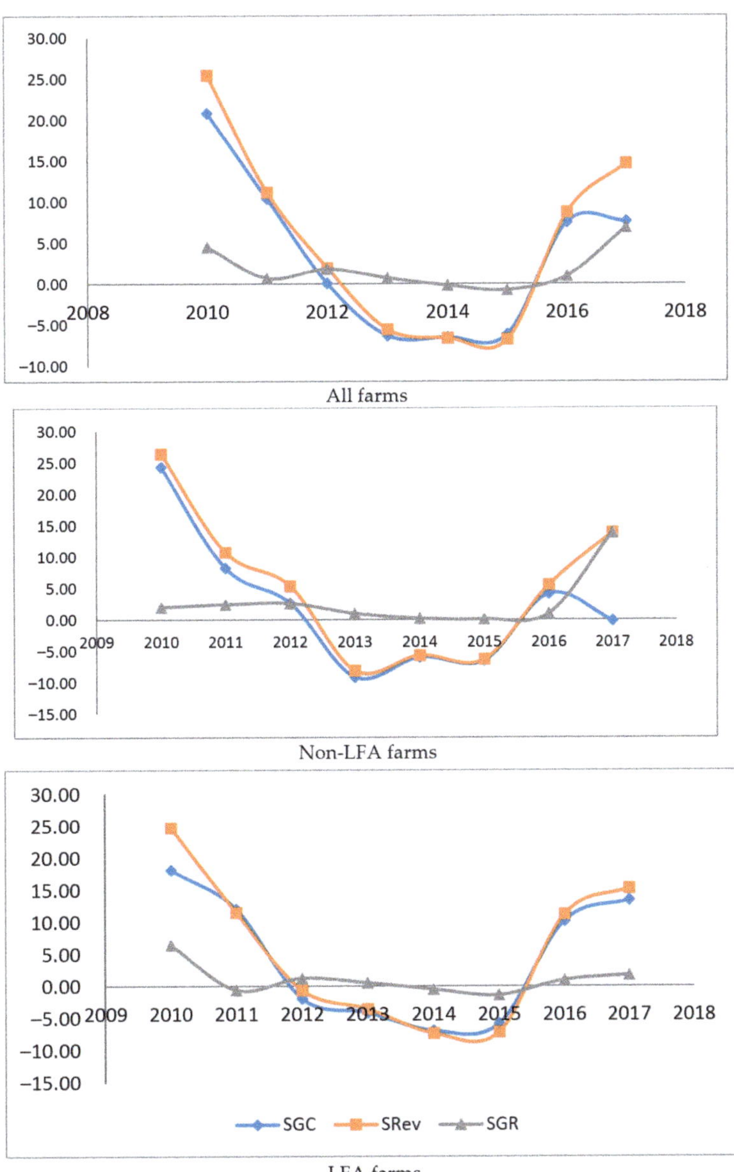

Figure 2. SGC and its components for non-LFA, LFA and all farms in 2010–2017. Source: own computation based on FADN data [38].

It should be added that the variability of SGC may be explained by boom/bust cycles in the agricultural sectors where the so-called price scissors effect may increase even after the 2004 expansion of the EU (Figure 3). As shown in Figure 3, the values of the agricultural price gap (index of the price relation of sold agricultural products to prices of goods and services purchased for current agricultural production and investments) were lower than 100 in the 2013–2015 period, in particular in 2014 (95.1).

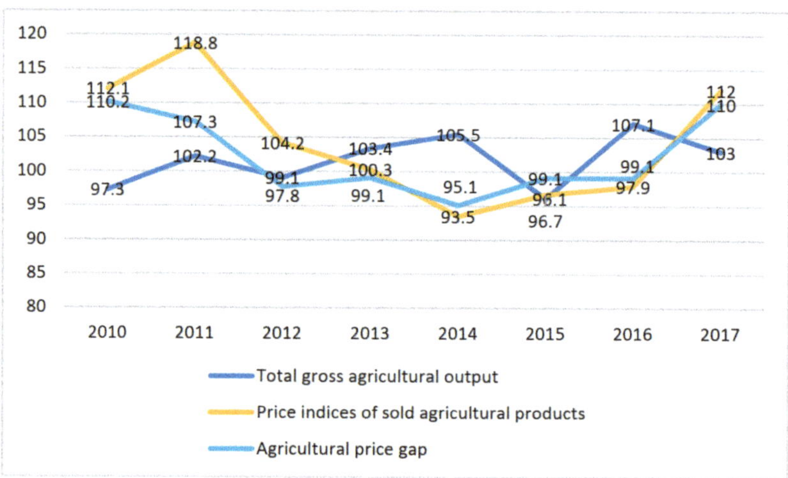

Figure 3. Agricultural boom/bust cycle in Poland. Note: Agricultural price gaps—index of the price relation of sold agricultural products to prices of goods and services purchased for current agricultural production and investments. Source: based on GUS (Central Statistical Office in Poland) [45].

As presented in Table 3, there are four patterns that can be distinguished by comparing their SGC components. Only in 2010 (non-LFA farms) and in 2011 (LFA farms) was the difference between SRev and SGR positive. In the subperiod 2012–2015, the values of SGC were negative as a result of the very low SGR. This shows problems of LFA farms with improvements in sales dynamics. It should be noted that the absolute value of SGC for LFA farms was close to zero in 2012. Nevertheless, the negative value of SGR indicates a higher sensitivity of LFA farms to market conditions, such as falling prices of agricultural products and increasing input costs (e.g., fuel, fertilisers). Taking into account the nature of the inequalities in the SGC model, several patterns could be identified: (1) SGC > SRev > SGR; (2) SRev > SGC > SGR; (3) SGR > SGC > SRev; (4) SGR > SGC > SRev. The same pattern was identified for both LFA and non-LFA farms only in 2013. We noted that in 2014 and 2015 the specific pattern (SGR > SGC > SRev) was only found on LFA farms. This means that ROE dynamics were slightly higher than total revenues dynamics in terms of the absolute values.

Results of the DuPont decomposition indicated that Polish farms generated positive incomes in the subperiods 2010, 2012, 2016–2017; values of ROE were positive in the aforementioned years. Table 4 presents descriptive statistics for components of DuPont expansion in the analysed period (presented in Appendix A—Table A2). This may be explained by external factors that determine the value of agricultural incomes. From the standpoint of agricultural policy, the complex assessment of ROE, its determinants and dynamics should be a rationale for changes in the agrarian structure. The higher rate of ROE for LFA farms may be attributed to mainly higher profit margins and asset turnover (AT). It should be added that the variability in profit margins may be explained by boom/bust cycles in the agricultural sector where the so-called price scissors effect may increase even after the 2004 expansion of the EU (Figure 3). As presented in Figure 3, the values of the agricultural price gap (an index of the price relation of sold agricultural products to prices of goods and services purchased for current agricultural production and investments) were lower than 100 in the 2013–2015 period, in particular in 2014 (95.1). As statistics show (see Table 4), the ROE of farm households located in LFAs is significantly lower than for farms located in areas without environmental handicaps (non-LFA). It should be added that the number of unprofitable farms in the case of LFAs is much higher, which is also confirmed by the median values. It can be observed that there is a short-term increase in farm productivity up to 2011—both in LFA and non-LFA farms. The tendencies are similar in both locations and are probably caused by external factors that are not strongly related to environmental handicaps.

Table 3. Patterns of relations between sustainable growth challenges (SGC), sustainable revenues (SRev) and the sustainable growth rate (SGR).

	SGC	SRev	SGR	Patterns
Non-LFA				
2010	24.36	26.42	1.99	SGC > Srev > SGR
2011	8.24	10.73	2.42	Srev > SGC > SGR
2012	2.64	5.33	2.58	Srev > SGC > SGR
2013	−9.13	−8.12	0.98	SGR > Srev > SGC
2014	−5.98	−5.70	0.19	SGR > Srev > SGC
2015	−6.49	−6.30	0.08	SGR > Srev > SGC
2016	4.09	5.50	0.96	Srev > SGC > SGR
2017	−0.16	13.85	13.84	Srev > SGC > SGR
LFA				
2010	18.15	24.72	6.48	Srev > SGC > SGR
2011	12.07	11.50	−0.60	SGC > Srev > SGR
2012	−1.91	−0.64	1.23	SGR > Srev > SGC
2013	−4.12	−3.56	0.52	SGR > Srev > SGC
2014	−6.93	−7.34	−0.51	SGR > SGC > Srev
2015	−5.82	−7.13	−1.40	SGR > SGC > Srev
2016	10.20	11.22	0.95	Srev > SGC > SGR
2017	13.45	15.26	1.74	SRev > SGC > SGR

Note: Negative values are in red. Means of indicators are presented.

It should be noted that non-LFA farms were less reluctant to use external financing, which was shown by lower values of the equity multiplier (e.g., in 2016 for LFA farm—1.059; non-LFA—1.047). This may be explained by the fact that more difficult environmental factors may decrease the creditworthiness of farm households.

Analysing the results of the Mann–Whitney U test (see Table 4), we note that in 2010–2017, location of farms in LFA significantly differentiates the financial efficiency of farms, as measured by the ROE and its drivers (excluding the EM in 2010 and AT in 2017) within the DuPont model expansion. Analysing median values in Table 5, we see that medians for non-LFA farms are higher than LFAs, which means that an LFA farm is less profitable than a non-LFA one. Based on the results presented in Table 4, we may confirm the validity of hypothesis H2 in Poland and state that the DuPont decomposition of LFA farms differs from non-LFA farms. As Balezentis et al. [44] (p. 13/15) state: "Decline in the profitability of Lithuanian family farms shows increasing extent with farm size." This means that the DuPont expansion should be analysed for particular classes according to economic size. Langemeier [46] has clearly explained that " … a small change in revenue or cost can have a significant impact on financial performance. Therefore (,) from managerial perspective (,) simulations on how an increase/decrease of the unit cost may affect financial performance are important for operationalisation of competitive financial strategy" [47].

Table 4. Components of DuPont expansion farms located on LFAs vs. farms located on non-LFAs (means and medians).

Year	Group	Statistics	Profit Margin	Asset Turnover	Equity Multiplayer	ROE
2010	Non-LFA	Mean	−0.137	0.118	1.061	0.95%
		Median	0.111	0.103	1.014	1.22%
	LFA	Mean	−0.179	0.112	1.061	0.62%
		Median	0.055	0.099	1.011	0.53%
	MWU	p	0.0000 ***	0.0000 ***	0.1669	0.0001 ***
2010	Non-LFA	Mean	−0.104	0.123	1.059	1.27%
		Median	0.145	0.109	1.012	1.60%
	LFA	Mean	−0.209	0.118	1.056	0.31%
		Median	0.052	0.105	1.006	0.56%
	MWU	p	0.0000 ***	0.0003 ***	0.0025 **	0.0000 ***
2012	Non-LFA	Mean	−0.113	0.123	1.062	1.39%
		Median	0.145	0.107	1.012	1.63%
	LFA	Mean	−0.265	0.113	1.058	0.17%
		Median	0.023	0.100	1.005	0.24%
	MWU	p	0.0000 ***	0.0000 ***	0.0082	0.0000 ***
2013	Non-LFA	Mean	−0.226	0.109	1.063	0.28%
		Median	0.066	0.094	1.008	0.66%
	LFA	Mean	−0.324	0.104	1.057	−0.24%
		Median	0.011	0.092	1.000	−0.11%
	MWU	p	0.0000 ***	0.0101 *	0.0003 ***	0.0000 ***
2014	Non-LFA	Mean	−0.356	0.102	1.062	−0.58%
		Median	0.023	0.089	1.007	−0.20%
	LFA	Mean	−0.450	0.097	1.055	−1.02%
		Median	0.106	0.086	1.000	−0.91%
	MWU	0.0000	0.0005 ***	0.0000 ***	0.0000 ***	0.0000
2015	Non-LFA	Mean	−0.369	0.099	1.063	−0.53%
		Median	0.025	0.082	1.004	−0.24%
	LFA	Mean	−0.582	0.092	1.053	−1.98%
		Median	0.214	0.078	1.000	−1.69%
	MWU	p	0.0000 ***	0.0000 ***	0.0000 ***	0.0000 ***
2016	Non-LFA	Mean	−0.204	0.101	1.059	0.61%
		Median	0.116	0.084	1.000	0.94%
	LFA	Mean	−0.271	0.096	1.047	0.14%
		Median	0.043	0.082	1.000	0.37%
	MWU	p	0.0000 ***	0.0046 **	0.0000 ***	0.0000 ***
2017	Non-LFA	Mean	−0.221	0.110	1.057	0.50%
		Median	0.084	0.093	1.000	0.76%
	LFA	Mean	−0.309	0.109	1.047	0.15%
		Median	0.042	0.095	1.000	0.39%
	MWU	p	0.001 ***	0.7008	0.0000 ***	0.040 *
2010–2017	Non-LFA	Mean	−0.220	0.110	1.061	0.46%
		Median	0.079	0.095	1.007	0.73%
	LFA	Mean	−0.327	0.105	1.054	−0.25%
		Median	0.013	0.092	1.000	−0.12%
			0.0000 ***	0.0000 ***	0.0000 ***	0.0000 ***

Note: Outliers were deleted. MWU, Mann–Whitney U test; significance levels are indicated by asterisks (* $p < 0.05$, ** $p < 0.01$, *** $p \leq 0.001$). Source: own computation based on FADN data [38].

Table 5. SGC vs. gender of farm operators—results of the Mann–Whitney U test.

Specification	SGC vs. Gender of Farm Operators		SGC vs. a Mobile Range of Age of Farm Operators		SGC vs. Agricultural-Oriented Profile of Educational Background of Farm Operators	
	LFA	Non-LFA	LFA	Non-LFA	LFA	Non-LFA
	Mann–Whitney U Test		Mann–Whitney U Test		Kruskal–Wallis Test	Kruskal–Wallis Test
	p	*p*	*p*	*p*	*p*	*p*
2010	0.8948	0.8553	0.9402	0.7780	0.0119	0.4671
2011	0.4437	0.8554	0.9042	0.6648	0.0002	0.0019
2012	0.0412	0.7572	0.0707	0.2208	0.7816	0.2892
2013	0.0171	0.0010	0.406	0.4809	0.0396	0.4548
2014	0.0964	0.5110	0.4952	0.1445	0.3250	0.5758
2015	0.0039	0.1074	0.004	0.0957	0.9418	0.4093
2016	0.2447	0.0094	0.5004	0.7303	0.0004	0.0511
2017	0.1896	0.5235	0.6737	0.9329	0.2850	0.0205
2010–2017	0.0429	0.0089	0.3951	0.0486	0.0001	0.0011

Source: own computation based on FADN data [38]. *p*-values < 0.05 were shaded.

In Table 5, we verified H3 hypothesis. We analysed the age, gender and education of the managing person and its influence on farm efficiency, as measured by the SGC factor (the main descriptive statistics for the aforementioned social demographic features are presented in the Appendix A—Tables A3–A6).

The gender of the managing person does not have a statistically significant influence on the differentiation of the SGC level for non-LFA and LFA farms. Based on descriptive statistics (Table A2), some trends can be noticed. In the analysed period, in non-LFA areas, more favourable values of the SGC index are achieved when the managing person is a man. For LFA sites, the difference between SGCs when managed by a woman or a man is not significant.

The age of the managing person was defined as mobile age (up to 44 years) and non-mobile age (over 44 years). This factor turned out to be statistically insignificant. However, looking at the descriptive statistics (Table A4), it can be noticed that in the case of non-LFA farms, the age of the managing person above 44 years has a positive effect on the farm's efficiency—the average SGC value in the analysed period is more favourable.

The factor differentiating SGC in non-LFA and LFA farms is the education of the managing person (specialised or other). In the case of farms located in non-LFA areas, agricultural education of the manager translates into better farm efficiency. Education influences effective management in the area of agricultural production and consequently also the financial results of the farm. In the case of farms located in the LFA areas, the education profile does not seem to be significant, as the SGC index has on average similar values in the analysed period (Table A3).

5. Discussion

By verifying H1, the behaviour of SGC indices and, indirectly, SR, Srev is significantly affected by factors connected with boom/bust cycles in agriculture (mainly, price scissors indices in agriculture). The observed slight differences in SGC and SRev values between LFA and non-LFA farms result from the system of subsidies to LFA farms—the compensation they receive for farming in areas with adverse environmental conditions. Generally, the impact of agricultural policies on LFA and non-LFA farms is noticeably significant and may weaken the effect on LFAs [4]. For example, the misbalanced structure of subsidies in LFAs and non-LFAs and between the 15 EU members and the Czech Republic has decreased the competitiveness of Czech agriculture in LFAs both among regions and compared with the agriculture of similar EU states [48]. Strengthening vertical and horizontal integration of the LFA farms within food chains may be important for the development of stronger market relations of these entities. The variability of SGC may be explained by boom/bust cycles in agricultural sectors where the price scissors effect may increase even after the

2004 expansion of the EU. LFA farms are strongly exposed to price risks. Nevertheless, price risk management in Polish agriculture (low interest in agricultural forwards and futures contracts, low market liquidity for this type of instruments on the Warsaw Stock Exchange) is not strongly developed.

By verifying H2, differences between values of the DuPont expansion indicators for LFA and non-LFA farms were statistically significant, with some exceptions. Asset turnover was higher in non-LFA farms. As Mishra, et al. (p. 60) [5] have stated, "low asset turnover ratios imply that the revenues generated from commercial agriculture are insufficient to justify the observed asset base" [5] (p. 60). This also relates to the case of LFA farms who suffer from overcapitalisation and the dominant role of farmland. It should be noted that the price of agricultural land on LFAs is slightly lower than non-LFAs because of the presence of natural environmental constraints (e.g., hills in the southern part of Poland). Nevertheless, the farmland asset base is relatively stable. One important recommendation for increasing asset turnover is to benefit from intermediate equipment that may be an important driver of the flexibility of farms [5] (p. 60). Furthermore, particular attention should be paid to strategies on how to increase profit margins on sales. This may be explained by the fact that selected regions in Poland with a dominant role of LFAs (e.g., the Małopolska and Podkarpackie voivodships/regions) may have historically weaker connections with the food industry. Conversely, a part of commodity-oriented farms in Wielkopolskie or Kujawsko-pomorskie voivodeships have been strongly integrated as the members of agricultural cooperatives (e.g., dairy cooperatives) or producer groups. The profit margin of LFA farms may be improved by a significant reduction in cost production, including through technological/marketing/business models innovations and vertical integration). Higher activity of agricultural extension may be helpful for the transfer of technological innovations to farms. The DuPont expansion should be analysed for particular classes according to economic size (see Balezentis et al. [44]) (p. 13/15). Furthermore, farmers should also include financial peculiarities related to the type of production (according to TF8 in FADN typology), which results in differences in needs for foreign capital or the length of the cash conversion cycle. Simulations on how an increase/decrease of the unit cost may affect financial performance are important at the farm level for operationalisation of a competitive financial strategy [47].

According to conclusions from H3 analysis, it is noticed that most factors are statistically insignificant.

Many studies highlighted the gender of farm operators as an important factor in farming. It has been argued that women have lower access to human capital, land and other assets that would allow them to be more efficient and enterprising [49]. Nevertheless, Gasson and Winter [50] have highlighted that women's independence and proactive nature may increase women's independent earnings and work experience, which influence their involvement in running their farm. Gender of farm operator does not significantly influence SGC values. However, it should be noticed that in Polish farm households, the gender of the managing person is very convenient because the decision-making model, especially in family farms, is collective, both in terms of gender and age of family members.

The age (mobile or not) of the managing person also does not statistically affect SGC. However, Gale [51] and Gardebroek et al. [52] found a negative relationship between age and farm growth. Other studies indicate that farm growth is less likely in the younger and older age groups of farm operators [53–56].

As a rule, farm managers of a mobile age, are characterised by greater flexibility in decision-making and less risk-taking, as well as by abandoning traditional agriculture. Younger farmers use innovations more often, both in terms of production technology and financing sources, which translates into the effectiveness of the farm. In the case of Polish farms, we rarely deal with a situation where a managing person is a young person in a mobile age, but this does not mean that they do not influence management. In the case of people of mobile age, there is a risk of abandoning farming activities and limiting involvement in farm work. Here, the influence of managers above the mobile age may be more important.

Level of education is a factor that often differentiates SGC values between LFA and non-LFA farms. The statistical significance of this factor was confirmed for a few selected years. As numerous studies suggest, education influences farm efficiency [57,58]. Weiss [59] found that households with a lower level of human capital (education, broader work experience) more often select a cooperative type of farming. In addition, it was shown that human capital may increase the earning capacity of a farm operator in the non-farm economy. This may improve farm survival if the operator were to put this income into a household and support agricultural production.

A higher level of education also increases awareness of subsidies from the second pillar of the Common Agricultural Policy and applications for financing from foreign capital; in addition, the effectiveness of financing with these funds increases. Moreover, the higher the education, the greater the interest in implementing innovative solutions for managing the networking capital on the farm. The interest in precision agriculture and agriculture 4.0 is also growing, which directly translates into better technological and financial efficiency of farms. The more specialised the education of a farmer, the deeper his or her ties with the links in the food chain, which generates opportunities to use innovations offered by suppliers of the means of production.

The level of education also affects the structure of financing sources of an agricultural holding. More specialised education, in particular in LFAs, is associated with a greater share of nonagricultural activities as an alternative source of agricultural income.

6. Conclusions

The financial behaviour of LFA farms was only partly significantly sustainable, as indicated by the value of the SGC of farm households. Higher subsidy rates of LFA farms (based on RDP LFA payments) may disturb the financial balance of these entities. One of the important recommendations is to strengthen the degree of vertical and horizontal integration of LFAs. This hinders the development of market relations within food chains.

It should be noted that since 2019, changes in LFA delimitation on a macro scale have been slight. Introducing a new classification (specific zones I and II) on the basis of local criteria resulted in the fact that the LFA areas covered both submontane/piedmont areas and other areas with difficult conditions not meeting the EU criteria. Some criteria (from previous delimitations) were very restrictive. Until 2018, criteria for lowland LFA areas (both for types I and II) and the lack of distinction of areas with high natural values, which favour landscape preservation and environmental biodiversity conducive to sustainable agriculture, were problematic. Such a division resulted from characteristics of the category of LFAs (including their natural and socioeconomic features). As a result, there were LFA areas with conditions that allowed for relatively stable agricultural production that did not require support. Unfortunately, the area of holdings covered by LFAs was significant. Considering such a large area qualified as LFA, the number of eligible farms should be reduced by setting boundary conditions (related to both production system and farm location). The payment per hectare could be multiplied by applying the maximum rates laid down in the EU regulations. The effect of changing the method of allocating support under the LFA would be the implementation of the objectives of the RDP, i.e., improving landscape preservation and environmental biodiversity [13,60].

The answer to the question whether the criteria selection should be more restrictive in order that the additional support is really targeted to those farms which really deserve it is affirmative. From Rural Development Programs perspective (RDP 2007-13 and RDP 2014–2020), aid could be targeted only to groups of farms requiring support. Moreover, the support mechanism could be changed so as to exclude farms not requiring subsidies from it. Currently, fulfilment of the eligibility criteria by land located on an agricultural plot is essential. In Poland, the main criteria included unfavourable soil structure and stoniness. The registered part of agricultural parcel ('obręb ewidencyjny') was classified as LFA when unfavourable conditions were found for 60% of the agricultural land in a given administrative unit. Thus, even 40% of agricultural land that did not meet the aforesaid criteria was eligible for LFA payments. Theoretically, support can be limited to agricultural parcels located on the registered parcels that meet LFA criteria. This would exclude land of higher production quality. This process would, however, be very costly at the level of LFA delimitation, monitoring of LFA areas, because the land parcels are divided or merged. Eligibility of LFA subsidies from a WTO perspective is controversial: LFA payments are included in the so-called Green Box, i.e., payments that do not affect production competitiveness. According to many experts, LFA area coverage should be more adapted to the implementation of environmental objectives, landscape conservation and promotion of traditional, environmentally sustainable agriculture [13,60].

We showed that differences between values of DuPont expansion indicators for LFA and non-LFA farms were statistically significant, with some exceptions. Asset turnover was higher in non-LFA farms. From the standpoint of agricultural policy, the complex assessment of the ROE, its determinants and dynamics should be a rationale for changes in the agrarian structure. The higher rate of ROE for LFA farms may be attributed to mainly higher profit margins and asset turnover. Additionally, the variability of profit margins may be explained by boom/bust cycles in the agricultural sector

Typical demographic factors do not significantly affect the SGC value, which results mainly from the specific characteristics of these farms. A significant part of farm households in Poland are family businesses, where all family members are involved in the process of managing agricultural production and its financing. In addition, agricultural communities are strongly integrated, hence some decisions result from the so-called effect of infection. This means that some decisions made on a given farm affect the management of neighbouring farms that readapt certain decisions in the areas of applied technologies or financing structures. This effect may appear with some delay, especially in the case of introducing innovative solutions, where imitation intensifies when a given technology becomes effective.

The observed trends in the value of the SGC index in the context of selected sociodemographic factors in non-LFA and LFA areas indicate that in the case of LFAs, the mean SGC values are similar, regardless of gender, age and education. Therefore, other factors determine the financial efficiency of these farms. It can be expected that the financial performance of LFA farms is largely due to the subsidies received. At the same time, the possibilities of farming in these areas are limited by their environment, which require the use of specific production models. Management decisions, including in the financial area, are also limited.

There are some important limitations to our study. First, we did not include behavioural determinants of farm profitability. Second, there are some factors related to planned and realised financial strategy that affect financial conditions of farms.

Further research should include the combination of both quantitative and qualitative factors that may determine proxies for the financial sustainability (e.g., sustainable growth challenges) of farm households. Survey data, including data based on questions on behavioural heuristics and biases, attitudes towards risk (i.e., a risk aversion assessment), may be incorporated into enhanced econometric models or statistical analyses.

Author Contributions: Conceptualization, R.P., M.S., M.J. and J.S.; methodology, R.P., M.S., M.J. and J.S.; formal analysis, R.P., M.S., M.J. and J.S.; investigation, R.P., M.S., M.J., J.S. and J.P.-T.; resources, J.P.-T.; data curation, M.S. and J.P.-T.; writing, R.P., M.S., M.J., J.P.-T. and J.S.; writing—review and editing, R.P., M.S., M.J. and J.S.; visualisation, M.S., M.J., J.S. and J.P.-T.; supervision, R.P. All authors have read and agreed to the published version of the manuscript.

Funding: This research received no external funding.

Institutional Review Board Statement: Not applicable.

Informed Consent Statement: Not applicable.

Data Availability Statement: Not applicable.

Conflicts of Interest: The authors declare no conflict of interest.

Appendix A

Table A1. The basic description of the FADN sample and farms located on LFAs.

Specification	The observation field of the PL FADN (n) *	The FADN sample (n) *	The number of farms located on LFA,	Lowlands (2)	Highlands (3)
2010	738,035	11,004	6208	6071	137
2011	738,038	10,890	6139	6019	120
2012	738,055	10,909	6126	6008	118
2013	730,905	12,117	6831	6736	95
2014	730,861	12,123	6784	6673	111
2015	730,895	12,105	6826	6719	107
2016	730,762	12,104	6905	6794	111
2017	730,904	12,103	6858	6756	102
Average for 2010–2017	733,557	11,669	6585	6472	113

Specification	UAA ** (hectares)	Total output (PLN)	Family Farm Income (w PLN)	Total output per 1 ha of UAA (PLN)	Family Farm Income per 1 ha of UAA (PLN)
2010	35.29	213,172.16	90,331.13	6040.15	2559.5
No-LFA farms (1)	34.9	226,037.57	94,318.72	6476	2702.24
Lowlands (2)	35.86	205,133.27	87,884.86	5720.39	2450.78
Highlands (3)	23.75	119,023.61	59,140.1	5011.04	2489.87
2017	35.04	246,963.87	58,868.57	7047.24	1679.84
No- LFA farms (1)	36.23	261,395.66	61,856.91	7214.26	1707.19
Lowlands (2)	34.27	237,570.03	56,755.86	6931.82	1656.02
Highlands (3)	25.01	127,061.59	45,138.91	5080.53	1804.87
Average no-LFA farms in years 2010–2017	36.02	255,036.41	97,901.24	7080.41	2717.97
Average of LFA farms—lowlands	35.53	235,220.42	89,446.41	6620.33	2517.49
Average of LFA farms—highlands	24.64	159,931.87	64,148.37	6490.74	2603.42

Note: * the number of farms, ** UAA, Utilised agricultural area. Source: own computation based on FADN data [38].

Table A2. Components of the DuPont expansion, descriptive statistics for the research period.

Specification	N	Mean	SD	Min	Median	Max
Profit Margin	117,644	2.448	978.0289	−2216.971	0.010	335,386.300
Asset Turnover	117,644	0.111	0.177	−1.043	0.091	41.064
Equity Multiplayer	117,644	1.082	2.642	−155.498	1.003	610.470
ROE_Du Ponta	117,644	0.005	1.674	−96.440	0.000	564.053

Source: own computation based on FADN data [37].

Table A3. SGC value in terms of the gender of farm operators—LFA vs. non-LFA farms, main descriptive statistics.

		Non-LFA				LFA			
Year	Groups	N	Mean	SD	Median	N	Mean	SD	Median
2010	Female	616	24.58	60.84	25.80	735	24.17	61.10	21.03
	Male	3968	24.32	51.59	24.57	5183	17.29	397.79	22.17
	Total	4584	24.36	52.92	24.72	5918	18.15	372.89	22.00
2011	Female	562	6.60	60.42	8.83	713	10.58	50.19	12.64
	Male	3763	8.48	49.87	8.24	4913	12.28	168.35	11.18
	Total	4325	8.24	51.36	8.27	5626	12.07	158.33	11.36
2012	Female	575	3.64	49.85	0.86	722	1.17	52.58	0.48
	Male	3756	2.49	50.19	1.22	4869	−2.36	48.15	−2.36
	Total	4331	2.64	50.14	1.19	5591	−1.91	48.75	−2.11
2013	Female	602	−6.15	53.36	−1.93	778	0.18	49.02	1.09
	Male	3854	−9.60	50.09	−10.65	4966	−4.80	50.23	−3.35
	Total	4456	−9.13	50.56	−9.59	5744	−4.12	50.10	−2.69
2014	Female	667	−4.83	58.21	−2.41	841	−10.38	60.65	−7.00
	Male	4250	−6.16	58.26	−2.89	5344	−6.39	56.85	−4.14
	Total	4917	−5.98	58.25	−2.86	6185	−6.93	57.40	−4.44
2015	Female	640	−5.26	73.30	−4.77	827	−1.28	66.93	−1.95
	Male	4214	−6.68	59.88	−7.05	5284	−6.54	63.77	−7.29
	Total	4854	−6.49	61.81	−6.65	6111	−5.82	64.23	−6.65
2016	Female	608	12.16	67.65	6.23	840	14.11	99.58	10.06
	Male	4082	2.89	61.49	2.72	5233	9.57	63.13	8.43
	Total	4690	4.09	62.39	3.12	6073	10.20	69.33	8.69
2017	Female	600	9.82	59.53	12.34	855	6.23	162.77	13.71
	Male	3933	−1.68	892.58	12.58	5242	14.63	57.13	15.62
	Total	4533	−0.16	831.69	12.54	6097	13.45	80.78	15.41
2010–2017	Female	4870	4.92	61.71	4.98	6311	5.34	86.08	6.42
	Male	31.820	1.62	318.11	3.57	41.034	4.21	160.99	4.79
	Total	36.690	2.06	297.10	3.78	47.345	4.36	153.14	4.95

Source: own computation based on FADN data [38].

Table A4. SGC value in terms of agricultural-oriented profile of the educational background of farm operators—LFA vs. non-LFA farms, main descriptive statistics.

		Non-LFA				LFA			
Year	Groups	N	Mean	SD	Median	N	Mean	SD	Median
2010	Unprofiled	1687	26.50	54.76	26.29	2455	10.91	575.81	21.76
	Profiled	2897	23.11	51.79	23.78	3463	23.28	50.66	22.21
	Total	4584	24.36	52.92	24.72	5918	18.15	372.89	22.00
2011	Unprofiled	1553	9.08	53.38	10.33	2322	15.57	240.17	11.91
	Profiled	2772	7.76	50.20	6.79	3304	9.61	46.31	11.04
	Total	4325	8.24	51.36	8.27	5626	12.07	158.33	11.36
2012	Unprofiled	1546	2.95	54.99	1.57	2282	−2.51	51.25	−1.62
	Profiled	2785	2.47	47.24	1.11	3309	−1.49	46.96	−2.39
	Total	4331	2.64	50.14	1.19	5591	−1.91	48.75	−2.11
2013	Unprofiled	1618	−9.32	51.68	−9.40	2350	−2.32	50.54	−1.00
	Profiled	2838	−9.02	49.91	−9.85	3394	−5.37	49.76	−3.82
	Total	4456	−9.13	50.56	−9.59	5744	−4.12	50.10	−2.69
2014	Unprofiled	1795	−4.90	55.12	−1.77	2536	−8.48	58.50	−5.07
	Profiled	3122	−6.60	59.97	−3.65	3649	−5.85	56.60	−3.91
	Total	4917	−5.98	58.25	−2.86	6185	−6.93	57.40	−4.44

Table A4. *Cont.*

Year	Groups	Non-LFA				LFA			
		N	Mean	SD	Median	N	Mean	SD	Median
2015	Unprofiled	1737	−4.99	66.11	−3.68	2506	−2.77	61.89	−5.43
	Profiled	3117	−7.33	59.27	−8.24	3605	−7.95	65.73	−7.81
	Total	4854	−6.49	61.81	−6.65	6111	−5.82	64.23	−6.65
2016	Unprofiled	1671	5.74	65.34	2.79	2471	12.17	56.48	9.95
	Profiled	3019	3.18	60.69	3.41	3602	8.85	76.89	7.66
	Total	4690	4.09	62.39	3.12	6073	10.20	69.33	8.69
2017	Unprofiled	1573	11.91	65.06	13.49	2444	14.27	58.66	16.21
	Profiled	2960	−6.57	1028.13	12.09	3653	12.91	92.69	14.73
	Total	4533	−0.16	831.69	12.54	6097	13.45	80.78	15.41
2010–2017	Unprofiled	13,180	4.49	59.63	4.81	19.366	4.56	226.75	5.80
	Profiled	23,510	0.70	368.45	3.15	27.979	4.22	64.00	4.35
	Total	36,690	2.06	297.10	3.78	47.345	4.36	153.14	4.95

Source: own computation based on FADN data [38].

Table A5. SGC value in terms of mobile age range of farm operators—LFA vs. non-LFA farms, main descriptive statistics.

	AGE	NonLFA				LFA			
		N	Mean	SD	Median	N	Mean	SD	Median
2010	Mobile	2088	24.97	52.58	25.32	2812	12.62	538.34	21.61
	Immobile	2496	23.85	53.21	24.36	3106	23.15	50.59	22.52
	Total	4584	24.36	52.92	24.72	5918	18.15	372.89	22.00
2011	Mobile	1893	8.63	50.75	8.93	2575	14.92	228.96	10.46
	Immobile	2432	7.93	51.84	8.06	3051	9.66	44.48	11.93
	Total	4325	8.24	51.36	8.27	5626	12.07	158.33	11.36
2012	Mobile	1872	2.07	49.26	0.04	2545	−3.37	51.60	−2.55
	Immobile	2459	3.08	50.80	2.08	3046	−0.68	46.22	−1.78
	Total	4331	2.64	50.14	1.19	5591	−1.91	48.75	−2.11
2013	Mobile	1864	−8.16	48.74	−10.16	2521	−4.34	51.63	−2.79
	Immobile	2592	−9.83	51.82	−9.18	3223	−3.95	48.87	−2.62
	Total	4456	−9.13	50.56	−9.59	5744	−4.12	50.10	−2.69
2014	Mobile	2073	−4.84	56.03	−1.98	2696	−6.08	57.43	−4.16
	Immobile	2844	−6.81	59.81	−3.19	3489	−7.58	57.37	−4.76
	Total	4917	−5.98	58.25	−2.86	6185	−6.93	57.40	−4.44
2015	Mobile	1998	−5.05	62.05	−6.08	2621	−3.38	65.19	−4.76
	Immobile	2856	−7.49	61.63	−7.41	3490	−7.66	63.45	−8.02
	Total	4854	−6.49	61.81	−6.65	6111	−5.82	64.23	−6.65
2016	Mobile	1931	4.09	63.75	2.72	2583	10.56	76.91	7.85
	Immobile	2759	4.10	61.44	3.38	3490	9.93	63.14	9.12
	Total	4690	4.09	62.39	3.12	6073	10.20	69.33	8.69
2017	Mobile	1910	12.45	56.05	12.19	2666	12.02	105.12	15.03
	Immobile	2623	−9.34	1092.29	12.85	3431	14.57	54.86	15.92
	Total	4533	−0.16	831.69	12.54	6097	13.45	80.78	15.41
2010–2017	Mobile	15,629	4.40	56.19	4.03	21.019	4.21	221.30	4.84
	Immobile	21,061	0.33	389.13	3.64	26.326	4.48	55.43	5.04
	Total	36,690	2.06	297.10	3.78	47.345	4.36	153.14	4.95

Source: own computation based on FADN data [38].

Table A6. SGC value in terms of educational background of farm operators—LFA vs. non-LFA farms.

		NonLFA				LFA			
	Education	N	Mean	SD	p50	N	Mean	SD	p50
2010	Primary	2095	25.36	53.57	25.08	3084	23.20	48.44	22.37
	Secondary	1978	22.90	51.64	24.42	2337	11.76	590.25	22.75
	Higher	511	25.92	55.07	25.84	497	16.81	55.04	17.31
	Total	4584	24.36	52.92	24.72	5918	18.15	372.89	22.00
2011	Primary	1946	10.37	48.57	10.06	2915	12.47	44.87	13.18
	Secondary	1891	7.52	52.79	7.85	2226	12.79	245.26	9.31
	Higher	488	2.47	55.98	4.13	485	6.38	51.64	8.11
	Total	4325	8.24	51.36	8.27	5626	12.07	158.33	11.36
2012	Primary	1911	2.87	49.05	1.64	2828	−2.09	44.92	−2.03
	Secondary	1933	1.72	51.03	0.19	2248	−1.59	50.71	−2.47
	Higher	487	5.40	50.77	3.95	515	−2.25	59.25	−0.39
	Total	4331	2.64	50.14	1.19	5591	−1.91	48.75	−2.11
2013	Primary	1919	−9.32	48.68	−9.81	2850	−3.24	49.77	−1.36
	Secondary	2018	−8.42	52.86	−8.75	2331	−5.15	49.83	−4.81
	Higher	519	−11.20	48.17	−11.67	563	−4.34	52.74	−2.22
	Total	4456	−9.13	50.56	−9.59	5744	−4.12	50.10	−2.69
2014	Primary	2026	−7.33	63.72	−3.11	3007	−5.12	54.90	−4.42
	Secondary	2277	−4.49	54.39	−2.44	2536	−7.95	57.96	−4.36
	Higher	614	−7.05	52.89	−3.51	642	−11.36	65.74	−5.35
	Total	4917	−5.98	58.25	−2.86	6185	−6.93	57.40	−4.44
2015	Primary	1981	−6.62	62.97	−6.18	2944	−6.24	61.33	−6.32
	Secondary	2242	−7.45	61.82	−7.92	2524	−5.84	64.75	−6.68
	Higher	631	−2.64	57.91	−5.52	643	−3.87	74.42	−8.51
	Total	4854	−6.49	61.81	−6.65	6111	−5.82	64.23	−6.65
2016	Primary	1867	6.14	57.28	4.19	2848	13.54	70.34	11.19
	Secondary	2195	3.34	64.69	3.02	2563	7.54	68.36	6.89
	Higher	628	0.65	68.40	0.33	662	6.15	68.07	3.88
	Total	4690	4.09	62.39	3.12	6073	10.20	69.33	8.69
2017	Primary	1780	13.17	45.39	12.22	2765	13.28	98.13	16.31
	Secondary	2122	−13.43	1214.44	13.40	2620	14.08	57.15	14.87
	Higher	631	6.87	59.74	7.66	712	11.83	80.50	13.34
	Total	4533	−0.16	831.69	12.54	6097	13.45	80.78	15.41
2010–2017	Primary	15,525	4.37	55.38	4.55	23.241	5.78	61.98	5.97
	Secondary	16,656	−0.13	436.67	3.57	19.385	3.18	227.17	4.06
	Higher	4509	2.22	57.69	1.53	4719	2.24	65.96	3.10
	Total	36,690	2.06	297.10	3.78	47.345	4.36	153.14	4.95

Source: own computation based on FADN data [38].

Table A7. Payment rates for farming in LFAs.

The payment rates for management in LFAs is calculated by type of area:

- LFA payment for mountain areas (mountain type)
  - 320 PLN/ha for beneficiaries continuing the 5-year LFA commitment undertaken under RDP 2007–2013
  - 450 PLN/ha for other beneficiaries
- Payment for lowland areas (lowland type LFA):
  - Lowland LFA type I: 179 PLN/ha
  - Lowland LFA type II: 264 PLN/ha
- Payment for specific areas (LFA specific type): 264 PLN/ha.

LFA payment is due to the area of agricultural land owned by a farmer on 31 May 2017, amounting to no more than 75 ha, and in cases of a 5-year commitment, amounting to no more than 300 ha.

The support within the framework of the measure, as before, will be granted in the form of an annual payment granted per hectare of agricultural land, which is a product of the rate established for a given type of LFA and the number of hectares declared by the farmer.

LFA payments will be subject to a degressive rate based on the total area of agricultural parcels or parts of them covered by the aid. Depending on this area, the payment will be granted as follows:

- from 1 to 25 ha: 100% payment
- from 25.01 to 50 ha: 50% payment
- from 50.01 to 75 ha: 25% of payments
- over 75 ha: payment will not be granted

Source: based on ARiMR (The Agency for Restructuring and Modernisation of Agriculture).

Table A8. Patterns of relations between sustainable growth challenges (SGC), sustainable revenues (SRev) and the sustainable growth rate (SGR) for LFA and non-LFA farm households.

Year	Group		SGC	SRev	SGR	Year	SGC	SRev	SGR
2010	Non-LFA	Mean	24.36	26.42	1.99	2014	−5.98	−5.70	0.19
		Median	24.72	26.88	1.41		2.86	−1.23	−0.19
	LFA	Mean	18.15	24.72	6.48		−6.93	−7.34	−0.51
		Median	22.00	23.80	0.76		4.44	−3.46	−0.98
	Total	Mean	20.86	25.46	4.52		−6.51	−6.61	−0.20
		Median	23.22	25.27	1.02		3.71	−2.51	−0.64
2011	Non-LFA	Mean	8.24	10.73	2.42	2015	−6.49	−6.30	0.08
		Median	8.27	11.71	2.02		6.65	−6.22	−0.16
	LFA	Mean	12.07	11.50	−0.60		−5.82	−7.13	−1.40
		Median	11.36	13.58	0.86		6.65	−7.07	−1.65
	Total	Mean	10.40	11.17	0.71		−6.12	−6.76	−0.75
		Median	10.17	12.79	1.34		6.65	−6.67	−0.96
2012	Non-LFA	Mean	2.64	5.33	2.58	2016	4.09	5.50	0.96
		Median	1.19	5.05	2.08		3.12	5.69	1.19
	LFA	Mean	−1.91	−0.64	1.23		10.20	11.22	0.95
		Median	2.11	−0.53	0.53		8.69	10.41	0.72
	Total	Mean	0.08	1.97	1.82		7.54	8.73	0.95
		Median	0.93	1.53	1.22		6.32	8.10	0.91
2013	Non-LFA	Mean	−9.13	−8.12	0.98	2017	−0.16	13.85	13.84
		Median	9.59	−7.13	0.81		12.54	14.62	1.04
	LFA	Mean	−4.12	−3.56	0.52		13.45	15.26	1.74
		Median	2.69	−1.36	0.07		15.41	17.18	0.63
	Total	Mean	−6.31	−5.56	0.72		7.65	14.66	6.89
		Median	5.49	−4.02	0.42		14.02	16.08	0.78
2014	Non-LFA	Mean	−5.98	−5.70	0.19	2010_2017	2.06	5.02	2.82
		Median	2.86	−1.23	−0.19		3.78	5.91	0.91
	LFA	Mean	−6.93	−7.34	−0.51		4.36	5.48	1.04
		Median	4.44	−3.46	−0.98		4.95	6.48	0.05
	Total	Mean	−6.51	−6.61	−0.20		3.36	5.28	1.82
		Median	3.71	−2.51	−0.64		4.42	6.23	0.46

Note: SRev = $\ln \frac{Revenue_t}{Revenue_{t-1}}$, SGR = ROE $\frac{Equity_{end}}{Equity_{beginning}}$. Source: own computation based on FADN data [38].

References

1. *Diagnoza sektora rolno-spożywczego i obszarów wiejskich w Polsce przygotowana dla potrzeb opracowania Krajowego Planu Strategicznego 2021–2027 (eng. Diagnosis of the agri-food sector and rural areas in Poland prepared for National Strategic Plan 2021–2027)*; Ministry of Agriculture and Rural Development in Poland: Warsaw, Poland, 2019.
2. Higgins, R.C. How Much Growth Can a Firm Afford? *Financ. Manag.* **1977**, *6*, 7–16. [CrossRef]
3. Escalante, C.; Turvey, C.; Barry, J.P. Farm-level evidence on the sustainable growth paradigm from grain and livestock farms. In Proceedings of the International Association of Agricultural Economists Conference, Gold Coast, Australia, 12–18 August 2006.
4. Escalante, C.; Turvey, C.; Barry, P.J. Farm business decisions and the sustainable growth challenge paradigm. *Agric. Financ. Rev.* **2009**, *69*, 228–247. [CrossRef]
5. Mishra, A.K.; Moss, C.B.; Erickson, K.W. Regional differences in agricultural profitability, government payments, and farmland values: Implications of DuPont expansion. *Agric. Financ. Rev.* **2009**, *69*, 49–66. [CrossRef]
6. Mishra, A.K.; Harris, J.M.; Erickson, K.W.; Hallahan, C.; Detre, J.D. Drivers of agricultural profitability in the USA: An application of the Du Pont expansion method 2012. *Agric. Financ. Rev.* **2012**, *72*, 325–340. [CrossRef]
7. Scerri, A.; James, P. Accounting for sustainability: Combining qualitative and quantitative research in developing 'indicators' of sustainability. *Int. J. Soc. Res. Methodol.* **2010**, *13*, 41–53. [CrossRef]
8. Definition of business sustainability. *Finacial Times*. Available online: http://lexicon.ft.com/Term?term=business-sustainability (accessed on 11 January 2015).
9. Baum, R. Sustainable development of agriculture and its assessment criteria. *J. Agribus. Rural Dev.* **2008**, *7*, 5–15.
10. Bodie, Z.; Merton, R.C. *Finanse*; PWE: Warsaw, Poland, 2003.
11. Moss, C. *Agricultural Finance*; Routledge: New York, NY, USA, 2013.
12. Delimitacja ONW Według Nowych Zasad UE (LFA Delimitation According to New EU Rules). Available online: https://www.gov.pl/web/rolnictwo/delimitacja-onw-wedlug-nowych-zasad-ue (accessed on 23 November 2018).
13. Płatności Dla Obszarów Z Ograniczeniami Naturalnymi Lub Innymi Szczególnymi Ograniczeniami (Płatność ONW) W Roku 2020-Podstawowe Informacje (Payments for Areas Facing Natural or Other Specific Constraints (LFA Payment) in 2020-Basic Information). Available online: https://www.arimr.gov.pl/pomoc-unijna/prow-2014-2020/dzialanie-13-platnosci-dla-obszarow-z-ograniczeniami-naturalnymi-lub-innymi-szczegolnymi-ograniczeniami-tzw-platnosc-onw-podstawowe-informacje/platnosci-onw-2020.html (accessed on 5 January 2021).
14. Regulation (EU) No 1305/2013 of the European Parliament and of the Council on Support for Rural Development by the European Agricultural Fund for Rural Development. Available online: https://www.legislation.gov.uk/eur/2013/1305/contents (accessed on 20 August 2020).
15. Proposal for a Regulation of the European Parliament and of the Council Laying Down Certain Transitional Provisions for the Support by the European Agricultural Fund for Rural Development (EAFRD) and by the European Agricultural Guarantee Fund (EAGF) in the Year 2021 and Amending Regulations (EU) No 228/2013, (EU) No 229/2013 and (EU) No 1308/2013 as Regards Resources and Their Distribution in Respect of the Year 2021 and Amending Regulations (EU) No 1305/2013, (EU) No 1306/2013 and (EU) No 1307/2013 as Regards Their Resources and Application in the Year 2021. Available online: https://www.europarl.europa.eu/doceo/document/A-9-2020-0101_EN.html (accessed on 20 August 2020).
16. CAP Transitional Rules for 2021 and 2022, AT A GLANCE, Plenary 2020. Available online: https://www.europarl.europa.eu/thinktank/en/document.html?reference=EPRS_ATA(2020)659387 (accessed on 20 August 2020).
17. Bigman, D. *Globalization and the Developing Countries. Emerging Strategies for Rural Development and Poverty Alleviation*; CAB International and ISNAR: Wallingford, UK, 2002.
18. Fan, S.; Chan-Kang, C. Returns to investment in less-favored areas in developing countries: A synthesis of evidence and implications for Africa. *Food Policy* **2004**, *29*, 431–444. [CrossRef]
19. Ruben, R.; Kuyvenhoven, A.; Hazell, P. Investing in poor people in less-favored areas: Institutions, technologies, and policies for poverty alleviation and sustainable resource use. In Proceedings of the International Conference on Staying Poor: Chronic Poverty and Development Policy, Manchester, UK, 7–9 April 2003; University of Manchester: Manchester, UK, 2003.
20. Kuyvenhoven, A. Creating an enabling environment: Policy conditions for less-favored areas. *Food Policy* **2004**, *29*, 407–429. [CrossRef]
21. Journeaux, P. The effect of environmental constraints on land prices. In Proceedings of the 2016 Conference, New Zealand Agricultural and Resource Economics Society, Nelson, New Zealand, 25–26 August 2016; AgFirst Waikato Ltd.: Hamilton, New Zealand, 2016.
22. Pender, J.; Hazell, P. Promoting sustainable development in less-favored areas. In *Focus 4*; International Food Policy Research Institute: Washington, WA, USA, 2000.
23. Kulawik, J. Rozwój finansowy a wzrost i rozwój ekonomiczny w rolnictwie. *Studia i Monografie* **1997**, *83*.
24. Latruffe, L.; Diazabakana, A.; Bockstaller, C.; Desjeux, Y.; Finn, J. Measurement of sustainability in agriculture: A review of indicators. *Stud. Agric. Econ.* **2016**, *118*, 123–130. [CrossRef]
25. Modigliani, F.; Miller, M.H. The cost of capital, corporation finance and the theory of investment. *Am. Econ. Rev.* **1958**, *48*, 261–297.

26. Myers, S.C. The capital structure puzzle. *J. Financ.* **1984**, *39*, 575–592. [CrossRef]
27. DeAngelo, H.; Masulis, R.W. Optimal capital structure under corporate and personal taxation. *J. Financ. Econ.* **1980**, *9*, 3–30. [CrossRef]
28. Jensen, M.C.; Meckling, W.H. Theory of the firm: Managerial behavior, agency costs and ownership structure. *J. Financ. Econ.* **1976**, *3*, 305–360. [CrossRef]
29. Barry, P.J.; Ellinger, P.N.; Hopkin, J.A.; Baker, C.B. *Financial Management in Agriculture*, 6th ed.; Interstate Publishers: Danville, CA, USA, 2000.
30. Lagerkvist, C.J.; Larsen, K.; Olson, K.D. Off–farm income and farm capital accumulation: A farm—Level analysis. *Agric. Financ. Rev.* **2007**, *67*, 241–257. [CrossRef]
31. Stiglitz, J.E. *The Global Financial Crisis Has Made Capital Account Management More Important than Ever*; Global Markets: London, UK, 2013.
32. Eggertsson, G.B.; Krugman, P. Debt, deleveraging, and the liquidity trap: A Fisher-Minsky-Koo approach. *Q. J. Econ.* **2012**, *127*, 1469–1513. [CrossRef]
33. Carolina Rezende de Carvalho Ferreira, M.; Amorim Sobreiro, V.; Kimura, H.; de Moraes Barboza, F.L. Asystematic review of literature about finance and sustainability. *J. Sustain. Financ. Invest.* **2016**, *6*, 112–147. [CrossRef]
34. Wauters, E.; De Mey, Y.; Van Winsen, F.; Lauwers, L. Farm household risk balancing: Implications for policy from an EU perspective. *Agric. Financ. Rev.* **2015**, *75*. [CrossRef]
35. Virra, A.-H.; Põder, A.; Värnik, R. The determinants of farm growth, decline and exit in estonia. *Ger. J. Agric. Econ.* **2013**, *62*, 52–64.
36. Balezentis, T.; Novickyte, L. Are lithuanian family farms profitable and financially sustainable? Evidence using DuPont model, sustainable growth paradigm and index decomposition analysis. *Transform. Bus. Econ.* **2018**, *17*, 237–254.
37. Lankoski, J.; Lehtonen, H.; Ollikainen, M.; Myyrä, S. Modelling policy coherence between adaptation, mitigation and agricultural productivity. In *OECD Food, Agriculture and Fisheries Papers 2018*; OECD Publishing: Paris, France, 2018.
38. FADN. Polish FADN Standard Results. Available online: https://fadn.pl/en/publications/standard-results/ (accessed on 20 August 2020).
39. Higgins, R.C. *Analysis for Financial Management*, 10th ed.; McGraw-Hill: New York, NY, USA, 2012.
40. Tigner, R. *Using the DuPont System to Increase Farm/Ranch Profitability*; Institute of Agriculture and Natural Resources: Lincoln, NE, USA, 2019.
41. Melvin, J.; Boehlje, M.; Dobbins, C.; Gray, A. The Dupont profitability analysis model: An application and evaluation of an e–learning tool. *Agric. Financ. Rev.* **2004**, *64*, 75–89. [CrossRef]
42. Nehring, R.; Gillespie, J.; Hallahan, C.; Michael Harris, J.; Erickson, K. What is driving economic and financial success of US cow-calf operations? *Agric. Financ. Rev.* **2014**, *74*, 311–325. [CrossRef]
43. Grashuis, J. A quantile regression analysis of farmer cooperative performance. *Agric. Financ. Rev.* **2018**, *78*, 65–82. [CrossRef]
44. Baležentis, T.; Namiotko, V.; Novickytė, L. Lithuanian family farm profitability: The economic dimension of sustainability: Scientific Study 2018. *Lith. Inst. Agrar. Econ.* **2018**, *104*. [CrossRef]
45. GUS. *Socio-Economic Situation of the Country in 2019*; Central Statistical Office: Warsaw, Poland, 2019.
46. Langemeier, M.R. Financial performance configurations. *Int. J. Agric. Manag.* **2011**, *1*, 1–3.
47. Shadbolt, N. Competitive strategy analysis of NZ pastoral dairy farming systems. *Int. J. Agric. Manag.* **2012**, *1*, 19–27.
48. Střeleček, F.; Lososová, J.; Zdeněk, R. Different farming conditions of agricultural holdings in the LFA and non-LFA. *J. Cent. Eur. Agric.* **2011**, *12*, 409–424. [CrossRef]
49. Blackden, C.M.; Bhanu, C. Gender, growth, and poverty reduction: Special program of assistance for Africa. *World Bank Technical Paper* **1999**, *428*. [CrossRef]
50. Gasson, R.; Winter, M. Gender relations and farm household pluriactivity. *J. Rural Stud.* **1992**, *8*, 387–397. [CrossRef]
51. Gale, H.F., Jr. Longitudinal analysis of farm size over the farmer's life cycle. *Rev. Agric. Econ.* **1994**, *16*, 113–123. [CrossRef]
52. Gardebroek, C.; Turi, K.N.; Wijnands, J.H. Growth dynamics of dairy processing firms in the European Union. *Agric. Econ.* **2010**, *41*, 285–291. [CrossRef]
53. Glauben, T.; Tietje, H.; Weiss, C.R. Intergenerational successionon family farms: Evidence from survey. *FE Work. Paper* **2002**, *0202*, 1–19.
54. Calus, M.; Van Huylenbroeck, G. The succession effect within management decisions of family farms. In Proceedings of the International Congress, Ghent, Belgium, 26–29 August 2008; European Association of Agricultural Economists: Ghent, Belgium, 2008.
55. Calus, M.; Van Huylenbroeck, G.; Van Lierde, D. The relationship between farm succession and farm assets on Belgian farms. *Sociol. Rural.* **2008**, *48*, 38–56. [CrossRef]
56. Väre, M. Spousal effect and timing of retirement. *J. Agric. Econ.* **2006**, *57*, 65–80. [CrossRef]
57. Rizov, M.; Mathijs, E. Farm survival and growth in transition economies: Theory and empirical evidence from Hungary. *Post-Communist Econ.* **2003**, *15*, 227–242. [CrossRef]
58. Breustedt, G.; Glauben, T. Driving forces behind exiting from farming in Western Europe. *J. Agric. Econ.* **2007**, *58*, 115–127. [CrossRef]

59. Weiss, C.R. Farm growth and survival: Econometric evidence for individual farms in Upper Austria. *Am. J. Agric. Econ.* **1999**, *81*, 103–116. [CrossRef]
60. Czapiewski, K.; Niewęgłowska, G.; Stolbova, M. *Obszary O Niekorzystnym Gospodarowaniu W Rolnictwie. Stan Obecny I Wnioski Na Przyszłość (Disadvantaged Areas in Agriculture. Current State of Play and Lessons to be Learned)*; Niewęgłowska, G., Ed.; Institute of Agricultural and Food Economics, National Research Institute: Warsaw, Poland, 2008.

 *sustainability*

Article

Predicting Behavioral Intention of Rural Inhabitants toward Economic Incentive for Deforestation in Gilgit-Baltistan, Pakistan

Saif Ullah [1], Ali Abid [1], Waqas Aslam [1], Rana Shahzad Noor [2], Muhammad Mohsin Waqas [3] and Tian Gang [1,*]

1   College of Economics and Management, Northeast Forestry University, Harbin 150040, China; ranasaif2014@outlook.com (S.U.); ali.abid@uaar.edu.pk (A.A.); aslam_waqas219@yahoo.com (W.A.)
2   Department of Agriculture, Biological, Environment and Energy Engineering, College of Engineering, Northeast Agricultural University, Harbin 150030, China; engr.rsnoor@uaar.edu.pk
3   Department of Agricultural Engineering, Khwaja Fareed University of Engineering and Information Technology, Rahim Yar Khan 64200, Pakistan; mohsinwaqas333@gmail.com
*   Correspondence: tiangang0451@nefu.edu.cn; Tel.: +86-189-4605-9819

Abstract: The conservation of forest in the northern areas of Pakistan is the major priority of the national environmental policy to fight against global warming. Despite the policy for the protection of forest, rural residents' behavior toward economic incentives for deforestation may undermine their conservation goals. Therefore, the purpose of this study was to understand the factors that affect the illegal behaviors related to deforestation in the northern areas of Pakistan. The present study applied the socio-psychological theory of planned behavior to predict the behavioral intention of rural residents toward economic incentives for deforestation. Correlations were explored between background factors toward motivations for deforestation based on positive and negative views through open-ended questions. Attitude and descriptive norm were found good predictors to perceive the behaviors. The findings of the study suggest that rural communities' support for compliance with policies is vital for the long-term efficacy and protection of the forest in the region. Further, change in the behaviors of inhabitants toward the ecosystem through training can be improved to manage the forest.

Keywords: forest conservation; forest management; rural residents; economic incentives; Pakistan

Citation: Ullah, S.; Abid, A.; Aslam, W.; Noor, R.S.; Waqas, M.M.; Gang, T. Predicting Behavioral Intention of Rural Inhabitants toward Economic Incentive for Deforestation in Gilgit-Baltistan, Pakistan. *Sustainability* **2021**, *13*, 617. https://doi.org/10.3390/su13020617

Received: 1 December 2020
Accepted: 6 January 2021
Published: 11 January 2021

Publisher's Note: MDPI stays neutral with regard to jurisdictional claims in published maps and institutional affiliations.

1. Introduction

Globally, forests have been receiving ever-growing attention to fight against global warming. Forest not only conserves biodiversity but also provides necessary ecosystem services to society. Nevertheless, recently several studies have demonstrated that restrictions on the use of natural resources have negative behavior among the rural residents who rely on the forest for their livelihood, which creates a lot of hurdles for the management of forests [1–5]. The mountainous rural area resident's major source of income is from the forest resources; therefore, the economic incentive from forests directs rural residents toward deforestation [6–12]. Therefore, comprehensive attention is needed to understand the relationship between rural residents, economic incentives, and forest conservation.

Scientific literature has explored the relationship between rural residents and deforestation [13–16]. Such literature has helped to identify the preferences and beliefs of residents on conservation issues [17]. Therefore, the evaluation of plausible relationships with intentions is important to predict the actions that influence and change the behavior of rural communities [18–21]. In the past, many scholars have suggested incorporating the beliefs of rural residents toward the conservation of natural resources in global forest policy [22–25].

According to the National Forest Policy of Pakistan 2015, the lawmakers have taken steps to manage forests and improve public awareness of the ecological and cultural values of forests [26–29]. Nevertheless, forests have been weakly preserved in recent years and it is a challenging task in northern areas of Pakistan to apply strict rules of preservation due to the high reliance of peoples on the natural resources of forests [30–32]. In the scarcely facilitated area of Gilgit-Baltistan, resident depends on natural resources for their livelihoods [32,33]. According to Reference [32], porters cut trees for cooking, walking sticks, and fuel. This is because people have no source of income in the region, and they used natural resources to fulfill their basic needs; therefore, policies regarding the conservation of forests adversely affect the behavior of rural inhabitants. In the northern areas of Pakistan, rural resident's relationships with the forest authorities and continuous illegal activities to use fuel-wood are of particular concern for deforestation. This is due to the existence of strong (informal) links between the authorities, influential groups, and timber mafia [34,35]. The increasing level of illegal activities also negatively affects the rest of the areas. Deforestation in the developed countries is considered due to expansion in agricultural land [36–42]. While in the case of developing countries like Africa, deforestation is continued for subsistence agriculture farming and wood production for local markets [43,44]. Agriculture and charcoal production are the main causes of deforestation in Tanzania [45] and fuel-wood in Senegal [46]. In Turkey, the main causes of deforestation are rapid urbanization and industrialization [47].

Previous studies related to deforestation in Pakistan are mainly focused on timber production, conversion to agriculture, roadbuilding, and human-caused fire [48]. According to the studies of Ullah et al. [14,49] and Ali et al. [14,49], deforestation in northern areas is due to the construction of roads and high population growth in the last few years. However, to the best knowledge of authors, no research has so far been conducted to examine the behavior of rural residents toward deforestation, especially the study related to the economic motivation of rural residents. Therefore, the purpose of this study is to investigate the intention of people toward economic incentives to control deforestation and help the government institutes to make effective policies for the livelihoods of rural residents as for as law enforcement agencies.

To perceive human behavior, Ajzen's theory of planned behavior, an extension of the theory of reasoned action, has been used [50]. Nonetheless, few studies on forest biodiversity have utilized such frameworks to analyze multiple predictors of behavior. [18]. According to the theory, the behavioral intentions arise from an individual's attitude (ATT), norms, and Planned Behavior Constructs (PBC), which can further be predicted by actual behavior in question [50]. Furthermore, a number of background factors including socioeconomic, knowledge, education, and past experience of the individual can influence the ATT, Normative (N), and PBC. [51]. Contextual aspects such as rules and legislation of government may often interfere and evaluate a behavior [52]. Attitude is defined as the extent to which an individual has a favorable or unfavorable view of a particular behavior. Descriptive normative (DN) is defined as the opinions of the people rather than what approve or disapprove by others [18,53]. PBC is the perception of how an individual feels ease or difficulty to perform a specific behavior [50,54,55]. Understanding behavioral intent toward, and factors that affect, illegal behaviors can help managers emphasize their actions to enhance people's compliance with laws and protect forest within those areas.

Therefore, our study investigated the ATT, DN, PBC, and Behavior Intention (BI) of the rural inhabitants toward their economic incentive for deforestation. By applying the theory of planned behavior, the purpose of the present study was to (a) identify background factors that may influence the intention of rural residents; (b) identify inhabitants' attitudes, descriptive norms, perceived behavioral control toward economic incentives for deforestation; and (c) identify rural inhabitant's illegal behavior toward the economic incentives for deforestation.

2. Theoretical Model

2.1. Theory of Planned Behavior

Many contemporary studies have used intentions as a key component to understand the behavior in question [50,56–60]. In psychology, the theory of planned behavior is an attempt to shape an individual's behavioral intentions with a combination of three factors: attitudes toward the behavior, norms, and perceived behavioral control [51,61,62] as given in Figure 1.

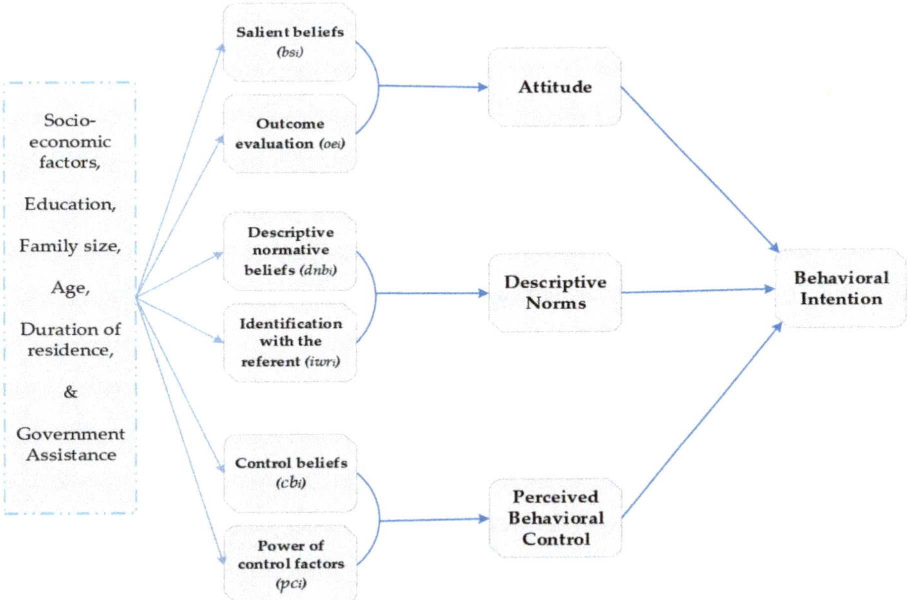

Figure 1. Factors affecting the Behavior Intention: Theory of Planned Behavior Model [50].

2.2. Expectancy-Value Model

The theory of planned behavior follows an expectancy-value model to predict the behavior of individual under question [50,51,63,64]. The beliefs-based measures are probably regarded to get a more accurate prediction of cognitive intention than its direct predictors alone.

The attitude is comprised of (silent beliefs -bs_i) and evaluation of the outcomes (oe_i) and it can be obtained according to the following formula:

$$ATT \propto \sum bs_i \, oe_i \tag{1}$$

Strength of behavioral belief (bs_i) is described as a possibility that can produce a particular outcome by performing a behavior (i) and the outcome evaluation (oe_i) can be termed as the utility obtained if the result (i) occurs [50,51,63,64].

The descriptive norm (DN) describes the whereas descriptive norms refer to perceptions that others are or are not performing the behavior. Normative beliefs can be explored by assessing a person's identification with the referent (iwr_i), multiplying the measures of descriptive normative beliefs (dnb_i) regarding given referents by the corresponding identity measures, and then summing the normative belief by identity products [65,66]. A belief-based measure of the descriptive norm (DN) can be obtained according to the following formula:

$$DN \propto \sum dnb_i \, iwr_i \tag{2}$$

Strength of descriptive norms is formed by considering multiple descriptive normative beliefs (dnb_i), or beliefs that behavior is normative for peers and individuals we look up to in social groups [65,66].

"Perceived behavioral control—This refers to a person's perception of the ease or difficulty to perform the behavior of interest. It consisted of personal control beliefs (cb_i) and the perceived strength of these specific control factors to facilitate or impede actions (power to affect—pc_i) [50,67,68].

$$PBC \propto \sum cb_i \, pc_i \tag{3}$$

The strength of each control belief (cb_i) is weighted by the perceived power (pc_i) of the control factor to perform a specific behavior [50,51].

Behavioral intention (BI) refers to "a person's subjective probability that he will perform some behavior". BI is the function of three antecedents, namely, attitude, norms, and PBC. By incorporating the belief-based measures, BI can be calculated according to the following formula:

$$BI = \beta_1 ATT + \beta_2 DN + \beta_3 PBC + \varepsilon \tag{4}$$

β_1, β_2, and β_3 are the coefficients to evaluate each component, and (ε) is an error term.

3. Materials and Methods

3.1. Study Area

The present study was conducted in the three districts (Skardu, Gilgit, and Astore) of Gilgit-Baltistan, Pakistan (Figure 2), where rural residents not only use forest resources to meet their livelihood but also generate substantial cash income through trade in forest products.

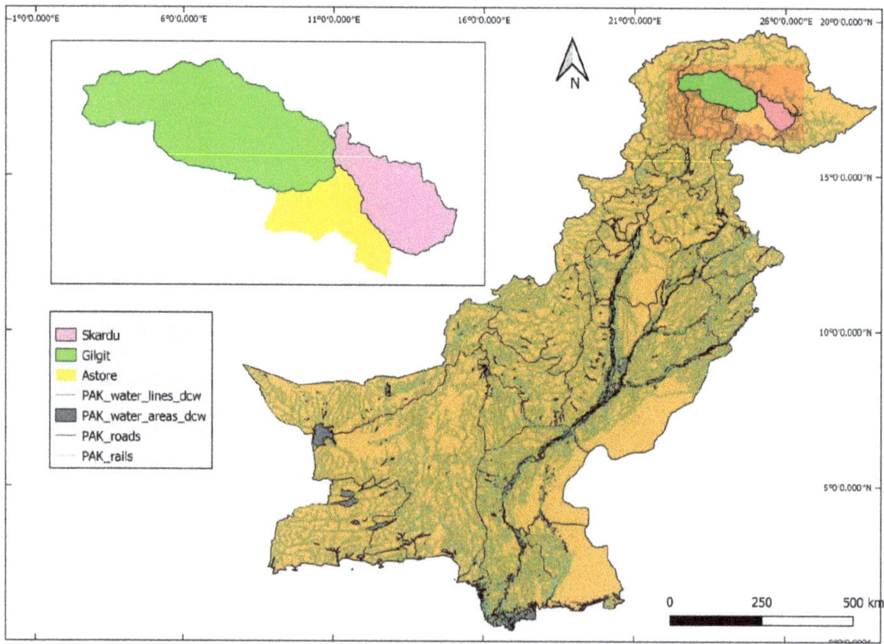

Figure 2. Study map.

3.2. Data Collection

For the present study, survey data were collected from January 2019 to April 2019 (Figure 3). Face-to-face interviews were conducted with the residents. All the meetings were scheduled with the consent of the participants and their privacy was ensured.

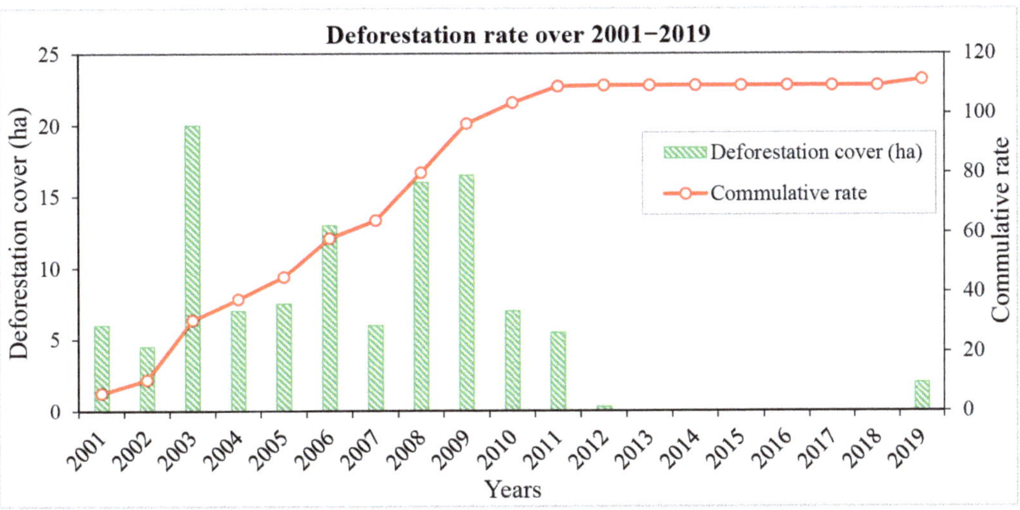

Figure 3. Continuous decline in forest cover area in northern Pakistan from 2001 to 2019. (Taken from https://www.globalforestwatch.org/dashboards/country/PAK).

From 2001 to 2019, Northern Areas lost 112 ha of tree cover, equivalent to a 0.23% decrease in tree cover since 2000. The survey also consisted of some close-ended questions of (a) socioeconomic factors: age, education, time duration to stay in the region, family size, and support from the government ("royalty" income offered to local residents for not utilizing the forest), (b) attitudes toward economic incentives for deforestation; descriptive norms defined in the study as the opinions of other people's behavior toward deforestation; perceived behavioral control as the respondent's views about the presence of law enforcement in the region; and behaviors toward economic incentives for deforestation. Moreover, an open-ended questionnaire has been used to explore the perceptions of the respondents on natural resources including the one where they lived. Categorization of these opinions was done according to the values of nature described by Kellert [69–71]. The opinions of the respondents were categorized into only two groups of natural values "(moralistic values, which represent a respect to the nature; and utilitarian values, which represents the material benefits that a person obtain from nature; Kellert, 2005)".

3.3. Measurement of Variables

Behavioral intention was defined for this research as the intention of rural residents toward the economic incentive of deforestation by replacing small-scale agriculture. The question was designed as follows: Do you have the intention to replace secondary forest with small-scale agriculture to get economic incentives? To evaluate the construct, a five-point bipolar Likert scale was used ranging from 4—very likely to 0—very unlikely.

Attitude was measured directly by utilizing five points "bipolar Likert scale ranging from very unlikely (0) to very likely (4)". For assessment of attitude, question was formed as follows: Do you think forest should be replaced with small-scale agriculture to get economic incentives? For the indirect assessment of belief-based items, (b) were measured on five points "Likert scale ranging from strongly disagree (0) to strongly agree (4)", while for (e) very important (4) to not very important (0). For the belief-based measures, three behavioral beliefs were presented to the respondents concerning that rural residents think by replacing forest with small-scale agriculture will provide food for the family, increase livelihood, and increase tourism (see Table 1).

Table 1. Items used for evaluation of Theory of Planned Behavior constructs.

Behavioral Intention	Do you have the intention to replace secondary forest with small-scale agriculture to get economic incentives?	
Attitude	Do you think forest should be replaced with small-scale agriculture to get economic incentives?	
Descriptive Norm	Do you think that rural residents in the area replacing the forest with small-scale agriculture for economic incentives?	
Perceived Behavioral Control	In my view, law enforcement is sufficient in the regions to control rural resident's activities?	
Indirect evaluation of Theory of Planned Behavior constructs		
	Salient beliefs (bs$_i$)	**Outcome evaluation (oe$_i$)**
	Rural residents think by replacing forest with small-scale agriculture will ...	
ATT $\alpha \sum bs_i oe_i$	... provide foods for family	For me, food for family is ...
	... increase livelihood	For me, revenue is ...
	... increase tourism	For me, tourism is ...
	Descriptive normative beliefs (dnb$_i$)	**Identification with the referent (iwr$_i$)**
DN $\alpha \sum dnb_i iwr_i$	In my view people deforest in the region.	With regards to deforestation, I am not similar to people.
	Control beliefs (cb$_i$)	**Power of control factors (pc$_i$)**
PBC $\alpha \sum cb_i pc_i$	I think legislation is insufficient to control people's activities in the region.	Without legislation, it is more difficult to control people's activities in the region.
	I think training of personnel is unsuited to control people's activities in the region.	Without proper training, it is more difficult to control illegal activities in the region.

Talking about the actions of others (descriptive norms) was found more comfortable for the respondent than perceived social pressure from others for an individual to behave in a certain manner (subjective norms). Descriptive norm was measured directly by utilizing a five-point unipolar Likert scale ranging from not at all (0) to a large extent (4). The item was designed as follows: Do you think that rural residents in the area replacing the forest with small-scale agriculture for economic incentives? The indirect assessment of belief-based items (dnb$_i$) was performed by using unipolar Likert scale, from 0—strongly disagree to 4—strongly agree, while for (iwr$_i$) "unipolar Likert scale was used ranging from 0—strongly disagree to 4—strongly agree".

For the measurement of perceived behavioral control, a five-point "unipolar Likert scale from (0) strongly disagree to (4) strongly agree" was used. The statement was formed as follows: In my view, law enforcement is sufficient in the regions to control rural resident's activities. For the indirect assessment of perceived behavioral control, control beliefs (cb$_i$) were assessed on a Likert scale ranging from (0) strongly disagree to (4) strongly agree, while p was measured on a scale ranging from (0) strongly disagree to (4) strongly agree.

3.4. Analysis Methods

Descriptive statistics were used to present the results, and "Pearson's correlation coefficients and regression analysis were chosen for interpretation". The Pearson correlation coefficients are used in statistics to measure how strong a relationship is between two variables. A value of 0 demonstrates that there is no correlation between the two variables. A value higher than 0 describes a positive relationship and a value less than 0 indicates a negative relationship. Regression analyses investigate the relationship between dependent and independent variables.

To determine the relationship between dependent variables (attitude, descriptive norms, perceived behavioral control, behavioral intention) and independent variables ($\sum bs_i oe_i$, $\sum dnb_i\ iwr_i$, and $\sum cb_i\ pc_i$), a linear relationship was presumed (as mentioned in the mathematical formulation of TPB). Therefore, a multiple linear regression technique was used to analyze the data. P-values were used to interpret the significance level of regression analysis and coefficients. β weights and t-values were used to interpret the results. R^2 was applied for the evaluation of the explanatory power of the regression analysis. F-test was used for the assessment of the overall significance level of the models. Data analysis was performed using "Microsoft Excel for Windows version 19 and SPSS ver. 25 software".

4. Results

4.1. Background Factors

The survey was conducted with 207 interviewers, 92 percent of whom were males and 8 percent were females. the detail of background factors is presented in Table 2.

Table 2. Demographic survey of rural residents in protected areas of northern Pakistan.

		Gilgit (n = 65)	Skardu (n = 73)	Astore (n = 69)	Total (n = 207)	Chi-Square Value (X^2)	p-Value
Age	Less than 22 years	11%	21%	14%	15%	$X^2 = 6.62$	0.577
	23 to 35 years	14%	16%	19%	16%		
	36 to 50 years	38%	25%	32%	32%		
	51 to 65 years	17%	18%	22%	19%		
	66 and Above	20%	21%	13%	18%		
Duration of Residence	0 to 20 years	15%	22%	19%	19%	$X^2 = 2.01$	0.732
	21 to 40 years	51%	51%	45%	49%		
	41 and above	34%	27%	36%	32%		
Education	Less than Primary	35%	24%	22%	27%	$X^2 = 19.33$	0.012 *
	Primary	16%	30%	12%	19%		
	High School	19%	13%	24%	18%		
	Middle	23%	18%	35%	26%		
	College	7%	15%	8%	10%		
Family Size	1–4 person	15%	15%	22%	17%	$X^2 = 4.21$	0.377
	5–10 person	60%	54%	45%	53%		
	11 and above	25%	31%	34%	30%		
Government Assistance	Yes	20%	36%	25%	27%	$X^2 = 4.55$	0.102
Profession	Small-scale Agriculture	83%	77%	83%	81%	$X^2 = 1.14$	0.565

* $p < 0.05$.

Based on the feedback of the open-ended question about the participants' opinions on deforestation, 86% of participants reported negative beliefs, including 42% people cannot work, 20% people claimed that government does not pay a royalty for the properties, 11% think an increase in unemployment, and 10% thinks that usage of firewood is prohibited. Only 14% mentioned just negative values. Respondents' beliefs categorized into negative values including restrictions and positive values including forest protection, relationship between education and profession, education and government assistance, Duration of Residence and government assistance, Age and government assistance, and found that the regression relationship is significant and multicollinearity is less than 5. So, these results indicate that there is an influence of these variables regarding deforestation if it is from the government sector or local communities [72].

4.2. Components of Theory of Planned Behavior

4.2.1. Components of Behavioral Intention

The majority of the participants (46%) had thought to carry out the behavior in question, while 23% of them declared that it is an illegal activity to deforest. Attitude toward performing behavior was 59%. At the same time, descriptive norms for the respondent's views toward other people were 42% that were engaging in illegal activities. As far as PBC is concerned, 47% of respondents were not satisfied with the performance of law enforcement.

4.2.2. Components of Attitude

Most of the participants (61%) believed that the replacement of forests with small-scale agriculture will provide food for the family (agree and strongly agree on the *"bipolar Likert scale"*). With regards to other important beliefs that categorized the respondents are livelihood (82%), however, respondents' beliefs toward replacing forest with tourism were adverse (Figure 4a). As far as outcome evaluation is concerned, 94% of the participants perceived that provide food to the family is important or very important *"3 and 4 on the bipolar Likert scale"*, while only 28% of the respondents stated that, for them, increase livelihood is of some importance.

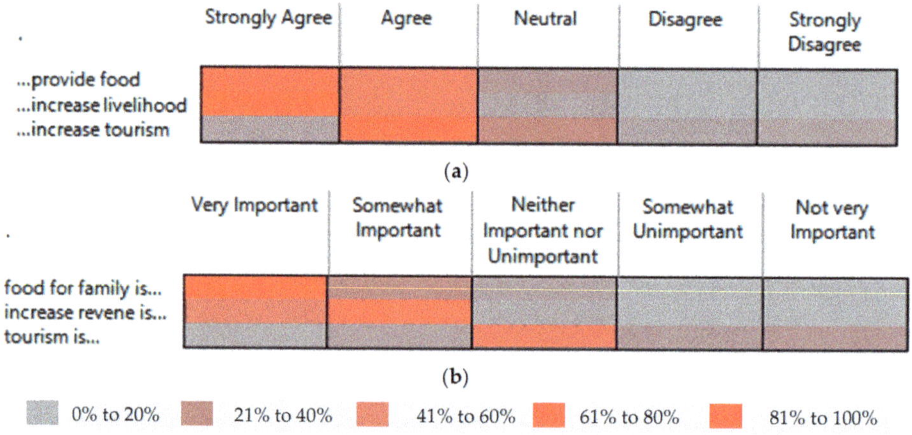

Figure 4. (a,b) Evaluating the percentage of components of attitude (behavioral belief and outcome evaluation).

The results obtained from Pearson correlation coefficients described that belief-assessment $(b_i e_i)$ had a relatively strong association with the direct measurement of attitude (the values of Pearson coefficient lie between 0.2 and 0.4), with the notable exception of increasing tourism (see Table 3). The regression analysis verified the relationship between endogenous variable attitude and the exogenous variables $b_i e_i$: the β-tourism coefficient was not found significant, whilst the β-coefficient for all other exogenous variables was significant.

Table 3. Regression analysis to predict Attitude (dependent variable) from beliefs—evaluation (independent variables) Pearson correlation coefficients between b_ie_i and Attitude (F = 14.671, R^2 = 0.178 and significance level = 0.000).

Beliefs × Outcome Evaluation (Biei)	Correlations Coefficients (r)	B—Coefficients	t-Values
Provide food × food for family	0.396 **	0.297 **	3.838
Increase livelihood × increase revenue is	0.327 **	0.138 **	1.767
Increase tourism × tourism is	0.223 *	0.069	0.984

** significant for $p \leq 0.01$. * significant for $p \leq 0.05$.

4.2.3. Components of Descriptive Norm

With regards to the perception of other people, it was perceived (78%, say Yes) rural inhabitants replace the secondary forest with small-scale agriculture. Respondents generally confirmed their most people (85%) have illegal activities in the forest (Figure 5).

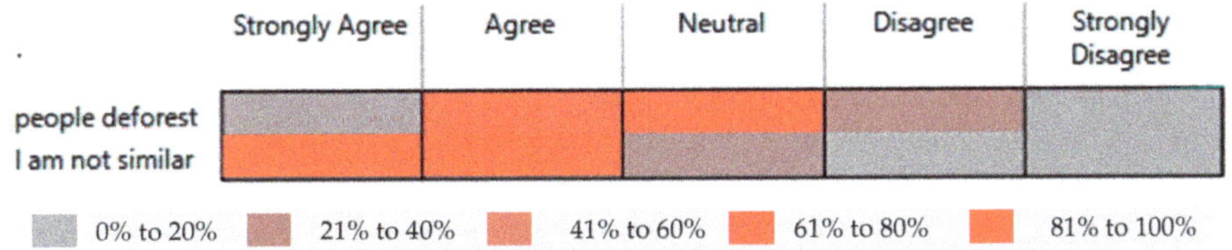

Figure 5. Percentage distribution of descriptive beliefs on the protection of forest.

By assessing correlation coefficients of descriptive normative beliefs (dnb_i) and identification with the referent (iwr_i), the descriptive norm was found correlated with these two products. The regression results demonstrated 20 percent of the variation in descriptive norm that confirmed illegal activities of the people in the region (Table 4).

Table 4. Multicollinearity among variables.

Model	B	t	Sig	R^2	Multicollinearity
Duration of Residence and Age	0.801	12.779	0.000 *	0.642	VIF < 1
Govt Assistance and Agriculture	0.789	12.252	0.000 *	0.623	VIF < 1
Education and Agriculture	0.661	8.411	0.000 *	0.437	VIF < 1
Education and Govt Assistance	0.663	8.445	0.000 *	0.439	VIF < 1
Duration of Residence and Govt Assistance	0.797	12.582	0.000 *	0.635	VIF < 1
Age and Govt Assistance	0.789	12.252	0.000 *	0.623	VIF < 1

* $p < 0.000$.

4.2.4. Components of PBC

The statement on unsuitable legislation was supported by a majority of respondents, while the power of this belief to impede law enforcement was also perceived as high. Majority of respondents agreed without suitable training illegal behavior of residents would not stop (Figure 6a,b).

With regards to the evaluation of control belief x perceived power to influence products (cb_ip_i), it was found that both products were correlated with the PBC (see Table 5). However, insufficient legislation demonstrated a major factor affecting the perceived power of control. Both the Pearson correlation coefficient and the regression analysis results demonstrated that insufficient legislation as well as unsuited training significantly predicted PBC.

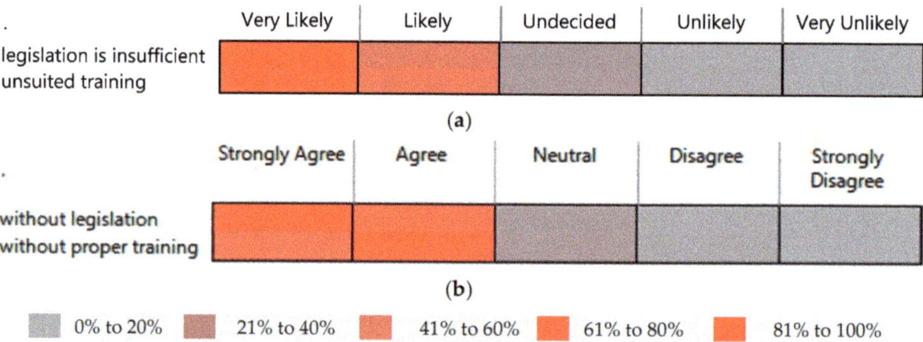

Figure 6. (**a**,**b**): Percentage distribution of control beliefs (**a**) and compliance with the beliefs (**b**) in law enforcement to protect the forest.

Table 5. Regression analysis to predict perceived behavior control (dependent variable) from beliefs—evaluation (independent variables) Pearson correlation coefficients between biei and attitude ($F = 37.679$, $R^2 = 0.270$ and significance level = 0.000).

Control Beliefs (cb_i) × Power of Control Factors (pc_i)	Correlations Coefficients (r)	B—Coefficients	t-Values
Legislations insufficient × without legislation	0.501 **	0.401 **	5.425
Training of personnel unsuited × without proper training	0.405 **	0.170 **	2.305

** significant for $p \leq 0.01$.

4.3. Factors Predicting the Rural Residents' Motivations Towards Deforestation

The components of the theory of planned behavior—attitude, DN, and perceived behavior control—had shown strong associations with behavioral intention. Furthermore, the sums of all three products ($\sum b_i e_i$, $\sum nb_i mc_i$, and $\sum cb_i p_i$) had a positive correlation with behavioral intention. Since all the variables had positive Pearson 's correlation coefficients with behavioral intention (significant level of all variables = $p \leq 0.01$), none of the variables was excepted from the linear regression analysis that was proposed to explain the behavioral intention (Table 6).

Table 6. TPB (Theory of planned behavior) model explaining the Pearson's correlation coefficients.

	Mean	Standard Deviation	BI	ATT	$\sum bs_i oe_i$	DN	$\sum dnb_i irw_i$	PBC	$\sum cb_i pc_i$
BI	3.32	0.740	1.000						
ATT	3.53	0.621	0.611 **	1.000					
$\sum bs_i oe_i$	11	2.388	0.276 **	0.413 **	1.000				
DN	3.14	0.770	0.511 **	0.326 **	0.222 **	1.000			
$\sum dnb_i irw_i$	10.00	4.211	0.508 **	0.359 **	0.333 **	0.455 **	1.000		
PBC	3.29	0.785	0.539 **	0.334 **	0.113 *	0.391 **	0.204 **	1.000	
$\sum cb_i pc_i$	11.442	3.067	0.458 **	0.291 **	0.310 **	0.197 *	0.381 **	0.508 **	1.000

** significant for $p \leq 0.01$. * significant for $p \leq 0.05$.

Through the evaluation of multiple linear regression, all possible models to explain behavioral intention were examined. (Table 7). "The standardized regression coefficients (and t-values) of so-called basic model, having as explanatory variables attitude, descriptive norm, and perceived behavioral control, demonstrated that all of three variables had high explanatory power in behavioral intention variation. The basic model was also statistically significant ($p \leq 0.01$) and explained 56% ($R^2 = 0.552$) of the total variation of the intention".

Table 7. Regression analysis explaining the economic motives of rural residents toward behavioral intention (t-values and β -coefficients).

Variables	β -Coefficients (t-Values)						Model Features
	ATT	DN	PBC	$\sum bs_i oe_i$	$\sum dnb_i irw_i$	$\sum cb_i pc_i$	
Basic model	0.436 ** (8.390)	0.256 ** (4.887)	0.295 ** (5.623)	-	-	-	F = 83.455 Significance Level = 0.000 $R^2 = 0.552$
Model 1	0.429 ** (8.333)	0.318 ** (6.339)	-	-	-	0.270 ** (5.447)	F = 82.201 Significance Level = 0.000 $R^2 = 0.548$
Model 2	0.452 ** (8.460)	-	-	-	0.259 ** (4.691)	0.228 ** (4.227)	F = 71.000 Significance Level = 0.000 $R^2 = 0.512$
Model 3	0.388 ** (7.592)	-	0.348 ** (7.153)	-	0.297 ** (6.047)	-	F = 91.905 Significance Level = 0.000 $R^2 = 0.576$
Model 4	-	0.321 ** (5.428)	0.395 ** (6.810)	0.160 * (2.927)	-	-	F = 49.282 Significance Level =0.000 $R^2 = 0.421$
Model 5	-	0.426 ** (7.554)	-	0.073 * (1.248)	-	0.351 ** (6.079)	F = 44.861 Significance Level = 0.000 $R^2 = 0.399$
Model 6	-	-	-	0.059 * (0.955)	0.375 ** (5.915)	0.297 ** (4.717)	F = 35.313 Significance Level = 0.000 $R^2 = 0.343$
Model 7	-	-	0.449 ** (8.553)	0.098 * (1.791)	0.384 ** (6.925)	-	F= 58.583 Significance Level = 0.000 $R^2 = 0.464$

** significant for $p \leq 0.01$. * significant for $p \leq 0.05$.

However, attitude was the most powerful component to explain the behavior intentions with highest β = 0.436, followed by descriptive norms β = 0.256 and perceived behavioral control β = 0.295 (Figure 7).

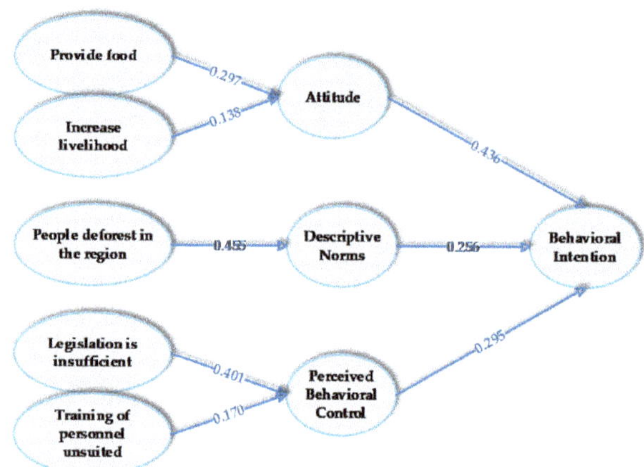

Figure 7. β-coefficient explaining the factors affecting behavioral intention.

Other analyzed models had the explanatory power in explaining the total variation of the behavioral intention (R^2 ranking between 0.576 and 0.343).

5. Discussion

Ajzen's theory of planned behavior suggests that attitude, norms, and perceived behavioral control are better predictors to explain an individual's behavior [51]. In this research, a preliminary exploration of behavior toward economic incentive for deforestation in the study region was conducted using the theory of planned behavior as a framework for structuring our analysis. The findings of the study suggest that the level of education influenced respondent's behavioral intention to deforest for economic incentive. The present study explored that attitude, descriptive norm, and perceived behavior control may be good predictors to investigate the behaviors.

The analysis of the regression models has demonstrated that attitude was the major factor to explain behavioral intention. Perceived behavioral control was followed closely by descriptive norms, which verified the high power of influence. The PBC generally has the characteristic of high power, which is confirming by the other studies [73–79]. Besides these components, intention to deforest with small-scale agriculture was also high because of unsuited law effort in the region.

With regards to belief-based evaluation, the major item explaining the attitude was food for family, followed by an increased livelihood. In the context of Pakistan, the influence of these two factors is due to rural residents' reliance on natural resources. Interestingly, *tourism* was not perceived as an important factor. This could be due to mostly residents want to focus on agricultural activities and especially on family food. Descriptive norms were highly influenced by the respondent's behavior toward other people. Unlike other studies that stated perceived behavioral control might be a good predictor of behavior (e.g., [80–83], we found that people's perception of law enforcement did not affect their behavior. Rural resident's view of law enforcement might not be sufficient to avoid negative behavior, as the activities carried out involve the use of resources vital to the livelihoods of local communities [9,84–86].

Regarding that most rural residents are involved in agrarian activities, their main complaint was that they were prohibited from substituting secondary forests for small-scale agriculture, which impeded their work and livelihood. This feeling was expressed in negative attitudes toward forest conservation, as well as negative behavior, as some of the residents replacing secondary forests with small-scale farming.

Deforestation by residents of the area is a great challenge for forest management because reconciling land use and preservation of ecosystems inside the region requires a preventative measure to ensure the protection of remaining fragments without affecting the livelihoods of inhabitants.

The current research did not deal with all of the theory of planned behavior components (particularly, exclude the actual behavior and SN and explore the general element of PBC and DN related to the economic incentive for deforestation), attempting to prevent the use of the basic theory of planned behavior framework and complete analysis of structural equations modeling or multivariate regression. However, it identifies areas to be addressed by forestry managers to change the behavior of residents in relation to important conservation issues of deforestation.

There should be a prosecution of corrupt government officials in charge of the forestry laws and policies along with illegal loggers. Environmental awareness should be made accessible to the general public about the devastating consequences of deforestation on people and society at large. The government should embark on a program of tree planting by enlightening the public to fathom that we have only one earth. Government, Non-governmental organizations, and spirited individuals should organize an enlightenment program on the impacts of climate change. The government should add more effort to the poverty eradication program, and the educated unemployed youths should be accorded employment. To curb the rate of deforestation, a skills training system should be coordinated for rural women dwellers and the uneducated youth. In conclusion, therefore, it is necessary to recognize and introduce successful ways of addressing the daily needs of the communities. The emphasis needs to be placed on seeking alternative energy sources, sustainable agricultural practices, diversifying income sources, and supporting rural development for young people and disadvantaged community members. In order to allow communities to engage actively in decision-making processes aimed at conserving the forest and improving the livelihoods of rural communities, forestry education and extension should be geared toward institutional strengthening at the local level.

6. Conclusions

The economic benefits include the provision of subsidies for forest products, an enhanced system of taxes on exploited forest goods, the procurement of well-monitored hunting licenses, alternative job opportunities, credit provision, and a limited ban on round log exports in northern areas of Pakistan. The study suggests that the level of education influenced the respondent's behavioral intention to deforest for economic incentive. The attitude, descriptive norm, and perceived behavior control may be good predictor to investigate the behaviors. Besides these components, the intention to deforest with small-scale agriculture was also high because of unsuited law efforts in the region. As far as outcome evaluation is concerned, 94% of the participants perceived that provide food to family is an important or very important component of livelihood. Socio-economic factors that affect the forest, such considerations are profoundly rooted in the everyday needs of the communities as regards forest products that meet the increasing population rather than knowledge of the degradation and its implications of forest resources.

However, insufficient legislation demonstrated a major factor that is affecting the perceived power of control. Both the Pearson correlation coefficient and the regression analysis demonstrated that insufficient legislation as well as unsuited training demonstrates that Human activities are environmentally hazardous in combination with our daily work and actions at home, in industry, and even in agriculture, endanger the stability of the climate and the ecological balance. All these human actions endanger nature. Adequate economic incentives can be an important tool for reducing deforestation.

Author Contributions: S.U., A.A., R.S.N. and M.M.W. conceptualized the idea of the study design; R.S.N. and W.A. contributed to data collection and formal analysis; S.U., A.A. and R.S.N. performed the statistical analysis and wrote the manuscript; S.U., R.S.N. and M.M.W. edited and reviewed the manuscript; T.G. supervised and gave the conceptual insight. All authors have read and agreed to the published version of the manuscript.

Funding: This study has not received any funding.

Acknowledgments: Authors acknowledge the efforts of Central Karakoram National Park and staff for helping in data collection.

Conflicts of Interest: The authors declared no conflict of interest.

References

1. Durand, L.; Lazos, E. The Local Perception of Tropical Deforestation and its Relation to Conservation Policies in Los Tuxtlas Biosphere Reserve, Mexico. *Hum. Ecol.* **2008**, *36*, 383–394. [CrossRef]
2. Duguma, L.A.; Atela, J.; Minang, P.A.; Ayana, A.N.; Gizachew, B.; Nzyoka, J.; Bernard, F. Deforestation and Forest Degradation as an Environmental Behavior: Unpacking Realities Shaping Community Actions. *Land* **2019**, *8*, 26. [CrossRef]
3. Alcocer-Rodríguez, M.; Arroyo-Rodríguez, V.; Galán-Acedo, C.; Cristóbal-Azkarate, J.; Asensio, N.; Rito, K.F.; Hawes, J.E.; Veà, J.J.; Dunn, J.C. Evaluating extinction debt in fragmented forests: The rapid recovery of a critically endangered primate. *Anim. Conserv.* **2020**, 1–12. [CrossRef]
4. Gogoi, B.; Nath, T.; Kashyap, D.; Sarma, S.; Kalita, R. Sustainable agriculture, forestry and fishery for bioeconomy. In *Current Developments in Biotechnology and Bioengineering*; Elsevier Science: Amsterdam, The Netherlands, 2020; pp. 349–371.
5. Osorio-González, C.S.; Suralikerimath, N.; Hegde, K.; Brar, S.K. Sustainability of Ecosystem Services (ESs). *Sustainability* **2020**, 277–294. [CrossRef]
6. Quang, N.V.; Noriko, S. Forest Allocation Policy and Level of Forest Dependency of Economic Household Groups: A Case Study in Northern Central Vietnam. *Small-Scale For.* **2008**, *7*, 49–66. [CrossRef]
7. Adhikari, B.; Di Falco, S.; Lovett, J.C. Household characteristics and forest dependency: Evidence from common property forest management in Nepal. *Ecol. Econ.* **2004**, *48*, 245–257. [CrossRef]
8. Luna, T.O.; Zhunusova, E.; Günter, S.; Dieter, M. Measuring forest and agricultural income in the Ecuadorian lowland rainforest frontiers: Do deforestation and conservation strategies matter? *For. Policy Econ.* **2020**, *111*, 102034. [CrossRef]
9. Zeb, A.; Armstrong, G.W.; Hamann, A. Forest conversion by the indigenous Kalasha of Pakistan: A household level analysis of socioeconomic drivers. *Glob. Environ. Chang.* **2019**, *59*, 102004. [CrossRef]
10. Yiwen, Z.; Kant, S.; Dong, J.; Liu, J. How communities restructured forest tenure throughout the top-down devolution reform: Using the case of Fujian, China. *For. Policy Econ.* **2020**, *119*, 102272. [CrossRef]
11. Puri, L.; Nuberg, I.; Ostendorf, B.; Cedamon, E. Locally Perceived Social and Biophysical Factors Shaping the Effective Implementation of Community Forest Management Operations in Nepal. *Small-Scale For.* **2020**, *19*, 291–317. [CrossRef]
12. Jannat, M.; Hossain, M.K.; Uddin, M. Socioeconomic factors of forest dependency in developing countries: Lessons learned from the Bandarban hill district of Bangladesh. *Am. J. Pure Appl. Sci.* **2020**, *2*, 77–84.
13. Sirivongs, K.; Tsuchiya, T. Relationship between local residents' perceptions, attitudes and participation towards national protected areas: A case study of Phou Khao Khouay National Protected Area, central Lao PDR. *For. Policy Econ.* **2012**, *21*, 92–100. [CrossRef]
14. Ullah, S.; Gang, T.; Rauf, T.; Sikandar, F.; Liu, J.Q.; Noor, R.S. Identifying the socio-economic factors of deforestation and degradation: A case study in Gilgit Baltistan, Pakistan. *GeoJournal* **2020**, 1–14. [CrossRef]
15. Brankov, J.; Jojić, G.T.; Pešić, A.M.; Petrović, M.D.; Tretiakova, T.N. Residents' Perceptions of Tourism Impact on Community in National Parks in Serbia. *Eur. Countrys.* **2019**, *11*, 124–142. [CrossRef]
16. Walde, J.; Huy, D.T.; Tappeiner, U.; Tappeiner, G. A protected area between subsistence and development. *Int. J. Commons* **2019**, *13*, 175. [CrossRef]
17. Badola, R.; Barthwal, S.; Hussain, S.A. Attitudes of local communities towards conservation of mangrove forests: A case study from the east coast of India. *Estuarine Coast. Shelf Sci.* **2012**, *96*, 188–196. [CrossRef]
18. John, F.A.V.S.; Edwards-Jones, G.; Jones, J.P.G. Conservation and human behaviour: Lessons from social psychology. *Wildl. Res.* **2011**, *37*, 658–667. [CrossRef]
19. Khan, W.; Hore, U.; Mukherjee, S.; Mallapur, G. Human-crocodile conflict and attitude of local communities toward crocodile conservation in Bhitarkanika Wildlife Sanctuary, Odisha, India. *Mar. Policy* **2020**, *121*, 104135. [CrossRef]
20. Al Idrus, A.; Syukur, A.; Zulkifli, L. The livelihoods of local communities: Evidence success of mangrove conservation on the coastal of East Lombok Indonesia. *AIP Conf. Proc.* **2019**, *2199*, 050010.
21. Garekae, H.; Lepetu, J.; Thakadu, O.T.; Sebina, V.; Tselaesele, N. Community Perspective on State Forest Management Regime and its Implication on Forest Sustainability: A Case Study of Chobe Forest Reserve, Botswana. *J. Sustain. For.* **2020**, *39*, 692–709. [CrossRef]

22. Waylen, K.A.; Fischer, A.; McGowan, P.J.K.; Thirgood, S.J.; Milner-Gulland, E.J. Effect of Local Cultural Context on the Success of Community-Based Conservation Interventions. *Conserv. Biol.* **2010**, *24*, 1119–1129. [CrossRef] [PubMed]
23. Dhakal, B.; Kattel, R.R. Effects of global changes on ecosystems services of multiple natural resources in mountain agricultural landscapes. *Sci. Total Environ.* **2019**, *676*, 665–682. [CrossRef] [PubMed]
24. Musavengane, R.; Kloppers, R. Social capital: An investment towards community resilience in the collaborative natural resources management of community-based tourism schemes. *Tour. Manag. Perspect.* **2020**, *34*, 100654. [CrossRef]
25. Avtar, R.; Tsusaka, K.; Herath, S. REDD+ Implementation in Community-Based Muyong Forest Management in Ifugao, Philippines. *Land* **2019**, *8*, 164. [CrossRef]
26. Shahbaz, B.; Ali, T.; Suleri, A.Q. A Critical Analysis of Forest Policies of Pakistan: Implications for Sustainable Livelihoods. *Mitig. Adapt. Strat. Glob. Chang.* **2007**, *12*, 441–453. [CrossRef]
27. Nandigama, S. Performance of success and failure in grassroots conservation and development interventions: Gender dynamics in participatory forest management in India. *Land Use Policy* **2020**, *97*, 103445. [CrossRef]
28. Sheikh, Y.; Ibrar, M.; Iqbal, J. Impact of Joint Forest Management on Rural Livelihoods in the Kalam and Siran Forest Divisions, Khyber Pakhtunkhwa Pakistan. *Glob. Reg. Rev.* **2019**, *4*, 225–237. [CrossRef]
29. Payne, D.; Snethlage, M.; Geschke, J.; Spehn, E.M.; Fischer, M. Nature and People in the Andes, East African Mountains, European Alps, and Hindu Kush Himalaya: Current Research and Future Directions. *Mt. Res. Dev.* **2020**, *40*, A1. [CrossRef]
30. Kabir, M.; Hameed, S.; Ali, H.; Bosso, L.; Din, J.U.; Bischof, R.; Redpath, S.; Nawaz, M.A. Habitat suitability and movement corridors of grey wolf (*Canis lupus*) in Northern Pakistan. *PLoS ONE* **2017**, *12*, e0187027. [CrossRef]
31. Hussain, M.; Butt, A.R.; Uzma, F.; Ahmed, R.; Rehman, A.; Ali, M.U.; Ullah, H.; Yousaf, B. Divisional disparities on climate change adaptation and mitigation in Punjab, Pakistan: Local perceptions, vulnerabilities, and policy implications. *Environ. Sci. Pollut. Res.* **2019**, *26*, 31491–31507. [CrossRef]
32. Hussain, J.; Zhou, K.; Akbar, M.; Khan, M.Z.; Raza, G.; Ali, S.; Hussain, A.; Abbas, Q.; Khan, G.; Khan, M.; et al. Dependence of rural livelihoods on forest resources in Naltar Valley, a dry temperate mountainous region, Pakistan. *Glob. Ecol. Conserv.* **2019**, *20*, e00765. [CrossRef]
33. Ali, N.; Hu, X.; Hussain, J. The dependency of rural livelihood on forest resources in Northern Pakistan's Chaprote Valley. *Glob. Ecol. Conserv.* **2020**, *22*, e01001. [CrossRef]
34. Shahbaz, B.; Mbeyale, G.; Haller, T. Trees, trust and the state: A comparison of participatory forest management in Pakistan and Tanzania. *J. Int. Dev.* **2008**, *20*, 641–653. [CrossRef]
35. Ali, J.; Benjaminsen, T.A. Fuelwood, Timber and Deforestation in the Himalayas. *Mt. Res. Dev.* **2004**, *24*, 312–318. [CrossRef]
36. Macdonald, K.; Rudel, T.K. Sprawl and forest cover: What is the relationship? *Appl. Geogr.* **2005**, *25*, 67–79. [CrossRef]
37. Jeon, S.B.; Olofsson, P.; Woodcock, C.E. Land use change in New England: A reversal of the forest transition. *J. Land Use Sci.* **2014**, *9*, 105–130. [CrossRef]
38. Wang, H.; Qiu, F. Investigating the Impact of Agricultural Land Losses on Deforestation: Evidence from a Peri-urban Area in Canada. *Ecol. Econ.* **2017**, *139*, 9–18. [CrossRef]
39. de Freitas, D.; van Eeden, T.S.; Christie, L. A psychographic framework for determining South African consumers' green hotel decision formation: Augmenting the Theory of Planned Behaviour. *J. Consum. Sci.* **2020**, *5*, 1–18. Available online: https://www.ajol.info/index.php/jfecs/article/view/195428 (accessed on 5 January 2021).
40. Pendrill, F.; Persson, U.M.; Godar, J.; Kastner, T. Deforestation displaced: Trade in forest-risk commodities and the prospects for a global forest transition. *Environ. Res. Lett.* **2019**, *14*, 055003. [CrossRef]
41. Pendrill, F.; Persson, U.M.; Godar, J.; Kastner, T.; Moran, D.; Schmidt, S.; Wood, R. Agricultural and forestry trade drives large share of tropical deforestation emissions. *Glob. Environ. Chang.* **2019**, *56*, 1–10. [CrossRef]
42. Murshed, M.; Ferdaus, J.; Rashid, S.; Tanha, M.M.; Islam, J. The Environmental Kuznets curve hypothesis for deforestation in Bangladesh: An ARDL analysis with multiple structural breaks. *Energy Ecol. Environ.* **2020**, 1–22. [CrossRef]
43. Curtis, P.G.; Slay, C.M.; Harris, N.; Tyukavina, A.; Hansen, M.C. Classifying drivers of global forest loss. *Science* **2018**, *361*, 1108–1111. [CrossRef] [PubMed]
44. De Sy, V.; Herold, M.; Achard, F.; Avitabilie, V.; Baccini, A.; Carter, S.; Clevers, J.G.P.W.; Lindquist, E.; Pereira, M.; Verchot, L. Tropical deforestation drivers and associated carbon emission factors derived from remote sensing data. *Environ. Res. Lett.* **2019**, *14*, 094022. [CrossRef]
45. Doggart, N.; Morgan-Brown, T.; Lyimo, E.; Mbilinyi, B.; Meshack, C.K.; Sallu, S.M.; Spracklen, D.V. Agriculture is the main driver of deforestation in Tanzania. *Environ. Res. Lett.* **2020**, *15*, 034028. [CrossRef]
46. Diouf, B.; Miezan, E. The Biogas Initiative in Developing Countries, from Technical Potential to Failure: The Case Study of Senegal. *Renew. Sustain. Energy Rev.* **2019**, *101*, 248–254. [CrossRef]
47. Günşen, H.B.; Atmiş, E. Analysis of forest change and deforestation in Turkey. *Int. For. Rev.* **2019**, *21*, 182–194. [CrossRef]
48. Khan, A.M.; Hussain, A.; Siddique, S.; Khan, S.M.; Iqbal, A.; Ahmad, T.; Suliman, M.; Ali, G. Effects of deforestation on socio-economy and associated insect pests in district Swat, Pakistan. *J. Entomol. Zool. Stud.* **2017**, *5*, 36–42.
49. Ali, J.; Benjaminsen, T.A.; Hammad, A.A.; Dick, Ø.B. The road to deforestation: An assessment of forest loss and its causes in Basho Valley, Northern Pakistan. *Glob. Environ. Chang.* **2005**, *15*, 370–380. [CrossRef]
50. Ajzen, I. The theory of planned behavior. *Organ. Behav. Hum. Decis. Process.* **1991**, *50*, 179–211. [CrossRef]

51. Ajzen, I.; Fishbein, M. The Influence of Attitudes on Behavior. In *The Handbook of Attitudes*; Albarracín, D., Johnson, B.T., Zanna, M.P., Eds.; Lawrence Erlbaum Associates Publishers: Hillsdale, NY, USA, 2005; pp. 173–221.
52. Stern, P.C. New Environmental Theories: Toward a Coherent Theory of Environmentally Significant Behavior. *J. Soc. Issues* **2000**, *56*, 407–424. [CrossRef]
53. Cialdini, R.B.; Petrova, P.K.; Goldstein, N.J. The hidden costs of organizational dishonesty. *MIT Sloan Manag. Rev.* **2004**, *45*, 67.
54. Ajzen, I. The theory of planned behavior: Frequently asked questions. *Hum. Behav. Emerg. Technol.* **2020**, *2*, 314–324. [CrossRef]
55. Si, H.; Shi, J.-G.; Tang, D.; Wu, G.; Lan, J. Understanding intention and behavior toward sustainable usage of bike sharing by extending the theory of planned behavior. *Resour. Conserv. Recycl.* **2020**, *152*, 104513. [CrossRef]
56. Miniard, P.W.; Cohen, J.B. Modeling Personal and Normative Influences on Behavior. *J. Consum. Res.* **1983**, *10*, 169–180. [CrossRef]
57. Daxini, A.; Ryan, M.; O'Donoghue, C.; Barnes, A.P. Understanding farmers' intentions to follow a nutrient management plan using the theory of planned behaviour. *Land Use Policy* **2019**, *85*, 428–437. [CrossRef]
58. Popa, B.; Niță, M.D.; Hălălișan, A.F. Intentions to engage in forest law enforcement in Romania: An application of the theory of planned behavior. *For. Policy Econ.* **2019**, *100*, 33–43. [CrossRef]
59. Pakravan, M.H.; Maccarty, N. An Agent-Based Model for Adoption of Clean Technology Using the Theory of Planned Behavior. *J. Mech. Des.* **2021**, *143*, 021402. [CrossRef]
60. Liang, Y. Toward a Sustainable Online Q & A Community Via Design Decisions Based on Individuals' Expertise: Evidence from Simulations. Ph.D. Thesis, Michigan State University, East Landing, MI, USA, 2020.
61. Ajzen, I.; Madden, T.J. Prediction of goal-directed behavior: Attitudes, intentions, and perceived behavioral control. *J. Exp. Soc. Psychol.* **1986**, *22*, 453–474. [CrossRef]
62. Rahman, U.H.F.B.; Zafar, M.K. Factors Influencing Uber Adoption in Bangladesh and Pakistan. *Open Econ.* **2020**, *3*, 86–97. [CrossRef]
63. Çoker, E.N.; van der Linden, S. Fleshing out the theory of planned of behavior: Meat consumption as an environmentally significant behavior. *Curr. Psychol.* **2020**, 1–10. [CrossRef]
64. Ateş, H. Merging Theory of Planned Behavior and Value Identity Personal Norm Model to Explain Pro-Environmental Behaviors. *Sustain. Prod. Consum.* **2020**, *24*, 169–180. [CrossRef]
65. Fishbein, M.; Ajzen, I. *Predicting and Changing Behavior: The Reasoned Action Approach*; Taylor & Francis Group: New York, NY, USA, 2011.
66. Nisson, C.; Earl, A. The Theories of Reasoned Action and Planned Behavior. *Wiley Encycl. Health Psychol.* **2020**, 755–761. [CrossRef]
67. Ajzen, I.; Driver, B.L. Application of the Theory of Planned Behavior to Leisure Choice. *J. Leis. Res.* **1992**, *24*, 207–224. [CrossRef]
68. Zhang, Y.; Li, L. Intention of Chinese college students to use carsharing: An application of the theory of planned behavior. *Transp. Res. Part F Traffic Psychol. Behav.* **2020**, *75*, 106–119. [CrossRef]
69. Kellert, S.R. *Nature and Childhood Development. Building for Life: Designing and Understanding the Human-Nature Connection*; Island Press: Washington, DC, USA, 2005.
70. Kylkilahti, E.; Berghäll, S.; Autio, M.; Nurminen, J.; Toivonen, R.; Lähtinen, K.; Vihemäki, H.; Franzini, F.; Toppinen, A. A consumer-driven bioeconomy in housing? Combining consumption style with students' perceptions of the use of wood in multi-storey buildings. *Ambio* **2020**, *49*, 1943–1957. [CrossRef]
71. Ullah, F.; Saqib, S.E.; Ahmad, M.M.; Fadlallah, M.A. Flood risk perception and its determinants among rural households in two communities in Khyber Pakhtunkhwa, Pakistan. *Nat. Hazards* **2020**, *104*, 225–247. [CrossRef]
72. Hair, J.F., Jr.; Hult, G.T.M.; Ringle, C.M.; Sarstedt, M. A Primer on Partial Least Squares Structural Equation Modeling (PLS-SEM). *J. Acad. Mark. Sci.* **2017**, *45*, 616–632. [CrossRef]
73. Fielding, K.S.; McDonald, R.; Louis, W.R. Theory of planned behaviour, identity and intentions to engage in environmental activism. *J. Environ. Psychol.* **2008**, *28*, 318–326. [CrossRef]
74. Karppinen, H. Forest owners' choice of reforestation method: An application of the theory of planned behavior. *For. Policy Econ.* **2005**, *7*, 393–409. [CrossRef]
75. Masud, M.M.; Al-Amin, A.Q.; Junsheng, H.; Ahmed, F.; Yahaya, S.R.; Ahmed, S.; Banna, H. Climate change issue and theory of planned behaviour: Relationship by empirical evidence. *J. Clean. Prod.* **2016**, *113*, 613–623. [CrossRef]
76. Judge, M.; Warren-Myers, G.; Paladino, A. Using the theory of planned behaviour to predict intentions to purchase sustainable housing. *J. Clean. Prod.* **2019**, *215*, 259–267. [CrossRef]
77. Oteng-Peprah, M.; De Vries, N.; Acheampong, M. Households' willingness to adopt greywater treatment technologies in a developing country—Exploring a modified theory of planned behaviour (TPB) model including personal norm. *J. Environ. Manag.* **2020**, *254*, 109807. [CrossRef] [PubMed]
78. Iranmanesh, M.; Mirzaei, M.; Hosseini, S.M.P.; Zailani, S. Muslims' willingness to pay for certified halal food: An extension of the theory of planned behaviour. *J. Islam. Mark.* **2019**, *11*, 14–30. [CrossRef]
79. Yuriev, A.; Dahmen, M.; Paille, P.; Boiral, O.; Guillaumie, L. Pro-environmental behaviors through the lens of the theory of planned behavior: A scoping review. *Resour. Conserv. Recycl.* **2020**, *155*, 104660. [CrossRef]
80. Mintzer, V.J.; Schmink, M.; Lorenzen, K.; Frazer, T.K.; Martin, A.R.; Da Silva, V.M.F. Attitudes and behaviors toward Amazon River dolphins (Inia geoffrensis) in a sustainable use protected area. *Biodivers. Conserv.* **2015**, *24*, 247–269. [CrossRef]
81. Zubair, M.; Garforth, C. Farm Level Tree Planting in Pakistan: The Role of Farmers' Perceptions and Attitudes. *Agrofor. Syst.* **2006**, *66*, 217–229. [CrossRef]

82. Mahmood, M.I.; Zubair, M. Farmer's Perception of and Factors Influencing Agroforestry Practices in the Indus River Basin, Pakistan. *Small-Scale For.* **2020**, *19*, 107–122. [CrossRef]
83. Tokede, A.; Banjo, A.; Ahmad, A.; Fatoki, O.; Akanni, O. Farmers' knowledge and attitude towards the adoption of agroforestry practices in Akinyele Local Government Area, Ibadan, Nigeria. *J. Appl. Sci. Environ. Manag.* **2020**, *24*, 1775–1780. [CrossRef]
84. De Boer, W.F.; Baquete, D.S. Natural resource use, crop damage and attitudes of rural people in the vicinity of the Maputo Elephant Reserve, Mozambique. *Environ. Conserv.* **1998**, *25*, 208–218. [CrossRef]
85. Miyamoto, M. Poverty reduction saves forests sustainably: Lessons for deforestation policies. *World Dev.* **2020**, *127*, 104746. [CrossRef]
86. Putraditama, A.; Kim, Y.-S.; Meador, A.J.S. Community forest management and forest cover change in Lampung, Indonesia. *For. Policy Econ.* **2019**, *106*, 101976. [CrossRef]

 *sustainability*

Article

Assessing the Contribution of Citrus Orchards in Climate Change Mitigation through Carbon Sequestration in Sargodha District, Pakistan

Ghulam Yasin [1,2], Muhammad Farrakh Nawaz [3], Muhammad Zubair [2], Ihsan Qadir [2], Aansa Rukya Saleem [4], Muhammad Ijaz [5], Sadaf Gul [6], Muhammad Amjad Bashir [7], Abdur Rehim [7], Shafeeq Ur Rahman [5,8,*] and Zhenjie Du [8,*]

1. Department of Forestry Range Wildlife Management, The Islamia University Bahawalpur, Bahawalpur 63100, Pakistan; yasin_2486@yahoo.com
2. Department of Forestry and Range Management, FAS &T, Bahauddin Zakariya University, Multan 61000, Pakistan; dikhan2000@hotmail.com (M.Z.); drbhabha@bzu.edu.pk (I.Q.)
3. Department of Forestry and Range Management, University of Agriculture Faisalabad, Faisalabad 38000, Pakistan; kf_uaf@yahoo.com
4. Department of Earth and Environmental Sciences, Bahria University, Islamabad 44000, Pakistan; arukya.buic@bahria.edu.pk
5. College of Agriculture, Bahauddin Zakariya University Multan, Bahadur Sub-Campus, Layyah 31200, Pakistan; muhammad.ijaz@bzu.edu.pk
6. Department of Botany, University of Karachi, Karachi 74200, Pakistan; sadafgpk@yahoo.com
7. Department of Soil Science, Faculty of Agricultural Sciences and Technology, Bahauddin Zakariya University, Multan 60800, Pakistan; Amjad.bashir941@gmail.com (M.A.B.); Abdur.rehim@bzu.edu.pk (A.R.)
8. Farmland Irrigation Research Institute, Chinese Academy of Agricultural Sciences, Xinxiang 453000, China
* Correspondence: malikshafeeq1559@gmail.com (S.U.R.); imdzj11@163.com (Z.D.)

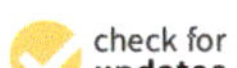

Citation: Yasin, G.; Farrakh Nawaz, M.; Zubair, M.; Qadir, I.; Saleem, A.R.; Ijaz, M.; Gul, S.; Amjad Bashir, M.; Rehim, A.; Rahman, S.U.; et al. Assessing the Contribution of Citrus Orchards in Climate Change Mitigation through Carbon Sequestration in Sargodha District, Pakistan. *Sustainability* **2021**, *13*, 12412. https://doi.org/10.3390/su132212412

Academic Editors: Marco Lauteri and Muhammad Sultan

Received: 26 September 2021
Accepted: 4 November 2021
Published: 10 November 2021

Publisher's Note: MDPI stays neutral with regard to jurisdictional claims in published maps and institutional affiliations.

Abstract: Adopting agroforestry practices in many developing countries is essential to combat climate change and diversify farm incomes. This study investigated the above and below-ground biomass and soil carbon of a citrus-based intercropping system in six sites (subdivisions: Bhalwal, Kot Momin, Sahiwal, Sargodha, Shahpur and Silanwali) of District Sargodha, Southeast Pakistan. Tree biomass production and carbon were assessed by allometric equations through a non-destructive approach whereas, soil carbon was estimated at 0–15 cm and 15–30 cm depths. Above and below-ground biomass differed significantly, and the maximum mean values (16.61 Mg ha^{-1} & 4.82 Mg ha^{-1}) were computed in Shahpur due to greater tree basal diameter. Tree carbon stock fluctuated from 6.98 Mg C ha^{-1} to 10.28 Mg C ha^{-1} among selected study sites. The surface soil (0–15 cm) had greater bulk density, organic carbon, and soil carbon stock than the subsoil (15–30 cm) in the whole study area. The total carbon stock of the ecosystem ranged from 25.07 Mg C ha^{-1} to 34.50 Mg C ha^{-1} across all study sites, respectively. The above findings enable us to better understand and predict the carbon storage potential of fruit-based agroforestry systems like citrus. Moreover, measuring carbon with simple techniques can produce trustworthy outcomes that enhance the participation of underdeveloped nations in several payment initiatives such as REDD+.

Keywords: agroforestry; allometric equations; biomass; carbon stock; organic carbon

1. Introduction

The rapid increase of greenhouse gases has been responsible for severe global warming throughout the world in the last few decades [1]. On an average basis, around 9.9 billion metric tons of CO_2 have been deposited into the atmosphere annually, causing significant threats to the global environment [2]. According to Stocker [3], severe combustion of fossil fuels and land cover change are the leading anthropogenic causes of this higher CO_2 content in the environment. In the background of this higher CO_2 amount and global warming risk, there is great interest in evolving every possible approach to diminish CO_2 concentration emitted through human activities to mitigate climate change [4].

In the terrestrial ecosystem, carbon sequestration is achieved by photosynthesis, which eliminates carbon from the environment and deposits in the biosphere [5,6]. Biomass is considered a vital carbon reservoir in the terrestrial ecosystem, thus playing a crucial role in the global carbon cycle [7]. Furthermore, vegetation biomass is greatly dependent on the growth pattern of its various components [8,9] and is strongly influenced by the management efforts [10]. Agroforestry is a well-managed system in which planting of woody trees is done along with crops on the same piece of land [11,12], is currently practised over more than one billion hectares in various parts of the globe and is acting as a major carbon sink around the world [13]. The carbon storage potential of various agroforestry systems is much inconsistent and ranges from 0.29 to 15.21 Mg ha^{-1} yr^{-1} around the globe [14].

Orchards are considered an important land-use type and cover approximately 22% of irrigated agricultural land across the globe [15]. Citrus orchards form almost 20% of global orchards, including both commercial (61%) and non-commercial (39%) types [16]. Citrus trees with medium-high canopy and shade indices have the potential to sequester 36.11-million-ton carbon in the Three Gorges Reservoir region of Chongqing, in which about 88.3% in soil and 11.7% in a citrus plant, and the economic value of that was more than 11.49 billion Yuan [17]. Various experimental fruit tree orchards, including citrus, has shown promising potential in fixing carbon [18]. The assimilatory activity by the photosynthetic leaves of these species accounts for a majority of carbon inputs [1,19]. Some fruit-based agroforestry systems in the tropics captured about 1.5 to 3. 5 Mg ha^{-1} yr^{-1} carbon [20].

Fruit tree-based agroforestry systems have been practiced throughout Pakistan, especially in irrigated plains. Along with crops on the same land unit, these systems provide constant and better output in income, food, and fruits to local dwellers [21]. Moreover, fruit-based agroforestry systems are more concerned and compatible as equated to crops [9]. Citrus is the prominent fruit crop and a chief constituent of agroforestry in Pakistan and is commercially interplanted along with the crops. Pakistan is ranked 12th among citrus-producing countries worldwide [22], with a total production of 2.4×10^5 t annually [23]. Out of the total country area under fruits, ~29.55% is under citrus, and ~60% is under *Citrus reticulata* Blanco, producing more than 75% of the citrus exports and providing labor days or full-time jobs for more than 75,000 people (about 57 million labor days in production and remaining in marketing sectors) [23].

Keeping in view the production, commercial benefits, and carbon capturing prospective, the citrus-based intercropping agroforestry systems have been studied and acknowledged by several scientists around the globe [21,24,25]. Although knowledge regarding the importance of fruit-based land-use practices in combating climate change is growing in the country, information about citrus-based agroforestry systems under local climatic conditions is limited. The present research work was conceived with the primary objective to inspect the current carbon stock and CO_2 mitigation potential of a citrus *reticulata* based planting on farmlands. The study was performed in six towns of District Sargodha, the hub of citrus *reticulata* based planting, to estimate the accurate dissemination of biomass and carbon stocks in above and below-ground components of the system.

2. Materials and Methods

2.1. Description of Study Sites and Sampling Methodology

The present research was carried out in six subdivisions of Sargodha, a district in the Punjab province of Pakistan. The study area is an agricultural region having an area of 2260 square miles, and citrus *reticulata* are commercially inter-planted in the fields along with farm crops across the whole study area. The alluvium of the area is highly fertile for citrus cultivation and is locally known as "Chaj Doab", as depicted in Figure 1. The whole district shares its boundary with salt range on the northern side, whereas the remaining three sides are adjacent to rivers Jhelum and Chenab [26]. The area experiences very long and hot summers with a maximum temperature of 50 °C, while winters are short and cold with minimum temperature falls below freezing point in winters at some places. Overall, the area has a 23.8 °C average temperature with 410 mm average annual precipitation (https://en.climate-data.org/location/2195/). Overall, 60 points were constructed for tree inventory across six sites of districts of Sargodha: Bhalwal, Kot Momin, Sahiwal, Sargodha, Shahpur, and Silanwali to estimate the biomass production and carbon storage both in woody biomass and in soil (Figure 1). Across the whole study area, a total of 300 plots (0.405 ha) with citrus orchards were randomly selected and measured by implementing the lottery method [9].

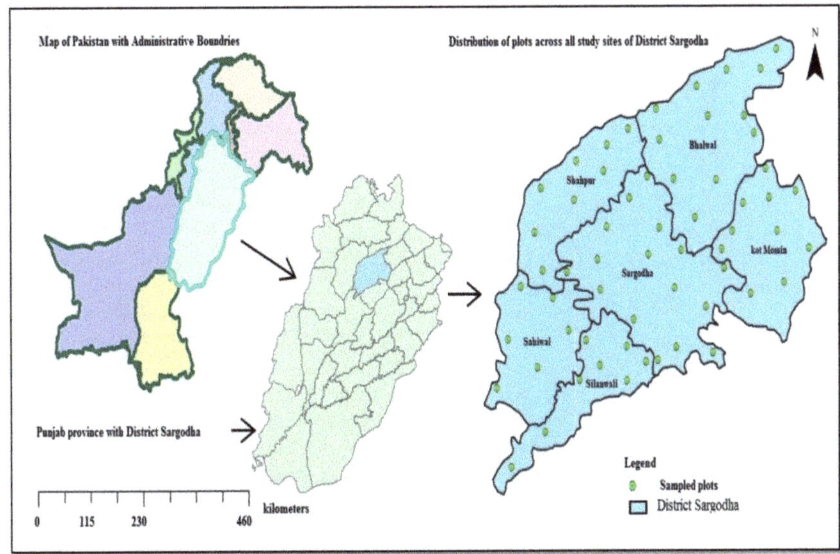

Figure 1. Map showing study sites with the distribution of sampling points of the citrus-based intercropping system in district Sargodha.

2.2. Tree Biomass and Carbon Estimation

For the collection of inventory data, field visits were carried out from April 2016 to August 2016. Sampling was performed in each citrus orchard (0.405 ha) by considering the method described by Pearson et al. [27]. An area of 20 m × 20 m was sampled from each plot. The stem basal diameter and height of each tree in the sampling plots were measured. The diameter was recorded at 30 cm above the ground to avoid grafted stem and tree forking at 130 cm. For trees with two or more branches below 30 cm, the basal diameter was measured individually, and an equivalent diameter was calculated at the end. Tree age ranged from 4 to 18 years of measured plots across the whole study area. Tree basal area for individual trees and plot was also computed from the measured basal diameter. Above-ground biomass kg per plot of citrus trees was assessed by different allometric equations converted to Mg ha^{-1} [28,29]. The results obtained from three equations were then averaged to increase the precision. Below-ground biomass was calculated by dividing the above-ground biomass by 0.24 [30,31]. The carbon stock per plot was estimated by multiplying the biomass by 0.48 [32].

2.3. Soil Sampling and Analysis

Samples of soil were taken randomly with the help of a soil auger from 10 plots of each site at two depths: 0–15 cm and 15–30 cm depths. Across all study sites, soil samples were obtained from three points in each plot, and a combined sample was made for each depth. Overall, 120 samples for both depths were taken from the whole study area. After collection, these samples of the same site and depth were separated and air-dried [33]. Soil bulk density was computed by the metal core of dimension 4×5 cm at both depths. All the soil samples were ground first and then sieved through a 0.25 mm sieve to determine the organic carbon. The wet oxidation method described by Walkley and Black [34] was used to estimate the organic carbon. Finally, the soil carbon contents per hectare for each depth was computed as follows

$$\text{Soil carbon stock} = \text{OC \%} \times \text{Bulk density (g cm}^{-3}) \times \text{sampling depth (cm)}$$

2.4. Statistical Analysis

Descriptive statistics were performed by using Statistics 8.1 statistical software package. Means of all the parameters were compared by one-way ANOVA, followed by the LSD method to test the difference across all study sites.

3. Results and Discussion

3.1. Citrus Biomass and Carbon Stock

The basal diameter and height of citrus trees showed some variations among study sites in district Sargodha (Table 1). However, the difference in inventory parameters across all study sites was not significant ($p > 0.05$). The total biomass production of citrus trees ranged from 14.55 Mg ha^{-1} to 21.43 Mg ha^{-1} across six study sites with maximum accumulation at Shahpur and minimum at Sahiwal. The above and below-ground biomass accumulation varied significantly ($p \leq 0.05$), and the distribution status of biomass amongst study sites was in the order of Shahpur > Sargodha > Silanwali > Bhalwal > Kot Momin > Sahiwal (Table 2). Based on biomass production, the above and below-ground carbon storage of citrus-based agroforestry system among all study sites was significantly different ($p \leq 0.05$). Overall, the ranking of carbon storage in the citrus intercropping system among study sites was in the order of Shahpur (10.28 Mg C ha^{-1}) > Sargodha (9.69 Mg C ha^{-1}) > Silanwali (8.65 Mg C ha^{-1}) > Bhalwal (8.43 Mg C ha^{-1}) > Kot Momin (7.77 Mg C ha^{-1}) > Sahiwal (6.98 Mg C ha^{-1}), (Figure 2). The carbon stock in Shahpur was 6.08%, 18.84%, 21.80%, 32.30% and 47.27% higher than that in Sargodha, Silanwali, Bhalwal, Kot Momin, and Sahiwal, respectively. The estimated basal area per plot showed a strong and positive linear relationship ($R^2 = 0.91$, $p \leq 0.05$) with total citrus carbon stock in the complete inventory plots of the study area (Figure 3).

Table 1. Growth parameters of the citrus-based intercropping system in district Sargodha.

Study Sites	# Plots (0.405 ha)	Basal Diameter (cm)	Height (m)
Bhalwal	60	13.21 ± 3.01	3.92 ± 0.71
Kot Momin	50	11.75 ± 2.54	3.34 ± 1.26
Sahiwal	30	10.67 ± 2.09	3.13 ± 0.59
Sargodha	80	13.42 ± 2.70	3.59 ± 0.98
Shahpur	40	13.97 ± 2.39	4.03 ± 0.71
Silanwali	40	12.64 ± 1.61	3.37 ± 0.51

Values are means $\pm$ standard deviation.

Table 2. Biomass production estimation of a citrus-based intercropping system in district Sargodha.

Study Sites	Above-Ground Biomass (Mg ha^{-1})	Below-Ground Biomass (Mg ha^{-1})	Total Biomass (Mg ha^{-1})
Bhalwal	$14.06^{\,b} \pm 2.62$	$3.52^{\,c} \pm 0.65$	$17.58^{\,cd} \pm 3.27$
Kot Momin	$12.94^{\,bc} \pm 3.51$	$3.27^{\,c} \pm 0.87$	$16.21^{\,cd} \pm 4.39$
Sahiwal	$11.48^{\,c} \pm 1.35$	$3.07^{\,c} \pm 0.35$	$14.55^{\,d} \pm 1.70$
Sargodha	$16.14^{\,a} \pm 2.61$	$4.04^{\,b} \pm 0.66$	$20.18^{\,ab} \pm 3.26$
Shahpur	$16.61^{\,a} \pm 1.85$	$4.82^{\,a} \pm 0.77$	$21.43^{\,a} \pm 2.42$
Silanwali	$14.42^{\,ab} \pm 1.76$	$3.60b^{\,c} \pm 0.44$	$18.02^{\,bc} \pm 2.19$

Values are means $\pm$ standard deviation; means followed by different letters are significantly different at 5% probability level.

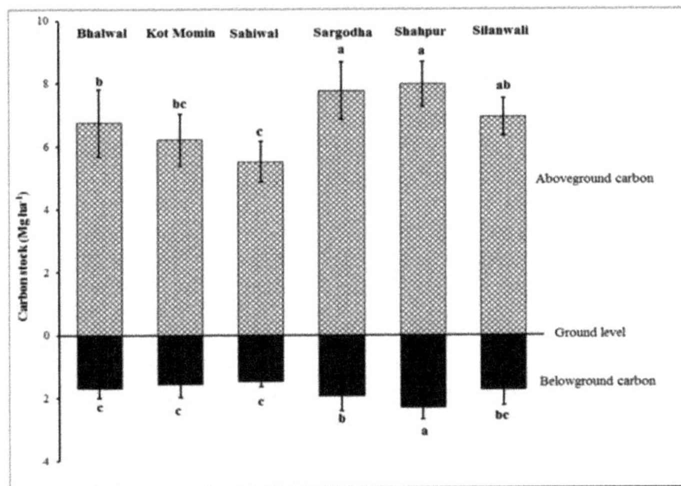

Figure 2. Carbon stock concentration (Mg ha^{-1}) in the citrus-based agroforestry system above- and below-ground parts. a–c represents the standard errors.

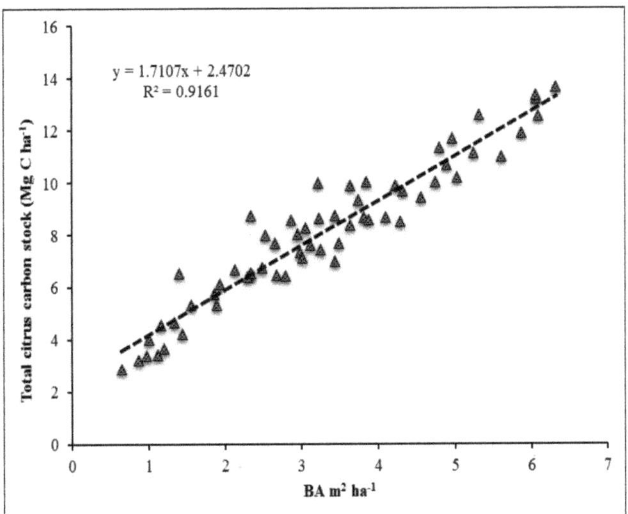

Figure 3. Correlation between citrus total carbon stock (Mg C ha^{-1}) and basal area (m^2 ha^{-1}) in whole plots of study sites.

In contrast to other tree species, there is some specific feature of the citrus-based agroforestry system concerning carbon-capturing ability to combat climate change. Because of their particular physiological features, evergreen plants like citrus play an important role in fixing atmospheric carbon [1], as earlier studies have demonstrated that citrus plants could be a suitable option for combating climate change on a sustained basis [18,21,35]. Moreover, trees are known as a vital part of biomass accumulation in agroforestry systems. Total plant biomass and carbon accumulation showed little variation among all sites, highest in Shahpur and Sargodha and lowest in Sahiwal. Total biomass in a citrus-based intercropping system varied between 14.55 and 21.43 Mg ha^{-1} in the current study. This variation of tree biomass between the sites is due to the difference between age, size, the density of trees along with management practices in the area as articulated by Ramachandran Nair, Mohan Kumar, and Nair [14]; Liu et al. [36]; Dash and Behera [37]. For instance, Shahpur has a greater biomass accumulation due to higher basal diameter and height than all other sites. This might be endorsed to a slight difference in climatic conditions along with awareness of agroforestry management practices across the study area. Similarly, Yadav, Bisht, and Pandey [21] and Yadav, Gupta, Bhutia, Bisht, Pattanayak, Meena, Choudhary, and Tiwari [10] have described similar differences in biomass accumulation in different fruit-based agroforestry systems. Like biomass, carbon concentration in a system is directly determined by several factors: environmental and socioeconomic and largely depends on the structure and function of the system [9,38]. The establishment of trees on farmlands enhances carbon sequestration ability both in soil and vegetation [39]. The above and below ground carbon stock of the current studied system ranged from 5.51 to 7.97 Mg C ha^{-1} & 1.47 to 2.31 Mg C ha^{-1}, respectively, among all sites of district Sargodha. Likewise, biomass, these small variations among carbon stocks are again dependent on the quality of the site, soil type, growth pattern of citrus trees on each site, age of trees, management practices in the area in combination with their relation to below-ground components of the system [37,40]. Similar trends of biomass carbon accumulation have been reported in various fruit-based agroforestry systems, especially in Indian Himalaya, e.g., Yadav, Bisht, and Pandey [21] have described the above-ground carbon accumulation (8.4 Mg C ha^{-1}) in the lemon + wheat system, a slightly greater than those presented in this study.

3.2. Soil Carbon Density

Organic carbon contents, bulk density, and soil carbon stock in the same soil depth varied among all study sites of the citrus intercropping system, and this difference was significant for both soil depths: 0–15 and 15–30 cm (Table 3). At surface soil: 0–15 cm, among six study sites, the maximum soil carbon concentration (0.58%) was computed in Sargodha and the lowest (0.44%) in Silanwali ($p \leq 0.05$). At 15–30 cm soil depth, Shahpur (0.52%) and Sargodha (0.49%) have significantly greater soil carbon contents than other study sites. The maximum values of soil bulk density were computed in Silanwali (1.49 & 1.55 g cm^{-3}), followed by Shahpur (1.47 & 1.54 g cm^{-3}) and Sargodha (1.42 &1.51 g cm^{-3}), which were significantly higher as compared to the other three sites: Bhalwal, Kot Momin and Sahiwal, respectively ($p \leq 0.05$). Due to soil pore space filled with eroded soil, porosity is reduced, and bulk density increases. Similar to carbon concentration and soil bulk density, there was some variation in the soil carbon stock between the study sites of district Sargodha. A higher amount of soil carbon was recorded in surface soil (0–15 cm) for all study sites than 15–30 cm soil layer. At 0–15 cm depth, soil carbon stock ranged from 9.42 Mg C ha^{-1} to 12.43 Mg C ha^{-1} across all study sites whereas, at 15–30 cm depth, higher soil carbon stock (12.06 Mg C ha^{-1}) was observed in Shahpur and was significantly different ($p \leq 0.05$) from all other sites. Overall, total soil carbon stock was greater in Shahpur (24.21 Mg C ha^{-1}) and Sargodha (23.65 Mg C ha^{-1}) as compared to other study sites at 0–30 cm depth (Table 4).

Table 3. Biomass production estimation of a citrus-based intercropping system in district Sargodha.

Study Sites	Above-Ground Biomass (Mg ha^{-1})	Below-Ground Biomass (Mg ha^{-1})	Total Biomass (Mg ha^{-1})
Bhalwal	14.06 [b] ± 2.62	3.52 [c] ± 0.65	17.58 [cd] ± 3.27
Kot Momin	12.94 [bc] ± 3.51	3.27 [c] ± 0.87	16.21 [cd] ± 4.39
Sahiwal	11.48 [c] ± 1.35	3.07 [c] ± 0.35	14.55 [d] ± 1.70
Sargodha	16.14 [a] ± 2.61	4.04 [b] ± 0.66	20.18 [ab] ± 3.26
Shahpur	16.6 [a] ± 1.85	4.82 [a] ± 0.77	21.43 [a] ± 2.42
Silanwali	14.42 [ab] ± 1.76	3.60b [c] ± 0.44	18.02 [bc] ± 2.19

Values are means ± standard deviation; means followed by different letters are significantly different at 5% probability level.

Table 4. Soil organic carbon, bulk density, and soil carbon stock of citrus-based intercropping system in district Sargodha.

Study Sites	Organic Carbon (%)		Bulk Density (g cm^{-3})		Soil Carbon Stock (Mg ha^{-1})	
	0–15 cm	15–30 cm	0–15 cm	15–30 cm	0–15 cm	15–30 cm
Bhalwal	0.53 [b] ± 0.05	0.47 [ab] ± 0.04	1.39 [cd] ± 0.04	1.46 [b] ± 0.05	11.28 [b] ± 1.04	10.33 [bc] ± 1.03
Kot Momin	0.50 [bc] ± 0.04	0.42 [bc] ± 0.05	1.34 [d] ± 0.03	1.39 [c] ± 0.06	10.13 [c] ± 1.03	8.93 [d] ± 1.54
Sahiwal	0.46 [cd] ± 0.06	0.39 [c] ± 0.05	1.35 [d] ± 0.03	1.43 [bc] ± 0.05	9.42 [c] ± 1.11	8.60 [d] ± 1.19
Sargodha	0.58 [a] ± 0.05	0.49 [a] ± 0.07	1.42 [bc] ± 0.07	1.51 [a] ± 0.06	12.43 [a] ± 1.19	11.21 [ab] ± 1.74
Shahpur	0.54 [ab] ± 0.02	0.52 [a] ± 0.04	1.47 [ab] ± 0.07	1.54 [a] ± 0.09	12.14 [ab] ± 0.94	12.06 [a] ± 1.26
Silanwali	0.44 [d] ± 0.07	0.40 [c] ± 0.06	1.49 [a] ± 0.06	1.55 [a] ± 0.05	9.81 [c] ± 1.48	9.49 [cd] ± 1.35

Values are means ± standard deviation; means followed by different letters are significantly different at 5% probability level.

Soil carbon stock is considered an important carbon pool in measuring carbon balance in different biomes throughout the globe [41]. The major factor of maximum total carbon loss to the atmosphere from agroecosystems is soil respiration. Soil respiration is highly variable in several plant species depending upon plant age, growth habits, and climatic conditions [16]. The present study's overall range of soil carbon stocks (0–30 cm) was 25.07 and 34.50 Mg C ha^{-1} between all study sites. The estimates of our study showed that soil carbon stocks among sites were more or less similar to those estimated in other citrus-based land-use systems [1,34]. However, Yadav, Gupta, Bhutia, Bisht, Pattanayak, Meena, Choudhary, and Tiwari [10] estimated soil carbon density between 54.9 to 59.5 t C ha^{-1} in four agroforestry-based land-use systems (agrisilviculture, agrihorticulture, agrihortisilviculture, and agrisilvihorticulture, etc.) in Indian Himalaya was higher than our estimates. These variations might be because of the differences with regard to space and time in physical components of the plants and their management. The higher concentration of soil carbon in various agroecosystems is associated with a greater amount of biomass reverted to soil resulting in higher soil organic matter stabilization and lower decomposition rates [42]. Apart from this, other factors, such as soil type and age, greatly affect and adjust the soil carbon amount in agroforestry land-use systems [43,44].

3.3. Total Carbon Stock (Biomass + Soil) of System

An overview of the findings (Table 5) indicates that the total carbon stock (biomass carbon + soil carbon) of the citrus intercropping system varied significantly across the study sites ($p \leq 0.05$). The maximum total carbon density was computed in Shahpur (34.50 Mg C ha^{-1}), followed by Sargodha (33.34 Mg C ha^{-1}), and was significantly higher as compared to other sites, while minimum (25.07 Mg C ha^{-1}) total carbon storage was computed in Sahiwal ($p \leq 0.05$). In Shahpur, the total carbon stock was 14.80%, 23.39%, 28.53% and 37.61% was higher when compared with Bhalwal, Silanwali, Kot Momin, and Sahiwal, respectively. Our findings illustrated that the soil carbon pool was prominent in the system's total carbon stock. Similar findings regarding total carbon storage have also been described [21,45] in different agroforestry systems. Likewise, Yasin, Nawaz, Siddiqui, and Niazi [9] and Nawaz, Shah, Gul, Afzal, Ahmad and Ghaffar [30] reported a similar range of total carbon stock in *P. deltoides* based bund planted agroforestry systems under semi-arid conditions and *E. camaldulensis* based agroforestry system under arid conditions on marginal lands. However, the total carbon stock in the agrisilvihorticulture system in Indian Himalayas was much greater (93 t C ha^{-1}) than our estimates [10]. Similarly, the present estimates are far below the estimates (12–228 Mg C ha^{-1}) of Krankina and Harmon [46] for agrisilvicultre systems of humid tropical regions of Southeast Asia and in different land uses varying from 51 to 448 t C ha^{-1} in Ethiopia [47]. However, the total carbon stocks of the present study are greater than the estimates (15–18 Mg C ha^{-1}) of Winjum et al. [48] for silvopastoral systems of low humid tropical regions of northern Asia.

Table 5. Total carbon stock (biomass + soil) of a citrus-based intercropping system in district Sargodha.

Sampling Sites	Total Biomass Carbon (Mg C ha^{-1})	Total Soil Carbon (Mg C ha^{-1})	Total Carbon Stock (Mg C ha^{-1})
Bhalwal	8.43 [cd] ± 1.57	21.61 [b] ± 1.64	30.05 [b] ± 1.62
Kot Momin	7.78 [cd] ± 2.10	19.06 [c] ± 1.62	26.84 [cd] ± 2.52
Sahiwal	6.98 [d] ± 0.76	18.02 [c] ± 1.78	25.07 [d] ± 1.37
Sargodha	9.68 [ab] ± 1.56	23.65 [a] ± 2.23	33.34 [a] ± 2.80
Shahpur	10.28 [a] ± 1.16	24.21 [a] ± 1.24	34.50 [a] ± 2.08
Silanwali	8.65 [bc] ± 1.05	19.31 [c] ± 1.97	27.96 [c] ± 2.02

Values are means ± standard deviation; means followed by different letters are significantly different at 5% probability level.

4. Conclusions

Citrus reticulata is widely interplanted across district Sargodha on a commercial basis. We concluded that the citrus-based planting showed a remarkable amount of carbon storage in tree biomass and soil. The system total carbon stocks varied across study sites due to differences among age, growth pattern of citrus trees, and their management across the study area. Thus, the ability and potential of fruit-based agroforestry systems make them a viable option to cope with climate change by sequestering a reasonable concentration of CO_2 from the atmosphere. Apart from this, these agroforestry systems provide livelihood security to local dwellers, especially in underdeveloped countries. Our findings suggest that the authorities promote such agroforestry systems to improve environmental services and increase farmers' income.

Author Contributions: Conceptualization, G.Y., S.U.R. and Z.D.; Data curation, I.Q., A.R.S., M.I., S.G., M.A.B. and A.R.; Formal analysis, G.Y., M.Z., I.Q., A.R.S., S.G., M.A.B. and A.R.; Funding acquisition, S.U.R. and Z.D.; Investigation, G.Y., M.Z., I.Q., A.R.S. and S.G.; Methodology, G.Y. and M.F.N.; Resources, M.F.N., A.R.S., M.I., S.G., M.A.B. and A.R.; Software, A.R.S., M.I., S.G. and M.A.B.; Supervision, A.R. and S.U.R.; Validation, S.U.R. and Z.D.; Visualization, A.R. and S.U.R.; Writing—review & editing, A.R.S., M.I., A.R. and S.U.R. All authors have read and agreed to the published version of the manuscript.

Funding: The authors are grateful to the funding agency (HEC) for providing the necessary funds and facilities to complete this study under NRPU Project #2459. The authors are also grateful to the National Natural Science Foundation of China (51779260), Henan province Key Point Research and Invention Program (192102110051), the Technical Innovation Program of the Chinese Academy of Agricultural Sciences Grant NO. ASTIP202101.

Institutional Review Board Statement: Not applicable.

Informed Consent Statement: Not applicable.

Data Availability Statement: All data required to support this research is already presented in this manuscript.

Acknowledgments: The authors are grateful to the funding agency (HEC) for providing the necessary funds and facilities to complete this study under NRPU Project #2459.

Conflicts of Interest: Authors declare that there is no conflict of interest.

References

1. Iglesias, D.J.; Quinones, A.; Font, A.; Martínez-Alcántara, B.; Forner-Giner, M.Á.; Legaz, F.; Primo-Millo, E. Carbon balance of citrus plantations in Eastern Spain. *Agric. Ecosyst. Environ.* **2013**, *171*, 103–111. [CrossRef]
2. Kumar, K.K.; Nagai, M.; Witayangkurn, A.; Kritiyutanant, K.; Nakamura, S. Above Ground Biomass Assessment from Combined Optical and SAR Remote Sensing Data in Surat Thani Province, Thailand. *J. Geogr. Inf. Syst.* **2016**, *08*, 506–516. [CrossRef]
3. Stocker, T. *Climate Change 2013: The Physical Science Basis: Working Group I Contribution to the Fifth Assessment Report of the Intergovernmental Panel on Climate Change*; Cambridge University Press: Cambridge, UK, 2014.

4. Parry, M.; Parry, M.L.; Canziani, O.; Palutikof, J.; Van der Linden, P.; Hanson, C. *Climate Change 2007-Impacts, Adaptation and Vulnerability: Working Group II Contribution to the Fourth Assessment Report of the IPCC*; Cambridge University Press: Cambridge, UK, 2007; Volume 4.
5. Chavan, B.; Rasal, G. Potentiality of Carbon Sequestration in six year ages young plant from University campus of Aurangabad. *Glob. J. Res. Eng.* **2011**, *11*, 7-C.
6. Victor, A.D.; Valery, N.N.; Louis, Z.; Aimé, V.B.T.; Aliou, S. Carbon Sequestration Potential and Economic Value in Agroforestry Parkland to *Tectona grandis* L. f.(Verbenaceae) in Central Africa: A Case Study to Department of Poli (Northern Region in Cameroon). *Adv. Res.* **2019**, *8*, 1–16. [CrossRef]
7. Chave, J.; Andalo, C.; Brown, S.; Cairns, M.A.; Chambers, J.; Eamus, D.; Fölster, H.; Fromard, F.; Higuchi, N.; Kira, T.; et al. Tree allometry and improved estimation of carbon stocks and balance in tropical forests. *Oecologia* **2005**, *145*, 87–99. [CrossRef]
8. Yadav, R.P.; Gupta, B.; Bhutia, P.L.; Bisht, J.K. Socioeconomics and sources of livelihood security in Central Himalaya, India: A case study. *Int. J. Sustain. Dev. World Ecol.* **2017**, *24*, 545–553. [CrossRef]
9. Yasin, G.; Nawaz, M.; Siddiqui, M.; Niazi, N. Biomass, carbon stocks and CO2 sequestration in three different aged irrigated populus deltoides bartr. Ex marsh. Bund planting agroforestry systems. *Appl. Ecol. Environ. Res.* **2018**, *16*, 6239–6252. [CrossRef]
10. Yadav, R.P.; Gupta, B.; Bhutia, P.L.; Bisht, J.K.; Pattanayak, A.; Meena, V.S.; Choudhary, M.; Tiwari, P. Biomass and carbon budgeting of sustainable agroforestry systems as ecosystem service in Indian Himalayas. *Int. J. Sustain. Dev. World Ecol.* **2019**, *26*, 460–470. [CrossRef]
11. Raj, A.; Jhariya, M.K.; Pithoura, F. Need of agroforestry and impact on ecosystem. *J. Plant Dev. Sci. Es* **2014**, *6*, 577–581.
12. Nawaz, M.; Yousaf, M.; Yasin, G.; Gul, S.; Ahmed, I.; Abdullah, M.; Rafay, M.; Tanvir, M.; Asif, M.; Afzal, S. Agroforestry status and its role to sequester atmospheric CO_2 under semi-arid climatic conditions in Pakistan. *Appl. Ecol. Environ. Res.* **2018**, *16*, 645–661. [CrossRef]
13. Mohan Kumar, B.; Ramachandran Nair, P. *Carbon Sequestration Potential of Agroforestry Systems*; Springer: Berlin, Germany, 2011.
14. Nair, P.K.R.; Kumar, B.M.; Nair, V.D. Agroforestry as a strategy for carbon sequestration. *J. Plant Nutr. Soil Sci.* **2009**, *172*, 10–23. [CrossRef]
15. Qubaja, R.; Yang, F.; Amer, M.; Tatarinov, F.; Yakir, D. Ecophysiology of an urban citrus orchard. *Urban For. Urban Green.* **2021**, *65*, 127361. [CrossRef]
16. Yixiang, W.A.N.G.; Boqi, W.E.N.G.; Na, T.I.A.N.; Zhong, Z.; Mingkuang, W.A.N.G. Soil organic carbon stocks of citrus orchards in Yongchun county, Fujian Province, China. *Pedosphere* **2017**, *27*, 985–990.
17. Xiaolian, W.; Yueqing, C.; Youjin, L.; Xia, C.; Yonghong, X. Carbon sequestration and storage of citrus orchard system in Three Gorges Reservoir region of Chongqing, Southwest China. *J. Agric. Sci.* **2014**, *27*, 693–698.
18. Liguori, G.; Gugliuzza, G.; Inglese, P. Evaluating carbon fluxes in orange orchards in relation to planting density. *J. Agric. Sci.* **2009**, *147*, 637–645. [CrossRef]
19. Iglesias, D.J.; Lliso, I.; Tadeo, F.R.; Talon, M. Regulation of photosynthesis through source: Sink imbalance in citrus is mediated by carbohydrate content in leaves. *Physiol. Plant.* **2002**, *116*, 563–572. [CrossRef]
20. Marti, B.V.; Estornell, J.; Cortés, I.L.; Martí-Gavilá, J. Calculation of biomass volume of citrus trees from an adapted dendrometry. *Biosyst. Eng.* **2012**, *112*, 285–292. [CrossRef]
21. Yadav, R.; Bisht, J.K.; Pandey, B.M. Above ground biomass and carbon stock of fruit tree based land use systems in Indian Himalaya. *Ecoscan* **2015**, *9*, 779–783.
22. Sharif, M.; Farooq, U.; Malik, W.; Bashir, M. Citrus Marketing in Punjab: Constraints and Potential for Improvement [with Comments]. *Pak. Dev. Rev.* **2005**, *44*, 673–694. [CrossRef]
23. Usman, M.; Ashraf, I.; Chaudhary, K.M.; Talib, U. Factors impeding citrus supply chain in Central Punjab, Pakistan. *Int. J. Agric. Ext.* **2018**, *6*, 01–05. [CrossRef]
24. Segura, M.; Kanninen, M.; Suárez, D. Allometric models for estimating above-ground biomass of shade trees and coffee bushes grown together. *Agrofor. Syst.* **2006**, *68*, 143–150. [CrossRef]
25. Akram, B.; Abbas, F.; Ibrahim, M.; Nawaz, M.; Zahra, S.; Salik, M.; Hammad, H. Above-ground carbon pools of citrus acreage in Pakistan. *JAPS J. Anim. Plant Sci.* **2017**, *27*, 1903–1908.
26. Mobeen, M.; Ahmed, H.; Ullah, F.; Riaz, M.O.; Mustafa, I.; Khan, M.R.; Hanif, M.U.; Muhammad, M. Impact of climate change on the precipitation pattern of district Sargodha, Pakistan. *Int. J. Clim. Chang. Strat. Manag.* **2017**, *9*, 21–35. [CrossRef]
27. Pearson, T.; Walker, S.; Brown, S. Sourcebook for Land Use, Land-Use Change and Forestry Projects. 2013. Available online: http://hdl.handle.net/10986/16491 (accessed on 10 October 2021).
28. Schroth, G.; D'Angelo, S.A.; Teixeira, W.G.; Haag, D.; Lieberei, R. Conversion of secondary forest into agroforestry and monoculture plantations in Amazonia: Consequences for biomass, litter and soil carbon stocks after 7 years. *For. Ecol. Manag.* **2002**, *163*, 131–150. [CrossRef]
29. Brown, S. *Estimating Biomass and Biomass Change of Tropical Forests: A Primer*; Food & Agriculture Org.: Roma, Italy, 1997; Volume 134.
30. Mehta, L.C.; Singh, J.; Chauhan, P.S.; Singh, B.; Manhas, R.K. Biomass accumulation and carbon storage in six-year old Citrus reticulata Blanco plantation. *Indian For.* **2016**, *142*, 563–568.
31. Nawaz, M.F.; Shah, S.A.A.; Gul, S.; Afzal, S.; Ahmad, I.; Ghaffar, A. Carbon sequestration and production of Eucalyp-tus camaldulensis plantations on marginal sandy agricultural lands. *Pak. J. Agric. Sci.* **2017**, *54*, 335–342.
32. Thomas, S.C.; Martin, A.R. Carbon Content of Tree Tissues: A Synthesis. *Forests* **2012**, *3*, 332–352. [CrossRef]

33. Arora, G.; Chaturvedi, S.; Kaushal, R.; Nain, A.; Tewari, S.; ALAM, N.M.; Chaturvedi, O.P. Growth, biomass, carbon stocks, and sequestration in an age series of Populus deltoides plantations in Tarai region of central Himalaya. *Turk. J. Agric. For.* **2014**, *38*, 550–560. [CrossRef]
34. Walkley, A.; Black, I.A. An examination of the Degtjareff method for determining soil organic matter, and a proposed modification of the chromic acid titration method. *Soil Sci.* **1934**, *37*, 29–38. [CrossRef]
35. Gratani, L.; Varone, L.; Catoni, R. Relationship between net photosynthesis and leaf respiration in Mediterranean ever-green species. *Photosynthetica* **2008**, *46*, 567–573. [CrossRef]
36. Liu, Z.-W.; Chen, R.-S.; Song, Y.-X.; Han, C.-T. Above-ground biomass and water storage allocation in alpine willow shrubs in the Qilian Mountains in China. *J. Mt. Sci.* **2015**, *12*, 207–217. [CrossRef]
37. Dash, M.C.; Behera, N. Carbon sequestration and role of earthworms in Indian land uses: A review. *Ecoscan* **2013**, *7*, 1–7.
38. Rajput, B.S.; Bhardwaj, D.R.; Pala, N.A. Factors influencing biomass and carbon storage potential of different land use systems along an elevational gradient in temperate northwestern Himalaya. *Agrofor. Syst.* **2017**, *91*, 479–486. [CrossRef]
39. Haile, S.G.; Nair, P.K.R.; Nair, V.D. Carbon Storage of Different Soil-Size Fractions in Florida Silvopastoral Systems. *J. Environ. Qual.* **2008**, *37*, 1789–1797. [CrossRef] [PubMed]
40. Jana, B.K.; Biswas, S.; Majumder, M.; Roy, P.K.; Mazumdar, A. Carbon sequestration rate and above-ground biomass carbon potential of four young species. *J. Ecol. Nat. Environ.* **2009**, *1*, 15–24.
41. Okuda, H.; Noda, K.; Sawamoto, T.; Tsuruta, H.; Hirabayashi, T.; Yonemoto, J.Y.; Yagi, K. Emission of N_2O and CO_2 and Uptake of CH4 in Soil from a Satsuma Mandarin Orchard under Mulching Cultivation in Central Japan. *J. Jpn. Soc. Hortic. Sci.* **2007**, *76*, 279–287. [CrossRef]
42. Kell, D.B. Large-scale sequestration of atmospheric carbon via plant roots in natural and agricultural ecosystems: Why and how. *Philos. Trans. R. Soc. B Biol. Sci.* **2012**, *367*, 1589–1597. [CrossRef] [PubMed]
43. Chiti, T.; Díaz-Pinés, E.; Rubio, A. Soil organic carbon stocks of conifers, broadleaf and evergreen broadleaf forests of Spain. *Biol. Fertil. Soils* **2012**, *48*, 817–826. [CrossRef]
44. Chauhan, S.K.; Sharma, R.; Singh, B.; Sharma, S.C. Biomass production, carbon sequestration and economics of on-farm poplar plantations in Punjab, India. *J. Appl. Nat. Sci.* **2015**, *7*, 452–458. [CrossRef]
45. Verma, A.; Kaushal, R.; Alam, N.M.; Mehta, H.; Chaturvedi, O.P.; Mandal, D.; Tomar, J.M.S.; Rathore, A.C.; Singh, C. Predictive models for biomass and carbon stocks estimation in Grewia optiva on degraded lands in western Himalaya. *Agrofor. Syst.* **2014**, *88*, 895–905. [CrossRef]
46. Krankina, O.N.; Harmon, M.E. The impact of intensive forest management on carbon stores in forest ecosystems. *World Resour. Rev.* **1994**, *6*, 161–177.
47. Bajigo, A.; Tadesse, M.; Moges, Y.; Anjulo, A. Estimation of carbon stored in agroforestry practices in Gununo Water-shed, Wolayitta Zone, Ethiopia. *J. Ecosyst. Ecography* **2015**, *5*, 1.
48. Winjum, J.K.; Dixon, R.K.; Schroeder, P.E. Estimating the global potential of forest and agroforest management practices to sequester carbon. *Water Air Soil Pollut.* **1992**, *64*, 213–227. [CrossRef]

MDPI

St. Alban-Anlage 66

4052 Basel

Switzerland

Tel. +41 61 683 77 34

Fax +41 61 302 89 18

www.mdpi.com

Sustainability Editorial Office

E-mail: sustainability@mdpi.com

www.mdpi.com/journal/sustainability